Nanoparticle Assembly: From Fundamentals to Applications

Indian Institute of Technology Bombay
Mumbai, India
7–9 January 2016

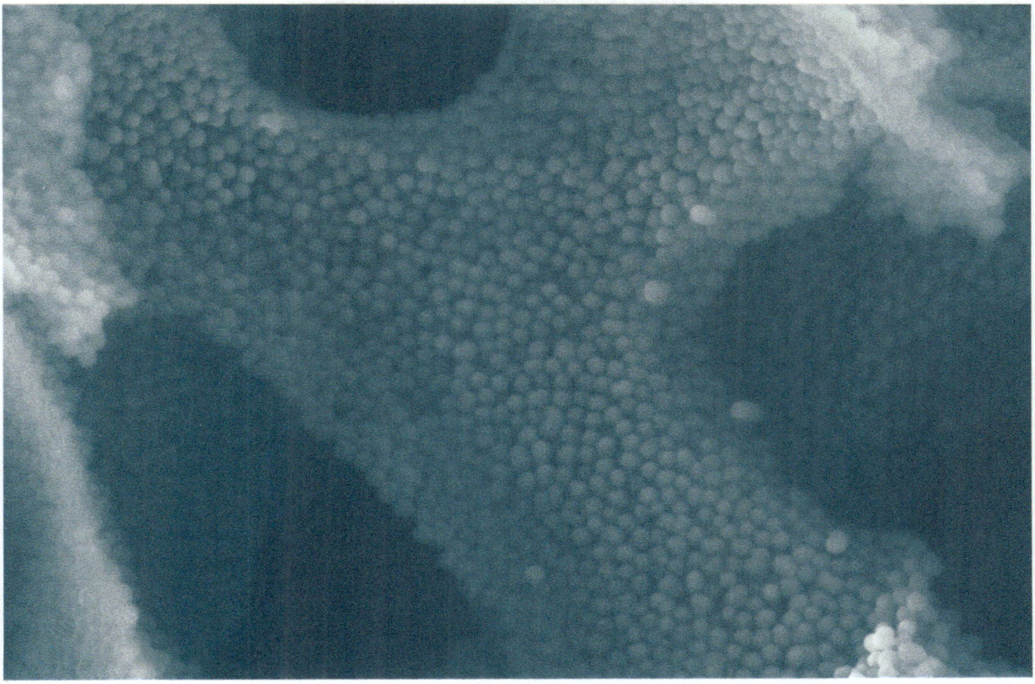

FARADAY DISCUSSIONS

Volume 186, 2016

ROYAL SOCIETY OF CHEMISTRY

In 2014 the Royal Society of Chemistry requested **Charusita Chakravarty** to organize this Faraday Discussion in India. Over time she recruited several of us - Sanat Kumar, Guruswamy Kumaraswamy, BLV Prasad, Rajdip Bandopadyay - to help with organizing and running this international meeting.

From the initial stages to the final highly successful meeting that was held in Mumbai in January 2016, this project has Charu's fingerprints all over it, including this final printed volume. No detail was small enough for her, from fine-tuning the program, interacting with chairs, stimulating discussion both during the meeting as well as during the social time, to sharing Indian culture with visitors. All the while she was fighting a debilitating disease, to which she succumbed on 29 March 2016.

It is this sense of commitment that we celebrate here. We, her friends, collaborators and colleagues, dedicate this issue to her memory. Her fighting spirit and her basic humanity is what we would like to remember her by.

Nanoparticle Assembly: From Fundamentals to Applications

Faraday Discussions

www.rsc.org/faraday_d

A General Discussion on Nanoparticle Assembly: From Fundamentals to Applications was held in Mumbai, India on the 7th, 8th and 9th of January 2016.

CONTENTS

ISSN 1359-6640; ISBN: 978-1-78262-470-7

Cover
Front cover image courtesy of Guruswamy Kumaraswamy, CSIR-National Chemical Laboratory.

INTRODUCTORY LECTURE

PAPERS AND DISCUSSIONS

The Faraday Division of the Royal Society of Chemistry, previously the Faraday Society, founded in 1903 to promote the study of sciences lying between Chemistry, Physics and Biology.

Faraday Discussions (Print ISSN 1359-6640, Electronic ISSN 1364-5498) is published 10 times a year by the Royal Society of Chemistry, Thomas Graham House, Science Park, Milton Road, Cambridge, UK CB4 0WF.

Volume 186 ISBN-13: 978-1-78262-470-7

2016 annual subscription price: print+electronic £929 US $1635; electronic only £883, US $1553. Customers in Canada will be subject to a surcharge to cover GST. Customers in the EU subscribing to the electronic version only will be charged VAT. All orders, with cheques made payable to the Royal Society of Chemistry, should be sent to RSC Order Department, Royal Society of Chemistry, Thomas Graham House, Science Park, Milton Road, Cambridge, CB4 0WF, UK.

Tel +44 (0) 1223 432398; E-mail **www.orders@rsc.org**

If you take an institutional subscription to any RSC journal you are entitled to free, site-wide web access to that journal. You can arrange access via Internet Protocol (IP) address at **www.rsc.org/ip**. Customers should make payments by cheque in sterling payable on a UK clearing bank or in US dollars payable on a US clearing bank.

Printed in the UK

MIX
Paper from responsible sources
FSC www.fsc.org FSC® C013604

Faraday Discussions are unique international discussion meetings that focus on rapidly developing areas of chemistry and its interfaces with other scientific disciplines.

CONCLUDING REMARKS

ADDITIONAL INFORMATION

MANTIS

MANTIS DEPOSITION LTD

Faraday Discussions

PAPER

Nanoparticle puzzles and research opportunities that go beyond state of the art

Ah-Young Jee,[a] Kai Lou,[a] Hyun-Sook Jang,[a] K. Hima Nagamanasa[a] and Steve Granick*[ab]

Received 24th February 2016, Accepted 29th February 2016
DOI: 10.1039/c6fd00024j

We present an overview of current progress and research challenges in the field of nanoparticle assembly, touching on the following topics: (1) historical perspective; (2) consideration of what is a nanoparticle; (3) contrast between nanoparticle self-assembly and top-down construction; (4) opportunities for nanoparticles with more intelligent sub-structures; (5) opportunities for nanoparticle systems cued to interact subtly in space and time. In this personal and subjective account, certain holy grails for nanoparticle science and technology are identified.

Introduction

Recent magnificent progress in nanoparticle science and nanotechnology, stimulated especially by pioneering government initiatives,[1] is accompanied nowadays by a sense of topicality and optimistic excitement that is shared internationally. Nanoparticles have become an interdisciplinary subject that transcends all conventional disciplines and provides opportunities both scientific and career related for many chemists, physicists, and engineers, not to speak of those interested in biology, drugs, and medicine. A quarter of a century since the first initiatives at the U.S. National Science Foundation,[1] it is fitting and proper to step back and reflect on where we have come from; where we are now; and where we as a field should set our sights to be going.

Where we come from

We stand on the shoulders of giants whose names it is well to remember as without their seminal contributions, the science-based approach to nanoparticles that we presently enjoy would be unthinkable. In our opinions expressed in this

[a]*Center for Soft and Living Matter, Institute for Basic Science (IBS), Ulsan 44919, Republic of Korea. E-mail: sgranick@ibs.re.kr*
[b]*Department of Chemistry, UNIST, Ulsan 44919, Republic of Korea*

highly personal account, four great scientists are worth remembering especially. One is P.-G. de Gennes, the "Newton of our times";[2] by brilliance and force of personality, he convinced a generation of physicists that it is possible to do honest, rigorous, penetrating physics on problems that, in prior years, had been considered too messy, unclean, and tainted by too much association with applications and technology. P.-G. de Gennes made an essential contribution culturally: he helped to make our discipline respectable.

If we did not understand surface forces, we could not understand how nanoparticles interact with their environments. Nanoscientists working today have probably almost all learned about this from the beautiful monograph of Jacob Israelachvili.[3] It is a gem: beautifully written and presented with a point of view and humor that give it personal character in addition to great scientific content. Thanks to it, we as a community are comfortable with analyzing nanoparticle stability and instability; we understand van der Waals forces, double layer forces, hydration and hydrophobic forces, and more; we know the limits of these concepts; and we are sensitive to how to deepen understanding of these concepts. Tending to take this knowledge for granted as part of the tools of our trade, we forget that there was a day when these ideas were known only in smaller niche communities, and also when the concepts had not yet been validated. Much of the scientific work to perform the experimental validation was performed by Israelachvili and the field of surface forces and AFM research that he helped to spawn.[3]

Perhaps equally as important, much related research was performed by a scientific titan in the former Soviet Union: Boris Derjaguin and the scientific community around him.[4] It is fascinating to survey the development of nanoscience in retrospect and to reflect on how ideas were developed and pursued independently in different corners of the world, back in the days when the scientific world was less networked than it is today. The source materials to which we just referred[2,3] contain numerous original references to the many other scientists, living and dead, who also contributed in essential ways.

Forces are just part of the story, however. We call especial attention to the intellectual contributions of Ludwig Boltzmann, whose shadow continues to be cast into the 21st century though he lived so long ago. Battles fought long ago by Boltzmann taught the world that nanoparticles and atoms exist. Nanoparticles are buffeted by thermal energy; enthalpic forces of interaction between them are countervailed by entropy; sometimes paradoxically so.[5] The essential role played by entropy in understanding nanoparticle order has been preached eloquently.[5]

In making these few references to specific scientists, we apologize to the numerous other scientists who deserve to be cited, if this were a history of the subject. But instead this is a short and personal essay, so we hope our brevity with citations can be forgiven.

What is a nanoparticle?

Like great art, we recognize nanoparticles but to agree on what they are, this is more difficult. We mean something more subtle than simply a material that is nm in dimension, as macromolecules are also nm in dimension: not only in the synthetic world of polymers and microgels but also in the biological world of proteins, RNA, and DNA. So: we cannot discriminate on the basis of size alone. However, it is helpful, in organizing our thinking, to exclude large objects the size of soccer balls, even though it is true that a Devil's Disciple might argue that a soccer ball is nm-sized relative to the size of a sun. For soccer balls, thermal energy = 0, but not so for nanoparticles.

Quantum mechanics is at the heart of the matter, when understanding a molecule. One needs to understand covalent bonding, aromaticity, molecular structure, and much more; and the whole is *not* the sum of its parts. Consider the caffeine molecule, for example. Information about its atomic composition, $C_8H_{10}N_4O_2$, is relatively uninformative as the organization of those atoms in space with quantum mechanical interactions between them is so relevant and interesting. But from the nanoparticle perspective, nanoparticle and colloidal "molecules" are a metaphor[6] for systems where often we are content to know the structural makeup: that the "molecule" is a dimer, trimer, or some other cluster. To a useful approximation, such clusters indeed are the sum of their parts. In the current understanding of nanoparticles, quantum mechanics is often secondary and the classical language of enthalpy and entropy are sufficient to describe the interactions between them.

Exceptions matter

The burgeoning field of plasmonic assemblies does rely on quantum mechanical interactions between nanoparticles, especially when the nanoparticle is a metal.[7] Furthermore, understanding of active matter is developing rapidly, but active matter does not involve a significant contribution from thermal energy.[8] An intriguing challenge for the future is that the importance of biological nanomachines ("molecular motors") is clear[9] but synthetic active nanomachines are at an early stage of development.

If the nanoparticle is sufficiently large, beyond roughly 20 nm in linear dimension, the span of surface forces is smaller than this, and hence interactions can be decomposed pairwise. But smaller nanoparticles often obey force interactions whose distance span actually exceeds the particle size. In those cases, numerous nanoparticles are mutually influenced by the same force interactions, and simplifications such as DLVO cease to be helpful to the point of becoming an oversimplification. It is not widely enough appreciated that the ratio of nanoparticle size to span of surface force interactions is so essential.

Nanoparticle self-assembly *versus* macroscopic top-down construction

Most of the attendees at this meeting arrived by automobile or airplane. Modern airliners are marvelous machines with millions of parts going into their construction; their cost is enormous, their size is enormous, they employ hundreds of thousands of workers for their construction and maintenance, and they contain millions of parts. It is fascinating to contrast such mega-machines with living cells: living cells reproduce themselves, they are tiny, their cost is low, and yet they too contain millions of parts, parts which we know by the name of "protein". We can find from company literature that a commercial airplane contains on the order of 2 million parts, while a living cell contains on the order of 2 million proteins per cubic micron. The airplane is created by a careful and deliberate top-down construction, but living cells just grow.

Nanoscience and nanotechnology are a long way from knowing how to achieve this synthetically but nature shows us, by example, that the concept is possible. A holy grail for our community is to learn how to assemble complex machines from the bottom-up and with nm-sized interacting parts. Whatever progress we make in this direction, it could become a step towards implementing one of the goals of this meeting: to understand how a fundamental understanding of the rapidly evolving field of nanoparticle self- and driven assembly can be translated into design principles useful for applications.

Nanoparticles with more intelligent substructures

In this exciting age of discovery, nanoscientists have contributed to understanding biology; as an example, a lovely recent work shows some principles of protein assembly and reveals a figurative "periodic table of protein complexes".[10] Conversely, we nanoscientists have much to learn from the example of nm-sized proteins, not just from their role as nanomachines[9] but even more generally so.

A limitation is that our nanoparticles often consist of frozen structures: for example, a gold or silver or silica nanoparticle, or a quantum dot, or some other construct whose size and structure is known. For these frozen structures, their size and surface chemical makeup are most of the information needed to understand the nanoparticle's physical behavior. This has carried the field a long way but increasingly the field is looking towards new functions that cannot be thus achieved. Might it be possible to design new nanoparticles whose function changes by binding to a function that is far away? If we could learn to do this, it would be the "allostery" effect known for proteins. In fact, proteins present, by their existence, proof that there can be a dynamic continuum of allosteric phenomena, with rigid-body motions of protein sub-elements on one extreme of relatively little disorder, and local unfolding or even intrinsically disordered nanoparticle structures on the other extreme of high disorder.[11] There is the possibility to engineer new kinds of synthetic nanoparticles whose structure will respond to the environment in a more subtle fashion than in conventional ways. Responses to pH changes, temperature changes and the like are already known and well understood[1] but there is scope for allosteric responses that would be significantly more subtle and complex, not with the goal of reproducing what nature has already achieved, but to find ways to learn from it by analogy and metaphor. To achieve such goals, our nanoscience community will need to invent fundamentally new ways to design nanoparticles with intelligent substructures.

Nanoparticle systems cued to interact subtly in space and time

Upon a survey of the literature, it is striking to notice that the interactions between nanoparticles are built so much on the classical forces of colloid science[3,4] without the specificity that nature displays in any living cell.

To a nanoscientist, one of the most gloriously surprising aspects of living cells is that the nm-sized proteins manage to coexist without too much interference with one another; they interact as needed, and also they *fail to interact*, as needed. This allows the presence of systems in which nm-sized proteins of many sorts coexist while allowing each to perform its function.[12] They follow rules of network interactions, consisting of "subnets" within larger networks, in which interacting elements are orthogonal in space and time, thus enabling functions which go magnificently beyond what can be designed using concepts of classical nanoscience and nanotechnology.[12] Nanoscience can learn thematically from this proof of concept too. Presently we know too little about how to signal interactions between selected nanoparticles, too little about how to switch interactions on and off, too little about how to design the structure of one spatio-temporal element of a nanoparticle assembly while leaving other portions of the sample unaffected. This should be possible to achieve in the synthetic nanoparticle world but our community does not know well enough how to design it. This could become a holy grail for nanotechnology of the future.

Acknowledgements

This work was supported by the Institute for Basic Science, project code IBS-R020-D1.

References

1 M. C. Roco, The long view of nanotechnology development: the National Nanotechnology Initiative at 10 years, *J. Nanopart. Res.*, 2011, **13**, 427–445.

2 The Nobel Prize in Physics, 1991, http://www.nobelprize.org/nobel_prizes/physics/laureates/1991/.

3 J. N. Israelachvili, *Intermolecular and Surface Forces*, Academic Press, 3rd edn, 2011.

4 N. V. Churaev, B. V. Derjaguin and V. M. Muller, *Surface Forces*, Springer, 1987.

5 D. Frenkel, Order through entropy, *Nat. Mater.*, 2015, **14**, 9–12.

6 A. van Blaaderen, Colloidal molecules and beyond, *Science*, 2003, **301**, 470.

7 C. X. Xi, P. F. Marina, H. B. Xia and D. Y. Wang, Directed self-assembly of gold nanoparticles into plasmonic chains, *Soft Matter*, 2015, **11**, 4562–4571.

8 S. Ramaswamy, The mechanics and statistics of active matter, *Annu. Rev. Condens. Matter Phys.*, 2010, **1**, 323–345.

9 M. Peplow, The tiniest Lego: a tale of nanoscale motors, rotors, switches and pumps, *Nature*, 2015, **525**, 18–21.

10 S. E. Ahnert, J. A. Marsh, H. Hernández, C. V. Robinson and S. A. Teichmann, Principles of assembly reveal a periodic table of protein complexes, *Science*, 2015, **350**, 1331.

11 H. N. Motlagh, J. O. Wrabl, J. Li and V. J. Hilser, The ensemble nature of allostery, *Nature*, 2014, **508**, 331–339.

12 R. Zaidel-Bar, S. Itzkovitz, A. Ma'ayan, R. Iyengar and B. Geiger, Functional atlas of the integrin adhesome, *Nat. Cell Biol.*, 2007, **9**, 858–867.

Faraday Discussions

PAPER

Tailoring structure formation and mechanical properties of particle brush solids *via* homopolymer addition

Michael Schmitt,[a] Chin Ming Hui,[b] Zachary Urbach,[a] Jiajun Yan,[b] Krzysztof Matyjaszewski*[b] and Michael R. Bockstaller*[a]

Received 17th August 2015, Accepted 26th October 2015
DOI: 10.1039/c5fd00121h

Recent progress in the area of surface-initiated controlled radical polymerization (SI-CRP) has enabled the synthesis of polymer-grafted colloids with precise control over the architecture of grafted chains. The resulting 'particle brush materials' are of interest both from a fundamental as well as applied perspective because structural frustrations (associated with the tethering of chains to a curved surface) imply a sensitive dependence of the interactions between brush particles on the architecture of surface-tethered chains that offers new opportunities to design hybrid materials with novel functionalities. An important prerequisite for establishing structure–property relations in particle brush materials is to understand the role of homopolymer impurities that form, for example, by thermal self-initiation. This contribution presents a detailed discussion of the role of homopolymer additives on the structure and mechanical properties of particle brush materials. The results suggest that the dissolution of homopolymer fillers follows a two-step mechanism comprised of the initial segregation of homopolymer to the interstitial regions within the array and the subsequent swelling of the particle brush (depending on the respective degree of polymerization of brush and linear chains). Addition of even small amounts of homopolymer is found to significantly increase the fracture toughness of particle brush assembly structures. The increased resistance to failure could enable the synthesis of robust colloidal crystal type materials that can be processed into complex shapes using 'classical' polymer forming techniques such as molding or extrusion.

Introduction

The tethering of polymeric chains to the surface of nano-sized particles or colloids has emerged as a versatile tool to modulate the interactions and assembly

[a]Department of Materials Science and Engineering, Carnegie Mellon University, 5000 Forbes Ave., Pittsburgh, PA 15213, USA. E-mail: bockstaller@cmu.edu

[b]Chemistry Department, Carnegie Mellon University, 4400 Fifth Ave., Pittsburgh, PA 15213, USA. E-mail: km3b@anrew.cmu.edu

behavior of particulate systems but also as a means towards nanocomposite materials with novel functionalities.[1–4] Examples include the formation of particle films and colloidal crystal structures with enhanced fracture resistance or the formation of 'one-component' hybrid materials with enhanced electric breakdown strength or novel phonon propagation characteristics.[5–9] A prerequisite to harness the opportunities that are afforded by polymer-tethered particulate materials is the ability to control the architecture, *i.e.* the particle radius as well as the degree of polymerization and density of grafted chains. Among the various synthetic techniques surface-initiated controlled radical polymerization (SI-CRP) has attracted particular interest for the synthesis of polymer-tethered particle systems because of the high level of control over molecular parameters such as degree of polymerization (N) or grafting density (σ) as well as the wide range of accessible chemistries and chain constitutions.[10–12] Surface-initiated atom transfer radical polymerization (SI-ATRP) has attracted particular interest due to the wide range of accessible monomer compositions and functionalities, its tolerance to impurities as well as compatibility with aqueous solvent environments.[13–17]

However, while SI-ATRP has proven to be a versatile synthetic tool for polymer-tethered particle materials one challenge is the possibility of side reactions such as radical termination or thermal self-initiation of monomer systems. The latter process presents a particular challenge because the formation of homopolymer impurities alters the physicochemical properties of product materials and hence complicates the elucidation of structure–property relations in polymer-tethered particle materials. To understand the conditions favoring the thermal self-initiation (TSI) of polymers during surface-initiated polymerization, our group recently evaluated the influence of reaction parameters on TSI for the particular case of poly(methyl methacrylate) (PMMA) and polystyrene (PS) tethered silica colloids.[18] It was found that the influence of TSI increased with reaction temperature and decreasing catalyst reactivity. For the case of PS-tethered particle systems a fraction of up to 23% of untethered chains was observed (depending on reaction conditions). Interestingly the presence of even small amounts of homopolymer impurities was found to result in significant toughening of particle film structures. Fig. 1 illustrates the general mechanism of TSI and also summarizes the toughening effect (measured in terms of the normalized mode-1 stress intensity factor) that was observed in TSI afflicted systems.

The pronounced toughening in the presence of TSI that is observed in Fig. 1 raises the prospect of homopolymer additives as 'functional fillers' to enhance the mechanical properties and processability of polymer-tethered particle materials and thus motivates a more detailed analysis of the effect of homopolymer fillers on the structure formation and mechanical properties in polymer-tethered particle materials. The objective of the present contribution is hence to establish the effect of homopolymer addition on the order formation and micromechanical properties of polymer-tethered particle model systems as a function of both the degree of polymerization of the tethered and linear chains. To limit the parameter space our focus is exclusively on densely polymer-tethered systems (that in the following will be called 'particle brushes'). The structure of the paper is as follows: in the first part, the miscibility of homopolymer fillers within particle brush monolayers as well as its effect on the 'degree of order' (as determined on the basis of Voronoi tessellation analysis) will be evaluated using electron imaging. In the second part, nanoindentation will be used to assess the effect of

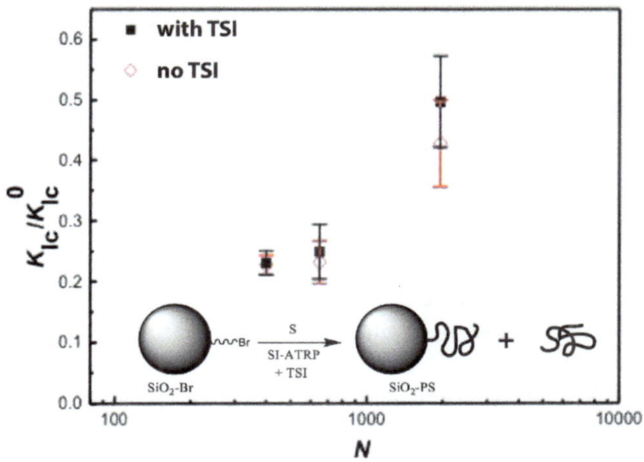

Fig. 1 Dependence of the (normalized) mode-1 stress-intensity factor (fracture toughness) of 57SiO$_2$-S*N* thick films on the degree of polymerization of tethered chains in the absence (red open triangles) and presence (filled black squares) of thermal self-initiation. The fracture toughness is normalized with respect to the respective value of high molecular polystyrene. The effect of thermal self-initiation (TSI) is found to increase with degree of polymerization of tethered chains. Inset illustrates mechanism of homopolymer formation by TSI. Adapted from ref. 18.

homopolymer addition on the elastic and fracture characteristics of bulk particle brush materials.

Experimental methods

Synthesis of polystyrene grafted silica particles

Particle brush synthesis was performed using surface-initiated atom transfer radical polymerization as described previously.[20,21] In a typical synthetic procedure, silica nanoparticles were modified with a tetherable initiator: 6-(triethoxysilyl)hexyl 2-bromo-2-isobutyrate for atom transfer radical polymerization (ATRP). The initiator was attached to the particle surface using NH_4OH as a catalyst in ethanol, using previously established procedures, prior to removal of excess initiator by washing with ethanol. For the polymerization, a mixture of initiator-modified silica nanoparticles (SiO_2–Br) and anisole was stirred in a Schlenk flask for 24 h to form a homogenous suspension. Subsequently, styrene, *N*,*N*,*N*′,*N*″,*N*″-pentamethyldiethylenetriamine (PMDETA) and $CuBr_2$ were added to the flask with a rare earth magnetic stir bar. The use of a sufficiently strong stir bar was required to prevent vitrification that can occur, especially at high monomer conversion. The solution underwent three freeze–pump–thaw cycles before being immersed in liquid nitrogen and then purged with nitrogen. Then, CuBr was added to the flask. The flask was sealed with a glass stopper and evacuated before being back-refilled with nitrogen three times. The reaction mixture was then warmed to room temperature and placed in an oil bath heated to 70 °C to initiate polymerization. The final molar ratios of reaction components in a typical reaction were $[\text{styrene}]_0 : [\text{SiO}_2\text{–Br}]_0 : [\text{CuBr}]_0 : [\text{CuBr}_2]_0 : [\text{PMDETA}]_0 =$

2000 : 1 : 2.5 : 0.25 : 2.75 with a volume fraction of non-reactive solvents of 5.4% dimethylformamide and 40% anisole in a 100 mL flask and stirred at approximately 1000 rpm. The polymerization was stopped by exposing the catalyst to oxygenated tetrahydrofuran after cooling under continuous stirring. The final product was dialyzed against tetrahydrofuran and methanol until the copper(II) catalyst was removed as evidenced by disappearance of its characteristic color.

Styrene (St, Aldrich, 99%) was purified by passing through a basic alumina column before use. Copper(I)bromide was prepared by reduction of an aqueous solution of $CuBr_2$ with an aqueous solution of ascorbic acid, and copper(I)chloride was prepared by reduction of $CuCl_2$ aqueous solution using an aqueous solution of sodium sulfite. Both copper halides were then sequentially filtered, washed with methanol, dried and stored under vacuum before use. Silica nanoparticles (SiO_2NP), 30% solution in isopropanol, effective diameter, $d_{TEM} \cong 113.2$ nm, were donated by Nissan Chemical Corporation and used as received. 5-Hexen-1-ol (98%), α-bromoisobutyryl bromide (98%), triethoxysilane (95%), ethyl 2-bromoisobutyrate (EBiB, 98%), 4,4′-dinonyl-2,2′-bipyridine (dNbpy, 99%), *N*,*N*,*N*′,*N*′′,*N*′′-pentamethyldiethylenetriamine (PMDETA, 99%), and anisole (99%) were purchased from Aldrich and used as received. All other chemicals and solvents were supplied by Aldrich and Acros Organics.

The dried product of the polymerization was redispersed in toluene, sonicated for 1 h, then stirred for 24 h before centrifugation. The centrifugation was carried out gently until the large particle brushes had sedimented at the bottom where they were collected. Removal of the free homopolymer was verified by monitoring the interparticle distance *via* TEM.

Molecular weight and dispersity were measured by size exclusion chromatography (SEC) using a Waters 515 pump and Waters 2414 differential refractometer using PSS columns (Styrogel 10^5, 10^3, and 10^2 Å) in THF as an eluent (35 °C, flow rate of 1 mL min^{-1}) with toluene and diphenyl ether used as internal references. A linear polystyrene (PS) standard was used for calibration. To perform SEC, chains were cleaved from particles by etching of particles in HF in a polypropylene vial for 20 h, neutralized with ammonium hydroxide, and dried with magnesium sulfate before running SEC. Hydrofluoric acid (50 vol% HF) was purchased from Acros Organics and used as received. THF was purchased from Aldrich and used as received.

Assessment of the grafting density and inorganic content of particles was made using weight fractions measured from thermogravimetric analysis (TGA) on a Q50 TGA analyzer from TA Instruments under nitrogen up to 850 °C. The grafting density was calculated by using the weight fractions measured with TGA converted to the number of polymer chains using the molar mass of the polymer chains (as determined using size exclusion chromatography) and surface area (using a silica density of 2.2 g cm^{-3}).

Transmission electron microscopy (TEM)

Approximately monolayer films were prepared by drop casting dilute solutions (~1 mg mL^{-1}) of particle brushes and homopolymer in xylene onto poly(acrylic acid) (PAA) films. PAA was obtained as a 25 weight percent solution in water from Sigma Aldrich. The use of a low vapor pressure solvent such as xylene allowed the films to better equilibrate during deposition. The films were thermally annealed at 120 °C for 24 h to fully equilibrate. The film was then placed onto a surface of water to allow the PAA substrate to dissolve. Residual particle brush films were suspended on the water surface and were lifted off onto copper grids for analysis. Particle film morphology was studied by transmission electron microscopy (TEM) with a JEOL EX2000 electron microscope operated at 200 kV. Phase and amplitude contrast mode images were taken using a Gatan Orius SC600 high-resolution camera.

Nanoindentation

Elastic modulus and hardness were measured *via* nanoindentation. Measurements were made with an MTS Nanoindenter XP with a Berkovich tip calibrated to a quartz standard under displacement control to no more than 10% of the nanocrystal film thickness using films (measured to be ~50 μm by profilometry). Experimental data for particle brush samples were obtained from at least 20 indentations per sample, and the standard deviation of the measurements was used as experimental error. The displacement rate during the indentation of the particle brush samples was 5 nm s^{-1} to maximum load followed by constant load indentation for 10 seconds.

Atomic force microscopy (AFM)

The residual indentations from nanoindentation were imaged using atomic force microscopy (AFM) on an NT-MDT SolverNEXT system in semi-contact mode with silicon cantilevers (300 kHz resonance frequency, 40 N m^{-1} force constant) of small tip radius (<10 nm). Samples were imaged in height, phase, and deflection imaging modes in order to clearly see radial cracks at the corners of the residual indents. Cracks were most clearly observed in deflection imaging and crack lengths were measured using these images.

Results and discussion

The systems in this study consist of a low dispersity polystyrene (PS) homopolymer with degrees of polymerization, $P = 85$, 442, and 905 (hereafter referred to as '10 kDa', '50 kDa', and '100 kDa', respectively) as well as densely polystyrene-tethered silica particles with a particle core radius $R_0 = 56.6 \pm 6$ nm. Homopolymers were used as purchased from Sigma-Aldrich. Silica particle brushes were synthesized using surface-initiated atom transfer radical polymerization (SI-ATRP) according to previously published protocols.[11] The molecular characteristics of the silica brush particles were respectively: $\sigma = 0.61$ nm^{-2}, $N = 130$, $Đ_M = 1.08$ (Sample ID: 57SiO_2-S130), and $\sigma = 0.49$ nm^{-2}, $N = 1000$, $Đ_M = 1.8$ (Sample ID: 57SiO_2-S1000) where σ is the density of the surface grafted chains, N is the degree of polymerization, and $Đ_M$ the molar-mass dispersity of the surface grafted chains. The grafting density of all particle brush systems in the present study was in the range $\sigma = 0.5$–0.6 nm^{-2}, hence, all systems were considered to be in the dense grafting regime.

Effect of homopolymer addition on structure formation in particle brush solids

To determine the solubility of the homopolymer additive in particle brush solids, the morphology of the thin films of the particle brush/homopolymer blends was evaluated by analysis of the electron micrographs of the thin films of the particle brush/homopolymer mixtures. This methodology was adopted due to the limitation of alternative techniques (such as small-angle X-ray scattering or cross-sectional imaging) that cannot be applied in the present case (due to the strong scattering of particle cores and interference of the large particles with the microtoming process, respectively). Substrate interactions might alter the distribution of components within the film; however, we expect the effect of such interactions to be of limited influence due to the identical chemical composition of brush and linear polymer. For each blend system, ten micrographs from different regions of the film (averaging about 500 μm^2) were evaluated to allow for representative sample analysis. Contrast gradient analysis was applied to differentiate homopolymer domains from void space that might form during film casting and hence to determine the miscibility of the particle brush/homopolymer blend systems. Fig. 2 depicts representative transmission electron micrographs of the structure formation of 57SiO_2-S1000/10 kDa (panels a–d), 57SiO_2-S1000/100 kDa (panels e–h), and 57SiO_2-S130/1000 kDa (i–l).

The micrographs reveal that in the case of the 57SiO_2-S1000 blend systems the particle array morphology is approximately maintained up to high concentrations of homopolymer ($\phi \sim 0.5$) while the formation of 'patch' structures in the case of 57SiO_2-S130/1000 kDa at low homopolymer concentration ($\phi > 0.26$) indicates

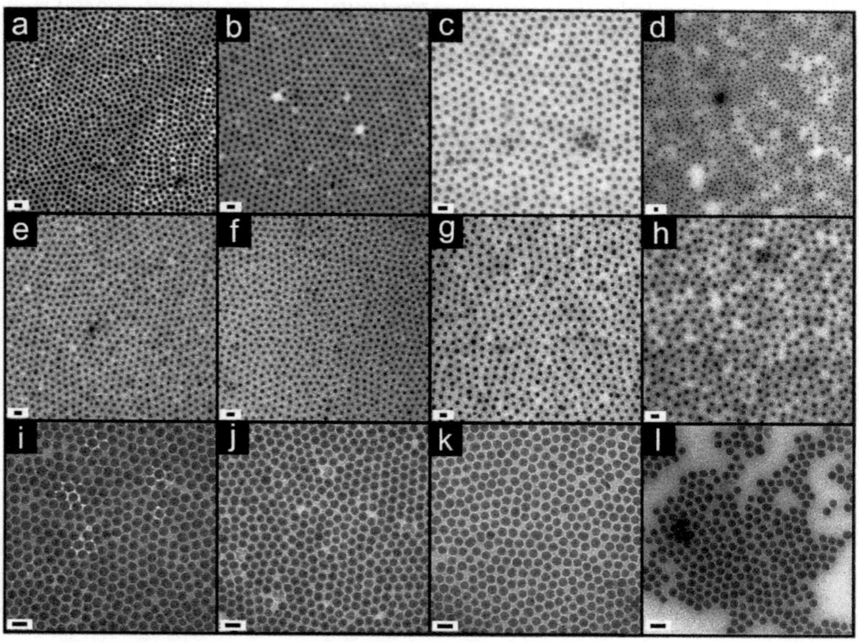

Fig. 2 Representative transmission electron micrographs of 57SiO_2-S1000 blends with (a–d) 10 kDa and (e–h) 100 kDa PS and 57SiO_2-S100 blends with (i–l) 100 kDa PS. Homopolymer volume fractions are (a and i) 0, (j) 0.15, (b, e and k) 0.26, (l) 0.35, (c and f) 0.41, (g) 0.51, and (d and h) 0.58. For the 57SiO_2-S1000 blend systems, a significantly higher solubility of homopolymer is observed as compared to the 57SiO_2-S130 based systems (see text for more details). All scale bars are 250 nm.

a reduced solubility in this latter case. The observed trend bears analogy to prior reports on the miscibility of particle brush fillers in polymer melts that has been found to primarily depend on the ability of matrix (melt) chains to wet (interpenetrate) the brush.[21–23] Pioneering work by Leibler and co-workers revealed the interaction between planar polymer brushes and melts is primarily dependent on two parameters: the degree of polymerization of the tethered chains (N) and matrix (P) chains.[24–26] Specifically, the authors determined (for the case of chemically identical brush and matrix polymers) that wetting of the brush by the matrix is subject to the condition $P < N$. Hence autophobic dewetting (and particle brush aggregation) is expected if the degree of polymerization of the tethered chains is less than the matrix. More recent theoretical studies have extended the Leibler model to account for the effect of surface curvature (such as in the case of particle brushes).[27] Harton and Kumar found that in the case of curved brushes the wetting condition follows the same general trend, however, the upper bound of the degree of polymerization of the matrix chains to facilitate wetting of the brush was found to increase with the curvature of the brush (*i.e.* with reduced particle size).[28] This latter effect was attributed to the reduced chain crowding in high curved (convex) brush systems. The role of brush and matrix degree of polymerization on the brush wetting characteristics is illustrated in Fig. 3.

Although the miscibility of the homopolymer in particle brush films (see Fig. 2) generally follows the predicted trend for particle brush-in-polymer dispersions, it is important to note that the solubility of 100 kDa homopolymer in the $57SiO_2$-S100 brush system up to concentrations of $\phi \sim 0.26$ (see Fig. 2k and l) is surprising since in this case $P \sim 10 \times N$ and hence strong autophobic dewetting is expected. The pronounced miscibility of high molecular homopolymer filler points to additional contributions to the free energy of mixing that are not considered in the classical brush/melt interaction model. In particular, we hypothesize that in particle brush array structures the surface energy associated with interstitial space contributes an additional driving force for mixing that is not accounted for in the classical wetting theory (since the latter has been designed to capture the wetting conditions of dilute particle brush-in polymer dispersions rather than the 'insertion' of polymer chains in a uniform particle brush phase). In a first approximation, the amount of void space in particle brush array structures can be estimated on the basis of a close-packed hard sphere model for which the amount of interstitial space is $\phi_{\text{void}} = 0.26$. The surface energy associated with the polymer/air interfaces across void space provides a strong energetic driving force for

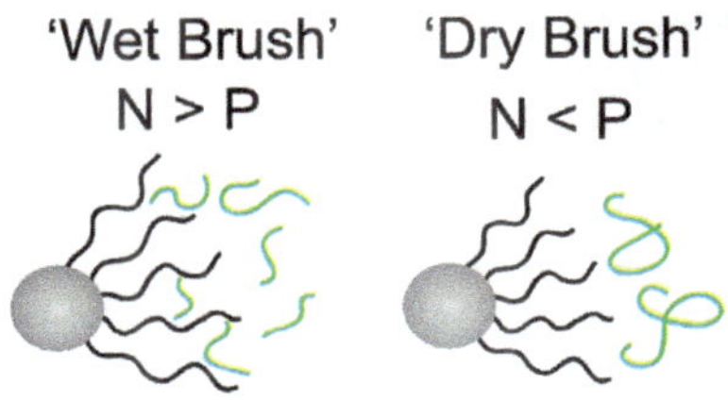

Fig. 3 Wet brush behavior, where the homopolymer easily interpenetrates the graft layer, is expected when the homopolymer chain length, P, is less than the graft length, N ($P < N$). Dry brush behavior, *i.e.* the expulsion of the matrix polymer from the brush is expected when the homopolymer chain length exceeds the graft length ($P > N$).

homopolymer filling of interstitial regions. For example, by considering the typical values for the air/PS interfacial energy ($\gamma_{\text{air-PS}} = 40$ mN m^{-1}) it follows that the surface energy penalty per chain amounts to approximately $10^5 k_B T$ (assuming $N \sim 1000$ and $R_0 = 60$ nm). In particle brush array structures the actual amount of void space will be reduced because of the stretching of tethered chains to fill the interstitial space. This latter process is energetically favored because the energy penalty associated with the reduction of chain conformational entropy during chain stretching amounts to only a small fraction of the energy gain due to the reduction of surface energy.[19] Chain stretching along the interstitial regions (that correspond to the corners of the corresponding Wigner–Seitz cell of the array) can indeed be inferred from the contrast distribution in electron micrographs or pristine particle brush arrays in Fig. 2 where interstitial regions appear as 'brighter regions' (indicating local thinning of the film). Despite this relief of the energy associated with void spaces it can be expected that the presence of 'interstitials' in particle brush arrays creates 'energy hot spots' that will drive the segregation of homopolymer within interstitial regions (to reduce either surface energy or the stored elastic energy in the case of chain stretching). This is supported by previous reports by Ojha *et al.* who evaluated the morphology of blends of asymmetric (*i.e.* large/small) particle brush mixtures and reported small particle brushes to decorate interstitial regions.[19] On the basis of the above arguments we rationalize the solubility of 100 kDa fillers within the 57SiO$_2$-S130 as a consequence of the initial 'segregation' of homopolymer within interstitial regions that is independent of brush/matrix interpenetration. Consistent with this argument we find that the solubility limit of 100 kDa PS is approximately 26% and hence of the same order as the estimated interstitial space in the particle brush lattice.

Support for the conclusion that chain relaxation across interstitial regions drives the distribution of homopolymer in the limit of small filling fractions can

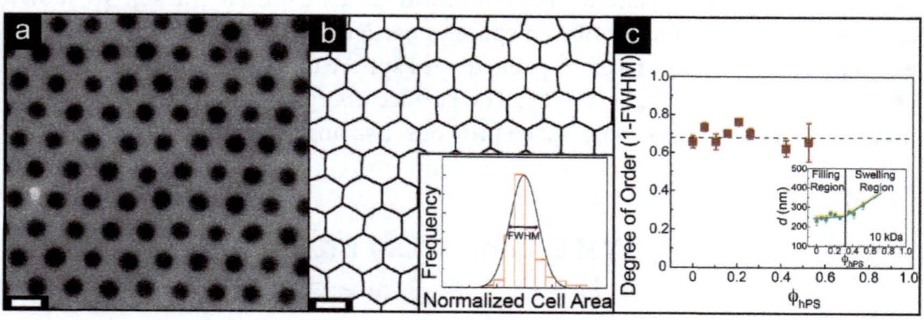

Fig. 4 (a) Representative electron micrograph of the 57SiO$_2$-S1000 array. (b) Corresponding Voronoi diagram used to calculate the degree of order. The degree of order is defined as 1-FWHM, where FWHM is the full-width half-maximum of a Gaussian fitted to the distribution of normalized Voronoi-cell areas calculated from the diagram (shown in inset). (c) Degrees of order for the range of soluble 57SiO$_2$-S1000 and 10 kDa PS composites showing that order is retained for all homopolymer volume fractions. Inset shows that while order is maintained in all systems, interparticle distance increases faster when the homopolymer volume exceeds the expected interstitial volume. Scale bars are 200 nm.

be derived from the analysis of the 'degree of order' of particle brush array structures that is shown in Fig. 4.

The 'degree of order' of the particle brush monolayers (Fig. 4a) was evaluated on the basis of the full-width-at-half-maximum (FWHM) of the normalized Voronoi cell area distribution (Fig. 4b and c) that was introduced by Choi *et al.* as a quality factor to compare order formation in brush particle systems.[29] Analysis of the quality factor (1-FWHM) in the case of the $57SiO_2$-S1000/10 kDa system reveals that order formation is unaffected by the presence of homopolymer filler for all experimental compositions thus confirming the solubility of the 10 kDa additive (consistent with $N \gg P$). Interestingly, the analysis of the nearest neighbor surface-to-surface distance (see inset in Fig. 4c) reveals that the particle distance is approximately constant up to a threshold concentration that corresponds approximately to the estimated volume fraction of the interstitial space beyond which the particle distance increases with filler concentration. This trend of particle distance is consistent with the process of homopolymer dissolution being governed by two steps: the initial segregation of homopolymer within the interstitial regions of the array structure and the subsequent swelling of the brush (provided that $N > P$). Fig. 5 illustrates the proposed two-step dissolution process.

Mechanical properties of the homopolymer filled particle brush solids

To evaluate the effect of homopolymer addition on the mechanical properties of the particle brush solids, particle brush/polymer blends were evaluated by nanoindentation. Nanoindentation has emerged as a versatile technique to determine the micromechanical properties of materials for which macroscopic

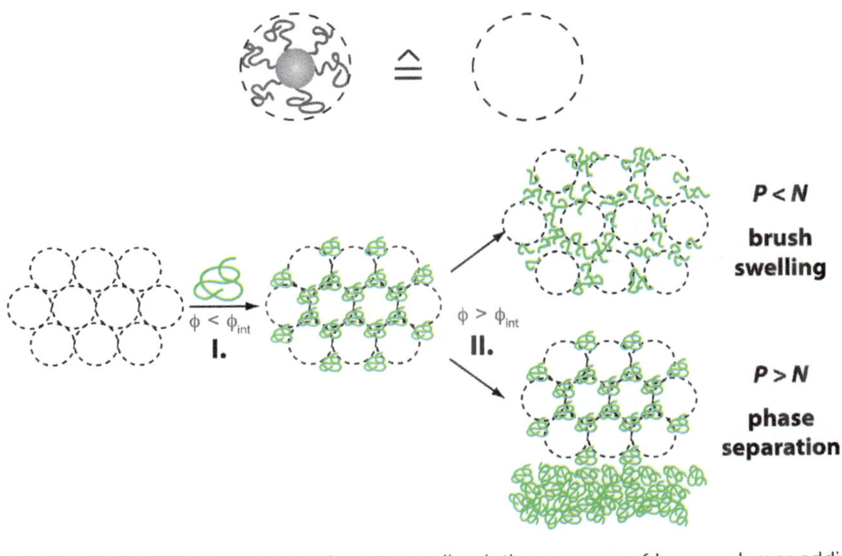

Fig. 5 Illustration of the proposed two-step dissolution process of homopolymer additives in particle brush array structures. In the first step, the homopolymer filler segregates to the interstitial regions of the array structure (independent of *P* and *N*). In the second step, as $\phi > \phi_{int}$ the homopolymer swells the brush (if $N > P$) or phase separates from the mixture (if $N < P$).

mechanical tests cannot be performed (for example because of insufficient sample amount).[30] Films of ∼50 μm thickness were indented to a depth of 2 μm using a Berkovich indenter at a rate of 5 nm s^{-1} (rates from 1–25 nm s^{-1} were tested to exclude the influence of viscoelastic effects on the measurement result). Maximum displacement was limited to less than 10% of the film thickness to preclude substrate effects, for select samples the measurements were performed at varying film thicknesses to ensure consistency of results (data not shown here). Note that for the present experimental conditions, the indentation test sampled a volume of about 50 μm^3 and hence the results are expected to be insensitive to microstructural defects. Fig. 6 depicts typical load–displacement behaviors for a 57SiO_2-S1000/50 kDa blend system.

Because the reliable evaluation of the mechanical properties of films requires sufficient mechanical stability of the films, only the 57SiO_2-S1000 particle brush system was chosen for nanoindentation analysis. Furthermore, 50 kDa PS was selected as the additive to facilitate both matrix/brush compatibility (*i.e.* wet brush conditions) as well as chain entanglement. From the nanoindentation measurements three mechanical parameters can readily be determined under the assumption of elastic deformation behavior: the hardness H (from the maximum load level during indentation), the elastic modulus E (from the initial slope of the unloading curve) as well as the stress intensity factor for mode-1 fracture K_{1c} (from the length of cracks that form at the indenter tip regions).[7,31] The latter provides a measure for the toughness of the material and hence describes the resistance of the material to fracture. Data analysis was performed on the basis of at least 20 indentations per sample to determine average values as well as experimental error. We note that the above analysis is based on the assumption of elastic deformation and hence only applicable to polymers well below the glass transition temperature. To exclude significant contributions from plastic

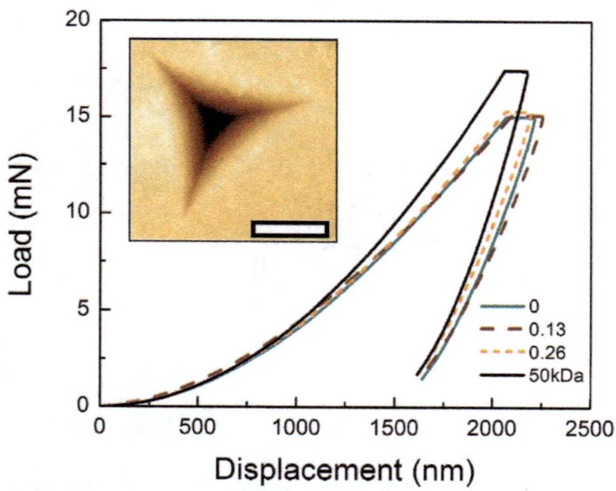

Fig. 6 Characteristic load–displacement curves for the 57SiO_2-S1000/50 kDa composites with different volume fractions of homopolymer (see legend) show similar behavior to the 50 kDa homopolymer filler. Inset shows a residual indent for the pristine 57SiO_2-S1000 system revealing the absence of pile up or plastic deformation. Scale bar is 5 mm.

deformation the morphology of the indents was evaluated by atomic force microscopy (AFM) – the inset in Fig. 6 depicts a representative micrograph of an indent. The lack of pile-up or sink-in confirms the absence of (significant) plastic deformation. To determine the stress intensity factor the length of the cracks emanating from the corners of residual indentation sites were analysed using the following equation first derived by Oliver and Pharr:[31]

$$K_{\mathrm{Ic}} = 0.0161\left(\frac{a}{l}\right)^{\frac{1}{2}}\left(\frac{E}{H}\right)^{\frac{2}{3}}\frac{P_{\max}}{c^{\frac{3}{2}}} \tag{1}$$

where K_{Ic} is the fracture toughness, a is the distance from the center of the residual indent to the corner, l is the length of the crack (from the indent corner to tip), c is the total length of the crack and center-to-corner distance ($a + l$), E is the Young's modulus, H is the hardness, and $P_{\max}$ is the maximum load.[31] The stress intensity factor was calculated based on measurements of at least 10 residual indentations. This analysis was previously used to calculate the fracture toughness of particle brushes with small particle sizes ($R_0 = 8$ nm).[7] All error bars reported are equal to one standard deviation in the measurement value above and below the mean value.

Fig. 7 depicts a summary of the mechanical properties of the $57SiO_2$-S1000/50 kDa blend systems that were determined by indentation experiments.

Fig. 7a reveals that Young's modulus remains approximately constant for all concentrations of added homopolymer and significantly below the respective value of the 50 kDa reference PS (black dashed line). This trend can be interpreted as a consequence of the elastic deformations (*i.e.* Young's modulus) being evaluated in the small strain limit where short-ranged dispersion interactions provide the major contribution to the restoring force. Since dispersion interactions depend on the molecular polarizability and both the tethered and linear polymer exhibit identical chemical composition, no significant change is expected in the limit of small amounts of added filler. We hypothesize that residual effects related to interstitial regions (such as void spaces) are responsible for the modulus of the particle brush/homopolymer films remaining below the corresponding value of

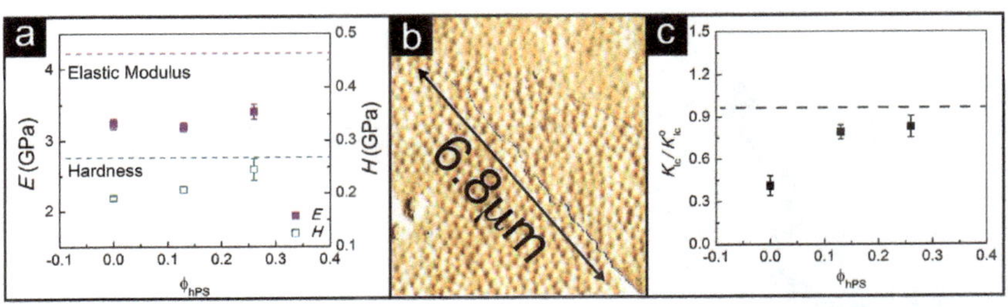

Fig. 7 (a) Young's modulus and hardness of $57SiO_2$-S1000 and 50 kDa composites as a function of homopolymer volume fraction. Modulus and hardness show only weak dependence on the amount of added homopolymer. (b) Fracture cracks propagating from corners of the residual indent as shown for the case of $57SiO_2$-S1000. Crack length is reduced upon homopolymer addition. (c) Normalized mode-1 stress intensity factor (fracture toughness, normalized with respect to the value of 100 kDa homopolymer (dashed line)) for particle brush/homopolymer blend systems. Dashed line corresponds to normalized fracture toughness of 50 kDa polystyrene.

the 50 kDa homopolymer. While elastic constants show only a weak dependence on the homopolymer content, a pronounced increase of the films' fracture toughness by about 300% is observed upon addition of even small amounts of homopolymer (see Fig. 7c). Here it should be noted that fracture toughness (*i.e.* the work required to fracture a material) – in contrast to small strain elastic deformations – sensitively depends on the presence of entanglements.[32,33] We thus rationalize that the pronounced increase of the material fracture toughness occurs with the increase of the materials' entanglement density as the homopolymer additive fills the interstitial regions in the array. The proposed toughening effect of 'decorated interstitial regions' is illustrated in Fig. 8.

An interesting experiment to validate the above proposition would be to evaluate the effect of low molecular polymer fillers for which no entanglement formation is expected (*i.e.* in the limit of $P < N_{entgl}$). Unfortunately, despite extensive experimentation no reliable data could be obtained on these small P systems during the course of the present study (due to the brittleness of films that rendered determination of K_{1c} difficult).

Conclusions

We have analyzed the effect of homopolymer addition on the structure formation and mechanical properties of particle brush solids. In the limit of small volume fraction, homopolymer additives are soluble within the particle brush solid even for 'dry brush' compositions. The solubility of high molecular weight polymers that are otherwise expected to be immiscible within the brush is interpreted as a consequence of a strong driving force to fill 'interstitial regions' within the particle brush solid that derives from either surface energy contributions (in the case of short-chain brushes) or the recovery of relaxed chain conformations (in the case of long-chain brushes in which chain stretching prevents the formation of void space in the interstitial region). The addition of small amounts of homopolymer corresponding to only a fraction of the interstitial volume significantly raises the toughness of particle brush solids indicating chain entanglement of polymer chains that are segregated within the interstitial regions. The ability to raise the material's fracture toughness at low concentrations of added

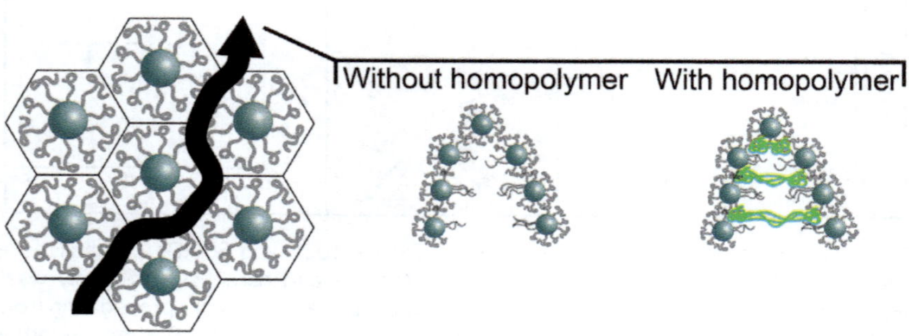

Fig. 8 Illustration of the homopolymer-induced toughening effect. As a crack (shown by the thick black line) propagates through the material, the presence of entangled homopolymer in the interstitial regions introduces microscopic flow processes (*i.e.* crazing) that stabilize the crack against further propagation.

filler (where microstructural order is unaffected by the filler addition) promises new opportunities for designing particle-brush based materials that combine the favorable ability of forming ordered assembly structures of dense brush particles with polymer-like processability and mechanical robustness. Future work will test alternative approaches to facilitate this goal, for example, by means of bimodal particle brush systems.[34]

Acknowledgements

M. R. B. and K. M. acknowledge financial support by the National Science Foundation (*via* DMR-1410845 and DMR-1501324) as well as the Department of Energy (*via* DE-EE0006702). M. S. gratefully acknowledges financial support from the John and Clare Bertucci Graduate Fellowship.

References

1 C. M. Hui, J. Pietrasik, M. Schmitt, C. Mahoney, J. Choi, M. R. Bockstaller and K. Matyjaszewski, *Chem. Mater.*, 2014, **26**, 745–762.
2 N. J. Fernandes, H. Koerner, E. P. Giannelis and R. A. Vaia, *MRS Commun.*, 2013, **3**, 13–29.
3 S. K. Kumar, N. Jouault, B. Benicewicz and T. Neely, *Macromolecules*, 2013, **46**, 3199–3214.
4 S. L. Hilburg, A. N. Elder, H. Chung, R. L. Ferebee, M. R. Bockstaller and N. R. Washburn, *Polymer*, 2014, **55**, 995–1003.
5 P. Voudouris, J. Choi, N. Gomopoulos, R. Sainidou, H. Dong, K. Matyjaszewski, M. R. Bockstaller and G. Fytas, *ACS Nano*, 2011, **5**, 5746–5754.
6 J. Choi, H. Dong, K. Matyjaszewski and M. R. Bockstaller, *J. Am. Chem. Soc.*, 2010, **132**, 12537–12539.
7 J. Choi, C. M. Hui, J. Pietrasik, H. Dong, K. Matyjaszewski and M. R. Bockstaller, *Soft Matter*, 2012, **8**, 4072–4080.
8 C. A. Grabowski, H. Koerner, S. Meth, A. Dang, C. M. Hui, K. Matyjaszewski, M. R. Bockstaller, M. F. Durstock and R. A. Vaia, *ACS Appl. Mater. Interfaces*, 2014, **6**, 21500–21509.
9 E. Alonso-Redondo, M. Schmitt, Z. Urbach, C. M. Hui, R. Sainidou, P. Rembert, K. Matyjaszewski, M. R. Bockstaller and G. Fytas, *Nat. Commun.*, 2015, **6**(8309), 1–8.
10 R. Barbey, L. Lavanant, D. Paripovic, N. Schuewer, C. Sugnaux, S. Tugulu and H.-A. Klok, *Chem. Rev.*, 2009, **109**, 5437–5527.
11 R. Advincula, *Adv. Polym. Sci.*, 2006, **197**, 107–136.
12 Y. Tsujii, K. Ohno, S. Yamamoto, A. Goto and T. Fukuda, *Adv. Polym. Sci.*, 2006, **197**, 1–45.
13 V. Coessens, T. Pintauer and K. Matyjaszewski, *Prog. Polym. Sci.*, 2001, **26**, 337–377.
14 K. Matyjaszewski, *Macromolecules*, 2012, **45**, 4015–4039.
15 K. Matyjaszewski and N. V. Tsarevsky, *Nat. Chem.*, 2009, **1**, 276–288.
16 K. Matyjaszewski and N. V. Tsarevsky, *J. Am. Chem. Soc.*, 2014, **136**, 6513–6533.
17 K. Matyjaszewski and J. Xia, *Chem. Rev.*, 2001, **101**, 2921–2990.
18 C. M. Hui, A. Dang, B. Chen, J. Yan, D. Konkolewicz, H. He, R. Ferebee, M. R. Bockstaller and K. Matyjaszewski, *Macromolecules*, 2014, **47**, 5501–5508.

19 S. Ojha, B. Beppler, H. Dong, K. Matyjaszewski, S. Garoff and M. R. Bockstaller, *Langmuir*, 2010, **26**, 13210–13215.

20 J.-S. Wang and K. Matyjaszewski, *J. Am. Chem. Soc.*, 1995, **117**, 5614–5615.

21 S. Ojha, A. Dang, C. M. Hui, C. Mahoney, K. Matyjaszewski and M. R. Bockstaller, *Langmuir*, 2013, **29**, 8989–8996.

22 S. K. Kumar, N. Jouault, B. Benicewicz and T. Neely, *Macromolecules*, 2013, **46**, 3199–3214.

23 J. F. Moll, P. Akcora, A. Rungta, S. Gong, R. H. Colby, B. C. Benicewicz and S. K. Kumar, *Macromolecules*, 2011, **44**, 7473–7477.

24 A. Gast and L. Leibler, *Macromolecules*, 1985, **19**, 686–691.

25 I. Borukhov and L. Leibler, *Macromolecules*, 2002, **35**, 5171–5182.

26 I. Borukhov and L. Leibler, *Phys. Rev. E: Stat. Phys., Plasmas, Fluids, Relat. Interdiscip. Top.*, 2000, **62**, R41–R44.

27 D. Meng, S. K. Kumar, J. M. D. Lane and G. S. Grest, *Soft Matter*, 2012, **8**, 5002–5010.

28 S. E. Harton and S. K. Kumar, *J. Polym. Sci., Part B: Polym. Phys.*, 2008, **46**, 351–358.

29 J. Choi, C. M. Hui, M. Schmitt, J. Pietrasik, S. Margel, K. Matyjazsewski and M. R. Bockstaller, *Langmuir*, 2013, **29**, 6452–6459.

30 A. C. Fischer-Cripps, *Nanoindentation*, Springer, New York, 3rd edn, 2011.

31 W. C. Oliver and G. M. Pharr, *J. Mater. Res.*, 2004, **19**, 3–20.

32 W. Graessley, *Adv. Polym. Sci.*, 1974, **16**, 1–179.

33 E. J. Kramer and L. L. Berger, *Adv. Polym. Sci.*, 1990, **91/92**, 1–68.

34 J. Yan, T. Kristufek, M. Schmitt, Z. Wang, G. Xie, A. Dang, C. Hui, J. Pietrasik, M. Bockstaller and K. Matyjaszewski, *Macromolecules*, 2015, **48**, 8208–8218.

Faraday Discussions

PAPER

Soft-shear induced phase-separated nanoparticle string-structures in polymer thin films

Ren Zhang,[a] Bongjoon Lee,[b] Michael R. Bockstaller,[b] Abdullah M. Al-Enizi,[c] Ahmed Elzatahry,[de] Brian C. Berry[f] and Alamgir Karim*[a]

Received 8th September 2015, Accepted 21st September 2015
DOI: 10.1039/c5fd00141b

Application of shear stress has been shown to unidirectionally orient the microstructures of block copolymers and polymer blends. In the present work, we study the phase separation of a novel nanoparticle (NP)–polymer blend thin film system under shear using a soft-shear dynamic zone annealing (DZA-SS) method. The nanoparticles are densely grafted with polymer chains of chemically dissimilar composition from the matrix polymer, which induces phase separation upon thermal annealing into concentrated nanoparticle domains. We systematically examine the influence of DZA-SS translation speed and thus the effective shear rate on nanoparticle domain elongation and compare this with the counterpart binary polymer blend behavior. Unidirectionally aligned nanoparticle string-domains are fabricated in the presence of soft-shear in confined thin film geometry. We expect this DZA-SS method to be applicable to various NP–polymer blends towards unidirectionally aligned nanoparticle structures, which are important to functional nanoparticle structure fabrication.

Introduction

Polymer–nanoparticle (NP) hybrid materials attract extensive research interest in nanotechnology since they offer a unique opportunity to engineer novel materials resulting from the combination of the nanoscale properties of NPs and the

[a]Department of Polymer Engineering, The University of Akron, 250 S Forge St, Akron, OH 44325, USA. E-mail: alamgir@uakron.edu; Tel: +1-330-972-8324

[b]Department of Materials Science and Engineering, Carnegie Mellon University, 5000 Forbes Ave, Pittsburgh, PA 152133, USA

[c]Chemistry Department, Faculty of Science, King Saud University, PO Box 2455, Riyadh 11451, Saudi Arabia

[d]Materials Science and Technology Program, College of Arts and Sciences, Qatar University, PO Box 2713, Doha, Qatar

[e]Polymer Materials Research Department, Advanced Technology and New Materials Research Institute, City of Scientific Research and Technology Applications, New Borg El-Arab City, Alexandria 21934, Egypt

[f]Department of Chemistry and Applied Science Program, University of Arkansas at Little Rock, College of Arts, Letters & Sciences, 2801 S University Avenue, Little Rock, AR 72204, USA

mechanical properties and processability of polymer materials. With various novel nanoparticle synthesis and functionalization techniques, precisely designed polymer grafted nanoparticles are promising building blocks towards versatile nanoparticle structures with advanced photonic, electronic, optical and magnetic properties.[1–5] Polymer grafted nanoparticles have been embedded in a wide range of polymer matrices and the distribution of NPs was successfully manipulated. For instance, nanoparticles were selectively segregated in phase-separated block copolymer templates[6,7] or supramolecular frameworks;[8] a "self-corralled" nanoparticle superlattice was generated by unfavorable enthalpic interactions between a polymer brush and a chemically dissimilar polymer matrix;[9] a jamming morphology was observed for NP filled phase-separated polymer blends.[10,11] One simple example of a NP–polymer blend is the case where the grafted polymer ligand is composed of the same chemical composition as the free polymer matrix chain, which is the so-called "athermal" blend. In this situation, the dispersion behavior of nanoparticles in a homopolymer matrix is determined by the NP size relative to the free polymer chains, grafting density, and the ratio of the degree of polymerization of the polymer matrix and brush.[12,13] Generally, the dispersion of nanoparticles is promoted when the nanoparticle size is smaller than the radius of the gyration of the free polymer chains because of the reduced elastic stretching of the polymer matrix chains and the enhanced translation entropy of the nanoparticles.[13,14]

Nanoparticle–polymer blend systems where the grafted polymer chains are of chemically dissimilar compositions from the polymer matrix are less studied, yet the enthalpic interactions between the grafted and free polymer chains can potentially be utilized to control the assembly structures of nanoparticles. As has been demonstrated in binary polymer blend films, diverse phase-separated morphologies were observed depending on the composition and annealing conditions.[15,16] More importantly, to fabricate novel functional materials such as photonic bandgap materials, nanostructured solar cells, and high-density magnetic storage media, it is essential to control nanoparticle assembled structures. We have previously demonstrated that a soft-shear dynamic zone annealing (DZA-SS) process is an efficient way to fabricate unidirectionally aligned structures of phase-separated block copolymers (BCPs) and NP–BCP blend films.[17,18] The shear stress during the DZA-SS process is generated by local thermal expansion and contraction of a confining soft elastomer capping layer (as it traverses the temperature gradient) at the polymer film surface, which is termed "soft shear". The more significant deformation of the elastomer capping layer defining the shear direction occurs in the direction of the DZA thermal gradient or sample motion direction rather than the orthogonal direction, which generates unidirectionally aligned phase separated BCP structures together with nanoparticles. In this study, we investigated a novel binary thin film system composed of densely grafted gold nanoparticles with polystyrene (PS, $M_{n,PS} = 11.5$ kg mol^{-1}) ligands (hereafter denoted as AuPS) and a low molecular mass poly(methyl methacrylate) (PMMA, $M_{n,PMMA} = 3.1$ kg mol^{-1}) homopolymer matrix. The average radius of the gold core (R_0) is 1.21 nm and the PS grafting density is 0.7 nm^{-2}. The unfavorable interaction between the PMMA matrix and the PS brush results in phase-separated morphologies where AuPS nanoparticles form discrete domains in the PMMA matrix while the well-dispersed state of the NPs is maintained within the domain structures. DZA-SS induced oscillatory soft-shear was applied to these AuPS/PMMA blend films at various conditions and compared with binary PS/PMMA blend films with $M_{n,PS} = 11.5$ kg mol^{-1}. The correlation between the translation speed (v) during the DZA-SS process and the deformation morphology was established considering a conceptual "effective shear rate" ($\dot{\gamma}_E$). Stabilization of the highly elongated AuPS "string" morphology was achieved due to the shear flow and finite size effect, beyond the expected balance of the elastic restoring interfacial tension of the nanoparticle domain against the elongating shear flow. Furthermore, prior to coalescence into lines, the nanoparticle domains line-up on average in straight-line fashion presumably to reduce frictional/hydrodynamic interactions between multi nanoparticle domains.

Results and discussion

The initial distribution state of AuPS nanoparticles in the PMMA matrix was examined by top-view transmission electron microscopy (TEM) as shown in Fig. 1a. This random in-plane dispersion of the nanoparticles induced by good solubility of the AuPS nanoparticles in toluene solvent provided a well-defined initial state. Upon vacuum oven annealing at 180 °C, which is significantly higher than the glass transition temperatures (T_g) of both PS ($T_{g,PS} \approx 90$ °C) and PMMA ($T_{g,PMMA} \approx 105$ °C), for 16 h the 20% AuPS/PMMA blend films demonstrated a phase separated morphology. As shown in Fig. 1b, the AuPS nanoparticles segregated together and formed discrete circular domains in the PMMA matrix. It is worth noting that even in these phase-separate domains, the nanoparticles are in a well-dispersed state without aggregation or crystallization. The surface morphology of the phase-separated AuPS/PMMA blend films was characterized by atomic force microscopy (AFM). As shown in Fig. 2a and b, both PMMA and AuPS phases were observed at the smooth film surface with root-mean-square roughness $R_{rms} \approx 0.76$ nm. By comparison, the phase-separated morphology of the 20% PS/PMMA blend films was studied where the molecular weight of the linear PS chains was the same as the grafted PS chains ($M_{n,PS} = 11.5$ kg mol^{-1}). Fig. 2c and d show the AFM height and phase images of these binary polymer blends where similar phase separated morphology was observed with circular PS domains in the PMMA matrix. The film surface is also smooth with $R_{rms} \approx 1.83$ nm. This

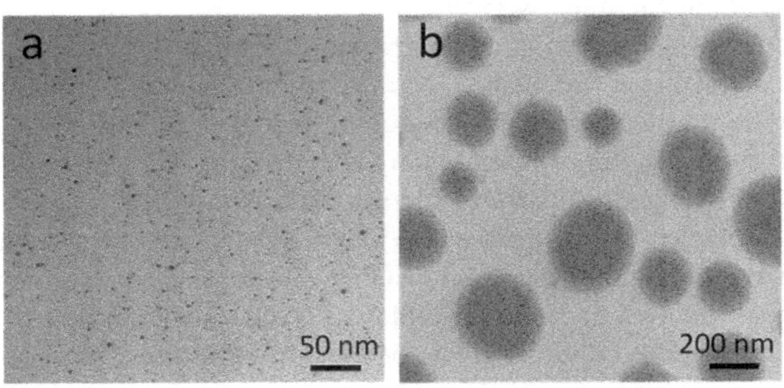

Fig. 1 Top-view TEM micrographs of (a) as-cast 20% AuPS/PMMA blend films and (b) phase-separated 20% AuPS/PMMA blend films after thermal annealing at 180 °C for 16 h. The dark circular domains are AuPS-rich domains while the light continuous phase is composed of PMMA.

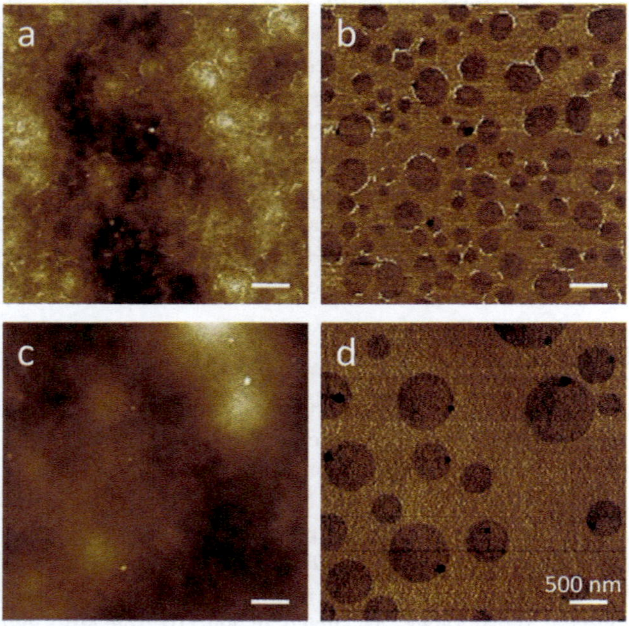

Fig. 2 Comparison of 20% AuPS/PMMA (a and b) and 20% PS/PMMA (c and d) blend film phase-separated morphologies after thermal annealing at 180 °C for 16 h. (a) and (c) are AFM height images; (b) and (d) are corresponding AFM phase images.

analogous phase separation behavior of the AuPS/PMMA blends and binary polymer blends stems from the common thermodynamic driving force. In both cases, it is the unfavorable enthalpic interaction (Flory–Huggins interaction parameter $\chi_{PS-PMMA} \sim 0.037$,[19] 180 °C) that drives the phase separation process, considering that the sufficiently high grafting density of the nanoparticles effectively shields the Au cores.[20] The major difference between the AuPS/PMMA and PS/PMMA phase-separated structures is that the phase separation scale and average domain size (S) of the binary polymer blends is significantly larger compared to NP–polymer blends, which might be correlated to the collective higher viscosity of the densely grafted AuPS nanoparticles[21] as well as configuration differences of the nanoparticle grafted PS chains.

To exert control over the phase-separated structures of these blend films and achieve symmetry breaking of the isotropic circular domains, shear stress was applied to these blend films though a soft-shear dynamic zone annealing (DZA-SS) process. As schematically illustrated in Fig. 3, the as-cast AuPS/PMMA blend film on a quartz substrate is laminated with a smooth poly(dimethyl siloxane) (PDMS) capping layer. This assembly is translated across the in-plane temperature gradient profile defined by cold–hot–cold zones. The translation speed (ν) ranging from 1 μm s^{-1} to 140 μm s^{-1} is prescribed by a programmed movable arm. The maximum temperature is controlled at $T_{max} \approx 180$ °C. When the PDMS confined blend film is translated across the thermal gradient field and the temperature rises above T_g, the expansion and contraction of the PDMS generates shear stress in the sample translation direction at the surface of the NP–polymer blend film.

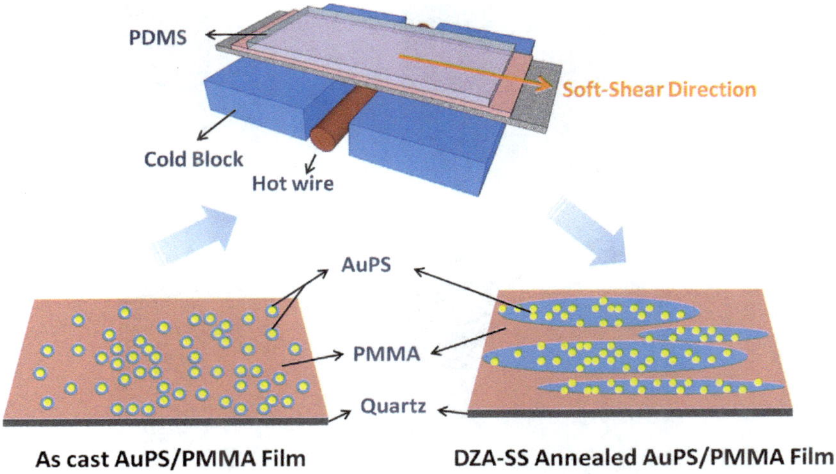

Fig. 3 Schematic of the soft-shear dynamic zone annealing (DZA-SS) process to generate unidirectionally aligned elongated AuPS nanoparticle domain structures.

This drives simultaneous phase separation and NP-domain elongation. As the confined blend films are subjected to temperature increases and decreases centered around $T_{\max}$, the shear stress increases and decreases accordingly, composing an oscillatory shear cycle. The shear rate ($\dot{\gamma}$) variation induced by a shear stress (τ) change is dependent on the thermal gradient field, which is controlled and the same for all the blend films in the current study. However, the final blend film morphology can be effectively tuned by changing the sample translation speed (v) utilizing the synergistic interaction between the phase separation kinetics and the oscillatory shear rate ($\dot{\gamma}$).

The influence of translation speed (v) during the DZA-SS process on the final phase-separated morphology of 20% AuPS/PMMA blend films is demonstrated in Fig. 4. Clearly, the AuPS domain deformation is enhanced in the soft-shear direction as v increases from 1 $\mu m\ s^{-1}$ to 60 $\mu m\ s^{-1}$. At a low translation speed of 1 $\mu m\ s^{-1}$, the circular AuPS domains were slightly deformed with an average aspect ratio (AR) of 1.43 (Table 1), with no spatial distribution preference of the nanoparticle domains. The domain shapes and in-plane arrangement were almost the same as in the vacuum oven annealed case (Fig. 1). As v increased to 20 $\mu m\ s^{-1}$, the average AuPS domain size decreased slightly ($S \approx 0.053\ \mu m^2$) and the elongation of nanoparticle domains became more pronounced with AR = 2.5. Interestingly, the AuPS domains were lined up in the soft-shear direction, forming a "necklace-like" structure. Further increase in speed ($v = 40\ \mu m\ s^{-1}$ and 60 $\mu m\ s^{-1}$) resulted in coexistence of small discrete drops ($S \approx 0.02\ \mu m^2$) and highly elongated nanoparticle domains in the soft-shear direction. This extremely elongated morphology under high shear rate is referred to as "strings".[22] The domains are also roughly aligned in straight lines on average, suggesting a minimization of the friction/hydrodynamic interactions between domains. In all the above cases, the initial high dispersion of the nanoparticles is retained after the DZA-SS process (Fig. 4b). Previous studies have shown the formation of a necklace of particles in viscoelastic fluid under shear flow, which was attributed

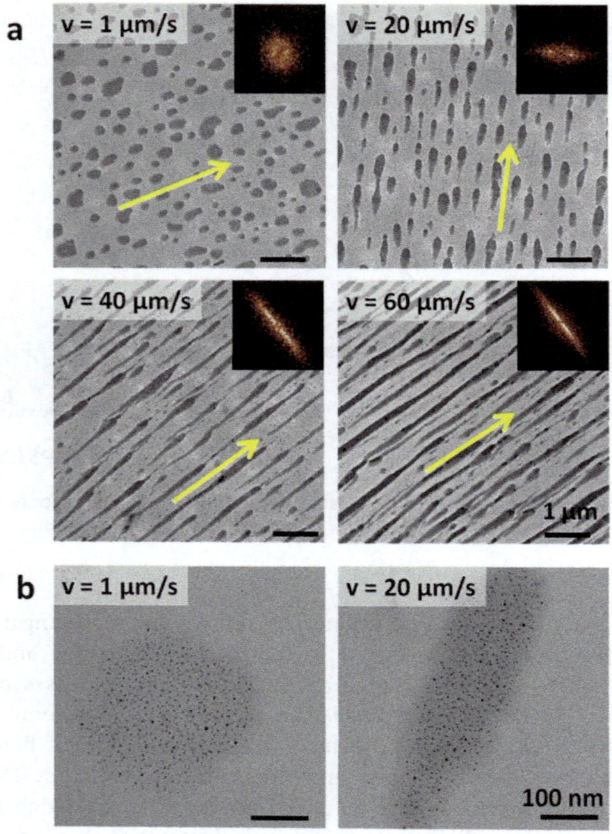

Fig. 4 (a) Top-view TEM micrographs for DZA-SS induced morphologies of 20% AuPS/PMMA blend films processed at $v = 1$, 20, 40 and 60 μm s^{-1}. The arrow denotes the soft-shear direction. The right corner insets are the corresponding FFT images. (b) Magnified TEM micrographs of 20% AuPS/PMMA films processed at $v = 1$ and 20 μm s^{-1}.

to normal stress effects.[23,24] Particle alignment was generated at low volume fractions, indicating enhanced hydrodynamic interactions in viscoelastic fluid. For Newtonian polymer blends, Migler *et al.*[25] also reported a "necklace" structure during a droplet-string morphology transition. Alignment of the phase-separated polymer blend domains was observed when the droplet size was comparable to the gap width between the shearing surfaces. They associated this "necklace"

Table 1 Geometric characteristic comparison of the DZA-SS processed blend films

Characteristics	AR		S (μm^2)		W (nm)
v (μm s^{-1})	1	20	1	20	60
20% AuPS/PMMA	1.43 ± 0.36	3.06 ± 1.11	0.064 ± 0.041	0.053 ± 0.033	114 ± 16
20% PS/PMMA	1.72 ± 0.66	2.03 ± 0.84	0.79 ± 0.45	0.15 ± 0.10	161 ± 46

structure, similar to the current morphology at $v = 20\ \mu m\ s^{-1}$, with a distortion of the velocity field in the presence of the walls. By comparison, binary polymer blend films (20% PS/PMMA) were also subjected to soft-shear dynamic zone annealing (DZA-SS) under the same conditions ($T_{max} \approx 180\ °C$, at $v = 1$, 20 and 60 $\mu m\ s^{-1}$). Fig. 5 displays the AFM height images of the PS/PMMA blend films upon shear annealing by the DZA-SS process. The PS phase was selectively removed by immersion into cyclohexane to enhance the contrast between the PS and PMMA phases. Similar to 20% AuPS/PMMA blend films, the binary polymer blends demonstrated a morphology transition from weakly deformed droplets, to a "necklace-like" pattern of squashed domains, to highly elongated "strings" in the soft-shear direction as the translation speed increased.

To quantitatively compare the effect of soft-shear on NP–polymer and binary polymer blend films, the average domain aspect ratio (AR), domain size (S), and "string" width (W) were calculated by binarizing the TEM or AFM images, as listed in Table 1. At low translation speed ranges ($v = 1$ and 20 $\mu m\ s^{-1}$), the discrete domains were squashed and AR increased with increasing speed. The deformation trend is more pronounced for NP–polymer blends. The average domain size and string width in AuPS/PMMA blends were significantly smaller than those in PS/PMMA blends, which might be attributed to higher viscosity and surface tension modification due to the presence of multiple PS chains surrounding an Au core. Assuming that the discrete elongated domains are ellipse-shaped, the average minor axis dimension (b) perpendicular to the soft-shear direction can be extracted from the S values and thereby also obtains the average domain widths ($W = 2b$). The correlation between W and translation speed is displayed in Fig. 6 for 20% AuPS/PMMA and 20% PS/PMMA blend films. Generally, the average domain width decreases as the speed increases and the larger PS domain width at low speeds gradually converges with the AuPS domain width at higher speeds. In addition, the domain width distribution becomes narrower as v increases. Furthermore, soft shear imposes a certain length scale on the highly stretched structure in the shear direction, which is the characteristic inter-domain spacing (L^*).[21] The translation speed dependence of the phase-separated morphology coarsening is quantitatively analyzed by 2D Fourier transformation (FFT) of the TEM and AFM images. The characteristic inter-domain spacing (L^*) is determined from the maximum intensity wavenumber of the radial power spectrum. For both blend systems, an increase in speed induced a decrease in L^* because the

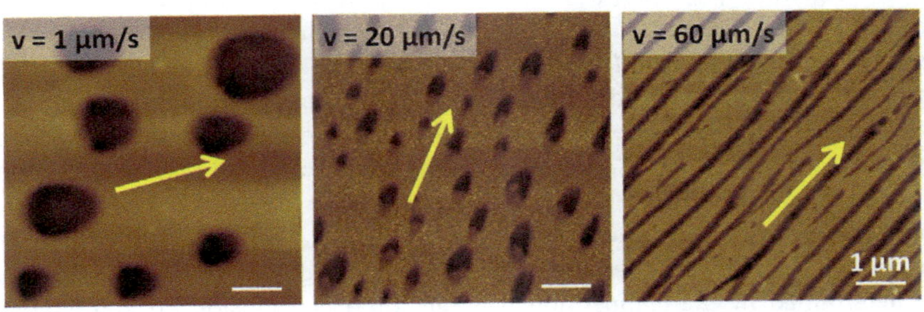

Fig. 5 AFM height images for DZA-SS induced morphologies of 20% PS/PMMA blend films processed at $v = 1$, 20 and 60 $\mu m\ s^{-1}$. The arrow denotes the soft-shear direction.

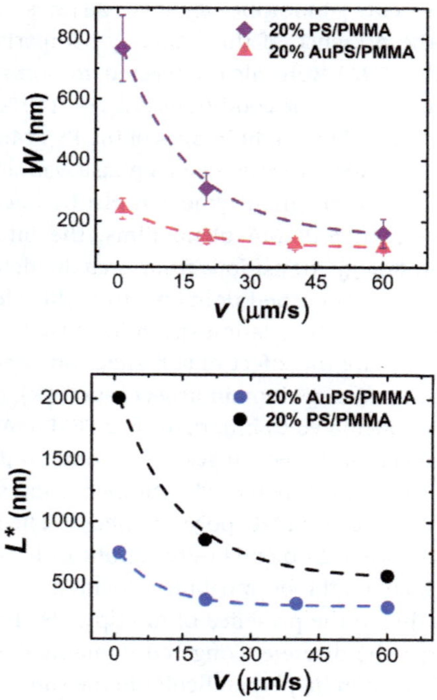

Fig. 6 DZA-SS translation speed dependence of the average domain width (*W*) orthogonal to the soft-shear direction and characteristic inter-domain spacing (*L**) for 20% AuPS/PMMA and 20% PS/PMMA blend films.

equivalent annealing time above T_g is shortened (Fig. 6). In accordance with the domain width transition trends, AuPS/PMMA blend films show a shallower decrease to an almost constant value of L^*, in contrast to the steep decline in PS/PMMA blend films as v increases.

To rationalize the soft-shear induced domain deformation transitions, the maximum shear rate ($\dot{\gamma}_{max}$) during one oscillatory shear period in the DZA-SS process is estimated by $\dot{\gamma}_{max} = \tau_{max}/\eta_m$, where τ_{max} is the maximum shear stress and η_m is the viscosity of the polymer matrix (*i.e.* PMMA). The maximum shear stress (τ_{max}) induced by a thermal expansion mismatch of PDMS layers and polymer blend films has been previously estimated as $\tau_{max} \approx 12 \times 10^4$ Pa based on the equation $\tau_{max} = G_{PDMS} \times \varepsilon$, where G_{PDMS} is the shear modulus of PDMS and ε is the maximum strain.[17] In the current blend systems, the molecular weights of the PS ligands ($M_n = 11.5$ kg mol^{-1}) and PMMA matrix chains ($M_n = 3.1$ kg mol^{-1}) are lower than the entanglement molecular weights, respectively.[26,27] Consequently, the viscosities can be evaluated by a zero shear rate viscosity, where $\eta_{PS\ (11.5k)} \approx 210$ Pa s and $\eta_{PMMA\ (3.1k)} \approx 2680$ Pa s at 180 °C.[27,28] Assuming that the viscous behavior of the blends is mainly responsible for the continuous matrix phase,[29] the maximum shear rate $\dot{\gamma}_{max} = \tau_{max}/\eta_{PMMA} \approx 45$ s^{-1}. When bulk immiscible blends are placed in a shear flow field, the deformation of the droplets is determined by the competition between interfacial tension (σ) and viscous shear stress (τ).[30] Interfacial tension tends to maintain spherical shaped droplets while shear stress exposes a deformation trend in the shear direction. The relative effect of viscous shear stress and interfacial tension is described by the capillary number, $Ca = \eta_m \dot{\gamma} R/\sigma$, where η_m, $\dot{\gamma}$, R, and σ denote the matrix viscosity, shear rate, droplet radius, and interfacial tension, respectively.[31] When Ca exceeds the critical capillary number (Ca_{cr}), the droplets become unstable and break up, and the critical stable droplet radius is determined to be $R^* = \eta_m \dot{\gamma}/Ca_{cr}\sigma$.[32] For the current PS/PMMA blends, the viscosity ratio of the dispersed phase to the matrix phase is $\eta_{PS\ (11.5k)}/\eta_{PMMA\ (3.1k)} \approx 0.078$, and thus the critical capillary number is estimated to be $Ca_{cr} \approx 0.67$.[31] Based on $\eta_{PMMA\ (3.1k)} \approx 2680$ Pa s, $\dot{\gamma}_{max} \approx 45\ s^{-1}$ and $\sigma_{PS/PMMA} = 1.26$ mJ m^{-2} at 180 °C,[33] the corresponding critical droplet radius is estimated to be $R^* \approx 7$ nm. However, the observed average elongated domain width is substantially larger than $2R^*$ (Fig. 6). This discrepancy (stabilization of strings) is attributed to both the shear flow and finite size effects. In the presence of shear flow, the varicose fluctuations of the elongated droplets can be convected away, thus suppressing the pressure differences and preventing Rayleigh instability growth.[34] Furthermore, thin film geometry also facilitates the formation of remarkably elongated "strings". Because the in-plane width of the elongated domains is larger than the film thickness (~90 nm), the 3D configuration of the domains is non-axisymmetric "ribbon-like" instead of axisymmetric "string-like". The varicose fluctuations induce a non-axisymmetric sinusoidal thread configuration, which requires a higher wavelength distortion in order to generate a net decrease in surface area.[35] Consequently, the highly elongated "ribbon" domains are stable with respect to capillary instability.

The above discussion only considered the influence of maximum shear rate ($\dot{\gamma}_{max}$) during the DZA-SS process. However, the final microstructures of the blend films are not determined only by $\dot{\gamma}_{max}$, but also by the full shear rate profile since the shear rate varies as the temperature changes across the entire thermal gradient profile. Above T_g, the shear rate increases and decreases as the temperature rises and falls, centered around T_{max} in response to the expansion and contraction of the PDMS capping layer.[17] The final morphology of the AuPS or PS domains is determined by the integrated shear rate over one oscillatory DZA-SS shear cycle, which is correlated to the translation speed v. If the translation speed is sufficiently low (*e.g.* $v = 20$ μm s^{-1}), the phase separated domains will deform accordingly as the shear rate increases followed by a symmetric decrease, and after a shear cycle, slightly elongated domain morphology is observed. Upon increasing the translation speed (*e.g.* $v = 60$ μm s^{-1}), the final morphology is reflective of a higher overall "effective" shear rate since the elongated structures get frozen as the blend film rapidly traverses the thermal profile. A further increase in the translation speed may even generate less deformed structures as the blend systems cannot adjust the phase organization kinetics as fast as the shear rate variation. We then conclude that the variation of translation speed of the DZA-SS process can influence the "effective shear rate" ($\dot{\gamma}_E$) at the blend film surfaces and thus the phase-separated morphologies. While mainly a conceptual concept, the effective shear rate is essentially the determinant shear rate for the final observed morphologies over the course of the oscillatory shear cycle. The increase of translation speed from 1 μm s^{-1} to 60 μm s^{-1} induces a droplet-to-string transition for 20% AuPS/PMMA blend films confirming an increase of $\dot{\gamma}_E$ with DZA-SS speed. To gain further insights into how higher NP blend compositions may be amenable to such anisotropic sheared morphologies, DZA-SS processed 50% AuPS/PMMA blend films are investigated as a function of translation speed (varying effective shear rate). As shown in Fig. 7a, a 50% loading of nanoparticles relative to the PMMA weight resulted in spinodal decomposition structures upon vacuum oven annealing at 180 °C for 16 h. The application of soft shear then generated deformed spinodal structures as shown in Fig. 7b. As v increased from 20 μm s^{-1} to 60 μm s^{-1}, the AuPS domains evolved into drastically elongated strings, indicating increasing $\dot{\gamma}_E$ with increasing translation speed. Simultaneously, the characteristic inter-domain spacing L^* decreased from ~1164 nm to ~332 nm. As v was raised to 120 μm s^{-1} and then to 140 μm s^{-1}, the unidirectionally aligned structures were distorted again accompanied with an increase in inter-domain spacing ($L^* \approx 703$ nm at $v = 120$ μm s^{-1}). This initially enhanced and then suppressed elongation of the AuPS domains implies an increase followed by a decrease in the effective shear rate. Furthermore, clearly, the effective shear rate variation dominates the phase separation scale instead of the corresponding annealing time, where higher $\dot{\gamma}_E$ induced a smaller inter-domain spacing L^*.[21]

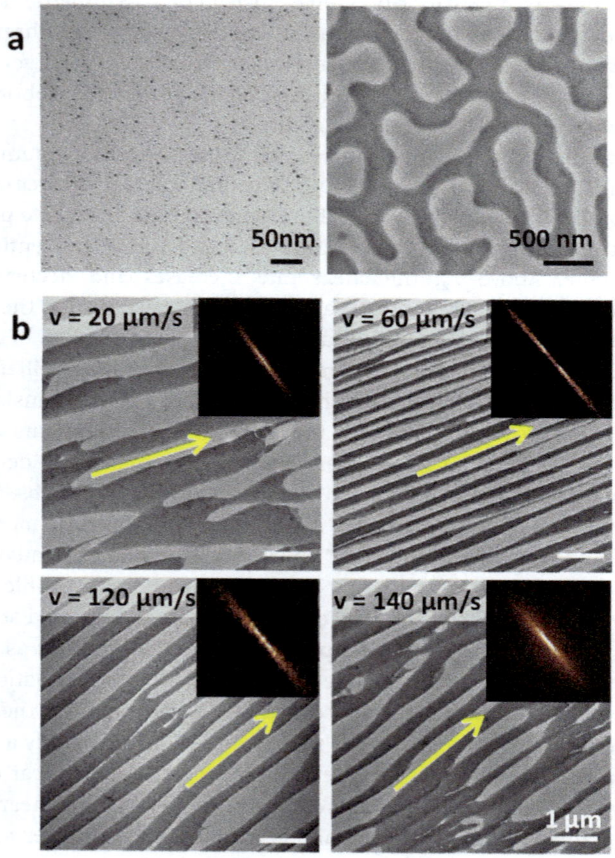

Fig. 7 (a) Top-view TEM micrographs for as-cast (left) and thermal annealed 50% AuPS/PMMA blend films at 180 °C for 16 h (right). (b) DZA-SS induced morphologies of 50% AuPS/PMMA blend films at v = 20, 60, 120 and 140 μm s^{-1}. The arrow denotes the soft-shear direction. The upper corner insets are the corresponding FFT images.

Conclusions

In summary, we studied simultaneous phase separation structure formation and deformation of a novel binary blend thin film system composed of AuPS nanoparticles and a PMMA matrix under dynamic thermal gradient field induced soft shear (DZA-SS). Progressive elongation of the AuPS domains was achieved by adjusting the translation speed (ν) and the correlation between the translation speed, the effective shear rate determinant factor and the final blend film morphology was established. Remarkably unidirectionally aligned nanoparticle strings were fabricated with the presence of soft shear in a confined thin film geometry. This soft shear directed nanoparticle assembly strategy can be easily adapted to other NP–polymer blend systems with roll-to-roll compatibility for production scale-up.

Methods

Poly(methyl methacrylate) (PMMA, $M_n = 3.1$ kg mol^{-1}, polydispersity = 1.09) and polystyrene (PS, $M_n = 11.5$ kg mol^{-1}, polydispersity = 1.25) were purchased from Polymer Source Inc. and used as obtained. The thiol-polystyrene (PS-SH) grafted gold nanoparticles (AuPS) were synthesized by phase transfer reduction of [$AuCl^4$] in the presence of thiol ligands.[36] The average radius of the gold core $R_0 = 1.21 \pm 0.42$ nm. The PS grafting density is 0.7 nm^{-2}. PMMA solutions (3 wt% in toluene) were admixed with appropriate amounts of the AuPS nanoparticles to generate blend solutions where the weight ratio of AuPS to PMMA is 20% and 50%. Separately, PS and PMMA solutions (both 3 wt% in toluene) were mixed together to generate 20% PS/PMMA solutions. The AuPS/PMMA and PS/PMMA solutions were flow coated into thin films on quartz substrates (G.M. Associates, Inc.). The quartz substrates were exposed to 1 h of ultraviolet-zone (UVO) treatment before flow coating. The film thickness ($\approx$90 nm) was measured by an interferometer (F-20 UV Thin Film Analyzer, Filmetrics, Inc.). AuPS/PMMA and PS/PMMA blend films were vacuum oven annealed at 50 °C for 6 h to remove residual solvent before further vacuum annealing or soft-shear dynamic zone annealing (DZA-SS). Upon vacuum oven annealing at 180 °C for 16 h, the AuPS nanoparticles experienced a subtle size increase ($R_0 = 1.35 \pm 0.48$ nm) due to the thermal instability of the thiol–Au bond.[18,37]

Sylgard-184, a two-part thermocurable poly(dimethyl siloxane) (PDMS) elastomer (Dow Corning Corp.), was used to make the PDMS capping layer. A mixture of PDMS elastomer and curing agent in 20 : 1 weight ratio was poured on glass slides and cured at 120 °C for 6 h. The PDMS was then peeled off the glass slide to obtain ~0.5 mm thick smooth films. Shear stress was applied to the blend film surface by soft-shear dynamic zone annealing (DZA-SS) with $T_{max} \approx 180$ °C at various translation speeds (ν = 1, 20, 60, 120 and 140 μm s^{-1}). The procedure for the DZA-SS experiments has been explained in detail in our previous papers.[17,38] After thermal annealing, the PDMS capping layer was removed and the blend film was further characterized. The nanoparticle distribution in these blend films was characterized with a JEOL JEM-1230 transmission electron microscope (TEM) at 200 kV. Specimens for TEM were prepared by pre-coating a thin layer ($\approx$10 nm) of aqueous poly(4-styrenesulfonic acid) (PSS; Sigma-Aldrich) solution on to the substrates prior to flow coating the blend films, annealing the multilayer films, and then floating the films by immersion in distilled water, followed by transferring them to copper grids. The surface topography of these blend films was imaged using a Dimension Icon atomic force microscope (AFM) (Bruker AXS) in tapping mode. Some of the annealed PS/PMMA blend films were immersed in cyclohexane for 30 min at room temperature to selectively remove the PS-rich domains.

Acknowledgements

We acknowledge the National Science Foundation *via* Grant NSF DMR-1411046 for support of this Cold Zone Annealing Soft-Shear study. MRB further acknowledges funding by the National Science Foundation *via* DMR-1410845. The authors extend their sincere appreciation to the Deanship of Scientific Research at King Saud University for its funding of the Prolific Research group (PRG-1436-14). We are grateful to Dr C. C. Han and Dr K. B. Migler for valuable discussions.

References

1 M. R. Bockstaller, R. A. Mickiewicz and E. L. Thomas, *Adv. Mater.*, 2005, **17**, 1331.
2 A. C. Balazs, T. Emrick and T. P. Russell, *Science*, 2006, **314**, 1107.
3 V. Ganesan and A. Jayaraman, *Soft Matter*, 2014, **10**, 13.
4 V. Ganesan, C. J. Ellison and V. Pryamitsyn, *Soft Matter*, 2010, **6**, 4010.
5 J. Kao, K. Thorkelsson, P. Bai, B. J. Rancatore and T. Xu, *Chem. Soc. Rev.*, 2013, **42**, 2654.
6 S. B. Darling, N. A. Yufa, A. L. Cisse, S. D. Bader and S. J. Sibener, *Adv. Mater.*, 2005, **17**, 2446.
7 M. Bockstaller and E. Thomas, *Phys. Rev. Lett.*, 2004, **93**, 166106.
8 J. Kao, P. Bai, J. M. Lucas, A. P. Alivisatos and T. Xu, *J. Am. Chem. Soc.*, 2013, **135**, 1680.
9 S. Gupta, Q. Zhang, T. Emrick and T. P. Russell, *Nano Lett.*, 2006, **6**, 2066.
10 S. Gam, A. Corlu, H.-J. Chung, K. Ohno, M. J. A. Hore and R. J. Composto, *Soft Matter*, 2011, **7**, 7262.
11 H.-j. Chung, K. Ohno, T. Fukuda and R. J. Composto, *Nano Lett.*, 2005, **5**, 1878.
12 S. K. Kumar, N. Jouault, B. Benicewicz and T. Neely, *Macromolecules*, 2013, **46**, 3199.
13 L. Meli, A. Arceo and P. F. Green, *Soft Matter*, 2009, **5**, 533.
14 M. E. Mackay, A. Tuteja, P. M. Duxbury, C. J. Hawker, B. Van Horn, Z. Guan, G. Chen and R. S. Krishnan, *Science*, 2006, **311**, 1740.
15 A. Karim, T. M. Slawecki, S. K. Kumar, J. F. Douglas, S. K. Satija, C. C. Han, T. P. Russell, Y. Liu, R. Overney, J. Sokolov and M. H. Rafailovich, *Macromolecules*, 1998, **31**, 857.
16 H. Wang and R. J. Composto, *Macromolecules*, 2002, **35**, 2799.
17 G. Singh, K. G. Yager, B. Berry, H.-C. Kim and A. Karim, *ACS Nano*, 2012, **6**, 10335.
18 R. Zhang, G. Singh, A. Dang, L. Dai, M. R. Bockstaller, B. Akgun, S. Satija and A. Karim, *Macromol. Rapid Commun.*, 2013, **34**, 1642.
19 T. P. Russell, R. P. Hjelm and P. A. Seeger, *Macromolecules*, 1990, **23**, 890.
20 P. F. Green, H. Oh, P. Akcora and S. K. Kumar, *Dynamics of Soft Matter*, 2012, p. 349.

21 D. Derks, D. G. A. L. Aarts, D. Bonn and A. Imhof, *J. Phys.: Condens. Matter*, 2008, **20**, 404208.
22 T. Hashimoto, K. Matsuzaka, E. Moses and A. Onuki, *Phys. Rev. Lett.*, 1995, **74**, 126.
23 J. Vermant and M. J. Solomon, *J. Phys.: Condens. Matter*, 2005, **17**, R187.
24 J. Feng and D. D. Joseph, *J. Fluid Mech.*, 1996, **324**, 199.
25 K. B. Migler, *Phys. Rev. Lett.*, 2001, **86**, 1023.
26 Z. Dobkowski, *Rheol. Acta*, 1995, **34**, 578.
27 K. M. O'Connor and K. M. Scholsky, *Polymer*, 1989, **30**, 461.
28 J.-C. Majeste, J. P. Montfort, A. Allal and G. Marin, *Rheol. Acta*, 1998, **37**, 486.
29 C. Wippler, *Polym. Bull.*, 1991, **25**, 357.
30 G. I. Taylor, *Proc. R. Soc. London, Ser. A*, 1932, 138.
31 C. L. Tucker III and P. Moldenaers, *Annu. Rev. Fluid Mech.*, 2002, **34**, 177.
32 J. A. Pathak and K. B. Migler, *Langmuir*, 2003, **19**, 8667.
33 C. J. Carriere, G. Biresaw and R. L. Sammler, *Rheol. Acta*, 2000, **39**, 476.
34 A. Frischknecht, *Phys. Rev. E: Stat. Phys., Plasmas, Fluids, Relat. Interdiscip. Top.*, 1998, **58**, 3495.
35 Y. Son, N. S. Martys, J. G. Hagedorn and K. B. Migler, *Macromolecules*, 2003, **36**, 5825.
36 M. Brust, M. Walker, D. Bethell, D. J. Schiffrin and R. Whyman, *J. Chem. Soc., Chem. Commun.*, 1994, 801.
37 X. Jia, J. Listak, V. Witherspoon, E. E. Kalu, X. Yang and M. R. Bockstaller, *Langmuir*, 2010, **26**, 12190.
38 G. Singh, K. G. Yager, D.-M. Smilgies, M. M. Kulkarni, D. G. Bucknall and A. Karim, *Macromolecules*, 2012, **45**, 7107.

Faraday Discussions

PAPER

Microtubule targeted therapeutics loaded polymeric assembled nanospheres for potentiation of antineoplastic activity

Radhika Poojari,* Rohit Srivastava and Dulal Panda*

Received 18th August 2015, Accepted 12th October 2015
DOI: 10.1039/c5fd00123d

Polymeric nanoassemblies represent an attractive strategy for efficient cellular internalization of microtubule targeted anticancer drugs. Using dynamic light scattering, zeta potential, transmission electron microscopy and scanning electron microscopy, the physical properties and surface morphology of microtubule-binding PEGylated PLGA assembled nanospheres (100–200 nm) were analyzed. The present approach leads to strong internalization as observed by confocal laser scanning microscopy and transmission electron microscopy in hepatocarcinoma cells. The effect of these nanoassemblies on microtubules and mitosis were explored using immunofluorescence microscopy. The effects of these nanoassemblies on cancer cell proliferation and cell death revealed their antitumor enhancing effects. Perturbation of the microtubule assembly, mitosis and nuclear modulations potentiated the antineoplastic effects delivered *via* nanospheres in hepatocarcinoma cells. The extensive biomolecular and physical characterizations of the synthesized nanoassemblies will help to design potent therapeutic materials and the present approach can be applied to deliver microtubule-targeted drugs for liver cancer therapy.

1. Introduction

Liver cancer is one of the most common causes of cancer-related mortality. The development of multidrug resistance, poor drug penetration into solid tumors, poor target selection and abnormal apoptosis machinery is found to decrease the efficacy of liver cancer chemotherapy.[1–3] Certain aberrantly expressed proteins in liver cancer lead to dysregulation in the proliferative and apoptotic signaling pathways, cell cycle anomalies and other pathways, which promote the growth of liver cancer cells.[4,5] Despite invasive chemotherapy, the survival rate of liver cancer patients is poor. Thus, nanotechnology based novel strategies can be helpful in designing new therapies against liver cancer. One such approach could

Department of Biosciences and Bioengineering, Indian Institute of Technology Bombay, Mumbai 400076, India. E-mail: drradhikapoojari@gmail.com; panda@iitb.ac.in; Tel: +91-22-25767838

be the use of nanoformulations to deliver microtubule-targeted drugs in solid tumors like liver cancer.

Microtubules (MTs) are formed by the polymerization of a heterodimeric protein tubulin. MTs display a dynamic state of instability comprising polymerization and depolymerization cycles.[6–9] The MT cytoskeleton is critical for cellular processes like mitosis, spindle formation, intracellular transport and locomotion. Killing cancer cells by disrupting the microtubules is one of the effective therapeutic strategies. Estradiol is a naturally occurring estrogen present in mammals, which is catalyzed from 2-hydroxyestradiol by catechol-*O*-methyltransferase (COMT) with no estrogen-related mitogenic activity.[10–12] Estradiol is a potent microtubule-targeted anticancer agent. It disrupts microtubule functioning, which then leads to cell cycle arrest at the G2/M-phase.[10–12] It has also been reported to possess anti-angiogenesis effects. Clinical drawbacks such as poor water solubility, poor pharmacokinetics, short half-life and adverse drug effects like systemic-neurotoxicity are responsible for the limited therapeutic window, which contributes to poor drug efficacy.[13–16] Nano drug delivery systems (NDDS) are considered to be one of the best effective modalities to circumvent the above problems.[17,18] NDDS have several advantages.[17,18] For example, NDDS protect the drug from inactivation and metabolism. Colloidal NDDS are stable and sustained release of the drug from the NDDS into the blood stream reduces the side effects. Furthermore, they are biocompatible and can accumulate at the tumor site by enhanced permeability and retention mechanisms.[17,18] Therefore, polymeric nanoassemblies are of considerable interest presently and represent one of the promising materials for cancer therapy as well as imaging.[19,20] An earlier study reported the preparation of 2-methoxyestradiol (2-ME) microsphere based poly(DL-lactide-*co*-glycolide) (PLGA) by an emulsion solvent extraction method for breast cancer therapy. The 2-ME PLGA microspheres with a particle size of 55.44 ± 12.21 μm exhibited controlled release over 46 days, and could efficiently inhibit the growth of MCF-7 cells.[21] Shen *et al.*[22] reported 2-ME nanosuspension preparation by a nanoprecipitation-high-frequency ultrasonication method with a particle size of 168.4 ± 3.2 nm. This 2-ME nanosuspension exhibited significant suppression of Lewis lung carcinoma indicating it as a possible delivery mode for lung cancer therapy. Estradiol loaded PLGA nanoparticles have been shown to improve oral bioavailability of the drug without any inflammatory response.[23,24]

In a quest to find an effective delivery modality with low toxicity, we synthesized a formulation of PEGylated PLGA assembled nanosphere-based estradiol delivery for potentiating the antineoplastic effects in human Huh7 hepatocarcinoma cells. We analyzed the particle size, zeta potential, encapsulation efficiency and release profile of the PEGylated PLGA assembled nanosphere encapsulated estradiol in Huh7 hepatocarcinoma cells. Furthermore, the *in vitro* cellular uptake, cytotoxicity and cell death proficiency of estradiol entrapped in the nanospheres were examined in Huh7 hepatocarcinoma cells. PEGylated PLGA assembled nanosphere encapsulated estradiol treatment strongly depolymerized microtubules in these cells. We address two key questions in this paper: how do the synthesized polymeric PEGylated PLGA assembled nanosphere entrapped microtubule binding drugs inhibit the proliferation of liver cancer cells and how do the polymeric PEGylated PLGA assembled nanospheres influence the antineoplastic activity.

2. Experimental

2.1 Materials

Poly(DL-lactide-*co*-glycolide) (PLGA, MW 17 000) was kindly provided by Purac Biomaterials, Netherlands. NH_2–PEG_{2000}–COOH was obtained from Jenkem Technology USA Inc., USA. Polyvinyl alcohol (PVA) was obtained from SD Fine-Chem. Ltd., Mumbai. The drug estradiol, the dyes Rhodamine B (RB), Sulforhodamine B (SRB) reagent and Hoechst 33258 were purchased from Sigma-Aldrich, USA. An Annexin V/PI apoptosis detection kit was procured from BD Pharmingen (San Diego, USA). All other reagents used in the experiments were of analytical reagent grade.

2.2 Synthesis of polymeric PEGylated PLGA assembled nanospheres

PEGylated PLGA was synthesized using NH_2–PEG_{2000}–COOH and 1-ethyl-3-(3-dimethylaminopropyl)carbodiimide hydrochloride (EDC)–sulfo-*N*-hydroxysuccinimide (Sulfo-NHS) conjugation chemistry as described previously.[25,26] Estradiol loaded PEGylated PLGA assembled nanospheres were prepared by an oil-in-water emulsion-solvent evaporation method. The organic phase was composed of PEGylated PLGA polymer (25 mg) and antimicrotubule agent estradiol (1 mg) dissolved in dimethylformamide. This organic phase in turn was emulsified with the aqueous phase containing 2.5% PVA as the surfactant as well as the stabilizer. This emulsion was sonicated for 2 min. The organic solvents used for emulsion preparation were removed through solvent-evaporation using a Rotavapor. The polymeric nanospheres were centrifuged at 20 000 × *g* for 20 min. The free drug estradiol in the supernatant was removed, the pellets were washed with deionized water and re-centrifuged for 20 min. The washing steps were performed three to four times for the removal of excess surfactant and drug. The pellets were resuspended in phosphate buffered saline and then lyophilized. Dye RB (50 μg mL^{-1}) loaded PEGylated PLGA assembled nanospheres and PEGylated PLGA assembled nanospheres were prepared similarly to the above method with the dye and in absence of a drug or a dye, respectively. Only PLGA nanoparticles were prepared in the absence of PEG. PLGA nanoparticles, PEGylated PLGA assembled nanospheres and PEGylated PLGA assembled nanosphere encapsulated estradiol were subjected to physico-chemical characterization.

2.3 Dynamic light scattering (DLS) and zeta potential measurements

The hydrodynamic particle size of PLGA nanoparticles, PEGylated PLGA assembled nanospheres and PEGylated PLGA assembled nanosphere encapsulated estradiol at 0.1 mg mL^{-1} dispersed in Milli Q water after brief sonication was measured by DLS using a 632 nm red laser (DLS 90Plus, Brookhaven Instruments Corporation, USA) at 25 °C (scattering angle of 90°). The surface charge of the PLGA nanoparticles, PEGylated PLGA assembled nanospheres and PEGylated PLGA assembled nanosphere encapsulated estradiol was measured using a Zeta-Plus Zeta-potential analyzer (Brookhaven Instruments Corporation, USA) in Milli Q water (pH 7.0) at room temperature. The physical stability of the PEGylated PLGA assembled nanosphere encapsulated estradiol on storage at 4 °C as a suspension in PBS (pH 7.4) was monitored for one week. The estradiol encapsulation efficiency in the PLGA nanoparticles and PEGylated PLGA assembled nanospheres was measured using a UV spectrophotometer (Perkin Elmer, Lambda 25) and calculated as follows; estradiol drug encapsulation efficiency (%) = (amount of loaded estradiol/amount of estradiol added) × 100%. All the measurements were repeated three times and their mean was calculated.

2.4 Transmission electron microscopy (TEM)

The size and surface morphology of the PEGylated PLGA assembled nanospheres and PEGylated PLGA assembled nanosphere encapsulated estradiol were investigated using a TEM (JEOL, JEM-2100 F, Japan) operating at 120 kV. The nanosphere suspension (0.1 mg mL^{-1}) was drop-cast on a carbon-coated copper grid, negatively stained with 1% phosphotungstic acid and then air dried.

For an ultrastructural cellular internalization study by TEM analysis, human Huh7 hepatocarcinoma cells were grown in Dulbecco's Modified Eagle Medium (DMEM) supplemented with 10% fetal bovine serum and 1% penicillin–streptomycin (HiMedia, Mumbai) at 37 °C under a 5% CO_2 atmosphere. Huh7 cells (>1 × 10^6 cells per mL) were then incubated with PEGylated PLGA assembled nanospheres and PEGylated PLGA assembled nanosphere encapsulated estradiol at 25 μg mL^{-1} for 24 h in 25 mm tissue culture flasks (Nunc, USA). Briefly, the untreated control cells and treated cells were centrifuged, fixed in 2.5% glutaraldehyde and washed with PBS, followed by fixation in 1% osmium tetroxide. The cells were then subjected to dehydration in a graded series of alcohol, embedded using a DER 332-732 embedding kit (EMS, USA) and polymerized for 48 h. These resin blocks with cells were sectioned using an ultramicrotome (Leica Ultracut UCT, Germany) and ultrathin sections (70 nm) were placed on 300 mesh Formvar and carbon coated copper grids (EMS, USA). The grids were then processed with uranyl acetate and lead citrate solutions as previously described to visualize the cellular ultrastructures.[27] Micrographs of internalization of the PEGylated PLGA assembled nanospheres, PEGylated PLGA assembled nanosphere encapsulated estradiol and untreated control cells were taken on a TEM (FEI TECNAI, 12 Bio-Twin, Netherlands) operating at 80 kV.

2.5 Scanning electron microscopy (SEM)

The morphology of the nanoparticles was observed using a SEM (JSM-7600 F, Japan). Prior to SEM application, the dried PEGylated PLGA assembled nanospheres and PEGylated PLGA assembled nanosphere encapsulated estradiol were placed on a stub and coated with gold using a sputter gold coater Auto Fine Coater (JEOL, Tokyo, Japan).

2.6 Drug release profiles

The release of estradiol from PLGA nanoparticles and PEGylated PLGA assembled nanospheres was analyzed by dialyzing (dialysis bags, molecular weight cut off of 12 kDa, HiMedia, Mumbai) the samples against phosphate buffered saline (PBS, pH 7.4) at 37 °C. An aliquot (1 mL) was taken from the release medium at predetermined intervals and was then replaced with an equivalent volume of fresh PBS to maintain the constant conditions. The percentage of estradiol content released as a function of time was quantified using a UV spectrophotometer.

2.7 Confocal laser scanning microscopy (CLSM) for analysis of cellular internalization and MT interactions

For the study of intracellular uptake by CLSM analysis, Huh7 cells (5×10^4 cells per well) were seeded on 12 mm glass cover-slips in 24 well plates (Nunc, USA). On reaching 70% confluency, the culture medium was replaced with PEGylated PLGA assembled nanospheres and PEGylated PLGA assembled nanosphere encapsulated RB at 50 μg mL^{-1} for 24 h of incubation, and then subjected to indirect immunocytochemistry CLSM analysis. Untreated cells were considered as the negative control. Images were acquired using CLSM (LSM 510 Meta, Carl Zeiss, Germany) equipped with a 63× oil immersion objective lens and Zeiss LSM Image Browser software.

For evaluating the effect of PEGylated PLGA assembled nanospheres on microtubules by immunocytochemistry analysis, Huh7 cells (5×10^4 cells per well) in 24 well plates were subjected to treatment with 5.6 μM IC_{50} concentrations of estradiol and PEGylated PLGA assembled nanosphere encapsulated estradiol for 48 h. Untreated cells and PEGylated PLGA assembled nanosphere treated cells (50 μg mL^{-1}) were the negative controls. The cells were then fixed in 3.7% formaldehyde at 37 °C for 30 min and subjected to immunostaining with primary antibody anti-α-tubulin mouse monoclonal antibody (Sigma Aldrich, USA), which was followed by incubation with secondary antibody anti-mouse IgG-Alexa 568 conjugate (Invitrogen, USA). Hoechst 33258 was used for the nuclear staining and the cells were observed under CLSM.[28,29]

2.8 Antiproliferative analysis

Cell viability was measured by an SRB assay.[30,31] Hepatocarcinoma Huh7 cells were incubated in 96-well plates (Nunc, USA) at 5×10^4 cells per well density and then subjected to treatment with PEGylated PLGA assembled nanospheres (50–500 μg mL^{-1}), free drug estradiol (1–7 μM) and PEGylated PLGA assembled nanosphere encapsulated estradiol (1–7 μM) for 48 h to evaluate their anti-proliferative activity. The absorbance was measured with a 96-well fluorescence plate reader (SPECTRAmax M2, Molecular Devices, USA) equipped with the SOFTmax Pro software (Molecular Devices), at a wavelength of 550 nm. The half maximal inhibitory concentration (IC_{50}) values of free drug estradiol *vs.* PEGylated PLGA assembled nanosphere encapsulated estradiol were analyzed by OriginPro 8.0 software.

2.9 Cell death analysis

The apoptosis activity in Huh7 cells was measured by an Annexin V and Propidium iodide (PI) staining detection kit.[28] Huh7 cells (1×10^5 cells per mL) were seeded in 6-well plates and treated with estradiol and PEGylated PLGA assembled nanosphere encapsulated estradiol at the IC_{50} concentration (5.6 μM) for 48 h. Untreated cells and PEGylated PLGA assembled nanosphere (50 μg mL^{-1}) treated cells were considered as the negative controls. Cells were washed twice with PBS and resuspended in 1× binding buffer (400 μL). Later, Annexin V-FITC (5 μL) and PI (5 μL) solutions were added to test samples, incubated in the dark for 15 min at room temperature, and subsequently analyzed using a flow cytometer (BD FACS ARIA special order system equipped with BD FACSDiva, Becton and Dickinson,

USA software program). For each test sample, the fluorescence of ~20 000 cells was counted. Data represent the mean of three independent experiments.

2.10 Statistical evaluation

The data represent mean ± SD. Statistical evaluation was carried out using OriginPro 8.0 software.

3. Results and discussion

3.1 PEGylated PLGA assembled nanosphere characterization – size, shape, charge and stability

A PEGylated PLGA based drug delivery system has a number of advantages over other polymers used in drug, gene delivery and various biomedical applications, including biodegradability, biocompatibility, non-antigenicity and easy scale-up for manufacturing, and is a FDA approved polymer for human use.[26,32] It can be used for site-specific targeting and controlled-release of encapsulated drugs. Also, PEG provides stealth surface effects, increases the stability and half-plasma life of the PEGylated PLGA nanoassemblies in order to prevent non-specific interactions minimizing the side effects, decreases the phagocytic uptake, improves the bioavailability of the drug, and accumulates in tumour tissues.[25,26,33] Herein, we report the preparation of PEGylated PLGA assembled nanospheres encapsulated with a microtubule-binding agent, estradiol by an emulsion-solvent evaporation method using PVA as the emulsifier (Fig. 1). PEGylation of these polymeric nanospheres provided steric stability to the delivery system.

The size and shape of the nanoparticles are known to greatly influence cellular internalization.[32,34,35] Spherical shaped nanoparticles have been reported to possess five times more uptake capacity than the rod shaped ones suggesting that the elongated nanoparticles required more time for the cellular endocytosis process.[34] The charge of the nanoparticles also influences their interaction with the cell membrane and thereby, affects the cellular internalization. The surface

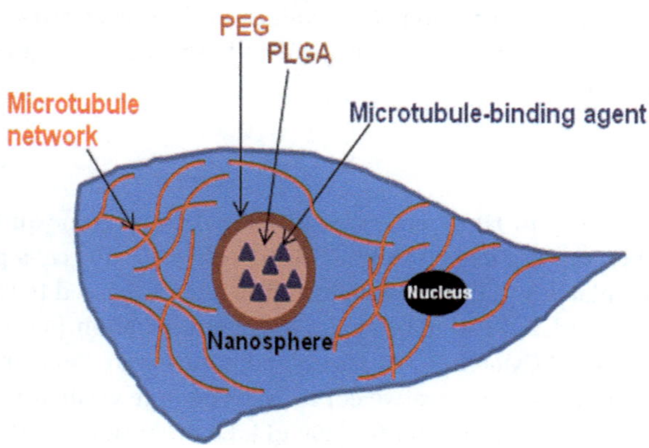

Fig. 1 An illustration depicting a microtubule-binding chemotherapeutic agent encapsulated inside a polymeric PEGylated PLGA assembled nanosphere to target liver cancer cells.

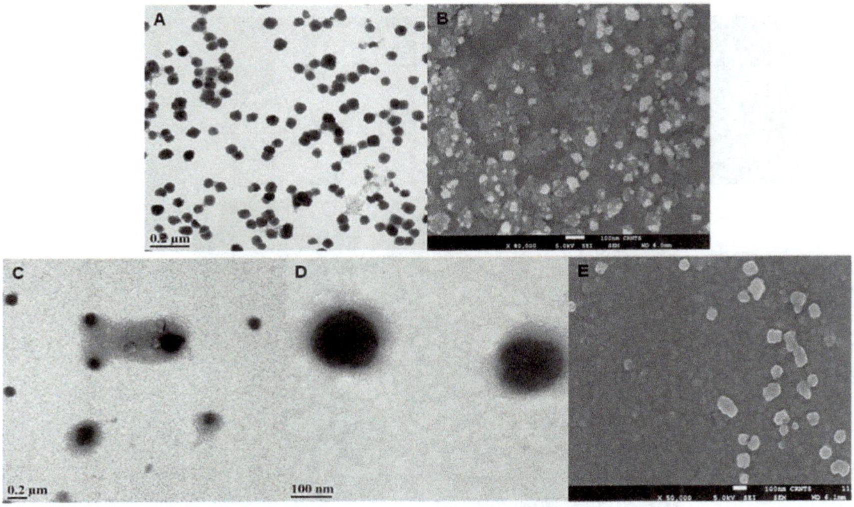

Fig. 2 (A) TEM and (B) SEM images of PEGylated PLGA assembled nanospheres. (C and D) TEM images at 200 nm and 100 nm scales, and (E) SEM image of PEGylated PLGA assembled nanosphere encapsulated estradiol.

charge on the particles can be explored to carry the cargo inside the cell.[32,35] DLS results of PLGA nanoparticles, PEGylated PLGA assembled nanospheres and PEGylated PLGA assembled nanosphere encapsulated estradiol depicted mean hydrodynamic diameters of 211 ± 8, 180 ± 6 and 200 ± 6 nm and zeta potentials of -18.7 ± 5.3, -7.4 ± 1.1 and 4.5 ± 2.1 mV, respectively. The mean polydispersity indices (PDI) of the PLGA nanoparticles, PEGylated PLGA assembled nanospheres

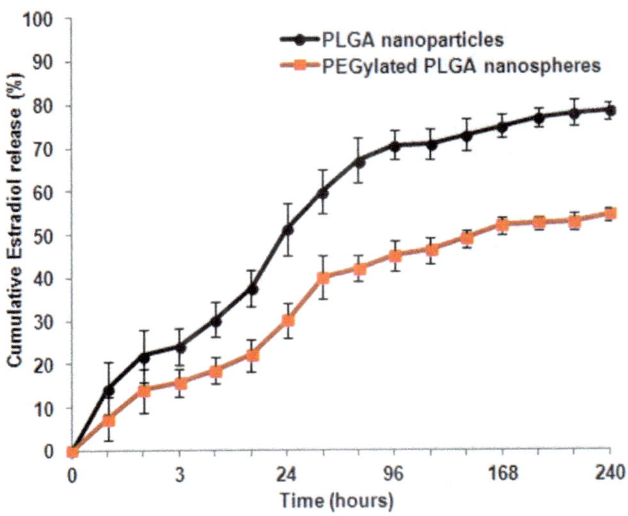

Fig. 3 The release profiles of estradiol from the PLGA nanoparticles and the PEGylated PLGA assembled nanospheres in PBS at pH 7.4.

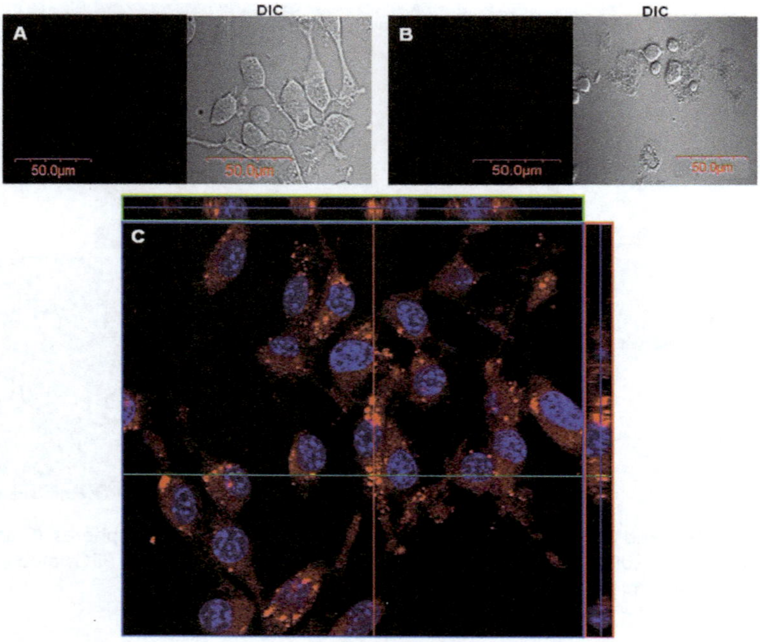

Fig. 4 Differential interference contrast (DIC) microscopy images of (A) control cells, (B) PEGylated PLGA assembled nanosphere treated cells, and (C) CLSM reconstructions of Huh7 cells showing the intracellular distribution of PEGylated PLGA assembled nanosphere encapsulated RB at 24 h. The PEGylated PLGA assembled nanosphere encapsulated RB was seen as an intense bright red fluorescence in the cells. Cell nuclei were stained with Hoechst 33258 (blue fluorescence).

and PEGylated PLGA assembled nanosphere encapsulated estradiol were 0.27 ± 0.06, 0.20 ± 0.01 and 0.13 ± 0.01, respectively. The PDI of 0.13 indicated a narrow size distribution of the estradiol loaded polymeric nanospheres, and good dispersion. The encapsulation efficiency was found to be 36 ± 5.1% for PLGA nanoparticles and 40 ± 3.4% for PEGylated PLGA assembled nanospheres by using UV absorption spectroscopy. The physical stability of the PEGylated PLGA assembled nanosphere encapsulated estradiol stored at 4 °C was monitored for 1 week. No significant change either in the hydrodynamic size (209 ± 4 nm) or in the zeta potential (5.2 ± 1.1 mV) was observed in the samples after 1 week, indicating the stealth effects of PEG.[26,33] TEM and SEM analysis of the synthesized PEGylated PLGA assembled nanospheres and PEGylated PLGA assembled nanosphere encapsulated estradiol indicated that the nanospheres were spherical in shape with a smooth surface (Fig. 2A–E).

3.2 Estradiol release profiles

The cumulative estradiol release from the PLGA nanoparticles and PEGylated PLGA assembled nanospheres in PBS (pH 7.4) at 37 °C has been shown in Fig. 3. PLGA nanoparticles showed a biphasic release pattern, which was characterized with an initial burst release of estradiol followed by a sustained release. At 24 h, the amount of estradiol released from the PLGA nanoparticles in the initial burst

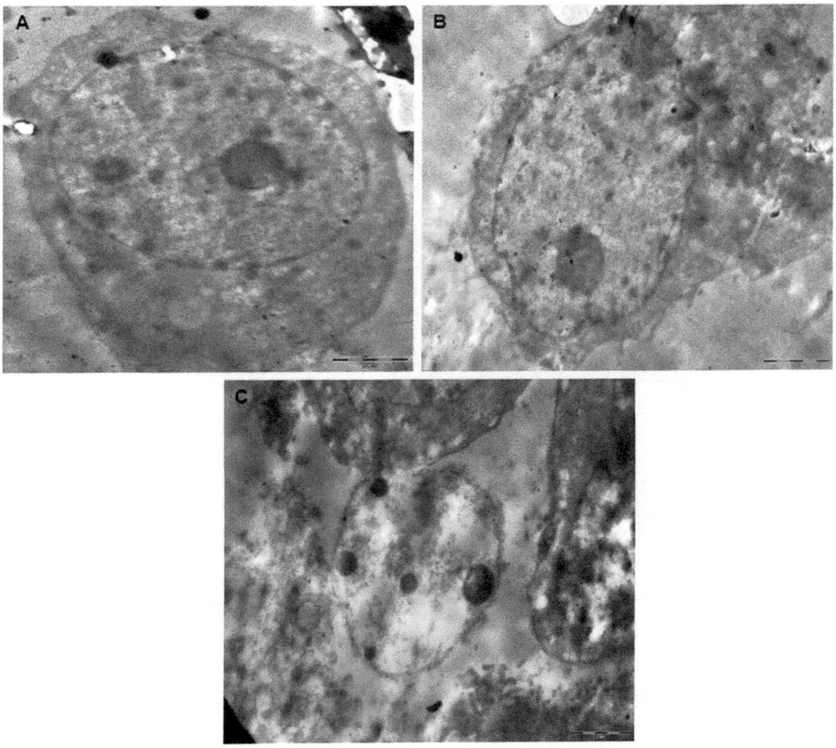

Fig. 5 Ultrastructural TEM images of Huh7 cells: (A) control cells, (B) cells treated with PEGylated PLGA assembled nanospheres and (C) cells treated with PEGylated PLGA assembled nanosphere encapsulated estradiol showing efficient cellular internalization at 24 h. Scale: 2 μm.

was 51% and ~67% of the drug was released at 72 h. No significant burst release of estradiol was observed in the initial stages in the case of the PEGylated PLGA assembled nanospheres showing about 30% release in 24 h and about 42% release in 72 h, indicating its slower diffusion than the PLGA nanoparticles. The cumulative drug release in 10 days was found to be 54% in the PEGylated PLGA assembled nanospheres whereas it was 78% for the PLGA nanoparticles, indicating that the sustained slower release of estradiol was due to the presence of

Table 1 Cell death evaluation of PEGylated PLGA assembled nanospheres in Huh7 cells

Population (%)	Control	PEGylated PLGA assembled nanospheres	Estradiol	PEGylated PLGA assembled nanosphere encapsulated estradiol
Live cells	99.8	99.7	80.5	49.5
Necrotic cells	0.1	0.1	10	8.5
Early apoptotic cells	0.1	0.2	6.5	12
Late apoptotic cells	0	0	3	30

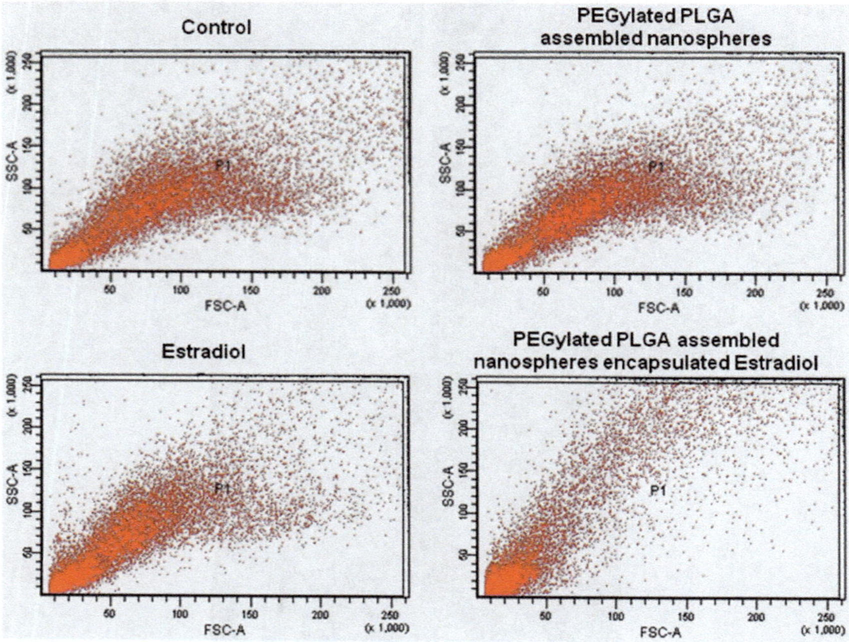

Fig. 6 PEGylated PLGA assembled nanosphere induced apoptosis in Huh7 cells. Cells were treated with PEGylated PLGA assembled nanospheres for 48 h and analyzed by flow-cytometry.

PEG, a stealth surface layer surrounding the PLGA (*i.e.* PLGA core–PEG shell) nanospheres. PEGylation is known to impart increased circulation of drug delivery carriers for controlled and optimal drug release.[36–38] This inferred that the PEGylated PLGA assembled nanospheres are suitable for a prolonged estradiol delivery system, which is in agreement with previous reports.[36–38]

3.3 Influence of internalization of the PEGylated PLGA assembled nanospheres

The delivery efficiency of our synthesized PEGylated PLGA assembled nanospheres was evaluated by the CLSM qualitative analysis of fluorescent dye RB taken up by the Huh7 cells. Cellular internalization is an important factor for the evaluation of different drug delivery therapeutic modalities in cancer.[17] CLSM studies showed that the PEGylated PLGA assembled nanosphere encapsulated RB (50 μg mL^{-1}) was readily internalized by Huh7 cells distributed in the cytoplasm (intense bright red fluorescence) in 24 h (Fig. 4C). The untreated control cells and those treated with PEGylated PLGA assembled nanospheres (50 μg mL^{-1}) showed no fluorescence signal (Fig. 4A and B).

The ultrastructural studies of untreated control cells showed normal cellular architecture (Fig. 5A). TEM analysis of the PEGylated PLGA nanosphere (25 μg mL^{-1}) treated Huh7 cells revealed the intact structure of the cell membrane and nucleus (Fig. 5B) whereas, disrupted cell membrane integrity, condensation of the nucleus and nuclear fragmentation were clearly observed in the treated PEGylated

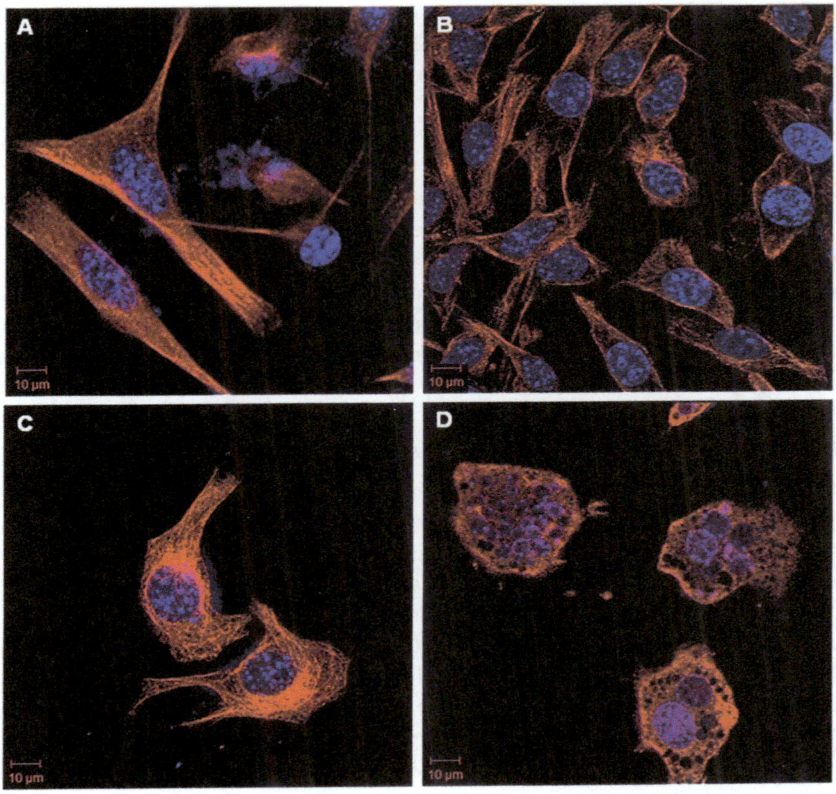

Fig. 7 Effects of the PEGylated PLGA assembled nanospheres and the free drug on the microtubule organization and nuclear morphology in Huh7 cells. (A) Control cells, (B) PEGylated PLGA assembled nanospheres (C) free drug estradiol and (D) PEGylated PLGA assembled nanosphere encapsulated estradiol treated cells for 48 h at 5.6 μM. The fixed cells were stained with anti-α-tubulin (red fluorescence) and the nucleus with Hoechst 33258 (blue fluorescence). Scale: 10 μm.

PLGA assembled nanosphere encapsulated estradiol (25 μg mL^{-1}) at 24 h (Fig. 5C). It has been reported that the PEGylation promotes cellular uptake due to an increase in the circulation time of the nanoparticles.[26,36,38] Hence, from our results it can be inferred that the PEGylated PLGA assembled nanospheres enhanced the RB cellular uptake and also enhanced the PEGylated PLGA assembled nanosphere encapsulated estradiol internalization in Huh7 cells, which is in agreement with previous reports.[26,36,38]

3.4 Influence of the PEGylated PLGA assembled nanospheres on cell proliferation

2-ME has been reported to induce cytotoxicity in prostate cancer cells.[39] Herein, the dose responses of PEGylated PLGA assembled nanospheres (50–500 μg mL^{-1}), free estradiol (1–7 μM) and PEGylated PLGA assembled nanosphere encapsulated estradiol (1–7 μM) in Huh7 cells were evaluated by quantification of the total metabolic activity after 48 h of exposure by the SRB assay.[30,31] The delivery of

PEGylated PLGA assembled nanosphere encapsulated estradiol increased the cell killing efficiency in Huh7 cells. Antiproliferative effects of free estradiol exhibited an IC_{50} of 5.6 ± 0.8 μM in comparison to PEGylated PLGA assembled nanosphere encapsulated estradiol with an IC_{50} of 2.2 ± 0.5 μM. These results indicated enhanced cell growth inhibition *via* PEGylated PLGA assembled nanosphere encapsulated estradiol in comparison to the free estradiol. The drug loaded polymeric nanospheres were able to better internalize into the cells and efficiently deliver estradiol in comparison to the free drug in the cells. The PEGylated PLGA assembled nanospheres at 50, 100, 250 and 500 μg mL^{-1} were non-toxic to cells with no significant difference in cell viabilities in comparison to the control cells, indicating their biocompatibility. Overall, the data suggested that PEGylated PLGA assembled nanospheres provided an efficient delivery modality for estradiol.

3.5 Influence of the PEGylated PLGA assembled nanospheres on cell death

A strategy to accelerate tumor cell death remains a crucial factor in cancer chemotherapy.[40,41] A previous study has shown 2-ME to inhibit growth and induce apoptosis in prostate cancer cells by arresting the cells in the G2/M phase.[39] Herein, the influence of a PEGylated PLGA assembled nanosphere treatment on cell death was evaluated in Huh7 cells using an Annexin V/PI assay at IC_{50} concentrations for 48 h by flow cytometry as shown in Table 1 and Fig. 6. The cancer cells die through both apoptosis and necrosis. The data revealed that PEGylated PLGA assembled nanosphere encapsulated estradiol induced greater cell death (apoptosis and necrosis) compared to that of the free drug at 5.6 μM. ~6.5% of early apoptosis (*i.e.* Annexin V stained positive cells) and ~3% of late apoptosis (*i.e.* PI stained cells) was observed after treatment with estradiol for 48 h in comparison to ~12% of early apoptosis and ~30% of late apoptosis being observed for PEGylated PLGA assembled nanosphere encapsulated estradiol in Huh7 cells. No difference in the apoptotic rate was observed between the treated PEGylated PLGA assembled nanospheres and the control cells, indicating that the polymeric nanospheres by themselves did not elicit significant apoptosis. Increased circulation time of the nanoparticles by PEGylation has been reported in enhanced cell uptake, and thereby its therapeutic effects.[26,36,38] Hence, from the results it can be inferred that PEGylated PLGA assembled nanosphere encapsulated estradiol could enhance apoptosis induction effectively due to the efficient internalization by these polymeric nanospheres in comparison to the free estradiol (Table 1 and Fig. 6).

3.6 Effects of the PEGylated PLGA assembled nanospheres on microtubule organization in Huh7 cells

Estradiol binds at the colchicine site of α-tubulin, interferes with the interphase and mitotic microtubules and causes mitotic arrest.[10,12,42] Immunocytochemistry revealed strong microtubule depolymerization, irregular cells and induction of multinucleation with PEGylated PLGA assembled nanosphere encapsulated estradiol (Fig. 7D) in comparison to the free drug (Fig. 7C) in Huh7 cells. Estradiol was able to elicit less activity wherein, the microtubule cytoskeleton was not strongly disrupted. Morphological evidence of the PEGylated PLGA assembled nanosphere encapsulated estradiol treated cells with respect to nuclear alterations in the form of fragmented nuclei and apoptotic bodies was visible in Hoechst 33258 stained cells. A strong induction of multinucleation of the cells indicated a well-disrupted mitotic division. Interestingly, morphologic alteration in the nuclei of the estradiol treated cells was not very effective. In the case of the PEGylated PLGA assembled nanosphere treated cells the microtubule cytoskeletal architecture was intact with no cell damage and the nuclear morphology was normal (Fig. 7A and B). The superior antineoplastic effect of PEGylated PLGA assembled nanosphere mediated estradiol delivery in Huh7 cells in comparison to the free estradiol reflects the efficient internalization *via* polymeric nanospheres leading to potent disruption of the mitotic spindle apparatus and apparently cell death induction.

4. Conclusions

We report a strategy to effectively deliver estradiol using PEGylated PLGA assembled nanospheres. The PEGylated PLGA assembled nanosphere encapsulated estradiol strongly depolymerized the microtubules, inhibited cell proliferation and induced cell death in Huh7 hepatocarcinoma cells in comparison to the free drug. Fluorescence and ultrastructural analyses suggested that the PEGylated PLGA assembled nanospheres enhanced the internalization process of RB or estradiol, respectively in Huh7 hepatocarcinoma cells. PEGylated PLGA assembled nanosphere encapsulated estradiol was found to be stable and potentiated the antineoplastic effects of the drug in hepatocarcinoma cells. Further scale-up of the PEGylated PLGA assembled nanosphere encapsulated estradiol and its evaluation in xenografted hepatocarcinoma animal models need to be explored for translational research. The results discussed here provide insights into the development of microtubule-binding polymeric nanoassemblies for the treatment of hepatic neoplastic disorders.

Acknowledgements

R. Poojari acknowledges a DBT-Post doctoral fellowship in Biotechnology and Life Sciences, Government of India. The study was supported by a DAE-SRC fellowship, Government of India, granted to D. Panda. The authors thank the Centre for Research in Nanotechnology and Science (CRNTS) for providing the characterization facility, S. Kini for help with the cell culture studies at IIT Bombay, India, and the Advanced Centre for Treatment, Research and Education in Cancer (ACTREC), India for providing the CLSM digital imaging facility.

Notes and references

1 R. Siegel, J. Ma, Z. Zou and A. Jemal, *Ca-Cancer J. Clin.*, 2014, **64**, 9–29.
2 H. El-Serag, *N. Engl. J. Med.*, 2011, **365**, 1118–1127.
3 J. M. Llovet, A. Burroughs and J. Bruix, *Lancet*, 2003, **362**, 1907–1917.
4 X. Bisteau, M. J. Caldez and P. Kaldis, *Cancers*, 2014, **6**, 79–111.
5 J. M. Llovet and J. Bruix, *Hepatology*, 2008, **48**, 1312–1327.
6 D. J. Needleman, M. A. Ojeda-Lopez, U. Raviv, H. P. Miller, Y. Li, C. Song, S. C. Feinstein, L. Wilson, M. C. Choi and C. R. Safinya, *Faraday Discuss.*, 2013, **166**, 31–45.

7 R. A. Stanton, K. M. Gernert, J. H. Nettles and R. Aneja, *Med. Res. Rev.*, 2011, **31**, 443–481.

8 M. A. Jordan and L. Wilson, *Nat. Rev. Cancer*, 2004, **4**, 253–265.

9 R. Mohan and D. Panda, *Cancer Res.*, 2008, **68**, 6181–6189.

10 E. Aizu-Yokota, K. Ichinoseki and Y. Sato, *Carcinogenesis*, 1994, **15**, 1875–1879.

11 J. Saussede-Aim, E. L. Matera, C. Ferlini and C. Dumontet, *Cell Motil. Cytoskeleton*, 2009, **66**, 378–388.

12 R. J. D'Amato and M. J. Folkman, *US Pat.*, 7,381,848 B2, 2008.

13 K. Pal, S. Pore, S. Sinha, R. Janardhanan, D. Mukhopadhyay and R. Banerjee, *J. Med. Chem.*, 2011, **54**, 2378–2390.

14 A. Gupta, P. Saha, C. Descôteaux, V. Leblanc, É. Asselin and G. Bérubé, *Bioorg. Med. Chem. Lett.*, 2010, **20**, 1614–1618.

15 M. R. Harrison, N. M. Hahn, R. Pili, W. K. Oh, H. Hammers, C. Sweeney, K. Kim, S. Perlman, J. Arnott, C. Sidor, G. Wilding and G. Liu, *Invest. New Drugs*, 2011, **29**, 1465–1474.

16 S. Verenich and P. M. Gerk, *Mol. Pharm.*, 2010, **7**, 2030–2039.

17 N. Kamaly, Z. Xiao, P. M. Valencia, A. F. Radovic-Moreno and O. C. Farokhzad, *Chem. Soc. Rev.*, 2012, **41**, 2971–3010.

18 M. Ferrari, *Nat. Rev. Cancer*, 2005, **5**, 161–171.

19 Z. Liu, Y. Wang and N. Zhang, *Expert Opin. Drug Delivery*, 2012, **9**, 805–822.

20 H. J. Lee, A. Ponta and Y. Bae, *Ther. Delivery*, 2010, **1**, 803–817.

21 X. Guo, Q. Mei, Y. Xing, L. Ye, H. Zhang, X. Shi and Z. Zhang, *Drug Delivery*, 2012, **19**, 143–148.

22 G. Shen, Q. Wang, Q. Zhang, H. Sun, Y. Zhao, Z. Zhang and B. Du, *Drug Delivery*, 2012, **19**, 327–333.

23 G. Mittal, D. K. Sahana, V. Bhardwaj and M. N. V. Ravi Kumar, *J. Controlled Release*, 2007, **119**, 77–85.

24 D. K. Sahana, G. Mittal, V. Bhardwaj and M. N. V. Ravi Kumar, *J. Pharm. Sci.*, 2008, **97**, 1530–1542.

25 J. Cheng, B. A. Teply, I. Sherifi, J. Sung, G. Luther, F. X. Gu, E. Levy-Nissenbaum, A. F. Radovic-Moreno, R. Langer and O. C. Farokhzad, *Biomaterials*, 2007, **28**, 869–876.

26 F. Gu, L. Zhang, B. A. Teply, N. Mann, A. Wang, A. F. Radovic-Moreno, R. Langer and O. C. Farokhzad, *Proc. Natl. Acad. Sci. U. S. A.*, 2008, **105**, 2586–2591.

27 A. M. Schrand, J. J. Schlager, L. Dai and S. M. Hussain, *Nat. Protoc.*, 2010, **5**, 744–757.

28 P. K. Gajula, J. Asthana, D. Panda and T. K. Chakraborty, *J. Med. Chem.*, 2013, **56**, 2235–2245.

29 A. Rai, T. K. Gupta, S. Kini, A. Kunwar, A. Surolia and D. Panda, *Biochem. Pharmacol.*, 2013, **86**, 378–391.

30 V. Vichai and K. Kirtikara, *Nat. Protoc.*, 2006, **1**, 1112–1116.

31 P. Skehan, R. Storeng, D. Scudiero, A. Monks, J. McMahon, D. Vistica, J. T. Warren, H. Bokesch, S. Kenney and M. R. Boyd, *J. Natl. Cancer Inst.*, 1990, **82**, 1107–1112.

32 J. Panyam and V. Labhasetwar, *Adv. Drug Delivery Rev.*, 2003, **55**, 329–347.

33 E. Nogueira, A. Loureiro, P. Nogueira, J. Freitas, C. R. Almeida, J. Härmark, H. Hebert, A. Moreira, A. M. Carmo, A. Preto, A. C. Gomes and A. Cavaco-Pauloa, *Faraday Discuss.*, 2013, **166**, 417–429.

34 B. D. Chithrani, A. A. Ghazani and W. C. W. Chan, *Nano Lett.*, 2006, **6**, 662–668.
35 K. Unfried, C. Albrecht, L. O. Klotz, A. V. Mikecz, S. G. Beck and R. P. F. Schins, *Nanotoxicology*, 2007, **1**, 52–71.
36 N. Vij, T. Min, R. Marasigan, C. N. Belcher, S. Mazur, H. Ding, K. T. Yong and I. Roy, *J. Nanobiotechnol.*, 2010, **8**, 22.
37 R. Gref, Y. Minamitake, M. T. Peracchia, V. Trubetskoy, V. Torchilin and R. Langer, *Science*, 1994, **263**, 1600–1603.
38 J. Park, P. M. Fong, J. Lu, K. S. Russell, C. J. Booth, W. M. Saltzman and T. M. Fahmy, *Nanomedicine*, 2009, **5**, 410–418.
39 T. M. LaVallee, X. H. Zhan, C. J. Herbstritt, E. C. Kough, S. J. Green and V. S. Pribluda, *Cancer Res.*, 2002, **62**, 3691–3697.
40 S. W. Fesik, *Nat. Rev. Cancer*, 2005, **5**, 995.
41 W. L. Scott and W. L. Athena, *Carcinogenesis*, 2000, **21**, 485–495.
42 J. L. Kipp and V. D. Ramirez, *Neuroendocrinology*, 2003, **77**, 258–272.

Faraday Discussions

PAPER

Colloidal assembly by ice templating†

Guruswamy Kumaraswamy,* Bipul Biswas and Chandan Kumar Choudhury

Received 23rd August 2015, Accepted 8th October 2015
DOI: 10.1039/c5fd00125k

We investigate ice templating of aqueous dispersions of polymer coated colloids and crosslinkers, at particle concentrations far below that required to form percolated monoliths. Freezing the aqueous dispersions forces the particles into close proximity to form clusters, that are held together as the polymer chains coating the particles are crosslinked. We observe that, with an increase in the particle concentration from about 10^6 to 10^8 particles per ml, there is a transition from isolated single particles to increasingly larger clusters. In this concentration range, most of the colloidal clusters formed are linear or sheet like particle aggregates. Remarkably, the cluster size distribution for clusters smaller than about 30 particles, as well as the size distribution of linear clusters, is only weakly dependent on the dispersion concentration in the range that we investigate. We demonstrate that the main features of cluster formation are captured by kinetic simulations that do not consider hydrodynamics or instabilities at the growing ice front due to particle concentration gradients. Thus, clustering of colloidal particles by ice templating dilute dispersions appears to be governed only by particle exclusion by the growing ice crystals that leads to their accumulation at ice crystal boundaries.

Introduction

When an aqueous solution or dispersion is frozen, the ice crystals that form exclude solutes and dispersed particles. These solutes or dispersed particles are spatially localized in the interstitial regions between ice crystals that form as water freezes. Subsequent removal of the ice, typically by freeze drying, results in the formation of a porous material as a negative replica of the ice crystals. This process, called ice templating or ice segregation-induced self-assembly (ISISA) has emerged as a powerful technique for the production of macroporous materials.[1-3] There are several advantages to ice templating: it uses water as a solvent and is therefore, environmentally friendly and economical. Furthermore, the processing

Complex Fluids and Polymer Engineering, Polymer Science and Engineering Division, CSIR-National Chemical Laboratory, Dr. Homi Bhabha Road, Pune 411008, India. E-mail: g.kumaraswamy@ncl.res.in; Fax: +91-20-2590-2618; Tel: +91-20-2590-2182

† Electronic supplementary information (ESI) available. See DOI: 10.1039/c5fd00125k

equipment required to practise ice templating are widely accessible since they are relatively inexpensive.

Ice templating is not a new technique: reports of usage of synthetic cryogels in the biomedical literature go back several decades.[4,5] Ice templating has been used to prepare porous monoliths from rubber latex,[6] from thermoset resorcinol–formaldehyde,[7] as well as from a wide variety of biocompatible polymers, including, collagen, polylactic acid, polyglycolic acid, gelatin, *etc.*[8–12] Freeze-drying induced partial crystallization of polymers, such as poly(vinyl alcohol), has also been used to generate porous structures.[13] Ice templating of ceramic slurries has been employed to freeze cast a variety of near net shape porous products. Recently, there has been tremendous interest in combining ice templating with directional freezing to produce oriented structures. This was pioneered by researchers from DuPont[14] who demonstrated that directional freezing of an aqueous solution of silicic acid and its subsequent polymerization resulted in the formation of oriented silica wires. In the last decade, researchers have extended directional ice templating to produce aligned titania fibers[15] and macroporous aligned honeycomb monoliths of silica or combinations of alumina and silica.[16,17] Aligned 1-D and 2-D structures comprising polymer or polymer–inorganic composites have also been produced by directional templating followed by freeze drying.[18] Nacre-like tough layered composites have been produced in a two-step process by forming alumina sheets through directional ice templating, followed by in-filling with either epoxy or metal[19,20] or even a glassy phase.[21] Oriented anisotropic polymer nanocomposite hybrids have also been reported by other researchers,[22–24] attesting to the versatility of directional ice templating.

Control over the properties of ice-templated materials requires a thorough understanding of the development of structures as solutes or dispersed particles are organized by growing ice crystals. Apart from porous materials synthesis, this understanding has implications for waste water treatment[25] and geophysical phenomena such as freezing of wet soil.[26] The interaction between dispersed particles and a solid–liquid interface was examined by Uhlmann and co-workers.[27] By balancing the hydrodynamic lubrication forces experienced by a suspended particle as it is approached by a growing solid interface, with the attractive interaction between the particle and the interface, they determined a critical velocity for the interface, above which the particles are overgrown by the solid medium and trapped within it. At velocities lower than the critical velocity, particles are unable to penetrate the growing solid front and are aggregated at the liquid–solid interface. Experimental studies[28] suggested that this critical velocity for particle trapping as water was frozen was higher in ice grain boundary grooves. The presence of a gradient of temperature or solute concentration at the ice–water interface can also influence ice growth. The ice front can become unstable as a consequence of such gradients and break up into cellular structures, characterized by a wavelength, l_c, that is predicted by the linear Mullins–Sekerka theory.[29–31] The presence of particles at the ice growth front also results in constitutional undercooling and depresses the freezing point of the water – similar to the case of atomic or molecular systems. Furthermore, as particles accumulate at the ice front, the transport properties change in a highly nonlinear manner – unlike the case for atomic or molecular systems. By carefully accounting for these, significant progress has recently been made in understanding ice structure formation during freezing of aqueous colloidal dispersions, and theoretical estimates of the conditions for the onset of growth instabilities are now available.[32,33] Following this approach, the group of Deville[34–37] has determined an experimental phase diagram for instabilities in ice growth in colloidal dispersions. Specifically, they have identified regions where particles are rejected by the ice and are concentrated at the ice–water interface, as well as metastable and unstable regions, where transverse ice lenses are likely to form. We note that, based on their observations, particles up to a micron in size are always rejected by the growing ice front for ice growth velocities of $\sim O(0.1–1\ \mu m\ s^{-1})$.

Thus, while experimental implementation of ice templating is simple, a detailed understanding of the effects of solutes and colloids on solidification is important for controlling the structure of the porous aggregate formed. Our understanding of structure development during freezing of aqueous systems containing both dispersed colloid and dissolved polymers is incomplete. For example, Zhang and co-workers[18] note that the presence of a small fraction of the polymer poly(vinyl alcohol) critically influences the shape of the ice-templated structures formed by directional ice-templating of silica colloidal dispersions. Plate like objects are formed in the absence of the polymer while aligned porous structures result when ice-templating polymer/colloid hybrids. Similarly, Shi and co-workers[22] have noted that the presence of a polymeric "mortar" (in their case, poly-*N*-isopropylacrylamide) was essential for the formation of anisotropic rod morphologies by direction ice templating of silica particle dispersions. The reasons that underly these observations are not fully understood at this time.

Recently, we have demonstrated[38] the formation of porous hybrid structures by ice templating an aqueous dispersion containing rigid colloidal particles, polymers and crosslinkers, followed by crosslinking of the hybrid structure in the frozen state. This process does not require freeze drying since the hybrid structure is covalently crosslinked. Thus, the ice can simply be thawed to recover a porous monolith. Remarkably, we observe that these monoliths are elastic and can be reversibly compressed to about 10% of their original size – even when the inorganic content exceeds 90% by weight. This behaviour was observed for porous materials prepared using various colloids (silica particles, polystyrene latex and hydroxyapatite particles), for different polymers (polyethyleneimine and gelatin) and for different crosslinking chemistries. We concluded that the mechanical response observed was a consequence of a unique microstructure in the particle–polymer hybrid that was created by crosslinking during ice templating.

In most of the previous investigations, the focus was on the synthesis of macroscopic assemblies or monoliths. Therefore, those studies used colloidal dispersions containing a relatively large volume fraction of colloids. In this work, we investigate the formation of colloidal aggregates by ice templating extremely dilute colloidal dispersions and by crosslinking the assemblies of polymer coated colloids in the frozen state. The dispersion concentrations used here are far below that required for a transition to percolated porous monoliths. We demonstrate that the colloidal assemblies exhibit a progression of shape from linear assemblies to tape-like structures to 2D aggregates. Simulations that rely on a simple kinetic model of particles being expelled by growing ice crystals are able to capture the qualitative features of the particle clusters formed.

Materials and methods

Materials

Poly(ethylene imine), PEI (50% w/v solution; molecular weight specified by supplier, $M_w = 750$ kg mol^{-1}) and polyethyleneglycol diepoxide (PEG-diepoxide, supplier specified $M_n = 500$ g mol^{-1}) were obtained from Sigma-Aldrich and were used as received. Fluorescent polystyrene (PS) latex particles, 1.04 μm in size with a polydispersity of 4%, were used as received from Microparticles GmbH, Germany. These particles were negatively charged with sulphonic acid groups on their surface and were supplied as 2.5% w/v aqueous dispersions. DI water was used from a Millipore system (pH = 7, resistivity = 18.2 MΩ cm).

Sample preparation

1 μl of PS latex dispersion (containing $\approx 2.96 \times 10^8$ colloidal particles) was added to 100 μl DI water. This was sonicated for one minute, followed by a minute of vortexing to prepare a uniform colloidal dispersion. 1.5 μl of a stock solution of PEI (prepared by adding 100 mg of as received PEI in 1 ml of DI water) was diluted by adding to 100 μl DI water and vortexing for 2 minutes. While the PEI solution was vortexed, the latex dispersion was added slowly. After addition of the latex, the sample was centrifuged at 3000 rpm for 8 minutes. The supernatant (corresponding to about 75% of the solution) was carefully removed using a micropipette. We note that there was almost no loss of particles in the washing step since only about 75% of the supernatant was removed and since care was taken to pipette it out gently. 200 μl DI water was added to the sedimented latex particles and they were redispersed by ultrasonication for two minutes and vortexing for one minute. Subsequently, the dispersion was further diluted by addition of 280 μl DI water. This washing step, *viz.* centrifugation and redispersion, was performed one more time. 3 μl PEG-diepoxide was diluted in 100 μl of water and vortexed for a minute. To this, the latex dispersion was added while continuously vortexing. After addition of the latex dispersion, the sample was transferred to a freezer maintained at −18 °C and stored for 12 hours, during which time the water froze, and crosslinking proceeded. A schematic of the ice-templating process is shown in Fig. 1. Subsequently, the sample was thawed at room temperature for 30–40 minutes and loaded in a microscope cavity slide and sealed. Particles and particle clusters in the sealed sample were allowed to settle over an hour before the sample was imaged. Similarly, we prepared samples at different particle concentrations: 5.41×10^7 particles per ml, 7.39×10^7 particles per ml, 3.39×10^6 particles per ml and 8.46×10^5 particles per ml.

Experimental methods and data analysis

Laser scanning fluorescence microscopy was performed using a Carl Zeiss Axio Observer with a 40× objective (NA = 0.75 and working distance = 0.71 mm). Laser light (wavelength = 543 nm) was used to excite the fluorescent PS particles and images were collected using a 610 nm emission filter. We collected over 200 images at each concentration and analysed these to characterize particle clustering. At the lowest particle concentration (8.46×10^5 particles per ml), images containing nearly 7500 particles were analysed while at the highest concentration

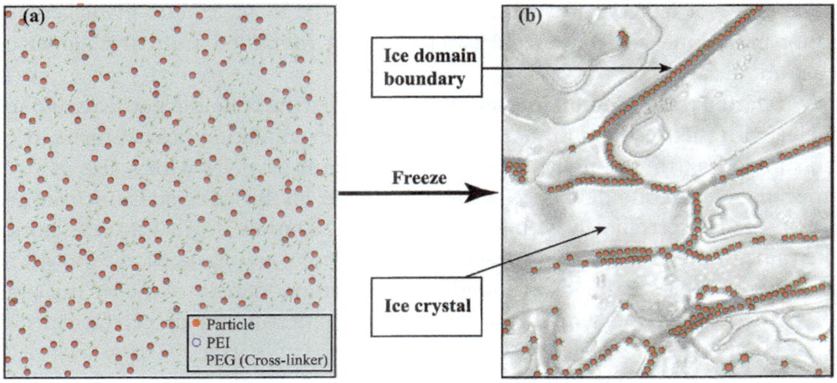

Fig. 1 Schematic of the ice templating process. On freezing a dilute dispersion (a), latex particles (in red), polymers (PEI, in blue) and crosslinker (in green) segregate to the boundaries of the ice crystals (b) and are crosslinked to form clustered aggregates.

(2.96×10^8 particles per ml), there were nearly 140 000 particles in the images analysed.

Image analysis was performed using Image J software. After subtraction of a constant background, the images were thresholded and converted into binary. We then employed the inbuilt particle analysis option in Image J to obtain the area and perimeter of each cluster in the image.

Computational details

In our simulation scheme to investigate aggregate formation during ice templating, we use Brownian dynamics to model the motion of individual colloids. We anticipate, based on our experiments, that colloid aggregation is essentially a kinetic phenomenon, driven by particle expulsion at the water–ice interface. Thus, we implement a minimal simulation scheme, in which we neglect hydrodynamic interactions between colloids. Colloidal particles are modelled as hard spheres, with a diffusion coefficient ($D = 0.2235\ \mu m^2\ s^{-1}$) that is estimated from the Stokes–Einstein relation for 1 μm colloids diffusing in water at a temperature, $T = 273.15$ K. Thus,

$$D = (k_B T)/(6\pi\eta a) \tag{1}$$

where, k_B is the Boltzmann constant, η is the viscosity of water at 273.15 K, and a (= 0.5 μm) is the radius of the colloidal particles.

We implement our simulation by updating the position of particles (x) as in eqn (2).

$$x = x + \sqrt{2D\mathrm{d}t}r \tag{2}$$

where r is a random number drawn from a normal distribution with mean = 0 and standard deviation = 1 and dt represents the simulation time step. The time step was set to 0.5 ms. The size of the colloidal particles was taken as unity. We allowed the colloidal particles to translate in three dimensions according to eqn (2). We carried out simulations for dilute dispersions of the colloidal particles in the absence of ice crystals and confirmed that the mean square displacement is linear in time and has a diffusion constant (obtained from the slope) of 0.22 $\mu m^2 s^{-1}$.

We started the ice templating simulations by placing colloidal particles and ice nuclei, with their centers randomly located in the simulation box. At the start of the simulation, the ice nuclei were treated as point particles with a size of 10^{-3} times that of the colloidal particles. In our scheme, all the ice crystals nucleate at $t = 0$ and grow isotropically with a growth rate of 100 nm s^{-1} (*viz.* 0.1 units per s). The value of the growth rate is based on experimental estimates for ice growth in our experiments (see the Results section). The number density of the ice nuclei is chosen to correspond with the average number density of ice crystals in our experiments.

As the ice crystals grow, the motion of the colloidal particles is restricted since the volume available to the particles decreases. We implement the following rules to dictate particle motion:

(a) If a particle overlaps with another particle during a move, then the move is rejected.

(b) If a particle overlaps with an ice crystal, the motion is reflected. Thus, if a move $x_p = x_p + \delta x$, results in overlap of the particle position with ice, then we implement $x_p = x_p - \delta x$ (where x_p is the position of the colloid and $|\delta x| = \sqrt{2D\mathrm{d}tr}$ is its translation, determined by eqn (2)). If the particle overlaps with an ice particle even after the move is mirrored, the move is rejected. Initially all systems were equilibrated for 2500 steps. These equilibrated coordinates were used to investigate the effect of ice growth on the aggregation of colloidal particles. The simulations were stopped when we visually observed aggregation of the colloidal particles, resulting in a cessation of their Brownian motion. The trajectory was written after every 500 ms (=1000 steps). The aggregation time was observed to vary with particle concentration. All simulations were carried out in triplicate, using three sets of initial coordinates.

We investigated systems with particle concentrations of 10^6, 10^7, 10^8, and 10^9 colloidal particles per ml. The number of particles, ice nuclei and box dimensions corresponding to these concentrations are tabulated in Table 1. We employed periodic boundary conditions in all three directions and used minimum image convention.[39]

Table 1 Simulation system details

Concentration of colloidal particles per ml	No. of colloidal particles in simulation box	No. of ice nuclei in simulation box	Total no. of particles in simulation box	Box dimensions (μm)
10^6	1000	5000	6000	1000.0 × 1000.0 × 1000.0
10^7	1000	500	1500	464.1 × 464.1 × 464.1
10^8	950	50	1000	215.0 × 215.0 × 215.0
10^9	995	5	1000	100.0 × 100.0 × 100.0

Results and discussion

The PS colloids used in these experiments are negatively charged and therefore, PEI chains adsorb on their surface. We have previously characterized the adsorption of PEI on these latices using zeta potential measurements.[40] In this work, a dilute aqueous dispersion of PEI coated PS colloids is frozen by storing at −18 °C. As the water forms ice crystals, particles, PEI and crosslinkers are expelled at the water–ice boundary. Recently, we have reported[38] the preparation of macroporous colloid/crosslinked PEI hybrid materials by ice-templating of particle dispersions (at particle concentrations that are significantly higher than in this work). These macroporous monoliths were characterized by approximately isotropic pores. Also, while the pores were not monodisperse, most pores were within a factor of 1.5 from the mean size. As the pores are templated by ice crystals, these observations suggest that our freezing protocol results in the formation of ice crystals that are approximately uniform in size and that are nucleated in a spatially random manner. We anticipate that this will also hold for ice crystals in the experiments reported here. In accordance with this expectation, optical microscopy on thin (~50–100 μm) samples of dilute latex dispersions, frozen in a cold stage, showed the formation of randomly nucleated ice crystals that had a monomodal size distribution. We can estimate the average growth rate of ice crystals from the time taken for the sample to freeze, assuming that the ice crystals formed are uniform. This gives a growth rate $\sim O(0.1\ \mu\text{m s}^{-1})$. Microscopic observations also indicate that the colloidal particles are accumulated at the growing ice–water front. This is consistent with recent calculations[33] and with a reported experimental "phase diagram".[34] As the water freezes, both the PEI coated colloids and the diepoxide crosslinkers are concentrated at the boundaries of the ice crystals. We use an excess of diepoxide crosslinker to ensure that the PEI coated particles that are in contact are covalently crosslinked. Covalent crosslinking of the PEI proceeds in the frozen state, resulting in the formation of a polymeric mesh around the particles.[38] Thus, colloids that are forced into proximity by the ice crystals form bound aggregates, connected by a crosslinked PEI shell. We allow crosslinking to proceed for 12 hours and subsequently, the sample is thawed. Samples are imaged using a fluorescence microscope to quantify cluster formation.

At the lowest particle concentration (8.46×10^5 particles per ml), we observe a large number of "monomers", *viz.* isolated particles and dimers as well as a few short (~4 to 5 particle long) linear strands and planar aggregates (Fig. 2, top row). As the concentration increases to 1.35×10^7 particles per ml (Fig. 2, middle row) and to the highest concentration investigated here, 2.96×10^8 particles per ml, we observe the formation of longer linear strands, going up to over 40 particles long, as well as elongated planar structures that are 2 particles wide (that we term "tapes") and larger planar structures. Interestingly, for the planar structures, we observe that the monodispersity of the colloids drives the formation of locally crystalline structures. Crystalline order has been observed when monodisperse colloidal dispersions undergo gravity-induced sedimentation characterized by a small Peclet number $\sim O(0.1)$.[41,42] We estimate the Peclet number (Pe) in our experiments based on the growth rate of ice crystals, v. Thus, $\text{Pe} = vR/D = 0.23$, where R is the colloid size and D is the diffusion coefficient of the colloids,

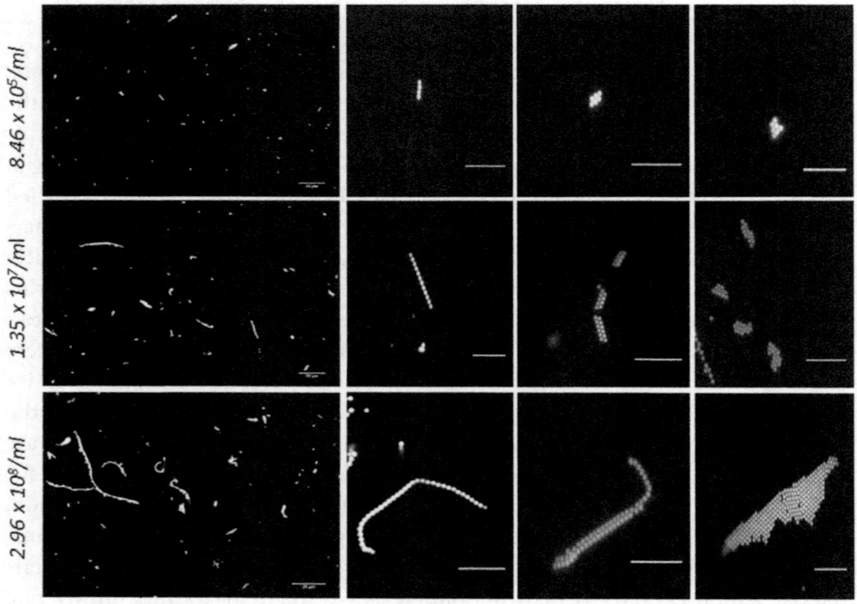

Fig. 2 Microscopic images of particle aggregates from ice templating dispersions at particle concentrations of (top) 8.46×10^5 particles per ml; (middle) 1.35×10^7 particles per ml and (bottom) 2.96×10^8 particles per ml. The scale bar for the low magnification images on the left is 20 μm. The scale bar for the magnified images of individual clusters (block of 3×3 images) on the right is 10 μm.

consistent with the observation of colloidal crystals. At all concentrations, we observe the formation of only a small fraction of three dimensional aggregates. The majority of the clusters are organized into planar assemblies.

The fraction of particles that form monomers or dimers, f_n ($n < 3$), is about 0.4 for the lowest concentration (8.46×10^5 particles per ml) and this decreases approximately logarithmically with concentration to about 0.15 at 2.96×10^8 particles per ml (Fig. 3). We term particle aggregates that comprise $n \geq 3$ particles as clusters and analyse the cluster size distribution as a function of the particle concentration in the ice-templated aqueous dispersions. We define P_n as the number fraction of clusters comprising n particles. We observe that at the lowest concentration, ice templating does not result in the formation of clusters larger than about 15 particles (Fig. 4). However, with an increase in concentration, a few clusters larger than 100 particles are observed to form. We note that all the experiments reported here are at concentrations that are significantly below that required for the particles to organize into a percolated network. We observe that, for all the particle concentrations investigated, $P_n \sim n^{-2}$, for clusters comprising $\approx n < 30$ particles (Fig. 4). For larger n, P_n deviates from n^{-2} to exhibit a tail that stretches to increasingly higher n with an increase in the particle concentration (Fig. 4).

Cluster size distributions have been reported to follow a power law in several aggregation processes. When large clusters form by diffusion limited aggregation of clusters, the time dependent cluster size distribution has been observed[43] to

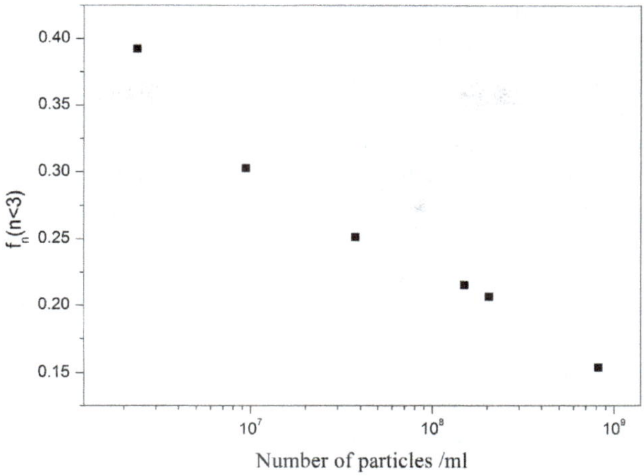

Fig. 3 Fraction of particles existing as unimers or dimers after ice templating.

follow, $P_n \sim t^{-w}n^{\tau}f(n/t^z)$, where t is the aggregation time and f is a function that asymptotes to 1 at long times. In such diffusion limited cluster aggregation, the power law exponent, $\tau \approx -0.75$. MD simulations of clustering of Lennard-Jones fluids below the percolation threshold have been reported[44] to give $P_n \sim n^{-2.1\pm0.1}$. For fluids with short range attractive and long range repulsive interactions, the cluster size distribution has been shown[45] to follow a power law distribution similar to that for random percolation, *viz.* $P_n \sim n^{-2.2}$. Glotzer[46] has reported that the distribution of non-percolated clusters of mobile particles as a glass forming Lennard-Jones fluid is cooled scales as $P_n \sim n^{-1.9}$. A similar power law exponent ($P_n \sim n^{-1.7}$) has been reported[47] for the distribution of crystalline clusters formed in a concentrated dispersion of monodisperse hard spheres. Thus, cluster size distributions that follow a power law, with a power law exponent close to −2 have

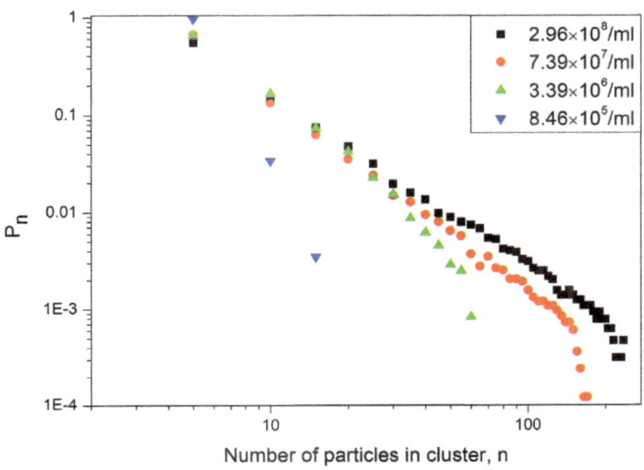

Fig. 4 Cluster size distribution obtained from ice templating.

been observed for a wide variety of particle aggregation processes. It is interesting that a similar behaviour is observed when particles aggregate during ice templating, and that this scaling is independent of concentration over a wide concentration range.

Of particular interest are clusters that are organized as connected linear strings of colloidal particles. Such systems represent colloidal analogues of linear polymers and represent promising models for studying fundamental problems in polymer physics. Therefore, there has been significant interest in realizing such linear architectures.[48–59] We plot the number fraction of linear chains as a function of their size ($P_n^{\,l}$ = number of linear chains of size *n*/total number of linear chains). We observe that 99% of the linear chains have a chain length less than 20 and that the number fraction of linear chains decreases to less than 10^{-3} as *n* approaches 40 (Fig. 5). Remarkably, $P_n^{\,l}$, does not appear to be a very strong function of the colloid concentration in the range 3.39×10^6 to 2.96×10^8 particles per ml.

Finally, we examine the effect of particle concentration on the shape of the colloidal clusters, focusing on planar assemblies that comprise the majority of the clusters formed. We define a connectivity parameter for a cluster, *C*, as the ratio of the cluster area to the square of the perimeter, *viz.* C = area/(perimeter)^2. An increase in *C* represents a transition in cluster shape from a linear aggregate towards a circular shape. We calculate the average *C* at each cluster size, *n*, for ice templated clusters at various dispersion concentrations (Fig. 6). It is difficult to attribute much significance to the variation in *C* with *n*. For example, it is easy to visualize that the difference in *C* between a linear chain and an isotropic planar cluster is small for $n = 5$, relative to that for larger *n*, say, $n = 50$. However, at a fixed cluster size, *n*, it is possible to comment on the relative values of *C* for clusters prepared starting from different particle concentrations. We note that there is a systematic increase in *C* with particle concentration for $n > 10$ (Fig. 6). This indicates that with an increase in concentration from 1.35×10^7 to 7.39×10^7 to 2.96×10^8 particles per ml, there is a transition from linear chains to tapes to planar clusters.

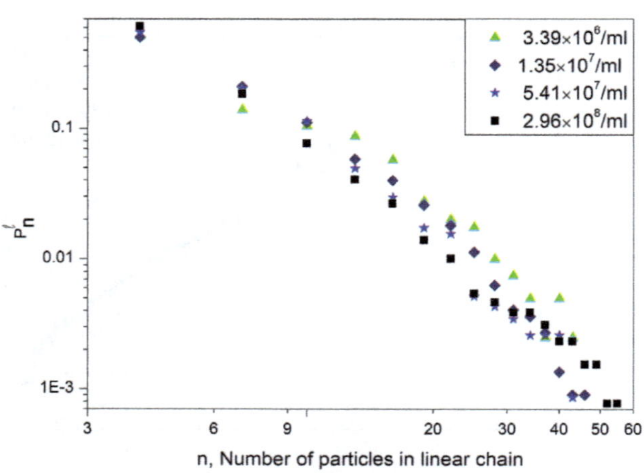

Fig. 5 Size distribution of linear aggregates (colloidal chains) obtained from ice templating.

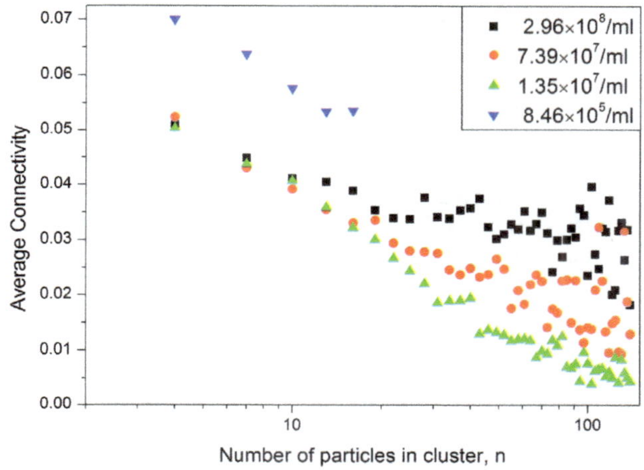

Fig. 6 Variation in the average connectivity parameter with cluster size for ice templated systems at different particle concentrations.

Finally, we note that we obtain qualitatively similar data for the cluster size distribution and the evolution of cluster shape when a dispersion of 1 μm silica particles is ice-templated using the same experimental protocol. As the specific gravity for the silica particles (≈ 2) is significantly higher than that for the polystyrene latex (≈ 1.05), this indicates that gravitational settling during the freezing process is unimportant in these experiments.

We employ simulations to obtain insights into the mechanism of cluster formation in our experiments. Our simulations incorporate several severe assumptions. Firstly, we completely neglect hydrodynamics in modelling the particle motions. Particles are modelled as hard sphere Brownian colloids with no hydrodynamic interactions. We also assume that all ice crystals are nucleated at the initial time, that their nuclei are fixed in space and that the ice grows as isotropic, spherical crystals until they spatially overlap. Furthermore, we assume that, at the low dispersion concentrations employed in our experiments, growth instabilities of the propagating ice front can be ignored. Finally, we assume that the particles experience repulsion when they encounter the growing ice crystals. The simulation is terminated when particles are confined at ice crystal boundaries and when their motions cease. These assumptions are premised on our intuition that particle aggregation during ice templating of such dilute colloidal dispersions is essentially a kinetic phenomenon. Thus, our simulations model a situation where the only physics that governs particle clustering is their exclusion from growing isotropic ice crystals that are randomly nucleated in space. We anticipate that the results of these simulations are sensitive to the nucleation density and growth rate of ice crystals and use simulation values for these parameters that are based on estimates from our experimental observations.

We present snapshots from the late stages of our simulations, where particle motion has ceased due to the particles being trapped between ice crystals (Fig. 7). The ice crystals are not shown in these images, so that the particle clusters can be easily visualized. At a low particle concentration (10^6 particles per ml), we observe

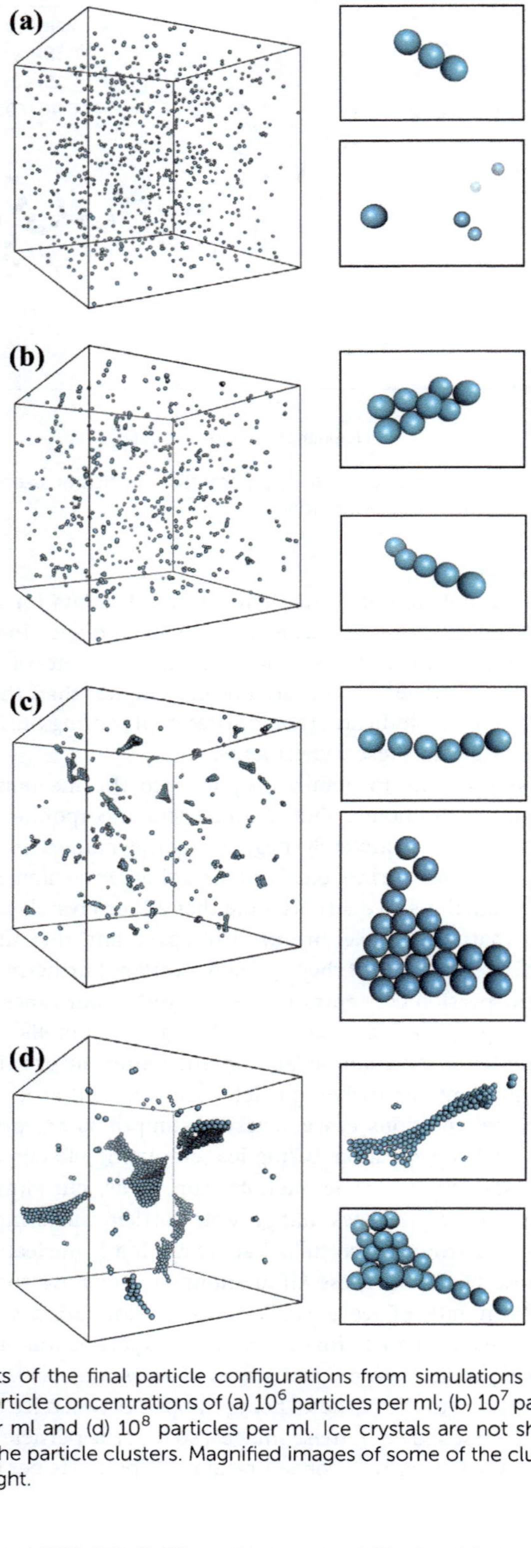

Fig. 7 Snapshots of the final particle configurations from simulations of ice templated dispersions at particle concentrations of (a) 10^6 particles per ml; (b) 10^7 particles per ml; (c) 10^8 particles per ml and (d) 10^8 particles per ml. Ice crystals are not shown for ease of visualization of the particle clusters. Magnified images of some of the clusters formed are shown on the right.

that most colloids exist as isolated monomeric objects. With an increase in the particle concentration, we see the emergence of small particle clusters (at 10^7 particles per ml) that transition to larger clusters (at 10^8 particles per ml) and finally, we observe the formation of several three dimensional structures at 10^9 particles per ml (Fig. 7). Visually, we observe a transition from isolated colloids to short linear structures to more isotropic planar aggregates and finally to large three-dimensional clusters with an increase in the particle concentration, in qualitative agreement with our experimental observations. As in our experiments, we observe local crystalline ordering in the planar aggregates and three dimensional clusters (Fig. 7, right). The number of particles that exist as monomers and dimers decreases, approximately logarithmically, with the dispersion concentration (ESI, Fig. S1†), in qualitative agreement with our experimental results (Fig. 3). However, the agreement with experiments is not quantitative, probably due to the large number of simplifying assumptions in our simulations.

We obtain a cluster size distribution (P_n) for the ice template aggregates as a function of the dispersion concentration. Our simulations sample a smaller number of clusters relative to our experiments. Therefore, while we observe the formation of a very small fraction of large colloidal aggregates in our experiments (Fig. 4), these are not observed in our simulations. Remarkably, our simulations show that for small n ($<\sim 20$), $P_n \sim n^{-2}$, independent of the ice-templated dispersion concentration from 10^7 to 10^9 particles per ml (Fig. 8). Beyond this power law region, there is a dispersion concentration-dependent deviation in P_n. The similarity in the trends in P_n values obtained from simulations with those obtained from experiments is striking. Similarly, the size distribution of linear colloidal chains, $P_n^{\,l}$ obtained from the simulations is independent of the dispersion concentration between 10^6 and 10^9 particles per ml (Fig. 9), in qualitative agreement with our experimental data (compare with Fig. 5).

Therefore, despite the numerous simplifying assumptions, our simulations are able to accurately capture qualitative trends in the size and shape distribution of particle clusters formed during ice templating. This suggests that, at low

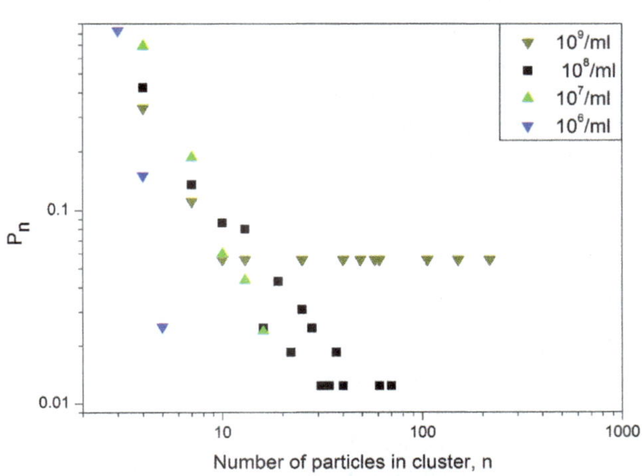

Fig. 8 Cluster size distribution obtained from ice templating simulations.

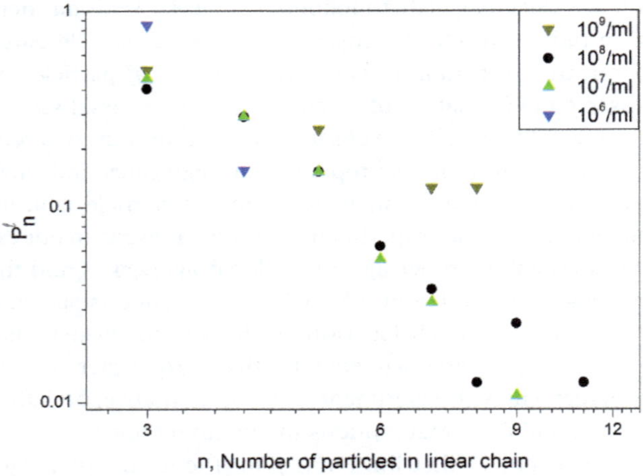

Fig. 9 Size distribution of linear aggregates obtained from ice templating simulations.

concentrations, particle clustering during ice templating is purely kinetic in origin.

Conclusions

Ice templating very dilute colloidal dispersions (containing $\approx 10^6$ to $\approx 10^8$ particles per ml) results in the formation of clusters. With an increase in particle concentration in the ice templated dispersion, the average cluster size increases and there is a transition from linear clusters to two-particle wide "tapes" to more isotropic two-dimensional shapes. We observe relatively few large three-dimensional clusters in this range of particle concentration. In this concentration range, the cluster size distribution exhibits a power law, $P_n \sim n^{-2}$, for clusters containing less than ≈ 30 particles, independent of the dispersion concentration. The distribution of linear colloidal clusters is also only weakly dependent on dispersion concentration. Simulations that ignore hydrodynamic interactions between colloidal particles and that model the exclusion of particles by randomly nucleated, isotropically growing ice crystals capture the essential features of our experimental observations. This suggests that, for low colloidal concentrations, clustering during ice templating is purely kinetic in origin.

Acknowledgements

We acknowledge funding from the Science and Engineering Research Board, Department of Science and Technology through grant number SB/S3/CE/072/2014. BKB acknowledges an Inspire fellowship from the Department of Science and Technology. We acknowledge the IUSSTF Joint Center for Nanoparticle Assemblies that made it possible to have discussions with Prof. Sanat Kumar, Columbia University, USA.

Notes and references

1 W. L. Li, K. Lu and J. Y. Walz, *Int. Mater. Rev.*, 2012, **57**, 38.
2 S. Deville, *Adv. Eng. Mater.*, 2008, **10**, 155.
3 M. C. Gutiérrez, M. L. Ferrer and F. Del Monte, *Chem. Mater.*, 2008, **20**, 634.
4 D. N. Ross, *Lancet*, 1962, **2**, 487.
5 T. M. Freyman, I. V. Yannas and L. J. Gibson, *Prog. Mater. Sci.*, 2001, **46**, 273.
6 J. A. Talalay, *US Pat.* 2,432,353, 1947.
7 S. R. Mukai, H. Nishihara, T. Yoshida, K. I. Taniguchi and H. Tamon, *Carbon*, 2005, **43**, 1563.
8 F. J. O'Brien, B. A. Harley, I. V. Yannas and L. Gibson, *Biomaterials*, 2004, **25**, 1077.
9 N. Dagalakis, J. Flink and P. J. Stasikelis, *J. Biomed. Mater. Res.*, 1980, **14**, 511.
10 G. Chen, T. Ushida and T. Tateishi, *Biomaterials*, 2001, **22**, 2563.
11 M. H. Ho, P. Y. Kuo, H. J. Hsieh, T. Y. Hsien, L. T. Hou, J. Y. Lai and D. M. Wang, *Biomaterials*, 2004, **25**, 129.
12 M. H Ho, D. M. Wang, C. E. Liu, C. H. Hsieh, H. C. Tseng and H. J. Hsieh, *Carbohydr. Polym.*, 2007, **67**, 124.
13 R. Ricciardi, F. Aureimma and C. De Rosa, *Macromol. Symp.*, 2005, **222**, 49.
14 W. Mahler and M. F. Bechtold, *Nature*, 1980, **285**, 27.
15 S. R. Mukai, H. Nishihara, S. Shichi and H. Tamon, *Chem. Mater.*, 2004, **16**, 4987.
16 S. R. Mukai, H. Nishihara and H. Tamon, *Chem. Commun.*, 2004, **116**, 874.
17 H. Nishihara, S. R. Mukai, Y. Fujii, T. Tago, T. Masuda and H. Tamon, *J. Mater. Chem.*, 2006, **16**, 3231.
18 H. Zhang, I. Hussain, M. Brust, M. F. Butler, S. P. Rannard and A. I. Cooper, *Nat. Mater.*, 2005, **4**, 787.
19 S. Deville, E. Saiz, R. K. Nalla and A. P. Tomsia, *Science*, 2006, **311**, 515.
20 E. Munch, M. E. Launey, D. H. Alsem, E. Saiz, A. P. Tomsia and R. O. Ritchie, *Science*, 2008, **322**, 1516.
21 F. Bouville, E. Maire, S. Meille, B. van de Moortèle, A. J. Stevenson and S. Deville, *Nat. Mater.*, 2014, **13**, 508.
22 Q. Shi, Z. An, C. K. Tsung, H. Liang, N. Zheng, C. J. Hawker and G. D. Stucky, *Adv. Mater.*, 2007, **19**, 4539.
23 C. A. L. Colard, R. A. Cave, N. Groissard, J. A. Covington and S. A. F. Bon, *Adv. Mater.*, 2009, **21**, 2894.
24 J. L. Vickery, A. J. Patil and S. Mann, *Adv. Mater.*, 2009, **21**, 2180.
25 P. Vesilind and C. Martel, *J. Environ. Eng.*, 1990, **116**, 854.
26 K. A. Jackson and B. Chalmers, *J. Appl. Phys.*, 1958, **29**, 1178.
27 D. R. Uhlmann, B. Chalmers and K. A. Jackson, *J. Appl. Phys.*, 1964, **35**, 2986.
28 J. Cisse and G. F. Bolling, *J. Cryst. Growth*, 1971, **10**, 67.
29 J. S. Langer, *Rev. Mod. Phys.*, 1980, **52**, 1.
30 W. W. Mullins and R. F. Sekerka, *J. Appl. Phys.*, 1964, **35**, 444.
31 M. F. Butler, *Cryst. Growth Des.*, 2001, **1**, 213.
32 S. S. L. Peppin, J. A. W. Elliot and M. G. Worster, *J. Fluid Mech.*, 2006, **554**, 147.
33 S. S. L. Peppin, J. S. Wettlaufer and M. G. Worster, *Phys. Rev. Lett.*, 2008, **100**, 238301.
34 S. Deville, E. Maire, G. Bernard-Granger, A. Lasalle, A. Bogner, C. Gauthier, J. Leloup and C. Guizard, *Nat. Mater.*, 2009, **8**, 966.

35 S. Deville, E. Maire, A. Lasalle, A. Bogner, C. Gauthier, J. Leloup and C. Guizard, *J. Am. Ceram. Soc.*, 2010, **93**, 2507.
36 A. Lasalle, C. Guizard, E. Maire, J. Adrien and S. Deville, *Acta Mater.*, 2012, **60**, 4594.
37 S. Deville, J. Adrien, E. Maire, M. Scheel and M. Di Michiel, *Acta Mater.*, 2013, **61**, 2077.
38 R. Rajamanickam, S. Kumari, D. Kumar, S. Ghosh, J. C. Kim, G. Tae, S. Sengupta and G. Kumaraswamy, *Chem. Mater.*, 2014, **26**, 5161.
39 D. Frenkel and B. Smit, *Understanding molecular simulation*, Academic, London, 1996.
40 K. P. Sharma, A. K. Ganai, S. Sengupta and G. Kumaraswamy, *Chem. Mater.*, 2011, **23**, 1448.
41 J. P. Hoogenboom, D. Derks, P. Vergeer and A. van Blaaderen, *J. Chem. Phys.*, 2002, **117**, 11320.
42 M. Holgado, F. García-Santamaría, A. Blanco, M. Ibisate, A. Cintas, H. Míguez, C. J. Serna, C. Molpeceres, J. Requena, A. Mifsud, F. Meseguer and C. López, *Langmuir*, 1999, **15**, 4701–4704.
43 T. Vicsek and F. Family, *Phys. Rev. Lett.*, 1984, **52**, 1669.
44 D. M. Heyes and J. R. Melrose, *Mol. Phys.*, 1989, **66**, 1057–1074.
45 J. C. F. Toledano, F. Sciortino and E. Zaccarelli, *Soft Matter*, 2009, **5**, 2390–2398.
46 S. C. Glotzer, *J. Non-Cryst. Solids*, 2000, **274**, 242–255.
47 C. Valeriani, E. Sanz, P. N. Pusey, W. C. K. Poon, M. E. Cates and E. Zaccarelli, *Soft Matter*, 2012, **8**, 4960–4970.
48 E. M. Furst, C. Suzuki, M. Fermigier and A. P. Gast, *Langmuir*, 1998, **14**, 7334.
49 E. M. Furst and A. P. Gast, *Phys. Rev. E: Stat. Phys., Plasmas, Fluids, Relat. Interdiscip. Top.*, 2000, **62**, 6916.
50 A. L. Hiddessen, S. D. Rodgers, D. A. Weitz and D. A. Hammer, *Langmuir*, 2000, **16**, 9744.
51 S. L. Biswal and A. P. Gast, *Anal. Chem.*, 2004, **76**, 6448.
52 Z. Nie, D. Fava, E. Kumacheva, S. Zou, G. C. Walker and M. Rubinstein, *Nat. Mater.*, 2007, **6**, 609.
53 K. Liu, Z. Nie, N. Zhao, W. Li, M. Rubinstein and E. Kumacheva, *Science*, 2010, **329**, 197.
54 S. Sacanna, W. T. M. Irvine, P. M. Chaikin and D. J. Pine, *Nature*, 2010, **464**, 575.
55 D. Li, S. Banon and S. L. Biswal, *Soft Matter*, 2010, **6**, 4197.
56 Y. Wang, Y. Wang, D. R. Breed, V. N. Manoharan, L. Feng, A. D. Hollingsworth, M. Weck and D. J. Pine, *Nature*, 2012, **491**, 51.
57 H. R. Vutukuri, A. F. Demiroers, B. Peng, P. D. J. van Oostrum, A. Imhof and A. van Blaaderen, *Angew. Chem., Int. Ed.*, 2012, **51**, 11249.
58 B. Bharti, G. H. Findenegg and O. D. Velev, *Sci. Rep.*, 2012, **2**, 1004.
59 M. B. Bannwarth, S. Utech, S. Ebert, D. A. Weitz, D. Crespy and K. Landfester, *ACS Nano*, 2015, **9**, 2720.

Faraday Discussions

PAPER

Highly stable noble metal nanoparticles dispersible in biocompatible solvents: synthesis of cationic phosphonium gold nanoparticles in water and DMSO

Yon Ju-Nam,*[a] Wanisa Abdussalam-Mohammed[a] and Jesus J. Ojeda[ab]

Received 28th August 2015, Accepted 30th October 2015
DOI: 10.1039/c5fd00131e

In this work, we report the synthesis of novel cationic phosphonium gold nanoparticles dispersible in water and dimethyl sulfoxide (DMSO) for their potential use in biomedical applications. All the cationic-functionalising ligands currently reported in the literature are ammonium-based species. Here, the synthesis and characterisation of an alternative system, based on phosphonioalkylthiosulfate zwitterions and phosphonioalkylthioacetate were carried out. We have also demonstrated that our phosphonioalkylthiosulfate zwitterions readily disproportionate into phosphonioalkylthiolates *in situ* during the synthesis of gold nanoparticles produced by the borohydride reduction of gold(III) salts. The synthesis of the cationic gold nanoparticles using these phosphonium ligands was carried out in water and DMSO. UV-visible spectroscopic and TEM studies have shown that the phosphonioalkylthiolates bind to the surface of gold nanoparticles which are typically around 10 nm in diameter. The resulting cationic-functionalised gold nanoparticles are dispersible in aqueous media and in DMSO, which is the only organic solvent approved by the U.S. Food and Drug Administration (FDA) for drug carrier tests. This indicates their potential future use in biological applications. This work shows the synthesis of a new family of phosphonium-based ligands, which behave as cationic masked thiolate ligands in the functionalisation of gold nanoparticles. These highly stable colloidal cationic phosphonium gold nanoparticles dispersed in water and DMSO can offer a great opportunity for the design of novel biorecognition and drug delivery systems.

1. Introduction

The ability to modify noble metal nanoparticles, such as gold nanoparticles with various organic thiolate ligands, makes them versatile systems and opens a range of possibilities for their use as drug delivery systems,[1–3] and biosensing[4–6]

[a]*Swansea University, College of Engineering, Engineering Central Building, Bay Campus, Crymlyn Burrows, Swansea, SA1 8EN, UK. E-mail: y.k.ju-nam@swansea.ac.uk*
[b]*Brunel University London, Institute of Materials and Manufacturing, Experimental Technique Centre, Uxbridge, UB8 3PH, UK*

applications. Even though the use of these nanoparticles in therapeutic purposes is promising, challenges still remain and when designing drug nanocarriers, the following considerations are usually taking into account: (i) possible drawbacks due to free drugs, (ii) tissue permeation and cellular uptake, and (iii) homogenous distribution within the target tissues and organs.[7–13] Therefore, for any biomedical application the main requirements are the affinity to aqueous media and compatibility with the immediate biological environment or appropriate receptor groups for sensing or diagnostic applications.

Ligands are introduced to the surface of the nanoparticle either to stabilise the particle and avoid uncontrolled aggregation, or to add functionality to the particle for biomolecular recognition, to enhance transport and anchoring properties.[7] Recent studies have shown that the size, shape and surface charge of nanoparticles, in general, play crucial roles in their entry and subsequent access of the nanoparticles into living cells.[6,13–16] For instance, cationic nanoparticles are more effective at penetrating into mammalian cells than anionic nanoparticles.[17] Nanoparticle surface functionalisation plays an important role in the cellular uptake and producing cellular responses.[18,19] However, information published in the literature on how functionalised nanoparticle surface characteristics, such as aromaticity and hydrophobicity, affect cellular internalisation processes is still unclear. Al-Hajaj *et al.* have shown that the functionalised quantum dots they produced with non-specified ligands were taken up by lipid host mediated endocytosis in human kidney and liver cells.[20] Designing and tailoring the surface of inorganic nanoparticles with suitable ligands may potentially offer long-term stability under a wide range of conditions such as high electrolyte concentration, a broad pH range, and biogenic or naturally occurring thiols.[3] Zhan *et al.* have shown that quantum dots modified with multidentate lipoic ligands containing a zwitterionic head group can be highly biocompatible nanomaterials,[21,22] making them attractive nanodevices for clinical *in vitro* and *in vivo* imaging applications.[23–25] More recently, Rotello *et al.* have studied the behaviour of gold nanoparticles with different hydrophobicities and the effect of surface functionality on hemolysis.[26] They observed a linear haemolytic behaviour with increasing hydrophobicity in the absence of serum media. McIntosh *et al.* synthesised gold nanoparticles capped with a mixture of octanethiol and 11-trimethylammonium-undecanethiol, and others functionalised with trimethyl ammonium cationic head groups on the surface can interact electrostatically and bind with the negatively charged phosphate backbone of 37mer duplex DNA.[27] As in these studies reported in the literature, the majority of the cationic species investigated and used are ammonium-based groups. Here in this work, we are proposing the use of cationic phosphonium-based groups as alternatives to the ammonium ones. The phosphonium groups can offer a range of advantages including the availability of a broad range of organic derivatives, which allows of design and production of a wide variety of functionalised gold nanoparticles. The ability of triphenylphosphonium head groups to travel across cell membranes is well-known.[28–30] There are also several studies on the use of phosphines and phosphine oxides to stabilise nanoparticles. However, there have been far fewer studies of the use of organophosphorus ligands.[31–33]

In addition to imparting specific functionalities to nanoparticles, their dispersion in water or any other biocompatible solvent is crucial for their use in biomedical or biological applications. Therefore, water should be the primary solvent used to disperse functionalised nanoparticles. However, in many cases, the ligands used to protect and provide functionality to nanoparticles, and the corresponding nanomaterials produced are not soluble or dispersible in water. It would be highly beneficial to have a range of other biocompatible solvents. One potential alternative to water may be dimethyl sulfoxide (DMSO). According to the FDA, DMSO is a solvent derived from wood, which has been the subject of considerable interest for its potential use as a drug. At present, the only human use for which DMSO has been approved is interstitial cystitis, a bladder condition. DMSO is also known as a "carrier" chemical; it could deliver substances into the bloodstream through the skin.[34] For all the reasons outlined here, DMSO was the chosen solvent to synthesise and disperse our nanoparticles.

Here in this paper we are reporting for the first time the synthesis of cationic gold nanoparticles by reducing gold(III) salt in the presence of cationic phosphonium ligands in DMSO. We also describe the synthesis of a new cationic phosphonium ligand, tri(*p*-tolyl)phosphoniopropylthiosulfate zwitterion, used to functionalise the gold nanoparticle surface in DMSO. Two previously reported phosphonium ligands, triphenylphosphoniopropylthiosulfate zwitterion[35] and (3-thioacetylpropyl)triphenylphosphonium bromide,[36] were used in this work to produce the cationic phosphonium gold nanoparticles in DMSO and compare them with the new ligand synthesised, tri(*p*-tolyl)phosphoniopropylthiosulfate.

2. Materials and methods

(a) General

All chemicals and solvents were purchased from Sigma-Aldrich and Fisher Scientific Ltd. For the characterisation of the ligands and functionalised gold nanoparticles several analytical techniques were used. TLC, ATR-FTIR, NMR and ESI-MS techniques were employed to characterise the phosphonium-containing ligands. UV-visible spectroscopy and TEM were used to characterise and obtain the size of the cationic phosphonium gold nanoparticles. The stabilities of the colloidal gold nanoparticle solutions were monitored using UV-visible techniques.

Thin layer chromatography (TLC). Analytical thin layer chromatography was carried out on Merck silica gel 60 F254 plates by using mixtures of methanol : dichloromethane (DCM) as the eluent system.

Attenuated total reflectance-Fourier transform infrared (ATR-FTIR) spectroscopy. IR spectrum acquisition was performed on solid samples using a Nicolet 5700 FT-IR spectrometer equipped with Omnic software (version 7.1) and a Smart Omni-Sampler (ATR cell with single reflectance germanium crystal). Each recorded spectrum was the average of 32 scans with a spectral resolution of 4 cm^{-1} from 400 to 4000 cm^{-1} on a dried sample, and a background spectrum was recorded before each analysis. Spectra were measured, analysed and fitted using Origin software (version 7.5) equipped with a Peak-Fitting Module (PFM).

Nuclear magnetic resonance spectroscopy (NMR). ^1H NMR spectra were acquired on a JEOL 400 MHz spectrometer. Data are reported as follows: s, singlet; d, doublet; t, triplet; q, quartet; m, multiplet. Appropriate solvents such as dimethyl sulfoxide, chloroform and diethyl ether were used to dissolve the samples to a concentration of 15 to 40 mg ml^{-1} for the ^1H experiments.

Electrospray ionisation-mass spectrometry (ESI-MS). ESI-MS analyses were generally performed on an Orbitrap XL using a nano-electrospray in positive or negative ionisation modes. The Orbitrap XL high resolution instrument gives accurate mass measurements over the full mass range in electrospray resolution: up to 100 000 (FWHM). Mass range: m/z 50–2000 or m/z 200–4000. Mass accuracy: <3 ppm RMS with external calibration, <2 ppm RMS with internal calibration.

All samples were initially solvated in 250 μl of DCM before approximately 10 μl of the solutions were added to 1 ml of methanol with NH_4OAc. A portion of these samples was added to a 96 well plate to be analysed by positive nano-electrospray.

Ultraviolet-visible absorption spectroscopy (UV-visible). UV-visible spectra were obtained on a HITACHI U-2900 spectrophotometer with a double beam principle system and data recording using a Vision software version on Windows XP/2000. Samples were analysed in a UV-quartz cuvette with a 1 cm optical path. Appropriate dilutions to the original samples were carried out using either water or DMSO.

Transmission electron microscopy (TEM). A JEOL2100 field emission gun transmission electron microscope (FEG TEM) set at 100 kV was used for the analysis of the particles and obtaining the TEM micrographs. The size distribution was obtained by counting and measuring over 150 gold nanoparticles. A deposition method was used to prepare the samples for TEM analyses. A drop of a diluted colloidal solution of the functionalised gold nanoparticles was placed and suspended on a Holey carbon coated copper grid. Then the grids were left to dry at room temperature.

(b) Synthesis of the cationic phosphonium ligands

Tri(*p*-tolyl)phosphoniopropylthiosulfate zwitterion synthesis and characterisation are reported for the first time in this work. The syntheses of the triphenylphosphoniopropylthiosulfate zwitterion and ω-thioacetylpropylphosphonium salt were carried out following previously reported methods.[35,36] The synthetic protocols used to produce the three ligands are outlined in this section.

Synthesis of the tri(*p*-tolyl)phosphoniopropylthiosulfate zwitterion. The synthesis of the tri(*p*-tolyl)phosphoniopropylthiosulfate zwitterion was carried out following the reactions shown in Scheme 1. In a round bottom flask with a reflux condenser a mixture of tri(*p*-tolyl)phosphine (**1**) (3.8 mmol) with an approximate amount of bromo-propanol (15 mmol) was refluxed for five hours in acetonitrile (0.024 mmol) and the solid precipitate was collected and recrystallized from ethanol (reaction 1, Scheme 1). The yield was 70%. The resulting salt (**2**) was dissolved in hydrobromic acid (0.18 mmol, 10 ml) (48%) in a round bottom flask under reflux for five hours (reaction 2, Scheme 1). After the reaction mixture was left to cool down, the compound appeared as a yellow oil (**3**). In order to prepare the tri(*p*-tolyl)phosphoniopropylthiosulfate zwitterion (**4**), a mixture of ω-bromopropyl-tri(*p*-tolyl)phosphonium bromide (**3**) (1 mmol) and sodium thiosulfate (1.5 mmol) was refluxed for five hours in aqueous ethanol (1 : 1, 10 ml) (reaction 3, Scheme 1). TLC was used in order to follow the progress of the reaction by using 20% : 80% methanol : DCM as the mobile phase. All compounds were purified further by liquid–liquid extraction of the reaction mixture using DCM and by trituration with dry diethyl ether.

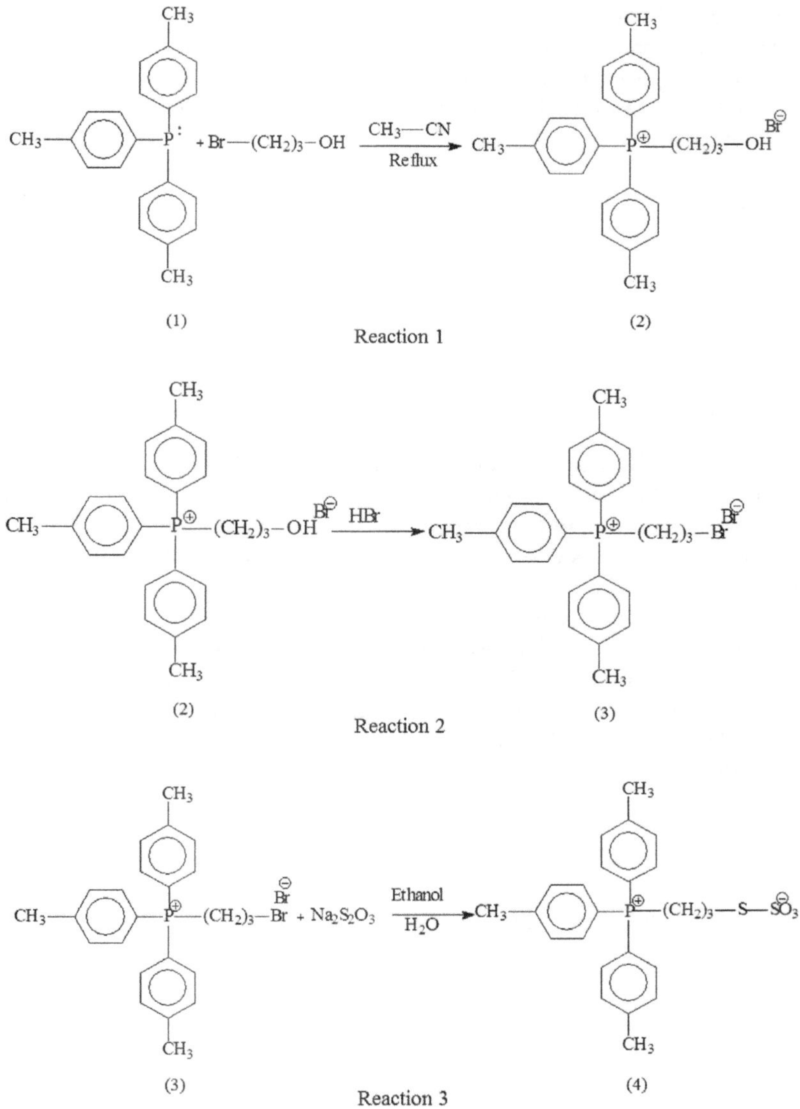

Scheme 1 Synthesis of the tri(*p*-tolyl)phosphoniopropylthiosulfate zwitterion (**4**).

Synthesis of triphenylphosphoniopropylthiosulfate. In order to obtain the triphenylphosphoniopropylthiosulfate zwitterion,[35] three main chemical reactions were carried out. The first stage involved the synthesis of the hydroxylpropylphosphonium salt and this was generated by a quaternation reaction of triphenylphosphine (4 mmol) and the corresponding bromo-propanol (15 mmol) in acetonitrile under reflux for four to five hours. The resulting salt was dissolved in hydrobromic acid (48%) and the reaction was carried out under reflux for five hours to obtain the ω-bromopropyltriphenylphosphonium bromide salt. The latter (1 mmol) was then treated with sodium thiosulfate (1.5 mmol) in ethanol/

water under reflux for five hours in order to produce the phosphoniopropylthiosulfate zwitterion. The progress of all the reactions was monitored by TLC using methanol : dichloromethane in a 1 : 4 ratio as the mobile phase. All the compounds were purified by dichloromethane extraction and re-crystallisation using diethyl ether.

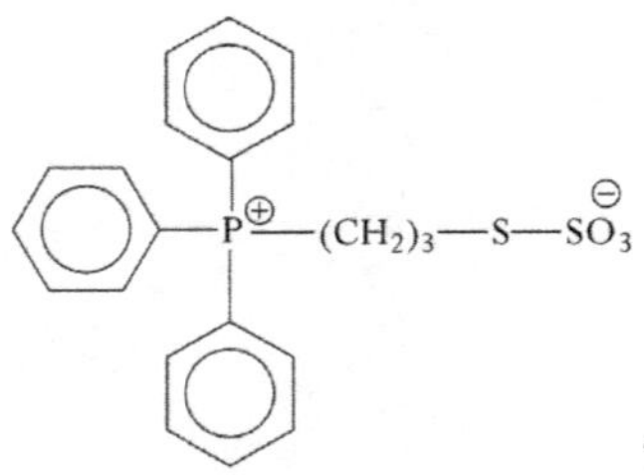

Triphenylphosphoniopropylthiosulfate zwitterion

Synthesis of (3-thioacetylpropyl)triphenylphosphonium bromide. In order to obtain (3-thioacetylpropyl)triphenylphosphonium bromide, hydroxypropylphosphonium and ω-bromopropyltriphenylphosphonium bromide salts were also synthesised as described in the synthesis of the triphenylphosphoniopropylthiosulfate zwitterion.[35,36] Once the bromide salt (2 mmol) was produced, it was reacted with potassium thioacetate (3 mmol) in a mixture of ethanol/water. The progress of the reaction was followed by TLC using a mobile phase of methanol : dichloromethane in a 1 : 4 ratio. (3-Thioacetylpropyl)triphenylphosphonium bromide was isolated and purified by liquid–liquid extraction using DCM as the extracting solvent, and re-crystallisation of the bromide salt using diethyl ether.

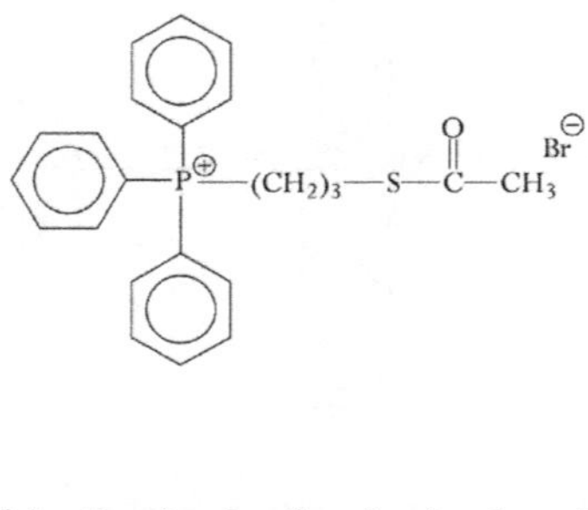

(3-Thioacetylpropyl)triphenylphosphonium bromide

(c) Synthesis of cationic phosphonium gold nanoparticles in DMSO and water

The method for the synthesis of colloidal cationic phosphonium gold nanoparticles in DMSO is reported for the first time in this work. These nanoparticles were obtained by a reduction of the gold salt in the presence of protecting ligands in DMSO. Colloidal cationic phosphonium gold nanoparticle solutions in water were also prepared for comparison purposes, following the principles involved in the method developed by Brust and Schifrin for a two-phase liquid–liquid system.[35] Variations were made to this protocol due to the solubility of the phosphonium ligands synthesised.

Cationic phosphonium gold nanoparticles dispersed in DMSO. In order to prepare the colloidal gold nanoparticles in DMSO, solutions of the phosphonium ligands, triphenylphosphoniopropylthiosulfate zwitterion, (3-thioacetylpropyl) triphenylphosphonium bromide and tri(*p*-tolyl)phosphoniopropylthiosulfate zwitterion were prepared in DMSO (0.25 mmol, 0.30 mmol, and 0.8 mmol, respectively). The volume used for these solutions was 20 ml. A solution of potassium tetrachloroaurate (0.12 mmol) in DMSO (10 ml) was also prepared. Then both the ligand and gold salt solutions were mixed and stirred. The reduction was carried out by adding 5 ml of freshly prepared aqueous solution of sodium borohydride (2.0 mmol) to the DMSO mixture. In order to remove the excess of ligands in each one of the gold colloidal solutions prepared, liquid–liquid extractions were carried out using diethyl ether.

Cationic phosphonium gold nanoparticles dispersed in water. Solutions of the phosphonium ligands (triphenylphosphoniopropylthiosulfate zwitterion and (3-thioacetylpropyl)triphenylphosphonium bromide) were prepared in 20 ml of dichloromethane (DCM) (0.25 mmol, 0.30 mmol, 0.8 mmol, respectively) and potassium tetrachloroaurate (0.12 mmol) was also added to the DCM solution.[35,36] This was vigorously stirred for 2 hours and until the potassium gold salt was totally dissolved. The reduction of gold(III) to gold(0) in the presence of phosphonium ligands was carried out by adding drop by drop a freshly prepared sodium borohydride solution in water (3 ml, 400 mM) with vigorous stirring. After 20 minutes of stirring, 15 ml of deionised water was then added to the mixture. The latter is kept stirring overnight. The initial DCM was removed from the aqueous phase, as the functionalised nanoparticles transferred to this phase. Then, three liquid–liquid extractions were carried out using DCM in order to purify the cationic phosphonium gold nanoparticles dispersed in water.

3. Results and discussion

(a) Synthesis of tri(*p*-tolyl)phosphoniopropylthiosulfate zwitterion

The synthesis of the tri(*p*-tolyl)phosphoniopropylthiosulfate zwitterion involved the preparation of two intermediates, which were obtained and purified in order to get the zwitterion. The first compound synthesised was the hydroxylpropylphosphonium salt and this was obtained by the quaternisation of tri(*p*-tolyl)phosphine with bromo-propanol as shown in Scheme 1. Hydroxypropyl-tri(*p*-tolyl)phosphonium bromide was then treated with HBr in order to obtain the ω-bromopropyl-tri(*p*-tolyl)phosphonium bromide. The latter is then reacted with $Na_2S_2O_3$ in aqueous ethanol. The resulting zwitterion was purified by trituration using diethyl ether. The final product had an oil appearance and was yellow coloured. When studied by ESI-MS in positive ion mode, an ion corresponding to $M + H^+$ was observed (459). The ^1H NMR showed a distinctive chemical shift at 2.3 ppm corresponding to the nine protons from the $-CH_3$ attached to the aromatic ring. This chemical shift is the main difference between the tri(*p*-tolyl)phosphoniopropylthiosulfate and the triphenylphosphoniopropylthiosulfate ^1H NMR spectra. It can be assumed that the sulfur–sulfur bond of tri(*p*-tolyl)phosphoniopropylthiosulfate undergoes cleavage with loss of the sulfite ion under reductive conditions as in the synthesis of gold colloidal nanoparticles, as was shown and reported in the case of triphenylphosphoniopropylthiosulfate. This cleavage will allow the sulfur of the cationic phosphonium ligand to interact with the gold nanoparticle surface. As previously mentioned, these thiosulfate and thioacetate ligands are usually known as masked thiol ligands. For triphenylphosphoniopropylthiosulfate zwitterion and (3-thioacetylpropyl)triphenylphosphonium bromide, similar results to those previously reported were observed when studied using IR, NMR and ESI-MS. Solubility tests were carried out for all three masked thiol ligands. They were soluble in methanol, ethanol, DMSO, chloroform and DCM.

(b) Cationic phosphonium gold nanoparticles dispersed in DMSO and water

Triphenylphosphoniopropylthiosulfate and (3-thioacetylpropyl)triphenylphosphonium bromide ligands have shown their ability to functionalise and stabilise gold nanoparticle surfaces in water. For this work, another biocompatible solvent, DMSO, was used in order to assist with the solubilisation of more hydrophobic ligands (in our case the tri(*p*-tolyl)phosphoniopropylthiosulfate zwitterion), its interaction with the gold nanoparticle surface, and then penetration into the lipophilic cell wall. These thiosulfate containing salts are also known as Bunte salts. Shon *et al.* were the first to show the production of alkanethiolate–gold nanoparticles from Bunte salts.[37] They used sodium *S*-dodecylthiosulfate and showed that the sulfur–sulfur bond of the thiosulfate salt underwent cleavage and acted as the free thiol.[35,37] Functionalised gold nanoparticles dispersed in water were obtained when triphenylphosphoniopropylthiosulfate and (3-thioacetylpropyl)triphenylphosphonium bromide ligands, and the two-phase liquid–liquid (DCM/water) method were used. However, when tri(*p*-tolyl)phosphoniopropylthiosulfate zwitterion was used as the protecting ligand with the same method, no transfer of these nanoparticles to the water phase was observed, and, therefore, no stable gold colloidal nanoparticles were obtained in water. Various attempts to synthesise phosphonium cationic gold nanoparticles were carried out using tri(*p*-tolyl)phosphoniopropylthiosulfate zwitterion as the stabilising ligand and the two phase DCM/water method. The concentration of this ligand was also increased to double (80 mM) and the same results were obtained. Therefore, in the case of the tri(*p*-tolyl)phosphoniopropylthiosulfate zwitterion, even increments in the concentration of the ligand did not help the formation of functionalised gold nanoparticles dispersed in water using the *p*-tolyl ligand.

DMSO was chosen as an alternative solvent to disperse the gold nanoparticles functionalised with tri(*p*-tolyl)phosphoniopropylthiolate ligand. Due to the properties described in the introduction, DMSO was used to synthesise gold nanoparticles functionalised with the tri(*p*-tolyl)phosphoniopropylthiosulfate zwitterion, and also allow us to use more hydrophobic phosphonium ligands to fabricate carriers with potential abilities to permeate and penetrate the lipophilic cell walls.

In order to prepare the colloidal gold nanoparticles in DMSO, a reduction of gold(III) to gold(0) was carried out in the presence of tri(*p*-tolyl)phosphoniopropylthiosulfate zwitterion in DMSO, using sodium borohydride as the reducing agent. With this reducing method using DMSO as the dispersing solvent, a ruby-coloured colloidal solution of gold nanoparticles functionalised with tri(*p*-tolyl)phosphoniopropylthiolate was obtained. Using the same method in DMSO, other colloidal solutions of gold nanoparticles functionalised using triphenylphosphoniopropylthiosulfate zwitterion and (3-thioactylpropyl)triphenylphosphonium bromide as protecting ligands were also successfully obtained. Stability studies were also carried out in order to monitor over time the potential aggregation in the colloidal solutions prepared following this method.

(c) UV-visible and TEM studies

Using a UV-vis spectroscopy technique, the formation of colloidal gold nanoparticles in DMSO (Fig. 1) and as previously mentioned, the stability of these solutions were monitored (Fig. 2–4).

It is well-known that the dark-red colour of gold colloids dispersed in water reflects the surface plasmon band (SPB), which is a broad absorption band in the visible region around 520 nm.[38] The SPB is due to the collective oscillations of the electrons at the surface of nanoparticles (6s electrons of the conduction band for AuNPs) that are correlated with the electromagnetic field of incoming light.[39] The maximum and bandwidth are also influenced by the particle size.[40–43] It has been shown that for gold nanoparticles of mean diameters of 9, 15, 22, 48, and 99 nm, the SPB maximum λ_{max} was observed at 517, 520, 521, 533, and 575 nm, respectively, in aqueous media.[40,44]

The method for the synthesis of colloidal gold nanoparticles in DMSO using triphenyl- and tri(*p*-tolyl)phosphoniopropylthiolate as protecting ligands is reported for the first time in this work. Evidence of the formation of gold nanoparticles using the reduction method in DMSO was shown by a UV-visible spectroscopy technique. Using DMSO as the blank, bands centred at 519, 519 and 529 nm were observed in the resulting UV spectra (Fig. 1) corresponding to the burgundy gold colloidal solutions prepared with triphenylphosphoniopropylthiosulfate (TPPTS-AuNPs), (3-thioacetylpropyl)triphenylphosphonium bromide (TPPTA-AuNPs) and tri(*p*-tolyl)phosphoniopropylthiosulfate (TTPTS-AuNPs), respectively. Their stability was also monitored by UV-vis spectroscopy and the DMSO solutions were shown to be stable for 12 weeks at least (Fig. 2–4). The colloidal solutions were stored in the dark and at room temperature.

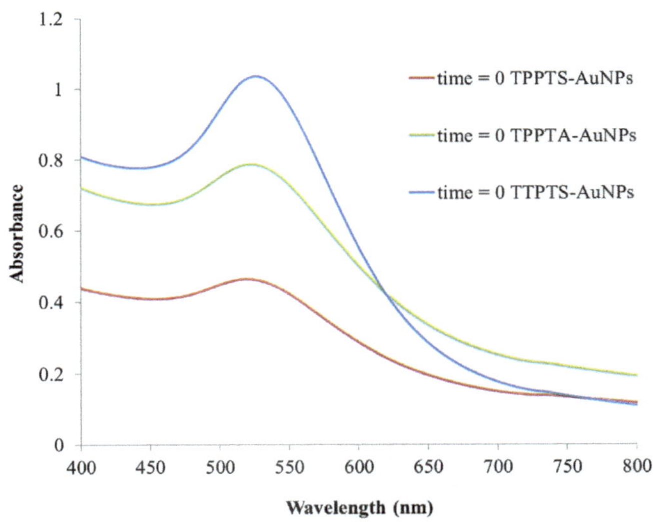

Fig. 1 UV-visible absorption spectra of the colloidal solutions of gold nanoparticles functionalised with triphylphosphoniopropylthiosulfate zwitterion (TPPTS-AuNPs), (3-thioacetylpropyl)triphenylphosphonium bromide (TPPTA-AuNPs), and tri(*p*-tolyl)phosphoniopropylthiosulfate zwitterion (TTPTS-AuNPs) in DMSO, time = 0.

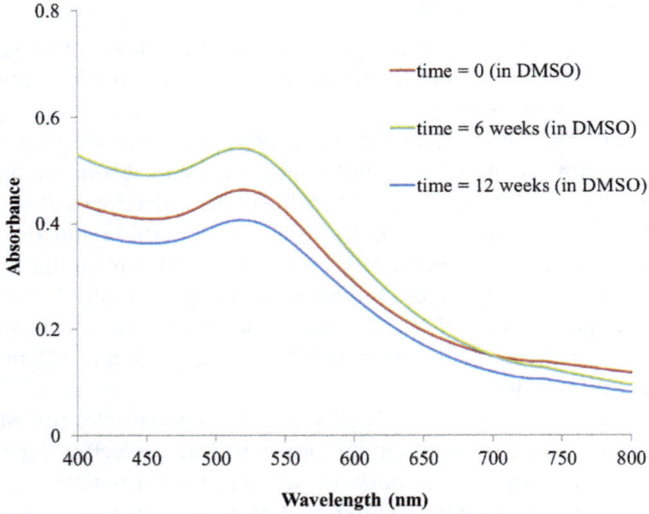

Fig. 2 UV-visible absorption spectra of the colloidal solutions of gold nanoparticles functionalised using triphylphosphoniopropylthiosulfate zwitterion as protecting ligand (TPPTS-AuNPs). The different UV-visible spectra represent these nanoparticles dispersed in DMSO at time = 0, 6 and 12 weeks.

As previously reported and repeated for this work for comparison purposes,[35,36,45] gold nanoparticles synthesised using triphenylphosphonio-propylthiosulfate and (3-thioacetylpropyl)triphenylphosphonium bromide as protecting ligands, dispersed in water, showed broad bands centred at 519 nm in

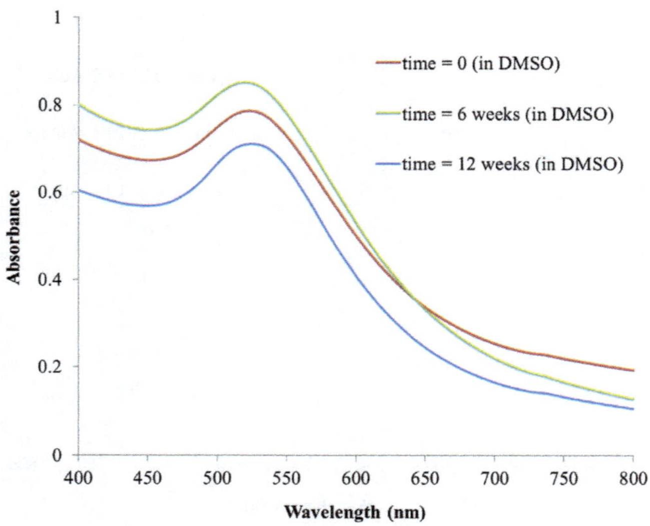

Fig. 3 UV-visible absorption spectra of the colloidal solutions of gold nanoparticles functionalised using (3-thioacetylpropyl)triphenylphosphonium bromide as protecting ligand (TPPTA-AuNPs). The different UV-visible spectra represent these nanoparticles dispersed in DMSO at time = 0, 6 and 12 weeks.

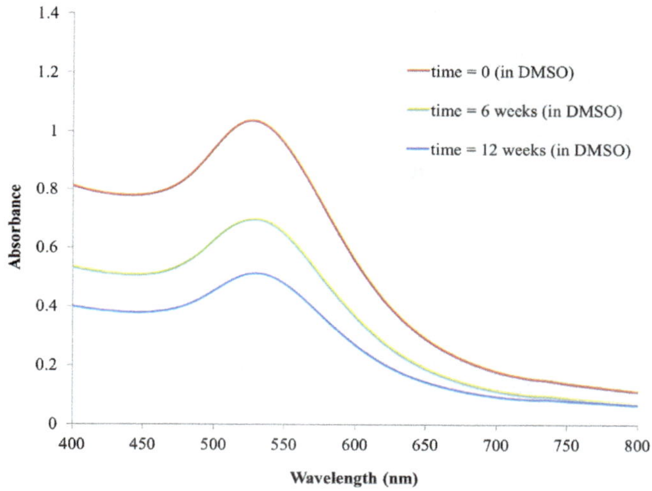

Fig. 4 UV-visible absorption spectra of the colloidal solutions of gold nanoparticles functionalised using tri(*p*-tolyl)phosphoniopropylthiosulfate zwitterion as protecting ligand (TTPTS-AuNPs). The different UV-visible spectra represent these nanoparticles dispersed in DMSO at time = 0, 6 and 12 weeks.

both cases in the UV-visible spectra. When compared with the UV spectra of the colloidal solutions of gold nanoparticles functionalised with triphenylphosphoniopropylthiolate in water and DMSO (time = 0), they did not show significant differences in the wavelengths. In our case the refractive index of the solvent DMSO has not been shown to induce a shift of the SPB of our nanoparticles as suggested in the literature.[44]

It is known that the ligand shell also alters the refractive index and causes either a red or blue shift.[41] This shift is especially significant with thiolate ligands, which are responsible for a strong ligand field interacting with a surface electron cloud.[41] Similar evidence can be observed when comparing the colloidal solutions in DMSO of gold nanoparticles stabilised with triphenylphosphoniopropylthiolate ligand with those corresponding to gold nanoparticles functionalised with tri(*p*-tolyl)phosphoniopropylthiolate. Differences in wavelengths (519 and 529 nm, respectively) can be observed when the ligands change in structure (see Fig. 2 and 4).

The colloidal solutions of gold nanoparticles functionalised using triphenylphosphoniopropylthiosulfate (Fig. 5, TEM1), (3-thioacetylpropyl)triphenylphosphonium bromide (Fig. 5, TEM2) and tri(*p*-tolyl)phosphoniopropylthiosulfate (Fig. 5, TEM3) as protecting ligands, were all analysed by TEM. Micrographs of all three gold nanoparticle samples showed spherical or semi-spherical shaped particles. Then, particle sizes for samples TEM1, TEM2 and TEM3 were obtained by analysing at least 150 particles per sample from several images taken.

All samples showed similar particle sizes, and mean values were obtained by calculating the averages and standard deviations. The calculated mean diameters were the following: 11 ± 1 nm (TEM1), 8 ± 1 nm (TEM2) and 10 ± 1 nm (TEM3). Samples TEM1 and TEM2 showed similar particle sizes to those reported and synthesised using the two-phase DCM/water method.[35,36] This evidence is confirmed by that found in the corresponding UV-visible spectra. Although the solvents used to disperse the nanoparticles were different, the principle of the reduction method used and also the concentrations used for both methods, two phase DCM/water and DMSO, were similar.

Gold nanoparticles in DMSO derived from triphenylphosphoniopropylthiosulfate zwitterion and (3-thioacetylpropyl)triphenylphosphonium bromide are slightly different in size and particle size distribution. This can be due to the differing passivation kinetics of the types of ligands used for this study, and the cleavage of the sulfur–sulfur and sulfur–carbon bonds in the thiosulfate and thioacetate species.[45] The mechanisms of these cleavages are still unclear. However, the presence of water traces in the solvent used for the synthesis of colloidal gold nanoparticles has been shown to be crucial for the hydrolysis of the thiosulfate and thioacetate ligands and formation of alkylthiolates.[46,47] These thiolate anions will lead to the formation of phosphonium-functionalised gold

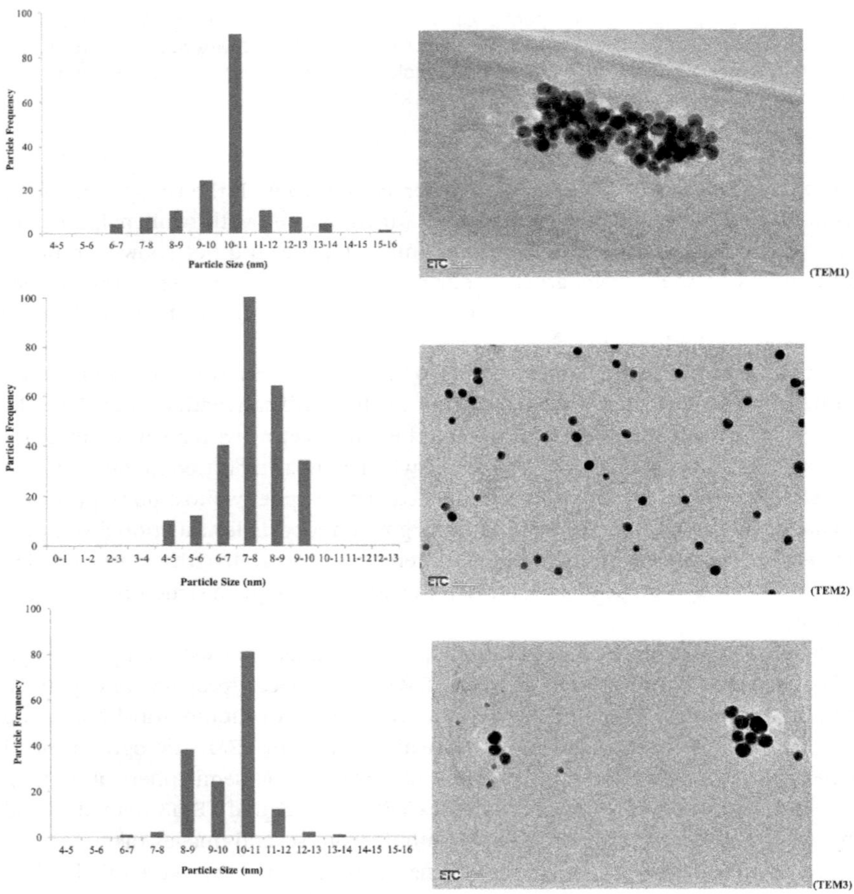

Fig. 5 Representative TEM micrographs of cationic phosphonium gold nanoparticles synthesised using triphenylphosphoniopropylthiosulfate (TEM1), (3-thioacetylpropyl)triphenylphosphonium bromide (TEM2), and tri(*p*-tolyl)phosphoniopropylthiosulfate (TEM3) as protecting ligands in DMSO, and corresponding particle size histograms.

nanoparticles. The presence of a reducing agent in the solution also influences the cleavage of sulfur–sulfur and sulfur–carbon bonds.[45]

In the case of the tri(*p*-tolyl)phosphoniopropylthiolate functionalised gold nanoparticles, they could only be produced using the DMSO method. The corresponding UV-visible spectrum shows that the SPB is centred at 529 nm (Fig. 4). With this evidence only, it could be assumed that the nanoparticles were slightly larger in size than those functionalised with the triphenylphosphoniopropylthiolate ligands, which show bands centred at 519 nm (Fig. 2 and 3). The slight blue shift observed in the UV-visible spectrum of the tri(*p*-tolyl)phosphoniopropylthiolate functionalised gold nanoparticles might be due to the ligand shell altering the refractive index and interacting more strongly with the electron cloud of the nanoparticle surface, and not due to the particle size. This evidence supports the experimental observations and results of Su *et al.*[41]

The agglomeration of nanoparticles can be observed in the TEM1 and TEM3 micrographs (Fig. 5). This recurrent experimental observation can be due to the use of the drop deposition method for sample preparation. This technique is widely used for TEM sample preparation, however, according to studies reported in the literature,[48] the use of this method can lead to agglomeration of particles at the substrate edges as the solvent evaporates. This accumulation will also depend on the sample concentration. No particle agglomeration can be observed in the case of gold nanoparticles prepared using (3-thioacetylpropyl)triphenylphosphonium bromide (TEM2, Fig. 5). We can speculate that in this case, the sample concentration was lower than those in the case of the solutions of gold nanoparticles prepared with the other two ligands.

4. Conclusions

The fabrication of cationic phosphonium gold nanoparticles using triphenylphosphoniopropylthiosulfate zwitterion, (3-thioacetylpropyl)triphenylphosphonium bromide and tri(*p*-tolyl)phosphoniopropylthiosulfate zwitterion as protecting ligands and a reduction method in DMSO is reported here for the first time; and also the synthesis of a new masked thiol, tri(*p*-tolyl)phosphoniopropylthiosulfate zwitterion. These Bunte salts and thioacetate ligands offer an advantage over the free thiol ligands as they showed better stability to air oxidation. The use of a biocompatible solvent such as DMSO to produce these nanoparticles can represent an advantage for their potential applications as novel therapeutic agents, especially when more hydrophobic ligands (in our case, tri(*p*-tolyl)phosphoniopropylthiosulfate zwitterion) are used to functionalised gold nanoparticles, as they may have more affinity with the lipophilic cell walls. We have shown that under reductive conditions in DMSO, the formation of these nanoparticles occurred in the presence of all three cationic phosphonium ligands synthesised for this work. We have shown that the thiosulfate and thioacetate compounds convert into the corresponding thiolate anions. We can assume that these anions interact with the surface of the gold nanoparticles produced *in situ* when a strong reducing agent is added to the DMSO solutions. The formation and the stability over time of triphenylphosphoniopropylthiolate and tri(*p*-tolyl)phosphoniopropylthiolate functionalised gold nanoparticles in DMSO were confirmed and followed by UV-visible spectroscopy. Bands centred at 519 and 529 nm, respectively, were observed in the UV-visible spectra, and did not change over a period of 3 months. We can also speculate that the changes in the ligand chemical structure can affect the refractive index of the colloidal solution and produce a shift to the left (blue shift). As previously mentioned, tri(*p*-tolyl)phosphoniopropylthiolate functionalised gold nanoparticles were successfully produced and dispersed in DMSO, however, these nanoparticles could not be obtained using the two-phase liquid/liquid (DCM/water) method and dispersed in water. The TEM analyses of cationic phosphonium gold nanoparticles functionalised using triphenylphosphoniopropylthiosulfate and tri(*p*-tolyl)phosphoniopropylthiosulfate zwitterions as protecting ligands showed similar particle sizes (11 nm). However, when (3-thioacetylpropyl)triphenylphosphonium bromide was used, the size of the nanoparticles obtained was different (8 nm). This might be due to different passivation kinetics of the thiosulfate and thioacetate containing ligands.

Visualisation of the cellular internalisation of our lipophilic cationic moieties containing gold nanoparticles by using a flow cytometry technique is planned for future. This is a biophysical technique usually employed in the diagnosis of various health disorders. The cells are normally suspended in a stream of fluid and passed by an electronic detection apparatus. In order to prepare the samples for cytometry analysis, our cationic phosphonium gold nanoparticles are planned to be internalised by incubation.

5. Characterisation of the cationic phosphonium ligands

(a) Hydroxypropyl-tri(*p*-tolyl)phosphonium bromide

The IR spectrum showed absorption bands in the regions 725–720 cm^{-1} (ν CH_2), 1600 cm^{-1} (ν C–O), 3100–3000 cm^{-1} (ν CH-aromatic), and 3344 cm^{-1} (ν C–OH). The ^1H NMR spectrum gave signals at δ 1.8 (9H, s, $2CH_3$), δ 2.0 (2H, m, $-CH_2-$), δ 2.6 (1H, s, OH), δ 3.4 (2H, m, P–CH_2), δ 3.6 (2H, t, OH–CH_2), and δ 7.3–7.5 (12H, m, aromatic H) ppm. ESI-MS 363 [M − Br], 364 [M − Br] + [H^+].

(b) ω-Bromopropyl-tri(*p*-tolyl)phosphonium bromide

The IR spectrum showed absorption bands in the regions 677 cm^{-1} (ν C–Br), 748 cm^{-1} (ν CH_2), 988 cm^{-1} (ν P–CH_2), and 3100–3000 cm^{-1} (ν CH-aromatic). The ^1H NMR spectrum gave signals at δ 2.3 (9H, s, $-CH_3$), δ 2.4 (2H, m, P–CH_2), δ 3.8 (2H, m, $-CH_2-$) and δ 7.4–7.9 (12H, m, aromatic H) ppm. ESI-MS 427 [M].

(c) Tri(*p*-tolyl)phosphoniopropylthiosulfate zwitterion

The IR spectrum showed absorption bands in the regions 725–720 cm^{-1} (ν CH_2), 1329 cm^{-1} (ν S=O), 1400 cm^{-1} (ν C–S), 3100–3000 cm^{-1} (ν CH-aromatic). ^1H NMR showed signals at δ 2.3 (s, 9H, $3CH_3$), δ 3.2 (t, 2H, P–$\underline{CH_2}$–CH_2–CH_2), δ 3.4 (t, 2H, P–CH_2–CH_2–$\underline{CH_2}$), δ 3.6 (m, 2H, P–CH_2–$\underline{CH_2}$–CH_2), δ 7.24 (1H, $CDCL_3$), and δ 7.3–7.6 (12H, m, aromatic H) ppm. ESI-MS 459 [M + H^+].

(d) Triphenylphosphoniopropylthiosulfate zwitterion

The IR spectrum showed absorption bands in the regions 725–720 cm^{-1} (ν CH_2), 1330 cm^{-1} (ν S=O), 1345 cm^{-1} (C–S–), and 3100–3000 cm^{-1} (ν CH-aromatic). The

^{1}H NMR spectrum gave signals at δ 2.2 (2H, m, P–<u>CH$_2$</u>–CH$_2$–CH$_2$), δ 3.3 (2H, m, P–CH$_2$–CH$_2$–<u>CH$_2$</u>), δ 4.0 (2H, m, P–CH$_2$–<u>CH$_2$</u>–CH$_2$) and at 7.6–7.8 (15H, m, aromatic H) ppm. ESI-MS 417 [M + H$^+$].

(e) (3-Thioacetylpropyl)triphenylphosphonium bromide

The IR spectrum showed absorption bands in the regions 735–740 cm^{-1} (ν CH$_2$), 1345 cm^{-1} (ν C–S), 1470–1450 cm^{-1} (ν C–H corresponding to bend-CH$_2$), 1678 cm^{-1} (ν C=O), and 3100–3000 cm^{-1} (ν CH–aromatic). ^{1}H NMR showed signals at δ 2.4 (3H, s, –CH$_3$), δ 2.9 (2H, m, P–<u>CH$_2$</u>–CH$_2$–CH$_2$), δ 3.8 (2H, m, P–CH$_2$–CH$_2$–<u>CH$_2$</u>), δ 4.1 (2H, m, P–CH$_2$–<u>CH$_2$</u>–CH$_2$), and δ 7.6–7.9 (15H, m, aromatic H) ppm. ESI-MS 380 [M + H$^+$].

Acknowledgements

We would like to thank the Libyan Embassy for funding Wanisa's PhD project and also to Ashley Howkins (ETC, Brunel University London) for the provision of the TEM micrographs.

References

1 M. Ferrari, *Nat. Rev. Cancer*, 2005, **5**, 161–171.
2 G. F. Paciotti, L. D. G. I. Kingston and L. Tamarkin, *Drug Dev. Res.*, 2006, **67**, 47–54.
3 G. Y. Tonga, D. F. Moyano, C. S. Kim and V. M. Rotello, *Curr. Opin. Colloid Interface Sci.*, 2014, **19**, 49–55.
4 N. L. Rosi and C. A. Mirkin, *Chem. Rev.*, 2005, **105**, 1547–1562.
5 M. M. C. Cheng, G. Cuda, Y. L. Bunimovich, M. Gaspari, J. R. Heath, H. D. Hill, C. A. Mirkin, A. J. Nijdam, R. Terracciano, T. Thundat and M. Ferrari, *Curr. Opin. Chem. Biol.*, 2006, **10**, 11–19.
6 K. Saha, S. S. Agasti, C. Kim, X. Li and V. M. Rotello, *Chem. Rev.*, 2012, **112**, 2739–2779.
7 P. Ghosh, G. Han, M. De, C. K. Kim and V. M. Rotello, *Adv. Drug Delivery Rev.*, 2008, **60**, 1307–1315.
8 A. Schroeder, D. A. Heller, M. M. Winslow, J. E. Dahlman, G. W. Pratt, R. Langer, T. Jacks and D. G. Anderson, *Nat. Rev. Cancer*, 2012, **12**, 39–50.
9 T. M. Allen and P. R. Cullis, *Science*, 2004, **303**, 1818–1822.
10 M. R. Dreher, W. Liu, C. R. Michelich, M. W. Dewhirst, F. Yuan and A. Chilkoti, *J. Natl. Cancer Inst.*, 2006, **98**, 335–344.
11 H. Maeda, J. Wu, T. Sawa, Y. Matsumura and K. Hori, *J. Controlled Release*, 2000, **65**, 271–284.
12 R. M. Sawant, J. P. Hurley, S. Salmaso, A. Kale, E. Tolcheva, T. S. Levchenko and V. P. Torchilin, *Bioconjugate Chem.*, 2006, **17**, 943–949.
13 E. C. Cho, Q. Zhang and Y. N. Xia, *Nat. Nanotechnol.*, 2011, **6**, 385–391.
14 C. S. Kim, B. Duncan, B. Creran and V. M. Rotello, *Nano Today*, 2013, **8**, 439–447.
15 E. C. Cho, L. Au, Q. Zhang and Y. N. Xia, *Nano Lett.*, 2009, **9**, 1080–1084.
16 E. C. Cho, L. Au, Q. Zhang and Y. N. Xia, *Small*, 2010, **6**, 517–522.

17 J. Dausend, A. Musyanovych, M. Dass, P. Walther, H. Schrezenmeier, K. Landfester and V. Mainlander, *Macromol. Biosci.*, 2008, **8**, 1135–1143.
18 H. Yang, S. Y. Fung and M. Y. Liu, *Angew. Chem., Int. Ed.*, 2011, **50**, 9643–9646.
19 E. Oh, J. B. Delehanty, K. E. Sapsford, K. Susumu, R. Goswami, J. B. Banco-Canosa, P. E. Dawson, J. Granek, M. Shoff, Q. Zhag, P. L. Goering, A. Huston and I. L. Medintz, *ACS Nano*, 2011, **5**, 6434–6448.
20 N. A. Al-Hajaj, A. Moquin, K. D. Neibert, G. M. Soliman, F. M. Winnik and D. Maysinger, *ACS Nano*, 2001, **5**, 4909–4918.
21 N. Zhan, G. Palui, M. Safi, X. Ji and H. Mattoussi, *J. Am. Chem. Soc.*, 2013, **135**, 13786–13795.
22 Z. J. Zhu, Y. C. Yeh, R. Tang, B. Yan, J. Tamayo, W. Vachet and V. M. Rotello, *Nat. Chem.*, 2011, **3**, 963–968.
23 B. Ballou, B. C. Lagerholm, L. A. Ernst, P. M. Bruchez and A. S. Waggoner, *Bioconjugate Chem.*, 2004, **15**, 79–86.
24 Y. He, Y. Zhong, Y. Su, Y. Lu, Z. Jiang, F. Peng, T. Xu, S. Su, Q. Huang, C. Fan and S. T. Lee, *Angew. Chem., Int. Ed.*, 2011, **50**, 5695–5698.
25 B. Dubertret, P. Skourides, D. J. Norris, V. Noireaux, A. H. Brivanlou and A. Libchaber, *Science*, 2002, **298**, 1759–1762.
26 K. Saha, D. F. Moyano and V. M. Rotello, *Mater. Horiz.*, 2014, **1**, 102–105.
27 C. M. McIntosh, E. A. Esposito, A. K. Boal, J. M. Simard, C. T. Martin and V. M. Rotello, *J. Am. Chem. Soc.*, 2001, **123**, 7626–7629.
28 M. P. Murphy and R. J. A. Smith, *Annu. Rev. Pharmacol. Toxicol.*, 2007, **47**, 629–656.
29 M. F. Ross, T. Da Ros, F. H. Blaikie, T. A. Prime, C. M. Porteous, I. I. Severina, V. P. Skulachev, H. G. Kjaergaard, R. A. J. Smith and M. P. Murphy, *Biochem. J.*, 2006, **400**, 199–208.
30 M. F. Ross, G. F. Kelso, F. H. Blaikie, A. M. James, H. M. Cocheme, A. Filipovska, T. Da Ros, T. R. Hurd, R. A. J. Smith and M. P. Murphy, *Biochemistry*, 2005, **70**, 222–230.
31 C. N. R. Rao, S. R. C. Vivekchand, K. Biswas and A. Govindaraj, *Dalton Trans.*, 2007, **34**, 3728–3749.
32 W. H. Binder, R. Sachsenhofer, C. J. Straif and R. Zirbs, *J. Mater. Chem.*, 2007, **17**, 2125–2132.
33 S. Pyrpassopoulos, D. Niarchos, G. Nounesis, N. Boukos, I. Zafiropoulou and V. Tzitizioe, *Nanotechnology*, 2007, **18**, 485604.
34 http://www.accessdata.fda.gov/cms_ia/importalert_169.html, accessed date: 01/10/2014.
35 Y. Ju-Nam, N. Bricklebank, D. W. Allen, P. H. E. Gardiner, M. E. Light and M. B. Hursthouse, *Org. Biomol. Chem.*, 2006, **4**, 4345–4351.
36 Y. Ju-Nam, D. W. Allen, P. H. E. Gardiner and N. Bricklebank, *J. Organomet. Chem.*, 2008, **693**, 3504–3508.
37 Y. S. Shon, S. M. Gross, B. Dawson, M. Porter and R. W. Murray, *Langmuir*, 2000, **16**, 6555–6561.
38 A. P. Alivisatos, *Science*, 1996, **271**, 933–937.
39 D. L. Feldheim, and A. F. Colby Jr, *Metal Nanoparticles-Synthesis, Characterization and Applications*, Marcel Dekker, New York, 2002.
40 S. Link and M. A. El-Sayed, *J. Phys. Chem. B*, 1999, **103**, 4212–4217.
41 K. H. Su, Q. H. Wei, X. Zhang, J. J. Mock, D. Smith and S. Schulz, *Nano Lett.*, 2003, **3**, 1087–1090.

42 S. Link, M. B. Mohamed and M. A. El-Sayed, *J. Phys. Chem. B*, 1999, **13**, 3073–3077.
43 B. Yan, Y. Yang and Y. Wang, *J. Phys. Chem. B*, 2003, **107**, 9159.
44 M. C. Daniel and D. Astruc, *Chem. Rev.*, 2004, **104**, 293–346.
45 Y. Ju-Nam, Y.-S. Chen, J. J. Ojeda, D. W. Allen, N. A. Cross, P. H. E. Gardiner and N. Bricklebank, *RSC Adv.*, 2012, **2**, 10345–10351.
46 R. J. Fealy, S. R. Ackerman and G. S. Ferguson, *Langmuir*, 2011, **27**, 5371–5376.
47 R. G. Pillai and M. S. Freund, *Langmuir*, 2011, **27**, 9028–9033.
48 K. J. Wilkinson, and J. R. Lead, *Environmental Colloids and Particles: Behaviour, Separation and Characterisation*, John Wiley and Sons, Chichester, 2007.

Faraday Discussions

PAPER

Large-scale dynamic assembly of metal nanostructures in plasmofluidic field†

Partha Pratim Patra, Rohit Chikkaraddy,‡ Sreeja Thampi, Ravi P. N. Tripathi and G. V. Pavan Kumar*

Received 24th August 2015, Accepted 15th October 2015
DOI: 10.1039/c5fd00127g

We discuss two aspects of the plasmofluidic assembly of plasmonic nanostructures at the metal–fluid interface. First, we experimentally show how three and four spot evanescent-wave excitation can lead to unconventional assembly of plasmonic nanoparticles at the metal–fluid interface. We observed that the pattern of assembly was mainly governed by the plasmon interference pattern at the metal–fluid interface, and further led to interesting dynamic effects within the assembly. The interference patterns were corroborated by 3D finite-difference time-domain simulations. Secondly, we show how anisotropic geometry, such as Ag nanowires, can be assembled and aligned in unstructured and structured plasmofluidic fields. We found that by structuring the metal-film, Ag nanowires can be aligned at the metal–fluid interface with a single evanescent-wave excitation, thus highlighting the prospect of assembling plasmonic circuits in a fluid. An interesting aspect of our method is that we obtain the assembly at locations away from the excitation points, thus leading to remote assembly of nanostructures. The results discussed herein may have implications in realizing a platform for reconfigurable plasmonic metamaterials, and a test-bed to understand the effect of plasmon interference on assembly of nanostructures in fluids.

Introduction

The manipulation and reversible assembly of a large number of nanoparticles in a fluid is of relevance in various aspects of microfluidic science and technology.[1–7] The assembly of nanoparticles, especially of noble metals, has also derived attention in self-assembled plasmonic metamaterials[8–14] that can control various parameters of light at sub-wavelength scales. An inherent challenge to overcome while handling nanoparticles in liquids is Brownian motion. Various possible methods have been developed to overcome this hindrance, of which optical trapping has attracted significant attention.[15–21] Although laser-based

Photonics & Optical Nanoscopy Laboratory, Division of Physics and Center for Energy Science, Indian Institute of Science Education and Research, h-cross, Pune-411008, India. E-mail: pavan@iiserpune.ac.in

† Electronic supplementary information (ESI) available. See DOI: 10.1039/c5fd00127g

‡ Current address: University of Cambridge.

conventional optical trapping has been very successful in trapping and manipulating sub-micron scale objects, it still needs powerful lasers to create gradient and scattering forces at a tightly focused location.[22–24] This requirement of large optical power is detrimental to trapping of soft nanomaterials, as it can severely damage the structure of the trapped object. An alternative method of trapping sub-micron scale objects with low power laser excitation is to utilize the evanescent field of surface plasmons at the metal–dielectric interface.[25–31] Surface plasmons – the collective oscillation of free electrons and light at the metal–dielectric interface – can localize optical fields to subwavelength scale and propagate optical signals beyond the diffraction limit.[32–37] This capability to localize and propagate light can be further harnessed to trap and manipulate large assemblies of nanoparticles, including resonant plasmonic nanoparticles,[29,38–40] at the metal–fluid interface. Such trapping methods which use the plasmonic field in fluidic environments are generally called plasmofluidic traps,[6] and can be harnessed to trap and manipulate nano and microscale objects in the presence of Brownian motion.

Recently, our group developed a method[41] to trap and manipulate plasmonic nanoparticle assemblies at an unstructured metal–fluid interface. We showed that, by using single evanescent optical excitation, our trapping mechanism could not only manipulate nanostructures in a fluid, but also detect single-molecule Raman scattering signatures in the created trap. The same method was also shown to work with dual evanescent wave excitation, which was harnessed to show the interaction between dynamic assemblies of nanoparticles. Our method, in essence, has expanded the capability of dynamic trapping and single-molecule assisted probing of nanoparticle assemblies using surface plasmons. Furthermore, there are still many interesting questions regarding our method that are to be addressed such as: (a) what will be the effect of multiple evanescent plasmon excitation on the dynamic assembly of plasmonic nanoparticles in fluids? Answering this question will lead to understanding how nanoparticle assemblies interact with each other at the metal–fluid interface, especially when the interaction is mediated by the near-field interference of surface plasmons. (b) Can we extrapolate our method to trap anisotropic geometries such as plasmonic nanowires? The answer to this question will open up a new possibility of manipulating and arranging nanowires in liquids, which can be further harnessed for nano-optical and nano-optoelectronic circuitry in liquids.

In this paper we present experimental results that address the above-mentioned questions for some special cases. We show (a) how three and four evanescent-optical excitations can lead to interesting assemblies of plasmonic nanoparticles at the metal–fluid interface; and (b) how plasmonic Ag nanowires can be trapped and aligned at an unstructured and structured metal–fluid interface.

Experimental section

Preparation of unstructured and structured plasmonic thin film

The unstructured gold thin films were prepared by a direct current (DC) plasma sputtering technique, already discussed in our previous work.[41] The thickness of the film was (50 ± 5) nm.

The structured gold thin films were produced by photolithography followed by Au deposition through plasma sputtering. First, a glass substrate was thoroughly cleaned and heated at 80–100 °C. Then, it was coated with the photo-resist at 4000 rpm for 3 min. The photo-resist coated glass substrate was baked at 85 °C. Thereafter, the patterns were written using a 405 nm laser line. After developing the glass substrate, 50 nm Au was deposited on to it by DC sputtering. Finally this was washed with acetone to remove the remaining photoresist to make ready for the assembly experiment. The geometry of the structured film was periodic strips; each strip was of length – 50 μm, width – 2 μm, thickness – 50(±5) nm and the gap between two consecutive strips was 2 μm.

Synthesis of silver nanoparticles and nanowires

In this work we used aqueous phase silver colloidal nanoparticles, synthesized by the citrated reduced method.[42] 45 mg of $AgNO_3$ was dissolved in 250 mL water and heated. 5 mL of boiling 1% sodium citrate aqueous solution was added and kept boiling for 1 h.

The silver nanowires were synthesized by polyol process.[32,43] 3 mL of each, 0.1 M $AgNO_3$ solution and 0.6 M polyvinylpyrrolidone (PVP, MW, 55 000) solution in ethylene glycol (EG), were mixed together at room temperature. This mixture was then injected drop wise into 5 mL of preheated EG at 160 °C and kept for 1 h. The nanowires were then washed in acetone and water. The final Ag nanowire suspension was made in water.

FDTD simulations

The simulation set up consisted of a 50 nm thin gold film placed on a glass slab and the medium above the metal film was modeled with a refractive index of 1.34 (water). A vertical dipole (polarization perpendicular to the plane of the metal film) of 532 nm was placed at the interface of metal and glass for the excitation of the SPPs on the metal film. A two-dimensional (2D) frequency domain power monitor was set at the metal–water interface to record the near-field image. Perfectly matched layer (PML) boundary conditions were applied in the far-field domain to avoid reflections from the boundaries. For multiple excitations, equivalent numbers of vertical dipoles were placed at the appropriate (*xy*-plane) position at the metal–glass interface.

To model the structured film (metallic strip) experiment, simulations were done for a vertical dipole excitation at the center of a single strip (dimensions matching with experimental parameters) with periodic boundary conditions in the *x*-direction and PML on the other two axes. The difference between the excitation methods between the experiments and simulation was justified since only the *z*-component of the electric-field (p-polarization) contributes to the SPP excitation.

Results and discussion

Plasmofluidic assembly of metal-nanoparticles

In our previous work[41] we used a plasmofluidic method to assemble plasmonic nanoparticles at the metal–fluid interface. Fig. 1a shows the assembly

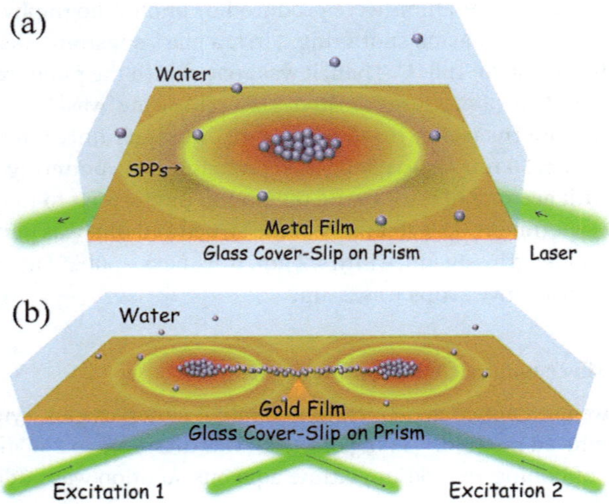

Fig. 1 (a) Schematic illustration of plasmonic assembly of nanoparticles. Black arrows indicate the direction of 532 nm laser (green beam) used for SPP excitation of the metal film deposited over glass coverslip. Grey spheres are the metal nanoparticles dispersed in the water medium. (b) Schematic representation of dual assembly of Ag nanoparticles on Au film coupled to a prism: green lines are the two laser beams of 532 nm introduced at two opposite ends of the prism. Black arrows indicate the direction of the laser beams. Reprinted with permission from *Nature Communications*, **5**, 4357 (2014).

configuration for single spot excitation, and Fig. 1b shows the dual spot configuration. In the current work, we have extrapolated this scheme to multiple excitations.

Multiple-spot-excited plasmofluidic assembly of Ag nanoparticles

We first addressed the issue of how to excite multiple evanescent-waves at the metal–fluid interface to assemble metal nanoparticles. In order to do so we chose to excite multiple surface plasmon waves on the metal film in Kretschmann geometry (Fig. 2a). The choice of nanoparticles, metal film and wavelength of excitation are crucial, so we utilized the optimized parameters from our previous study.[41]

The plasmon excitation set-up (Fig. 2a) consists of a 532 nm continuous-wave frequency-doubled Nd:YAG laser (maximum power up to 200 mW) coupled with a Dove-prism (N-BK7, refractive index, 1.519). A glass coverslip coated with a gold thin film (50 ± 5 nm) was used as the substrate. The p-polarized light was used at an incidence angle of 72.7° for the assembly process. The multiple beams with equal power were created by using 50–50 beam splitters (BS), polarizing beam-splitters (PBS) and half-wave plates ($\lambda/2$). Four weakly focusing lenses of the same focal length were used to focus the light. In such a geometry the power density at the excitation spot is approximately 100 W cm^{-2} which is very low compared to the conventional objective lens based excitation ($\sim 10^6$ W cm^{-2}). To capture the real time assembly process, objective lenses (OBL) with magnifications of $\times 4$ and $\times 60$ were used. To capture bright field images a white light source was combined

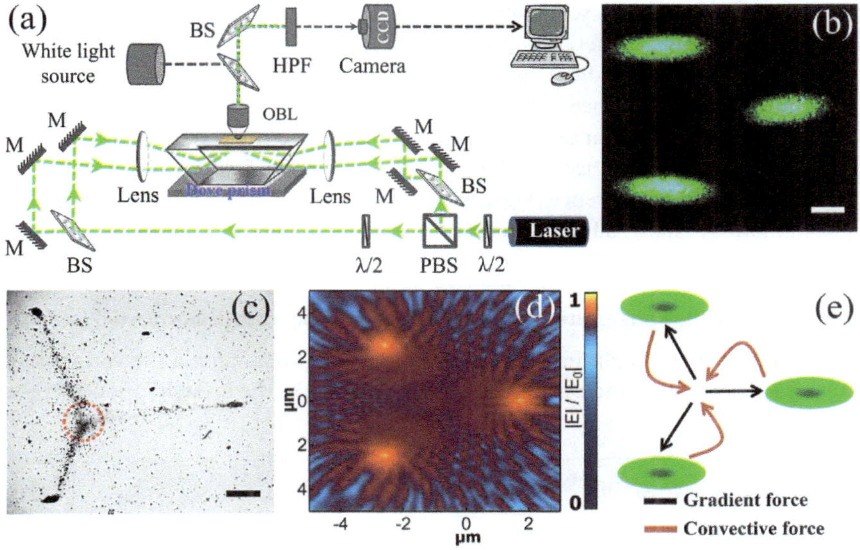

Fig. 2 (a) Schematic of the evanescent-wave multi-trap optical set-up used for plasmofluidic assembly of plasmonic nanoparticles. Acronyms: λ/2, half wave plate; PBS, polarizing beam splitter; BS, 50–50 beam splitter; M, mirror; OBL, microscopic objective lens; HPF, high pass filter. (b) Optical dark-field image of the three, evanescently-excited 532 nm laser spots on a gold film (50 nm thick) Optically coupled to the Dove prism. Scale bar is 200 μm. We call this assembly the triple-trap configuration; (c) Optical bright-field image of plasmofluidic assembly of Ag nanoparticles at gold film–water interface due to excitation shown in (b). The excitation laser was filtered before imaging. Scale bar is 200 μm. Note that the assembly of nanoparticles is denser at the excitation points and at the orthocenter of the three excitation spots indicated by dotted circle. (d) Simulated electric near-field distribution of the three laser spots at the gold–water interface. The geometry of the simulation was identical to the experimental configuration. (e) Schematic of the forces involved in the plasmofluidic trapping process.

with the imaging system and a high pass filter (HPF) was introduced before the CCD camera to block the excitation beam. In Fig. 2b the dark-field image of the triple-spot-excited gold thin film is shown.

Onto this film 200 μL of Ag colloidal solution was drop-casted and the assembly process was recorded in real time (movie-1 in ESI†). Fig. 2c is a bright-field snap shot of the assembly process taken after 12 minutes of excitation. We observed that the nanoparticles were assembled at the excitation spots as well as at the orthocenter where there was no direct optical-excitation. In our earlier study,[41] we elaborately explained the role of hydrodynamic and optical forces in the assembly of nanoparticles at the single excitation spot. Here, a convoluted effect of multiple excitations resulted in the formation of nanoparticle assemblies at the orthocenter and excitation points.

To understand the assembly pattern, we mapped the electric fields at the metal–fluid interface through full-wave 3-D FDTD simulations. The near-field $|E|$ map is shown in Fig. 2d. We see that the electromagnetic fields from the propagating plasmons of the metal film interact with each other. The collective interference of SPP waves from three excitation points results in a unique pattern of nearfield at the metal–fluid interface, which in turn governs both optical and hydrodynamic forces on the nanoparticles. A schematic of the interaction of the forces is shown in Fig. 2e. At the orthocenter, two predominant forces get involved but in opposite directions: one is a gradient force due to laser excitation and the other is the convective force due to heating by the same.[44–46] Thus, the assembly at the orthocenter is due to the balance between two counter steering forces in the plasmofluidic environment. To explain the nanoparticle chain formation, we primarily attribute the SPP interference (Fig. 2d) at the metal–fluid interface, though in such plasmofluidic systems all the effects are codependent and thus complex in nature.

Kinetics of triple-spot-excited plasmofluidic assembly of Ag nanoparticles

To obtain further insight into the assembly process, we analyzed the variation of nanoparticle density and position with time. In Fig. 3a the growth kinetics of the

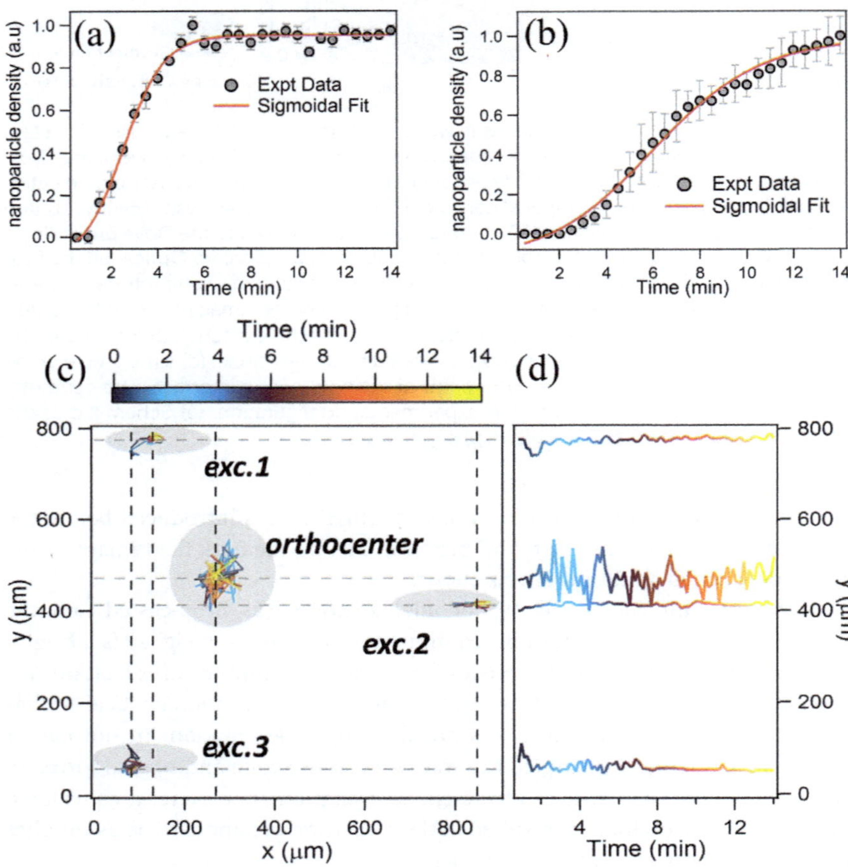

Fig. 3 Kinetics of the nanoparticle assembly in the triple-trap configuration. (a) Variation of nanoparticle-density at an excitation point as a function of time. (b) Variation of the nanoparticle density at the orthocenter of the triple-trap. (c) Spatio-temporal variation of the center of nanoparticle assembly at four different locations: excitations 1, 2 and 3, and at the orthocenter. (d) Spatial variation of the same centers along the y-axis as a function of time. Note that the fluctuation at the orthocenter is more pronounced compared to other locations.

nanoparticle assembly at the excitation points is shown. The growth plot fits well with a sigmoidal curve. It took only a few minutes (4–5 min) for the nanoparticles to assemble and after a particular time the assembly process seems to be saturated. This type of assembly kinetics was reported in our previous paper.[41] However, the scenario at the orthocenter is quite different as we envisage distinct forces involved in this assembly process. Fig. 3b represents the growth kinetics at the orthocenter. It is evident from the plot that saturation of the nanoparticle assembly had not been achieved within the experimental time limit *i.e.* ~15 minutes. As the assembly process at the orthocenter depends on the three excitation points and the balance between two counter steering forces, the assembly process is complicated. The effect of the complex nature of this process is perceived from the dynamics of the assembly process, (movie-2 in ESI†). It is also important to notice that a significant number of nanoparticles escape out from the orthocenter at regular intervals. Furthermore, tracing the central position of the nanoparticle-assembly shows (Fig. 3c) that, the assembly position at the orthocenter is not static, unlike at the excitation spots. The variations in position of the nanoparticles are compared in Fig. 3d and it is apparent that at the orthocenter the nanoparticle assembly fluctuates over a 100 μm diameter whereas the assembly at the excitation points is very compact. So this analysis helped us to understand at least the basic mechanism behind the multiple-excitation plasmofluidic assembly process. Though the whole process is complex to realize, still one can be curious enough to introduce one more excitation into our experimental configuration.

Quadruple-spot-excited plasmofluidic assembly of Ag nanoparticles

Thus, we performed the assembly experiment with four-spot-excitation geometry. Fig. 4a shows the dark-field image of the excitation geometry. In Fig. 4b the FDTD simulation reveals the near-field distribution due to excitation of four spots. We can see that the SPP interference forms a unique pattern on the plasmonic film, which in turn governs the formation of the nanoparticles chain. We would like to

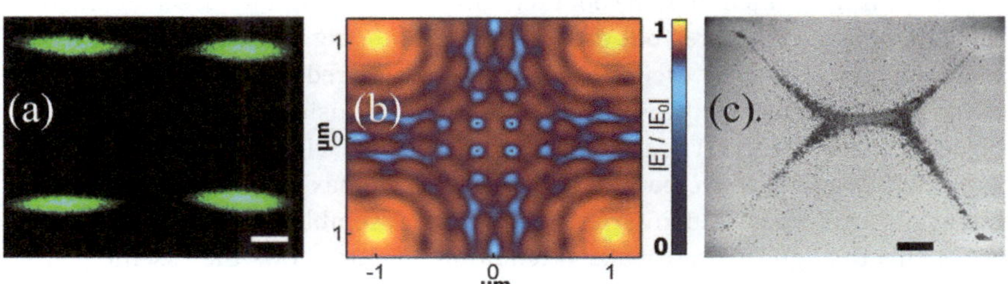

Fig. 4 Plasmofluidic assembly of nanoparticles using quadruple-trap configuration. (a) Optical dark-field image of the four, evanescently-excited 532 nm laser spots on a gold film (50 nm thick) optically coupled to the Dove prism. Scale bar is 200 μm. (b) Simulated electric near-field distribution of the four laser spots at the gold–water interface. The geometry of the simulation was identical to the experimental configuration. Plasmonic interference patterns can be clearly observed. (c) Optical bright-field image of plasmofluidic assembly of Ag nanoparticles at gold film–water interface due to excitation shown in (a). Scale bar is 200 μm. The nanoparticle densities at the corners of the geometry are different due to small variance in the laser powers.

clarify that the length scales of the simulation (around 2 μm × 2 μm) and the experiments (200 μm scale bar in Fig. 4) were different. A large, millimeter scale 3D simulation is computationally expensive, and currently beyond our resources. The simulation results that we have shown are indicative of the kind of interference that we can expect when we have four-spot illumination. It is not the exact description of the field profile in the experiment.

In the context of the experiments, there are two components of the electric field. One is the localized plasmon field at the location of illumination, and the other is the propagating surface plasmon polariton field that is delocalized. The propagating fields originate from the four different excitations, and interfere with each other. This interaction leads to a complex field pattern, which can facilitate potential wells that are much deeper than the ones produced at the location of excitation.

Fig. 4c presents the bright-field image of the nanoparticle assembly with the excitation geometry shown in Fig. 4a. For such geometry the assembly process followed the same basic pattern of creating two types of assembly: one at the excitation point and another at the central part. We emphasis here that there is some accumulation of nanoparticles at the point of excitation, but the in-plane field gradient is such that the majority of nanoparticles tend to assimilate at the location where the potential well is steeper. The difference in the nanoparticle density at the four excitation spots occurs due to the slight inequality in power.

Extrapolation from three-spot-excitation to four-spot-excitation geometry in the plasmofluidic field resulted in interesting patterns of nanoparticle-assemblies and can be further tailored by changing the excitation points. Relevant studies can be pursued in this direction.

Aligned assembly of Ag nanowires in plasmofluidic field

Next we asked whether this method could be extrapolated to anisotropic nanogeometries. To test the versatility of our method, we implemented this technique to assemble one-dimensional nanostructures. Herein, we used Ag nanowires (prepared by polyol process[43]) suspended in an aqueous medium; the assembly process was performed in dual spot excitation geometry and we observed not only the assembly of nanowires but their alignment too. Fig. 5a shows the dark field image of the dual spot excitation of the unstructured thin film. The portion marked by a dotted box was closely monitored by using a ×60 magnification objective lens. The FDTD simulation for similar excitation geometry reveals a typical SPP interference pattern with consecutive maxima and minima, shown in Fig. 5b. This type of SPP interference not only assembles the Ag nanowires, but also helps to align in a particular orientation. Fig. 5c shows the assembly of Ag nanowires on an unstructured film in dual excitation geometry. Thus, we see that the same SPP interference which was attributed to the formation of a nanoparticle-chain can be utilized to manipulate much heavier nanowires at the metal–fluid interface. Although the nanowires assembled around the illumination points, their alignment was not satisfactory.

This motivated us to repeat the experiment with a structured plasmonic film. The gold thin film was patterned as strips of 2 μm width each and the gap between two consecutive strips was 2 μm (see Experimental section for more details). In this experiment only a single-spot excitation was used. Fig. 5d shows the optical

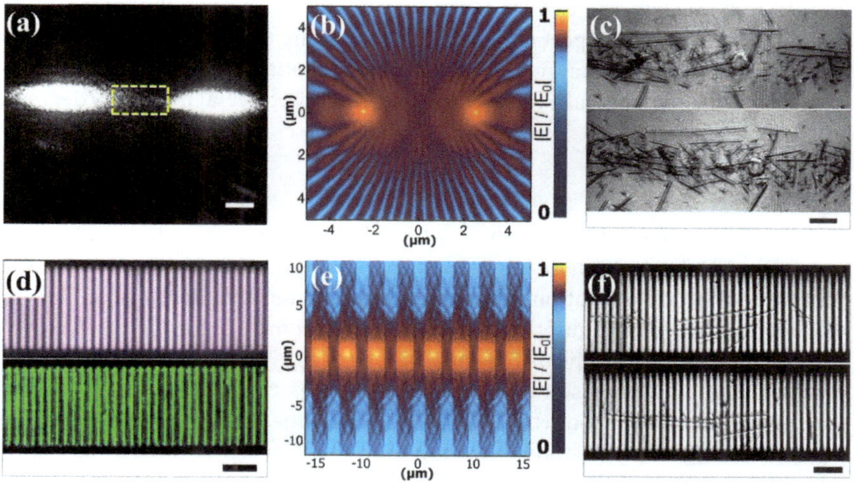

Fig. 5 Assembling Ag nanowires using unstructured and structured plasmonic thin films. (a) Optical dark-field image of dual-trap configuration. The region marked with a yellow dashed box is the region of interest in our experiments. Scale bar is 200 μm. (b) Simulated electric near-field distribution of the two laser spots at the gold–water interface. Plasmonic interference patterns can be clearly observed. (c) Optical bright field images at the Ag nanowire assembly at two different instances. We found weak alignment of the longer nanowires. (d) Optical bright-field (top) and dark-field (bottom) images of structured gold film evanescently-excited using a single 532 nm laser beam. The length of each Au strip was 50 μm, width was 2 μm and the thickness was 50 nm; and the gap between two consecutive strips was 2 μm. (e) Simulated electric near-field distribution of a single laser spot exciting the gold strip–water interface. (f) Optical bright-field snap-shots of the aligned nanowires at two different instances. We found the trap to be stiffer and confined to a smaller area compared to the dual-trap configuration. All scale bars are 20 μm long. Note that experiments in the first row of the (a) were performed using dual evanescent-excitation, whereas experiments in the second row were performed using a single evanescent excitation.

bright and dark-field images of the structured substrate. The near-field distribution was simulated in FDTD (Fig. 5e) which revealed a pattern of consecutive field maxima and minima which actually aided the nanowire alignment. The bright field image of the assembled Ag nanowires is shown in Fig. 5f. The clearly defined field pattern created by the structured thin film assembled and aligned the heavier nanowires with greater precision (movie-3 in ESI†). The power density used for this experiment was approximately 100 W cm^{-2}. The two important aspects of this aligned assembly are reversibility and ease. Thus, one can recreate this alignment of nanowires for multiple cycles of experiments with a very low power excitation.

An interesting question to ask is why do nanowires align along the direction perpendicular to the patterned gold strips? The reason is that surface plasmon polaritons can be excited only with p-polarized light, especially in the Kretschmann configuration that we use (see ref. 36, page numbers 27 and 42). This restricts the in-plane electric field wavevectors of the surface plasmons along the horizontal direction. The optical potential follows the field distribution. Therefore, the nanowires align mainly along the horizontal direction.

If by some means, one can excite the surface plasmons along the length of the nanowire, then we can expect nanowires to align along the vertical direction. This may be achieved in two ways: (a) by using an orthogonal excitation configuration to the one used in this study; (b) by tilting the structured metal film by 90 degrees.

Conclusion

The motivation behind this study was twofold. First, we wanted to explore the possibility of plasmonic-nanoparticle assemblies due to three and four evanescent-wave plasmon excitation at the metal–fluid interface. Our results showed that for triple-trap configuration, nanoparticles assembled not only at the excitation points but also at the orthocentre of the triangle formed between the three excitation spots. The reason for such assembly was mainly due to the surface plasmon interference from three different excitations. However, the convective forces resulting from plasmonic heating may also contribute towards assembly formation. Kinetic studies based on particle-tracking at various locations in the assembly indicated a dynamic equilibrium, with the assembly at the orthocentre exhibiting greater fluctuation in its position compared to the excitation spots. For the four-spot configuration, we observed interesting assembly patterns which were mainly governed by the plasmon interference patterns. The second motivation of our study was to explore the possibilities of assembling anisotropic plasmonic geometries, such as Ag nanowires, using our method. We found that a dual-trap configuration can indeed assemble and weakly align nanowires in liquid, and the directionality of the alignment was governed by interference pattern of counter-propagating surface plasmons at the metal–fluid interface. Interestingly, we also found that nanowires can also be assembled and aligned, with better efficiency, using structured metal films and a single evanescent-wave excitation of surface plasmons. All these experiments indicated that plasmon excitation and interference can be harnessed for micron to millimetre scale reversible-assembly of nanoparticles and elongated nanostructures at the metal–fluid interface. There are new directions that our work can lead to such as reconfigurable plasmonic metamaterials at the metal–fluid interface, where control of various parameters of light – intensity, polarization, phase, *etc.* – can be controlled in fluids by tailoring the geometrical arrangements of nanoparticle assemblies. Such reconfigurable optical platforms can be further harnessed for nano and micro-optical circuits and devices in fluidic environments. In addition to that, nanoparticle assemblies may be harnessed as test-beds of various plasmofluidic effects where understanding the interaction between surface plasmons and the surrounding fluids is of fundamental interest.

Acknowledgements

This research work was partially funded by DTS Nanomission Grant (SR/NM/NS-1141/2012) and DST Nanoscience Unit Grant (SR/NM/NS-42/2009).

References

1 A. Rida and M. A. M. Gijs, *Anal. Chem.*, 2004, **76**, 6239–6246.

2 A. Jahn, W. N. Vreeland, M. Gaitan and L. E. Locascio, *J. Am. Chem. Soc.*, 2004, **126**, 2674–2675.
3 H. Lee, A. M. Purdon, V. Chu and R. M. Westervelt, *Nano Lett.*, 2004, **4**, 995–998.
4 L. Huang, S. J. Maerkl and O. J. Martin, *Opt. Express*, 2009, **17**, 6018–6024.
5 H. Hwang and J.-K. Park, *Adv. OptoElectron.*, 2011, **2011**, 13.
6 J. Kim, *Lab Chip*, 2012, **12**, 3611–3623.
7 O. D. Velev and K. H. Bhatt, *Soft Matter*, 2006, **2**, 738–750.
8 B. Gong, X. Zhao, Z. Pan, S. Li, X. Wang, Y. Zhao and C. Luo, *Sci. Rep.*, 2014, **4**, 4713.
9 N. I. Zheludev, *Opt. Photonics News*, 2011, **22**, 30–35.
10 J. Dintinger, S. Mühlig, C. Rockstuhl and T. Scharf, *Opt. Mater. Express*, 2012, **2**, 269–278.
11 S. Mühlig, C. Rockstuhl, V. Yannopapas, T. Bürgi, N. Shalkevich and F. Lederer, *Opt. Express*, 2011, **19**, 9607–9616.
12 J. A. Fan, C. Wu, K. Bao, J. Bao, R. Bardhan, N. J. Halas, V. N. Manoharan, P. Nordlander, G. Shvets and F. Capasso, *Science*, 2010, **328**, 1135–1138.
13 S. Mühlig, A. Cunningham, J. Dintinger, T. Scharf, T. Bürgi, F. Lederer and C. Rockstuhl, *Nanophotonics*, 2013, **2**, 211–240.
14 Q. Liu, Y. Cui, D. Gardner, X. Li, S. He and I. I. Smalyukh, *Nano Lett.*, 2010, **10**, 1347–1353.
15 G. Volpe, R. Quidant, G. Badenes and D. Petrov, *Phys. Rev. Lett.*, 2006, **96**, 238101.
16 A. Ashkin, J. M. Dziedzic, J. E. Bjorkholm and S. Chu, *Opt. Lett.*, 1986, **11**, 288–290.
17 R. Agarwal, K. Ladavac, Y. Roichman, G. Yu, C. Lieber and D. Grier, *Opt. Express*, 2005, **13**, 8906–8912.
18 O. M. Marago, P. H. Jones, P. G. Gucciardi, G. Volpe and A. C. Ferrari, *Nat. Nanotechnol.*, 2013, **8**, 807–819.
19 D. G. Grier, *Nature*, 2003, **424**, 810–816.
20 A. S. Urban, S. Carretero-Palacios, A. A. Lutich, T. Lohmuller, J. Feldmann and F. Jackel, *Nanoscale*, 2014, **6**, 4458–4474.
21 J. R. Moffitt, Y. R. Chemla, S. B. Smith and C. Bustamante, *Annu. Rev. Biochem.*, 2008, **77**, 205–228.
22 Z. Yan, M. Pelton, L. Vigderman, E. R. Zubarev and N. F. Scherer, *ACS Nano*, 2013, **7**, 8794–8800.
23 Z. Yan, J. E. Jureller, J. Sweet, M. J. Guffey, M. Pelton and N. F. Scherer, *Nano Lett.*, 2012, **12**, 5155–5161.
24 O. Brzobohatý, M. Šiler, J. Trojek, L. Chvátal, V. Karásek, A. Paták, Z. Pokorná, F. Mika and P. Zemánek, *Sci. Rep.*, 2015, **5**, 8106.
25 K. Wang and K. B. Crozier, *ChemPhysChem*, 2012, **13**, 2639–2648.
26 W. Zhang, L. Huang, C. Santschi and O. J. F. Martin, *Nano Lett.*, 2010, **10**, 1006–1011.
27 Y. Zhang, J. Wang, J. Shen, Z. Man, W. Shi, C. Min, G. Yuan, S. Zhu, H. P. Urbach and X. Yuan, *Nano Lett.*, 2014, **14**, 6430–6436.
28 K. Wang, E. Schonbrun, P. Steinvurzel and K. B. Crozier, *Nat. Commun.*, 2011, **2**, 469.
29 C. Min, Z. Shen, J. Shen, Y. Zhang, H. Fang, G. Yuan, L. Du, S. Zhu, T. Lei and X. Yuan, *Nat. Commun.*, 2013, **4**, 2891.
30 M. L. Juan, M. Righini and R. Quidant, *Nat. Photonics*, 2011, **5**, 349–356.

31 V. Garcés-Chávez, R. Quidant, P. J. Reece, G. Badenes, L. Torner and K. Dholakia, *Phys. Rev. B: Condens. Matter Mater. Phys.*, 2006, **73**, 085417.
32 R. Chikkaraddy, D. Singh and G. V. Pavan Kumar, *Appl. Phys. Lett.*, 2012, **100**, 043108.
33 G. Kumar, D. Singh, P. Patra and A. Dasgupta, *Pramana*, 2014, **82**, 59–70.
34 A. Dasgupta, D. Singh, S. Tandon, R. P. N. Tripathi and G. V. Pavan Kumar, *Nanoparticle*, 2013, **8**, 083899.
35 D. K. Gramotnev and S. I. Bozhevolnyi, *Nat. Photonics*, 2010, **4**, 83–91.
36 S. A. Maier, *Plasmonics: fundamentals and applications*, Springer Science & Business Media, 2007.
37 A. V. Zayats, I. I. Smolyaninov and A. A. Maradudin, *Phys. Rep.*, 2005, **408**, 131–314.
38 K. C. Toussaint, M. Liu, M. Pelton, J. Pesic, M. J. Guffey, P. Guyot-Sionnest and N. F. Scherer, *Opt. Express*, 2007, **15**, 12017–12029.
39 M. Dienerowitz, M. Mazilu, P. J. Reece, T. F. Krauss and K. Dholakia, *Opt. Express*, 2008, **16**, 4991–4999.
40 T. Shoji and Y. Tsuboi, *J. Phys. Chem. Lett.*, 2014, **5**, 2957–2967.
41 P. P. Patra, R. Chikkaraddy, R. P. Tripathi, A. Dasgupta and G. P. Kumar, *Nat. Commun.*, 2014, **5**, 4357.
42 P. C. Lee and D. Meisel, *J. Phys. Chem.*, 1982, **86**, 3391–3395.
43 Y. Sun, Y. Yin, B. T. Mayers, T. Herricks and Y. Xia, *Chem. Mater.*, 2002, **14**, 4736–4745.
44 J. S. Donner, G. Baffou, D. McCloskey and R. Quidant, *ACS Nano*, 2011, **5**, 5457–5462.
45 P. M. Bendix, S. N. S. Reihani and L. B. Oddershede, *ACS Nano*, 2010, **4**, 2256–2262.
46 N. J. Hogan, A. S. Urban, C. Ayala-Orozco, A. Pimpinelli, P. Nordlander and N. J. Halas, *Nano Lett.*, 2014, **14**, 4640–4645.

Faraday Discussions

PAPER

2D-patterning of self-assembled silver nanoisland films

Semen Chervinskii,*[ab] Igor Reduto,[abc] Alexander Kamenskii,[bc] Ivan S. Mukhin[cd] and Andrey A. Lipovskii[bc]

Received 24th August 2015, Accepted 27th October 2015
DOI: 10.1039/c5fd00129c

The paper is dedicated to the recently developed by the authors technique of silver nanoisland growth, allowing self-arrangement of 2D-patterns of nanoislands. The technique employs silver out-diffusion from ion-exchanged glass in the course of annealing in hydrogen. To modify the silver ion distribution in the exchanged soda-lime glass we included the thermal poling of the ion-exchanged glass with a profiled electrode as an intermediate stage of the process. The resulting consequence consists of three steps: (i) during the ion exchange of the glass in the $Ag_xNa_{1-x}NO_3$ ($x = 0.01$–0.15) melt we enrich the subsurface layer of the glass with silver ions; (ii) under the thermal poling, the electric field displaces these ions deeper into the glass under the 2D profiled anodic electrode, the displacement is smaller under the hollows in the electrode where the intensity of the field is minimal; (iii) annealing in a reducing atmosphere of hydrogen results in silver out-diffusion only in the regions corresponding to the electrode hollows, as a result silver forms nanoislands following the shape of the electrode. Varying the electrode and mode of processing allows governing the nanoisland size distribution and self-arrangement of the isolated single nanoislands, pairs, triples or groups of several nanoislands—so-called plasmonic molecules.

Introduction

Plasmonic nanostructures have recently attracted significant attention due to their potential for applications employing light–matter interactions. Silver, copper and gold nanoparticles, their complexes, and various nanostructures of these materials demonstrate surface plasmon resonance (SPR) in the visible spectral range and, significant enhancement of the local electric field under irradiation with the resonant wavelength.[1] These resonant properties have been employed in surface-enhanced Raman spectroscopy,[2] biosensing,[3] light emission and fluorescence

[a]*University of Eastern Finland, P. O. Box 111, Joensuu, 80101 Finland. E-mail: semen.chervinskii@uef.fi; semen.chervinsky@gmail.com*
[b]*Peter the Great St. Petersburg Polytechnic University, 29 Polytechnicheskaya, St. Petersburg, 195251 Russia*
[c]*St. Petersburg Academic University, 8/3 Khlopina, St. Petersburg, 194021 Russia*
[d]*ITMO University, 49 Kronverksky, St. Petersburg, 197101 Russia*

enhancement,[4,5] catalysis, photovoltaics and solar cells.[6–8] The wavelength of the SPR depends on the material, geometry and size of a nanostructure as well as on the dielectric permittivity of its environment. Particular interest has focused on different ways of arranging plasmonic nanostructures from separate nanoparticles.[9–11] Control over the shape and size of the nanoparticles and their interactions allows governing resonant properties of the whole nanostructure.[12,13] For example, arranging plasmonic nanoparticles with small gaps between them allows achieving extremely high electric field intensity in these narrowings at the SPR wavelength.[12] At the same time, single nanoparticles are promising for sensing as their characteristics can be affected by analytes even more drastically then those of their complexes.[14–16] Formation of intrinsically disordered self-assembled nanostructures has also gained interest due to the resonant properties of disordered media.[17–19]

To fabricate separated nanoparticles, chemical[20] or, more recently, laser ablation and other techniques[21,22] are in use. Also, different sputtering techniques combined with thermal annealing are standard in manufacturing metal island films (MIFs)[23,24] that are ensembles of metal nanoislands, *i.e.* nanoparticles. Fabrication of groups of nanoparticles is usually based on deposition of a metal film and nanostructuring of this film using lithographic, preferably electron or nanoimprint lithography, processing.[25,26] Other techniques include constructing plasmonic structures from commercially available nanoparticles using lithographic or organic templates.[27,28] Processing MIFs with lithographic techniques could be a good candidate to make given groups of nanoparticles in a faster way, however the fragility of MIFs does not allow this.

Recently we developed a plasmonic nanoparticle growth technique based on out-diffusion of metals from ion-exchanged glasses in the course of their annealing in a reducing atmosphere.[29] The advantages of this technique are its easiness and thus cheapness, since it can be performed using very simple equipment. Ion exchange is a well-known method to enrich the subsurface layer of a glass with metal ions from the molten salt of the metal *via* replacement of the initial alkali ions of the glass.[30] After that, these ions are subjected to reduction, for that we heat up the samples in a reducing (hydrogen) atmosphere, the reduced atoms then cluster and can form nanoparticles either on the glass surface (MIF) or in the subsurface layer of the glass (glass–metal nanocomposite, GMN).[31,32]

In this paper, we discuss our latest results on the growth of 2D-patterned silver nanoisland films, the ensemble of small groups of nanoislands. To provide the 2D-patterning we adjusted the silver ion distribution in the glass after ion exchange. For that we used thermal poling (heat treatment under DC electrical field) of the ion-exchanged glasses profiled by a submicron scale anodic electrode. The resulting structures of silver nanoislands followed the shape of the electrode surface. We demonstrated that depending on the shape and processing conditions it is possible to obtain a variety of nanostructures, from areas filled with randomly distributed nanoislands to single nanoislands and groups of two, three or more nanoislands ("plasmonic molecules").

Experimental

Ion exchange

We used commercially available Menzel soda-lime glass slides with compositions as shown in Table 1.[33] The pre-cleaned with acetone slides of $1.0 \times 26 \times 76$ mm^3

were subjected to silver–sodium ion exchange in a $Ag_xNa_{1-x}NO_3$ ($x = 0.01$–0.15) melt at a temperature of 325 °C for durations of 5 minutes to an hour (Fig. 1-I). For better uniformity of the processing the samples were periodically shaken to avoid formation of bubbles on the glass surface resulting from minor decomposition of the melt. During this step, silver ions from the melt replaced the sodium ions of the glass which resulted in silver-enriched layers under immersed surfaces of the glass. The final ion distribution in such a layer depends on the properties of the glass, the silver concentration in the initial melt, the ion exchange temperature, and the duration.[34] Under typical conditions of $x = 0.05$ and 20 min duration of the ion exchange at 325 °C, about 50% of the sodium ions at the glass surface were replaced with silver ones, the descending concentration profile of which decayed by a depth of 10 microns with the 50% concentration level corresponding to a 5 micron distance from the surface.[35] These conditions of ion exchange are default for the experiments presented in this paper, if others are not stated.

Annealing

Ion exchanged glasses were annealed in a reducing atmosphere (mainly hydrogen, but air was also employed) at 75–350 °C for durations of 30 seconds to 3 hours (Fig. 1-II). During this step hydrogen (in the case of air it comes from water vapor decomposition at the glass surface[36]) diffuses into the subsurface region of the glass and reduces the silver ions to neutral atoms *via* replacing silver in the bonds with nonbridging oxygen:[37]

$$\equiv \mathrm{Si}-\mathrm{O}-\mathrm{Ag}^{+}+\frac{1}{2}\mathrm{H}_2 \rightarrow \equiv \mathrm{Si}-\mathrm{O}-\mathrm{H}^{+}+\mathrm{Ag}^{0} \tag{1}$$

The reduction results in the formation of a solid solution of neutral silver in the glass matrix, which tends to decompose because of the low solubility of atomic silver in glasses. The decomposition causes the formation of silver clusters and nanoparticles, in particular, silver self-assembled nanoisland films, on the glass surface, which are a strong sink for silver atoms.[32] These out-diffused nanoisland films have demonstrated applicability in surface enhanced Raman scattering.[37,38] Generally, the above description can be employed to grow nanoparticles of all noble metals. In other experiments, we used this for the formation of copper nanoparticles.

Poling and annealing

To grow predetermined groups of silver nanoislands we modified the silver ion distribution in the glass by thermal poling[39] of the ion-exchanged glass slides with

Table 1 Menzel glass chemical composition, wt% (ref. 33)

Silicon dioxide	SiO_2	72.20%
Sodium oxide	Na_2O	14.30%
Potassium oxide	K_2O	1.20%
Calcium oxide	CaO	6.40%
Magnesium oxide	MgO	4.30%
Aluminium oxide	Al_2O_3	1.20%
Ferric oxide	Fe_2O_3	0.03%
Sulfur trioxide	SO_3	0.30%

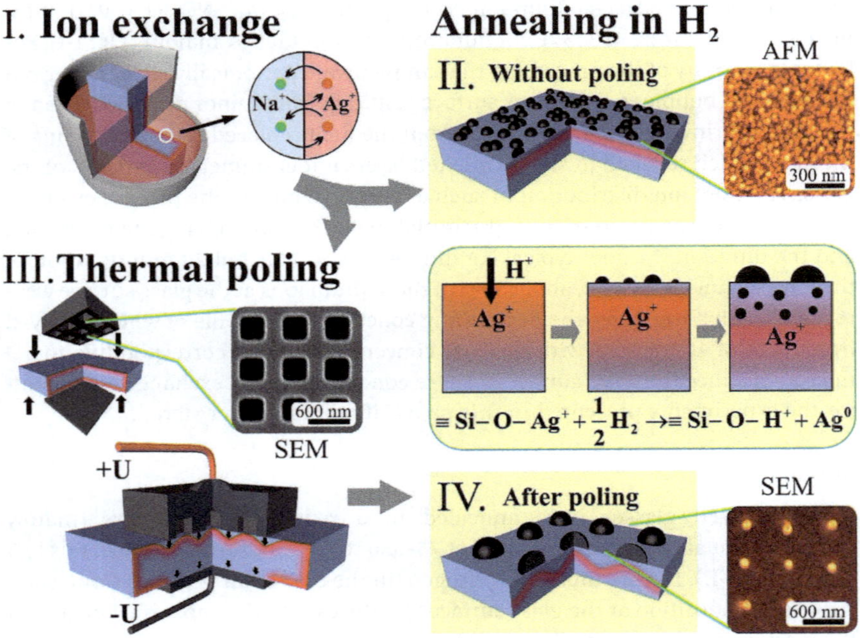

Fig. 1 Schematic of the process of silver nanoisland film and structure growth. (I) During ion exchange, the soda-lime glass slide is immersed in a $Ag_xNa_{1-x}NO_3$ ($x = 0.01$–0.15) melt (325 °C, 5 minutes to hour). (II) After annealing of the ion-exchanged glass slide in a hydrogen atmosphere (75–350 °C, 30 seconds to 3 hours) the silver nanoisland film grows. (III) Thermal poling of the ion-exchanged glass using a profiled anode (200–350 °C, DC 500–600 V, 30 seconds to 30 minutes) allows adjustment of the silver ion distribution. (IV) After annealing of the poled ion-exchanged glass, silver nanoislands grow only where the anodic electrode did not contact with the surface during poling. Inset: hydrogen penetrating the glass reduces the silver ions to atoms; in the course of diffusion and self-arrangement silver forms nanoparticles either on the glass surface or both on the surface and in the bulk of the glass.

a profiled anodic electrode with submicron hollows in the surface. The electrode was manufactured using electron lithography followed by ion etching, as described in the next section, while the nickel cathodic electrode was flat. DC voltage applied for the poling was of 500–600 V. The poling was performed at temperatures of 200–350 °C for durations of 30 seconds to 30 minutes; in some cases a preheating process for several minutes before applying the DC field was used. The DC field moved positively charged silver ions under the positive electrode deeper in the glass, and the deepening under the hollows in the electrode was less than under the places where the electrode was in contact with the glass surface (Fig. 1-III). It is worth noting that this difference in deepening decreases with decreasing size of the hollows in the anodic electrode.[40]

After proper annealing in hydrogen using the procedure described in the previous section, this modification of ions and, respectively, reduced silver distribution, controlled the out-diffusion of neutral silver. This resulted in the formation of silver nanoisland films, groups of nanoislands or isolated ones—depending on the electrode surface profiling and parameters of poling and annealing—only in the areas which had been under the hollows in the electrode used during poling (Fig. 1-IV). Since there are no available data on the diffusivity of silver atoms and hydrogen molecules in unpoled or poled glasses, the choice of both poling and hydrogen annealing parameters was mainly empirical.

TiO_2 coating

To protect the silver nanoislands from contamination from the atmosphere, in some cases we coated them with amorphous TiO_2 layers of several nanometers in thickness using an atomic layer deposition technique. TiO_2 was deposited at 120 °C in a Beneq TFS-200 reactor (Beneq, Espoo, Finland) using titanium tetrachloride ($TiCl_4$) and water (H_2O) as precursors; the growth rate was 0.07 nm per cycle; between deposition cycles nitrogen was purged to remove extra precursor materials from the reactor chamber. It has also been demonstrated in experiments[37] and theoretically,[41] that these coatings can be employed to tune the surface plasmon resonance wavelength by modifying the effective dielectric permittivity of surrounding media around the nanoislands due to the high (n = 2.3–3 in a wavelength range of 300–1300 nm) refractive index of TiO_2. In our experiments the thickness of the film varied from 3 to 100 nm.

Electron lithography

To make the profiled anode electrode from the polished glassy carbon ($R_a < 50$ nm, plates of $3 \times 100 \times 100$ mm^3 were obtained from Svensk Specialgrafit AB and cut into pieces of $3 \times 20 \times 25$ mm^3) we employed electron beam lithography. A chromium layer of 50 nanometers was e-beam sputtered as a mask on the glassy carbon using a Kurt J. Lesker Company Lab 18 thin film deposition system, after it the sample was spin-coated with electron resist, baked and exposed in Vistec Lithography Ebeam EBPG5000. Development after the exposure was performed using a SSE OPTIspin SST 120 setup. The chromium layer was etched through developed resist in Plasmalab 100, after which the final glassy carbon was etched through a chromium mask in Plasmalab 80plus. The chromium mask was then removed, and the profiled glassy carbon was cleaned consequently in acetone and isopropanol using an ultrasonic bath for 5–20 minutes. A resulting relief height of 50–400 nm was produced depending on the etching parameters.

Optical spectroscopy

To characterize the optical properties of the silver nanoisland films we used a Specord 50 spectrophotometer, the absorption spectrum was measured firstly for the sample with the MIF, after this the MIF was removed using cotton pads with acetone or isopropanol, and the spectrum was measured for the second time. The difference between these two spectra gave us the spectrum of the nanoisland film itself. Examples of measured and differential spectra are shown in Fig. 2.

Atomic force and electron microscopy

Structural characterization of manufactured electrodes, nanoisland films, groups and single nanoislands was performed by means of a Veeco Dimension-3100 atomic force microscope with Bruker TESPA/TESP-SS tips in tapping mode and a Zeiss Supra 25 scanning electron microscope. For SEM in some cases the glass

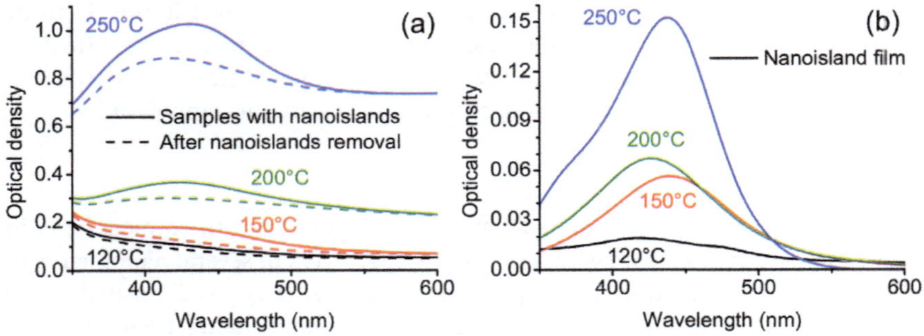

Fig. 2 Spectra of samples of ion-exchanged glasses, annealed in hydrogen for 10 minutes at various temperatures (marked near the spectra) and spectra of samples after removal of the grown MIF (a), and differential spectra of the nanoisland film (b).

samples were coated with 10 nanometers of gold using an Emitech K675X sputterer. A cross-section image of the ion-exchanged glass after annealing was made using a JEOL JEM 2100-F transmission electron microscope.

Results and discussion

Silver nanoisland films

Using the basic techniques of silver ion-exchange and annealing in a reducing atmosphere we succeeded in the growth of metal island films consisting of silver islands from ones to tens of nanometers in size with separations of a few nanometers (Fig. 3). The silver nanoislands demonstrated an almost hemispherical shape in the AFM (Fig. 4). Depending on the processing conditions, percolated and continuous films could also be grown. For the latter, more silver is needed to be reduced to obtain the nanoisland film, therefore ion exchange should be performed in the melt with a higher concentration of silver ions, and the annealing temperature and duration are also higher than for the nanoislands. We tried different annealing modes to reveal the nanoisland size distribution dependencies on the processing conditions (Fig. 4). In general, nanoisland self-assembly occurred in agreement with the recently developed model[42,43]—longer annealing as well as higher temperatures resulted in a decreasing concentration and increasing size of the nanoislands.[29]

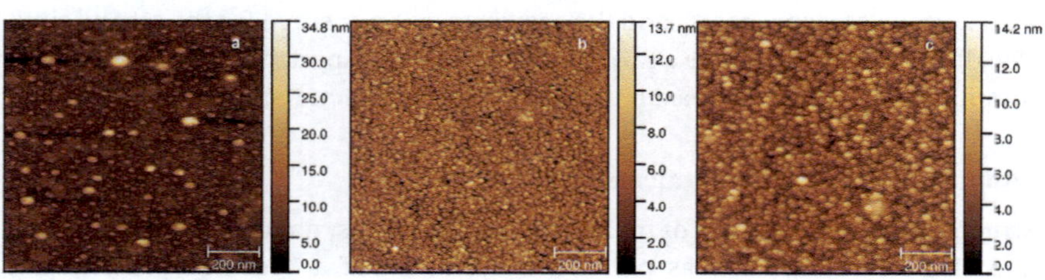

Fig. 3 Silver nanoisland films grown after annealing of ion-exchanged glass at different temperatures for 10 minutes: (a) 120 °C, (b) 150 °C, (c) 200 °C.

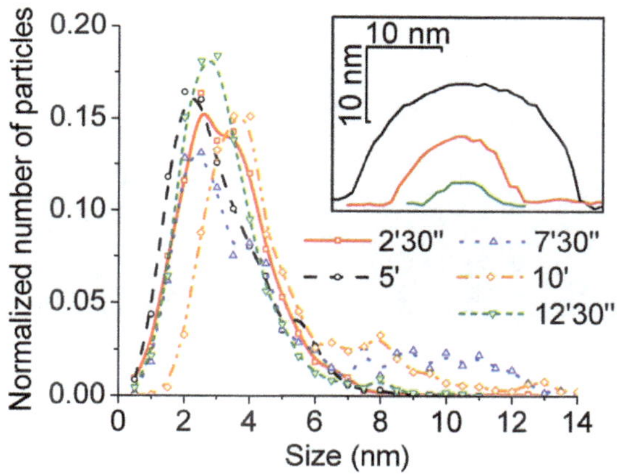

Fig. 4 Size distributions of films, grown after annealing of ion-exchanged glass at 120 °C for different durations of 2.5 minutes to 12.5 minutes. Inset: profiles of different nanoislands from the same sample. Reproduced with permission from (ref. 29). Copyright 2013, AIP Publishing LLC.

We also studied the relation of the optical properties of the nanoisland films and their processing parameters. Predictably, absorption (intensity of surface plasmon resonance) was proportional to the amount of reduced silver, that is to the annealing duration and temperature (Fig. 2). The ion exchange parameters (duration and amount of Ag^+ in the ion-exchange bath) determine the silver ion distribution before annealing, therefore also affecting the film growth. However since the reduction of silver ions takes place in a layer thinner than the ion-exchanged region of the glass, the duration of the ion exchange influences nanoisland formation only in the case of very short durations. Similarly, since almost all sodium ions near the glass surface are replaced by silver ones at $x = 0.15$ in the $Ag_xNa_{1-x}NO_3$ melt,[45] a less significant change in MIF formation was observed after processing of the slides in higher concentrated melts.

It should also be mentioned that at certain annealing parameters (higher temperatures and longer times) hydrogen reduces the silver deeply enough to form nanoparticles in the bulk of the glass, not only the surface. The typical diameter of these mostly spherical nanoparticles is below 10 nanometers (Fig. 5). Their amount also grows with the duration of hydrogen processing (compare Fig. 5a and b). Studies of the formation and plasmonic properties of these particles are out of scope of this paper and have been reported elsewhere,[31,32] however, they should be taken into account in certain experiments with out-diffused silver nanoislands.

Poling of silver ion-exchanged glass

Spatially non-uniform thermal poling was used to adjust the silver ion distribution in the glass in order to spatially control the growth of silver nanoislands. In poling, a DC electric field moves all the cations bonded before the poling to the non-bridging oxygen atoms of the glass network, in particular silver ions, away from the anodic surface of the glass. Using this effect, we had to work out a regime which produced deepening of the silver ions sufficient to prevent nanoisland

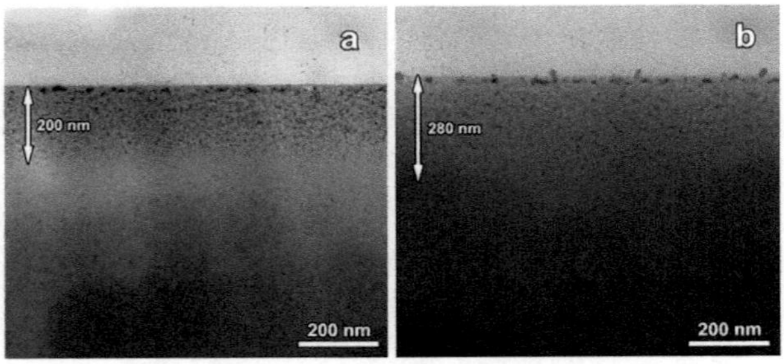

Fig. 5 TEM image of silver nanoparticles, formed in ion-exchanged glass in the course of hydrogen annealing for 7.5 (a) and 15 minutes (b) at 300 °C.

formation everywhere except the desirable positions after following annealing of the poled glass slides in hydrogen. For a flat anodic electrode the deepening increases with applied voltage and duration of the poling.[45–47] Generally speaking, nanosized hollows in the anodic electrode used in poling should provide a spatially non-uniform electric field which is weaker under the hollows, while the maximum field intensity and thus silver ion deepening is provided between these hollows. However, it should be noted that according to the existing model[40] lateral components of the electric field generated by structured electrodes induce lateral motion of positive ions. Thus, the difference in deepening under the hollows in the anodic electrode and between the hollows is less for smaller hollows. Moreover, the time dependence of this difference goes through a maximum and decays for longer poling,[40] that is, the optimal mode of the processing depends on the anodic electrode topology. Precise modeling of this problem can hardly be performed because of the unknown mobilities/diffusion coefficients of several species participating in the formation of the nanoislands: molecular hydrogen, neutral silver and hydronium, H_3O^+, which is the actual carrier for a hydrogen ions.[44] Moreover, since these species diffuse in the poled region of the glass, the composition and structure of which essentially differs with ones of virgin glass[48,49] this should modify their diffusivity in comparison with the virgin glass. Similarly, since the resulting 3D silver ion distribution could also hardly be directly measured, we evaluated it based on the shape and distribution of the nanoislands grown after the annealing. A comparison of the results of nanoisland growth obtained after poling of the ion-exchanged glasses with electrodes of different shape confirms that although the silver ion distribution follows the shape of the anodic electrode, it is always smoother than the electrode rectangular shape. In general, the poling mode controls the silver ion deepening which either allows or forbids formation of the silver nanoislands on the glass surface.

It is worth noting that the deepening results not only in a delay of silver reduction and out-diffusion because of the time necessary for hydrogen to go through the silver ion depleted layer and for silver atoms to go towards the glass surface, but the other effect is a decreasing of silver atom flux to the surface from bigger depths. This is because if the reduction takes place farther from the glass surface the strength of the surface sink for arisen silver atoms decreases, and the probability of out-diffusion and growth of silver nanoislands drops. Thus, the optimal poling regime should provide sufficient difference in the deepening of silver ions under the flat surface of the anodic electrode and under the hollow. This is necessary to withdraw formation of silver nanoislands in stronger poled regions between the hollows in the electrode, while the nanoislands under the hollows should grow.

2D-patterned self-assembled silver nanoisland films

Using annealing of ion-exchanged glass poled with a 2D-patterned electrode we succeeded in the growth of structures from nanoislands, following the shape of the electrode. It was demonstrated that silver nanostructures with different distributions of nanoislands can be grown by varying the poling and annealing conditions (Fig. 6), while the electrode shape provides one more variable to design these nanostructures (Fig. 7). Using patterns in the micron scale we obtained areas with nanoisland films inside which were quite similar to ones grown without poling, the modified ion distribution influenced the growth of the nanoislands on the edges of these areas only. This influence resulted in bigger nanoislands on the edges of the pattern, which can be explained by a lower concentration of silver ions near the edges and, respectively, the bigger size of critical nuclei.[50] Definitely, the same relates to nanoparticles of any material in the case of their diffusion-governed growth. Patterns with dimensions of 1 μm and less, down to 100 nm, affected the growth of the nanoisland structures very much, allowing the nanoislands to grow almost an order bigger than in the non-patterned films, up to 150–200 nm in diameter. Due to this, the small hollows in the electrode (300×300 nm^2 or less) allowed us to grow isolated single particles or groups of a few particles—so-called plasmonic molecules—depending on processing conditions.[42,43] Fig. 8 illustrates that decreasing the size of the hollows (keeping all other conditions the same) resulted not only in an obvious decrease of the number of nanoislands, but also in an increase in the size of the biggest islands in the "molecule"—the possible reasons for this are the difference in ion distribution at the edges of the poled areas and the decrease in the quantity of growing nanoparticles and the decrease in the quantity of growing nanoparticles, which lowers their competition for silver atoms. It is essential that the same electrodes under different poling and annealing conditions can be used to form arrays of either isolated single nanoislands (Fig. 9), pairs (dimers) or triples (Fig. 10). It is worth noting that by choosing an appropriate width of the electrode hollow to fit only one particle (at certain processing conditions) self-arranged chains of nanoislands can be obtained (Fig. 11 and 7c). Similarly, wider elongated hollows could provide more complicated chains containing several nanoparticles placed in a transverse direction. It is worth noting that all these chains possess some disordering both in the longitudinal and transverse directions.

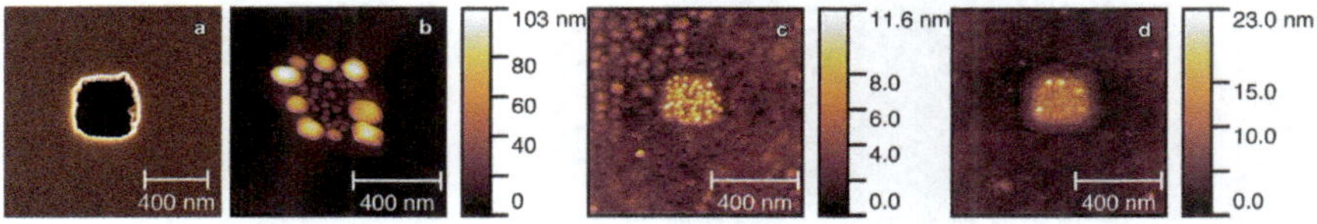

Fig. 6 Electrodes with hollows of 400×400 nm^2 (a) and structures, grown after poling with this electrode (b–d). Ion exchanged glasses were poled with a 500 V DC field at 300 °C for 60 seconds, and annealed in hydrogen for 10 minutes at 300 (b), 250 (c) and 200 °C (d).

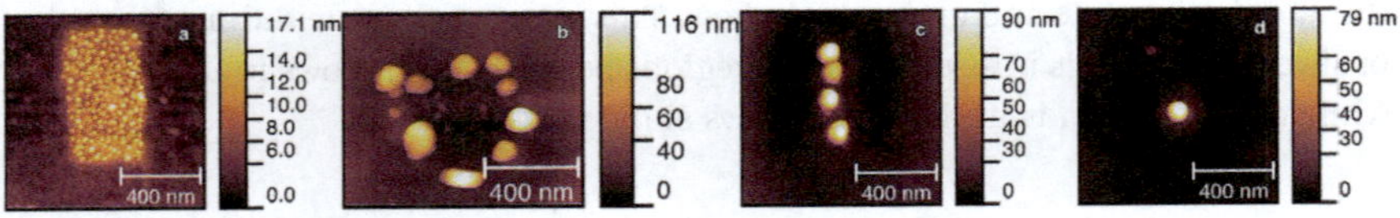

Fig. 7 Examples of silver nanoisland structures obtained using the developed 2D-patterning method with different electrodes, poling and annealing conditions.

Stability of out-diffused silver nanoislands

Although silver is known to react with oxygen/sulphur dioxide in the air, in the case of the out-diffused nanoislands these reactions did not eliminate silver, they just passivated the nanoislands. We tested the stability of the grown nanoisland films by studying the optical spectra evolution over time and the spectra remained almost the same for 6 months (Fig. 12). Minor changes can, however, be explained not only by continuing degradation, but also the slow reduction of silver in the ion-exchanged glass even in air at room temperature which can take place.[51]

To additionally protect the samples and avoid both effects in some cases we covered them with 3–5 nanometers of amorphous TiO_2 by means of atomic-layer deposition. Using this coating, a plasmon resonance shift by 20–50 nanometers to longer wavelengths should be taken into account. It should be noted that this shift could be less if a coating material of lower refractive index, for example Al_2O_3, is used.

Stability of glassy-carbon electrode

We tested different materials to make the anode and found glassy carbon to be the best option, as it is solid and does not suffer from anodic bonding (compared to silicon, for example), can be patterned in the nanoscale, and is commercially available with a perfectly flat surface. It is worth mentioning that after 90 cycles of

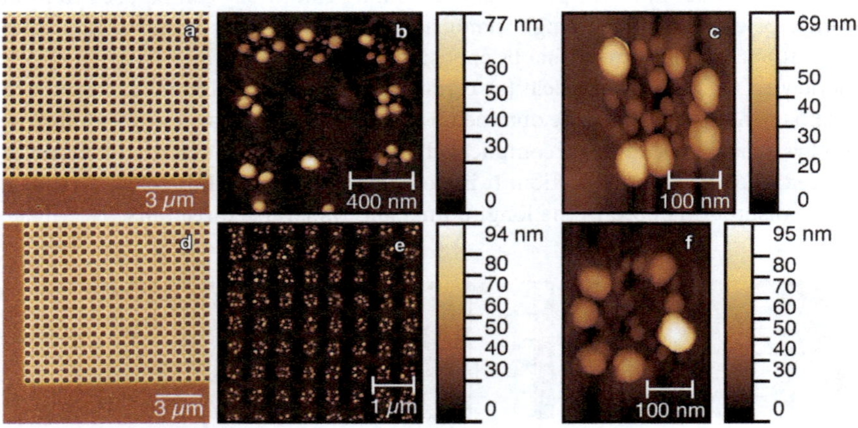

Fig. 8 SEM images of the electrodes with 200 × 200 (a) and 300 × 300 nm^2 hollows (d). AFM images of silver nanostructures grown in the course of 3 minutes annealing at 315 °C, before that glass was consequently ion-exchanged and poled for 10 seconds with DC 500 V at 300 °C using these electrodes—(a) for (b and c) and (d) for (e and f), respectively.

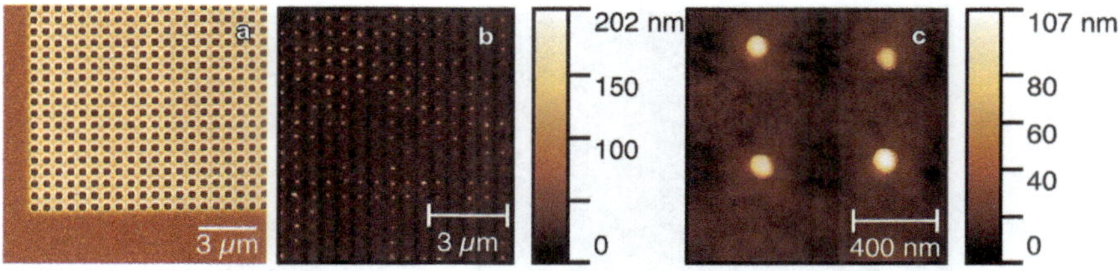

Fig. 9 SEM image of the electrode with 300 × 300 nm^2 hollows (a). AFM images of silver nanostructures (b and c) grown in the course of 5 minutes annealing at 300 °C, before that glass was consequently ion-exchanged and poled using shown in (a) electrode for 10 seconds with DC 500 V at 300 °C.

poling and cleaning, the etched hollows in the anodic electrode enlarged only by 50 nanometers at each side which characterizes the stability of the structure. However, the basic cleaning technique used in the laboratory—mechanical cleaning using cotton pads with acetone—was found to cause minor scratches on the electrode surface as well as leaving some acetone remaining affected the contact between the electrode and glass in the poling. The best option was consequent ultrasonication in acetone and isopropanol.

Conclusions

Finally, we have designed a technique to control the self-assembled growth of silver nanoislands in predefined areas of a glass surface. This is provided by the

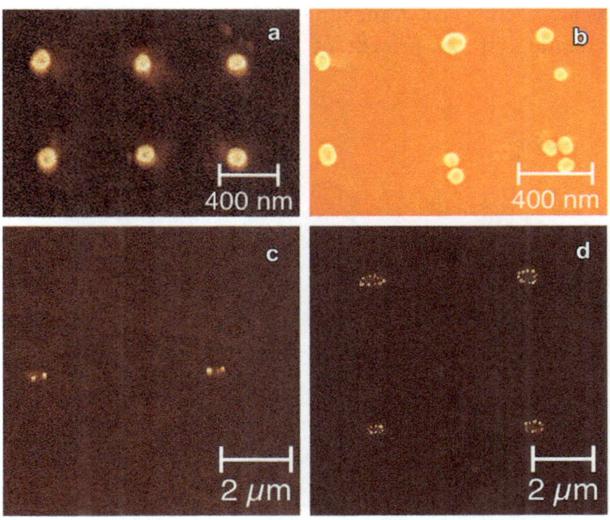

Fig. 10 SEM images of silver nanostructures grown in the course of annealing for 10 minutes at 350 °C (a,c), for 30 seconds at 300 °C (b), and for 30 minutes at 300 °C (d), before that glass was consequently ion-exchanged and poled for 60 seconds with DC 500 V at 350 °C (a,c) and 300 °C (b,d).

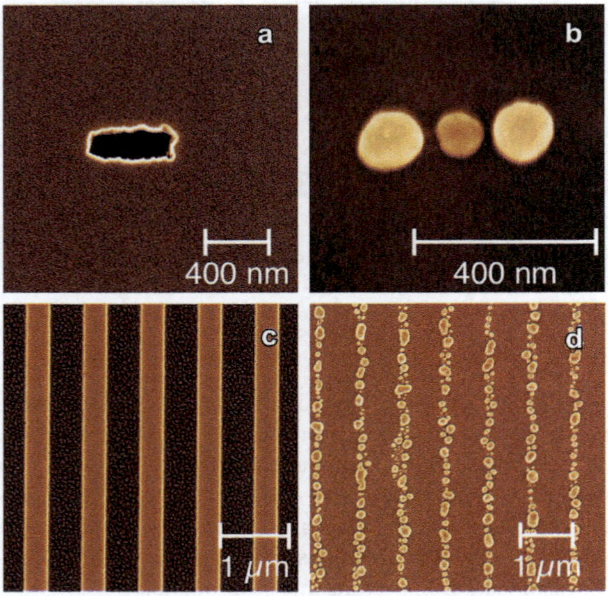

Fig. 11 SEM images of the electrode with 200 × 600 nm² (a) and 1 μm stripe (c) hollows. SEM images of silver nanostructures grown in the course of annealing for 10 minutes (b) and for 60 seconds (d) at 300 °C, before that glass was consequently ion-exchanged and poled for 60 seconds with DC 500 V at 300 °C.

thermal poling of a silver-exchanged glass with a profiled anodic electrode, which allows the nanoscale tuning of the spatial distribution of silver ions in the subsurface region of the glass. Annealing of glasses with altered ion distributions in hydrogen results in the growth of silver nanoislands and groups of the

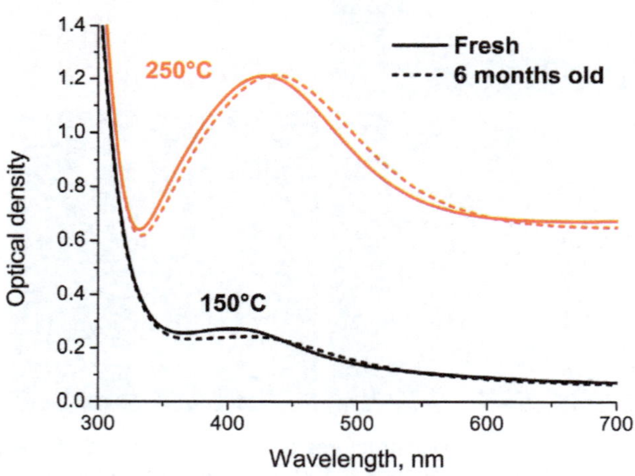

Fig. 12 Optical absorption spectra of silver nanoisland films after preparation (solid) and after 6 months in air at room temperature (dashed). Films were grown in the course of 10 minutes annealing in hydrogen at the temperatures shown in the graph.

nanoislands in less-poled areas, which correspond to hollows in the anode used in poling of the glass. Governing the 3D distribution of silver ions in the glass by the conditions of the poling and annealing enables control of neutral silver out-diffusion. The latter allows varying the structures of the nanoislands which can self-arrange in the less-poled regions—from areas with densely placed small silver nanoislands of a few nanometers in size to the arrays of single islands as big as 150–200 nm in diameter, arrays of isolated dimers and triplets of nanoislands, as well as chains and circles of islands. It has been shown that the use of glassy-carbon as the anodic electrode material provides decent stability of the nano-profiled anode on a run of tens of high voltage poling and solvent cleaning cycles. The optical properties of the manufactured silver nanoislands are stable after 6 months exposure in the air. The developed technique is capable of easily fabricating differing in their spectral response structures consisting of several metal nanoislands for plasmonics and photonics applications.

Author contributions

S. Ch. made the nanoprofiled electrodes and prepared the main part of the manuscript text. S. Ch. and I. R. developed the 2D-patterning technique. S. Ch., I. R. and A. K. made samples, measured spectra, and analysed all experimental data. I. R. drew the 3D-schematic of the process. I. R. and I. S. M. performed SEM measurements. A. K. made AFM scans. A. A. L. supervised the whole work.

Acknowledgements

The work was supported by the EU (Marie Curie FP7 project #269140), Academy of Finland (grant #288151), and Russian Science Foundation (grant no. 14-22-00136). The authors thank Pavel Brunkov and Alexander Baklanov from the Ioffe Institute for their help in the AFM studies. S. Ch. is thankful to Janne Laukkanen, Mikhail Petrov and Denis Karpov from the University of Eastern Finland for their support in electron beam lithography. The AFM and TEM studies were performed using equipment at the Joint Research Centre ‘Material science and characterization in advanced technology’ (Ioffe Institute, St. Petersburg, Russia).

Notes and references

1 S. A. Maier, *Plasmonics: Fundamentals and Applications*, Springer, 2007.
2 E. C. le Ru and P. G. Etchegoin, *Principles of Surface-Enhanced Raman Spectroscopy*, Elsevier, 2009.
3 J. N. Anker, W. P. Hall, O. Lyandres, N. C. Shah, J. Zhao and R. P. van Duyne, *Nat. Mater.*, 2008, 7(6), 442–453.
4 V. A. Podolskiy, P. Ginzburg, B. Wells and A. V. Zayats, *Faraday Discuss.*, 2015, **178**, 61–67.
5 M. Gaio, M. Castro-Lopez, J. Renger, N. van Hulstbc and R. Sapienza, *Faraday Discuss.*, 2015, **178**, 237–252.
6 P. Xu, L. Kang, N. H. Mack, K. S. Schanze, X. Han and H. L. Wang, *Sci. Rep.*, 2013, **3**, 2997.
7 C. Clavero, *Nat. Photonics*, 2014, **8**, 95–103.

8 J. Kim, H. Choi, C. Nahm and B. Park, *Electron. Mater. Lett.*, 2012, **8**(4), 351–364.

9 H. Wang, D. W. Brandl, P. Nordlander and N. J. Halas, *Acc. Chem. Res.*, 2007, **40**(1), 53–62.

10 N. Zohar, L. Chuntonov and G. Haran, *J. Photochem. Photobiol., C*, 2014, **21**, 26–39.

11 C. Lee, M. Tame, J. Lim and J. Lee, *Phys. Rev. A*, 2012, **85**, 063823.

12 V. Klimov, *Nanoplasmonics*, Pan Stanford, 2007.

13 R. Esteban, A. Zugarramurdi, P. Zhang, P. Nordlander, F. J. Garcıa-Vidal, A. G. Borisov and J. Aizpurua, *Faraday Discuss.*, 2015, **178**, 151–183.

14 E. Ringe, B. Sharma, A.-I. Henry, L. D. Marks and R. P. van Duyne, *Phys. Chem. Chem. Phys.*, 2013, **15**, 4110–4129.

15 A. D. McFarland and R. P. van Duyne, *Nano Lett.*, 2003, **3**(8), 1057–1062.

16 B. de Nijs, R. W. Bowman, L. O. Herrmann, F. Benz, S. J. Barrow, J. Mertens, D. O. Sigle, R. Chikkaraddy, A. Eiden, A. Ferrari, O. A. Scherman and J. J. Baumberg, *Faraday Discuss.*, 2015, **178**, 185–193.

17 I. M. Lifshits, S. A. Gredeskul and L. A. Pastur, *Introduction to the theory of disordered systems*, Wiley, 1988.

18 V. Markel and A. Sarychev, *Phys. Rev. B: Condens. Matter Mater. Phys.*, 2007, **75**, 085426.

19 M. Petrov, *Phys. Rev. A*, 2014, **91**(2), 023821.

20 S. Panigrahi, S. Kundu, S. Ghosh, S. Nath and T. Pal, *J. Nanopart. Res.*, 2004, **6**(4), 411–414.

21 V. Amendola and M. Meneghetti, *Phys. Chem. Chem. Phys.*, 2009, **11**, 3805–3821.

22 M. V. Gorokhov, V. M. Kozhevin, D. A. Yavsin, A. V. Ankudinov, A. A. Sitnikova and S. A. Gurevich, *Tech. Phys.*, 2012, **57**(6), 868–873.

23 K. Aslan, Z. Leonenko, J. R. Lakowicz and C. D. Geddes, *J. Fluoresc.*, 2005, **15**(5), 643–654.

24 A. Schaub, P. Slepička, I. Kašparkova, P. Malinsky, A. Mackova and V. Švorčik, *Nanoscale Res. Lett.*, 2013, **8**, 249.

25 S. Madsen, P. Kempen and R. Sinclair, *Microsc. Microanal.*, 2013, **19**(2), 1554–1555.

26 B. Radha, S. H. Lim, M. S. M. Saifullah and G. U. Kulkarni, *Sci. Rep.*, 2013, **3**, 1078.

27 X. Zhang, A. V. Whitney, J. Zhao, E. M. Hicks and R. P. van Duyne, *J. Nanosci. Nanotechnol.*, 2006, **6**, 1–15.

28 R. McCaffrey, H. Long, Y. Jin, A. Sanders, W. Park and W. Zhang, *J. Am. Chem. Soc.*, 2014, **136**(5), 1782–1785.

29 S. Chervinskii, V. Sevriuk, I. Reduto and A. Lipovskii, *J. Appl. Phys.*, 2013, **114**, 224301.

30 F. Gonella, *Ceram. Int.*, 2015, **41**(5), 6693–6701.

31 G. de Marchi, F. Caccavale, F. Gonella, G. Mattei, P. Mazzoldi, G. Battaglin and A. Quaranta, *Appl. Phys. A*, 1996, **63**(4), 403–407.

32 Y. Kaganovskii, A. Lipovskii, M. Rosenbluh and V. Zhurikhina, *J. Non-Cryst. Solids*, 2007, **353**, 2263–2271.

33 Menzel-Glaser: Microscope slides (accessed 22.08.2015) http://menzel.de/Microscope-Slides.687.0.html?&L=1.

34 J. Linares, D. Sotelo, A. A. Lipovskii, V. V. Zhurikhina, D. K. Tagantsev and J. Turunen, *Opt. Mater.*, 2000, **14**(2), 145–153.
35 V. V. Zhurikhina, M. I. Petrov, K. S. Sokolov and O. V. Shustova, *Tech. Phys.*, 2010, **55**(10), 1447–1452.
36 P. L. Inácio, B. J. Barreto, F. Horowitz, R. R. B. Correia and M. B. Pereira, *Opt. Mater. Express*, 2013, **3**(3), 390–399.
37 S. Chervinskii, A. Matikainen, A. Dergachev, A. Lipovskii and S. Honkanen, *Nanoscale Res. Lett.*, 2014, **9**, 398.
38 V. V. Zhurikhina, P. N. Brunkov, V. G. Melehin, T. Kaplas, Y. Svirko, V. V. Rutckaia and A. A. Lipovskii, *Nanoscale Res. Lett.*, 2012, **7**, 676.
39 C. M. Lepienski, J. A. Giacometti, G. F. L. Ferreira, F. L. Freire and C. A. Achete, *J. Non-Cryst. Solids*, 1993, **159**(3), 204–212.
40 K. Sokolov, V. Melehin, M. Petrov, V. Zhurikhina and A. Lipovskii, *J. Appl. Phys.*, 2012, **111**(10), 104307.
41 S. A. Scherbak, O. V. Shustova, V. V. Zhurikhina and A. A. Lipovskii, *Plasmonics*, 2015, **10**(3), 519–527.
42 A. Redkov, S. Chervinskii, A. Baklanov, I. Reduto, V. Zhurikhina and A. Lipovskii, *Nanoscale Res. Lett.*, 2014, **9**, 606.
43 A. Redkov, S. Chervinskii, A. Baklanov, I. Reduto, V. Zhurikhina and A. Lipovskii, *Nanoscale Res. Lett.*, 2015, **10**, 201.
44 V. V. Zhurikhina, M. I. Petrov, K. S. Sokolov and O. V. Shustova, *Tech. Phys.*, 2010, **55**(10), 1447–1452.
45 R. H. Doremus, *Appl. Phys. Lett.*, 2005, **87**, 232904.
46 A. A. Lipovskii, V. V. Rusan and D. K. Tagantsev, *Solid State Ionics*, 2010, **181**(17–18), 849–855.
47 M. I. Petrov, A. V. Omelchenko and A. A. Lipovskii, *J. Appl. Phys.*, 2011, **109**(9), 094108.
48 M. Dussauze, V. Rodriguez, A. Lipovskii, M. Petrov, C. Smith, K. Richardson, T. Cardinal, E. Fargin and E. I. Kamitsos, *J. Phys. Chem. C*, 2010, **114**(29), 12754–12759.
49 A. V. Redkov, V. G. Melehin and A. A. Lipovskii, *J. Phys. Chem. C*, 2015, **119**(30), 17298–17307.
50 A. V. Redkov, A. A. Lipovskii and V. V. Zhurikhina, *J. Non-Cryst. Solids*, 2013, **376**, 152–157.
51 I. Reduto, S. Chervinskii, A. Matikainen, A. Baklanov, A. Kamenskii and A. Lipovskii, *J. Phys.: Conf. Ser.*, 2014, **541**, 012073.

Faraday Discussions

DISCUSSIONS

Synthesis of Nanoparticle Assemblies: general discussion

Madhura Som, Sristi Majumdar, Nirmalya Bachhar, Guruswamy Kumaraswamy, G. V. Pavan Kumar, Vinothan N. Manoharan, Sanat Kumar, Madivala G. Basavaraj, Siddharth Kulkarni, Ranjini Bandyopadhyay, Sudeep Punnathanam, Himani Medhi, Ajeet Srivastav, Daan Frenkel, Mukta Tripathy, Erika Eiser, Lola Gonzalez-Garcia, Priyadarshi Roy Chowdhury, Jayant Singh, Vimala Sridurai, Alison Edwards, B. L. V. Prasad, Amit Kumar Singh, Michael Bockstaller, Neena S. John, Jyoti Seth, Mayank Misra, Charusita Chakravarty, Vandana Shinde, Rajdip Bandyopadhyaya, Jacques Jestin, Radhika Poojari, Nicholas Kotov, Oleg Gang, Alamgir Karim, Yon Ju-Nam, Steve Granick, Semen Chervinskii and Andrea Tao

DOI: 10.1039/C6FD90001A

Nicholas Kotov opened a discussion of the introductory lecture by Steve Granick: I much appreciated the historical perspective of the particle self-assembly studies. It is important to know where the current research came from. The connection between the nanoscale matter and biological matter is also significant and represents the growth point of the field. From your studies it is clear that the colloidal systems become particularly interesting when particles acquire anisotropy which is dipolar anisotropy in your works. What are the other degrees of isotropies that you can engineer into these particles and what self-assembly effects would you expect from them?

Steve Granick responded: Your important question encapsulates some of the most urgent and pressing questions of this research field. Anisotropy can generate important and interesting shapes, which we as a community know well, starting from the proud tradition of studying surfactant self-assembly. Increasingly recognized is that anisotropy can also generate important and interesting shapes out of equilibrium, as we know in nature from the flocking of birds and the generation of schools of fish, and perhaps also in the social interactions of people. Many people around the world are exploring the implications.

Vinothan N. Manoharan remarked: In response to Prof. Kotov's question about designing other types of anisotropy in nanoparticles: one point to take away from Prof. Granick's lecture is that we should focus not only on spatial anisotropy, but also temporal anisotropy – that is, particles with interactions that change in time or in response to an event that happens within the dispersion. With such interactions, we could design systems that evolve in complex ways.

Nicholas Kotov further remarked: The great advantage of micro and mesoscale particles is the possibility to observe them in optical microscopes. This is often not the case for nanoparticles. What aspects of nanoparticle behavior can we learn from the dynamics of micron scale particles? Can the nanoscale particles behave in the same way?

Steve Granick replied: Beyond differences in convenience of observation, it is important to recognize that nanoparticles can differ fundamentally from micron-sized particles when the (nano) range of their interactions becomes comparable to the (nano) size, which cannot happen for colloidal-sized particles. But progress is being made in observing nanoparticle behavior directly in microscopes. One approach is fluorescence spectroscopy in the "super-resolution" mode such that the resolution is less than what previously was considered to be a diffraction-based limit. A second approach is the mounting progress in employing electron microscopy for direct observation using liquid cells.

Nirmalya Bachhar addressed Steve Granick and Nicholas Kotov: Since, motion of Janus particles are oriented by their respective dipole, would it be appropriate to have several such patches on particles, as multiple such patches may nullify the effect of a single dipole which will lead to random collisions?

Nicholas Kotov responded: The randomization of the patches will result in reduction of anisotropy and lead to randomization of the assembly. However, if the number of the gold spots is increased whereas the fidelity of surface indexing of their placement remains high, it can also lead to an increase in complexity of the assemblies. Collective effects might also be an important factor that will guide their self-organization behavior.

Steve Granick replied: If I understand you correctly, you are correct in your intuition that too much seasoning can spoil the broth: indeed, too many specific patchy interactions can have the effect of smearing everything out so that it appears homogeneous or nearly so.

Sudeep Punnathanam questioned: You have used electrical energy to impart motion of Janus particles. Can you achieve the same through the use of a chemical reaction? Are there examples in the literature where energy from a chemical reaction has been used to create active matter of nanoparticles?

Steve Granick responded: Indeed, your idea has been explored in the literature. An excellent recent review is ref 1.

1 C. Bechinger, R. Di Leonardo, H. Löwen, C. Reichhardt, G. Volpe and G. Volpe, *Active Brownian Particles in Complex and Crowded Environments*, 2016, http://arxiv.org/abs/1602.00081.

Mukta Tripathy asked: In your talk, you have presented movies of Janus particles moving under the influence of an alternating current. Their motion at low densities evolves into a rotating ring-like motion at long times. However, at high densities the motion is more chaotic. Is there a way to characterize the motion of these particles at high densities?

It also seems from the movies that the crossover in the dynamics of the particles may have been a result of the collision rate (which would increase with particle density). Would you agree?

Also, it seemed from your answer to a previous question, that in this particular experimental setup, there is limited control over the speed of the particles. Are there other experimental systems where the speeds can be more minutely controlled, in order to properly check if it is indeed the collision rate that is dictating the crossover in the dynamics?

Steve Granick replied: You have identified a number of interesting problems. We are working on them.

Charusita Chakravarty remarked: One of the ways you showed of activating colloidal particles seemed to depend on imposing macroscopic flows that can couple to particle motion. Is this indeed the case and can it be mimicked at the nanoparticle scale?

Steve Granick responded: Rather than simply seek ways to activate particles, I prefer to begin with considering the purpose for which one wishes to do so. Once one has a desired function of some sort in mind, all sorts of methods can be dreamed up to accomplish it.

Nicholas Kotov said: As a comment, I would highlight the fact that the studies at the intersection of nanoscale matter and biology become more and more fascinating every day. I would point out that the similarity with biological phenomena that we can trace in the case of nanoparticles and some micro/mesoscale particles are not accidental. The nanoparticles can be considered as simplified and less structurally controlled versions of the nanoscale species known from biology. From my perspective, the studies of (nano)particle assemblies have a lot to give to biology because these studies provide better understanding where biology became biology. Self-assembly is a particularly good example of that because it is a unifying phenomenon between physical chemistry and biochemistry. If we dig deeper, we may one day realize that the smartness of biology can be replicated by simple particle models.

Madhura Som addressed Steve Granick and Nicholas Kotov: Most of the work Steve Granick presented has a general theme on how we can be inspired from nature/biology to know more about nanoparticle assemblies. What about the other way round? Are we now headed in a direction where we can think of contributing to the fundamental knowledge we have in biology today using what we know about nanoparticle assembly or nanotechnology in general?

Nicholas Kotov responded: Yes, we can contribute to fundamental knowledge in biology using the data on self-assembly of nanoparticles. These data help us identify essential components of molecular and nanoscale interactions necessary for self-assembly phenomena to occur. To some degree, the data on nanoparticle self-assembly allow us to simplify and demystify self-organized systems known from biology.

Steve Granick answered: I am a humble man and feel it is not my place to speak for what we can contribute to another field. I am sure that we as a community would be thrilled to have the broadest impact possible, but until that impact is achieved, I am more comfortable to simply do the best we can, and let others praise our contributions. We can surely contribute to biology in the sense of elucidating intermolecular and surface forces, and also some of the principles of self-assembly, but I am not sure that these are the most important questions of the 21st century biology.

Himani Medhi asked Steve Granick: Biology to cosmology showed chaotic movement of particles which reminds me of asteroids and how planets form, eventually most materials form a large body of mass or vortex. Do you expect your system to form small vortices in a systematic manner?

Steve Granick replied: This is an intriguing question. Thank you.

Oleg Gang commented: It might be interesting to simulate an artificial temperature *via* tailored broad spectrum field excitation.

Michael Bockstaller questioned: Many thanks for the inspiring contribution. My question is with regard to the results that were presented in the context of the statement "Crowd Transcends Death" – *i.e.* the absence of a defined equilibrium state. In particular, is this statement based on theoretical analysis or is it an experimental assertion? Could it be related to the impracticality of experiments that a "final state" cannot be observed? Could you comment on the role of symmetry (of colloidal building blocks as well as aggregates) on the structure evolution process? Finally, the video suggested that circular aggregate structures did not translate and showed only a rotary motion. Does this not suggest that perhaps an "array" of circular aggregates represent a "stable state"?

Steve Granick answered: Your intriguing questions are in part philosophical, in part system-specific. To the philosophical question, whether a defined equilibrium state exists in an active system, of course the answer is no; life is out of equilibrium! Regarding the system-specific questions, we are currently working on these matters.

Ajeet Srivastav commented: It seems that the most stable system is the most unstable system provided activity is confined in a particular geometrical structure. How to correlate or model such active systems to their geometrical structure?

Steve Granick responded: You have posed an intriguing challenge for this field.

Priyadarshi Roy Chowdhury communicated: In regards to your opinion discussed between colloids and nanoparticles, express your views with respect to the optical density and anisotropic property between them. Please discuss your views on the structural-engineering aspects of the colloids and nanoparticles.

Steve Granick communicated in reply: You are correct in calling attention to this grand challenge of our field.

Priyadarshi Roy Chowdhury communicated: Which method do you consider, the best method for nanoparticle synthesis at present?

Steve Granick communicated in reply: I prefer to live in a world of variety. Which method is best, probably this will depend on the type of nanoparticle one wishes to synthesize, on the purpose to do so, and on the resources available. The human race has existed for thousands and thousands of years, but still we have not found the best way to raise a child.

Priyadarshi Roy Chowdhury communicated: How do activated colloids behave in comparison to the nanoparticles; does the former exhibit some contrasting behaviour in comparison to the nanoparticle with respect to its directionality, randomness, density, flow-property and band gap-engineering?

Steve Granick communicated in reply: Active nanoparticles have been less explored than active colloids, and comprise one of the challenges to resolve, for which more research is needed. A survey of recent understanding can be found in Ref. 1.

1 C. Bechinger, R. Di Leonardo, H. Löwen, C. Reichhardt, G. Volpe and G. Volpe, *Active Brownian Particles in Complex and Crowded Environments*, 2016, http://arxiv.org/abs/1602.00081.

Guruswamy Kumaraswamy opened a discussion of the paper by Michael Bockstaller: Why is it important for the grafted polymers to be in an extended brush configuration? For nanoparticles with sparsely grafted polymers in a mushroom configuration too, one can obtain repulsive interparticle interactions that allow them to crystallize. An advantage of this strategy is that the chains can interpenetrate, resulting in enhanced toughness at high particle content. What then, is the advantage of preparing particles densely grafted with chains with bimodal size distributions in a brush configuration?

Michael Bockstaller responded: The advantage of densely grafted particle systems is that core–core interactions are fully screened, which is not guaranteed in the case of sparsely grafted particles. Core–core interactions can sensitively affect the miscibility of particle brush systems. For example, a diverse range of particle assembly structures that has been observed for sparsely grafted particles admixed to a homopolymer matrix (of equal chemical composition to the tethered chains) has been attributed to the superposition of core–core, core–matrix and graft–matrix interactions as well as entropic driving forces.[1] The role of core–core interactions in sparsely grafted systems is difficult to predict and is expected to sensitively depend on material composition and geometry and hence significantly complicates the interpretation of results. In contrast, for densely tethered brush systems our group has shown previously that interactions can be tuned from hard sphere-type to soft (polymer)-type by tailoring of the degree of polymerization of tethered chains.[2] Even in the limit of small degree polymerizations the role of core–core van der Waals interactions is predicted to be negligible.[3] Finally, it is important to note that our choice of dense particles has been for strategic (rather than fundamental) reasons. First, to reduce the number of unknowns in the system but also to establish whether assembly structures formed by dense particle brush materials – that are of interest as photonic materials – can be toughened by homopolymer addition. The formation of ordered superlattice structures is specific to dense particle brushes interacting by (approximately) hard sphere-type interactions. We concur that a broader understanding of the governing parameters that control structure and interactions in particle brush assembly structures should include sparse particle brush systems. In fact, the recent development of synthetic strategies towards bimodal brush systems should enable the elucidation of the role of grafting density in the absence of core–core interations.[4]

1 S. K. Kumar, N. Jouault, B. Benicewicz and T. Neely, *Macromolecules*, 2013, **46**, 3199–3214.
2 J. Choi, C. M. Hui, M. Schmitt, J. Pietrasik, S. Margel, K. Matyjazsewski and M. R. Bockstaller, *Langmuir*, 2013, **29**, 6452–6459.
3 H. Yockell-Lelievre, J. Desbiens and A. M. Ritcey, *Langmuir*, 2007, **23**, 2843–2850.
4 J. Yan, T. Kristufek, M. Schmitt, Z. Wang, G. Xie, A. Dang, C. M. Hui, J. Pietrasik, M. R. Bockstaller and K. Matyjaszewski, *Macromolecules*, 2015, **48**, 8208–8218.

Jyoti Seth questioned: Polymer chains contained in thin films along with silica particles would experience a strain larger than the applied bulk strain. How does this affect the observations for modulus and hardness?

Michael Bockstaller replied: The constraints imparted by surface tethering fundamentally alter the conformation of chains in films and are certain to alter the deformation behavior of particle brush materials. Contrasting the behavior of particle brush systems with corresponding binary nanocomposite materials would be a very interesting research proposition. We are not aware of a detailed study on this interesting subject, however, we want to highlight that Brillouin scattering experiments have revealed characteristic differences in the mechanical anisotropy of thin films of particle brushes and binary nano composites that could be interpreted as a consequence of altered chain conformations.[1] Again, this would be an interesting topic for future research.

1 P. Voudouris, J. Choi, N. Gomopoulos, R. Sainidou, H. Dong, K. Matyjaszewski, M. R. Bockstaller and G. Fytas, *ACS Nano*, 2011, **5**, 5746–5754.

Jacques Jestin asked: Could the absence of autophobic dewetting be attributed to strong repulsive interactions between particles due to high particle loading?

Michael Bockstaller responded: Many thanks for this interesting question. Since the question is similar in scope as the previous question asked by Guruswamy Kumaraswamy, we refer to our previous answer.

Nicholas Kotov enquired: It seems to me that the observation about change of mechanical properties and order is important. The reorganization of the material is influential in determining toughness and stiffness. What are the deformation patterns in nanoscale and mesoscale?

Michael Bockstaller replied: Many thanks for this interesting question about the relationship between order and deformation behavior of the material. This is certainly an area that merits future research and, as indicated in the previous question asked by Guruswamy Kumaraswamy, interesting observations have been made that suggest new (or rather: unexpected) physics could be at work. In our previous answer to Guruswamy Kumaraswamy we summarize our current understanding of this aspect.

Sanat Kumar asked: Can you please comment on the order–disorder transition and how it affects properties, in this case, in particular, the toughness? We have found, in our own work, that the ordering of grafted nanoparticles into hexatic phases results in chain conformations that are frustrated because surface tension forces them to fill all the interstitial spaces. Moreover the chains from adjacent nanoparticles tend not to entangle. You may want to look at chains in the ordered phase *vs.* disordered state to see why you go from low toughness to a high toughness state.

Michael Bockstaller answered: Thank you for this interesting comment and question – in the following we will first provide a brief direct response to the question followed by a brief summary of our previous observations regarding the interrelationship between order and toughness of particle brush materials. The short answer to the question is that we have not explicitly tested the effect of the "degree of order" in homopolymer-free particle brush superlattice structures on mechanical properties. It is important to note that our study entails "large" particle brush systems in which the ordering kinetics in the melt state is very slow. In fact, we were not able to detect changes in the "degree of order" ("measured" by evaluating the structural uniformity for particle multilayers) of cast films during thermal annealing for periods of one week. Thermal annealing in our studies was therefore performed to relax frozen stresses rather than to allow brush rearrangement and large scale coarsening of particle brush solids. Consistent annealing conditions were applied to all material systems. The comment raises an interesting idea, *i.e.* an increasing order formation could result in the embrittlement of particle brush solids caused by chain stretching to fill the corner regions of Wigner Seitz cells and the associated reduction of chain entanglement. This is an intriguing proposition that would be somewhat counter intuitive since the packing fraction typically increases with the degree of order and hence the (average) entanglement density would be expected to increase with increasing order formation. Testing this proposition would, however, not be trivial since one has to consider that plastic deformation in ordered materials in general proceeds by different mechanisms as compared to their amorphous analogs. For example, the periodic packing in crystalline materials enables plastic deformation by slip processes that have a lesser activation barrier and hence reduce the resistance to plastic deformation. Hence crystalline materials are found to generally exhibit lower hardness and yield strength than amorphous homologues. The extent to which slip is relevant in particle brush based crystals is unknown and will likely depend on particle size as well as degree of polymerization of tethered chains. The experimental testing of this hypothesis would therefore have to not only establish a quantitative relationship between "structural order" and sample toughness but also differentiate the mechanisms involved in plastic deformation. It is also worthwhile to note that the results presented in our paper do not reveal a relation between order and toughness. For example, comparison of the data depicted in Figures 4c and 7c in our paper reveal that while the "degree of order" (in particle brush monolayers) is retained upon addition of homopolymer up to a filling fraction of approximately 30% a strong increase in material toughness is observed. We therefore interpret the increase of toughness to be predominantly a consequence of the added homopolymer rather than induced structural changes. In the following we briefly summarize some of our previous observations on the mechanical properties of particle brush systems. By systematic analysis of the elastic modulus, hardness and toughness of particle brush systems as a function of particle size and graft composition we found that the mechanical properties of small particle systems significantly increase with the graft length when the length of the grafted chains exceeds the expected transition from stretched to relaxed chain conformation.[1] The increase of toughness was attributed to chain entanglement that was evidenced by craze formation in thin films undergoing mode-I fracture.[2] Our recent work has shown that such trends hold true for even larger ($R_0 = 60$ nm) particle sizes. While mechanical properties of grafted particle assemblies increase with increasing chain length, quantification of short-range order in close-packed monolayers of grafted particles shows that order actually decreases with increasing chain length for all particle sizes.[3] Of course, in this case, the loss of order is caused by the increase in the degree of polymerization of surface grafted chains.

While it remains possible for the degree of order (or a possible disorder to order transition) to influence the mechanical properties of grafted particle assemblies in the case of entangled systems, we are currently limited by our ability to quantitatively assess the order of bulk (3D) assemblies of grafted particles. Early efforts to obtain information on the bulk ordering using X-ray tomography and small angle X-ray scattering did not yield sufficiently clear data to allow for quantification. This is at least partially due to the mesoscale size range (100–250 nm) of the large particles. Moderate thermal annealing, solvent annealing, or at least controlled solvent evaporation may influence the degree of order in these systems, but without an effective means to ascertain how much, if at all, the degree of order is increasing, a meaningful relationship between order and mechanical properties cannot be made. This presents opportunity for future study.

1 J. Choi, C. M. Hui, J. Pietrasik, H. Dong, K. Matyjaszewski and M. R. Bockstaller, *Soft Matter*, 2012, **8**, 4072–4080.
2 J. Choi, H. Dong, K. Matyjaszewski and M. R. Bockstaller, *J. Am. Chem. Soc.*, 2010, **132**, 12537–12539.
3 J. Choi, C. M. Hui, M. Schmitt, J. Pietrasik, S. Margel, K. Matyjazsewski and M. R. Bockstaller, *Langmuir*, 2013, **29**, 6452–6459.

Priyadarshi Roy Chowdhury communicated: What precautionary steps did you follow during the tailoring structure formation with regards to the synthesis of polymers mentioned?

Michael Bockstaller communicated in reply: In our work, we utilize surface-initiated atom transfer radical polymerization (SI-ATRP). Atom transfer radical polymerization (ATRP) is one of the most significant controlled radical polymerization techniques being employed today and, as such, a significant body of literature has been developed that defines a wide parameter space that can be used for designing polymer architectures. ATRP (as well as SI-ATRP) allows for precise control over many characteristics of the polymer chain, such as degree of polymerization, molecular weight distribution, and monomer chemistry to name a few. Setup of any SI-ATRP reaction requires careful attention in a couple areas. As an example, in order to achieve the desired architecture, precise molar ratios of the reactants must be obtained. In general, this requires an accurate estimate for the particle size given that the amount of initiator present in the system is inferred from the mass of the particles added to the reaction. Furthermore, the presence of non-inert gases in the solution from improper purification can cause early termination of the radical species leading to much shorter chains and higher chain molecular weight dispersity than desired. We refer you to our recent review in Chemistry of Materials for further details on SI-ATRP.[1]

1 C. M. Hui, J. Pietrasik, M. Schmitt, C. Mahoney, J. Choi, M. R. Bockstaller and K. Matyjaszewski, *Chem. Mater.*, 2014, **26**, 745–762.

Priyadarshi Roy Chowdhury communicated: What spectroscopic technique do you think is the most advanced/versatile in the present day for characterization of homo as well as co-polymers?

Michael Bockstaller communicated in reply: The appropriate spectroscopic technique for probing polymeric systems is dependent on which properties of the system one is interested in investigating as well as the relevant length-scales of the material. Many open questions remain about the dependence of the structure of such particle assemblies on the graft architecture as well as the dynamics of particles during assembly. Particularly promising – based on our own studies – have been the use of ultra-small angle X-ray scattering and X-ray tomography for determination of structure as well as light scattering to determine dynamics and interactions. If particles could be designed to be fluorescent this could provide new opportunities for microscopic investigation (such as confocal) or fluorescence correlation spectroscopy that could provide new insights into the dynamics of particle brush systems. Similarly, neutron scattering using isotopic labeled materials would be of interest. Mechanical spectroscopy – to reveal the role of particle brush architecture on the mechanical relaxation processes – would be similarly intriguing to perform.

Priyadarshi Roy Chowdhury further communicated: Did you do Chi-square fitting for your characteristic load-displacement curve, as depicted in Fig. 6 in your paper?

Michael Bockstaller communicated in reply: The plots in Fig. 6 are generated using a least squares fitting of the raw data from the nanoindenter, which is output at a sampling rate of 5 Hz. The elastic modulus is calculated based on the slope of the unloading curve, which is estimated using a finite difference method.

Priyadarshi Roy Chowdhury communicated: Do you consider Markov-Chain Monte Carlo (MCMC) to be an ideal theoretical approach in relation to the work you mentioned in this paper? Please discuss and relate if possible between the experimental and theoretical work.

Michael Bockstaller communicated in reply: Our group has not engaged in the theoretical modeling of particle brush systems, however, many other groups are involved in such efforts and interest in this area is growing. The prevailing methods employed during early efforts to model the behavior of particle brushes were self-consistent field theories (SCFT) and molecular dynamics (MD). A lot of this work has focused on understanding what relevant parameters determine particle dispersion states in polymer composites containing grafted particles. Recently, Monte Carlo (MC) methods have been employed in a variety of studies, but have found limitations in simulating particle movement in the matrix and are thus better suited for studies involving stationary particles.[1] We refer you to a recent review by Ganesan and Jayaraman that summarizes nicely the work done to simulate grafted particle behavior including current use of MC methods.[1]

1 V. Ganesan and A. Jayaraman, *Soft Matter*, 2014, **10**, 13–38.

Jacques Jestin opened a discussion of the paper by Alamgir Karim: From the analysis of the shape of the domains, could you say something about the chain deformation and about the influence of the particles on the chain deformation?

Alamgir Karim responded: An excellent question! We do not think there is much chain deformation of the grafted polymer chains attached to the nanoparticles. Reason is the TEM images show that the inter-nanoparticle distance is nearly the same in quiescent phase separated domains *versus* the highly sheared domains. We suspect that the relatively high interfacial tension at the boundary of the nanoparticle domain "absorbs" the shear energy; something like, Work of dynamic shear ~ Interfacial Tension (F/L) x rate change of boundary length (Delta) L/Time x Flexural/Bending Modulus of interface. The point is that the shear force does get significantly transmitted to the individual nanoparticle chains. So we expect the Daoud–Cotton model like 2 regimes near the particle to be a stretched dense grafting regime and a loose outer brush regime. In general, brushes on a curved surface may apply, *e.g.* see our paper published in *Macromolecules*.[1] Shear may deform the matrix chains, which is something that SANS *in situ* shear studies could tell in future. Also see Ref. 2.

1 J. M. R. d'Oliveira, J. M. G. Martinho, R. Xu and M. A. Winnik, *Macromolecules*, 1995, **28**, 4750–4752.
2 N. Singh, A. Karim, F. S. Bates, M. Tirrell and K. Furusawa, *Macromolecules*, 1994, **27**, 2586–2594.

Mayank Misra commented: I did some simulations for the system mentioned in your paper. I used hard sphere interactions for colloids and SRD particles to mimic a solvent. In no shear conditions, if the hydrodynamic radius is increased while keeping the core size the same, the probability of getting linear structures increases. So it looks like hydrodynamic forces play a critical role in maintaining linear structures. Further, when I apply high shear to the system, I find the strings to be aligned orthogonally to the shear direction, contrary to what you observed.

Alamgir Karim commented in reply: Excellent to know that! Please contact me regarding those simulations separately. I can say that at high nanoparticle concentrations than reported, we find orthogonal fractal like branches that develop on the main sheared axis of the domain under modest to high shear. Possibly, the high concentration affects the hydrodnamics of the system and there is a better coupling with shear forces and the orthogonal wings develop?

Amit Kumar Singh enquired: How are you introducing the localized defects to promote dewetting in a string-like fashion?

Alamgir Karim replied: Actually those are sheared phase separated nanoparticle string like domains caused by soft shear. However, the break-up into strings is by an instability mechanism similar to the dewetting droplet break-up.

Priyadarshi Roy Chowdhury communicated: In Fig. 1(b) and Fig. 2 of your paper, I could not find the average particle size distribution (histogram plot), calculated from the TEM images. If possible I request you to incorporate it in the ESI of your paper.

Alamgir Karim communicated in reply: Thanks, I will try to provide that, or in detail in a later paper on the topic if not possible in the present one.

B. L. V. Prasad remarked: It has been observed that the initial AuPS nanoparticles, that are well dispersed in the PMMA matrix, form discrete circular domains upon vacuum oven annealing at 180 °C for 16 h. What determines the size of these circular domains? Why don't they coalesce and become one big domain?

Alamgir Karim answered: The unfavorable interaction parameter between PS and PMMA controls the rate or driving force for phase separation domain growth. With increasing time the domains will grow and then pin. The circularity of the domains is driven by surface tension, but composition will also affect the domain shape.

Jyoti Seth enquired: The Au-PS structures align in response to soft shear but not the PS structures. Yet the critical capillary number, as expressed, does not account for the presence of particles in the matrix. Must there be an appropriate correction to the capillary number to explain the alignment occurring only in the presence of Au particles?

Alamgir Karim responded: Yes, the capillary number estimation needs to be renormalized in some way for this system. Probably by a suitable viscosity correction for the densely grafted nanoparticle in comparison to the homopolymer chains. This is indeed an interesting subject for further investigation. Theoretical insights into the problem would be useful as well.

G. V. Pavan Kumar asked: Since the inter-particle distances are small, they may show interesting plasmonic behavior. Have you explored them?

Alamgir Karim replied: Yes indeed, I would be very interested in pursuing this, but we don't have the instrument to measure this. I believe the distance is around 4–5 nm. The inter-particle distance can be further controlled by brush chain length and to some extent by grafting density.

Siddharth Kulkarni questioned: Depending upon the translational speed or speed of shearing, were the authors able to see any presence of particles inside the domains? Was the partitioning efficiency also determined?

Alamgir Karim responded: Our Faraday paper depicts the dense packing of the nanoparticles inside the clustered sheared domains. The PS-grafted-gold nanoparticle clusters behave like a disordered fluid with a finite interfacial tension with the PMMA matrix. The partitioning efficiency is near 100% so they are well-phase separated, but this is because PS and PMMA do not for the higher molecular weights. We have a paper in preparation showing that reducing the molecular weights can make this partitioning coefficient an intermediate value.

Ranjini Bandyopadhyay enquired: Are the string-like structures reversible under strain ramps? The experiments are understood in terms of an effective strain rate. Would it not be better to describe the results in terms of the shear strain rate profile?

Alamgir Karim answered: They are very good questions. Yes, if we turn off the shear, and anneal the distorted NP structures above Tg of PS and PMMA, we get isotropic (circular) polymer grafted gold nanoparticle cluster domains back in the PMMA matrix. We are able to observe the sheared nanoparticle clusters at the end of a non-uniform shear strain rate profile. Thus our profile being non-uniform is the challenge, but there are not too many methods for shearing ultrathin polymer blend films.

Sristi Majumdar opened a general discussion of the paper by Radhika Poojari: What is the shelf life of the amorphous nanoparticles and what is the probable effect of nanoparticles encapsulated Estradiol on normal tissue/cell?

Radhika Poojari responded: We have checked the physical stability of the pegylated polymeric nanoparticles which was found to be stable for 1 week without any aggregation. The stability of the nanoparticles was analyzed by DLS and zeta potential which did not show any significant change in the nanoparticle size and charge. We have checked the biocompatibility of the polymeric nanoparticle materials (50–500 $\mu g\ ml^{-1}$) on normal L929 and NIH 3T3 cells which showed no cytotoxicity. In the present study we have conducted the cell viability studies of the synthesized nanoparticles on hepatocellular carcinoma Huh7 cells to analyze its efficacy.

Priyadarshi Roy Chowdhury communicated: Please discuss the human side-effects associated with this particular therapeutic agent, which you have mentioned in your work.

Radhika Poojari communicated in reply: Some of the human side-effects involved with Estradiol are systemic-neurotoxicity, hypertension, breast pain, diarrhea, nausea, vomiting, hair thinning and dermatological toxicities.

B. L. V. Prasad said: 1. In Figure 3 in your manuscript the cumulative Estradiol release is shown as % amount. What was the actual amount (on wt/wt basis) of estradiol loaded into the nanoparticles? 2. How much of this (again on weight basis) was actually getting released? 3. The authors conclude microtubule targeting by confocal imaging? Is this sufficient? Are these conclusions supported by any other bio-chemical assays?

Radhika Poojari responded: The encapsulation efficiency of estradiol in nanoparticles was 40%. The release percentage of drug has been shown in the results. As you suggested we will definitely look into the parameter on weight basis. Microtubule targeting by confocal imaging is self-sufficient and proven. We have also done the western blotting of α-tubulin expression analysis which have not been included here.

Nicholas Kotov enquired: You presented a technique to encapsulate the drugs, which has both novelty and importance. My impression is often that the problem with cancer treatment is related to accurate targeting of the capsules and particles. Do you anticipate some kind of targeting and delivering mechanism to cancer cells and if yes, what are the advantages of your encapsulation for this aspect of cancer treatment?

Radhika Poojari responded: Targeted nanoparticles can deliver drugs at a sustained rate to cancer cells, which may provide better efficacy and lower toxicity for treating primary and advanced metastatic tumors. Other than antibodies one can use nucleic acid ligands aptamers, DNA or RNA oligonucleotides and peptides for targeting. One needs to calculate the number of targeting moieties binding to the receptor for accuracy. The sustained drug release from the nanoparticles will be beneficial in reducing the side effects of drugs, frequency of drug administration and thus improving patient compliance.

Nicholas Kotov commented: Multiple studies reported that antibody targeting is imperfect. The particles carrying antibodies are distributed into many organs as well as into healthy cells besides cancerous ones. Furthermore, even if the particle arrives at the cancer target, it needs to go deep inside the tumor, which is often prevented by antibodies that cause the drug to accumulate at the external part of the tumor. Did you consider these issues? Any ideas how to combat them?

Radhika Poojari replied: In my opinion targeting efficiency with antibody can be achieved when a specific cancer cell is over-expressing a particular protein using ligand–receptor mediated nanoparticle targeting.

Daan Frenkel remarked: Antibodies can bind very strongly to receptors that are over-expressed on the surface of target cells. However, they also bind strongly to the same receptors that may be expressed in lower concentrations on healthy cell surfaces. As a consequence, using antibodies to target cyto-toxic drug delivery may lead to undesired side effects.

In simple model-systems, a high discrimination between cells with high and low receptor densities may be achieved using many weakly binding ligands, rather than a smaller number that binds strongly (see *e.g.*: Ref. 1 and Ref. 2). I wonder whether a similar approach might be practical in the present context.

We always have to be careful to expand ideas from physical sciences to biology, particularly when targeting the delivery to specific cells, as antibodies bind very strongly. If target cells are only cells that express receptors, that is fine. However, if they don't only express receptors, this is not fine, as it is much better to have a large number of weakly binding receptors. Is that relevant in this specific case?

1 F. J. Martinez-Veracoechea and D. Frenkel, *Proc. Nat. Acad. Sci.*, 2011, **108**, 10963–10968.
2 G. V. Dubacheva, T. Curk, B. M. Mognetti, R. Auzely-Velty, D. Frenkel and R. P. Richter, *J. Am Chem. Soc.*, 2014, **136**, 1722–1725.

Radhika Poojari replied: It is based on selective targeting which over-expresses certain receptors in tumor cells. One needs to calculate the number of targeted nanoparticles binding to the receptors. Using nanoparticles functionalized with PEG protects from systemic clearance, macrophages, non-specific protein uptake in comparison to the non-targeted cells, and improves the drug pharmacokinetics with no undesirable side-effects. See for example Ref. 1–3.

1 J. M. Chan, L. Zhang, R. Tong, D. Ghosh, W. Gao, G. Liao, K. P. Yuet, D. Gray, J. W. Rhee, J. Cheng, G. Golomb, P. Libby, R. Langer and O. C. Farokhzad, *Proc. Nat. Acad. Sci.*, 2010, **107**, 2213–2218.
2 F. Gu, L. Zhang, B. A. Teply, N. Mann, A. Wang, A. F. Radovic-Moreno, R. Langer and O. C. Farokhzad, *Proc. Nat. Acad. Sci.*, 2008, **105**, 2586–2591.
3 O. C. Farokhzad, J. Cheng, B. A. Teply, I. Sherifi, S. Jon, P. W. Kantoff, J. P. Richie and R. Langer, *Proc. Nat. Acad. Sci.*, 2006, **103**, 6315–6320.

Ranjini Bandyopadhyay opened a discussion of the paper by Guruswamy Kumaraswamy by asking: How sensitive are the results to the rate of the temperature quench?

Guruswamy Kumaraswamy responded: For the ice-templating experiments with low particle concentration dispersions reported here, we have not systematically explored the effects of changing the cooling rate. However, we have done this for ice templating of higher concentration dispersions that we have reported earlier. We observe that when we freeze the sample by rapidly immersing it in liquid nitrogen, we observe a decrease in the pore size in the final macroporous composite. We attribute this to the enhanced nucleation density of ice crystals obtained for this cooling protocol.

Vinothan N. Manoharan said: The observation of linear chains and flat sheets suggests that the aggregation is happening within the Plateau borders and films of a foam-like structure. Since the underlying ice structure doesn't change with the concentration, the proportion of linear chains should be independent of the concentration of particles, just as you observed. Is this argument consistent with your understanding of the ice structure? Also, does your simulation yield a foam-like structure of the growing domains at long times?

Guruswamy Kumaraswamy replied: At the low concentrations that we use in this work, we do not observe any influence of particle concentration on the nucleation and growth of ice crystals. Therefore, as you have suggested, the distribution of cluster sizes and of linear chains reflect how particles are assembled by the (same) underlying ice crystal structure, independent of the particle concentration of the ice templated dispersion. We do not run our simulations at particle concentrations that are sufficiently high to form percolated structures. However, our simulations indicate the progression in cluster shape as a function of particle concentration. As we increase the particle concentration, we observe the formation of linear strands, tape-like structures and eventually 3D structures. The 3D structures that form in our simulations are indicative of the foam-like structures that we have observed in our previous experiments.[1]

1 R. Rajamanickam, S. Kumari, D. Kumar, S. Ghosh, J. C. Kim, G. Tae, S. Sengupta and G. Kumaraswamy, *Chem. Mater.*, 2014, **26**, 5161–5168.

Michael Bockstaller asked: Many thanks for the fascinating contribution. The crystallization driven "expulsion" of particles to grain periphery reminds of the process of "impurity segregation to grain boundaries in inorganic crystalline materials". There it is commonly observed that distinct grain boundaries (owing to their distinct energy and geometrical constraints) exhibit distinct affinity to filler segregation. Could you comment on whether a similar relationship holds here and whether it could be used to control the "fine structure" of the segregation process?

Guruswamy Kumaraswamy replied: The large (micron-sized) colloids that are the ice-template in our experiments are unlikely to be sensitive to the energetics of different ice crystal grain boundaries. However, such effects might be much more important for ice templating of smaller nanoparticles or other moieties that might specifically bind to specific ice crystal facets. Thank you for suggesting this fascinating possibility – to the best of my knowledge, this has not been explored in the literature.

Jayant Singh enquired: The question is related to the structure formation at the interface of ice crystals. Does the structure obtained have any signature of confinement effects? The other question I have is related to the initial state of the solution. How important is the dispersion state to start with for ice templating?

Guruswamy Kumaraswamy answered: As the ice crystals impinge, colloidal particles are confined to the spaces between the crystals. Therefore, the shape of the colloidal clusters reflect the geometry of the regions where the particles are confined. However, we do not observe confinement effects such as those observed when, say, a polymer chain is confined to a thin film. In both the experiments and the simulations, we always start with a well dispersed state. I would imagine that the dispersion of colloidal particles in the initial dispersion is important in determining the structure of the clusters. For example, if we had large pre-formed clusters in the initial dispersion, these would strongly influence the cluster size distribution in the final aggregated state.

Alison Edwards commented: In your paper, you make the point that the compounds are compressible to 10% of their starting volume and that on release of pressure they recover their starting volume. Given that the compounds are highly anisotropic, is the compressibility isotropic or anisotropic, and on the release of pressure is the recovered volume of the same shape as well as volume?

Guruswamy Kumaraswamy commented in reply: The macroporous scaffolds that we prepared using ice-templating[1] are isotropic at large length scales (*viz.* larger than the characteristic pore size). Therefore, the compressibility of these scaffolds is isotropic. The scaffolds recover their shape and volume on release of pressure after compression.

1 R. Rajamanickam, S. Kumari, D. Kumar, S. Ghosh, J. C. Kim, G. Tae, S. Sen Gupta and G. Kumaraswamy, *Chem. Mater.*, 2014, **26**, 5161–5168.

Nirmalya Bachhar asked: Since, ice freezes from the surface to the center, we logically should get a higher aggregate structure towards the center. Do you see such effect in your system? If not would you like to comment on the possible reasons?

Guruswamy Kumaraswamy answered: In our ice-templating experiments using high concentration particle dispersions to form macroporous foams,[1] we do not observe significant differences in the pore structure from the edge of the sample to the center. In these macroscopic ~*O*(1 cm) samples, complete freezing takes in the order of 10 min. It is possible that the temperature gradient (from the walls of the vial to the center) sets up convective flows that reduce spatial temperature variations. I would also like to note that in a previous work where we froze samples by immersing them in liquid nitrogen, we have reported variations in the pore structure from the walls of the vial to the center.[2]

1 R. Rajamanickam, S. Kumari, D. Kumar, S. Ghosh, J. C. Kim, G. Tae, S. Sen Gupta and G. Kumaraswamy, *Chem. Mater.*, 2014, **26**, 5161–5168.
2 M. Kar, P. Minois, K. Sharma, G. Kumaraswamy and S. Sen Gupta, *Langmuir* 2011, **27**, 12124–12133.

Rajdip Bandyopadhyaya questioned: (a) This work is about distribution of the size of colloidal aggregates (like, linear particle-chains, small or large clusters *etc.*). However, have you looked at shape and size of the ice crystals, which are available in the TEM images? One can use the idea of the Voronoi diagram to model the experimental ice-domain size distribution.

(b) What is the thickness of the interface between ice crystals, after thawing is over? Can a force balance at the interface in presence of cross-linked particles give the possibility of the formation of different aggregate-shapes observed in experiments? Is that a way to model this system?

Guruswamy Kumaraswamy replied: The shape and size distribution of ice crystals that form when water is frozen are well known from the literature. Also, there are previous reports on modelling the ice-interface/colloidal particle interactions as the colloid approaches the growing ice/water interface (see, for example: Ref. 1–4). What I find remarkable is that our simulations are able to capture the size distribution of colloidal aggregates without resorting to any such details. We model the ice crystals as growing out radially at a constant growth rate, model the particles as hard spheres that cannot overlap with the ice and ignore hydrodynamics. And yet, we are able to capture the size distribution. What this suggests to me, therefore, is that the specific shape of the ice crystals or their interface are probably not so important in determining the colloidal assemblies.

1 K. A. Jackson and B. Chalmers, *J. Appl. Phys.*, 1958, **29**, 1178.
2 D. R. Uhlmann, B. Chalmers and K. A. Jackson, *J. Appl. Phys.*, 1964, **35**, 2986.
3 M. F. Butler, *Cryst. Growth Des.*, 2001, **1**, 213.
4 H. Zhang, I. Hussain, M. Brust, M. F. Butler, S. P. Rannard and A. I. Cooper, *Nat. Mater.*, 2005, **4**, 787.

Mukta Tripathy said: In the simulations of ice crystal formation, ice nuclei were positioned randomly in the simulation box, and grew with a growth rate estimated from experiments. The particles interacted with the ice crystal by bouncing back if an overlap is set to occur. The ice crystals that are illustrated by the manuscript grow anisotropically. This seems to imply that the ice crystals were also effected by the particles. If this implication is correct, what was the nature of this interaction in the simulations?

Guruswamy Kumaraswamy responded: The growth of the ice crystals in our work is isotropic. Ice crystals are nucleated at the beginning of the simulation and grow out isotropically until they impinge.

Oleg Gang enquired: What is the major effect of solvent composition?

Guruswamy Kumaraswamy replied: We disperse polymer-coated colloids and crosslinkers in water and then freeze the water to drive the formation of colloidal assemblies. Crosslinking of the polymer coating around the particles locks the colloidal assemblies in place. The addition of small quantities of co-solvents changes the freezing point of water. If the freezing point is highly depressed, then the kinetics of the crosslinking reaction is very slow at the freezing temperature. Thus, the colloids assembled during freezing disperse on thawing. In addition, it is possible that the presence of the co-solvents, such as acetonitrile, inhibits the efficiency of the crosslinking reaction.

Daan Frenkel commented: From the theoretical work of van Dongen and Ernst,[1] it is known that the cluster-size distribution, that results from a wide class of cluster aggregation scenarios, results in a cluster-size distribution of the form $n^{-2} f(k/n)$, where $f(x)$ is a scaling function. Usually, this function is considered in the limit of large n. Clearly, in the regime where $f(x)$ is nearly constant, the n^{-2} scaling observed in your experiments would result naturally.

However, I do not know if the assumption that $f(x)$ is constant would apply in the case that you report and I would be interested to know if you think that such a theoretical approach would be profitable.

1 P. G. J. van Dongen and M. H. Ernst, *Phys. Rev. Lett.*, 1985, **54**, 1396–1399.

Guruswamy Kumaraswamy commented in reply: The work of van Dongen and Ernst[1] uses the Smoluchowski coagulation equation to estimate the cluster size distribution for a wide variety of gelling and non-gelling systems. The expression from van Dongen and Ernst provides the cluster size distribution in terms of the mean cluster size, s, as $P_n \sim s^{-2} f(k/s)$. This does not imply that $P_n \sim n^{-2}$ in all cases. This work followed computer experiments[2] that investigated aggregation of diffusing clusters of particles. Here, it was observed that $P_n \sim t^{-w} n^{-\tau} f(n/t^z)$. This can be written in the alternate form: $P_n \sim n^{-2} g(n/t^z)$. In finite size systems, all the particles will aggregate at long times to form one cluster. Therefore, this expression is valid only for an infinitely large system. However, in our experiments, the cluster size distribution is "frozen" when the ice crystals impinge. Therefore, there is no change in cluster size distribution after that time. In the work[2] of Vicsek and Family, $g(x) \ll 1$ when $x \gg 1$, *viz.* at large time, and $P_n \sim g(x) n^{-2}$ (corresponding to the case that Prof. Frenkel mentions). For $x \ll 1$, $P_n \sim t^{-1.75} n^{-0.75}$ – unlike what we observe in our experiments. Therefore, it does not appear that simulations of cluster formation by aggregation of randomly diffusing clusters capture the phenomenology of ice templating. In ice templating, the space available for diffusion of particles shrinks as the ice crystals grow. Our simulations suggest that clustering happens only at the late stages of the experiment, when the ice crystals are close to impinging. Therefore, it is not surprising that these models of aggregation by random diffusion are unable to capture the physics in the ice templating experiments.

1 P. G. J. van Dongen and M. H. Ernst, *Phys. Rev. Lett.*, 1985, **54**, 1396–1399.
2 T. Vicsek and F. Family, *Phys. Rev. Lett.*, 1984, **52**, 1669–1672.

Nicholas Kotov said: Did you see particle sheets as well as chains when the nanoparticle assembly is templated by ice? Have you, by any chance, tried other types of particles? They should also assemble under these conditions, I think.

Guruswamy Kumaraswamy responded: We see a transition from linear chains to 2–3 particle wide tapes and to sheets with an increase in the suspension concentration when we ice-template colloids. We never observed the formation of branched chains – therefore, the transition on increasing colloid concentration appears to happen through "widening" of the chains to form tapes and sheets (and then, three dimensional aggregates and finally a percolated 3D macroporous structure). We have tried this with PS latex and with silica particles. We observe the same behaviour in both these systems. Therefore, our simulation model does not incorporate any specifics of the particle chemistry and postulates a very simple mechanism of exclusion by the ice. We agree that this ought to work for other types of particles as well. Others in the literature have reported several types of particles (and polymers) – though, the literature does not explore ice templating of suspensions with such low particle concentrations.

Vimala Sridurai enquired: The concept of ice templating is very interesting to me, especially as I am working on gelation. In the microscopy images (Fig. 2 in your manuscript), the colloids form a linear chain like pattern for specific concentrations. If we compare this scenario with the gels formed by low molecular weight organo gelators, in which the gelator forms a three dimensional network within which the solvent medium is confined, do you think ice templating will favour formation of linear chains of gelator instead of a 3D network?

Guruswamy Kumaraswamy replied: One must be cautious when drawing comparisons between gelation of low molecular weight compounds and colloidal cluster formation during ice templating. In our simulations, we model the colloids as interacting through a hard sphere potential. We believe that in our experiments as well, structure formation is not driven through interparticle interactions. The formation of colloidal clusters is driven by the ice crystal templates and not by interparticle interactions. This is very different from the gel systems that you describe that are essentially driven by intermolecular interactions. Also, it is not obvious that we can extend what we have learnt from the ice templating of particles to ice templating of aqueous solutions of small molecules.

Rajdip Bandyopadhyaya asked: What happens when the structure is alternatively crystallized and thawed again repeatedly? Is the colloidal linear chain or other aggregates reversible under the thermal cycle? What happens to the polymer structure? Is it a rigid polymeric network?

Guruswamy Kumaraswamy answered: In our experiment, the polymer-coated colloids and the small molecule crosslinker are assembled at the boundary of the ice crystals. Crosslinking takes place over a day. If the system is thawed shortly after crystallization (*viz.* <6 h after ice formation), there is insufficient time for crosslinking and permanent colloidal assemblies do not form. Stable assemblies form only after completion of crosslinking. Formation of the colloidal assemblies is reversible when the freeze–thaw cycles are conducted before the crosslinking proceeds. It is not clear how assemblies organize on repeated thaw–freeze cycles once crosslinked aggregates are formed. In previous work,[1] we had shown that elongated anisotropic particles orient parallel to the edge of the templating domain. However, for flexible chains such as those prepared in our current work, it is not clear if alignment of the anisotropic structures will proceed in the same manner. Therefore, we cannot comment on whether crosslinked colloidal chains or aggregates or the colloidal 3D structure will survive repeated freeze–thaw cycles without a change in structure.

1 K. P. Sharma, A. K. Ganai, D. Sen, B. L. V. Prasad and G. J. Kumaraswamy, *Phys. Chem. B*, 2013, **117**, 12661–12668.

Jacques Jestin said: Did you see the influence of the particle size? Especially when decreasing the particle size to change the kinetic ratio between particle diffusion and the crystal growth rate?

Guruswamy Kumaraswamy responded: The group of Deville[1] have reported a phase diagram (as a function particle size and ice crystal growth velocity) for the structure formed when particles are ice templated. However, this phase diagram does not relate to the ordering of particles in the aggregates. For the particles that we use here, we observe that the particles form ordered aggregates. This is consistent with the Peclet number (that is related to the ratio of the crystal growth rate with particle diffusivity). We did not systematically study the influence of particle size on ordering – however, the Peclet number arguments would suggest that less ordered assemblies would result for larger crystal growth rates or larger particle sizes (*viz.* lower diffusivities).

1 S. Deville, E. Maire, G. Bernard-Granger, A. Lasalle, A. Bogner, C. Gauthier, J. Leloup and C. Guizard, *Nat. Mater.*, 2009, **8**, 966–972.

Priyadarshi Roy Chowdhury communicated: Which simulation model do you follow in this particular research work and why do you think it is a better simulation model than other statistical models?

Guruswamy Kumaraswamy communicated in reply: We use a very simple Brownian Dynamics simulation scheme that makes a large number of simplifying assumptions. For example, we completely ignore hydrodynamics. Even with this simple scheme, we are able to capture the qualitative results observed in our experiments, such as, for example the cluster size distribution for colloidal aggregates. Therefore, we are confident that we are able to capture the essential physics that governs cluster formation with our simple model.

Priyadarshi Roy Chowdhury communicated: What precautions do you take during compression of ice-crystals?

Guruswamy Kumaraswamy communicated in reply: We use ice crystals only as templates to generate the colloidal chains or macroporous assemblies. These are cross-linked in the presence of the ice crystals. Subsequently, the sample is thawed and there are no ice crystals that are present when we compress our samples.

Priyadarshi Roy Chowdhury further communicated: In designing your experiments, which parameters do you vary, so that your experiments could yield products of desired shape and morphology?

Guruswamy Kumaraswamy communicated in reply: The main control parameter is the concentration of colloids in the initial aqueous dispersion. The effect of varying the particle concentration in the dispersion is described in detail in our manuscript. What is interesting is that the distribution of colloidal clusters shows the same power law dependence over the concentration range explored (over a 100-fold variation in dispersion concentration). Of course, once we access much higher concentrations, the clusters percolate and we form a large three-dimensionally connected object.

Madivala G. Basavaraj opened a general discussion of the paper by Yon Ju-Nam: In several plots in the manuscript, you attribute the stability of particles to the fact that the peak position in the UV *viz* measurements does not change with time. However, there is a vertical shift, which does indicate to the decrease in concentration of particles either due to sedimentation or aggregation followed by sedimentation. Therefore, these data have to be interpreted by taking into account these issues.

Yon Ju-Nam replied: The differences in absorbance in the stability studies are due to the concentrations of the same colloidal sample analysed. They were prepared at different concentrations. In this case, these differences do not mean aggregation in the colloidal samples as there were no shifts observed in the surface plasmon bands.

Alison Edwards asked: I commend the inclusion of what appears to be the full measured UV-visible spectrum in Figures 2, 3 and 4 in your manuscript, which allows consideration of the long wavelength region in which it appears, in the 700–750 nm region, that there are some features in some spectra which may require consideration and/or interpretation. Although most instruments operate linear in wavelength, this is largely for historical reasons which ought perhaps to be reviewed. Presentation of the spectra linear in wavelength does not readily facilitate the examination of the spectra where subtle changes in absorbance can hide the presence of well-defined peaks which are quite apparent when the spectra are plotted linear in energy instead. Have you considered plotting your spectra in this alternative (preferable) mode, and if so, do these suggestions of broad peaks at high wavelength transform into useful experimental observations (peaks)? I note that comparing calculated UV-visible spectra with observed spectra plotted in this conventional wavelength *vs.* absorbance format can obscure the fact that the correct relative energy values may have been calculated and that a simple shift of origin (along the energy axis) for the calculated spectrum may result in a good match between calculation and experiment.

Yon Ju-Nam replied: The authors find this comment particularly useful, as the participant is suggesting an alternative way to show the UV spectra. For future works, the UV spectra will be plotted linear in energy.

Radhika Poojari questioned: What is the percentage of DMSO used? DMSO is known to be cytotoxic. In particular, what are the DMSO concentrations used for the synthesis of the nanoparticles?

Yon Ju-Nam responded: The cationic phosphonium gold nanoparticles with more hydrophobic organic layer can be dispersed in a solution of 1 : 5 DMSO : H2O.

B. L. V. Prasad remarked: The authors describe the preparation of gold nanoparticles in DMSO. But their ultimate aim is to use these nanoparticles for bio-applications. So have they checked the stability of these particles by adding water? Have they tested the stability of these particles by varying the pH and by adding electrolytes?

Yon Ju-Nam replied: Among the future works planned to continue investigating the stability of these cationic gold nanoparticles, the authors will be carrying out stability studies by adding electrolytes (varying pH) using potentiometric titration technique. Results from these experiments could potentially provide information about the pH range in which the cationic phosphonium nanoparticles are stable.

Rajdip Bandyopadhyaya remarked: You mentioned that the gold nanoparticles synthesized are highly stable, but the only data given is the UV-visible spectrum as a function of time. Have you looked at any scattering experiments to track particle evolution? Or any TEM imaging of the nanoparticles after 6 or 12 weeks, as is shown in the paper with UV-visible spectra after 6 or 12 weeks?

Yon Ju-Nam answered: TEM analyses were carried out to obtain the particle size distribution soon after the cationic gold nanoparticles were synthesised. No TEM analysis was carried out to after 12 weeks. We have done stability studies using the XPS technique in the past. However, the participant's suggestion of using TEM is good, as possible aggregations can be visualised by using this electron microscopy technique. We have done DLS analysis of our nanoparticles but they were not included in this *Faraday Discussions* paper.

Lola Gonzalez-Garcia enquired and commented: Do you know how the ligand is attached to the particle surface and how thick the ligand layer is? IR or Raman spectroscopy could clarify the first question while DLS measurements could give you information about the hydrodynamic radius of the synthesized nanoparticles.

Yon Ju-Nam responded: The best technique to see if the ligand is attached to the AuNP surface is XPS. No XPS experiments were carried out for this paper, however, in other papers published by the same authors in the past, XPS data and information about the chemistry of the surface of the nanoparticles were shown.[1] In this particular paper, the high resolution S(2p) XPS spectrum of the AuNP generated using the phosphonium zwitterion was obtained. The spectrum displayed a peak with a binding energy of 162.9 eV, corresponding to S(2p3/2). This value is consistent with values reported in the literature regarding sulfur species bound to the surface of the gold as a thiolate, which typically occurs in the range of 162.0 to 162.9 eV. XPS experiments are planned, as future work, to complement the results of this *Faraday Discussions* paper. The thickness of the ligand layer has not been measured, however, this can be an additional experiment to do as the participant is commenting.

1 Y. Ju-Nam, Y.-S. Chen, J. J. Ojeda, D. W. Allen, N. A. Cross, P. H. E. Gardiner and N. Bricklebank, *RSC Adv.*, 2012, **2**, 10345–10351.

Priyadarshi Roy Chowdhury communicated: Did you perform comparative analysis of your experiments by changing solvents other than DMSO? If so, what are the advantages of using DMSO for your experiments?

Yon Ju-Nam communicated in reply: Water and DMSO were the two solvents used to synthesise the cationic phosphonium gold nanoparticles. However, when the hydrophobicity of the ligands was increased, the functionalised nanoparticles could not be obtained in water. This was one of the reasons using DMSO. These colloidal solutions in DMSO can be diluted down in water, due to the miscibility of both solvents.

Priyadarshi Roy Chowdhury communicated: Did you perform XPS for surface and depth profile analysis of novel cationic phosphonium cationic gold nanoparticles?

Yon Ju-Nam communicated in reply: This question is similar to the one previously asked by Lola Gonzalez-Garcia. Therefore, please refer to my previous response.

Vandana Shinde communicated: Which actual information is obtained from the NMR and MS spectroscopy analyses of these phosphonium gold nanoparticles?

Yon Ju-Nam communicated in reply: The NMR and the MS data shown in the paper are the ones corresponding to the ligands. No NMR and MS analyses of the cationic phosphonium gold nanoparticles were carried out.

Madivala G. Basavaraj asked: Have you thought about the mechanism of long-term stability? A discussion of this would be useful.

Yon Ju-Nam replied: We haven't thought about studying the mechanism of long-term stability. It is a valuable suggestion of the participant. We believe that this can be carried out using surface techniques such as XPS and also TEM for at a least 6 months, analysing samples at least every month.

Sanat Kumar opened a discussion of the paper by G. V. Pavan Kumar: When you change the length of the triangle (formed by the three traps), does the fluctuation of the centroid vary like the square root of the side of the triangle? Does this variation happen over a modest range of triangle sizes and what defines these ranges?

G. V. Pavan Kumar responded: We do see the fluctuation at the orthrocenter changing when we change the length of the triangle. But we are yet to probe the nature of this fluctuation and its connection to the length of the triangle. In reference to the second question, what we have observed until now is that up to a critical distance between the excitation points, we obtain an assembly at the orthocenter. As we bring the excitation points close to each other, we see that the nanoparticles from the two excitation locations directly interact with each other along the line joining the excitation points.

Erika Eiser asked: Concerning Figure 4c in your manuscript, presented as a movie in the discussion, you mention you can track the gold nano particles. These are smaller than the diffraction limit of the laser light used for the trapping. What do you really track? In this context, you mention you measure the density of nano particles as function of time in these optical traps (Figure 3 in your manuscript). How is this possible, as these centres appear clearly black immediately after the traps are switched on?

G. V. Pavan Kumar replied: What we track in our experiments is not an individual nanoparticle, but a sub-micron sized aggregate formed at the location of the assembly. Our method is still constrained by the diffraction limit of light, so the fluctuations that we track are from nanoparticle assembly sizes in the order of the wavelength of light we use in our experiments. In reference to the question related to particle-density measurement, first of all, we earmark a specific area of interest in our field-of-view, and then ensure that the camera pixels are not saturated within the time period of the assembly experiment. With this setting, we track the decreasing brightness of the region to obtain the density. This is an imaging method that we have essentially borrowed from existing particle-tracking techniques.

Erika Eiser further enquired: How do you know that you don't have local temperature changes that may induce thermophoretic or convective flows? You irradiate the metal nano-particles for quite a long time – we observe severe changes due to radiation damage as the metal particles will heat up, leading to irreversible nano-particle aggregation. Do these possible changes contribute to the dynamics presented in your paper?

G. V. Pavan Kumar responded: The assembly that we have are mainly driven by two effects: one is the electric-field due to plasmon-polariton and the other is the thermal field created by the plasmon.

The second effect does contribute towards convective flows. In fact, that is the model that we have used to explain the assembly process (see Ref. 1). An important point to note is that the laser-excitation is evanescent in nature. This means that a small amount of power at the excitation location is confined to the metal-fluid interface, and the rest is out-coupled from the prism. At the powers we are operating in, they are not so high that they can damage the nanoparticles. This we confirm by switching off the excitation field, and observing the presence of nanoparticle aggregates, if any. We do not find any such irreversible nanoparticle aggregates.

1 P. P. Patra, R. Chikkaraddy, R. P. Tripathi, A. Dasgupta and G. V. Kumar, *Nat. Commun.*, 2014, **5**, 4357.

Neena S. John asked: What is the possibility of arresting the nanoparticle assemblies even after the removal of the optical field so that they can be examined in detail under electron microscopes?

G. V. Pavan Kumar replied: That is feasible. One can dry the solvent, in our case water, and then interrogate the arrested nanoparticle assembly under electron microscope. We intend to do that.

Rajdip Bandyopadhyaya questioned: Your experiments showed movement of particles under a field. Did you write equations of motion (like Navier–Stokes type) for the particles? Have you tried to model movement of a collection of particles to predict their trajectory? Or to predict their final equilibrium position, like the centroid of the triangle, as seen in your experiments?

G. V. Pavan Kumar responded: We have given a Navier–Stokes type model to understand dynamic effects in our experiments (see equation 1 in Ref. 1). This partly explains the process of nanoparticle assembly. However, predicting the trajectory is still far off. The reason is that the assembly has contributions not only from the thermal fields, but also from the plasmon wave interference. This combination collectively gives rise to observed effects, and modelling them will be interesting, and equally challenging.

1 P. P. Patra, R. Chikkaraddy, R. P. Tripathi, A. Dasgupta and G. V. Kumar, *Nat. Commun.*, 2014, **5**, 4357.

Guruswamy Kumaraswamy commented: It is well known that Rayleigh-Benard convection cells are formed when there is a temperature gradient in a fluid film, such that the lower part of the film is hotter than the top. I imagine that this would be true in your system due to heating by the laser spots. How important is this phenomenon for the pattern formation that you observe? Pattern formation by Rayleigh Benard convection cells ought to be dependent on the height of the liquid film, consistent with the previous experiments that you suggested.

G. V. Pavan Kumar replied: Yes, I agree. Rayleigh-Benard convection does play an important role in this phenomenon. The heating effect of laser on the plasmonic film is critical in our experiments. The overall pattern formation has collective contribution from heating of the film and the near-field of the plasmons. Probing the individual contribution in the absence of the other would be interesting, and equally challenging.

Alison Edwards stated: I note that you are using a synthesis in which silver nitrate is employed as the silver source, followed by reduction by citrate. I would then observe that the two syntheses of the AG_{44}thiolate nanoparticles[1,2] differ in the choice of silver source, with the citrate starting material reported in Nature Communications yielding what might be inferred to be the cleaner product based on the superior crystallizability evidenced by the determination of structures of 3 compounds to a very high standard.

Have you considered the possibility of using silver citrate as your silver source which may result in a synthetically more robust and reproducible procedure?

1 H. Yang, Y. Wang, H. Huang, L. Gell, L. Lehtovaara, S. Malola, H. Häkkinen and N. Zheng, *Nat. Commun.*, 2013, 4, 2422.
2 A. Desireddy, B. E. Conn, J. Guo, B. Yoon, R. N. Barnett, B. M. Monahan, K. Kirschbaum, W. P. Griffith, R. L. Whetten, U. Landman and T. P. Bigioni, *Nature*, 2013, **501**, 399–402.

G. V. Pavan Kumar responded: Thank you for an important suggestion. I agree with you that there is a room for improvement in the quality of the nanoparticle one can prepare. We have used the conventional Lee and Meisel method to prepare our nanoparticles. This method usually results in polydisperse Ag nanoparticles, but they do show strong plasmon resonance and SERS effect. An important criteria, at least for the SERS experiments, is that the capping agents should not interfere in the coupling between the molecule and the Ag nanoparticle. We had to take this criteria into our design, and the citrate-reduced Ag nanoparticle catered to this requirement. In future, we intend to use nanoparticles which are monodisperse, of single shape and SERS active. We are working towards this.

Michael Bockstaller commented: Many thanks for this very interesting contribution. I wanted to comment on the potential application of the trapping process in particle separation. I agree with the author that research along these lines has been published and thus might not be looked at as "conceptually new". However, to my knowledge, a persistent challenge has been the scalability of the process. Here, the "enhancement effect of a plasmic grating structure" could proof very useful.

G. V. Pavan Kumar commented in reply: Thank you for an interesting comment. I agree with you. Nanoparticle sorting is a very interesting direction to pursue. Several groups, to my knowledge, are exploring this opportunity.

I need to add that one can also sort optically resonant nanostructures using plasmofluidic methods. For example, Dholakia and coworkers[1] and Genet and coworkers[2] have reported some very interesting experiments on these lines.

1 M. Ploschner, T. Čižmár, M. Mazilu, A. Di Falco and K. Dholakia, *Nano Lett.*, 2012, **12**, 1923–1927.
2 A. Cuche, A. Canaguier-Durand, E. Devaux, J. A. Hutchison, C. Genet and T. W. Ebbesen, *Nano Lett.*, 2013, **13**, 4230–4235.

Ajeet Srivastav asked: What is the reason behind nanoparticles getting accumulated at the orthocenter of the three excitation spots indicated by the dotted circles (Fig. 2c in your manuscript)?

G. V. Pavan Kumar replied: There are two contributions towards the accumulation at the orthocenter. First is the interaction between the excitation light sources that generate plasmons at three different locations on the film, and these plasmons further interfere with each other. Second is the local heat generated by plasmon-polaritons which creates a plasmofluidic convection and further leads to an optical potential that traps the nanoparticles.

We know the contibutory factors, but we still do not know how they individually contribute towards the assembly at the orthocenter. If one considers the effect only due to plasmon interference or only due to thermal effects, one may not be able to explain the assembly process. A better understanding of the interactions between the film-plasmon, particle-plasmon, and the fluid is necessary. We are working towards this.

Oleg Gang asked: Can you estimate the local heating effects, *i.e.* on the nanoparticle and its proximity? What effect do they have on the particle's motion.

G. V. Pavan Kumar answered: In principle one can measure the local heating effects in the proximity of the nanoparticles, but such measurements will have to be done in a dynamic situation. A possible method is to coat the nanoparticles with temperature-sensitive dye molecules, and track their optical response at high spatial- and temporal-resolutions. Such measurements will equally serve the purpose of tracking the particle motion, aggregation and stratification, if any. In practice, implementing this would require a lot of optimization.

B. L. V. Prasad enquired: Is it possible to assemble the nanoparticles first using the optical techniques and then chemically crosslink them using the techniques described by Dr Guruswamy Kumaraswamy?

G. V. Pavan Kumar answered: It may be feasible. Chemically cross linking after assembly is an interesting suggestion. In what conditions such experiments work is yet to be explored.

Priyadarshi Roy Chowdhury communicated: In the simulation part of your paper, you have mentioned about the scale of simulation. If the scale is varied, what distortions would likely occur?

G. V. Pavan Kumar communicated in reply: We have performed FDTD simulation to show plasmonic interference effects. The scales at which the simulations are performed are small compared to the experimental situation. A larger scale simulation will be computationally expensive, but may show some long range interaction effect between the plasmonic waves, which may not be evident in the current simulations.

Priyadarshi Roy Chowdhury communicated: Will the distribution pattern of your simulation results change on changing the scales of the simulation?

G. V. Pavan Kumar communicated in reply: The field distribution pattern may show changes in the context of intensity. Plasmon waves decay as they propagate. So, one should expect some changes if the scales of the simulations are changed. Note that thermal effects are not part of the simulations. Therefore, one can obtain some interesting patterns if the local temperature is also taken into consideration.

Priyadarshi Roy Chowdhury communicated: Can particle tracking be made from the top only, or is there any special procedure and technique involved?

G. V. Pavan Kumar communicated in reply: We have tracked the assembly from the top. This was a convenient configuration in our experimental set-up. However, one can use different optical schemes to track the assembly, such as orthogonal imaging, but this will add to the optics hardware.

Priyadarshi Roy Chowdhury communicated: Please discuss the effect of thin-film interference on the nanoparticles mentioned in this work.

G. V. Pavan Kumar communicated in reply: We are using a plasmonic film, which is of around 50 nm in thickness. Conventionally, such films mainly support plasmon-polaritons rather than interference effects. These propagating polaritons further interact with localized plasmons of the nanoparticles, and this interaction is of direct interest to our work. When there is more than one excitation source on the film, then there will be interference between the propagating plasmon waves on the surface.

Priyadarshi Roy Chowdhury communicated: Mention the advantages and disadvantages of capping.

G. V. Pavan Kumar communicated in reply: Advantages: 1) they act against aggregation and provide stability to nanoparticles. 2) They play a role in defining the size and shape of the nanoparticle.

Disadvantage: 1) they reduce the electromagnetic coupling between the nanoparticle and the molecule of interest.

B. L. V. Prasad addressed Guruswamy Kumaraswamy and G. V. Pavan Kumar: When assembling the nanoparticles using light and ice templating, what is the size limit to which these techniques can be used? Is it possible to assemble 1 nm size particles, say in one line?

Guruswamy Kumaraswamy answered: In the work that we have presented, we demonstrate ice templating of 1 micron colloidal particles. Here, these particles are "pushed" along as an ice front propagates into water. Our simulations indicate that specific details of the ice front–particle or particle–particle interactions do not play a major role in determining particle clustering. This is unlikely to be true for much smaller nanoparticles. Work from the group of Deville[1] has suggested a "phase diagram" for the structures formed when particles are segregated at growing ice fronts. There is no data for particles as small as 1 nm – however, it is likely that interparticle interactions and particle–ice front interactions will play a much larger role in determining the structure of the colloidal clusters. Therefore, it might not be easy to translate our learnings from the ice templating of larger colloidal particles to this situation.

1 S. Deville, E. Maire, G. Bernard-Granger, A. Lasalle, A. Bogner, C. Gauthier, J. Leloup and Christian Guizard, *Nat. Mater.*, 2009, **8**, 966–972.

G. V. Pavan Kumar replied: In principle, our method based on plasmofluidic field should work for any nanoparticle, even those which are less than 10 nanometers. However, controlling the number of nanoparticles in a line is not yet achieved.

In our experiments, the excitation fields are microscopic in size, that is, the excitation profiles are few microns in length and width. The assembly of nanoparticle happens within this range. Precise control over the assembled nanoparticle would be a excellent goal to work towards.

Nicholas Kotov opened a general discussion of the paper by Semen Chervinskii: Can you speculate about the high stability of silver clusters?

Semen Chervinskii replied: Our main hypothesis is that silver could react with oxygen/sulphur in the air immediately after nanoisland preparation. By that a thin passivation layer forms over the nanoislands, protecting them from further reactions. In the beginning of our studies we expected that these reactions can degrade metal silver and used additional coatings to protect the samples, but later it was demonstrated, that nanoislands were stable as is.

Priyadarshi Roy Chowdhury communicated: What preventive measures did you take during the formation of silver nanoisland films?

Semen Chervinskii communicated in reply: During the silver nanoislands formation we performed only measurements mentioned in the paper parameters of the process. Authors will be grateful for any suggestions of *in situ* measurements on such small and fragile structures as these nanoislands.

Priyadarshi Roy Chowdhury communicated: If possible, insert 3D AFM height profile micrographs with explanation in the ESI of this paper. I feel it would enable higher standards to this wonderful piece of research work.

Madivala G. Basavaraj remarked: One of the alternate methods to make these silver islands is to use CVD followed by temperature induced dewetting. Does the technique you propose offer any advantages – for instance – do silver islands adhere better to the substrate or is there a better control of either size or their distribution on the substrate?

Semen Chervinskii responded: We suppose that the proposed technique provides the best adhesion of the nanoislands to the surface, as they grow "from the inside". However, the main advantage comparing to CVD is that the 2D-patterning technique allows to control the positions of the nanoislands/structures on the substrate.

Alison Edwards remarked: The two questions regarding this paper correspond to previous questions I have asked previous speakers in this session: the first, question regarding whether there might be any useful observations available if the UV-visible spectra were instead plotted linear in frequency? And the second question rgarding whether the choice of silver salt employed as the silver source has been varied?

Semen Chervinskii responded: Thank you for the remark about another way of spectra plotting. For these particularly nanoislands and films discussed in the paper there is no significant difference between plotting linear in frequency or in wavelength. In case of more complex spectra in UV-region plotting linear in frequency should definitely be useful.

Nitrate salts of silver and sodium are traditional for ion exchange processes because both are liquid and stable at certain temperature below the glass softening point. Other options of silver doping of glasses generally are not connected with ion exchange in the melt[1] and were not employed during this study.

1 F. Gonella, *Ceramics Int.*, 2015, **41**, 6693–6701.

G. V. Pavan Kumar commented: In solid state emission studies, researchers use energy scale to display the emission wavelength, and there are some advantages of using such a scale, especially in understanding energy distribution. Maybe, one can use such a method to display data on plasmon resonance of nanoparticles.

Alison Edwards also commented: I agree that it is preferable to use a scale which is linear in energy to depict the UV-visible spectrum of nanoparticles, in particular when the "plasmon resonance" is being discussed. The depiction of such spectra linear in wavelength can obscure potentially important features which are present in the recorded spectra, and also do not facilitate proper comparison with calculations. When depicted linear in energy, a calculation which yields correct relative energy can readily be compared with a measured spectrum also presented linear in energy and any zero-offset required to match the two is evident. This is not necessarily the case when wavelength is the framework within which the comparison is made.

Rajdip Bandyopadhyaya addressed Semen Chervinskii and Guruswamy Kumaraswamy and Andrea Tao: (a) Is there a way to correlate defect densities in a patterned or templated structure? How does the size of the building block influence it? Any other important factor?

(b) For downstream technological use, I believe defects in a regular pattern are more important than just showing a patterned structure is achieved in experiments. Is a phase diagram possible on defect densities as a function of building block type and number?

Guruswamy Kumaraswamy answered: We observe very regular packing/ordering of the spherical colloidal particles in our aggregates. We believe, based on the literature, that the Peclet number for the process (*viz.* based on the speed of the water/ice interface and based on the particle size/diffusivity) determines this ordering. Our data is consistent with the literature. We do not have data to show that increasing the speed of ice front propagation or that increasing the size of the particle results in defects in the colloid packing (though, this is anticipated from the Peclet number calculation). I agree that defects in patterned structures have important technological implications, for example, in optical multiplexers made from photonic bandgap materials. However, to the best of our knowledge, there is no phase diagram for defect densities as a function of building block type and density.

Semen Chervinskii replied: (a) We did not perform special analysis of defect densities, considering the absence of nanoislands or dimers/groups of island formation instead of single isolated ones (compare Fig. 10a and b of our paper) as defects. This correlation can be established through analysis of big arrays of nanoislands. Roughly, probability of defect formation increases with increase in size of the building block (hollow in the stamp) or in its aspect ratio. Another important factor is the stamp quality, as a hollow shape also affects this probability.

(b) Generally, a 3D phase diagram with coordinates (surface diffusion length, stamp hollow width, and stamp hollow length) can be built after the analysis of big arrays of nanoislands. This diagram will allow determining regions with single nanoislands and with defects (absence of nanoislands, dimers, *etc.*).

Andrea Tao responded: We've tried to measure defect densities using both quantitative image analysis from electron microscopy images of our assemblies, as well as using spectroscopic measurements to gauge differences between our measured optical responses *versus* the optical response for perfectly arranged nanoparticles in a periodic lattice. These defects seem to result almost directly from inhomogeneities in the as-made colloidal nanoparticles, *e.g.* defects in nanoparticle size or shape. For example, larger particles give rise to a larger number of nanoparticle binding sites, resulting in the formation of large clusters (in the case of our dimer work) *vs.* branch points (in the case of phase segregation work). I suspect that graft density at the surface of the nanoparticles plays an important role as well – but graft densities can be difficult to accurately measure.

I agree that phase diagrams are probably more utilitous for large-scale or industrial applications – phase diagrams not only of morphologies/patterns but also of functionality. The development of phase diagrams through experimental methods alone would probably be overwhelming and may not provide exhaustive information regarding certain assembly "phase spaces". I think this is where theory and modeling are currently playing a critical role in guiding experimental efforts.

Faraday Discussions

PAPER

Dewetting dynamics of a gold film on graphene: implications for nanoparticle formation

Sadanandam Namsani and Jayant K. Singh*

Received 11th August 2015, Accepted 7th September 2015
DOI: 10.1039/c5fd00118h

The dynamics of dewetting of gold films on graphene surfaces is investigated using molecular dynamics simulation. The effect of temperature (973–1533 K), film diameter (30–40 nm) and film thickness (0.5–3 nm) on the dewetting mechanism, leading to the formation of nanoparticles, is reported. The dewetting behavior for films ≤5 Å is in contrast to the behavior seen for thicker films. The retraction velocity, in the order of ~300 m s^{-1} for a 1 nm film, decreases with an increase in film thickness, whereas it increases with temperature. However at no point do nanoparticles detach from the surface within the temperature range considered in this work. We further investigated the self-assembly behavior of nanoparticles on graphene at different temperatures (673–1073 K). The process of self-assembly of gold nanoparticles is favorable at lower temperatures than at higher temperatures, based on the free-energy landscape analysis. Furthermore, the shape of an assembled structure is found to change from spherical to hexagonal, with a marked propensity towards an icosahedral structure based on the bond-orientational order parameters.

I. Introduction

The directed assembly of functional nanomaterials, using top-down and bottom-up techniques for controlling the patterns of the nanomaterials, is extremely useful in many micro and nanoscale electronic, optical and biological applications.[1] The patterning of nanoparticles, apart from fabrication techniques, relies on the selected substrate and the interaction of nanoparticles with the substrate. One of the popular techniques to form metallic nanoparticles on substrates is *via* dewetting of metallic thin films, which has gained much attention due to their novel physical and chemical properties.[2–4] Recent studies have used solid or liquid phase dewetting of the metal films for the directed self-assembly of metal nanoparticles to demonstrate novel functional properties, for example, plasmonics,[5–7] optical,[8] and magnetic[9] properties. Rack *et al.*[10] used pulse laser

Department of Chemical Engineering, Indian Institute of Technology Kanpur, Kanpur-2018016, India. E-mail: jayantks@iitk.ac.in

irradiation to study the dewetting and directed assembly process of nickel films in different initial shapes such as circles, triangles, and squares. The authors showed that the laser pulse induces instabilities in the edge of the nickel film, which can be modulated to direct assembly of the materials. Similarly, Wu *et al.*[11] have shown the formation of ordered nanoparticle arrays due to pulse-laser-induced dewetting of Cu rings. The directed assembly of bimetallic nanoparticles such as Cu–Ni has also been demonstrated by dewetting induced through irradiation of laser pulses on nanoscale Cu–Ni metallic films.[12]

In the last few years patterning of metal nanoparticles on graphene has gained increased attention due to their high-grade catalytic and electronic properties.[13–16] The chemical inertness, high conductivity and large surface area of graphene make it a novel platform for composites.[17] Decoration of the graphene surface with nanotubes,[18,19] nanowires,[20,21] and nanoparticles[22,23] has shown to induce superior properties compared to individual components. Hence, the understanding of decorated particles on the graphene surface is of immense interest. For example, Hupalo *et al.* [24] used experiments and density functional theory (DFT) to study graphene–Pb and graphene–Dy interactions in nanomaterials that have applications in gas sensors.

Recently, Fuentes-Cabrera *et al.*[25] studied the dewetting behavior of Cu nanodisks and nanowires on graphite and graphene surfaces using molecular dynamics (MD) simulations. They showed that the nanodisks and wires dewet the surface and form nanodroplets, which in turn jump off the surface with a velocity of 140 m s^{-1}. Moreover, the effect of surface roughness on the dewetting behavior of Cu nanodisks on a nanopillared graphene surface was studied using MD simulations by Li *et al.*[26] These simulation studies showed that the carbon nanopillars on graphene promote the dewetting behavior of the Cu film compared to that of the bare graphene. The dewetting and detachment behavior of Pt nanofilms on the graphitic surface was also studied using MD.[27] Dewetting of metal thin films leading to active metal nanoparticles on the graphene surface generated a new kind of SERS substrates known as graphene-mediated SERS (G-SERS),[28] which has created enhanced interest in the study of graphene–metal nanohybrids. Recently, gold–graphene nanohybrids have also gained attention because of their potential in SERS,[28] sensors[29] and graphene-based plasmonic device applications. For example, Zhou *et al.*[30,31] annealed a gold film in a furnace at different temperatures (973–1533 K) to generate gold nanoparticles on a graphene surface. The authors demonstrated that the resulting hybrid has superior SERS properties compared to graphene–gold film hybrids. Iyer *et al.*,[32] on the other hand, annealed a gold film on graphene at different temperatures (373–673 K) to form plasmonic nanostructures, and reported an increased SERS enhancement with temperature. Habenicht *et al.*[4] used a laser pulse technique to study the dewetting behavior of liquid triangle-shaped gold nanoparticles on graphene, where detachment of the gold particles was observed at 3129 K. The shape and size of the nanoparticles formed after annealing of the gold film on the graphene surface are shown to depend on temperature and film thickness. Zhou *et al.*[31] showed that the hexagonal and triangular-shaped nanoparticles exhibit superior SERS properties compared to hexagonal-shaped nanoparticles alone on the graphene surface. This indicates that the control of the size and shape of nanoparticles is also important, as they affect the properties of the nanoparticle–graphene hybrid material. The interactions at the metal–graphene interface in the metal–graphene hybrid system play a major role in structure control and composite performance. As the properties of metal–graphene hybrids are highly sensitive to the type of metal contact with the graphene surface, it is essential to understand the growth of metal nanostructures on graphene surfaces. SERS properties are dependent on the changes in the substrate structure. However, the formation of gold nanoparticles on graphene is not well understood. Typically on graphene, as observed by Iyer *et al.*,[32] heating of the gold film leads to the formation of nanoparticles, and graphene wraps around these nanoparticles to create wrinkles at the edges of the nanoparticles. Iyer *et al.*[32] clearly pointed out that the localized structure of graphene around the nanoparticle has a direct effect on the SERS properties. In particular, gold nanoparticles within the wrapped wrinkled shape of the graphene are shown to exhibit enhanced SERS activity. Electron-beam-lithography-based deposition of a serial array of gold nanoparticles on graphene shows SERS enhancement in the resulting system.[33] However, it was also observed that nanoparticles diffuse from their deposited location and subsequently agglomerate with neighboring nanoparticles, which is also not well understood.

As evident from the above, the formation of gold nanoparticles and their self-assembly on graphene are not well studied. Thus, we wish to evaluate gold nanostructures on graphene surfaces, and their agglomeration behavior during the annealing process. To this end, we present a detailed all-atom MD study at different temperatures to explain the nanoparticle formation and dewetting behavior of gold films of different thicknesses.

II. Model and methodology

All MD simulations in this work are performed using the LAMMPS package.[34] Metal–graphene systems are non-trivial systems, which require a suitable force field for a realistic simulation. In order to capture the behavior of the metal, an accurate metal–graphene interaction is required. One approach is to use a modified Lennard-Jones potential model (or any suitable model) such that some macroscopic properties (such as contact angles) match with experimental values. While this is more popular for organic or aqueous systems on graphene, such an approach is difficult to implement due to non-availability of experimental data for metal–graphene systems. However, a simple mixing rule has been used for metal–graphene systems, as recently employed by Fuentes-Cabrera *et al.*[25] for a Cu–graphene system. Another approach is deriving a force field using the data generated from an accurate DFT study. In this study, Au–Au and C–C interactions are modeled using the embedded-atom method (EAM)[35] and optimized Tersoff[36] potentials, respectively, which are well-tested models for the respective systems. The Au–C interactions are described using Buckingham potential based on our recent DFT work:[37]

$$U_{ij} = A_{ij}\exp\left(\frac{r_{ij}}{\rho_{ij}}\right) - \frac{C_{ij}}{{r_{ij}}^6}, \tag{1}$$

where A_{ij}, ρ_{ij}, and C_{ij} are the parameters. The A_{ij}, ρ_{ij}, and C_{ij} parameter values are 804.7145 eV, 0.3483 Å and 91.0355 eV Å^6 respectively. Since few DFT studies[38–43] and experiments[44] have shown that the charge transfer between gold and graphene is very small, the coulombic interactions between graphene and gold are neglected. Hence, the charges on gold and carbon atoms are taken as zero when performing the MD simulations. A cutoff of 12 Å is used for all the nonbonding interactions.

In order to understand the effect of the size and thickness of the film on the dewetting behavior of gold on graphene, we have considered circular disks of gold films of different diameters (30 and 40 nm) and variable thickness (5–30 Å). The graphene sheet of size 45 × 45 nm is considered in this work. The number of atoms in the simulation systems of 30 and 40 nm-diameter films on the graphene surface are 164 696 and 510 925, respectively. Initially, bulk gold is simulated at different temperatures (973, 1073, 1400 and 1533 K) using an isobaric and isothermal (NPT) ensemble at 1 atm. Subsequently, gold nanodisks of desired diameter and thickness are generated from the final structure of bulk gold obtained from NPT simulations. All the generated nanodisks are placed on the surface of the graphene at a vertical distance of 0.3 nm and positioned in the middle of the graphene sheet. The top and side views of a typical initial structure used in simulations are shown in Fig. 1(a) and (b), respectively. To study the dewetting dynamics of the deposited gold film on graphene, NVT ensemble simulations are performed at 973, 1073, 1400 and 1533 K.

The dewetting-induced density profile and retraction velocity of the gold film are calculated for gold nanodisks of different diameters and thicknesses at different temperatures. The retraction velocity of the film is defined as the

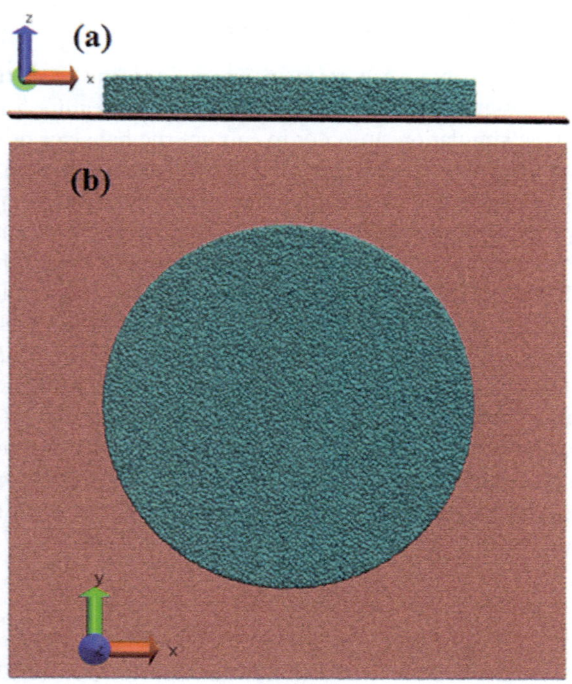

Fig. 1 (a) Top and (b) side views of a typical initial structure considered in the simulations. The gold disk is placed in the middle of the graphene surface. The diameter and thickness of the gold disk are 30 nm and 10 Å, respectively.

dewetting-induced change in the film radius per unit time. We have also reported the velocity of the center of mass (CM) of the gold film in the z direction, perpendicular to the surface, which is defined as $v_z(t) = \frac{dZ_{CM}}{dt}$, where Z_{CM} is the distance travelled by the CM of the gold film in the z direction. Inertia plays a dominant role in the dewetting process of liquid metals as they have low viscosity.[2,4] The liquid retraction velocity at the boundary of the liquid thin film is described by Brochard-Wyart:[45]

$$v = (|S|\rho^{-1}d^{-1})^{0.5}, \tag{2}$$

where v is the dewetting velocity, $|S|$ is the absolute value of the effective spreading coefficient, and ρ and d are the density and thickness of the film, respectively.

To understand the stability and self-assembly of gold nanoparticle structure formation on the graphene surface, we have calculated the free energy profiles for the gold-assembled structures. Upon heating the gold nanoparticle–graphene system at a particular temperature, the gold nanoparticles diffuse on the surface and form aggregates of different sizes. We use the population of definite cluster sizes of the assembled structures to calculate the free energy of the particular assembly. The free energy of the assembled structure of size k can be calculated using the following equation:[46,47]

$$F(k) = -k_B T \ln\left(\frac{kn_k}{\sum_{k=1}^{N} kn_k}\right), \tag{3}$$

where n_k is the number of assembled structures of size k. Here, the free energy of a single nanoparticle cluster ($k = 1$) is taken as the reference to calculate the free energy profiles, *i.e.*, $F(1) = 0.0$. To this end, we have considered a system of 25 gold nanoparticles of diameter 2 nm on a graphene surface to study the nanoparticle assembly behavior at different temperatures. We have considered 10 different initial arrangements of 25 nanoparticles on a graphene surface to generate an average free energy profile for the assembled nanostructures on the graphene surface. NVT simulations of the considered system are performed at different temperatures (673–1073 K) for 24 ns and the trajectories obtained from the simulations are used to generate the free energy profiles. In all the MD simulations the temperature and the pressure are maintained using the Nosé–Hoover thermostat[48,49] and barostat.[50] A time step of 1 fs is used to integrate Newton's equation of motion during the simulations.

The bond orientation order parameter method is used identify the nanoparticle structure.[51] For a considered bond r (a vector joining two neighboring atoms) the local bond order parameters are

$$Q_{lm}(r) \equiv Y_{lm}(\theta(r), \phi(r)), \tag{4}$$

where $\theta(r)$ and $\phi(r)$ are the polar and azimuthal angles for the bond with respect to a fixed coordinate frame, and $Y_{lm}(\theta(r), j(r))$ is the spherical harmonics. Global bond order parameter Q_l is defined as follows:

$$Q_l \equiv \sqrt{\frac{4\pi}{2l+1} \sum_{m=-1}^{l} |\overline{Q}_{lm}|^2}, \tag{5}$$

where $\bar{Q}_{lm}$ is the average of $Q_{lm}(r)$ over all the bonds in the nanoparticle. The third order invariants are defined as

$$W_l = \sum_{\substack{m_1,\ m_2,\ m_3 \\ m_1+m_2+m_3=0}} \begin{pmatrix} l & l & l \\ m_1 & m_2 & m_3 \end{pmatrix} \overline{Q}_{lm_1} \overline{Q}_{lm_2} \overline{Q}_{lm_3}, \tag{6}$$

where the coefficients (...) are the Wigner 3j symbols. The normalized quantity is defined as:

$$\hat{W} \equiv \frac{W_l}{\left(\sum_m |Q_{lm}|^2\right)^{3/2}}. \tag{7}$$

We have used Q_4, Q_6, $\hat{W}_4$ and $\hat{W}_6$ values to identify the crystal structure of the nanoparticle formed upon annealing the gold film on the graphene surface. This analysis is useful in comparing the nanoparticle structures on the graphene surface with that of free nanoparticles.

III. Results and discussions

We start the discussion with the dewetting behavior of a 5 Å thick and 30 nm diameter gold film at 1073 K, which is shown in Fig. 2. It is well known that the

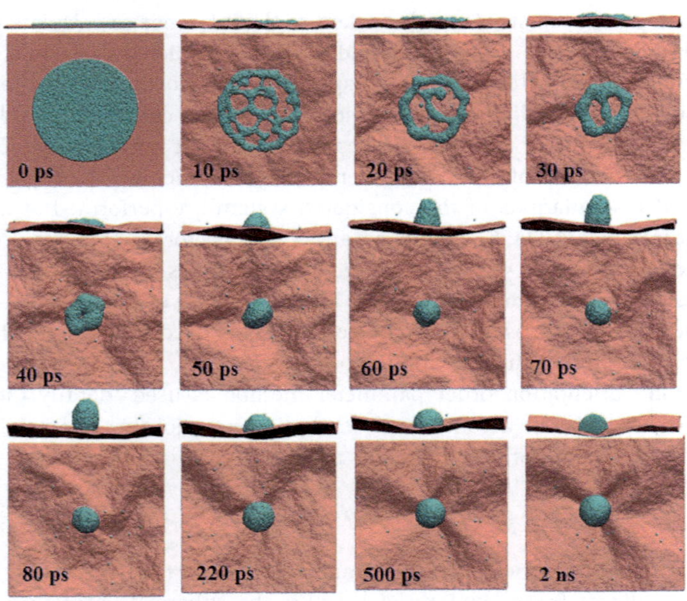

Fig. 2 MD simulation snapshots of the 30 nm diameter gold film of thickness 5 Å on a graphene surface at 1073 K.

dewetting begins at the edge of the film, where the surface tension is high.[4,10] Upon heating the film breaks within 10 ps to form holes of different sizes in the middle of the film, leading to a cage-like structure. Subsequently, the growth of holes occurs, whereby the interior regions (holes) coalesce either with the interior gold patch or with the outside ring, leading to an increase in the thickness of the rim of the cage. The propagation and coalescence of holes are clearly seen during the period 20–30 ps; however, simultaneously the film also contracts toward its CM. With increase in time, the film further contracts on the surface (decay of holes), and the holes finally disappear at ~40 ps. At ~200 ps a stable spherical gold nanodrop forms on the graphene surface. During the contraction of the film, it is noted that the CM of the film moves perpendicularly away from the graphene surface (*z* direction) until 90 ps. Subsequently, the CM moves towards the surface. After 200 ps the gold nanodrop settles on the surface with no change in the CM. Similar behavior was also observed for a Cu film on a graphene surface[25,26] and a Pt nanofilm on a graphitic surface.[27]

The morphology and sequence of evolution depend on the nature of the surface, surface–film interactions and film thickness, which are well known for polymeric systems.[52] To understand the effect of the thickness of the gold film on the dewetting process, we increase the film thickness from 5 to 10 Å. Fig. 3 shows the snapshots at different timesteps. It is apparent that the behavior is completely different from that seen for the 5 Å case. In contrast to the 5 Å film, upon heating at 1073 K the 10 Å film does not undergo rupture. Therefore, no holes are observed during the dewetting of the film as observed for the case of the 5 Å film. Instead, dewetting sets in through direct contraction of the film on the surface.

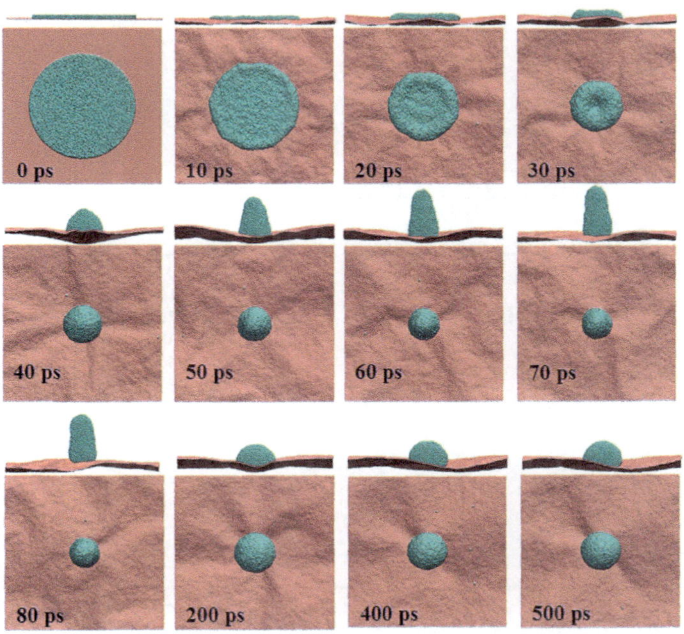

Fig. 3 MD simulation snapshots of the 30 nm diameter gold film of thickness 10 Å on a graphene surface at 1073 K.

Similar dewetting behavior, *i.e.*, hole formation in the 5 Å film and direct contraction in films of thickness >5 Å, were also observed in the dewetting process of Cu on graphene[25] and pillared graphene[26] surfaces, and Pt on a graphitic surface.[27] Despite different behavior for the 10 Å film, the film CM movement in the *z* direction is akin to that seen for the case of the 5 Å film. Interestingly, the CMs of both the films, 5 and 10 Å, move away from the surface up to 70 ps. Subsequently the film spreads, leading to a shift in the CM towards the surface. For both cases we observe a stable nanoparticle on the surface after 250 ps.

In order to further compare the behavior of the gold film for two different thicknesses we analyze the variation in the atomic surface density along the *x* direction with simulation time at 1073 K, which is shown in Fig. 4(a) and (b), for 5 Å and 10 Å films, respectively. The decrease in the length of the film in the *x* direction shows the contraction of the film during the dewetting stage. The spread of the film on the surface decreases with time, and the density of the film rim increases due to retraction of the edge atoms towards the CM of the film during contraction. The fluctuations in Fig. 4(a) are indicative of the formation of holes, which diminishes by 40 ps due to decay of holes, as reflected in the two small peaks. Subsequently, the density increases due to film contraction up to ~70 ps, as indicated by the single peak, signifying the formation of a nanodrop with a

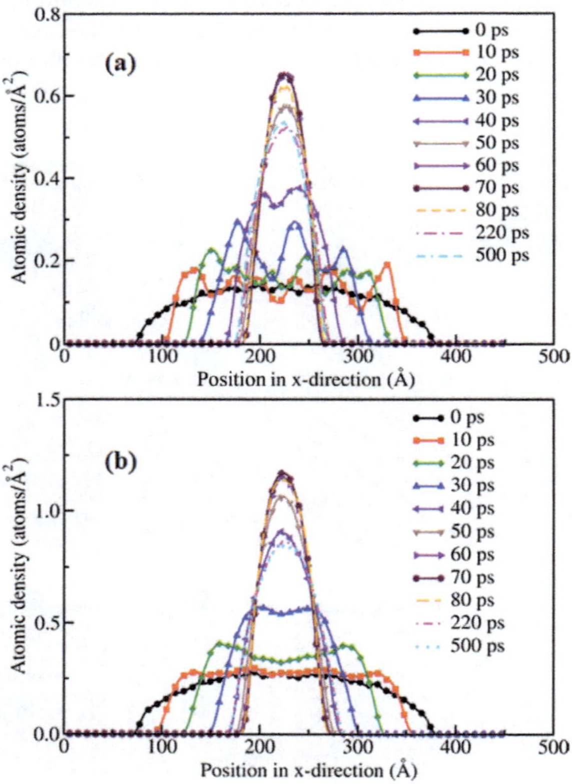

Fig. 4 Surface atomic density profiles of the (a) 5 Å and (b) 10 Å gold films along the *x* direction at 1073 K.

rod-like shape. After 70 ps the density of the film decreases with time; however, the base radius of the film on the surface increases, which indicates that the nanodrop spreads back on the surface. The spreading and rearrangement of the gold atoms continues until a stable nanoparticle forms at ~250 ps.

The dewetting behaviors of films at higher thicknesses, 15–30 Å, are akin to that seen for the 10 Å thick film, *i.e.*, the nanodrop formation takes place due to the direct contraction of the film, and no hole formation is observed in all these cases. For all the thicknesses considered, 5, 10, 15, 20, 25 and 30 Å, the CM of the film moves away from the graphene surface up to ~70 ps. Subsequently, the nanodrop returns to the surface and arranges itself to form a stable nanodrop at ~250 ps. Based on the above discussion, it is evident that the sequence of evolution is dependent on the thickness, as hole formation ceases to exist for film thicknesses >5 Å. However, the formation of gold nanoparticles and their final structure on the graphene surface are independent of the film thickness. Similar behavior was also observed for the dewetting process of a mercury film on a glass slab.[53]

In the case of a Cu nanofilm on a graphene surface at 2200 K, which is significantly higher than its melting temperature 1358 K, the nanoparticles formed due to dewetting are shown to jump off the surface at a constant speed of 140 m s^{-1}.[25] However, the temperature considered in this work is close to the melting point of the gold. Fig. 5 presents a typical CM velocity profile of the gold film in the *z* direction at 1073 K, for the case of a 10 Å film. The CM velocity profile, along with the corresponding MD snapshots of the system, clearly shows that the velocity of the CM increases up to 90 ps and subsequently decreases. It is evident from the snapshots shown in Fig. 5 that the nanoparticle formed on the surface is in contact with the surface all the time during the dewetting process. The contraction of the gold film towards the CM of the film resulted in the formation of a rod-like shape at ~70 ps, with an asphericity of ~0.3 (lower than the perfect cylinder's value of 0.5). As the nanodrop moves towards the surface (from ~70 ps

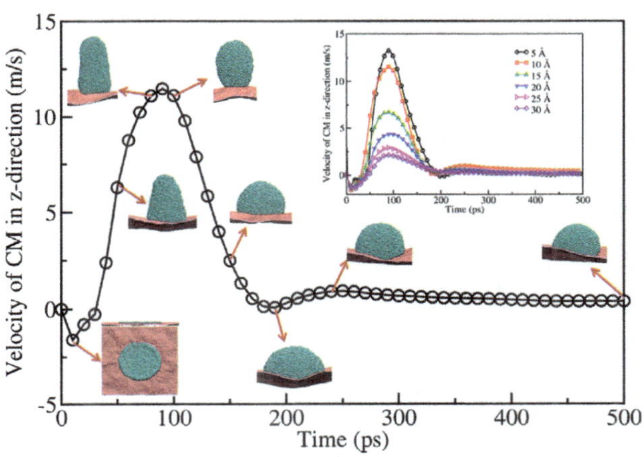

Fig. 5 Gold film CM velocity profile in the *z* direction along with the snapshots of MD simulation performed at 1073 K. The diameter and thickness of the film are 30 nm and 10 Å. Inset shows the velocity of the CM for different thicknesses of film.

to ~250 ps) the shape gradually changes from a rod-like shape to a spherical crown-like shape. At ~200 ps the nanodrop starts spreading on the surface. However, after 250 ps there is no change in the velocity, which shows a stable nanodrop formation on the surface, with an almost spherical shape (with an asphericity value of ~0.026).

The dewetting behavior seen for 15, 20, 25, and 30 Å thick films at 1073 K is similar to that seen for the case of a 10 Å thick film. The film CM velocity profiles for 5, 10, 15, 20, 25, and 30 Å are shown as an inset in Fig. 5. The maximum velocity of the CM, for all the thicknesses, occurs within ~80–90 ps. However, the magnitude of the velocity decreases with the increase in thickness. Moreover, for all the considered thicknesses of the film, a stable nanodrop forms at ~250 ps. This shows that the timescale of nanoparticle formation is independent of the thickness of the film. However, the maximum CM velocity, where the spread of the nanodrop is at a minimum, decreases with the film thickness. Nevertheless, we do not observe any detachment of the gold nanoparticles at any instant from the graphene surface within the temperature range considered in this study.

The dewetting-induced nanodrop formation involves three stages: contraction of the film, formation of a rod-like nanodrop and transformation of the rod-like nanodrop to a stable spherical nanodrop (see Fig. 5). To characterize the contraction of gold films on the surface we have calculated the dewetting area of the film. The dewetting area *vs.* time plot is shown in Fig. 6 for the case of a 15 nm diameter film on the graphene surface at 1073 K. The dewetted area (up to ~40 ps), corresponding to the contraction stage of the process, is observed to vary with t^{α}, where α is 0.87, 0.86, 0.96, 1.06 and 0.96 for 10, 15, 20, 25, 30 Å thickness films, respectively. This shows that the dewetted area of the film increases almost linearly with time during the contraction stage, and this is more prominent for thicker films. Similar linear variation of the dewetted area with time is also observed in the case of dewetting of thin polymer films.[54] After ~40 ps we do not observe a notable change in the dewetted area of the film. This indicates that

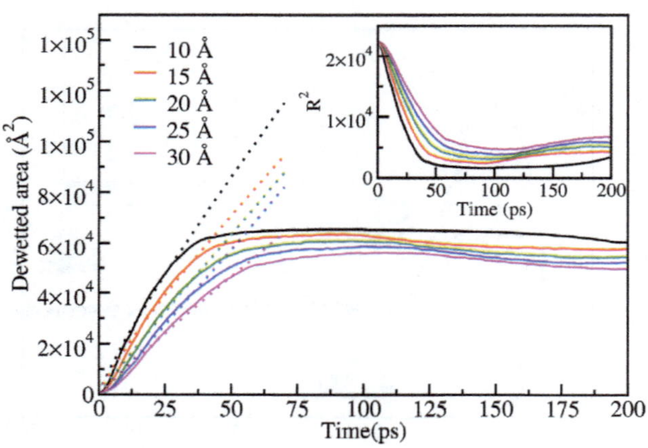

Fig. 6 Dewetted area $A(t)$ as a function of time at 1073 K for different thicknesses of the gold film on a graphene surface. The dotted lines show the linear fit to the variation of the dewetted area with time during the contraction stage. The inset figure shows the variation of R^2 with time (R is the radius of the film in Å).

during the process of rod-like shape formation and further transformation to spherical nanodrops the dewetted area does not change significantly. The inset of Fig. 6 shows a corresponding decreasing trend of the square radius of the film with time, as expected from the behavior of the dewetted area. Similar behavior for the dewetted area or radius of the film is also observed at other temperatures (973 K, 1400 K and 1533 K).

In order to investigate the retraction behavior for a gold film on a graphene surface, we calculate the retraction velocity (change in radius of the film per unit time),[53] for all the thicknesses considered in this study. The retraction velocity profiles are similar in nature to those of the CM velocity. The variation of maximum retraction velocity with film thickness is shown in Fig. 7. The retraction velocity increases with increase in temperature. On the other hand, the maximum retraction velocity decreases with the increase in film thickness. Similar behavior of decreasing retraction velocity with film thickness is also observed in the case of the Cu nanofilm[25] and the triangle shaped-gold film[4] dewetting on a graphene surface. However, at a temperature of 1533 K the retraction velocity of the 30 nm gold film of thickness 10 Å is ~320 m s^{-1}, which is significantly more than the maximum retraction velocity of 233 m s^{-1} [25] for the corresponding size of Cu film on a graphene surface at 2200 K. The higher retraction velocity of the gold film even at lower temperature may be attributed to its lower viscosity. It should be noted that the detachment of a Cu drop on graphene is observed at a temperature ~1.6 times its melting temperature. Although the Au–graphene binding energy is lower (−0.886 eV) than that of Cu–graphene (−0.909 eV), the temperature considered for gold in this work is around its melting temperature, as motivated by the experimental conditions.[55] We believe higher temperature would detach the gold drop, as seen for other metal–graphene cases.

Further, to study the system size effect on the dewetting behavior we have considered a gold disk of diameter 40 nm with different thicknesses (5, 10, 15, 20, 25, and 30 Å) on the graphene sheet. The dewetting behavior for the case of a

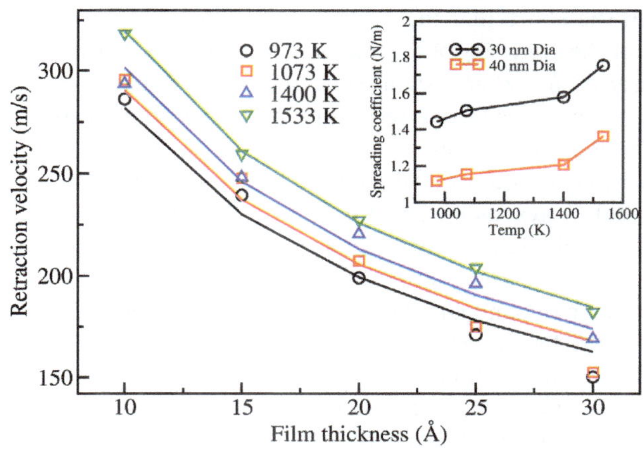

Fig. 7 Variation of the retraction velocity with film thickness for a 30 nm film at different temperatures. Solid lines represent the theoretical velocity based on the fit of eqn (2) to the simulation data. The inset shows the spreading coefficient variation with temperature.

40 nm diameter film is more or less the same as that seen for the 30 nm diameter film. At a thickness of 5 Å, the 40 nm film dewets the surface forming nanoholes, which is not seen for thicker films, in line with the behavior seen for the 30 nm diameter films. However, its retraction behavior leading to nanoparticle formation is slightly different. The CM of the film moves in the z direction up to ~160 ps in contrast to the 30 nm diameter film case. Moreover, the film forms a stable nanodrop on the surface at ~450 ps. The time taken for the 40 nm film to form a stable nanodrop on the surface is more than the time taken for the case of the 30 nm gold film. This shows that the dewetting process is dependent on the diameter of the film. The retraction velocity profiles for the case of the 40 nm diameter films also follow a similar trend to that seen for the 30 nm film. This retraction velocity data also fits well with the theoretical velocity equation (2). Retraction velocity profiles for both the cases (30 and 40 nm diameter films) clearly show that the retraction velocity increases with increase in temperature.

Now, we turn our attention towards the spreading behavior of the nanodrop, which can be analyzed based on the variation of the spreading coefficient, $|S|$, with temperature. The spreading coefficient is the driving force for the dewetting process, and takes into account the capillary force contribution.[53] In order to obtain the spreading coefficient, the theoretical velocity as shown in eqn (2) is fitted using the simulation data and is shown in the inset of Fig. 7. The absolute value of the spreading coefficient is found to increase with temperature, which is expected for the dewetting process. At low temperatures, $|S|$ linearly increases with temperature until the melting temperature. Subsequently, the change in $|S|$ is more drastic above the melting temperature, as seen for 1400 and 1533 K, indicating increased tendency for dewetting at higher temperature, which is also observed for the Cu-nanopillared surface system.[56] The spreading values in this work are lower than that reported by Habenicht *et al.*,[4] which is expected as the experiments performed by the authors were at a significantly higher temperature, 3129 K. It is also noted that an increase in the diameter of the film decreases the spreading coefficient, *i.e.*, dewetting is slow for the 40 nm film compared to the smaller diameter film. This is primarily due to the increased radius of curvature for the 40 nm film edges, leading to the reduction in the surface tension forces. It is clear from the above that the increase in temperature and decrease in film diameter favor the dewetting of the film on the graphene surface. Similar behavior was also observed for other systems such as for a Cu-nanopillared surface[26] and Pt–graphite,[27] where increase in temperature also led to increase in the detachment velocity of the nanodrop.

The dewetting of liquid metal leads to different shapes and sizes of nanoparticles on substrates, as observed by various researchers. For example, Zhou *et al.*[31] observed different shapes and sizes of gold nanoparticles on graphene surfaces at different temperatures. Moreover, the substrate can promote or suppress diffusion of nanoparticles. For example, in the case of a silica substrate gold nanoparticles have much less diffusion compared to on a graphene substrate due to the weaker interactions in the gold–graphene system. Hence agglomeration behavior is promoted on graphene, as observed by Schedin *et al.*[33] Such agglomeration behavior of nanoparticles can lead to extremely interesting structures, as seen in different systems.[57,58] It is now well known that the shape and size of the gold nanoparticle greatly affects the SERS properties.[31,32] Hence, to understand the aggregation behavior of gold nanoparticles on graphene, we consider 10 independent simulations, at different temperatures, with 25 nanoparticles of 2 nm diameter on a graphene surface. The nanoparticles diffuse and merge, leading to bigger agglomeration as the simulation runs. We track the agglomeration of nanoparticles using simple geometrical criteria, and using cluster size analysis we generate the free energy of the assembled structures as per eqn (3).

Fig. 8 shows the free energy profiles of different assembled structures calculated at 673, 773, 873, 973 and 1073 K. These profiles clearly show that there is a free energy barrier for the formation of an assembled structure of gold nanoparticles on a graphene surface, for all temperatures considered in this work. At 673 K the free energy barrier is about 15 kJ mol^{-1}, and the critical cluster size is $\sim k = 7$. This indicates that ~15 kJ mol^{-1} of energy is required to form an assembled structure of size 7. Once the critical cluster size is formed, subsequent growth is favorable to form a fully assembled structure. However, the largest cluster size obtained within 24 ns is 21, which is more stable than the single unassociated units by about 27 kJ mol^{-1}. In contrast, at 773 K the free energy barrier is higher, ~23 kJ mol^{-1}, corresponding to a cluster size of 8. The final assembled structure of size 20, within 24 ns, is more stable than the single unassociated units by ~23 kJ mol^{-1}. Furthermore, the assembly of nanoparticles is unfavorable at 773 K by 8 kJ mol^{-1} when compared to the process at 673 K. The energy barrier for the formation of critical clusters is found to increase with increasing temperature. The corresponding energy barriers at 873, 973, and 1073 K are ~26, ~32, and ~40 kJ mol^{-1}, respectively. In other words, the self-assembly processes at 873, 973, and 1073 K are unfavorable by 11, 17 and 25 kJ mol^{-1}, respectively, in comparison to the 673 K process. This analysis shows that the assembly of gold nanoparticles on the graphene surface is more favorable at low temperatures. The final assembled structures formed at 873, 973, and 1073 K, on the other hand, are more stable than the single unassociated nanoparticles by ~23, 24 and 16 kJ mol^{-1}, respectively.

Fig. 9 displays the assembled structures formed on the graphene surface after annealing of one arrangement of nanoparticles on the graphene surface, at

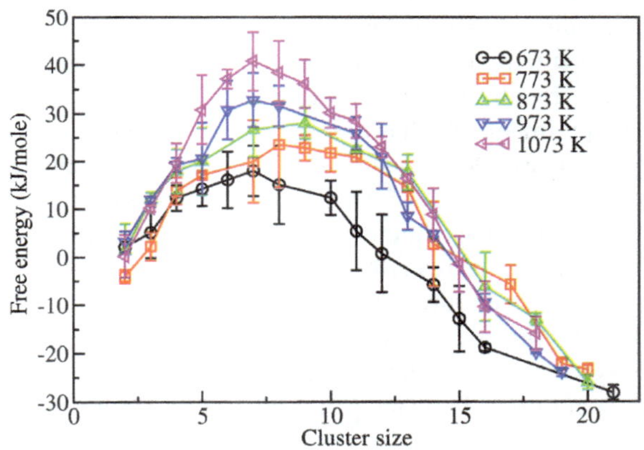

Fig. 8 Free energy profiles at different temperatures for gold-nanoparticle-assembled structures of different sizes. The error bars represent the standard deviation of the free energy value obtained form 10 independent simulations.

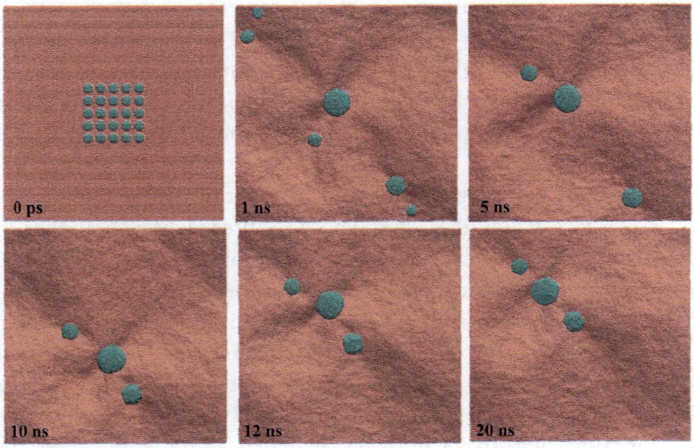

Fig. 9 MD simulation snapshots showing nanoparticle aggregation to form different sized nanoparticles. The simulation is performed at a temperature of 1073 K. Initially, the nanoparticles are kept in a square arrangement, with a nearest-neighbor distance of 1 nm.

1073 K. The final structure, however, does not change upon changing the initial structure (with the nearest neighbor nanoparticle–nanoparticle distance varying from 1 to 7 nm). Upon heating, the gold nanoparticles diffuse on the surface and form different sizes of aggregates on the surface. With time the size of the aggregates increases and the number density of clusters decreases. The aggregate structure however does not change beyond 5 ns, as apparent from Fig. 9. The nanoparticles formed at $\sim$5 ns on the graphene surface are spherical in shape. Between 5 and 12 ns we observe transformation of nanoparticles from spherical to hexagonal. Subsequently, the aggregates remain in the hexagonal shape, as seen in our simulations up to 24 ns. The behavior is similar in nature at different temperatures. The gold nanoparticles mainly in the hexagonal form were also observed in an experimental study by Zhou *et al.*[31] However, the authors also reported the formation of triangular-shaped nanoparticles, which was seen to be temperature-dependent. In particular, at 873 and 973 K hexagonal particles were seen in the experiment, in line with our simulation work. At a higher temperature of 1073 K, a mixture of triangular and hexagonal-shaped nanoparticles was observed in the experiment, which is not seen in our simulations. We believe such structures may be metastable in nature, which we plan to investigate in our future work. In the experimental work of Zhou *et al.*,[31] it was observed that the SERS enhancement was high at lower temperatures (873, 973 and 1073 K) compared to that seen at higher temperature (1100 K). This is due to well-faceted nanoparticle formation at lower temperatures. Although our study, based on classical MD simulations, does not provide direct reference to SERS data, we observe enhanced probability (lower energy barrier) of faceted nanoparticle formation with decreasing temperature, implying higher SERS enhancement at lower temperatures.

The bulk gold has an fcc crystal structure, but the gold nanoparticle is known to have a different crystal structure because of the different surface energy associated with the surfaces.[59,60] In order to understand the change in the structure of

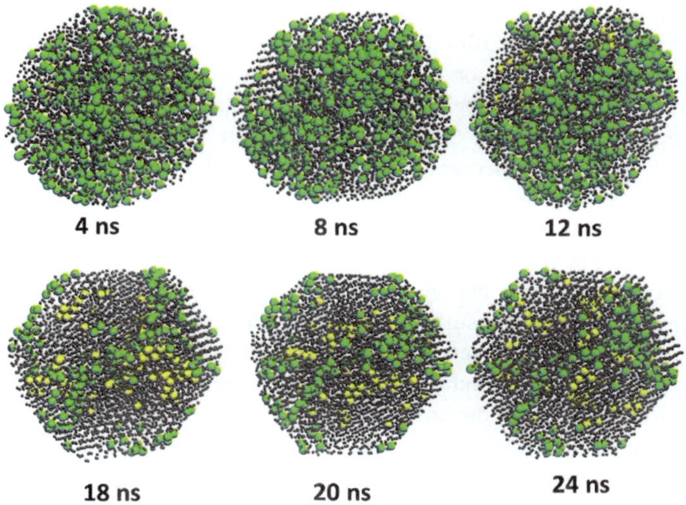

Fig. 10 Crystal structure analysis of the gold nanoparticles on the basis of local bond order parameter analysis. Green and yellow colors represent atoms with hcp and fcc structures, respectively. Gray color represents atoms in the disordered state.

the aggregated nanoparticle during the annealing process, we have used the bond orientation order analysis as per eqn (4)–(7). Bond orientation order parameter calculations are performed for the larger cluster containing 3984 gold atoms selected from the 1073 K annealing simulation. The four bond order parameters Q_4, Q_6, $\hat{W}_4$ and $\hat{W}_6$ are used to identify the crystal structure. Using the analysis of Wang *et al.*,[61] the atoms with $Q_6 > 0.3$ and $\hat{W}_6 < -0.12$ are considered to be icosahedral in structure. For $Q_4 > 0.16$ and $\hat{W}_4 < -0.13$ the atoms are considered to have the fcc structure; whereas, for $Q_4 < 0.12$ and $\hat{W}_4 > 0.10$ the atoms are considered to have the hcp structure.

Fig. 10 shows that at 4 ns there are two types of atoms, *viz.*, a few atoms in the hcp structure and others without any ordered structure. Moreover, the structure at 4 ns is spherical in shape. With time the structure of the nanoparticle deviates from its spherical shape, and at 8 ns we observe fcc-based arrangement for atoms. With increase in time the number of atoms with the fcc structure increases, while the number of atoms with the hcp structure decreases. Overall, the values of the bond order parameters are in line with the work of Wang *et al.*,[61] indicating propensity towards an icosahedral structure.

IV. Conclusions

We have studied the dewetting behavior of different diameter gold films with different thicknesses at different temperatures. At a lower thickness of the gold film of 0.5 nm, with a 30 nm diameter, dewetting sets in through the formation of nanoscale holes randomly in the middle of the film. These holes increase in size, but undergo coalescence with time along with contraction of the film. Hence, the holes disappear finally to form a drop on the surface. However, at higher thicknesses (>0.5 nm) drop formation is *via* a direct contraction mechanism without any hole formation. The nanoscale hole formation at lower thickness is also seen for increased diameters of film. For a given film diameter, the time taken to form a stable nanodrop does not change with increase in thickness of the film. However, the retraction velocity is found to depend on the film thickness and temperature of the system. The retraction velocity increases with increase in temperature, whereas it decreases with increasing film thickness. The dewetted area during the contraction stage, for different thickness of the film, is found to increase almost linearly with time. This study shows that increase in temperature favors dewetting of the gold film and nanostructure formation on the graphene surface. Finally, the free energy landscape analysis for the gold nanoparticle assembly on the graphene surface demonstrated that the self-assembly of gold nanoparticles is favorable at lower temperatures, as the free energy barrier increases with the temperature. The final assembled structures at various temperatures are thermodynamically more favorable than the unassociated single nanoparticles. We also present the bond orientation order parameter analysis of the nanoparticles formed on the graphene surface, which indicate inclination for an icosahedral structure.

References

1 L. Kondic, J. A. Diez, P. D. Rack, Y. Guan and J. D. Fowlkes, *Phys. Rev. E: Stat., Nonlinear, Soft Matter Phys.*, 2009, **79**, 026302.

2 J. Bischof, D. Scherer, S. Herminghaus and P. Leiderer, *Phys. Rev. Lett.*, 1996, **77**, 1536–1539.

3 S. Herminghaus, K. Jacobs, K. Mecke, J. r. Bischof, A. Fery, M. Ibn-Elhaj and S. Schlagowski, *Science*, 1998, **282**, 916–919.

4 A. Habenicht, M. Olapinski, F. Burmeister, P. Leiderer and J. Boneberg, *Science*, 2005, **309**, 2043–2045.

5 C.-H. Lin, L. Jiang, J. Zhou, H. Xiao, S.-J. Chen and H.-L. Tsai, *Opt. Lett.*, 2010, **35**, 941–943.

6 Y. Wu, J. D. Fowlkes and P. D. Rack, *J. Mater. Res.*, 2011, **26**, 277–287.

7 R. Sachan, V. Ramos, A. Malasi, S. Yadavali, B. Bartley, H. Garcia, G. Duscher and R. Kalyanaraman, *Adv. Mater.*, 2013, **25**, 2045–2050.

8 H. Reinhardt, H.-C. Kim, C. Pietzonka, J. Kruempelmann, B. Harbrecht, B. Roling and N. Hampp, *Adv. Mater.*, 2013, **25**, 3257.

9 H. Krishna, A. K. Gangopadhyay, J. Strader and R. Kalyanaraman, *J. Magn. Magn. Mater.*, 2011, **323**, 356–362.

10 P. D. Rack, Y. Guan, J. D. Fowlkes, A. V. Melechko and M. L. Simpson, *Appl. Phys. Lett.*, 2008, **92**, 223108.

11 Y. Wu, J. D. Fowlkes, P. D. Rack, J. A. Diez and L. Kondic, *Langmuir*, 2010, **26**, 11972–11979.

12 J. D. Fowlkes, Y. Wu and P. D. Rack, *ACS Appl. Mater. Interfaces*, 2010, **2**, 2153–2161.

13 I. Gierz, C. Riedl, U. Starke, C. R. Ast and K. Kern, *Nano Lett.*, 2008, **8**, 4603–4607.

14 E. Yoo, T. Okata, T. Akita, M. Kohyama, J. Nakamura and I. Honma, *Nano Lett.*, 2009, **9**, 2255–2259.

15 Z. Li, P. Zhang, K. Wang, Z. Xu, J. Wei, L. Fan, D. Wu and H. Zhu, *J. Mater. Chem.*, 2011, **21**, 13241–13246.

16 C. Xu, X. Wang and J. Zhu, *J. Phys. Chem. C*, 2008, **112**, 19841–19845.
17 P. V. Kamat, *J. Phys. Chem. Lett.*, 2010, **1**, 520–527.
18 D. H. Lee, J. E. Kim, T. H. Han, J. W. Hwang, S. Jeon, S.-Y. Choi, S. H. Hong, W. J. Lee, R. S. Ruoff and S. O. Kim, *Adv. Mater.*, 2010, **22**, 1247–1252.
19 V. C. Tung, L.-M. Chen, M. J. Allen, J. K. Wassei, K. Nelson, R. B. Kaner and Y. Yang, *Nano Lett.*, 2009, **9**, 1949–1955.
20 S. Biswas and L. T. Drzal, *Chem. Mater.*, 2010, **22**, 5667–5671.
21 D. Choi, M.-Y. Choi, W. M. Choi, H.-J. Shin, H.-K. Park, J.-S. Seo, J. Park, S.-M. Yoon, S. J. Chae, Y. H. Lee, S.-W. Kim, J.-Y. Choi, S. Y. Lee and J. M. Kim, *Adv. Mater.*, 2010, **22**, 2187–2192.
22 S.-M. Paek, E. Yoo and I. Honma, *Nano Lett.*, 2009, **9**, 72–75.
23 H. Zhang, X. Lv, Y. Li, Y. Wang and J. Li, *ACS Nano*, 2010, **4**, 380–386.
24 M. Hupalo, X. Liu, C.-Z. Wang, W.-C. Lu, Y.-X. Yao, K.-M. Ho and M. C. Tringides, *Adv. Mater.*, 2011, **23**, 2082–2087.
25 M. Fuentes-Cabrera, B. H. Rhodes, J. D. Fowlkes, A. López-Benzanilla, H. Terrones, M. L. Simpson and P. D. Rack, *Phys. Rev. E: Stat., Nonlinear, Soft Matter Phys.*, 2011, **83**, 041603.
26 X. Li, Y. He, Y. Wang, J. Dong and H. Li, *Sci. Rep.*, 2014, **4**, 1–6.
27 Y. Li, C. Tang, J. Zhong and L. Meng, *J. Appl. Phys.*, 2015, **117**, 064304.
28 W. Xu, X. Ling, J. Xiao, M. S. Dresselhaus, J. Kong, H. Xu, Z. Liu and J. Zhang, *Proc. Natl. Acad. Sci. U. S. A.*, 2012, **109**, 9281–9286.
29 A. Gutes, B. Hsia, A. Sussman, W. Mickelson, A. Zettl, C. Carraro and R. Maboudian, *Nanoscale*, 2012, **4**, 438–440.
30 H. Zhou, C. Qiu, Z. Liu, H. Yang, L. Hu, J. Liu, H. Yang, C. Gu and L. Sun, *J. Am. Chem. Soc.*, 2010, **132**, 944–946.
31 H. Zhou, F. Yu, M. Chen, C. Qiu, H. Yang, G. Wang, T. Yu and L. Sun, *Carbon*, 2013, **52**, 379–387.
32 G. R. S. Iyer, J. Wang, G. Wells, S. Guruvenket, S. Payne, M. Bradley and F. Borondics, *ACS Nano*, 2014, **8**, 6353–6362.
33 F. Schedin, E. Lidorikis, A. Lombardo, V. G. Kravets, A. K. Geim, A. N. Grigorenko, K. S. Novoselov and A. C. Ferrari, *ACS Nano*, 2010, **4**, 5617–5626.
34 S. Plimpton, *J. Comput. Phys.*, 1995, **117**, 1–19.
35 G. Grochola, S. P. Russo and I. K. Snook, *J. Chem. Phys.*, 2005, **123**, 204719.
36 L. Lindsay and D. A. Broido, *Phys. Rev. B: Condens. Matter Mater. Phys.*, 2010, **81**, 205441.
37 S. Namsani, J. Schulte, M. Böhm and J. K. Singh, Unpublished work.
38 A. Martin, S. Biplab, E. Olle and V. S. Natalia, *J. Phys.: Condens. Matter*, 2011, **23**, 205301.
39 O. V. Yazyev and A. Pasquarello, *Phys. Rev. B: Condens. Matter Mater. Phys.*, 2010, **82**, 045407.
40 G. M. Wang, J. J. BelBruno, S. D. Kenny and R. Smith, *Phys. Rev. B: Condens. Matter Mater. Phys.*, 2004, **69**, 195412.
41 R. Varns and P. Strange, *J. Phys.: Condens. Matter*, 2008, **20**, 225005.
42 S. S. Carara, R. J. C. Batista and H. l. Chacham, *Phys. Rev. B: Condens. Matter Mater. Phys.*, 2009, **80**, 115435.
43 K. S. Subrahmanyam, A. K. Manna, S. K. Pati and C. N. R. Rao, *Chem. Phys. Lett.*, 2010, **497**, 70–75.

44 L. Liu, Z. Chen, L. Wang, E. Polyakova, T. Taniguchi, K. Watanabe, J. Hone, G. W. Flynn and L. E. Brus, *J. Phys. Chem. B*, 2013, **117**, 4305–4312.

45 F. Brochard-Wyart, *J. Phys. II*, 1993, **3**, 21–26.

46 J. P. Mahalik and M. Muthukumar, *J. Chem. Phys.*, 2012, **136**, 135101.

47 T. K. Patra and J. K. Singh, *Soft Matter*, 2014, **10**, 1823–1830.

48 W. G. Hoover, *Phys. Rev. A: At., Mol., Opt. Phys.*, 1985, **31**, 1695–1697.

49 S. Nosé, *J. Chem. Phys.*, 1984, **81**, 511–519.

50 G. J. Martyna, M. E. Tuckerman, D. J. Tobias and M. L. Klein, *Mol. Phys.*, 1996, **87**, 1117–1157.

51 P. J. Steinhardt, D. R. Nelson and M. Ronchetti, *Phys. Rev. B*, 1983, **28**, 784–805.

52 J. Singh and A. Sharma, *J. Adhes. Sci. Technol.*, 2000, **14**, 145–166.

53 N. Péron, F. Brochard-Wyart and H. Duval, *Langmuir*, 2012, **28**, 15844–15852.

54 L. Liu, Z. Chen, L. Wang, E. Polyakova, T. Taniguchi, K. Watanabe, J. Hone, G. W. Flynn and L. E. Brus, *J. Phys. Chem. B*, 2013, **117**, 4305–4312.

55 A. Martin, L. Sébastien, E. Olle and V. S. Natalia, *J. Phys.: Condens. Matter*, **23**, 395001.

56 X. Li, Y. He, Y. Wang, J. Dong and H. Li, *Sci. Rep.*, 2014, **4**, 3938.

57 T. K. Patra, P. Katiyar and J. K. Singh, *Chem. Eng. Sci.*, 2015, **121**, 16–22.

58 B. Radha and G. U. Kulkarni, *Cryst. Growth Des.*, 2011, **11**, 320–327.

59 S. Iijima and T. Ichihashi, *Phys. Rev. Lett.*, 1986, **56**, 616–619.

60 C. L. Cleveland, W. D. Luedtke and U. Landman, *Phys. Rev. B: Condens. Matter*, 1999, **60**, 5065–5077.

61 Y. Wang, S. Teitel and C. Dellago, *J. Chem. Phys.*, 2005, **122**, 214722.

Faraday Discussions

PAPER

Chiral self-assembly of helical particles

Hima Bindu Kolli,†[a] Giorgio Cinacchi,*[b] Alberta Ferrarini*[c] and Achille Giacometti*[d]

Received 1st September 2015, Accepted 15th October 2015
DOI: 10.1039/c5fd00132c

The shape of the building blocks plays a crucial role in directing self-assembly towards desired architectures. Out of the many different shapes, the helix has a unique position. Helical structures are ubiquitous in nature and a helical shape is exhibited by the most important biopolymers like polynucleotides, polypeptides and polysaccharides as well as by cellular organelles like flagella. Helical particles can self-assemble into chiral superstructures, which may have a variety of applications, *e.g.* as photonic (meta) materials. However, a clear and definite understanding of these structures has not been entirely achieved yet. We have recently undertaken an extensive investigation on the phase behaviour of hard helical particles, using numerical simulations and classical density functional theory. Here we present a detailed study of the phase diagram of hard helices as a function of their morphology. This includes a variety of liquid-crystal phases, with different degrees of orientational and positional ordering. We show how, by tuning the helix parameters, it is possible to control the organization of the system. Starting from slender helices, whose phase behaviour is similar to that of rodlike particles, an increase in curliness leads to the onset of azimuthal correlations between the particles and the formation of phases specific to helices. These phases feature a new kind of screw order, of which there is experimental evidence in colloidal suspensions of helical flagella.

I. Introduction

Self-assembly refers to a process where multiple components spontaneously organize to form a large entity.[1] It is viewed as one of the most promising routes to

[a]*Dipartimento di Scienze Molecolari e Nanosistemi, Università Ca' Foscari di Venezia, Campus Scientifico, Edificio Alfa, via Torino 155, 30170 Venezia Mestre, Italy*

[b]*Departamento de Física Teórica de la Materia Condensada, Instituto de Física de la Materia Condensada (IFIMAC) and Instituto de Ciencias de Materiales "Nicolás Cabreras", Universidad Autónoma de Madrid, Campus de Cantoblanco, 28049 Madrid, Spain. E-mail: giorgio.cinacchi@uam.es*

[c]*Dipartimento di Scienze Chimiche, Università di Padova, via F. Marzolo 1, 35131 Padova, Italy. E-mail: alberta.ferrarini@unipd.it*

[d]*Dipartimento di Scienze Molecolari e Nanosistemi, Università Ca' Foscari di Venezia, Campus Scientifico, Edificio Alfa, via Torino 155, 30170 Venezia Mestre, Italy. E-mail: achille.giacometti@unive.it*

† Present address: Department of Chemistry, University of Oslo, Postboks 1033 Blindern, 0315 Oslo, Norway. E-mail: h.b.kolli@kjemi.uio.no

a "bottom-up" engineering of new nanomaterials. Recent developments, both numerical[2] and experimental,[3,4] include the possibility of obtaining a variety of complex structures using building blocks of essentially any possible shape. While the idea of exploiting the shape of the building blocks as a way to direct the self-assembly process towards desired architectures has been vigorously pursued in the last few years,[5,6] the idea of also exploiting their *chirality* has so far received less attention.

Yet the control of the chiral organization of materials is of major interest for application purposes. The promising field of chiral photonics aims at exploiting the unique optical properties of chiral (meta)materials, such as giant gyrotropy, unusually large optical activity and circular dichroism, with associated negative refractive index, or strong nonlinear chiroptical effects (for recent reviews see *e.g.* ref. 7–9). The early observation of an optical effect in a twisted structure was the detection of rotation of the plane of polarization of microwaves traversing a jute bundle by Bose in 1898.[10] In the last few years there has been a renewed interest in this topic and increasing efforts have been devoted to the design, synthesis and characterization of nano-engineered materials, made of chiral building blocks that may be viewed as "artificial chiral molecules". In this context, a special place is deserved for helical particles: as early as 1920 Lindman showed the polarization rotation of the microwave radiation transmitted through a system of randomly oriented metallic helices,[11] in analogy with the behavior exhibited by chiral molecules. This can be taken as the origin of helix-based chiro-optical metamaterials: special chiro-optical properties orders of magnitude stronger than those of molecular systems were recently discovered in ordered arrays of helical particles.[12] Further, thanks to progress in nanofabrication, arrays of plasmonic helices active in the visible wavelength region could be obtained.[13,14] Helically nanostructured materials are also attracting considerable attention because of their unique electrical and mechanical properties (see ref. 15–17 for recent reviews). The control of the morphology of individual helices and their 3D organization, which determines the interaction with electromagnetic fields[12] and may affect the electrical and mechanical response of the (meta)materials, is a major step towards the optimization of structures for different applications.

The importance of the helix in nature need not be stressed: a helical shape is exhibited by the most important biopolymers like polynucleotides, polypeptides and polysaccharides, as well as by biological particles like filamentous viruses and cellular organelles like flagella. These systems generally exhibit one or more liquid-crystal phases above a given density. The most common is the cholesteric (N*), which is a twisted nematic phase. It is nematic since the centers of mass of the helices are randomly distributed in space, but the helical axes are locally preferentially aligned to one another; it is twisted because the alignment direction (the director $\hat{\mathbf{n}}$) rotates in a helical way around a perpendicular axis. Typical periodicities (the cholesteric pitch $\mathscr{P}$) may range from micro- to milli-meters, depending on the specific system, but in all cases are orders of magnitude longer than the pitches (p) of the constituent helices. For some systems, like DNA,[18] in a restricted density range just below the N* phase, blue phases[19] have also been found. These can be described as fluid lattices of defects, with cubic symmetry and lattice periods of the order of the wavelength of visible light. They are locally chiral, since directors are locally arranged in double-twist cylinders. On the upper boundary of the N* phase, smectic phases (Sm) can be found, which, beside orientational order, exhibit partial positional order: the particles form layers, but within a layer the positional order is short-range or absent altogether. In polydisperse systems the formation of Sm phases is inhibited and columnar phases may form, where arrays of close packed columns are formed by axially stacked particles, with continuous translational symmetry along the axial dimension.

All these phases are not specific to helices and can be found in other chiral or non-chiral systems, and it was tacitly assumed that nothing special was associated with the helical shape of particles. However, this belief was challenged by a recent study of dense colloidal suspensions of helical flagella, which are strongly curled helical filaments, with a well defined pitch $p \sim 1$–2 μm.[20] To explain the results of polarizing and differential interference contrast microscopy experiments, a novel chiral nematic organization was proposed, where helical filaments would be preferentially aligned along a common direction, intercalated and in-phase with each other without positional order. This picture is compatible with the observation of a striped birefringent pattern with periodicity equal to the helix pitch p, since in such a structure the local tangents to helices undergo a conical rotation with period p around the alignment direction. Then, a number of questions arise: why is there no evidence of this phase in other helical systems? Are there special requirements for helices to form this new phase or is it only a matter of detection? In fact, whereas a periodicity in the micrometer range is immediately evidenced by polarization microscopy, analogously direct methods are not currently available in the nanometer range, within which are the helical pitches of most (bio) polymers. A more general question is whether self-assembly of helical particles may lead to other special phases, not yet discovered or overlooked.

Clearly, both fundamental and application questions call for a comprehensive insight into the self-assembly of helical particles. This issue has remained scarcely explored, which is surprising in view of the paradigmatic role of the helical shape. To fill this gap, we recently undertook a study of self-assembling hard rigid helices, using Monte Carlo simulations and classical density functional theory (DFT) at the second-virial (Onsager) and third-virial level.[21–24] The choice of hard particles, thus interacting through a purely repulsive short-range potential, was dictated by the intent of exploring the effects determined solely by the helical shape. We found a rich polymorphism, with the presence of chiral liquid-crystal phases not previously observed for other chiral systems. A remarkable result[22] was the detection of a chiral phase, called screw-nematic,[25] which has the same features of the phase detected in dense suspensions of helical flagella.[20] On increasing densities more, we found smectic phases, generally chiral, which are also specific to helical particles. These are followed, at very high density, by chiral compact phases.

In our studies, we used the helix model shown in Fig. 1 (left): each particle is composed of 15 partially fused hard spheres of diameter D, rigidly arranged in a helical fashion with a fixed contour length $L = 10D$. Hence, different helix morphologies can be achieved upon an independent change of the radius r and the pitch p. Here the bead diameter is a scaling parameter and the phase behaviour is determined by the values of r and p relative to D: for $p < 2D$, helices cannot intrude into each other's grooves and, irrespective of r, resemble threaded rods. The mutual helical character of particles clearly emerges only for (r, p) bigger than the sphere diameter. Hereafter, the bead diameter D is taken as the unit of length. In previous Monte Carlo simulations, we focused on three different

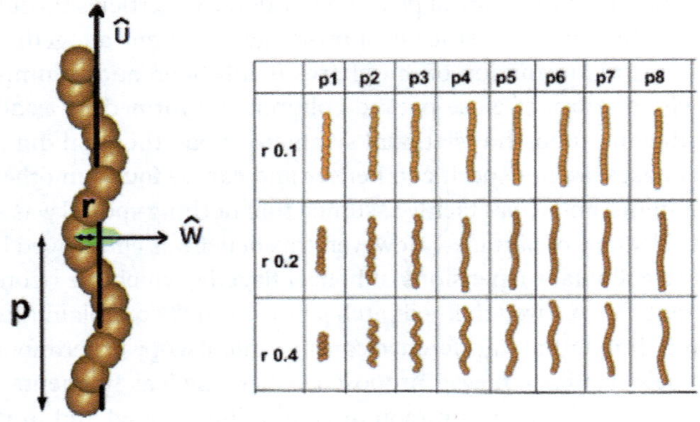

Fig. 1 (Left) Model helical particle with prescribed helix radius r and pitch p. Unit vectors $\hat{u}$ and $\hat{w}$ are parallel to the helical axis and to the two-fold symmetry axis, respectively. (Right) Morphology of the helix models used in this work: r is the helix radius and p is the pitch. Different combinations of (r, p) then give rise to different shapes and aspect ratios.

morphologies ($r = 0.2, p = 8$; $r = 0.2, p = 4$; $r = 0.4, p = 4$), all with clear helical character, and we identified the possible phases formed by the helical particles.[22,24] Here, with a considerable computational effort, we have extended our investigation to a wider range of (r, p) values, to get a more detailed insight into the phase behaviour as a function of the helical morphology. The differences in terms of combinations of radii and pitches is rather remarkable, as illustrated in Fig. 1 (right). The two extreme cases are represented by $r = 0.1$, where helices are rather tube-like for any pitch value, and by $r = 0.4$, where different pitches lead to significantly different morphologies and aspect ratios. The low chirality, $r = 0.1$, systems will allow us to investigate a region that remained hitherto unexplored, which however deserves to be considered, as representative of the behavior of systems like (bio)polymers. The case $r = 0.2$ presents intermediate phenomenologies. The analysis of the phase diagram will show how the organization of the systems can be controlled by tuning the helical character of the building blocks.

The remainder of the paper is organized as follows. In Section II we briefly recall the methods used in our study, whereas in Section III we review the definitions of the order parameters and correlation functions necessary to identify the various phases. Section IV contains the results for the phase diagrams and Section VI the conclusions and possible perspectives.

II. Methods

Isobaric Monte Carlo simulations were exploited to compute the equation of state. The interested reader is referred to previous works for details.[21,22,24] Here we shall recall only the few points necessary for the sake of completeness.

A variable number N, ranging between 900 and 2000, of hard helices, were used depending on the helix morphologies and the required densities. In all cases, the initial conditions were taken as an ordered compact configuration generated using the isopointal search method described in ref. 24, but the stability of the obtained results with respect to a change in the initial conditions was explicitly tested for. A typical Monte Carlo cycle consisted of $N/2$ translation moves, $N/2$ rotation moves, as well as one volume move, with a typical equilibration time of the order of 3 to 4 × 10^6 Monte Carlo steps, as well as production runs of additional 1 to 2 × 10^6 Monte Carlo steps.

In all simulations periodic boundary conditions (PBC) were used, with a floppy (*i.e.* shape adaptable) triclinic computational box. Such conditions are fully compatible with the existence of helical order with small periodicity, comparable with the particle length. However they do not allow the development of a cholesteric organization when this is characterized by a periodicity that is orders of magnitude longer than the scale of particle chirality.[26] For this reason, when using PBC in our simulations we could only find N phases, with a uniform rather than twisted $\hat{\mathbf{n}}$ director (*i.e.* $\mathscr{P} \rightarrow \infty$). To ascertain the possible existence of a director twist, we subsequently evaluated the pitch $\mathscr{P}$ for all state points in the putative N* phases, using DFT within the Onsager second-virial[28] framework with Parsons–Lee correction.[29] The method presented earlier[23,30] was adopted and calculations were performed according to the following procedure: (i) the order parameters at a given density were determined by minimization of the free energy of the untwisted N phase; (ii) these order parameters were used to calculate the chiral strength k_2 and the twist elastic constant K_{22}; (iii) the cholesteric pitch was determined as $\mathscr{P} = -k_2/K_{22}$. The elastic constant is a positive quantity, whereas the chiral strength, and thus the cholesteric pitch $\mathscr{P}$, are signed quantities (with opposite signs for enantiomers). Positive and negative $\mathscr{P}$ correspond to a right-handed and a left-handed cholesteric phase, respectively.

III. Order parameters and correlation functions

We used a combination of order parameters and correlation functions to fully characterize the various phases encountered in this study. Here we will briefly recall the main definitions, referring to ref. 22 and 24 for details.

The orientation of a helix in space is univocally identified by the unit vectors $\hat{\mathbf{u}}$ and $\hat{\mathbf{w}}$, parallel to the main axis and to the C_2 symmetry axis, respectively (see Fig. 1). Nematic order is identified by the order parameter

$$\langle P_2 \rangle = \left\langle \frac{3}{2}(\hat{\mathbf{u}}\cdot\hat{\mathbf{n}})^2 - \frac{1}{2} \right\rangle, \tag{3.1}$$

which describes the average (non polar) alignment of the $\hat{\mathbf{u}}$ axes of helices to the $\hat{\mathbf{n}}$ director and vanishes in the isotropic state. Here and henceforth the angular brackets denote ensemble averages. The $\hat{\mathbf{n}}$ director and the $\langle P_2 \rangle$ order parameter are computed according to the Veilliard-Baron procedure.[31]

Transversal polar order is quantified by the order parameter

$$\langle P_{1,c} \rangle = \langle \hat{\mathbf{w}}\cdot\hat{\mathbf{c}} \rangle, \tag{3.2}$$

which measures the average alignment of the $\hat{\mathbf{w}}$ axes of helices along a common direction ($\hat{\mathbf{c}}$). In the case of screw order the $\hat{\mathbf{c}}$ axis rotates in a helical way on moving along the $\hat{\mathbf{n}}$ director, with pitch $\mathscr{P}$ equal to the particle pitch p.[22,24]

Smectic order, with the formation of layers perpendicular to the director $\hat{\mathbf{n}}$, is monitored by using the order parameter

$$\langle \tau_1 \rangle = \left| \left\langle e^{i2\pi \frac{\mathbf{R}\cdot\hat{\mathbf{n}}}{d}} \right\rangle \right|, \tag{3.3}$$

where $\mathbf{R}$ is the position of the particle centre of mass and d the layer spacing. The most common smectic phases are those denoted as A (Sm_A) and B (Sm_B), which are both uniaxial and differ in their in-plane positional order, which is absent in the former, whereas in the latter there is short-range in-plane hexagonal order. This can be probed by using the hexagonal order parameter

$$\langle \psi_6 \rangle = \left\langle \frac{1}{N} \sum_{i=1}^{N} \left| \frac{1}{n(i)} \sum_{j=1}^{n(i)} e^{6i\theta_{ij}} \right| \right\rangle. \tag{3.4}$$

where $n(i)$ is the number of in-layer nearest-neighbors of the i-th helix, and θ_{ij} is the angle between the i, j intermolecular vector and a reference axis in the plane perpendicular to $\hat{\mathbf{n}}$, as illustrated in Fig. 2. Then $\langle \psi_6 \rangle$ takes on a non-zero value in the presence of short-range hexagonal order, and vanishes otherwise.

In ref. 24, new smectic phases were identified for helical-particle systems. Screw-smectic A ($Sm_{A,s}$) is a uniaxial smectic phase with screw order within the planes, quantified by a non-vanishing $\langle P_{1,c} \rangle$ value. The smectic B polar ($Sm_{B,p}$) phase has the same organization as the standard Sm_B phase with the additional feature of having azimuthal in-plane correlations, planes behaving independently from one another. This is called *polar* to highlight this specific correlation, and is unscrewed ($\langle P_{1,c} \rangle = 0$). Conversely, the screw-smectic B phase ($Sm^{*}_{B,s}$) is identified by a non vanishing value of the screw order parameter $\langle P_{1,c} \rangle$.

Table 1 summarizes the possible phases, as well as the corresponding set of order parameters that identify each of them.

We can finally complement and enrich the information provided by the above order parameters with specific correlation functions that are designed to highlight the various trends observed. The existence of the nematic director $\hat{\mathbf{n}}$ suggests the following decoupling of the vector $\mathbf{R}_{ij} = \mathbf{R}_j - \mathbf{R}_i$ in components parallel $\mathbf{R}^{\parallel}_{ij}$ and perpendicular $\mathbf{R}^{\perp}_{ij}$ to $\hat{\mathbf{n}}$

$$\mathbf{R}_{ij} = \mathbf{R}^{\parallel}_{ij} + \mathbf{R}^{\perp}_{ij} = (\mathbf{R}_{ij}\cdot\hat{\mathbf{n}})\hat{\mathbf{n}} + |\mathbf{R}_{ij} \times \hat{\mathbf{n}}|\hat{\mathbf{R}}^{\perp}_{ij}. \tag{3.5}$$

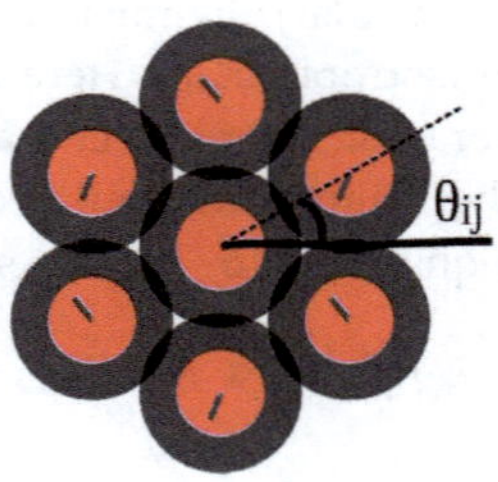

Fig. 2 Top view representation of short-range hexagonal ordering in helices. The dashed line joining the (i, j) nearest-neighbour helices makes an angle θ_{ij} with a reference axis (solid line). Helices are projected onto a plane perpendicular to the nematic director $\hat{\mathbf{n}}$. Red circles denote the radius of the helix, while the thick short line inside each circle indicates the orientation of the helix azimuthal direction $\hat{\mathbf{w}}$ associated with the C_2 symmetry axis.

Then, assuming L_z as the box length parallel to $\hat{\mathbf{n}}$ and L_x, L_y as the other two, we can introduce the parallel

$$g_{\parallel}(R_{\parallel}) = \frac{1}{N}\left\langle \frac{1}{\rho L_x L_y} \sum_{i=1}^{N} \sum_{j \neq i}^{N} \delta(R_{\parallel} - \mathbf{R}_{ij} \cdot \hat{\mathbf{n}}) \right\rangle \tag{3.6}$$

and perpendicular

$$g_{\perp}(R_{\perp}) = \frac{1}{2\pi R_{\perp} N}\left\langle \frac{1}{\rho L_z} \sum_{i=1}^{N} \sum_{j \neq i}^{N} \delta\left(R_{\perp} - \left|\mathbf{R}_{ij} \times \hat{\mathbf{n}}\right|\right) \right\rangle \tag{3.7}$$

positional correlation functions, that provide quantitative information on the spatial distribution of particles in the direction along $\hat{\mathbf{n}}$ and in planes perpendicular to it. Here, $\rho = N/V$ is the number density and $V = L_x L_y L_z$ is the box volume. The onset of screw ordering is signalled by an oscillatory behavior of a special correlation function

$$g_{1,\parallel}^{\hat{\mathbf{w}}}(R_{\parallel}) = \left\langle \frac{\sum_{i=1}^{N} \sum_{j \neq i}^{N} \delta(R_{\parallel} - \mathbf{R}_{ij} \cdot \hat{\mathbf{n}})(\hat{\mathbf{w}}_i \cdot \hat{\mathbf{w}}_j)}{\sum_{i=1}^{N} \sum_{j \neq i}^{N} \delta(R_{\parallel} - \mathbf{R}_{ij} \cdot \hat{\mathbf{n}})} \right\rangle. \tag{3.8}$$

that measures the azimuthal correlation between the $\hat{\mathbf{w}}$ axes of helices as a function of their distance along the $\hat{\mathbf{n}}$ director.

Typical behaviors of all the above quantities have been reported in ref. 22 and 24 and will not be reproduced here.

IV. Phase diagrams

In the following section we will present the calculated phase diagrams. In all cases, right-handed helical particles were studied. The same phase diagrams could be obtained for left-handed helices, with the only difference being that the

Table 1 Summary of the different phases identified in the MC simulations of hard helices, along with the corresponding non-vanishing order parameters

Phase	Code	Order parameter
Nematic	N	$\langle P_2 \rangle$
Screw-nematic	N_s^*	$\langle P_2 \rangle$, $\langle P_{1,c} \rangle^a$
Smectic A	Sm_A	$\langle \tau_1 \rangle$
Screw-smectic A	$Sm_{A,s}^*$	$\langle \tau_1 \rangle$, $\langle P_{1,c} \rangle^a$
Smectic B	Sm_B	$\langle \tau_1 \rangle$, $\langle \psi_6 \rangle$
Polar smectic B	$Sm_{B,p}$	$\langle \tau_1 \rangle$, $\langle \psi_6 \rangle$, $\langle P_{1,c} \rangle^b$
Screw-smectic B	$Sm_{B,s}^*$	$\langle \tau_1 \rangle$, $\langle \psi_6 \rangle$, $\langle P_{1,c} \rangle^a$

[a] $\hat{\mathbf{c}}$ rotates in helical way around $\hat{\mathbf{n}}$. [b] $\hat{\mathbf{c}}$ is defined for each layer, without inter-layer correlation.

handedness of the chiral phases would be reversed. Scaled quantities are used throughout, with the bead diameter D taken as the unit of length. Since the only interactions are hard-core repulsions, these systems are athermal and a single variable is sufficient to control the phase behaviour. Henceforth, the packing fraction η will be used, which is defined as $\eta = \rho v_0$, where v_0 is the volume of a helix.[21]

We start our analysis with the simplest cases of very slender helices with $r = 0.1$. Fig. 3 shows the phase diagram obtained for this case: irrespective of the particle pitch, on increasing density the sequence $\mathrm{I} \rightarrow \mathrm{N} \rightarrow \mathrm{Sm_A} \rightarrow \mathrm{Sm_B}$ is found, as for hard spherocylinders having the same aspect ratio, $L/D \sim 7.3$–7.4.[32] This example demonstrates that, if the helical character of particles is not sufficiently pronounced, which could be translated in terms of cavities that can be explored by neighboring particles, none of the distinct helix phases can be found. The only qualitative difference from the phase diagram of hard spherocylinders is in the low-density nematic phase, which here is cholesteric rather than uniform nematic. This is shown by the values of $\mathscr{P}$ obtained from DFT calculations (Fig. 4 and 5). Cholesteric phases are not special to helices. For all our right-handed helices with $r = 0.1$ a left-handed N* organization is predicted ($\mathscr{P} < 0$), with a pitch of the order of a few hundreds for helices with $p > 2$, which shows a dependence on density and tends to increase with decreasing p. A noticeable increase in the magnitude of $\mathscr{P}$, *i.e.* a decrease of the phase chirality, is found in the case of $p = 2$. This can be explained by the fact that in such helices the grooves are so tight that neighboring helices cannot easily lock into one another. This is a general behavior, which has already been pointed out:[23] hard helices with very low pitch p tend to form N* phases with large periodicity $\mathscr{P}$, whose magnitude and sign are particularly sensitive to the packing fraction.

The effects of the particle chirality on the phase diagram clearly appear when the helix radius increases to $r = 0.2$, as shown in Fig. 4. Considerable differences with the case $r = 0.1$ are visible with new phases, almost all of which are chiral. On increasing density, for all values of p, a transition from the isotropic to a cholesteric phase is found. This, according to DFT calculations, is left-handed, with a density-dependent $\mathscr{P}$ in the range 200–400 for $p > 2$, and about an order of magnitude longer for $p = 2$ (Fig. 4 and 5). These findings are not very different

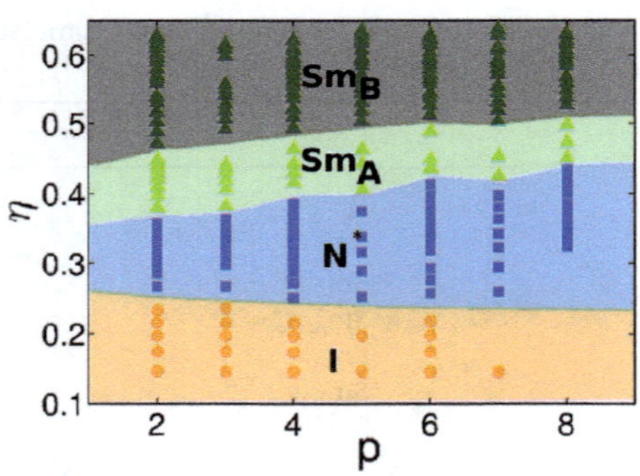

Fig. 3 Phase diagram in the plane packing fraction η *versus* particle pitch p for helices with radius $r = 0.1$.

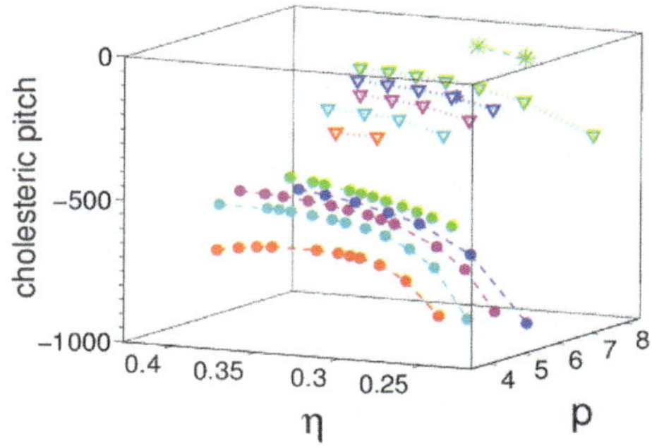

Fig. 4 Cholesteric pitch $\mathscr{P}$, calculated for helices with $p > 3$, in the N* phase of the phase diagrams in Fig. 3 ($r = 0.1$, full circles), Fig. 6 ($r = 0.2$, open triangles) and Fig. 9 ($r = 0.4$ asterisks). η is the packing fraction and p is the helix pitch: $p = 4$ (red), $p = 5$ (cyan), $p = 6$ (magenta), $p = 7$ (blue), $p = 8$ (green).

from those obtained for the case $r = 0.1$. However, the N* phase has now a narrower stability range, especially for small p, and is superseded, at higher density, by a screw-nematic phase N_s^* phase. This is characterized by a helical rotation of the C_2 symmetry axis of helices ($\hat{\mathbf{w}}$ vector in Fig. 1) around the main director $\hat{\mathbf{n}}$, which is the alignment direction of the helical axes $\hat{\mathbf{u}}$.

Unlike the cholesteric, the N_s^* phase has the same handedness (right) and the same pitch of the helical particles, essentially independent of the packing fraction. The physical mechanism of the transition to the N_s^* phase was clearly identified in ref. 22 and 24 as originating from a coupling between the rotation of each helix around its own main axis (the unit vector $\hat{\mathbf{u}}$) and a translation along the nematic director $\hat{\mathbf{n}}$. This coupling reduces the azimuthal freedom, and thus the rotational entropy, but such a loss is compensated by the increase of translational entropy allowed by the mutual sliding along $\hat{\mathbf{n}}$ of neighboring parallel helices.

For helices with $r = 0.2$ the screw-nematic N_s^* phase is followed, at higher density, by a screw-smectic A phase ($Sm_{A,s}^*$) which has screw order within layers. At

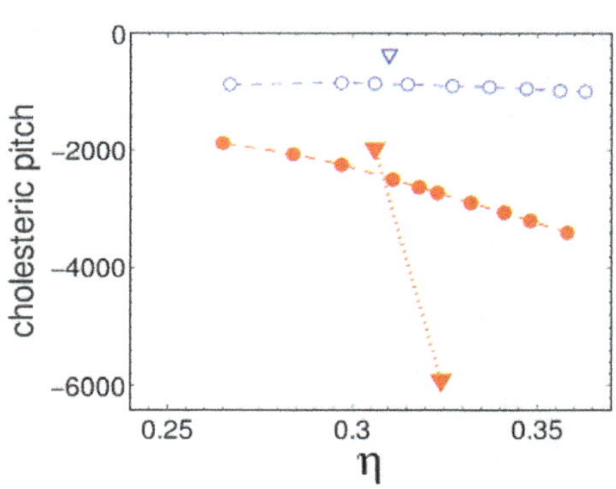

Fig. 5 Cholesteric pitch $\mathscr{P}$ as a function of the packing fraction η, calculated for state points in the N* phase of the phase diagrams in Fig. 3 ($r = 0.1$, circles) and Fig. 6 ($r = 0.2$, triangles), for helices with pitch $p = 2$ (full symbols) and $p = 3$ (open symbols).

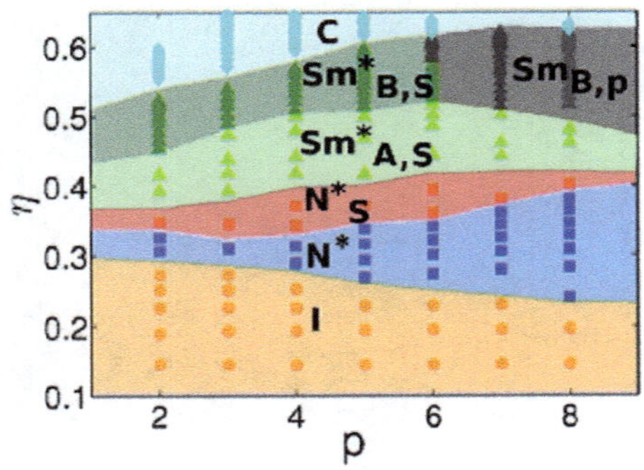

Fig. 6 Phase diagram in the plane packing fraction η *versus* particle pitch p for helices with radius $r = 0.2$.

even higher density two other smectic phases specific to helices appear, which have been denoted as screw-smectic B ($Sm^*_{B,s}$) and smectic B polar ($Sm_{B,p}$). The three smectic phases meet at a triple point located at $p \approx 6$ and $\eta \approx 0.5$.

To give a visual insight of the structure of the different phases, we show in Fig. 7 and 8 snapshots[33] taken from MC trajectories for helices with $r = 0.2$, $p = 4$ and $p = 8$, respectively. In the figures, color is coded according to the projection of the local tangent to helices onto a plane perpendicular to the director $\hat{\mathbf{n}}$. So, the stripes that can be seen in the N^*_s phase indicate the onset of azimuthal correlations between helices. The stripes persist in the $Sm^*_{A,s}$ phase, where there is the additional formation of layers. The regular spacing of the stripes indicates the presence of correlations between the layers: the helix $\hat{\mathbf{w}}$ axes rotate coherently in a screw-like way across the layers. This visual indication can be translated in quantitative terms by computing the periodicity of $g^{\hat{\mathbf{w}}}_{1,\parallel}(R_\parallel)$ given in eqn (3.8).

In the case of helices with $p = 4$ the correlation between layers is maintained in the subsequent $Sm^*_{B,s}$ phase (Fig. 7), which differs from $Sm^*_{A,s}$ due to the presence of short-range hexagonal ordering of the centers of mass of helices within the layers. Conversely, in the high density smectic phase of helices with $p = 8$ (Fig. 8), the intra-layer correlation is lost, in favour of the possibility for layers to freely slide and rotate with respect to one another, in order to achieve the optimal packing, still maintaining the in-plane short range hexagonal ordering of the centers of mass, as well as the azimuthal coupling of neighbouring helices. The resultant smectic B phase $Sm_{B,p}$ is named "B polar" to indicate the existence of in-plane hexagonal ordering (type B) and in-plane azimuthal coupling (polar).

At higher densities, a more compact configuration generically labelled "C" is found. Its characteristics are expected to be rather different from the crystal phases in the $r = 0.1$ case (not shown in Fig. 3), the latter probably being similar to those for hard spherocylinders. This is interesting in view of the possibility of forming crystals of specific shapes and chirality.

As the radius of the helix is further increased up to $r = 0.4$, the screw-nematic phase N^*_s widens its stability range at the expense of the N* phase, as illustrated in Fig. 7. This is clearly due to the increased curliness of the $r = 0.4$ case with respect to the $r = 0.2$ counterpart (Fig. 1), that favours interlock intrusion of neighboring helices, thus promoting the azimuthal correlations responsible for the onset of

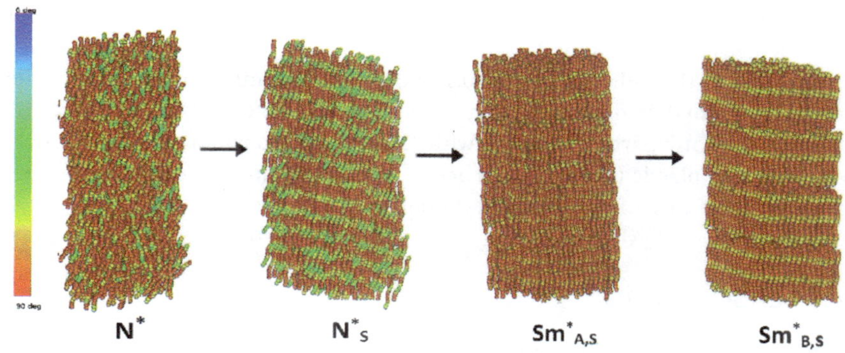

Fig. 7 Snapshots from MC trajectories in the different liquid-crystalline phases for helices with $r = 0.2$ and $p = 4$. Color is coded according to the projection of the local tangent to helices onto a plane perpendicular to the director $\hat{n}$.[33]

the screw-nematic phase. The disappearance of the cholesteric phase at sufficiently small particle pitch, $p < 3$, is a consequence of the small aspect ratio and reflects the lack of a nematic phase for short hard spherocylinders.[32] More interesting is the disappearance of the smectic B polar phase $\mathrm{Sm_{B,p}}$ in favour of the screw-smectic B phase $\mathrm{Sm^*_{B,s}}$, for any particle pitch above a certain packing fraction. This too can be rationalized in terms of the increased curliness of the helices that favours the screw order.

For helices with $p = 7$ and $p = 8$, a narrow cholesteric phase is predicted by DFT, which is left-handed as for the helices $r = 0.1$ and $r = 0.2$, but has a much tighter pitch. This is another consequence of the pronounced curliness of particles, which means high microscopic chirality. The predicted pitch $\mathscr{P}$, of the order of one hundred, is about 20 times the length scale of the particle chirality, thus much lower than typical ratios measured in cholesteric phases formed by molecular species, whether of low- or high-molar mass. Indeed, the high degree of microscopic chirality featured by helices with $r = 0.4$ seems out of reach for

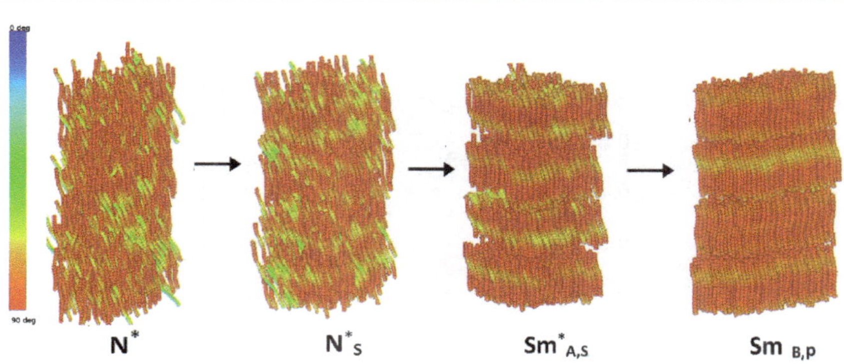

Fig. 8 Snapshots from MC trajectories in the different liquid-crystalline phases for helices with $r = 0.2$ and $p = 8$. Color is coded according to the projection of the local tangent to helices onto a plane perpendicular to the director $\hat{n}$.[33]

molecular systems, also because of their intrinsic flexibility, which has the effect of reducing the net chirality.[34]

Similarly to the helices with $r = 0.1$ and $r = 0.2$ examined above, also in the case $r = 0.4$ higher density configurations are characterized by compact "C" structures, at any particle pitch p, with the notable exceptions of low particle pitches where plastic (rotor) phases are to be expected in view of the low aspect ratio, as in the case of hard spherocylinders.[32]

V. Discussion

In molecular systems, well-defined and rigid helices like our model particles seem challenging and the phase behaviour can be affected by molecular flexibility.[35] If we would like to compare our predictions with experimental systems, then the best candidates appear to be colloidal suspensions of helical rigid particles. Indeed, screw-nematic order has been found in concentrated suspensions of helical flagella isolated from *Salmonella typhimurium*.[20] While flagella with a rodlike shape exhibit a nematic phase, filaments with a pronounced helical character have been found to undergo a direct transition from the isotropic to the N_s^* phase. The latter is the behaviour shown in Fig. 9 for most of the helices with $r = 0.4$. The experimental system did not exhibit any smectic phase, likely due to length polydispersity, as also suggested in the original work. Length polydispersity is expected to further promote the stability of the N_s^* phase against the N* phase, as helices of different lengths with the same p can all fit in the same N_s^*. While potentially beneficial to the stability of the N_s^* phase, length polydispersity is not essential to its formation, as our results on monodisperse systems show. This is further supported by recent results of a parallel study on monodisperse hard helix systems without employing PBC[36] that confirm the stabilisation of the N_s^* phase over the N* phase at high density and the absence of the latter in favour of the former for very curly helices, contrary to what was alluded elsewhere.[37]

The features of the N_s^* phase found in our simulations and the fact that this is the same phase described in ref. 20 were discussed in detail in ref. 22 and 24. We call this phase screw-nematic in line with ref. 25. This was denoted as *conical* in

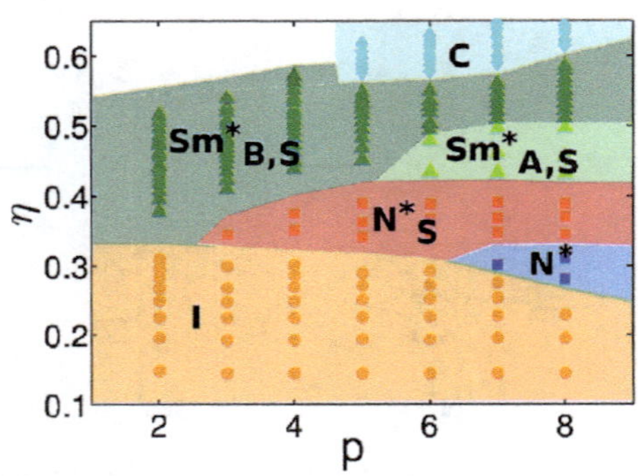

Fig. 9 Phase diagram in the plane packing fraction η *versus* particle pitch p for helices with radius $r = 0.4$.

ref. 20, to account for the conical arrangement of the local director, identified with the local tangent to the helical particles, as detected by polarization microscopy. The experimental observations are fully consistent with our description of screw order in the N_s^* phase, with the helix axes ($\hat{\mathbf{u}}$) preferentially aligned along the same direction (the main director $\hat{\mathbf{n}}$) and the two-fold symmetry axes of helices $\hat{\mathbf{w}}$ spiralling around $\hat{\mathbf{n}}$. We continue with "screw-nematic" though, to distinguish this phase from the *conical* or *heliconical* state induced in cholesterics by an applied electric or magnetic field parallel to the twist axis. In such a state, predicted almost 50 years ago[38] and observed only recently,[39] the director $\hat{\mathbf{n}}$ performs a helical rotation around an axis at an oblique angle, whence the name heliconical. Unlike the N_s^* phase, the heliconical state does not require the helical shape of constituents; moreover, the periodicity of the heliconical state is similar to that of the underlying N* phase, *i.e.* orders of magnitude longer than the molecular size, and can be controlled by the applied field.[39]

Related but still different is the *twist-bend* nematic (N_{TB}) phase, which was predicted for bent mesogens by theory[40] and simulations,[41] and then experimentally observed only a few years ago.[42–44] Like the N_s^*, the N_{TB} phase has a small pitch helical modulation ($\mathscr{P}$ of the order of few molecular lengths), but it does not require helical, nor even chiral constituents.

VI. Conclusions and perspectives

The aim of the present study was to obtain a detailed account of the phase behaviour of hard helices in terms of their morphology. For the model systems shown in Fig. 1 that resemble helical tubes of diameter D, the control parameters are the scaled radius r/D and pitch p/D. Using MC simulations, supplemented by DFT-Onsager second-virial calculations, we obtained phase diagrams as a function of these parameters, thus providing a detailed picture of the structures that are formed as density increases.

The phase diagrams are rich and exhibit a significant variety, even at a qualitative level, as the concavity of particles, which brings about the possibility of azimuthal correlations, becomes more pronounced. This is remarkable, especially in view of the absence of attractive interactions. Since only hard-core repulsions have been assumed, all effects originate solely from changes in the morphology of the helices and the screw phases, not found in the corresponding hard rod counterpart, are entropically driven.

For very slender helices, we find the sequence is $I \rightarrow N^* \rightarrow Sm_A \rightarrow Sm_B$ as in the phase diagram of hard spherocylinders,[32] with the only difference being that the uniform nematic phase is replaced here by the chiral cholesteric phase, due to the chiral nature of the particles. This shows that the helical character of the particles has to be pronounced enough to promote the onset of phase organizations specific to helices, which are characterized by azimuthal correlations between particles. From the thermodynamic point of view, this means that only if the helical character is sufficiently pronounced, will the entropy decrease due to the additional orientational order existing in such helix-specific phases be compensated by the increase in translational entropy allowed by the roto-translational coupling. Based on these results, helical (bio) polymers, which in most cases can be assimilated to very slender helices, do not appear suitable to form helix-specific phases. This might explain why no experimental evidence of such phases have been given thus far for polymeric systems. However, the possibility that additional interactions (*e.g.* electrostatic) might enhance the helical character of real systems cannot be excluded. Note also that the detection of screw distinct order in helical polymeric systems is made difficult by the very small characteristic length scale that is of the order of the particle pitch.

Upon increasing the radius of helices from $r = 0.1$ to $r = 0.2$, a richer polymorphism is found and new phases appear. For $p \leq 6$, we find the sequence I $\rightarrow$ N* $\rightarrow$ N_s^* $\rightarrow$ $Sm_{A,s}^*$ $\rightarrow$ $Sm_{B,s}^*$, whereas for $6 < p \leq 8$ $Sm_{B,s}^*$ is replaced by a $Sm_{B,p}$ phase, indicating the existence of a triple point between the $Sm_{A,s}^*$, $Sm_{B,s}^*$, and $Sm_{B,p}$ phases. This trend is continued at $r = 0.4$, with the direct sequence I $\rightarrow$ $Sm_{B,s}^*$ for $p \leq 3$, I $\rightarrow$ N_s^* $\rightarrow$ $Sm_{B,s}^*$ for $3 < p \leq 5$, I $\rightarrow$ N_s^* $\rightarrow$ $Sm_{A,s}^*$ $\rightarrow$ $Sm_{B,s}^*$ for $5 < p \leq 6$, and finally I $\rightarrow$ N* $\rightarrow$ N_s^* $\rightarrow$ $Sm_{A,s}^*$ $\rightarrow$ $Sm_{B,s}^*$ above that value, with several triple points in between. This very interesting phase behaviour requires highly helical particles and seems difficult to achieve in molecular systems. However, colloidal suspensions of natural or artificial particles would be very good candidates to verify the theoretical predictions. We hope that our results can stimulate experimental work in this direction.

In all cases that we have examined here, right-handed helices were found to form a left-handed cholesteric phase and right-handed phases, either nematic or smectic, with screw ordering. In ref. 37 examples of the N* phase were reported, including a few cases examined here, exhibiting a change of handedness from left to right with increasing density. However that study was limited to the cholesteric phase and did not account for the existence of the N_s^* phase nor any positionally ordered phase, either by DFT[45] or numerical simulation. Consideration of the whole phase diagram suggests that the change in N* handedness, although in principle possible, is fairly unlikely because of the relatively small range of stability of the N* phase. For very curly helices, on increasing density, the N* phase is superseded by the N_s^* phase before the inversion of cholesteric handedness has a chance to take place. Such inversion is more likely to be observed for large pitch cholesteric phases formed by slightly curly helices,[23] provided the density is sufficiently low that positionally ordered phases do not compete.

One interesting issue that originates from the present study is the analysis of the possible (if any) crystal phases of hard rigid helices, that is a detailed characterization of the part of the phase diagram that we have here generically labelled as "C". In general, given a hard non-spherical particle model, it is not straightforward to predict the structure of the crystal phases. This holds particularly true if the particles are chiral. In a very recent study,[47] this problem has been addressed by using an approach for shape design, where chiral entropic forces were tuned by either shape rounding or by an increase in depletant concentration. The use of helices where chiral entropic forces can be controlled in a very natural way by the particle morphology, may represent a more efficient way to control the onset of a chiral crystal structure from the particle chirality. We plan to address this issue in future work, possibly extending the methodology implemented in ref. 24 and combining it with insights stemming from ref. 47. Equally of interest would be to address the more general question of the dense, ordered or disordered, packings of hard helices.

Acknowledgements

G. C. thanks the Government of Spain for the award of a Ramón y Cajal research fellowship and the financial support under the grant FIS2013-47350-C5-1-R. This work was also supported by MIUR PRIN-COFIN2010-2011 (contract 2010LKE4CC). The use of the SCSCF multiprocessor cluster at the Università Ca' Foscari Venezia is greatfully acknowledged.

References

1 G. M. Whitesides and M. Boncheva, *Proc. Natl. Acad. Sci. U. S. A.*, 2002, **99**, 4769; G. M. Whitesides and B. Grzybowski, *Science*, 2002, **295**, 2418.
2 P. F. Damasceno, M. Engel and S. Glotzer, *Science*, 2012, **337**, 453.
3 S. Sacanna, M. Korpics, K. Rodriguez, L. Colón-Meléndez, S. Kim, D. Pine and G. Yi, *Nat. Commun.*, 2013, **4**, 1688.
4 S. Sacanna, D. J. Pine and G.-R. Yi, *Soft Matter*, 2013, **9**, 8096.
5 S. C. Glotzer and M. J. Solomon, *Nat. Mater.*, 2007, **6**, 557.
6 L. Cademartiri, K. J. M. Bishop, P. W. Snyder and G. A. Ozin, *Philos. Trans. R. Soc., A*, 2012, **370**, 2824.
7 C. M. Soukoulis and M. Wegener, *Nat. Photonics*, 2011, **5**, 523.
8 A. Guerrero-Martínez, J. L. Alonso-Gómez, B. Auguié, M. M. Cid and L. M. Liz-Marzán, *Nano Today*, 2011, **6**, 381.
9 A. Baev, P. N. Prasad, H. Ågren, M. Samoc and M. Wegener, *Phys. Rep.*, 2015, **594**, 1.
10 A. J. C. Bose, *Proc. R. Soc. London, Ser. A*, 1898, **63**, 146.
11 K. F. Lindman, *Ann. Phys.*, 1920, **63**, 621.
12 K. J. Gansel, M. Thiel, M. S. Rill, M. Decker, K. Bade, V. Saile, G. von Freymann, S. Linden and M. Wegener, *Science*, 2009, **325**, 1513.
13 G. K. Larsen, Y. He, J. Wang and Y. Zhao, *Adv. Opt. Mater.*, 2014, **2**, 245.
14 M. Esposito, V. Tasco, M. Cuscuná, F. Todisco, A. Benedetti, I. Tarantini, M. de Giorgi, D. Sanvitto and A. Passaseo, *ACS Photonics*, 2015, **2**, 105.
15 M. Yang and N. A. Kotov, *J. Mater. Chem.*, 2011, **21**, 6775.
16 J. Gibbs, A. G. Mark, T.-C. Lee, S. Eslami, D. Schamel and P. Fischer, *Nanoscale*, 2014, **6**, 9457.
17 L. Liu, L. Zhang, S. M. Kim and S. Park, *Nanoscale*, 2014, **6**, 9355.
18 F. Livolant and A. Leforestier, *Prog. Polym. Sci.*, 1996, **21**, 1115.
19 D. C. Wright and N. D. Mermin, *Rev. Mod. Phys.*, 1989, **61**, 385.
20 E. Barry, Z. Hensel, Z. Dogic, M. Shribak and R. Oldenbourg, *Phys. Rev. Lett.*, 2006, **96**, 018305.
21 E. Frezza, A. Ferrarini, H. B. Kolli, A. Giacometti and G. Cinacchi, *J. Chem. Phys.*, 2013, **138**, 164906.
22 H. B. Kolli, E. Frezza, G. Cinacchi, A. Ferrarini, A. Giacometti and T. S. Hudson, *J. Chem. Phys.*, 2014, **140**, 081101.
23 E. Frezza, A. Ferrarini, H. B. Kolli, A. Giacometti and G. Cinacchi, *Phys. Chem. Chem. Phys.*, 2014, **16**, 16225.
24 H. B. Kolli, E. Frezza, G. Cinacchi, A. Ferrarini, A. Giacometti, T. S. Hudson, C. De Michele and F. Sciortino, *Soft Matter*, 2014, **10**, 8171.
25 F. Manna, V. Lorman, R. Podgornik and B. Zeks, *Phys. Rev. E: Stat., Nonlinear, Soft Matter Phys.*, 2007, **75**, 030901.

26 The spontaneous formation of a cholesteric phase in a system subject to PBC would be possible provided that the size of the system investigated is large enough. Unfortunately, this cannot usually be the case if *realistic* chiral interactions are considered. However, by making the particles interact with strong chiral interactions, a cholesteric phase was observed in a Monte Carlo numerical simulation study employing PBC.[27]

27 R. Memmer, *J. Chem. Phys.*, 2001, **114**, 8210.
28 L. Onsager, *Ann. N. Y. Acad. Sci.*, 1949, **51**, 627; J. P. Straley, *Phys. Rev. A: At., Mol., Opt. Phys.*, 1976, **14**, 1835.
29 J. D. Parsons, *Phys. Rev. A: At., Mol., Opt. Phys.*, 1979, **19**, 1225; S. D. Lee, *J. Chem. Phys.*, 1987, **87**, 4972.
30 F. Tombolato and A. Ferrarini, *J. Chem. Phys.*, 2005, **122**, 054908.
31 J. Veilliard-Baron, *Mol. Phys.*, 1974, **28**, 809.
32 P. Bolhuis and D. Frenkel, *J. Chem. Phys.*, 1997, **106**, 666.
33 A. Gabriel, T. Meyer and G. Germano, *J. Chem. Theory Comput.*, 2008, **4**, 468.
34 N. Katsonis, E. Lacaze and A. Ferrarini, *J. Mater. Chem.*, 2012, **22**, 7088.
35 G. Cinacchi and L. de Gaetani, *Phys. Rev. E: Stat., Nonlinear, Soft Matter Phys.*, 2008, **77**, 051705.
36 G. Cinacchi, A. Ferrarini, A. Giacometti, C. Greco, H. B. Kolli and A. Tani, unpublished, 2015.
37 S. Dussi, S. Belli, R. van Roij and M. Dijkstra, *J. Chem. Phys.*, 2015, **142**, 074905.
38 R. B. Meyer, *Appl. Phys. Lett.*, 1968, **12**, 281; R. B. Meyer, *Appl. Phys. Lett.*, 1969, **14**, 208; P. G. de Gennes, *Solid State Commun.*, 1968, **6**, 163.
39 J. Xiang, S. V. Shiyanovskii, C. Imrie and O. D. Lavrentovich, *Phys. Rev. Lett.*, 2014, **6**, 217801.
40 I. Dozov, *Europhys. Lett.*, 2001, **56**, 247.
41 R. Memmer, *Liq. Cryst.*, 2002, **29**, 483.
42 M. Cestari, S. Diez-Berart, D. A. Dunmur, A. Ferrarini, M. R. de la Fuente, D. J. B. Jackson, D. O. Lopez, G. R. Luckhurst, M. A. Perez-Jubindo, R. M. Richardson, J. Salud, B. A. Timimi and H. Zimmermann, *Phys. Rev. E: Stat., Nonlinear, Soft Matter Phys.*, 2011, **84**, 031704.
43 D. Chen, J. H. Porada, J. B. Hooper, A. Klittnick, Y. Shen, M. R. Tuchband, E. Korblova, D. Bedrov, D. M. Walba, M. A. Glaser, J. E. Maclennan and N. A. Clark, *Proc. Natl. Acad. Sci. U. S. A.*, 2013, **110**, 15931.
44 V. Borshch, Y.-K. Kim, J. Xiang, M. Gao, A. Jkli, V. P. Panov, J. K. Vij, C. T. Imrie, M. G. Tamba, G. H. Mehl and O. D. Lavrentovich, *Nat. Commun.*, 2013, **4**, 2635.
45 Note that a study of positionally ordered liquid-crystal phases, like smectic and columnar, can also be addressed by similar DFTs, *e.g.* in ref. 46.
46 G. Cinacchi, L. Mederos and E. Velasco, *J. Chem. Phys.*, 2004, **121**, 3854.
47 P. F. Damasceno, A. S. Karas, B. A. Schultz, M. Engel and S. C. Glotzer, *Phys. Rev. Lett.*, 2015, **115**, 158303.

Faraday Discussions

PAPER

Enhancement of nucleation of protein crystals on nano-wrinkled surfaces

Praveen K. Bommineni and Sudeep N. Punnathanam*

Received 12th August 2015, Accepted 21st October 2015
DOI: 10.1039/c5fd00119f

The synthesis of high quality protein crystals is essential for determining their structure. Hence the development of strategies to facilitate the nucleation of protein crystals is of prime importance. Recently, Ghatak and Ghatak [*Langmuir* 2013, **29**, 4373] reported heterogeneous nucleation of protein crystals on nano-wrinkled surfaces. Through a series of experiments on different proteins, they were able to obtain high quality protein crystals even at low protein concentrations and sometimes without the addition of a precipitant. In this study, the mechanism of protein crystal nucleation on nano-wrinkled surfaces is studied through Monte Carlo simulations. The wrinkled surface is modeled by a sinusoidal surface. Free-energy barriers for heterogeneous crystal nucleation on flat and wrinkled surfaces are computed and compared. The study reveals that the enhancement of nucleation is closely related to the two step nucleation process seen during protein crystallization. There is an enhancement of protein concentration near the trough of the sinusoidal surface which aids in nucleation. However, the high curvature at the trough acts as a deterrent to crystal nucleus formation. Hence, significant lowering of the free-energy barrier is seen only if the increase in the protein concentration at the trough is very high.

1 Introduction

Proteins are in general hard to crystallize. Since the determination of protein structure is essential for many biotechnology applications such as drug design, a lot of effort has gone into discovering conditions and developing methods for facilitating protein crystallization. In 1997, ten Wolde and Frenkel[1] showed that under certain conditions, the barrier to crystal nucleation can be lowered substantially. Protein molecules, especially those of globular proteins, have a range of attraction that is typically much smaller than the size of the molecule. As a result, in such systems, the vapor–liquid coexistence curve becomes metastable and lies inside the two-phase solid–fluid region of the phase diagram. The study by ten Wolde and Frenkel[1] showed that if protein crystallization is carried out at temperatures close to the metastable critical point, the free-energy barrier

Department of Chemical Engineering, Indian Institute of Science, Bangalore – 560012, Karnataka, India. E-mail: sudeep@chemeng.iisc.ernet.in

for homogeneous nucleation is lower than elsewhere, even though the classical nucleation theory predicts the same barrier. The explanation for this lowering of the free-energy barrier is as follows. Close to the metastable critical point, nucleation follows a two-step mechanism. In the first step, large density fluctuations due to the proximity to the critical point result in spontaneous formation of dense liquid like clusters. The work of formation of these high density clusters is very small. In the next step, the formation of the protein nucleus takes place inside such high density clusters. Thus, the overall free energy barrier to nucleation gets reduced substantially.

Typically, heterogeneous nucleation is faster and more prevalent than homogeneous nucleation. Hence, there has been an ongoing search for a "universal nucleant"[2] that can facilitate formation of protein crystals. Initial attempts have been to use epitaxial nucleation through the introduction of mineral surfaces.[3,4] This requires a match between the lattice of the substrate and the nucleating protein crystal. Such interactions by their very nature are highly specific. Non-specific interactions have been introduced through the use of charged substrates.[5,6] Since proteins in solution have charges, such substrates can attract the protein molecules causing an enhancement in the concentration near the surface leading to nucleation.

An important discovery towards the development of "universal nucleants" for facilitating protein nucleation was the use of porous substrates by Chayen and co-workers.[7–9] They were able to nucleate seven kinds of proteins when their solutions were seeded with porous substrates. They also observed that if the pore sizes were unsuitable, *i.e.*, very small compared with protein diameters, no nucleation was observed. This led them to conclude that confined spaces, such as those found inside the pores, induced protein nucleation. To understand the role of confined spaces, simulations of protein nucleation have been performed by Sear and co-workers.[10,11] Page and Sear[10] studied nucleation of a two-dimensional Ising model inside the pores. From their study they concluded that the nucleation rate is orders of magnitude higher at the pores than at flat surfaces. The rate is also dependent on the size of the pore and there exists an optimal size for the maximum nucleation rate. Simulations of the crystal nucleation of globular proteins inside nanoscale pits were carried out by van Meel *et al.*[11] The simulations were carried out at temperatures slightly lower than the metastable critical point and at densities close to the saturated metastable vapor. Under these conditions, the simulations revealed that the nanoscale pits get filled with protein molecules in a phenomenon similar to capillary condensation. Nucleation then takes place inside these nanoscale pits where the high concentration of protein molecules facilitates a lower free energy barrier to protein crystal nucleation. The use of porous substrates facilitated the spontaneous formation of high density protein clusters not seen in homogeneous nucleation, thereby increasing the range of conditions under which protein crystallization can be performed.

Recently, Ghatak and Ghatak[12,13] crystallized several proteins on a nano-wrinkled PDMS surface. They observed that the presence of nano-wrinkles on the surface enhanced the nucleation rate of these proteins. In the absence of nano-wrinkles, precipitants were required to crystallize proteins. However, the presence of nano-wrinkles on the surface caused the crystals to nucleate at very low protein concentrations; sometimes even without the addition of any precipitant. They also concluded that both the large curvature and large spread in curvature on the surface diminished the free-energy barrier for nucleation. The effect of curvature of the surface on heterogeneous nucleation was studied through simulations by Cacciuto *et al.*[14] They simulated heterogeneous nucleation of hard spheres on curved surfaces. Crystallization was found to take place more readily on convex surfaces than concave surfaces. Also curvature on the surface tended to cause strain in the growing nuclei which could cause it to detach from the surface.

In this study, we perform Monte Carlo molecular simulations of heterogeneous crystal nucleation of globular proteins on nano-wrinkled surfaces in order to uncover the mechanism behind the increase in the nucleation rates. The system chosen is a simple model system that mimics the phase behavior of globular proteins such as lysozyme.

2 Model

The protein molecules are modeled by a spherically symmetric potential, $u(r)$, which is the same as the one used by ten Wolde and Frenkel.[1]

$$u(r) = \begin{cases} u^*(r) - u^*(r_c) & r \le r_c \\ 0 & r > r_c \end{cases} \tag{1}$$

$$u^*(r) = \begin{cases} \infty & r < \sigma \\ \dfrac{4\varepsilon}{\alpha^2}\left[\dfrac{1}{\left[\left(\frac{r}{\sigma}\right)^2 - 1\right]^6} - \dfrac{1}{\left[\left(\frac{r}{\sigma}\right)^2 - 1\right]^3}\right] & r \ge \sigma \end{cases} \tag{2}$$

where σ is the hard-core diameter of the particles, r is the interparticle distance, ε is the well depth and r_c is the cut-off distance at which the potential is truncated and shifted. The value of α is kept equal to 50 and that of r_c is equal to 2.0. A schematic of the potential is shown in Fig. 1.

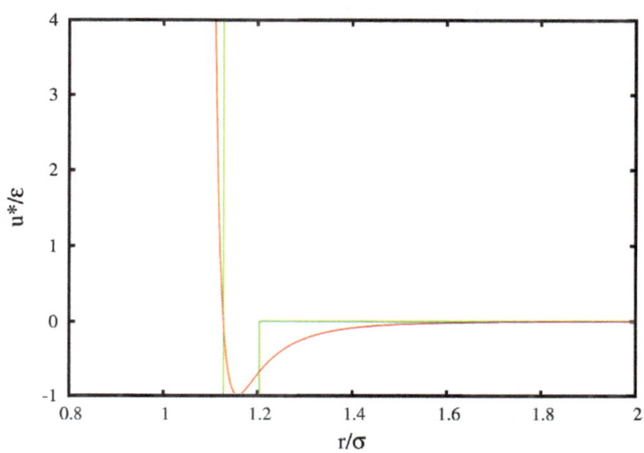

Fig. 1 Schematic of the inter-molecular potential between two protein molecules (red line). Also shown is a square well potential (green line) that has the same 2nd virial coefficient as the original potential at $T/T_c = 0.89$.

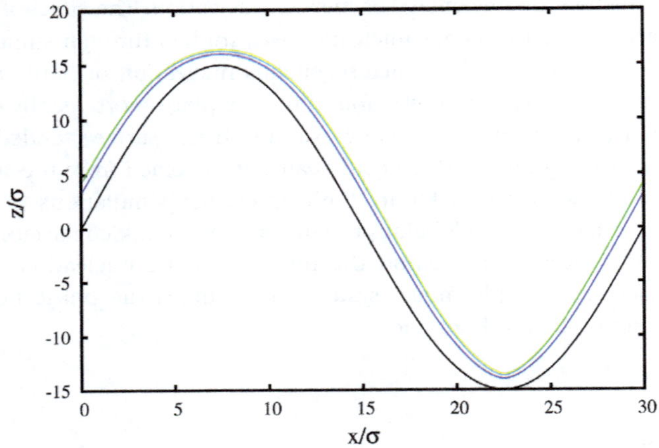

Fig. 2 Schematic of the wrinkled surface model by a sinusoidal curve having a wavelength of 30σ and amplitude of 15σ (black line). Also shown are the loci of points that are at a distance of σ (blue line) and $(1 + \lambda)\sigma$ (green line) from the surface. A value of $\lambda = 0.3$ is used in this schematic.

The nano-wrinkled surface is modeled by a single sinusoidal wave. The interaction between the protein molecule and the surface is modeled by a square-well potential.

$$u(d) = \begin{cases} \infty & d < \sigma \\ -\varepsilon & \sigma \leq d < (1+\lambda)\sigma \\ 0 & d \geq (1+\lambda)\sigma \end{cases} \quad (3)$$

where $\lambda\sigma$ is the width of the square well and d is the distance from the particle to the surface along the direction which is perpendicular to it at the point of contact (see Fig. 2). In this study, the surface is modeled by a sine wave of wavelength 30σ and amplitude 15σ. These values are chosen to approximately match the ratio of the dimensions of the steepest wrinkled substrate used in Ghatak and Ghatak[13] to

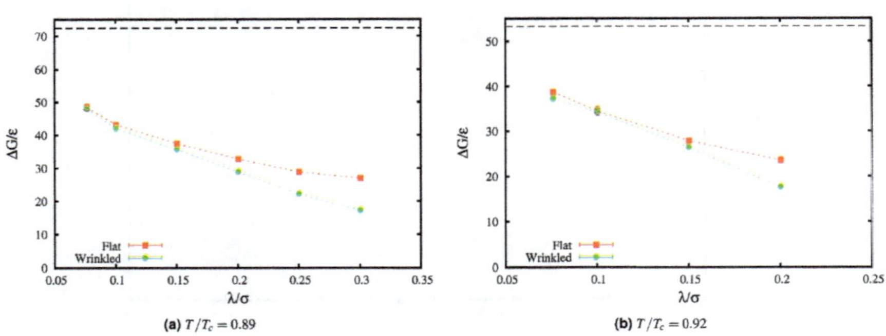

Fig. 3 Variation of the free-energy barrier for heterogeneous nucleation on flat and wrinkled surfaces as a function of the width of the square well interaction. The black dashed line denotes the free-energy barrier for homogeneous nucleation. The size of the error bars equals one standard deviation.

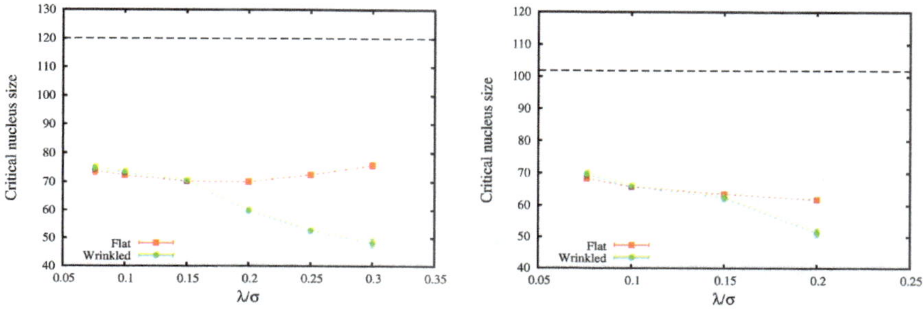

(a) $T/T_c = 0.89$ (b) $T/T_c = 0.92$

Fig. 4 Variation of size of the critical nucleus for heterogeneous nucleation on flat and wrinkled surfaces as a function of width of the square well interaction. The black dashed line denotes the size of the critical nucleus for homogeneous nucleation. The size of the error bars equals one standard deviation.

the diameter of the lysozyme protein. In order to compute the molecule–substrate interaction, the loci of points at a distance of σ and $(1 + \lambda)\sigma$ from the surface were determined at the start of the simulation and fitted to a Fourier series. During the simulations, the Fourier series was used to determine the square-well interaction of the molecule with the wrinkled substrate.

3 Simulation of crystal nucleation

The probability of observing the formation of the critical nucleus in a normal Monte Carlo simulation is vanishingly small. Hence, the simulations are biased with the umbrella sampling technique of Torrie and Valleau.[15] The order parameters used to define a crystal nucleus are described below.

3.1 Order parameters

The bond order parameters as defined by Steinhardt *et al.*[16] are used to identify the critical nucleus. The procedure is briefly described as follows. For every particle i, a $2l + 1$ dimensional complex vector $\mathbf{q}_l(i)$ with components is defined as

Table 1 Average value of q_6 for the critical clusters at various conditions. The value of q_6 for an FCC crystal is 0.575

	$T/T_c = 0.89$		$T/T_c = 0.92$	
λ	Wrinkled	Flat	Wrinkled	Flat
0.076	0.562(3)	0.562(2)	0.572(1)	0.565(4)
0.10	0.562(3)	0.561(2)	0.575(3)	0.562(2)
0.15	0.560(3)	0.562(2)	0.575(4)	0.563(4)
0.20	0.568(1)	0.561(1)	0.576(3)	0.562(3)
0.25	0.566(2)	0.561(1)		
0.30	0.568(4)	0.561(2)		

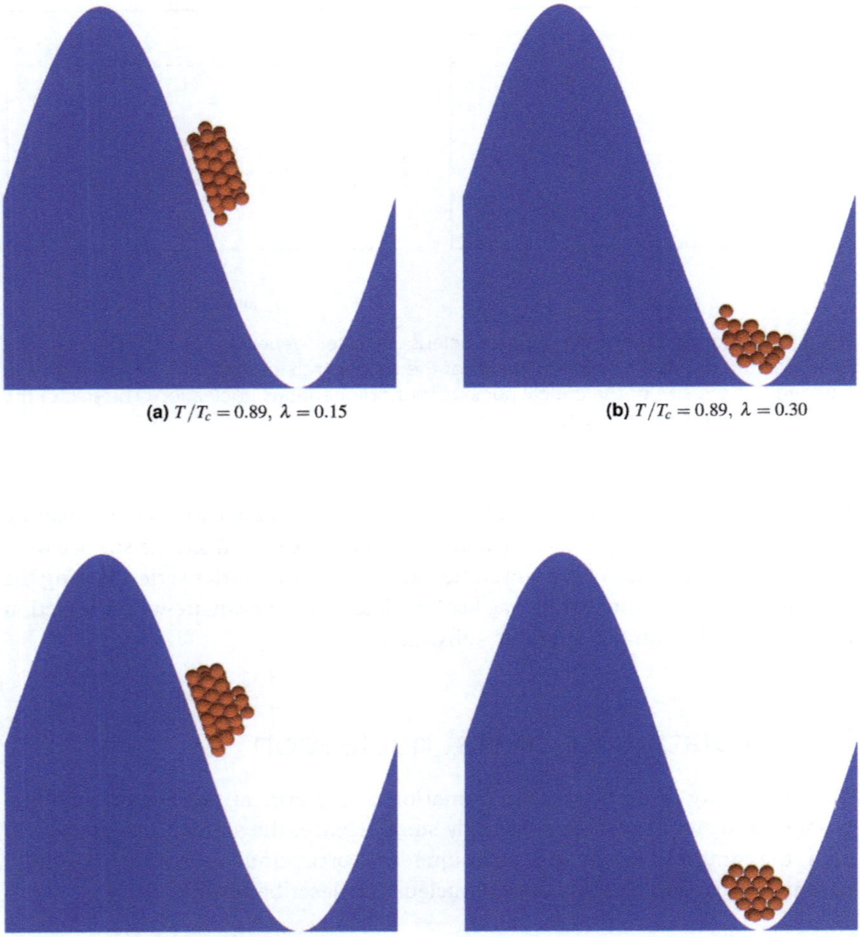

(a) $T/T_c = 0.89$, $\lambda = 0.15$ (b) $T/T_c = 0.89$, $\lambda = 0.30$

(c) $T/T_c = 0.92$, $\lambda = 0.10$ (d) $T/T_c = 0.92$, $\lambda = 0.20$

Fig. 5 Snapshots showing the location of the critical nucleus on the wrinkled surface. The protein molecules in the vapor phase are not shown for clarity.

$$q_{lm}(i) = \frac{1}{N_b(i)} \sum_{j=1}^{N_b(i)} Y_{lm}(\hat{\mathbf{r}}_{ij}), \tag{4}$$

where N_b is the number of neighboring particles and Y_{lm} is the spherical harmonics for the normalized direction vector $\hat{\mathbf{r}}_{ij}$ between particles i and j. Any other particle j is defined as a neighbor if it lies within a given cutoff radius r_c. For the calculations in this paper, values of $r_c = 1.35\sigma$ and $l = 6$ are used. In a crystalline environment, these vectors are aligned with each other but in a liquid-like environment, they are randomly oriented. To obtain a measure of the alignment, we calculate the correlation between the vectors $\mathbf{q}_6$ of neighboring particles i and j,

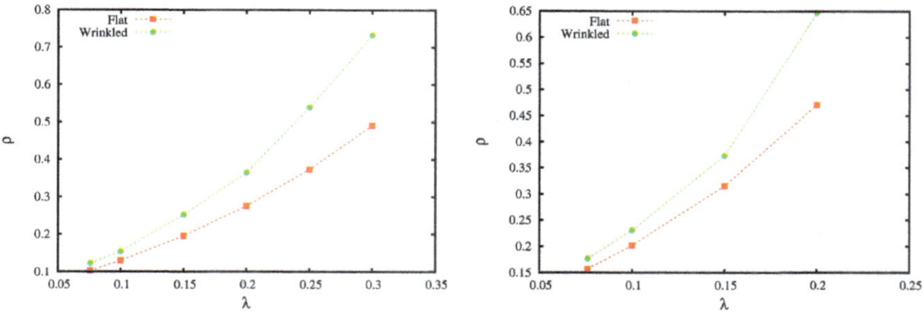

(a) $T/T_c = 0.89$ (b) $T/T_c = 0.92$

Fig. 6 Variation of fluid density near the flat and wrinkled surfaces as a function of λ. In the case of wrinkled surfaces, the values denote density at the trough.

$$\mathbf{q}_6(i)\cdot\mathbf{q}_6(j) = \sum_{m=-6}^{6} \tilde{q}_{6m}(i)\tilde{q}_{6m}^{*}(j), \tag{5}$$

where

$$\tilde{q}_{6m}(i) = \frac{q_{6m}(i)}{|q_{6m}(i)|}. \tag{6}$$

The asterisk in eqn (5) denotes the complex conjugate. Eqn (5) is a dot product between two normalized vectors and gives the cosine of the angle between them. A value of 0.55 is chosen to determine whether two neighboring particles are aligned or not. Particles are classified as "solid-like" if the ratio of the number of aligned neighbors to the total number of neighbors is greater than 0.5 and "liquid-like" otherwise. All the "solid-like" particles within a distance of r_c are determined to belong to the same cluster. The reaction coordinate for tracking the progress of nucleation is the number of particles belonging to the largest cluster.

3.2 Biasing potential

The simulations are biased with an external potential such that the sampling is weighted toward configurations that contain crystal nuclei. Instead of using a single biasing potential to sample all the relevant crystal nuclei sizes, it is more efficient to perform a series of simulations with different biasing potentials that sample a certain window of nuclei sizes. The biasing potentials are chosen such that the windows overlap with one another. Similar to previous studies[17–19] using umbrella sampling for crystal nucleation, the form of the biasing potential is

$$\omega(\mathbf{r}^N) = \frac{k}{2}\left[n(\mathbf{r}^N) - n_0\right]^2 \tag{7}$$

where $n(\mathbf{r}^N)$ is the number of particles in the largest cluster with a configuration $\mathbf{r}^N$, and k and n_0 are parameters in the biasing potential. A value of $k = 0.0561$ is used for all the simulation windows. The values of n_0 are different for each window and kept equal to 10, 20,… and so on until they reach sizes slightly larger than the critical nucleus size. During simulations, the stacking rearrangements in the cluster are very slow. In order to overcome the barriers towards stacking arrangements, all the windows are simulated in parallel using the parallel tempering technique of Geyer and Thompson.[20] The procedure for calculating the free energy of formation of the crystal nuclei from these umbrella sampling simulations is explained in Auer and Frenkel.[17,21]

3.3 Simulation details

The Monte Carlo simulations of crystal nucleation are performed in the grand canonical ensemble. For simulations of homogeneous nucleation and heterogeneous nucleation on flat surfaces, the dimensions of the simulation box were $22\sigma \times 22\sigma \times 22\sigma$. For heterogeneous nucleation on wrinkled surfaces, the simulation box dimensions were $30\sigma \times 20\sigma \times 40\sigma$. Periodic boundary conditions are implemented across all three dimensions for homogeneous nucleation and along two dimensions for heterogeneous nucleation. For simulations involving wrinkled surfaces, a hard wall is placed on one side and the sinusoidal surface is placed on the other side along the third dimension. The Monte Carlo moves consist of particle displacements, and insertions and deletions of particles. During simulations, the sampling of various properties is generally made after every Monte Carlo move. The quantity that we are interested in is the cluster size distribution. The calculations involved in the identification of the cluster and determination of the cluster size are expensive. Hence cluster size distributions are calculated after a fixed number of Monte Carlo moves called trajectories. We have defined a trajectory as a series of Monte Carlo moves consisting of on average 20 displacement, 1 insertion and 1 deletion moves per particle. Prior to sampling the properties, each window in the parallel tempering simulation was equilibrated. The equilibration run is performed in conjunction with the exchanges of configurations among windows. Initially, the acceptance rate for the exchange of configurations is very poor. The configurations for each window are considered to be equilibrated if the acceptance rate reaches a sufficiently high value such that each window is able to sufficiently sample all the cluster sizes. The number of trajectories required for equilibration was at least 1 000 000 trajectories. The production run consisted of 400 000 trajectories. For each state point studied, three independent simulations were performed.

4 Results and discussion

The phase diagram of protein solutions whose interactions are described by the potential shown in eqn (1) and (2) were computed by ten Wolde. The metastable critical temperature for this particular system was computed to be $T_c = 0.42\varepsilon/k_B$. In this study, simulations of protein crystal nucleation were carried out at two different temperatures, namely, (i) $T/T_c = 0.89$ and (ii) $T/T_c = 0.92$. The choice of these temperatures was motivated by previous heterogeneous nucleation studies on the same system by van Meel *et al.*[11] The values of the activity, $z = e^{-\beta\mu}/\Lambda^3$, used in the grand canonical simulations were equal to 0.024 and 0.0345 for temperatures $T/T_c = 0.89$ and $T/T_c = 0.92$, respectively, which gave bulk vapor densities of $0.04\sigma^{-3}$ and $0.066\sigma^{-3}$, respectively. These conditions were chosen such that the estimates for the free-energy barrier for homogeneous crystal nucleation calculated using the classical nucleation theory are equal at these two temperatures.

The effects of attractive interactions between the protein molecule and the surface were modeled by varying the width of the square well interaction, $\lambda\sigma$. The lowest value of λ studied is equal to 0.076. For this value, a system of square well particles with diameter 1.1276σ have the same 2nd virial coefficient at $T/T_c = 0.89$ as a protein system interacting *via* eqn (1) and (2). The value of 1.1276σ is the interparticle distance at which the potential given by eqn (1) and (2) goes to zero. A comparison of the square well potential with the protein potential is shown in Fig. 1. A surface interacting with this value of λ is assumed to approximate a substrate made up of protein molecules. The highest value of λ used was 0.3 for $T/T_c = 0.89$ and 0.2 for $T/T_c = 0.92$. At higher values of λ, the system spontaneously nucleated the protein crystals.

Fig. 3 shows the variation of the heterogeneous nucleation free-energy barrier with λ for flat and wrinkled surfaces. Also shown is the barrier for homogeneous nucleation. As expected, the barrier for heterogeneous nucleation on an attractive surface is lower than that for homogeneous nucleation and increasing the value of λ causes a decrease in the free-energy barrier. More importantly, these results show that the presence of wrinkles on the surface has an effect on the free-energy barrier. At lower values of λ the free-energy barriers on both surfaces are comparable. However, with increasing values of λ there is a pronounced lowering of the free-energy barrier due to the presence of wrinkles which in turn leads to faster nucleation of the protein crystals.

In Fig. 4, the number of particles in the critical nuclei is plotted against the value of λ. The average values of the bond orientational parameter, q_6, for the various critical nuclei are shown in Table 1. The values indicate that the nucleus has an FCC structure. Fig. 4 shows that, for both temperatures, the number of particles in the critical nuclei is nearly the same for the flat and wrinkled surfaces for values of λ smaller than or equal to 0.15. For higher values of λ, however, there is a marked difference between the sizes of the critical nuclei formed on both surfaces. To understand this dependence on λ, we determined the location of the critical nucleus on the wrinkled surface. In Fig. 5, a snapshot of the location of the critical nucleus is shown. At lower values of λ, nucleation occurs near the flat portion of the sinusoidal curve. Under these conditions, the regions of the surface with high curvature, *i.e.*, at the crest and the trough, are not favored for the formation of crystal nuclei. The primary reason is that the surface curvature induces stresses on the growing nuclei. Hence the preferred locations of the nuclei are regions on the surface with zero curvature. Since the local environments at these regions are similar to those found on a flat surface, there is little difference in the mechanism of heterogeneous nucleation on wrinkled and flat surfaces. At high values of λ, however, one begins to see a difference in the mechanism of nucleation. At these conditions, the formation of the crystal nuclei occurs in the trough of the sinusoidal surface. This shows that there are other countervailing factors that aid the formation of crystal nuclei near regions of high curvature. In Fig. 6, the density of the protein molecules near the surface is plotted against λ. The density values corresponding to the wrinkled surface are those located on the trough of the sinusoidal curve. These results show that the density near the trough is significantly enhanced compared to that found near a flat surface. This enhancement is similar to the occurrence of capillary condensation in pores as observed by van Meel *et al.*[11] The formation of the crystal nuclei at the trough is caused by this enhancement in the local density. The presence of the metastable critical point is necessary and its close proximity to the crystallization conditions is critical for enhancement of the nucleation rate.

5 Conclusions

In this study, molecular simulations were used to investigate the mechanism of heterogeneous nucleation of protein crystals on wrinkled surfaces. The aim of the study was to explain the enhancement in the heterogeneous nucleation of protein crystals on wrinkled surfaces seen in experiments. The simulations reveal that the free energy barrier to heterogeneous nucleation is lowered with the addition of wrinkles on the surface provided that the attractions of the surface with the protein molecules are sufficiently large. The enhancement of nucleation is caused by an enhancement of the local density near the troughs caused by the wrinkles. The formation of nuclei occurs in these troughs. The enhancement also requires that nucleation be carried out at conditions close to the metastable vapor–liquid critical point. If the above conditions are not met then the presence of wrinkles does not have an effect on the nucleation and the nuclei mostly form in regions of low curvature.

Acknowledgements

The authors acknowledge Prof. Animangsu Ghatak of the Department of Chemical Engineering, Indian Institute of Technology, Kanpur, India for the helpful discussions. The financial support for this work has been provided by grants sanctioned by the Department of Science and Technology, Government of India.

References

1 P. R. ten Wolde and D. Frenkel, *Science*, 1997, **277**, 1975–1978.
2 E. Saridakis and N. E. Chayen, *Trends Biotechnol.*, 2009, **27**, 99–106.
3 A. McPherson and P. Shlichta, *Science*, 1988, **239**, 385–387.
4 T. E. Paxton, A. Sambanis and R. W. Rousseau, *J. Cryst. Growth*, 1999, **198/199**, 656–660.
5 A. Sanjoh and T. J. Tsukihara, *J. Cryst. Growth*, 1999, **196**, 691–702.
6 A. Sanjoh, T. J. Tsukihara and S. Gorti, *J. Cryst. Growth*, 2001, **232**, 618–628.
7 N. E. Chayen, E. Saridakis, R. El-Bahar and Y. Nemirovsky, *J. Mol. Biol.*, 2001, **312**, 591–595.
8 N. E. Chayen, E. Saridakis and R. P. Sear, *Proc. Natl. Acad. Sci. U. S. A.*, 2006, **103**, 597–601.
9 S. Stolyarova, E. Saridakis, N. E. Chayen and Y. A. Nemirovsky, *Biophys. J.*, 2006, **91**, 3857–3863.
10 A. J. Page and R. P. Sear, *J. Am. Chem. Soc.*, 2009, **131**, 17550–17551.
11 J. A. van Meel, R. P. Sear and D. Frenkel, *Phys. Rev. Lett.*, 2010, **105**, 205501.
12 A. S. Ghatak and A. Ghatak, *Ind. Eng. Chem. Res.*, 2011, **50**, 12984–12989.
13 A. S. Ghatak and A. Ghatak, *Langmuir*, 2013, **29**, 4373–4380.
14 A. Cacciuto, S. Auer and D. Frenkel, *Nature*, 2004, **428**, 404–406.
15 G. M. Torrie and J. P. Valleau, *Chem. Phys. Lett.*, 1974, **28**, 578.
16 P. J. Steinhardt, D. R. Nelson and M. Ronchetti, *Phys. Rev. B: Condens. Matter Mater. Phys.*, 1983, **28**, 784.

17 S. Auer and D. Frenkel, *Nature*, 2004, **409**, 1020–1023.
18 S. Punnathanam and P. A. Monson, *J. Chem. Phys.*, 2006, **125**, 024508.
19 S. R. Ganagalla and S. N. Punnathanam, *J. Chem. Phys.*, 2013, **138**, 174503.
20 C. J. Geyer and E. A. Thompson, *J. Am. Stat. Assoc.*, 1995, **90**, 909.
21 S. Auer and D. Frenkel, *J. Chem. Phys.*, 2004, **120**, 3015–3029.

Faraday Discussions

PAPER

Analyzing a fractal gel of charged oblate nanoparticles in a suspension using time-resolved rheometry and DLVO theory

Shweta Jatav and Yogesh M. Joshi*

Received 24th August 2015, Accepted 21st September 2015
DOI: 10.1039/c5fd00128e

The disk-like nanoparticles of LAPONITE® are known to self-assemble to form a fractal gel within hours after a sufficiently large concentration of LAPONITE® is dispersed in water containing salt. The concentration of sodium counterions associated with LAPONITE® particles, however, continues to increase over a period of days, suggesting that delamination of LAPONITE® disks from stacks is sluggish and/or dissociation of counterions is slow. In either case, spontaneous self-assembly of LAPONITE® particles occurs even though delamination and/or counterion dissociation has not reached its equilibrium state. In order to determine the nature of the fractal gel as the extent of delamination and/or dissociation progresses towards equilibrium, we subject the LAPONITE® suspension to a freezing–defrosting cycle, which interestingly reinitiates the gelation process in suspension afresh. Application of time-resolved rheometry to a defrosted suspension shows that iso-frequency loss tangent curves intersect at an identical point, validating the Winter–Chambon criterion for a critical fractal gel state. Interestingly, while the time required to form a critical gel is observed to decrease with increased time elapsed since preparation, at which freezing–defrosting is carried out, the fractal dimension of the critical gel is observed to remain unaffected. We also solve DLVO theory for free energy interactions between the negatively charged LAPONITE® particle faces and analyze the observed phenomena.

1 Introduction

Colloidal particles suspended in aqueous media arrange themselves in a variety of microstructures depending upon their concentration, the nature of the charges present on the particles, and the nature and concentration of electrolytes.[1] Colloidal suspensions are used in a variety of applications, including in medicines, soaps and detergents, inks and paints, lotions and pastes, as a variety of food items, *etc.* The microstructure of a colloidal suspension on the one hand determines its application, while on the other hand it also determines its long-

Department of Chemical Engineering, Indian Institute of Technology, Kanpur, Kanpur-208016, India. E-mail: joshi@iitk.ac.in

term stability against sedimentation, creaming or thixotropic cake formation. At the limit of low concentration colloidal suspensions usually remain in the liquid state for a long time.[2] However, with increase in concentration their tendency to self-assemble increases. Sometimes the self-assembled structures are out of thermodynamic equilibrium and evolve with time in search of more stable states.[3] The presence of an anisotropic shape and uneven charges add further to the intricate interparticle interactions. In this work we study the self-assembly of nanoparticles of LAPONITE®, a synthetic clay mineral, which has a disk-like shape with uneven charge distribution when dispersed in aqueous media. LAPONITE® in aqueous media has been used in a variety of industries, including pharmaceutical, paper, paint, home care, personal care, petrochemical, agrochemical, *etc.*, primarily as a rheology modifier.[4–6] In academics, due to its spectacular time-dependent physical properties, aqueous LAPONITE® suspension has been termed a model soft glassy material.[7] In the literature, this system has been proposed to show a rich array of microstructures. In this work we employ rheological tools to gain further insight into its microstructure after subjecting it to a freezing–defrosting cycle.

LAPONITE® clay, with the unit cell chemical formula $Na_{+0.7}[(Si_8Mg_{5.5}Li_{0.3})O_{20}(OH)_4]_{-0.7}$, has a disk-like shape with a thickness of 1 nm and a 25 ± 2 nm diameter.[8] LAPONITE® belongs to a family of 2 : 1 polysilicates wherein one octahedral layer of magnesium oxide is sandwiched between two tetrahedral layers of silica. In the middle layer some magnesium atoms are replaced by lithium, which gives a negative charge on the faces of the LAPONITE® particle. This charge is compensated by sharing sodium atoms in the interlayer spacing between the particles of LAPONITE®. In a dry state LAPONITE® particles are present in the form of stacks. When LAPONITE® powder is dispersed in water, it enters the interlayer gallery, causing the stacks to swell.[9,10] During this process stacks of LAPONITE® particles with a variety of sizes are present in water.[9] On the one hand these stacks undergo swelling, while on the other hand LAPONITE® particles on the surface of the stacks delaminate as a function of time.

A suspension of LAPONITE® in water is known to show a spectacular increase in elastic modulus as a function of time, eventually acquiring a soft solid-like consistency.[11–14] It is typically observed that the ionic conductivity of the suspension increases very slowly over a timescale of days after preparation.[11,15,16] The increase in ionic conductivity is due to dissociation of Na^+ ions from the faces of LAPONITE®. Since significant dissociation of Na^+ ions occurs only after delamination of individual LAPONITE® particles from the stacks, sluggishness of either or both of the processes could lead to slow enhancement of ionic conductivity. The possibility of both processes being slow has been speculated in the literature.[11,17] The edges of the LAPONITE® particles where the crystal structure is broken primarily contain Mg–OH groups, whose charge depends upon the pH. The point of zero charge for Mg–OH is around 12.5,[18] below which the edges acquire positive charge due to protonation. Overall, negatively charged faces, positively charged edges and the anisotropic shape of LAPONITE® particles lead to complex interparticle interactions, which include edge-to-face attraction as well as face-to-face repulsion.[11,19,20] The presence of externally added salt in a suspension increases the concentration of ions, which are known to shield the charges on the particles.[11,21]

Owing to the presence of uneven charges, the microstructure of a LAPONITE® suspension is influenced by repulsive as well as attractive interactions present in between the LAPONITE® particles. Many versions of the LAPONITE® suspension phase diagram as a function of salt (NaCl) and LAPONITE® concentration have been proposed in the literature.[19,22–24] The LAPONITE® suspension shows phase separation below around 1 wt% or in the limit of very high salt concentrations (>20 mM).[19,24] Furthermore, while there is a consensus that between 1 and 2 wt% aqueous suspensions of LAPONITE® form a colloidal gel, the microstructure above a concentration of 2 wt% is a matter of debate. Ruzicka and coworkers,[25] by carrying out small angle X-ray scattering (SAXS), proposed that for suspensions with a concentration greater than 2 wt% LAPONITE® particles form a glass state in which particles are jammed in space, remaining untouched by other particles.[25] Interestingly Ruzicka and coworkers[26] indicated that while repulsion dominates the early stages after preparation of a LAPONITE® suspension, attraction among the particles plays a significant role as a suspension ages. Ruzicka and coworkers also carried out Monte Carlo simulations, wherein disk-like particles with charged patches (three positive potentials on the edge and one negative potential on each of the two faces) interact with each other.[27,28] Interestingly the structure factor obtained from simulations showed similarity with that obtained from SAXS measurements.

In an alternate proposal Mongondry and coworkers[24] proposed a LAPONITE® suspension to exist in a gel state, wherein particles form a space spanning network by forming edge-to-face attractive bonds. Interestingly Joshi and coworkers,[11,29,30] through a series of papers on rheological, dynamic light scattering and electrochemical experiments, showed that the rate of structural build-up in a LAPONITE® suspension increases with a decrease in repulsive screening length related to the LAPONITE® particle faces. Consequently, they claimed the resultant structure cannot be merely repulsion-dominated, but the attractive interactions must have a profound influence on the structure over all timescales since preparation. It should be noted that gel formation in aqueous media is not just limited to LAPONITE®. Other types of clays such as bentonite, montmorillonite, nontronite, *etc.*, which are structurally similar to LAPONITE®, have also been reported to form attractive gels when suspended in water.[10,31,32] Clays modified by hydrophobic chains are rendered compatible with organic solvents. Such clay dispersions have also been observed to undergo sol–gel transition.[33,34] However, this topic is beyond the scope of the present work.

A gel state and a glass state are distinguished from each other by a length scale above which the density of gel or glass becomes homogeneous.[3] In glasses, this length scale is of the order of an average interparticle distance, while in the case of gels this length scale is much greater than an average interparticle distance. This length scale can usually be obtained by scattering as well as microscopic techniques.[19,35] Rheologically, on the other hand, there are unique signatures to recognize the point of glass transition compared to that of the critical gel state. The point of glass transition, wherein a liquid undergoes ergodicity breaking, depends on the timescale of observation. Consequently, rheological properties at glass transition depend on frequency.[36] Typically at the point of glass transition $\tan\delta$, which is the ratio of G'' (viscous modulus) to G' (elastic modulus), shows a maximum. For molecular glasses, the glass transition temperature (associated with a maximum in $\tan\delta$) is observed to increase with increase in frequency

applied to monitor the evolution.[37] The critical gel state, on the contrary, is the state when particles just form a space-spanning percolated network. The point at which such a network is formed is independent of the timescale of observation or applied frequency.[38] Winter and Chambon[39] in a seminal paper proposed that at the critical gel point, elastic (G') and viscous (G'') modulus show identical frequency (ω) dependence given by:

$$G' = G'' \cot(n\pi/2) = \frac{\pi S}{2\Gamma(n)\sin(n\pi/2)}\omega^n, \tag{1}$$

where $\Gamma(n)$ is an Euler gamma function of power law exponent n and S is the gel strength. Consequently at the critical gel state $\tan\delta = \tan(n\pi/2)$ is independent of frequency. Therefore, rheologically the critical gel state is identified as the point at which iso-frequency $\tan\delta$ curves meet.[38] Beyond the critical gel state, a gel consolidates wherein interparticle bonds become stronger along with alteration of the network.[40]

The critical gel state that validates the Winter–Chambon criterion has been observed for a large number of crosslinking polymeric materials[41–44] and colloidal particulate materials undergoing gelation.[45–47] Interestingly the knowledge of n also leads to obtaining the fractal dimension of the critical gel.[48] Remarkably fractal dimensions obtained from independent small angle X-ray scattering measurements on polymeric[49] as well as colloidal[50] gels closely match those obtained from the rheology experiments. Interestingly, the dependence of G' on ω given by eqn (1) yields a relaxation time spectrum, given by $H(\tau) \sim \tau^{-n}$, where τ is relaxation time. Very recently Winter[51] proposed that the behavior of the relaxation time spectrum at the point of glass transition shows a positive slope when $H(\tau)$ is plotted as a function of τ on a double logarithmic scale. This is contrary to the behavior observed at the critical gel state, thereby indicating an important criterion to distinguish both the states.

In order to check the validity of the Winter–Chambon criterion given by eqn (1), a system evolving as a function of time is usually subjected to time-resolved rheometry, wherein a frequency sweep ranging from lower to higher values is applied in a continuous succession. Cocard and coworkers[52] were the first to investigate an aqueous suspension of LAPONITE® using time-resolved rheometry to validate the Winter–Chambon criterion. More recently Jatav and Joshi[53] validated the presence of a critical gel state for a freshly prepared 2.8 wt% LAPONITE® suspension with 3 mM NaCl, which is expected to consist of a network formed by bonds between positive edges and negative faces.[10] Interestingly, while the critical gel state is observed within a few hours of mixing LAPONITE® in water, slow dissociation of Na^+ ions continues over a timescale of days.[11,15] Consequently, the critical gel state observed for a freshly prepared LAPONITE® suspension is associated with only partial dissociation of Na^+ ions and/or partial delamination of particles.

As mentioned before, the negative potential on the faces of LAPONITE® particles is a consequence of dissociation of Na^+ ions.[11,30] Furthermore, the concentration of Na^+ ions in the electric double layer surrounding a particle also influences the effective width of the particle, which strongly influences interparticle attractive as well as repulsive interactions.[1,11] Discussion in previous paragraphs therefore suggests that at the time of critical gel formation either delamination of LAPONITE® particles and/or the dissociation of Na^+ ions from the faces of LAPONITE® particles has not reached equilibrium. As a result, the observed critical gel state is associated with that of only partly evolved delamination/dissociation dynamics. It is possible that the nature of the gel state could have been different if sufficient time had been given first for the delamination/dissociation process to reach equilibrium followed by gelation of the particles. However, the process of colloidal gelation, such as in the case of the LAPONITE® suspension, unlike chemical gelation,[54] cannot be stopped or paused at any point by killing the bond formation process.

It has been observed that application of a strong deformation field induces gel-to-liquid transition in a colloidal gel.[3] Subsequent to cessation of deformation, the rheological properties of the LAPONITE® suspension evolve as a function of time, whose dynamics are qualitatively similar to those of evolution in a freshly prepared suspension.[7,13] It may, therefore, appear that the nature of the critical gel state could be studied by shear melting a LAPONITE® suspension after the dissociation/delamination process has reached equilibrium. However it has been observed that shear melting does not obliterate all the interparticle bonds formed during the process of gelation.[11,53] Consequently, post shear melting restructuring of a LAPONITE® suspension does not sample the critical gel state.[53] The deviation from the critical gel state becomes so severe with age, at which the shear melting is applied, such that for a sufficiently old shear melted LAPONITE® suspension, iso-frequency tan δ curves do not even cross each other.[46,53]

The objective of the present work is to explore the possibility of determining the nature of the critical gel state at various concentrations of dissociated Na^+ ions until the equilibrium value, if reached, which also guarantees delamination to be nearly complete. In order to achieve this, we freeze a LAPONITE® suspension below 0 °C and after holding it there for some time, defrost it. Since at 0 °C the relative permittivity of ice is smaller than that of liquid water by over a factor of 25, defrosted samples are in a completely collapsed state. The gelation of such samples leads to characterization of the critical gel point at various intermediate states of partial dissociation and/or delamination.

2 Sample preparation and experimental protocol

The LAPONITE® XLG used in this study was obtained from BYK Additives, Ltd. The protocol to prepare a 2.8 wt% LAPONITE® suspension with 3 mM NaCl in water at pH 10 is the same as that reported by Jatav and Joshi,[53] which involves vigorous mixing of the mentioned ingredients with oven dried LAPONITE® using a Turrax drive for 30 min. The freshly prepared suspension was stored in air-tight polypropylene bottles in a constant temperature chamber maintained at 30 °C. The samples were stored for various rest times (t_R), and rheological, conductivity and pH measurements were carried out on them. The suspension was observed to remain chemically and physically stable over the duration of all the experiments, wherein no dissolution of magnesium ions and no phase separation were observed.[55]

Rheological experiments were carried out in the concentric cylinder geometry (outer diameter of 30 mm and gap of 1 mm) of a DHR3 rheometer (TA instruments). The suspension was loaded into the shear cell using a dropper, and an oscillatory stress of 25 Pa at 0.1 Hz frequency was applied for 5 min, during which the sample underwent shear melting, leading to a steady value of complex viscosity and strain magnitude. Subsequently the suspension was frozen to −10 °C, and after maintaining it at that temperature for 90 min it was defrosted to 30 °C. After reaching thermal equilibrium at 30 °C, the sample was again subjected to 25 Pa oscillatory stress at 0.1 Hz frequency for 5 min to assess the effect of freezing. Next oscillatory stress of 0.1 Pa with frequencies varying in the range: 0.5–25 rad s^{-1} was applied to the sample in continuous succession. The ionic conductivity was measured using a DS-71 conductivity meter with a 2 pole cell electrode, while pH was measured using an F-71 pH meter with a standard glass electrode (both instruments from HORIBA Scientific). Before measuring ionic conductivity and pH, the suspension was mechanically rejuvenated to convert it to a liquid, which facilitates unhindered ionic mobility.

3 Results and discussion

The ionic conductivity (σ) of the suspension measured before freezing on various days since preparation (t_R) is plotted in Fig. 1(a). It was observed that freezing–defrosting has practically no effect on σ. It can be seen that just after preparation value of σ is small but increases rapidly with increase in t_R. However beyond $t_R = 50$ h increase in σ becomes weak as a function of t_R and shows signs of reaching a constant value. As mentioned before, increase in σ over a prolonged period could arise from delayed delamination of LAPONITE® plates, and/or sluggish dissociation of Na^+ counter ions. The ionic conductivity is directly related to mobility and concentration of various ionic species in a suspension and is given by:[56]

$$\sigma = e\sum_i \mu_i c_i, \tag{2}$$

where e is electron charge (1.602×10^{-19} C), while μ_i and c_i are respectively the mobility and concentration of ith ionic specie. In the aqueous LAPONITE® suspension studied in the present work the various ionic species present are: Na^+, Cl^- and OH^- ions. The concentration of OH^- ions (c_{OH}) can be calculated from

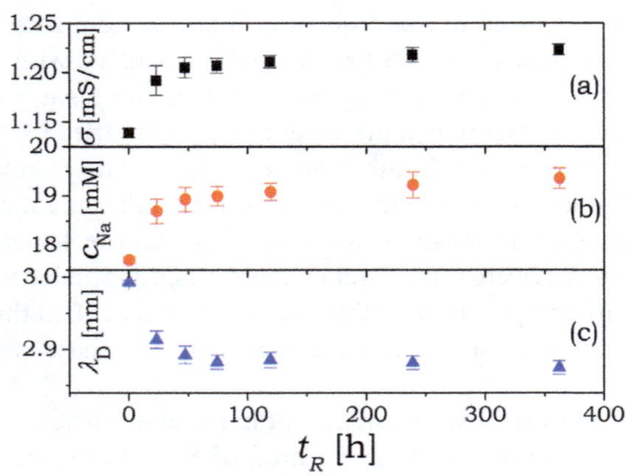

Fig. 1 (a) The ionic conductivity (σ), (b) concentration of Na^+ ions (c_{Na}) and (c) the Debye screening length (λ_D) is plotted as a function of rest time (t_R) of the suspension. It can be seen that after 100 h all three variables attain the equilibrium value.

the pH of the suspension, while the concentration of Cl^- ions (c_{Cl}) is the same as the concentration of externally added NaCl (3 mM). In principle H^+ ions are also present, however since the pH of the suspension is always around 10.3 over the explored duration, the contribution of H^+ ions is negligibly small. With knowledge of the μ_i values of the mentioned ionic species, the concentration of Na^+ ions (c_{Na}) can be easily calculated from eqn (2). In Fig. 1(b), c_{Na} is plotted as a function of t_R. It can be seen that, as expected, c_{Na} follows the same trend as σ.

Knowledge of c_i also enables us to estimate the Debye screening length (λ_D) associated with the electric double layer on the faces of LAPONITE® particles, and is given by:

$$\lambda_D = \left\{ \frac{\varepsilon_0 \varepsilon_r k_B T}{e^2 \sum_i z_i^2 c_i} \right\}^{1/2}, \tag{3}$$

where ε_0 is the permittivity of free space, ε_r is the relative permittivity of water, k_B is the Boltzmann constant, T is temperature, and z_i is the ionic charge of the ith specie. In Fig. 1(c) λ_D is plotted as a function of t_R. It can be seen that λ_D decreases strongly over the first 50 h followed by a weak decrease to reach a constant value. The decrease in λ_D can be attributed to an increase in the concentration of Na^+ ions in the bulk of the suspension, which compresses the electrostatic double layer around the particle faces.

The energy of the interactions between two clay particles approaching each other in a parallel fashion as a function of normal distance between the two faces ($2d$) is obtained by solving DLVO (Derjaguin, Landau, Verwey and Overbeek) theory.[1] In DLVO theory, the total free energy (V_T) is obtained by adding contributions from electrostatic (usually repulsive, V_E) and van der Waals (attractive, V_{vdW}) interactions between the faces separated by distance $2d$, and is given by:

$$V_T = V_E + V_{vdW} = (64 c_{Na} k_B T \lambda_D)\tanh^2(z/4)\exp\left(\frac{-2d}{\lambda_D}\right) - \frac{A_H}{48\pi d^2}. \tag{4}$$

In eqn (4) the terms on the right hand side are respectively the expressions of free energy for electrostatic and van der Waals interactions. Parameter z is a dimensionless number related to the surface charge density (ς) given by:

$$\sinh(z/2) = (8 c_{Na} \varepsilon_0 \varepsilon_r k_B T)^{-1/2} \varsigma, \tag{5}$$

while A_H is the Hamaker constant whose value is 1.06×10^{-20} J for LAPONITE®.[57] In eqn (5) ς is obtained by knowing the number of Na^+ ions dissociated per face of LAPONITE® particle, and is given by:

$$\varsigma = e(c_{Na} - c'_{Na})/(2 A_L n_p), \tag{6}$$

where c'_{Na} is the Na^+ ion concentration due to external sources such as NaCl and NaOH added to the suspension. Furthermore, the surface area of one face of a LAPONITE® particle is given by A_L, while n_p is the number density of LAPONITE® particles, so that the denominator of eqn (6) is the total surface area of all the faces of LAPONITE® particles, considering these to be completely delaminated. However, as mentioned above this may not be the case, particularly in the initial period after preparation. In such a case eqn (6) will provide a lower bound of ς so that the actual value will always be greater than or equal to the estimated value of ς, and the difference between them will decrease with t_R. Moreover, in the literature it has been proposed that the negative charge on the faces of clay particles exposed by dissociation of counterions attracts H^+ ions from the bulk.[58] However, since the suspension pH is around 10.3, the concentration of H^+ ions in the bulk is extremely small. We therefore expect alteration of surface charge density due to adsorption of H^+ ions from bulk to be negligible.

In Fig. 2, we plot the total interaction energy V_T obtained from eqn (4) as a function of d for the two values of t_R. It can be seen that owing to like charges on the faces of LAPONITE®, when particles approach each other face-to-face by remaining parallel, repulsion increases with decrease in d, and reaches a maximum at small values of d, before decreasing due to van der Waals interactions. From eqn (4) and (5) it is apparent that the height of the repulsive energy barrier is proportional to ς and $c_{Na}{}^{0.5}$ (as λ_D is proportional to $c_{Na}{}^{-0.5}$), and consequently the height of the repulsive energy barrier increases with t_R, as shown in Fig. 2. As mentioned above, the estimated value of ς associated with the faces of the particles is a lower estimate. In a limiting assumption that the dissociation of Na^+ ions is instantaneous, the upper bound of ς will always be constant. However, even in such limit c_{Na} would increase with t_R. Consequently for all the scenarios, with the controlling step being delamination, dissociation or a combination of both, the maximum height of the repulsive energy barrier will always increase with t_R. The width of the decaying repulsive barrier with increase in interplate distance, on the other hand, is governed by $\exp(-2d/\lambda_D)$, and therefore faster decay is observed with increase in t_R, as shown in the inset of Fig. 2.

The repulsive interaction shown in Fig. 2 is associated with face-to-face parallel arrangement of plates. However, in a suspension the particles may approach each other in any orientation. With increased electronegativity of the faces, decreased width of repulsive barrier, and possibly increased number density of LAPONITE® particles due to delamination, the rate of positive-edge-to-negative-face bond formation is expected to increase at higher values of t_R. Consequently gelation is expected to be faster at higher values of t_R. Moreover, all the above mentioned effects are also expected to influence the structure of the fractal gel.

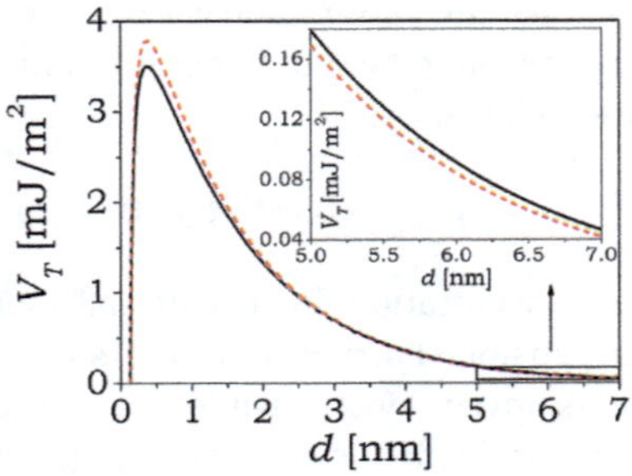

Fig. 2 The free energy of interactions obtained from DLVO theory, when two LAPONITE® particles move toward each other face-to-face, is plotted as a function of half of the normal distance between the two particles for two rest times: full line $t_R = 0$ h and dashed line $t_R = 362$ h. The inset shows the enlarged view of a tail.

We study the nature of the critical gel using time-resolved rheometry. A sample, which was aged over rest time t_R, was loaded into the shear cell, and after attaining thermal equilibrium at 30 °C, subjected to shear melting as mentioned before. Owing to aging over the duration of t_R, the material acquires high viscosity and elastic modulus. However during shear melting the complex viscosity (η^*) of the suspension decreases while the magnitude of oscillatory strain (γ_0) increases, and both the variables eventually reach a plateau. In Fig. 3, the plateau values of η^* and γ_0 are plotted as a function of t_R. It can be seen that the plateau value of η^* increases while that of γ_0 decreases with increase in t_R. This is because the shear melting cannot completely break the structure formed during aging that occurs over rest time (t_R).[11] The extent of unbroken structure increases with t_R, causing respectively increase and decrease in the plateau values of η^* and γ_0.

Subsequent to shear melting, the temperature of the sample was decreased to −10 °C and held there for 90 min, after which the temperature was raised to 30 °C. The temperature of −10 °C was chosen because the presence of LAPONITE® and salt decrease the freezing point of the suspension to −8 °C. The values of η^* and γ_0 of defrosted samples are also plotted in Fig. 3, which clearly show respectively very small and large values for η^* and γ_0 that are almost independent of t_R. A tremendous drop in the viscosity of the defrosted state compared to that of the completely shear melted state of the same sample suggests complete collapse of the structure during freezing–defrosting cycle, which could not be achieved by shear melting. We believe that this due to the significant difference in the relative permeabilities of ice and water. Upon increasing the temperature of the frozen LAPONITE® suspension, in the vicinity of the melting point ice surrounding the LAPONITE® particles undergoes a phase transformation from solid to liquid. Since the relative permittivity of liquid water at freezing point is 88 while that of ice is 3.2, the presence of crystallites along with water obstructs the electrostatic interactions, which is the very backbone of the structure. Consequently, after defrosting, the suspension acquires a liquid state of low viscosity with LAPONITE® particles undergoing Brownian motion, as shown in Fig. 3. The viscosity of the defrosted suspension being independent of t_R clearly suggests that obliteration of the structure has indeed occurred. This behavior is in contrast with

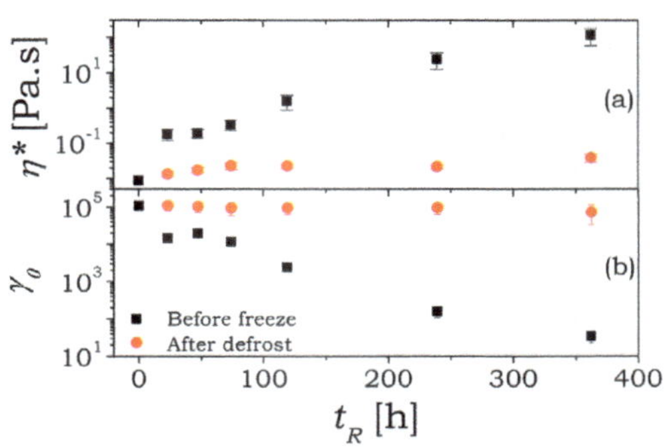

Fig. 3 The steady state values of (a) complex viscosity (η^*) and (b) magnitude of strain (γ_0) under a magnitude of oscillatory stress of 25 Pa at 0.1 Hz frequency are plotted as a function of rest time (t_R) before freezing (squares) and after defrosting (circles).

that observed during shear melting, wherein progressive lesser obliteration occurs with increase in t_R.

Subsequent to defrosting, we subject the sample to cyclic frequency sweep in the linear viscoelastic domain. The corresponding evolution of G', G'' and tan δ is plotted in Fig. 4 as a function of time elapsed since defrosting (t_w) for four different t_R values. After defrosting $G' \ll G''$ irrespective of frequency, suggesting the sample to be in a liquid state. Both the moduli increase as a function of time, with G' increasing at a faster rate than G'', so that they eventually cross each other. However with a decrease in frequency the crossover point ($G' = G''$) gets delayed. Rapid increase in G' compared to G'' leads tan δ to decrease with time, but with less intensity with increase in frequency. Consequently all the iso-frequency tan δ curves intersect with each other, but importantly at an identical point. Such a unique intersection point irrespective of frequency is a characteristic feature of the critical gel state, as proposed by the Winter–Chambon criterion.[54] We also subjected samples older than 20 days to the freezing–defrosting cycle. However, upon defrosting, the suspensions with $t_R > 20$ days, while undergoing liquid-to-solid transition, do not show a critical gel state. In this case all the decreasing iso-frequency tan δ curves simply approach each other as a function of time without crossing each other. We believe that such behavior could arise from sintering of some of the clay particles with each other over a long time, but these cannot be broken down to individual particle level upon freezing–defrosting, leading to the observed behavior.

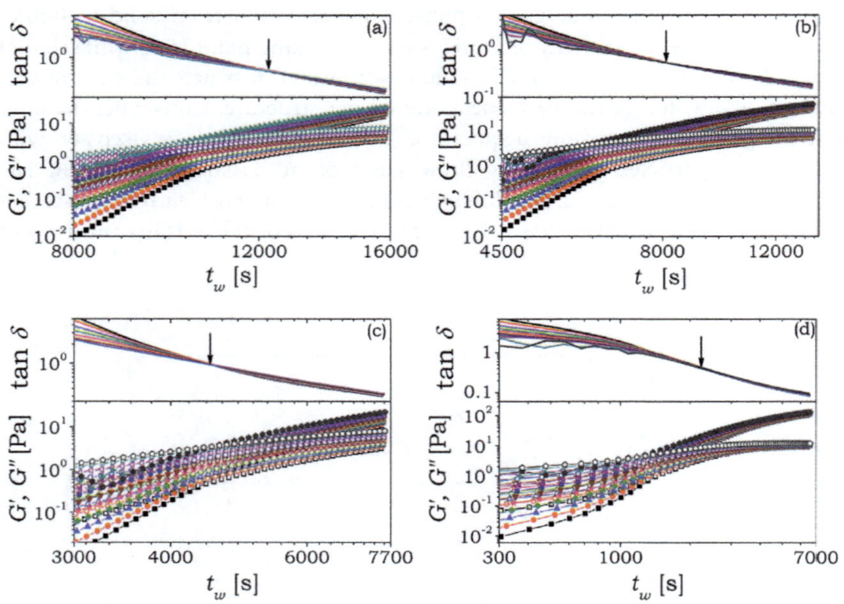

Fig. 4 The elastic modulus (G', filled symbols), viscous modulus (G'', open symbols) and loss tangent (tan δ) are plotted as a function of time elapsed since defrosting (t_w) for different rest times (t_R) = (a) 0 h, (b) 47 h, (c) 119 h and (d) 362 h. In all the G' and G'' plots frequency increases from bottom to top (0.5, 0.8, 1.25, 2, 3.15, 5, 8, 12.5, 20, and 25 rad s^{-1}) and for tan δ increase in frequency from top to bottom is applicable before the critical gel point. The critical gel point is shown with a vertical arrow.

For polymeric gels, the rate of change in G' and G'' near the gel point is observed to show a power law dependence on frequency (ω) given by:[59]

$$\left(\frac{\partial \ln G'}{\partial t}\right)_{t=t_g} \cong C\left(\frac{\partial \ln G''}{\partial t}\right)_{t=t_g} \sim \omega^{-\kappa}, \tag{7}$$

where C is the proportionality constant and κ is the dynamic critical exponent. Interestingly, for a variety of polymer gels (polymer gels containing different polymer–solvent pairs, as well as within these a polymer–solvent pair variation of chain length, concentration and stoichiometry) κ is observed to be within a range: 0.22 ± 0.02.[46,59–62] Even for physical gels the value of κ is observed to be in a similar range.[46] In addition, the proportionality constant C has been reported to be around 2 for various chemical as well as physical gel systems.[46,59] In Fig. 5, we plot κ and C as a function of t_R. Remarkably C is indeed observed to be around 2, while κ calculated from the rate of change of G'' is observed to be between 0.2 and 0.25, and κ calculated from rate of change of G' is around 0.15. Fig. 5, therefore indicates a very close match between the values of κ and C observed for various chemical and physical gels and the defrosted aqueous suspension of LAPONITE® studied in this work.

Fig. 4 clearly shows that the critical gel point occurs at earlier times with increase in t_R. In Fig. 6 we plot time to gelation (t_g) as a function of t_R, which shows a sharp decrease with increase in t_R for $t_R < 100$ h. However, at higher t_R the decrease in t_g becomes weaker. From eqn (1) a value of tan δ associated with the critical gel point leads to critical relaxation exponent n. The critical gel is known to have a fractal structure, and in a seminal contribution Muthukumar[48] obtained a relationship between exponent n and fractal dimension f_d by accounting for excluded volume interaction, and for a d dimensional system it is given by:[48]

$$f_d = \frac{(d+2)(2n-d)}{2(n-d)}. \tag{8}$$

We obtain f_d from eqn (8) for $d = 3$, which is also plotted in Fig. 6 as a function of t_R. Interestingly it can be seen that, unlike time to gelation (t_g), fractal dimension f_d remains practically constant in the vicinity of 2.2 over the explored range of t_R.

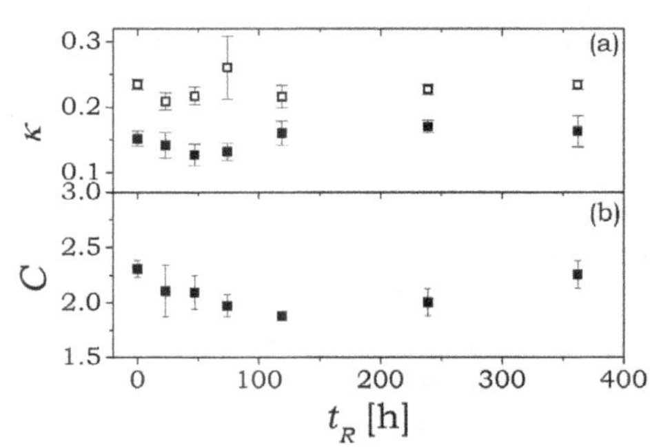

Fig. 5 The parameters (a) dynamic critical exponent (κ) (filled symbols obtained from G' and open symbols obtained from G'') and (b) proportionality constant (C), as defined by eqn (7), are plotted as a function of rest time (t_R). Both the parameters show remarkably a close match to those of critical gels of polymeric materials.

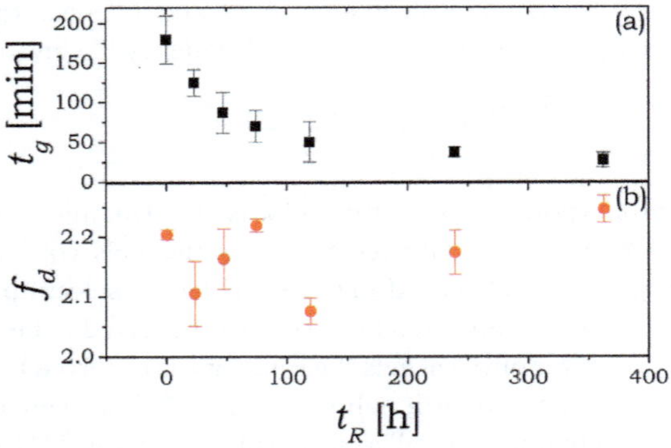

Fig. 6 (a) The time to gelation (t_g) and (b) fractal dimension (f_d) are plotted as a function of rest time (t_R). While t_g decreases, f_d remains constant within an uncertainty window.

The results of the rheological study in the light of DLVO theory lead to an interesting scenario. The results from DLVO theory clearly show that the height of the repulsive energy barrier (and/or negative charge on the faces) increases while its width ($\sim\lambda_D$) decreases with t_R. This trend clearly suggests that at higher t_R values LAPONITE® particles are more prone to form bonds when they approach each other in a non-parallel fashion, leading to faster gel formation and resulting in a decrease in t_g. However, faster dynamics of critical gel formation with possibly more delaminated particles and/or increase in concentration of dissociated counterions does not seem to affect the fractal dimension of the critical gel. It is generally observed that faster gel formation, because of fewer attempts for interparticle bond formation as typically is the case in diffusion-limited cluster aggregation (DLCA), leads to a more open structure (low fractal dimension in the range 1.7 to 1.9).[63] On the other hand, more attempts to form interparticle bonds, when the repulsive barrier is significant enough, leads to reaction-limited cluster aggregation (RLCA), which leads to a compact structure with a higher fractal dimension (typically above 2). However, in the present case decrease in t_g by a factor 6 does not influence the fractal dimension, with all the values remaining in the window of 2.1 to 2.3. This suggests that particles still need more attempts to form interparticle bonds, which can be attributed to a repulsive barrier among the particles, leading to compact structure formation through RLCA.

4 Conclusions

In this work we investigated an aqueous suspension of 2.8 wt% LAPONITE® with 3 mM NaCl, which spontaneously self-assembles to form a fractal gel within a few hours of preparation. On the other hand, the delamination of stacks of LAPONITE® and/or dissociation of counterions from the faces of the LAPONITE® progress over a timescale of days. Since delamination of particles and dissociated Na^+ counterions is expected to have a profound effect on the nature of the fractal gel, we subjected the LAPONITE® suspension to a freezing–defrosting cycle, which reinitiates the gelation process afresh but at a greater state of delamination/dissociation. Application of a cyclic frequency sweep to the defrosted suspension shows the formation of a critical fractal gel. Interestingly, while the time required to form the critical gel decreases with increased rest time, at which it is subjected to the freezing–defrosting cycle, the fractal dimension of the critical gel remains unaffected. With the knowledge of the concentrations of various ionic species in the suspension we determined the Debye screening length on the faces of the particles. We also solved DLVO theory between the faces of the LAPONITE® particles when they approach each other in a parallel fashion. DLVO theory suggests that the particles must overcome a repulsive energy barrier whose height grows but whose width shrinks with rest time. Consequently formation of the fractal gel, whose building block is considered to be an edge-to-face bond, gets accelerated, leading to faster gel formation, as observed in the rheology experiments. However, even though the increase in rest time causes greater delamination of stacks of LAPONITE® particles and/or dissociation of Na^+ ions, the fractal dimension of the critical gel is observed to remain unaffected.

Acknowledgements

We acknowledge the financial support from the Department of Atomic Energy–Science Research Council (DAE-SRC), Government of India.

References

1 J. N. Israelachvili, *Intermolecular and Surface Forces*, Academic Press, London, 3rd edn, 2010.
2 R. G. Larson, *The Structure and Rheology of Complex Fluids*, Clarendon Press, Oxford, 1999.
3 Y. M. Joshi, *Annu. Rev. Chem. Biomol. Eng.*, 2014, **5**, 181.
4 M. Ghadiri, H. Hau, W. Chrzanowski, H. Agus and R. Rohanizadeh, *RSC Adv.*, 2013, **3**, 20193–20201.
5 K. Sun and S. R. Raghavan, *Langmuir*, 2010, **26**, 8015–8020.
6 N Herrera, J.-M. Letoffe, J.-L. Putaux, L. David and E. Bourgeat-Lami, *Langmuir*, 2004, **20**, 1564–1571.
7 A. Shahin and Y. M. Joshi, *Langmuir*, 2012, **28**, 5826–5833.
8 M. Kroon, W. L. Vos and G. H. Wegdam, *Phys. Rev. E*, 1998, **57**, 1962–1970.
9 Y. M. Joshi, *J. Chem. Phys.*, 2007, **127**, 081102.
10 H. van Olphen, *An Introduction to Clay Colloid Chemistry*, Wiley, New York, 1977.
11 A. Shahin and Y. M. Joshi, *Langmuir*, 2012, **28**, 15674–15686.
12 A. K. Atmuri, G. A. Peklaris, S. Kishore and S. R. Bhatia, *Soft Matter*, 2012, **8**, 8965–8971.
13 A. S. Negi and C. O. Osuji, *Phys. Rev. E: Stat., Nonlinear, Soft Matter Phys.*, 2010, **82**, 031404.
14 A. S. Negi and C. O. Osuji, *Phys. Rev. E: Stat., Nonlinear, Soft Matter Phys.*, 2009, **80**, 010404.
15 R. Bandyopadhyay, D. Liang, H. Yardimci, D. A. Sessoms, M. A. Borthwick, S. G. J. Mochrie, J. L. Harden and R. L. Leheny, *Phys. Rev. Lett.*, 2004, **93**, 228302.
16 H. Tanaka, S. Jabbari-Farouji, J. Meunier and D. Bonn, *Phys. Rev. E: Stat., Nonlinear, Soft Matter Phys.*, 2005, **71**, 021402.
17 S. Ali and R. Bandyopadhyay, *Langmuir*, 2013, **29**, 12663–12669.

18 W. Stumm, *Chemistry of the solid–liquid interface*, John Wiley & Sons, New York, 1992.
19 B. Ruzicka and E. Zaccarelli, *Soft Matter*, 2011, **7**, 1268–1286.
20 H. Z. Cummins, *J. Non-Cryst. Solids*, 2007, **353**, 3891–3905.
21 T. Nicolai and S. Cocard, *Langmuir*, 2000, **16**, 8189–8193.
22 A. Mourchid, E. Lecolier, H. van Damme and P. Levitz, *Langmuir*, 1998, **14**, 4718–4723.
23 S. Jabbari-Farouji, H. Tanaka, G. H. Wegdam and D. Bonn, *Phys. Rev. E: Stat., Nonlinear, Soft Matter Phys.*, 2008, **78**, 061405–061410.
24 P. Mongondry, J. F. Tassin and T. Nicolai, *J. Colloid Interface Sci.*, 2005, **283**, 397–405.
25 B. Ruzicka, L. Zulian, R. Angelini, M. Sztucki, A. Moussaid and G. Ruocco, *Phys. Rev. E: Stat., Nonlinear, Soft Matter Phys.*, 2008, **77**, 020402–020404.
26 B. Ruzicka, L. Zulian, E. Zaccarelli, R. Angelini, M. Sztucki, A. Moussaïd and G. Ruocco, *Phys. Rev. Lett.*, 2010, **104**, 085701.
27 R. Angelini, E. Zaccarelli, F. A. de Melo Marques, M. Sztucki, A. Fluerasu, G. Ruocco and B. Ruzicka, *Nat. Commun.*, 2014, **5**, 4049.
28 B. Ruzicka, E. Zaccarelli, L. Zulian, R. Angelini, M. Sztucki, A. Moussaïd, T. Narayanan and F. Sciortino, *Nat. Mater.*, 2011, **10**, 56–60.
29 A. Shahin and Y. M. Joshi, *Langmuir*, 2010, **26**, 4219–4225.
30 D. Saha, R. Bandyopadhyay and Y. M. Joshi, *Langmuir*, 2015, **31**, 3012–3020.
31 E. Paineau, I. Bihannic, C. Baravian, A. M. Philippe, P. Davidson, P. Levitz, S. S. Funari, C. Rochas and L. J. Michot, *Langmuir*, 2011, **27**, 5562–5573.
32 C. Abrahamsson, L. Nordstierna, J. Bergenholtz, A. Altskar and M. Nyden, *Soft Matter*, 2014, **10**, 4403–4412.
33 S. Letaief and E. Ruiz-Hitzky, *Chem. Commun.*, 2003, 2996–2997.
34 H. J. M. Hanley, C. D. Muzny, D. L. Ho, C. J. Glinka and E. Manias, *Int. J. Thermophys.*, 2001, **22**, 1435–1448.
35 K. Sun, R. Kumar, D. E. Falvey and S. R. Raghavan, *J. Am. Chem. Soc.*, 2009, **131**, 7135–7141.
36 M. T. Shaw and W. J. MacKnight, *Introduction to Polymer Viscoelasticity*, Wiley, New York, 3rd edn, 2005.
37 G. W. Becker, *Kolloid-Z.*, 1955, **140**, 1–32.
38 H. H. Winter and M. Mours, *Adv. Polym. Sci.*, 1997, **134**, 167.
39 H. H. Winter and F. Chambon, *J. Rheol.*, 1986, **30**, 367–382.
40 S. Manley, B. Davidovitch, N. R. Davies, L. Cipelletti, A. E. Bailey, R. J. Christianson, U. Gasser, V. Prasad, P. N. Segre, M. P. Doherty, S. Sankaran, A. L. Jankovsky, B. Shiley, J. Bowen, J. Eggers, C. Kurta, T. Lorik and D. A. Weitz, *Phys. Rev. Lett.*, 2005, **95**, 048302.
41 H. Winter and M. Mours, *Adv. Polym. Sci.*, 1997, **134**, 165–234.
42 B. J. Adzima, H. A. Aguirre, C. J. Kloxin, T. F. Scott and C. N. Bowman, *Macromolecules*, 2008, **41**, 9112–9117.
43 M. Kaushal and Y. M. Joshi, *Macromolecules*, 2014, **47**, 8041–8047.
44 K. M. Schultz, A. D. Baldwin, K. L. Kiick and E. M. Furst, *Macromolecules*, 2009, **42**, 5310–5316.
45 S.-h. Hsu and A. M. Jamieson, *Polymer*, 1993, **34**, 2602–2608.
46 A. S. Negi, C. G. Redmon, S. Ramakrishnan and C. O. Osuji, *J. Rheol.*, 2014, **58**, 1557–1579.

47 A. Ponton, A. Bee, D. Talbot and R. Perzynski, *J. Phys.: Condens. Matter*, 2005, **17**, 821–836.

48 M. Muthukumar, *Macromolecules*, 1989, **22**, 4656–4658.

49 T. Matsumoto, M. Kawai and T. Masuda, *Macromolecules*, 1992, **25**, 5430–5433.

50 A. Ponton, S. Barboux-Doeuff and C. Sanchez, *Colloids Surf., A*, 2000, **162**, 177–192.

51 H. H. Winter, *Macromolecules*, 2013, **46**, 2425–2432.

52 S. Cocard, J. F. Tassin and T. Nicolai, *J. Rheol.*, 2000, **44**, 585–594.

53 S. Jatav and Y. M. Joshi, *J. Rheol.*, 2014, **58**, 1535–1554.

54 F. Chambon and H. H. Winter, *J. Rheol.*, 1987, **31**, 683–697.

55 S. Jatav and Y. M. Joshi, *Appl. Clay Sci.*, 2014, **97–98**, 72–77.

56 W. Benenson, J. W. Harris, H. Stöcker and H. Lutz, *Handbook of Physics*, Springer-Verlag, New York, 2002.

57 P. B. Laxton and J. C. Berg, *J. Colloid Interface Sci.*, 2006, **296**, 749–755.

58 *Handbook of Clay Science*, ed. F. Bergaya and G. Lagaly, Elsevier, Oxford, 2013.

59 J. C. Scanlan and H. H. Winter, *Makromol. Chem., Macromol. Symp.*, 1991, **45**, 11–21.

60 S. Mortimer, A. J. Ryan and J. L. Stanford, *Macromolecules*, 2001, **34**, 2973–2980.

61 J.-P. Eloundou, J.-F. Gerard, D. Harran and J. P. Pascault, *Macromolecules*, 1996, **29**, 6917–6927.

62 R. H. Colby, J. R. Gillmor and M. Rubinstein, *Phys. Rev. E: Stat. Phys., Plasmas, Fluids, Relat. Interdiscip. Top.*, 1993, **48**, 3712–3716.

63 P. J. Lu and D. A. Weitz, *Annu. Rev. Condens. Matter Phys.*, 2013, **4**, 217–233.

Faraday Discussions

PAPER

Effects of co-ordination number on the nucleation behaviour in many-component self-assembly

Aleks Reinhardt, Chon Pan Ho and Daan Frenkel*

Received 2nd September 2015, Accepted 14th September 2015
DOI: 10.1039/c5fd00135h

We report canonical and grand-canonical lattice Monte Carlo simulations of the self-assembly of addressable structures comprising hundreds of distinct component types. The nucleation behaviour, in the form of free-energy barriers to nucleation, changes significantly as the co-ordination number of the building blocks is changed from 4 to 8 to 12. Unlike tetrahedral structures – which roughly correspond to DNA bricks that have been studied in experiments – the shapes of the free-energy barriers of higher co-ordination structures depend strongly on the supersaturation, and such structures require a very significant driving force for structure growth before nucleation becomes thermally accessible. Although growth at high supersaturation results in more defects during self-assembly, we show that high co-ordination number structures can still be assembled successfully in computer simulations and that they exhibit self-assembly behaviour analogous to DNA bricks. In particular, the self-assembly remains modular, enabling in principle a wide variety of nanostructures to be assembled, with a greater spatial resolution than is possible in low co-ordination structures.

1 Introduction

Materials that can be formed by self-assembly have over time become increasingly more complex.[1] Furthermore, in the last few years, the field has seen something of an explosion in the number of self-assembling materials which exhibit not only structural complexity, but which are 'addressably' complex,[2] in the sense that the individual building blocks making up these structures are all distinct. Such self-assembled materials are not only interesting from the point of view of fundamental science, but are thought to hold considerable promise for applications in many aspects of nanotechnology,[3] especially since the addressable nature of the building blocks should allow the structures to be functionalised with sub-nanometre-scale resolution.

Recent experiments have demonstrated that it is possible to assemble structures comprising thousands of distinct modular building blocks into well-formed

Department of Chemistry, University of Cambridge, Lensfield Road, Cambridge, CB2 1EW, UK. E-mail: df246@cam.ac.uk

target structures by making use of single-stranded DNA molecules – termed 'DNA bricks' – designed to have an obligate set of hybridisation partners, in the sense that those parts of the DNA molecules that are designed to be bonded in the target structure have complementary sequences.[4–8] In the past few years, several theoretical and computational studies have also been undertaken, probing the intriguing self-assembly behaviour exhibited by such systems.[9–18]

We have previously shown that DNA brick self-assembly is made possible by the interplay between self-assembly and growth. In particular, as a system of DNA bricks is cooled, at some temperature the free-energy barrier to nucleation becomes small enough that nucleation can occur, but nucleation events remain sufficiently rare so that any clusters that do form do not interact significantly with one another, and monomers are not initially significantly depleted,[10,16] which enables these clusters to grow in an essentially error-free manner as the temperature is decreased. However, such behaviour only occurs over a very narrow window of temperatures: if the experiment is performed at a low temperature from the outset, misassembled aggregates dominate instead.[10] Nucleation thus plays an important role in enabling structures of this type to self-assemble successfully.

Whilst DNA bricks have been shown to self-assemble reliably, our previous theoretical work has indicated that the co-ordination number of the particles that form self-assembling structures determines their nucleation behaviour in both two and three dimensions.[16] In particular, the larger the co-ordination number, the more classical-looking the free-energy barrier to nucleation becomes. Yet one of the key aspects that seemed to enable the lower co-ordinate structures to form successfully was the non-classical nucleation barrier. Specifically, for tetrahedrally co-ordinated building blocks, the critical cluster size was found to be largely insensitive to the nature of the target structure, and the nucleation barrier was significant but surmountable at the point at which a large, nearly fully assembled cluster of the designed structure is thermodynamically stable. By contrast – and in agreement with the predictions of classical nucleation theory – for higher co-ordination number structures, the free-energy barrier to nucleation changes with temperature and is considerably larger than that for tetrahedral structures at the same supersaturation.[16] This suggests that, in order to overcome the free-energy barrier to nucleation, the driving force for growth must increase, for example by increasing the monomer concentration, reducing the temperature or increasing the bond strengths by choosing a different set of DNA sequences. Such approaches, however, would make competing structures in which monomers have not assembled as designed ever more stable, and our previous theoretical work thus suggests that, as the co-ordination number increases, the structures should become more and more difficult to form.

However, in order to create more varied target structures in an addressable way, we may well need to move to a system with a higher co-ordination number, as this should in principle allow us to construct structures with finer small-scale features due to the considerably greater spatial resolution of the system than we can achieve using tetrahedrally co-ordinated particles. Moreover, a greater degree of bonding can help to stabilise such structures, which may also be important in practical applications.

Although DNA bricks are tetrahedrally co-ordinated,[4] there are many possible ways in which addressable structures with higher co-ordination numbers might be experimentally realised. For example, one can envisage that colloidal particles with carefully positioned DNA strands grafted onto the particle in the correct geometry might be possible to assemble in the near future, perhaps similar to the experiments of Wang *et al.*[19] or Lu *et al.*,[20] but with each particle functionalised with a unique set of DNA strands. Alternatively, DNA Holliday junctions and multi-arm motifs[21] can be synthesised to correspond to high co-ordination number structures. Of course in practice, producing structures of this type in experiments may be non-trivial because, in our examples, each colloidal particle would have to be created with a unique set of grafted DNA strands, and each DNA junction with a different sequence would have to be pre-assembled. It is therefore important that future experiments focus on strategies that are likely to be successful. It is with this in mind that we have carried out the simulations presented here: if structures of this kind cannot be assembled on a computer with a toy model, then it may be risky to attempt to do so experimentally in the light of the significant cost and effort likely to be involved.

2 Simulation methods

We perform canonical ensemble simulations on a lattice, with periodic boundary conditions, using a Metropolis Monte Carlo[22] scheme. To determine the free-energy barriers as a function of the size of the largest crystalline cluster in the system, we use umbrella sampling with adaptive weights[23] in a time-step separated[24] Monte Carlo scheme. We use 'virtual moves'[25] to allow for realistic dynamics of cluster motion. In our simulations, clusters are randomly translated or rotated on a lattice, with 24 permissible orientations per particle, corresponding to all the possible neighbour interactions on a cubic lattice.[10]

Each particle in the system is hard in the sense that dual occupancy of lattice sites is not permitted, and each particle has n 'patches', where n is the co-ordination number. Every patch is assigned a DNA sequence such that, in the fully assembled target structure, adjacent patches have a complementary sequence, but otherwise these sequences are randomly assigned (subject to the rules identified by Wei *et al.*[5]).† Particles that are adjacent to each other interact with a slightly repulsive energy $\varepsilon_{\text{init}}/k_B = 100$ K,[10] to which we add the hybridisation free energy of the longest complementary sequence match between the nearest pair of patches, calculated using a standard thermodynamic model.[26] As in the experimental work of Ke *et al.*,[4] the outermost particles in the target structure are assigned a poly-T sequence to minimise any misbonding.

Particles which have 4 and 8 patches have a minimum interparticle distance of $a\sqrt{3}$, where a is the lattice parameter,[10] whilst particles with 12 patches have a minimum interparticle distance of $a\sqrt{2}$ to be able to accommodate the additional neighbours. This means that the effective densities are not strictly comparable, as the lower co-ordination structures have a greater excluded volume.

In grand-canonical simulations, we introduce particle addition and removal moves in addition to the canonical (virtual move) translations and rotations. Particles to be added or removed are chosen at random. A particle addition move

† It is by no means essential for particles in our system to interact *via* DNA hybridisation; it is sufficient that they have specific, designed interactions. In practice, however, we anticipate at this stage that DNA is the most likely candidate for an experimental realisation of such systems, and we have chosen to parameterise our model accordingly.

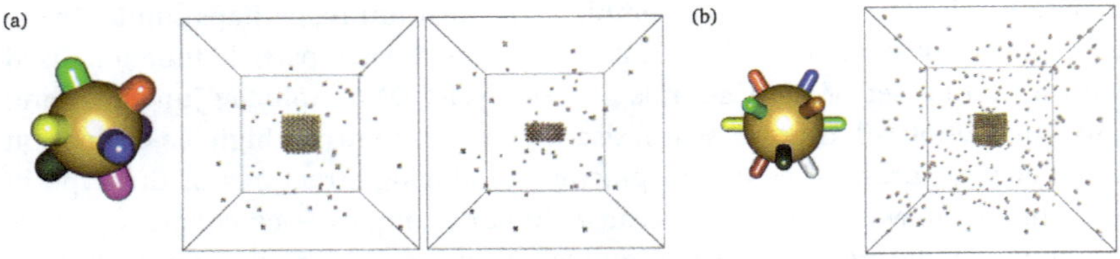

Fig. 1 A single monomer and snapshots towards the end of the nucleation process of some structures assembled from a vapour of monomers for the (a) 8 (332 K) and (b) 12 (345 K) co-ordinate monomers. The target structures in each case were simple rectangular parallelepipeds. In the simulation snapshots, correctly bonded clusters are shown in the same colour, but each particle, and each patch, is in fact distinct.

in which a particle of type i has been placed at a random position and with a random orientation in the simulation box is accepted with probability‡[27]

$$P_{\text{acc}}^{\text{add}} = \min\left[1, \frac{Vz_i}{N_i+1}\exp(-\Delta E/k_{\text{B}}T)\right], \tag{1}$$

where V is the volume of the simulation box, N_i is the current number of particles of type i in the system, ΔE is the trial change in the system's potential energy, and z_i is the fugacity of particles of type i, *i.e.* $z_i = \exp(\mu_i/k_{\text{B}}T)$, where μ_i is the particle's chemical potential. The ideal chemical potential is given by $\mu_{\text{id}} = k_{\text{B}}T\ln\rho$, where ρ is the number density; in the absence of interactions, the fugacity thus determines the target number density. An analogous acceptance probability holds for particle removals,

$$P_{\text{acc}}^{\text{rem}} = \min\left[1, \frac{N_i}{Vz_iN_{\text{types}}}\exp(\Delta E/k_{\text{B}}T)\right], \tag{2}$$

where N_{types} is the number of types of particle in the system. This accounts for the fact that when we add a particle, we choose its type uniformly at random, whereas when we remove a particle, we choose the particle at random: in order to obey detailed balance, we must account for the probability of choosing a particle of each type.

3 Results and discussion

We have previously considered tetrahedral co-ordination as applicable to DNA bricks. Here, we investigate the self-assembly behaviour of structures with co-ordination numbers of 8 (giving bcc-like target structures) and 12 (giving fcc-like target structures). The corresponding building blocks and sample target structures are shown in Fig. 1. The sequences associated with each patch for the structures we have studied are provided as supporting data.§ In the simulations reported here, the number of distinct particles in the target structures was 396 for the 4-, 403 for the 8- and 256 for the 12-co-ordinated structures.

‡ The de Broglie thermal wavelength is subsumed into the chemical potential, and cancels out in the case of an ideal chemical potential. For convenience, we have therefore set it to unity.

§ It is important to bear in mind that, if these sequences are chosen randomly, the temperatures at which nucleation and growth occur can change by a few degrees in identical conditions.[10] The temperatures we quote in the text refer to these specific DNA sequences. While the numerical values change with sequence choice, the qualitative behaviour does not.

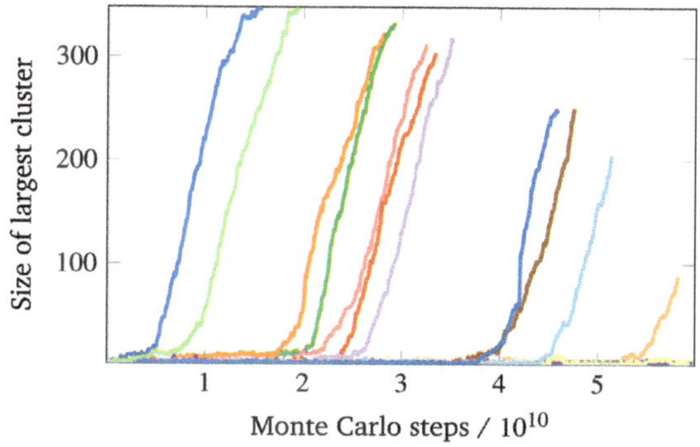

Fig. 2 The size of the largest cluster in the system as a function of Monte Carlo time for a canonical simulation with a total of 403 distinct particles with a co-ordination number of 8. $T = 332$ K, $\rho = 1/(62a)^3$. The different colours correspond to individual Monte Carlo trajectories starting from an equilibrated vapour of monomers. These trajectories were run for a fixed real-time; since virtual moves make simulations of larger clusters slower, simulations in which nucleation occurred later could run for a larger number of Monte Carlo steps.

Contrary to expectation,[16] brute-force simulations starting from a vapour of one copy of each of the monomers required to assemble a single target structure can, within a narrow temperature window, result in the successful self-assembly of the target structures shown in Fig. 1. Moreover, as evident from Fig. 2, this process is stochastic: under identical thermodynamic conditions, systems can exhibit drastically different lag times before any significant growth occurs. This is indicative of the presence of a free-energy barrier to nucleation, whereby a cluster of a sufficient size must form spontaneously before further growth is thermodynamically favoured. Since monomers coming together to form such a cluster lose a significant amount of translational and orientational entropy, this happens infrequently: there is a free-energy barrier associated with nucleation. Using umbrella sampling, we have calculated this free-energy barrier¶ for the two target structures shown in Fig. 1 at a number of temperatures, as shown in Fig. 3, where we also show a free-energy barrier for a reference tetrahedral system. Of course higher co-ordination structures are more stable at higher temperatures, since such structures entail many more bonds, and so the temperature scale at which nucleation occurs depends on the co-ordination number. Figs 2 and 3 indicate that the process is indeed nucleation-initiated for both the 8- and 12-co-ordinated target structures. However, whilst the process remains nucleation-initiated, there are significant differences in the systems' behaviours relative to the self-assembly of tetrahedral particles.

In particular, tetrahedrally co-ordinated structures, which include the experimentally studied DNA bricks, exhibit a free-energy barrier with a distinct jagged

¶ The order parameter used as a collective variable, *i.e.* the number of particles in the largest cluster, is a convenient choice consistent with classical nucleation theory. However, because each particle is different in these simulations, any *particular* cluster that forms can behave rather differently from this averaged behaviour. This is especially important if the cluster under consideration forms near a face or an edge of the target structure, where the average environments are different from those at the centre of the structure.

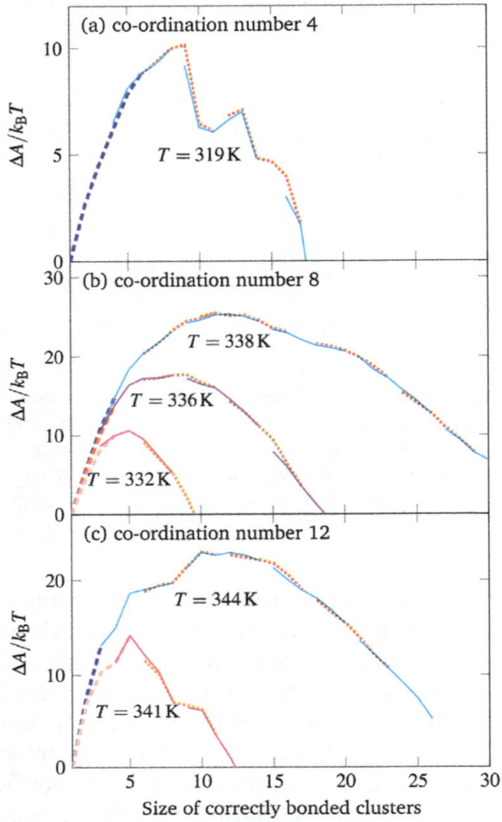

Fig. 3 The free-energy profile for cluster growth of particles with a co-ordination number of (a) 4, (b) 8 and (c) 12. Simulation results from different umbrella sampling windows are depicted in alternating styles to show their overlap. The thick dashed line corresponds to brute-force simulations.[10] In each case, there was one copy of each particle in a simulation box of dimensions $62a \times 62a \times 62a$.

appearance. This is not an artefact of the simulation technique used or a lack of equilibration, but rather reflects the fact that as clusters grow, there is a competition between the entropy loss associated with monomers losing their translational and orientational degrees of freedom when they are attached to a larger cluster and the energy gain associated with the formation of 'designed' interactions, which are, by construction, highly favourable. Tetrahedral structures grow in a very predictable fashion, with steps at which clusters can form closed cycles, for which the entropic penalty is compensated by not one, but two designed bonds forming, having a considerably lower free energy than other steps do.[10,15,16] The critical cluster for tetrahedrally co-ordinated structures is typically bicyclic or tricyclic (adamantane-like) with a single particle missing,[10,15,16] *i.e.* the size of the critical cluster is typically 8 or 9, and this cluster size appears to be essentially temperature independent in the regime where nucleation can occur. By contrast, the free-energy profiles shown in Fig. 3, in agreement with our theoretical prediction,[16] are considerably smoother, and the temperature greatly affects the size of the critical cluster. The reason for this behaviour is that there are a considerably larger number of possibilities of forming different clusters comprising the same number of building blocks;[16] this makes the nucleation considerably more classical, affecting both the smoothness and the dependence of the critical nucleus size on temperature.

However, despite this quite different behaviour at high temperatures, the systems behave in a less divergent manner at temperatures where the nucleation barrier is sufficiently small compared to the thermal energy that nucleation can reasonably be expected to occur. The degree of supercooling required in order to observe a nucleation event is not significantly different amongst the structures we have studied: if we deem the temperature at which a pre-formed target structure fully 'melts' to be an effective melting point, nucleation becomes sufficiently fast to observe in brute-force simulations at a supercooling of approximately 2% for all target structures considered. The point at which mass aggregation occurs is also similar, at roughly 4% supercooling. These results indicate that a more optimistic view of the possibility of assembling high co-ordination number structures is perhaps warranted.

Nevertheless, one difference in the behaviour observed is noteworthy. At temperatures at which there is a reasonably small free-energy barrier to nucleation, the driving force for growth is considerably larger for higher co-ordination number structures. One proxy for this is the gradient of the free-energy profile at post-critical cluster sizes: this gradient has roughly the same value ($\sim -1.1k_BT$ per particle) in the tetrahedral case where the critical free-energy barrier height is approximately $10k_BT$, and in the 8-co-ordinate structure at 338 K with a critical free-energy barrier height of $25k_BT$. As the temperature is decreased, the effective supersaturation increases: at 332 K, the large-cluster gradient of the free energy is already $-5k_BT$ per particle. This means that the conditions in which the 8- and 12-co-ordinate structures grow are considerably harsher than in the tetrahedral case, which is likely to lead to more mistakes during assembly.[11]

In simulations where only one particle of each component is present, the increased supersaturation may not interfere with correct self-assembly, since competing structures are less likely to occur. Of course, in experiments, many copies of each building block are present. To investigate whether higher co-ordination number structures can still form in circumstances where competition from additional monomers and clusters is possible, we have also simulated the self-assembly process in the grand-canonical ensemble. We have run simulations at a fugacity corresponding to the same ideal number density as in the canonical simulations, starting from an empty simulation box of various volumes, and we observe successful self-assembly to completion at a number of temperatures for both the 8- and 12-co-ordinate structures.|| Correctly assembled clusters grow one-by-one in such simulations: at sufficiently high temperatures, nucleation remains a rare event and the clusters grow essentially to completion before additional clusters nucleate. This observation supports the conclusion from canonical simulations that nucleation helps to prevent cluster interactions. These

|| In the grand ensemble, the stability of the target structure at temperatures at which nucleation occurs changes with the co-ordination number: for the tetrahedral structures, partially formed structures dominate, whilst for high co-ordination structures, essentially fully formed structures result at the end of the self-assembly process.

grand-canonical simulations also confirm that the lack of competition from monomers and clusters in solution is not the principal reason why self-assembly can succeed in the canonical ensemble, and the self-assembly process is surprisingly robust.

Despite this apparent success, the prediction that the greater supersaturation leads to more defects does hold. If we compare the largest assembled structures in the grand ensemble at the highest temperature at which nucleation was found to occur for co-ordination numbers of 4 and 12 (319 K and 338 K, respectively), the high co-ordination number structures typically have one or two incorrect particles embedded in the structure, and one or two vacancies, whilst the tetrahedral structure is entirely error free. The error rates would, moreover, be expected to be higher still if we implemented a 'kinetic constraint' to prevent the change of state for any particle wholly within the solid structure to account for the relative slowness of the relaxation dynamics within a solid structure:[28] this would, in particular, prevent vacancies from being filled when the rest of the structure has already formed around them. While the number of defects in absolute terms is not large even for the high co-ordination number structures, it is worth bearing in mind that incorrect particles on the surface of the cluster can lead to additional undesired clustering as the temperature is lowered and the clusters are allowed to undergo diffusion for long periods of time.

One way in which the driving force for nucleation can be changed is by strengthening or weakening the average bond energy between particles. When using DNA bases, this can be achieved by varying the proportion of G and C bases at the expense of A and T: the larger the GC content, the stronger on average the hybridisation between two complementary strands will be.[26] We have therefore simulated the self-assembly of the same target structures, but with differently chosen patch sequences. These are still chosen randomly, but with an appropriate bias towards either GC or AT base pairs.** Because the DNA hybridisation free energy itself depends strongly on the temperature, changing the bond strengths in this way is not equivalent to simply shifting the temperature scale. We show in Fig. 4 some additional free-energy barriers calculated for a system with stronger average interactions. While the basic behaviour remains unchanged, the different temperatures at which nucleation becomes feasible do affect the driving force for growth and thus the likelihood of defects occurring during the process. For example, if we compare the curves corresponding to $T = 352$ K and $T = 344$ K in Fig. 4, the system with weaker bonds has a less negative large-cluster gradient of the free energy as a function of the largest cluster size ($\sim -1.1k_BT$ per particle compared to $\sim -1.5k_BT$ per particle) and thus has a weaker driving force for growth, even though the nucleation free-energy barrier is considerably smaller ($23k_BT$ compared to 29.5 k_BT). Moreover, the system with stronger bonds appears to grow with more defects in a grand-canonical simulation, with typically three or four incorrect particles bonded in the final structure. A judicious choice of DNA sequences can thus significantly affect the probability that high co-ordination number structures in particular can grow in a reasonably error-free manner.

One of the main advantages of the work on DNA bricks has been their modularity, in the sense that a large range of target structures have been

** Terminal poly-T sequences are ignored in the GC content calculation.

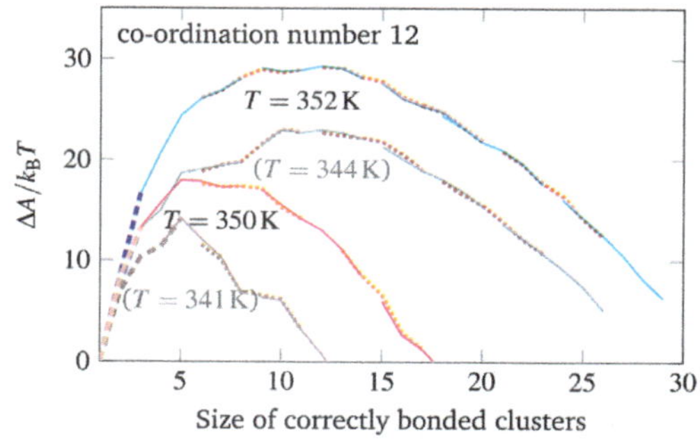

Fig. 4 The free-energy profile for cluster growth of particles with a co-ordination number of 12 and a GC content of 68%. The free-energy profiles of Fig. 3 are reproduced in a greyed-out hue.

assembled from essentially the same building blocks: the cubic target structures considered so far can be thought of as a 'molecular canvas'.[4,5] It is possible in experiment to construct more intricate structures simply by excluding the undesired bricks from the assembly pot, although in practice, poly-T DNA strands were used at every non-bonded position to minimise undesired interactions. To verify that this modularity continues to be a feature of target structures with a higher co-ordination number, we have run grand-canonical simulations with certain building blocks simply missing. This results in the self-assembly of more complex target structures, exactly as expected. We show two structures that have formed in such conditions in Fig. 5: a 'top hat' style structure and a cube with a cavity. The self-assembly of these target structures from a cubic canvas confirms that the modularity of the building blocks remains a feature in these high-co-ordination number structures.

Moreover, we have run simulations in which the target number density of the undesired building blocks is not set to zero, but rather to a finite but small number. In principle, one would expect that the undesired building blocks need not be completely absent from the reaction mixture, but must simply be vastly outnumbered by the correct building blocks. Our simulations suggest that this is indeed possible, but the fugacities (and hence the solution number densities) of the undesired particles must be set to very low values in order to form the target structure reproducibly. The precise value of the required fugacity depends on the environment of the undesired particles in the underlying canvas structure. For example, for structures with a co-ordination number of 12, if the target structure is a 'top hat' (Fig. 5(b)), most of the undesired particles are outlying particles with relatively few bonds connecting them to the remaining structure. It thus proves possible to form the desired target structure reliably when the undesired particle fugacities are set to approximately 0.5% of the desired particle fugacities of $z_{\rm des} = 2/(78a)^3$ (where $78a$ is the length of the simulation box in lattice units). Larger 'undesired' fugacities result in considerable attachment of the undesired particles over time. However, if the target structure is the central cavity structure of Fig. 5(c), most of the undesired particles are at the centre of the cubic canvas and any undesired bonding that does occur is rather stable; therefore an even lower

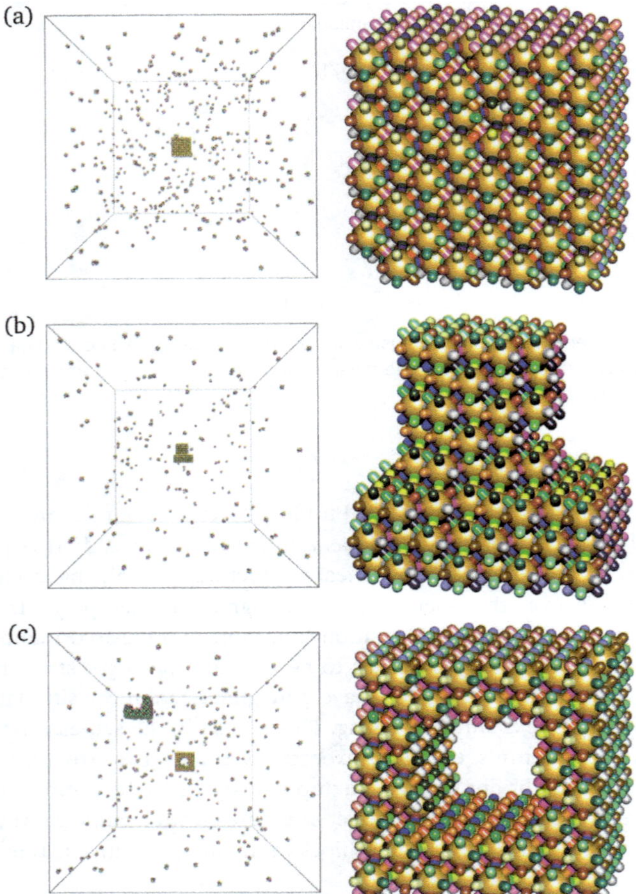

Fig. 5 Snapshots from grand-canonical ensemble simulations of 12-co-ordinate particles with a 68% GC content. $T = 344$ K. The fugacity of all 'desired' particle types is set to $z_{des} = 2/(78a)^3$, where $78a$ is the length of the simulation box in lattice units. All simulation snapshots shown here were obtained from the same building blocks, but with the fugacity of particles not part of the 'desired' structure set to zero. In each case, the whole simulation box and a close-up of the largest cluster are shown. (a) Original cubic target structure. (b) 'Top hat' structure. (c) Central cavity structure.

concentration of undesired particles is required in order to be able to self-assemble the target structure robustly.†† Whilst in theory, designed structures can form in a modular way even when the solution concentration of undesired particles is non-zero, if the target structures are not passivated as they are in experiment (with a poly-T sequence assigned to outlying non-bonding portions of the single-stranded DNA), there is always the chance that at least some undesired

†† In addition, such a structure is considerably more difficult to nucleate than the full cube, since the nucleus that forms must be near the edges of the target structure and has, of necessity, fewer bonds and is thus less stable.

particles will attach to the structure, either during growth or once the desired structure is already fully assembled. In this sense, the experimental strategy of passivating the outer surfaces appears to be very important and permits the desired structure to be assembled even in slightly unclean environments.

4 Conclusions

We have shown that, using a simplified computational model for addressable self-assembly, we are able to self-assemble structures with co-ordination numbers as high as 12. This was a somewhat unexpected result, because we had previously predicted that such structures will exhibit free-energy barriers to nucleation very different – and less conducive to self-assembly – from those previously determined for tetrahedrally co-ordinated structures. Our theoretical work suggested that the nucleation barriers would be less jagged in appearance and much more classical in shape. We predicted that this indicated that self-assembly would be considerably more challenging, because the supersaturation required for nucleation free-energy barriers to be surmountable would need to be greater: so great, we hypothesised, that competition from misassembled structures would dominate and it would be impossible for high co-ordination number structures to be assembled spontaneously in high yield.[16]

Indeed, the theoretical predictions we made about the free-energy barrier are borne out in simulations, but the hypothesis that such structures would be impossible to form is not. We have shown that the free-energy barriers do indeed become less jagged, the critical cluster size is considerably more temperature-dependent and it is more difficult to find mild conditions under which error-free self-assembly can occur. However, we have shown that despite this, it is still possible to find conditions under which the nucleation free-energy barrier is large enough that nucleation is rare, but sufficiently small that it can nonetheless sometimes occur, in conditions under which the stable structure lies along the pathway towards the formation of a fully assembled and designed target structure. This is very good news, because it gives us some confidence that higher co-ordination number structures, which are expected to be of considerable interest in nanotechnology, may indeed be possible to assemble using only a simple protocol.

We have also shown that the design process is modular in much the same way as it is for DNA bricks and that the designed structures self-assemble reproducibly in computer simulations. However, it is necessary to qualify these successes of the simulation method. The computational model we have used to study these effects is very crude and neglects a number of aspects that are likely to be important in any experimental realisation. Notwithstanding the molecular-level mechanisms of DNA hybridisation that have been coarse-grained away, one particular limitation of the model we have used is that it is a lattice model, which over-constrains the geometry of the growing structures and favours their successful assembly. This geometric constraint may be a significant issue in experimental work, perhaps especially so if DNA multi-arm motifs rather than coated colloidal particles were used in the assembly process, as they are themselves not very stiff, and the resulting poor geometry of the growing cluster may significantly retard the growth process. Such additional geometric considerations may cause difficulties not only during the nucleation stage itself, where the additional loss of entropy of the monomers required to form a compact structure would likely increase the height of the free-energy barrier, but because of the time involved in the reorganisation of the monomer structure when bonding to the growing clusters, they may also reduce the ratio of the rate of cluster growth relative to cluster diffusion. This may make it more likely for different clusters in the system to meet and interact, frustrating their correct assembly. It would be useful in future work therefore to characterise more fully the effect of the cooling protocol on addressable self-assembly.

These considerations may mean that not all possible experimental approaches to many-component building blocks may result in successful self-assembly, and so experimental success is far from guaranteed. It is likely to be the case that an experimental realisation of such building blocks may involve a significant investment of time, effort and not least money. Nevertheless, since we have shown that high co-ordination number self-assembly is computationally feasible, this indicates that the underlying physics does not preclude such structures from self-assembling: we hope this will help to stimulate experimental efforts to achieve similar complexity.

Acknowledgements

We thank William M. Jacobs and David M. Smith for helpful discussions. This work was supported by the Engineering and Physical Sciences Research Council [Programme Grant EP/I001352/1]. Research was carried out in part at the Center for Functional Nanomaterials, Brookhaven National Laboratory, which is supported by the US Department of Energy, Office of Basic Energy Sciences, under Contract No. DE-SC0012704. Supporting data are available at https://www.repository.cam.ac.uk/handle/1810/250582.

References

1 S. Whitelam and R. L. Jack, The statistical mechanics of dynamic pathways to self-assembly, *Annu. Rev. Phys. Chem.*, 2015, **66**, 143–163, DOI: 10.1146/annurev-physchem-040214-121215.

2 D. Frenkel, Order through entropy, *Nat. Mater.*, 2015, **14**, 9–12, DOI: 10.1038/nmat4178.

3 (*a*) V. Linko and H. Dietz, The enabled state of DNA nanotechnology, *Curr. Opin. Biotechnol.*, 2013, **24**, 555–561, DOI: 10.1016/j.copbio.2013.02.001; (*b*) J.-P. Sobczak and H. Dietz, DNA–nanoparticle crystals: exploiting shape complementarity, *Nat. Mater.*, 2015, **14**, 752–754, DOI: 10.1038/nmat4372.

4 Y. Ke, L. L. Ong, W. M. Shih and P. Yin, Three-dimensional structures self-assembled from DNA bricks, *Science*, 2012, **338**, 1177–1183, DOI: 10.1126/science.1227268.

5 B. Wei, M. Dai and P. Yin, Complex shapes self-assembled from single-stranded DNA tiles, *Nature*, 2012, **485**, 623–626, DOI: 10.1038/nature11075.

6 Z. Zhang, J. Song, F. Besenbacher, M. Dong and K. V. Gothelf, Self-assembly of DNA origami and single-stranded tile structures at room temperature, *Angew. Chem., Int. Ed.*, 2013, **52**, 9219–9223, DOI: 10.1002/anie.201303611.

7 B. Wei, M. Dai, C. Myhrvold, Y. Ke, R. Jungmann and P. Yin, Design space for complex DNA structures, *J. Am. Chem. Soc.*, 2013, **135**, 18080–18088, DOI: 10.1021/ja4062294.

8 Y. Ke, L. L. Ong, W. Sun, J. Song, M. Dong, W. M. Shih and P. Yin, DNA brick crystals with prescribed depths, *Nat. Chem.*, 2014, **6**, 994–1002, DOI: 10.1038/nchem.2083.

9 J. D. Halverson and A. V. Tkachenko, DNA-programmed mesoscopic architecture, *Phys. Rev. E: Stat., Nonlinear, Soft Matter Phys.*, 2013, **87**, 062310, DOI: 10.1103/physreve.87.062310.

10 A. Reinhardt and D. Frenkel, Numerical evidence for nucleated self-assembly of DNA brick structures, *Phys. Rev. Lett.*, 2014, **112**, 238103, DOI: 10.1103/physrevlett.112.238103.

11 L. O. Hedges, R. V. Mannige and S. Whitelam, Growth of equilibrium structures built from a large number of distinct component types, *Soft Matter*, 2014, **10**, 6404–6416, DOI: 10.1039/c4sm01021c.

12 J. Zenk and R. Schulman, An assembly funnel makes biomolecular complex assembly efficient, *PLoS One*, 2014, **9**, e111233, DOI: 10.1371/journal.pone.0111233.

13 Z. Zeravcic, V. N. Manoharan and M. P. Brenner, Size limits of self-assembled colloidal structures made using specific interactions, *Proc. Natl. Acad. Sci. U. S. A.*, 2014, **111**, 15918–15923, DOI: 10.1073/pnas.1411765111.

14 A. Murugan, Z. Zeravcic, M. P. Brenner and S. Leibler, Multifarious assembly mixtures: systems allowing retrieval of diverse stored structures, *Proc. Natl. Acad. Sci. U. S. A.*, 2015, **112**, 54–59, DOI: 10.1073/pnas.1413941112.

15 W. M. Jacobs, A. Reinhardt and D. Frenkel, Communication: theoretical prediction of free-energy landscapes for complex self-assembly, *J. Chem. Phys.*, 2015, **142**, 021101, DOI: 10.1063/1.4905670.

16 W. M. Jacobs, A. Reinhardt and D. Frenkel, Rational design of self-assembly pathways for complex multicomponent structures, *Proc. Natl. Acad. Sci. U. S. A.*, 2015, **112**, 6313–6318, DOI: 10.1073/pnas.1502210112.

17 S. Whitelam, Hierarchical assembly may be a way to make large information-rich structures, *Soft Matter*, 2015, **11**, 8225–8235.

18 J. Madge and M. A. Miller, Design strategies for self-assembly of discrete targets, *J. Chem. Phys.*, 2015, **143**, 044905, DOI: 10.1063/1.4927671.

19 Y. Wang, Y. Wang, D. R. Breed, V. N. Manoharan, L. Feng, A. D. Hollingsworth, M. Weck and D. J. Pine, Colloids with valence and specific directional bonding, *Nature*, 2012, **491**, 51–55, DOI: 10.1038/nature11564.

20 F. Lu, K. G. Yager, Y. Zhang, H. Xin and O. Gang, Superlattices assembled through shape-induced directional binding, *Nat. Commun.*, 2015, **6**, 6912, DOI: 10.1038/ncomms7912.

21 (*a*) R. Holliday, A mechanism for gene conversion in fungi, *Genet. Res.*, 1964, **5**, 282–304, DOI: 10.1017/s0016672300001233; (*b*) D. M. J. Lilley, Structures of helical junctions in nucleic acids, *Q. Rev. Biophys.*, 2000, **33**, 109–159; (*c*) C. Zhang, M. Su, Y. He, X. Zhao, P. Fang, A. E. Ribbe, W. Jiang and C. Mao, Conformational flexibility facilitates self-assembly of complex DNA nanostructures, *Proc. Natl. Acad. Sci. U. S. A.*, 2008, **105**, 10665–10669, DOI: 10.1073/pnas.0803841105.

22 N. Metropolis, A. W. Rosenbluth, M. N. Rosenbluth, A. H. Teller and E. Teller, Equation of state calculations by fast computing machines, *J. Chem. Phys.*, 1953, **21**, 1087–1092, DOI: 10.1063/1.1699114.

23 (*a*) G. M. Torrie and J. P. Valleau, Nonphysical sampling distributions in Monte Carlo free-energy estimation: umbrella sampling, *J. Comput. Phys.*, 1977, **23**, 187–199, DOI: 10.1016/0021-9991(77)90121-8; (*b*) M. Mezei, Adaptive umbrella sampling: self-consistent determination of the non-Boltzmann bias, *J. Comput. Phys.*, 1987, **68**, 237–248, DOI: 10.1016/0021-9991(87)90054-4.

24 B. Hetényi, K. Bernacki and B. J. Berne, Multiple "time step" Monte Carlo, *J. Chem. Phys.*, 2002, **117**, 8203–8207, DOI: 10.1063/1.1512645.

25 (*a*) S. Whitelam and P. L. Geissler, Avoiding unphysical kinetic traps in Monte Carlo simulations of strongly attractive particles, *J. Chem. Phys.*, 2007, **127**, 154101, DOI: 10.1063/1.2790421; (*b*) S. Whitelam and P. L. Geissler, Erratum: "Avoiding unphysical kinetic traps in Monte Carlo simulations of strongly attractive particles", *J. Chem. Phys.*, 2008, **128**, 219901, DOI: 10.1063/1.2931944.

26 J. SantaLucia Jr and D. Hicks, The thermodynamics of DNA structural motifs, *Annu. Rev. Biophys. Biomol. Struct.*, 2004, **33**, 415–440, DOI: 10.1146/annurev.biophys.32.110601.141800.

27 D. Frenkel and B. Smit, *Understanding molecular simulation: from algorithms to applications*, Elsevier Academic Press, London, 2nd edn, 2002.

28 R. V. Mannige and S. Whitelam, Predicting the outcome of the growth of binary solids far from equilibrium, 2015, e-print: arXiv:1507.06267.

Faraday Discussions

PAPER

How do polydisperse repulsive colloids crystallize?

Robert Botet,*[a] Bernard Cabane,[b] Lucas Goehring,[c] Joaquim Li[c] and Franck Artzner[d]

Received 14th October 2015, Accepted 4th November 2015
DOI: 10.1039/c5fd00145e

A modified version of the Gibbs-ensemble Monte-Carlo method reveals how polydisperse charged colloidal particles can build complex colloidal crystals. It provides general rules that are applicable to this fractionated crystallization that stems from size segregation. It explains the spontaneous formation of complex crystals with very large unit-cells in suspensions of nanoparticles with a broad size distribution.

1 Introduction

Fundamental physics requires experiments to observe, and theory to explain. In principle, nothing more is needed for a complete understanding of phenomena. However, in some situations, this sharing of the work is not sufficient, either because the experimental data are counter-intuitive (disagreement with the theory) or because the theory itself is impracticable. If one of these cases arises, numerical simulations turn out to be the essential tool to complement our ideas through direct 'observation' of the theory or through artificial changes of the microscopic rules (as in a Maxwell's demon-like approach to analyze cause-to-effect mechanisms).

In the present article, we discuss results of numerical simulations dedicated to the comprehension of recent experiments in which a dispersion of polydisperse repulsive nanoparticles was compressed through osmosis. Details of the experiments will be published elsewhere.[1] A suspension of ordinary colloidal silica particles (Ludox, polydispersity 14%, average radius 8 nm), was compressed through dialysis. This caused a spontaneous segregation of the particles, fractionated crystallization and multiple-phase coexistence (involving various crystal structures). For instance, the low-compression part of the phase sequence was seen to be a colloidal liquid at the lowest volume fractions, then nucleation of the

[a]*Laboratoire de Physique des Solides, CNRS UMR8502, Univ. Paris-Sud, Université Paris-Saclay, 91405 Orsay Cedex, France. E-mail: robert.botet@u-psud.fr; Fax: +33 1 6915 6086; Tel: +33 1 6915 6925*
[b]*LCMD, CNRS UMR 8231, ESPCI, 10 rue Vauquelin, 75231 Paris Cedex 05, France*
[c]*Max Planck Institute for Dynamics and Self-Organization (MPIDS), 37077 Göttingen, Germany*
[d]*Institut de Physique, CNRS UMR 6626, Univ. Rennes, 35042 Rennes, France*

BCC crystalline phase occurred at a silica volume fraction $\phi > 0.19$, and then there was coexistence of the BCC + Laves AB_2 crystal phases for $\phi > 0.22$. The complete phase diagram was found to be much richer than expected according to current theories.

It is important to note that we do not describe here ordinary fractionated crystallization caused by different solubilities of coexisting substances, but a new kind of fractionation that organizes populations of particles with different sizes, and makes the best use of available space. Without any theory at hand to explain the formation of such uncommon structures as the Laves AB_2 crystals,[2] we used the Gibbs-ensemble Monte-Carlo method[3] to understand the mechanisms used by the system to sort the various particle sizes and to put them in the correct places.

Actually, we encountered many unanswered questions from the experimental data. The numerical simulations helped us to answer some of them. We replicate hereafter this particular state of mind, dividing our comprehension process in a series of short questions and discussing the answers given by the numerical study. The reader interested in a quantitative comparison between the experimental data and the numerical simulations should refer to ref. 1.

To close this Introduction, let us give a warning which might be important when comparing experimental data with numerical simulations: the regular Monte-Carlo method does not consider real kinetic processes, but is an efficient tool to search for the equilibrium state using the physical ground of free energy minimization. Therefore, the results presented below are linked to the equilibrium state. If this is not the case (in experiments, in kinetic numerical simulations), other intermediate crystalline structures may appear for some time as unstable phases.

2 The system under study

We consider a dispersion of spherical hard particles. The surface of each particle is electrically charged with $0.4e$ per nm^2. The particle radius is denoted: a, and its value is Gaussian-distributed with the average value $\langle a \rangle = 8$ nm and standard deviation σ, truncated to the range: [4 nm; 12 nm] (that is: no radius is smaller than 4 nm or larger than 12 nm). The polydispersity, δ, of this population is defined as: $\delta^2 = \langle a^2 \rangle / \langle a \rangle^2 - 1$. Then, the value of δ coincides with $\sigma / \langle a \rangle$ for the small values of σ, and has the limiting value $1/\sqrt{12} \simeq 29\%$ for the infinite value of σ (rectangular distribution). Throughout the article, we shall take the value $\sigma = 1.2$ realizing the polydispersity $\delta = 15\%$ of the population.

If large polydispersity (say $\delta > 25\%$) has to be considered, the influence of moments of order larger than 2 should be studied, that is to investigate the various distributions – such as log-normal or Schulz distributions – with the same values of $\langle a \rangle$ and δ. In related problems, the shape of the radius-distribution is known to be unessential.[4]

We consider here the DLVO theory,[5] in which the interaction energy between two charged particles, labelled i and j, is generally of the Yukawa (screened-Coulomb) form:

$$U_{ij}/k_{\mathrm{B}}T = Z_i Z_j \frac{l_{\mathrm{B}}}{r_{ij}} \mathrm{e}^{-\kappa r_{ij}}, \quad \text{for } r_{ij} \geq a_i + a_j \tag{1}$$

where Z_i is the effective charge of the particle i, and r_{ij} the distance between their centers. $l_B = 0.7$ nm is the Bjerrum length, and $\kappa = 1/l_D$ is the screening parameter, with $l_D = 3$ nm the Debye screening length in our experimental conditions. The values of l_B and l_D are expected to be constant throughout the system.

Two definitions have to be stated:

• The energy at contact (*i.e.* using (1) for $r_{ij} = a_i + a_j$), traditionally written: $\beta\varepsilon$, is a central parameter for discussion. Its averaged value is $\beta\varepsilon = 21.6$ for the present set of parameters.

• The potential energy per particle at equilibrium is here denoted generically: E. It is defined as the sum of the pair-potentials (1) between all the couples of particles forming the equilibrated system, divided by the total number of particles, after having minimized the system free-energy using the Monte-Carlo method.

At small or intermediate volume fractions – namely: $\phi \lesssim 0.41$ for $\delta = 15\%$ – the Wigner–Seitz radius is larger than $\langle a \rangle + l_D$. In these conditions, the attractive van der Waals interaction between silica spheres in water is smaller than the thermal energy, k_BT, and thus can be neglected. In the same way, we do not discuss here controversial additional weak long-ranged attractive pairwise potentials such as those appearing in the Sogami–Ise theory.[6] Within our approach, such interactions would result in the same results as in the case of the purely repulsive DLVO potential, with an effective value of the osmotic pressure slightly larger than the experimental pressure.[7]

3 Q1: which of the FCC or BCC crystalline structures should appear for a system made of a polydisperse distribution of repulsive particles at the thermodynamic equilibrium?

A1: systems of monodisperse particles interacting through repulsive Yukawa potentials, form either FCC or BCC crystal structures in the equilibrium solid phase.[8] The BCC phase appears when $\phi < 0.5$ and the Debye screening length l_D is larger than a threshold depending on ϕ. In all other cases, the FCC phase is the solid structure at equilibrium.

In the case of polydisperse particles, the question is more complicated. Let us suppose first that the solid phase can only be FCC or BCC, without coexistence (note that, for the parameters as chosen, only the FCC crystalline phase can appear in the monodisperse case, since the value of $l_D/\langle a \rangle$ 0.38 is below the liquid–BCC–FCC triple critical point). We performed random exchanges of couples of particles in the system composed of either a single FCC phase or a single BCC phase, according to the standard Metropolis Monte-Carlo algorithm, until thermodynamic equilibrium was reached. The evolution of the difference of energy at equilibrium per particle, $(E_{FCC} - E_{BCC})/k_BT$, *versus* the volume fraction is plotted in Fig. 1, and the monodisperse case is also plotted for comparison. Unlike the monodisperse case, the polydisperse population is seen to undergo a clear change in the stability of the BCC/FCC structure: when $\phi > 0.3$, the BCC crystal is energetically more likely to appear than the FCC structure. Moreover this trend increases with the volume fraction.

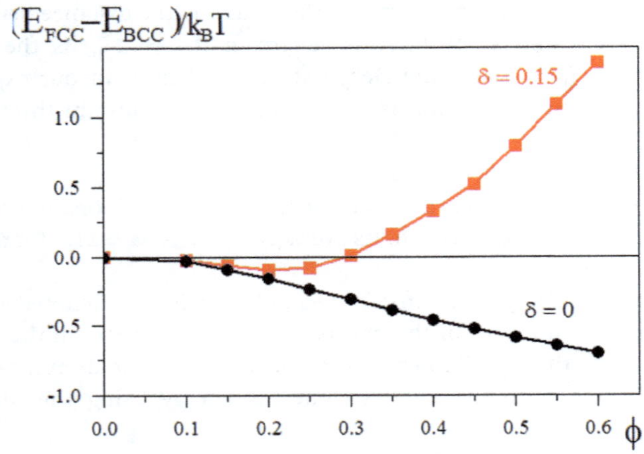

Fig. 1 Differences between the energies per particle of the FCC crystal and of the BCC crystal, at equilibrium, *versus* the system volume fraction, ϕ, for particles of average radius $\langle a \rangle = 8$ nm, Debye length $l_D = 3$ nm, density of charge $= 0.4$ nm^{-2}. Black circles are the exact calculation for the monodisperse case ($\delta = 0$); red squares are from the Metropolis Monte-Carlo method, 500 000 random particle-exchanges for each of the 50 realizations of the random radius-distribution with polydispersity $\delta = 0.15$, on the 2662 sites of the BCC crystal, or the 2916 sites of the FCC crystal. Positive values of the energy difference indicate that the BCC structure should be the stable phase at equilibrium. Then, polydispersity favors the BCC structure for the high volume fractions.

This is a sign of the correlated organization of the charges in the system, decreasing the energy of the BCC crystal more efficiently than for the FCC structure.

4 Q2: how do the polydisperse particles distribute in a system with the BCC crystal phase coexisting with FCC crystal phase?

A2: to understand why BCC becomes the stable phase when polydispersity is high enough, we consider the problem of the 2-phase system (coexistence): let us then suppose that the population of the polydisperse particles is arranged in part as a BCC lattice and in part as a FCC lattice, with proportions ρ_{BCC} and $\rho_{FCC} = 1 - \rho_{BCC}$, respectively.

To know the thermodynamic equilibrium of the system, we use a variant of the Gibbs ensemble Metropolis random exchanges of particles, introduced to study the thermodynamic evolution of different phases coexisting in a liquid.[9] Particles are exchanged randomly between the phases according to the usual Metropolis Monte-Carlo scheme, keeping constant throughout the total number of particles, the total volume of the system and the temperature. In this method, the homogeneous coexisting phases are simulated in separated boxes without interfaces between the phases, and each box has periodic boundary conditions. Particle exchanges occur either inside a single phase, or from different phases.

To make the approach more efficient, we sampled the configuration space into a finite number of configurations, for example taking the 19 configurations corresponding to the coexistence of one cubic BCC crystal and one cubic FCC crystal, the total sum of the sites being between 3500 and 4500 (in order to compare systems of about the same total sizes). When thermodynamic equilibrium is reached in every configuration, the total energy, E, of the system is evaluated for each configuration. The program then scanned, little by little, the energy landscape of (a sampling of) the configuration space, and the approximate partition function $Z = \sum_{\text{conf}} \exp(-E/k_BT)$ gives access to all thermodynamic quantities at equilibrium. The average value of the proportion of the BCC phase at equilibrium is estimated through:

$$\langle \rho_{\text{BCC}} \rangle = \frac{1}{Z} \sum_{\text{conf}} \rho_{\text{BCC}} \; e^{-E/k_BT} \tag{2}$$

and similarly for $\langle \rho_{\text{FCC}} \rangle$.

Let us discuss the results for the value $\phi = 0.3$ realizing the equality of the pure BCC and pure FCC energies, as seen in Fig. 1. As could be expected, $\langle \rho_{\text{BCC}} \rangle = \langle \rho_{\text{FCC}} \rangle = 0.5$ is obtained in this case because of the choice $E_{\text{BCC}} = E_{\text{FCC}}$ for the pure phases. However, the actual value of the energy per particle for BCC and FCC crystals of the same size, is found to be $E = 22.2\ k_BT$, significantly lower than the value for the single phase $E_{\text{BCC}} = E_{\text{FCC}} = 23.0\ k_BT$. This means that the system put different particle distributions in the different phases. This occurrence is exemplified in Fig. 2 where the at-equilibrium radius distributions in the two phases are compared for the configuration corresponding to a BCC crystal of 2000 sites, and a FCC crystal of 2048 sites. The radius-distribution in the FCC phase has approximately a Gaussian shape with a polydispersity of 6.5%, that is much smaller then the polydispersity (15%) of the total distribution including all the

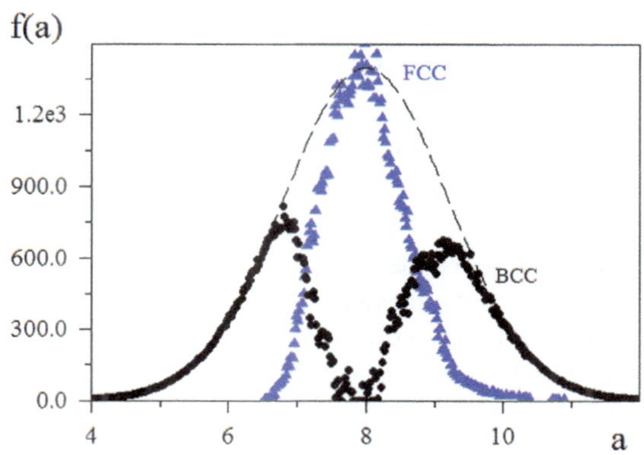

Fig. 2 At-equilibrium radius-distributions of the particles located in the phases of a system made of a FCC finite lattice of 2048 sites and of a BCC finite lattice of 2000 sites, both with periodic boundary conditions. The Gibbs Monte-Carlo scheme was applied with 300 000 random particle exchanges and 50 random realizations of the total radius-distribution (plotted as the dashed curve) with polydispersity $\delta = 0.15$, for the system volume fraction $\phi = 0.3$, all other parameters as listed in the caption of Fig. 1. The FCC phase takes all the particles with radii close to the average value $\langle a \rangle$, letting the wings of the distribution be managed by the BCC lattice. The polydispersity of the population of particles making the FCC lattice is 0.065, much smaller than the polydispersity of the total distribution, $\delta = 0.15$.

particles in the system. Consequently, both wings of the initial Gaussian distribution distribute in the BCC phase.

5 Q3: why does the BCC crystal phase accept handling two modes of particle radius while FCC sticks to one mode?

A3: the detailed management of the initial radius-distribution by the system results from the different abilities of the phases to accept particles of different radii (that is here, of different electric charges). Following this idea, one can define a parameter of tolerance as the maximum relative decrease of the system energy when changing the charges of a couple of particles, keeping the total charge unchanged. More precisely, let us define the tolerance, τ, by the expression:

$$\tau = \min\left\{-\frac{\Delta E}{U_o}\right\} \tag{3}$$

in which ΔE is the variation of the system energy when the charges of two particles are changed from the initial values Z to: $2Z$ and 0, respectively, and $U_o = Z^2\kappa l_B$ is a reference energy for the system. The minimum value in (3) is taken over all the possible positions of the couple of particles with modified charge. This concept supposes that the system is able to detect the best (from the point of view of the energy) relative positions of the two concerned particles. Then the repulsive interactions are expected to be long-ranged – as the Yukawa potential is – even if the characteristic lengths are finite. For a regular lattice with only one sort of site, the value of τ is given by the expression:

$$\tau = \frac{e^{-\kappa r_o}}{\kappa r_o} \tag{4}$$

where r_o is the value of the distance between the nearest neighbours.

For the BCC lattice of volume fraction ϕ, one has $r_o/a = \left(\pi\sqrt{27}/3\phi\right)^{1/3}$, and for the FCC lattice at the same volume fraction: $r_o/a = \left(\pi\sqrt{32}/3\phi\right)^{1/3}$. Then, $\tau_{BCC} > \tau_{FCC}$ for all values of ϕ (for $l_D = 3$ nm and $a = 8$ nm considered here: $\tau_{BCC}/\tau_{FCC} = 1.3$, for $\phi = 0.3$). This indicates that the FCC crystal phase is less tolerant than BCC, or, equivalently, that changing the charges of two particles as follows: $Z \rightarrow 2Z$, and $Z \rightarrow 0$, results in a larger decrease of the BCC lattice energy than in the FCC case. This is the reason why, from the point of view of the energy, the system prefers to fill the FCC lattice with particles of almost the same charges, and to arrange the rest of the particles in the BCC lattice.

As a conclusion, we propose here a 'tolerance rule' of the form: in the case of coexistence of several phases, the less tolerant phase prevails over all the other phases in the distribution of the radii.

6 Q4: what information can the method give about experiments of compression of a system of charged polydisperse particles?

A4: as explained in the Introduction, we recently explored the colloidal crystallization of polydisperse Ludox HS silica nanoparticles dispersed in water through near-equilibrium dialysis to achieve very slow compression of the polydisperse suspension.[1] We demonstrated that, in charge-stabilized colloidal dispersions at intermediate compressions, the system exhibits two crystalline phases, namely: a BCC phase and a Laves AB_2 phase, in coexistence with the liquid. Moreover, the FCC phase is missing. The Laves AB_2 crystalline system is known to be relevant for the opal structure,[10] and more generally for packing of bidisperse particles[11] with a radius ratio of $\sqrt{2/3} \simeq 0.82$. The tolerance of this phase, as defined in (3), is such that: if $\tau_{AB_2}/\tau_{BCC} = 3.55$, in the experimental conditions, then the Laves lattice is much more tolerant than BCC to accepting uneven particles. Therefore, an argument similar to the one developed in Section B, leads to the conclusion that the BCC phase should collect most of the particles close to the mode of the radius-distribution, while the wings of the distribution should fill the AB_2 phase.

To be more quantitative, the Gibbs ensemble Monte-Carlo method was used for the system made of a BCC phase in coexistence with an AB_2 phase. As previously, the ensemble average of each crystalline phase proportion is evaluated after sampling the configuration space. This results in the ensemble-average values: $\langle \rho_{BCC} \rangle = 0.52$; $\langle \rho_{AB_2} \rangle = 0.48$ corresponding to the respective volume proportions of the phases at equilibrium. Then, considering a system close to these values, that is: a BCC crystal of $9 \times 9 \times 9$ unit cells, and a Laves AB_2 crystal of $5 \times 5 \times 5$ unit cells, the resulting respective radius-distributions are plotted in Fig. 3.

Here too, we find that the less tolerant phase (the BCC phase in this case) collects most of the particles with radii close to the average radius. The final polydispersity of the BCC phase particles is 7.5% in this example, emphasizing the role of the repulsive interactions in the process, since this polydispersity significantly exceeds the terminal value 5.7% of the polydispersity of the crystal phase made of hard sphere colloidal particles precipitating from a fluid phase at equilibrium.[12,13] Moreover, the AB_2 phase is pretty well filled with particles with a particle-radius ratio of about 0.8, the bigger particles being located in the octahedral sites, and the smaller ones in the tetrahedral sites.

This suggests a simple "rule of thumb" for the respective proportions of the two phases: suppose that the radii of the particles filling the less tolerant phase are distributed in the Gaussian shape with polydispersity smaller than the overall polydispersity δ, and almost all the particles of radius $\langle a \rangle$ belong to this phase. Calculating the two modes of the remaining radius-distribution of the particles in the high-tolerant phase is then a simple exercise. If we constrain the ratio between the two radius modes to be $\sqrt{2/3}$ for the system to efficiently fit the AB_2 phase,† then one finds the relation:

$$\rho_B \simeq \frac{\delta^*}{\delta} \tag{5}$$

with ρ_B the proportion of the less-tolerant phase at equilibrium, and $\delta^{\star} = 5 - \sqrt{24} \simeq 10\%$. In (5), we approximated: $2 \ln \rho_B/(1 - 1/{\rho_B}^2) \simeq \rho_B$.

Interpretation of eqn (5) is as follows: when $\delta < \delta^*$ only the less-tolerant phase (here: BCC) can exist; when $\delta > \delta^*$, the proportion of the BCC phase is a decreasing

† In principle, one should consider the ratio of effective radii based on the energy density. This does not change eqn (5) but the value of the limit of polydispersity δ^* which generally depends on the system volume fraction.

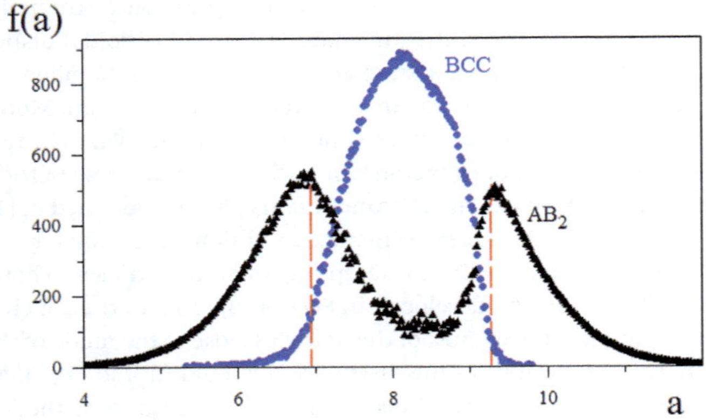

Fig. 3 At-equilibrium radius-distributions of the particles located in the phases of a system made of a BCC crystal of 1458 sites and of a Laves AB_2 crystal of 1500 sites, both with periodic boundary conditions. The Gibbs Monte-Carlo scheme was applied with 300 000 random particle exchanges and 50 random realizations of the total Gaussian radius-distribution with $\langle a \rangle = 8$ nm, polydispersity $\delta = 0.15$, for the system volume fraction $\phi = 0.3$, all other parameters were as listed in the caption of Fig. 1. The BCC phase takes most of the particles with radii close to the average value $\langle a \rangle$, letting the wings of the distribution be managed by the AB_2 lattice. The polydispersity of the population of particles making the BCC lattice is 7.5% compared with the polydispersity of the total distribution, $\delta = 15\%$. The two modes of the AB_2 radius-distribution are the vertical red segments, respectively at 6.9 nm for the tetrahedral sites, and at 9.3 nm for the octahedral sites. The ratio between these two values is 0.74, close to the value 0.82 for the AB_2 packing of hard spheres. Compared with Fig. 2, one can easily understand that the system prefers the present configuration for which all the modes are nearly at the correct place (unlike the BCC phase in Fig. 2, which had to be filled with a bi-disperse distribution).

function of the polydispersity, δ, of the system, that is: polydispersity favors the search for more tolerant phases.

It is worth noting that the appearance of the Laves phase in this context can be related to a work on bidisperse particles.[14] Indeed, such complex crystal structures were found to be spontaneously stabilized in the case of mixtures of spherical particles with two different diameters, interacting *via* simple repulsive potentials. This was shown by numerical simulations using the thermodynamic integration technique. However, our case is *a priori* much more surprising since there is not, at the beginning, any 'magic' radii ratio.

7 Q5: how long-ranged should the interaction be between the particles?

A5: the physical parameters that we have considered up to now are rather typical of short-range interactions, namely: an exponential decrease in (1) with a characteristic length $l_D = 3$ nm, much smaller than the nearest-neighbor distances, which are of the order 20 nm. Then, it is natural to ask the question whether only the nearest-neighbors are important or if the farther neighbors play a role too.

A quick answer is given considering the monodisperse case for the BCC or FCC lattices. In these cases, one knows the analytical formula for the energy per

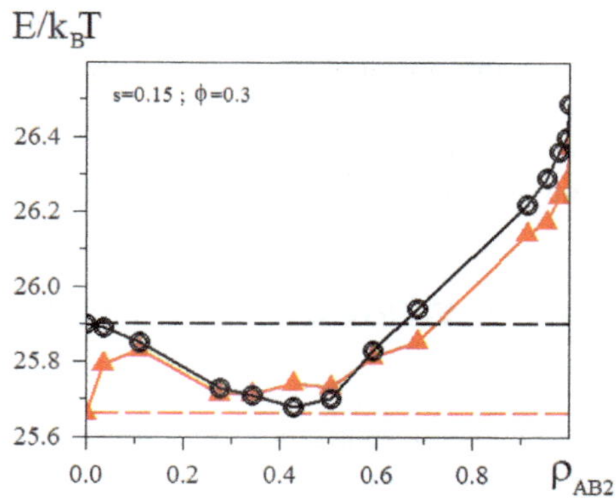

Fig. 4 Energies at equilibrium per particle of the system of the BCC crystal phase in the proportion $1 - \rho_{AB_2}$ coexisting with a Laves AB_2 crystal phase in the proportion ρ_{AB_2}. Equilibrium is reached after 300 000 random particle exchanges and 50 random realizations of the Gaussian radius-distribution with polydispersity $\delta = 0.15$, for the system volume fraction $\phi = 0.3$ and total number of sites between 2500 and 3500, all other parameters as listed in the caption of Fig. 1. The open circles are the total Yukawa interaction, while the red triangles are the Yukawa interaction truncated at the nearest-neighbors (the respective black and red continuous line are guides for the eyes). The black (respectively red) dashed line is the energy of the pure BCC phase considering all the neighbors (respectively only the first-neighbors). The statistical error bars are of the order $0.1k_BT$. When all the neighbors are included, the system energy reaches the minimum value for $\rho_{AB_2} \simeq 0.42$ (to be compared to the average value $\langle\rho_{AB_2}\rangle \simeq 0.48$ found in Section 6 using the partition function). On the other hand, all the results including only the nearest-neighbors are found to be above the energy of the pure BCC lattice, indicating that the system should not consider the AB_2 phase if the Yukawa interaction was truncated. As a matter of fact, the difference between the nearest-neighbors/all-neighbors is essentially important only for the less tolerant crystalline structure in this case (*i.e.* the BCC phase).

particle as a series in the order of the neighbors, then it is easy to compare the energy when only the nearest-neighbors are considered. For the same parameters as in the caption of Fig. 1 and $\phi = 0.3$, only the first-neighbors give the results: $E^{(1)}_{BCC}/k_BT = 24.98 < E^{(1)}_{FCC}/k_BT = 29.79$ which seems to show that the BCC crystal is more stable than the FCC crystal. But including the second-neighbors turns round that conclusion, as: $E^{(2)}_{BCC}/k_BT = 30.47 > E^{(2)}_{FCC}/k_BT = 30.32$. Taking into account all the neighbors confirms the latter conclusion: $E^{(\infty)}_{BCC}/k_BT = 30.82 > E^{(\infty)}_{FCC}/k_BT = 30.51$. Actually, this discussion was about the point $\delta = 0$; $\phi = 0.3$; $(E_{FCC} - E_{BCC})/k_BT = -0.31$ of Fig. 1. Then, even if $l_D/r_o \ll 1$, one must consider more than only the nearest-neighbors to get the correct conclusion about the stability of the crystals.

To check the relevance of this 'long-range *versus* short-range' question in the case of the coexistence of the BCC and AB_2 phases of polydisperse particles, we cut the Yukawa potential (1) right after the first nearest-neighbors (three kinds of sites are to be considered: one for the BCC lattice, the tetrahedral site and the octahedral site of the AB_2 lattice). Then, the second- and farther neighbors of a given particle do not interact with it any more. The main results are conveniently visualized in Fig. 4, after plotting the energies per particle of the system in phase coexistence *versus* the composition of the system (characterized by: $\rho_{BCC} = 1 - \rho_{AB_2}$ and ρ_{AB_2} as the respective volume proportions).

The conclusion is that the number of neighbors considered is relevant in the energy profile *versus* the composition of the system. In the example in Fig. 4, the system is probably a pure BCC crystal phase if the Yukawa interaction is cut at the first-neighbors, while coexistence of the BCC and AB_2 phases is clear when all the neighbors are taken into account.

8 Conclusions – what did you learn from the method?

In the case of the fractionated crystallization, the Gibbs ensemble Monte-Carlo method does not predict the nature of the phases at equilibrium. However, it can provide essential help for a quantitative description of the crystals in coexistence (such as: the respective volumes of the phases, the particle size-distributions in each phase, *etc.*) as well as the fundamental mechanisms leading to the final composition at equilibrium, within the framework of the DLVO theory. In the two cases presented here (the FCC + BCC and the BCC + AB_2 systems), spontaneous self-assembly processes sort out particles from a broad polydisperse population and build different sets of colloidal crystals with structures that are adapted to a best use of this population. One process selects the most popular particles to build a first set of crystals (the less tolerant crystal) and "burns" a hole in the original population of particle sizes. The residual particle population then resembles a bi-disperse mixture, and a second process makes use of the remaining sub-populations on either side of the "hole" to construct other crystals (more tolerant). This process of sieving is probably general and leads naturally to a sequence of fractionated colloid crystallization governed by the tolerance rule. On the basis of what we know about binary nanoparticle systems,[11,15] we can expect still more complex superlattices to be built spontaneously that way for still larger polydispersities.

This description of the mechanisms is close to the idea by Pusey[16] in a different context – an idea which has been validated much later through numerical simulations[17] and theoretical developments[13] – that suspensions of highly polydisperse hard spheres might crystallize by splitting the broad overall distribution into a number of narrower distributions, each of which being accommodated within a single crystalline phase.

We shall close this presentation with three discussions about the limitations of the numerical method.

• The method does not consider the real kinetics of the mechanisms, then reliability of the method for large volume fractions may be questionable since movements of the particles can be hindered. However, for systems of spherical particles, repulsive interactions act as lubrication between the particles and generate ergodicity. In other words, local rearrangement of the particles (related to the system capability to reach thermodynamic equilibrium) is facilitated by the repulsive interactions, and more precisely by the value of the energy at contact. This is not the case in the opposite situation of attractive interactions, resulting generally in non-equilibrium patterns.[18]

• In principle, the method cannot predict the phases at equilibrium, since it supposes to have *a priori* information – or to guess – about the phases in coexistence. However, one should note that if large unit crystal cells are being formed in the system, no numerical kinetic method (*e.g.* molecular dynamics) can nowadays predict the phases, due to essential computing limitations.

• A last but fundamental question is the role of the liquid phase in coexistence with the crystals. In principle, the same Gibbs ensemble Monte-Carlo method for a system of two crystal phases plus a liquid phase can be used if a fast-computing free energy method is at hand for the liquid phase. The optimized random-phase approximation (ORPA) is a good candidate for a systematic study. This approach could help to solve the similar problem with large polydispersities. Indeed, in this case, the wings of the charge-distribution function are probably too wide to be incorporated in the only Laves AB_2 phase, and the system should choose the liquid phase as a reservoir of unfitting particles (too large or too small), expressing the high tolerance of the liquid phase. However, experimental data for such systems with large polydispersity are scarce, and it is not clear whether such a system will choose intensive use of the liquid phase (*i.e.* melting the phases) or if it will find another crystalline structure with a still larger crystal unit cell.

Acknowledgements

Part of this work was realized in the framework of the GDR AMC2 "Approches Multi-Physiques pour les Colloïdes Concentrés", CNRS. We thank Christophe Labbez and Guillaume Bareigts for interesting comments about the Monte-Carlo numerical method used in this work.

References

1 B. Cabane, J. Li, F. Artzner, R. Botet, C. Labbez, G. Bareigts, M. Sztucki and L. Goehring, Hiding in plain view: Colloidal self-assembly from polydisperse populations, 2015, ArXiv: 1511.02493v1

2 F. Laves, Crystal Structure and Atomic Size, in *Theory of Alloy Phases*, The American Society of Metals, Cleveland, OH, 1956, pp. 124–198.

3 A. Z. Panagiotopoulos, in *Observation, prediction and simulation of phase transitions in complex fluids*, ed. M. Baus, L. R. Rull and J. P. Ryckaert, Kluwer Academic Publishers, Dordrecht, The Netherlands, NATO ASI Series C, 1995, vol. 460, pp. 463–501.

4 E. Zaccarelli, S. M. Liddle and W. C. K. Poon, *Soft Matter*, 2015, **11**, 324–330.

5 E. J. Verwey and J. T. G. Overbeek, *Theory of the Stability of Lyophobic Colloids*, Elsevier, Amsterdam, The Netherlands, 1948.

6 I. Sogami and N. Ise, *J. Chem. Phys.*, 1984, **81**, 6320–6332.

7 *Ordering and Phase Transitions in Charged Colloids*, ed. A. K. Arora and B. Tata, Wiley-VCH, New York, 1996.

8 M. O. Robbins, K. Kremer and G. S. Grest, *J. Chem. Phys.*, 1988, **88**, 3286–3312.

9 The Monte-Carlo Method in Condensed Matter Physics, *Topics in Applied Physics*, ed. K. Binder, Springer-Verlag Berlin Heidelberg, New York, 1992, vol. 71.

10 M. J. Murray and J. V. Sanders, *Philos. Mag. A*, 1980, **42**, 721–740.

11 Z. Chen and S. O'Brien, *ACS Nano*, 2008, **2**, 1219–1229.

12 D. A. Kofke and P. G. Bolhuis, *Phys. Rev. E: Stat. Phys., Plasmas, Fluids, Relat. Interdiscip. Top.*, 1999, **59**, 618–622.

13 M. Fasolo and P. Sollich, *Phys. Rev. Lett.*, 2003, **91**, 068301.

14 M. D. Eldridge, P. A. Madden and D. Frenkel, *Nature*, 1993, **365**, 35–37.

15 L. Filion and M. Dijkstra, *Phys. Rev. E: Stat., Nonlinear, Soft Matter Phys.*, 2009, **79**, 046714.

16 P. N. Pusey, *J. Phys.*, 1987, **48**, 709–712.

17 P. Bartlett, *J. Chem. Phys.*, 1998, **109**, 10970–10975.

18 G. Foffi, C. de Michele, F. Sciortino and P. Tartaglia, *J. Chem. Phys.*, 2005, **122**, 224903.

Faraday Discussions

PAPER

Magnetic filament brushes: tuning the properties of a magnetoresponsive supracolloidal coating

Pedro A. Sánchez,*[a] Elena S. Pyanzina,[b] Ekaterina V. Novak,[b] Joan J. Cerdà,[c] Tomàs Sintes[c] and Sofia S. Kantorovich[ab]

Received 1st September 2015, Accepted 28th October 2015
DOI: 10.1039/c5fd00133a

We present a theoretical study on the design of a supramolecular magnetoresponsive coating. The coating is formed by a relatively dense array of supracolloidal magnetic filaments grafted to a surface in a polymer brush-like arrangement. In order to determine and optimise the properties of the magnetic filament brush, we perform extensive computer simulations with a coarse-grained model that takes into account the correlations between the magnetic moments of the particles and the backbone crosslinks. We show that the self-assembly of magnetic beads from neighbouring filaments defines the equilibrium structural properties of the complete brush. In order to control this self-assembly, we highlight two external stimuli that can lead to significant effects: temperature of the system and an externally applied magnetic field. Our study reveals self-assembly scenarios inherently driven by the crosslinking and grafting constraints. Finally, we explain the mechanisms of structural changeovers in the magnetic filament brushes and confirm the possibility of controlling them by changing the temperature or the intensity of an external magnetic field.

1 Introduction

Self-assembly—the reorganisation of the components of a system into patterns or defined structures driven by their interactions—is a ubiquitous mechanism in many physical systems, as well as a key tool of increasing importance in the bottom-up design of materials and nanotechnologies.[1] As the knowledge of the self-assembly properties of simple systems grows, more complex components and strategies can be explored. In particular, the creation of sophisticated structures by means of the hierarchical self-assembly of their components—or building blocks—has experienced great development in recent years.[2] In this strategy, the self-assembly of the final structure takes place in multiple steps. Each step

[a]*University of Vienna, Sensengasse 8, 1090 Vienna, Austria. E-mail: pedro.sanchez@univie.ac.at*
[b]*Ural Federal University, Lenin av. 51, 620000 Ekaterinburg, Russia*
[c]*Instituto de Física Interdisciplinar y Sistemas Complejos (CSIC-UIB), E-07122 Palma de Mallorca, Spain*

produces building blocks of increasing complexity, which are able to self-assemble into new structures in the next step. One of the current leading goals of hierarchical self-assembly techniques, particularly in the context of soft matter systems, is the creation of stimuli-responsive materials, *i.e.*, materials that experience a controlled change in their properties as a response to an external drive.[3]

Among the different materials with relevant self-assembly properties available, magnetic colloids have attracted great interest in recent years due to their broad potential as basic building blocks of novel responsive materials and nanotechnologies.[4–6] Under adequate conditions, simple spherical magnetic colloids dispersed in a carrier fluid self-assemble into chains, rings and branched structures driven by the non-central character of the interparticle magnetic dipole–dipole interaction that favours the head-to-tail orientation of the neighbouring magnetic moments.[7,8] The self-assembly into chains with the head-to-tail orientation of the dipoles can also be stimulated by an applied external magnetic field. These magnetically-driven self-assembly processes lead to significant changes in the macroscopic behaviour of these systems: their overall magnetic response, rheology and thermodynamics depend strongly on the microstructure of magnetic soft matter.[9–11]

Besides their direct self-assembly, magnetic colloids play an increasingly important part in the development of new responsive and functional nanocomposite materials, particularly in combination with polymer molecules.[12,13] In general, the coating of colloidal particles with polymers is a technique frequently used to stabilise and/or functionalise their surfaces. Such polymeric coatings can also be used to permanently crosslink the individual colloids that form a self-assembled cluster, stabilising the overall structure and preventing their disaggregation even when the system is no longer under the appropriate self-assembly conditions. In the case of magnetic colloids, their self-assembled head-to-tail chains have been made permanent by polymer crosslinking for more than a decade.[14] The result of this crosslinking is a semi-flexible chain with a polymer-like supracolloidal structure, with the magnetic colloids playing the role of polymer monomers. The control of the properties of these systems, known as magnetic filaments, has experienced great progress in recent years thanks to the enhancement of crosslinking techniques.[15–18]

Recently, we proposed the use of magnetic filaments in order to create a magnetoresponsive thin film surface.[19] The basic idea of our approach is to replace, with magnetic filaments, the densely end-grafted polymer chains that form a polymer brush.[20,21] In general, polymer brushes are widely used for the creation of stimuli-responsive surfaces[3,22] but, except for few compounds at very extreme conditions, polymer molecules are unable to provide significant responses to external magnetic fields. The control of the properties of a surface by means of magnetic fields is particularly interesting in soft matter systems, where other parameters—like pH or electric field—have strong and complex interactions with most substances of interest, making the control of their effects more difficult. For these reasons, efforts have been recently made to create magnetoresponsive polymer brush structures by embedding magnetic colloids within the polymer array.[23,24] On the other hand, magnetic filaments with one end grafted to a surface, either individually or in sparse arrays, have largely been studied as micro- and nanofluidic actuators, pumpers or propellers.[25–30] In our case, we analyse the result of increasing the grafting density of tethered magnetic filaments to form a polymer brush-like structure at a supracolloidal scale. In a preliminary study,[19] we showed that this system shows a significantly more pronounced structural equilibrium response to an external field than former systems of polyelectrolyte brushes with extended electric dipoles,[31] and than the aforementioned polymer brushes with embedded magnetic colloids.[23,24] Our preliminary results also indicated that the origin of such a strong structural response should be attributed to the particular self-assembly properties of the crosslinked colloids placed in the highly constrained environment of the brush structure.

Here, we aim to pinpoint the intrinsic mechanisms responsible for the overall equilibrium structural properties of magnetic filament brushes. This understanding will provide the opportunity to control the thickness and the local density of the brush—the two most important features of a coating—with external stimuli. In contrast to our previous work, in the present study we focus on the influence of temperature and external field on the self-assembly of magnetic beads from neighbouring filaments. Self-assembly proves to be the key to manipulating the responses of the filament brushes, because it is the topology and the size of the formed clusters which define the local magnetic properties and reflect the balance between the entropy and the dipolar correlations.

2 Model and methods

2.1 Magnetic filament brush model

In our polymer brush-like system of supracolloidal magnetic filaments, we consider the latter to be formed by spherical single-domain ferromagnetic colloids that keep the orientation of their magnetic moments—described as simple point dipoles—fixed in one of the directions of the magnetic easy axis of the particle. These particles are considered to have sufficiently high internal anisotropy and sufficiently large volume to be above the superparamagnetic regime. The polymer crosslinkers that maintain the chain-like structure of these filaments are assumed to be attached to the surface of the colloids at the projection points of such easy axes. This combination corresponds to experimental magnetic filaments created by crosslinking single-domain colloids of strong ferromagnetic materials after their self-assembly into straight chains, with a head-to-tail arrangement of their magnetic moments. Therefore, the equilibrium structures of these filaments will tend to keep a nearly head-to-tail orientation of the dipoles between close neighbours along the chain. In order to model these types of magnetic filaments, we recently introduced a coarse-grained, bead-spring model that takes into account the coupling between the orientation of the magnetic dipoles and the chain backbone, led by the crosslinkers.[32,33] In this model, the magnetic colloids are represented as soft spherical beads, with a characteristic diameter d. The soft-core steric repulsion between such spheres is given by a WCA potential,[34]

$$U_{\mathrm{WCA}}(r) = \begin{cases} U_{\mathrm{LJ}}(r) - U_{\mathrm{LJ}}(r_{\mathrm{cut}}), & r < r_{\mathrm{cut}} \\ 0, & r \geq r_{\mathrm{cut}} \end{cases}, \tag{1}$$

where r is the centre-to-centre distance between the spheres, $U_{\mathrm{LJ}}(r)$ is the conventional Lennard-Jones potential,

$$U_{\rm LJ}(r) = 4\varepsilon_{\rm s}[(d/r)^{12} - (d/r)^6], \tag{2}$$

and $r_{\rm cut} = 2^{1/6}d$ is a shift that makes the potential purely repulsive. In addition, each soft sphere carries a point magnetic dipole at its centre, $\vec{\mu}$, which leads to a conventional magnetic dipole–dipole interaction between the beads:

$$U_{\rm dd}(i, j) = \frac{\vec{\mu}_i \cdot \vec{\mu}_j}{r^3} - \frac{3\left(\vec{\mu}_i \cdot \vec{r}\right)\left(\vec{\mu}_j \cdot \vec{r}\right)}{r^5}, \tag{3}$$

where $\vec{r} = \vec{r}_i - \vec{r}_j$ is the centre-to-centre displacement vector between the beads i and j, which have dipolar moments $\vec{\mu}_i$ and $\vec{\mu}_j$ respectively. If a uniform external magnetic field, $\vec{H}$, is applied to the system, each magnetic dipole $\vec{\mu}_i$ will experience an interaction that tends to align it with the field. The energy of such a dipole–field interaction has the form:

$$U_{\rm H}(i) = -\vec{\mu}_i \cdot \vec{H}. \tag{4}$$

In order to mimic the coupling between the orientation of the dipoles and the chain backbone introduced by the crosslinkers, we use the following attractive potential between any pair of crosslinked neighbours, which depends on both the centre-to-centre displacement vector between them, $\vec{r}$, and on the relative orientation of their dipoles:

$$U_{\rm S}(i, j) = \frac{1}{2}K_{\rm S}\left(\vec{r} - (\hat{\mu}_i + \hat{\mu}_j)\frac{d}{2}\right)^2, \tag{5}$$

where $\hat{\mu}_i = \vec{\mu}_i/|\vec{\mu}_i|$ and $\hat{\mu}_j = \vec{\mu}_j/|\vec{\mu}_j|$ are the unitary vectors parallel to each associated dipole moment. Expression (5) corresponds essentially to a harmonic spring potential that connects the surfaces of a pair of bonded particles at the projection points of the dipole head of one of them and the dipole tail of the other. The strength of the attraction is defined by the prefactor $K_{\rm S}$. The combination of potentials (1) and (5) provides the actual bonding interaction between the neighbouring beads in the filament. Fig. 1a shows a sketch of the described magnetic filament model.

Finally, in order to create a polymer brush-like structure, we arrange a relatively dense array of magnetic filaments—formed by N identical beads—tethered by one of their ends to a flat sterically repulsive surface in the XY plane. In particular, the grafting of each filament is achieved by fixing the position of its tethered end bead next to the surface, with the corresponding dipole permanently pointing perpendicular to the latter. The grafting positions are distributed in a square lattice arrangement with a separation constant a. This gives a number grafting density—*i.e.*, the number of grafting points per unit of surface area—of $\sigma = a^{-2}$. In the initial configuration of the system, the filaments are placed as completely straight chains perpendicular to the grafting surface. Fig. 1b shows an example of the initial configurations of magnetic filament brushes used in this study. The steric repulsion produced by the grafting surface on the filament beads is given by a truncated shifted 9–3 Lennard-Jones potential,[35] which is obtained by applying expression (1) to

$$U_{\rm LJ}{}^{9-3}(r) = \frac{3\sqrt{3}}{2}\left[\left(\frac{d}{2r}\right)^9 - \left(\frac{d}{2r}\right)^3\right], \tag{6}$$

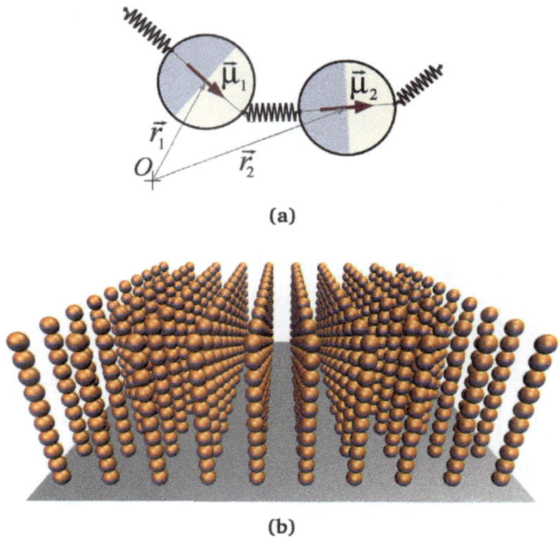

Fig. 1 (a) Sketch of the magnetic filament model. The magnetic beads are represented as two-colour spheres, indicating the orientation of the dipoles. The crosslinkers are represented as springs. (b) Example of a magnetic filament brush used as an initial configuration in our computer simulations, corresponding to a grafting density of $\sigma = 0.111$ of filaments formed by $N = 10$ beads.

where r is the distance between the centre of the bead and the surface. This potential is the result of integrating a conventional 12–6 Lennard-Jones potential over an infinite plane.

2.2 Simulation and analysis approaches

Our characterisation of the equilibrium properties of supracolloidal magnetic filament brushes is based on two main methods: first, we performed extensive computer simulations with the phenomenological model described above; second, we analysed the simulation results by means of a combination of conventional parameters—in the context of studies on polymer brush systems—and a detailed characterisation of the internal structure of the brush obtained by means of graph-theory techniques.

The simulation method we used is molecular dynamics in the canonical ensemble. In particular, we used a Langevin thermostat in order to implicitly treat the effects of the thermal fluctuations of the background fluid on the filament beads.[36] A pseudo-infinite brush system was mimicked by imposing lateral periodic boundaries. Efficient calculations of the long-range magnetic interactions under these conditions is rather difficult. To address this issue, we used the approximate dipolar-P^{3}M method,[37] in combination with the dipolar layer correction method,[38] that takes into account the slab geometry of the system. All the simulations were carried out with the ESPResSo 3.2.0 simulation package.[39,40] Further details of the simulation method can be found in our introductory work on this system.[19]

The analysis of simulation results for the equilibrium configurations of the filament brushes was performed in two different ways. First, we calculated two

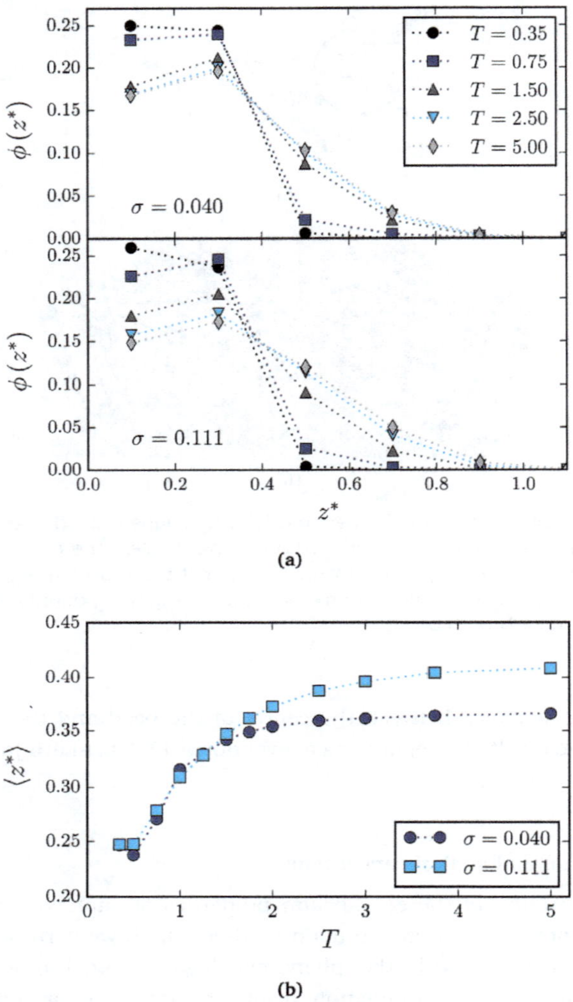

Fig. 2 Density profiles (a) and average brush heights (b) for different temperatures, grafting densities and zero external field.

traditional parameters used to characterise the overall structure of polymer brush-like systems: the vertical density profile, $\phi(z^*)$, and the average height, $\langle z^* \rangle$. $\phi(z^*)$ is simply defined as the number density distribution of the centres of the filament beads as a function of the reduced distance to the grafting surface, $z^* = z/d(N-1)$, where $d(N-1)$ is the contour length of the filaments measured in units of the characteristic diameter of the bead. The average height is obtained as the first moment of the density profile:

$$\langle z^* \rangle = \frac{1}{d(N-1)} \frac{\int_0^\infty \phi(z) z \mathrm{d}z}{\int_0^\infty \phi(z) \mathrm{d}z}. \tag{7}$$

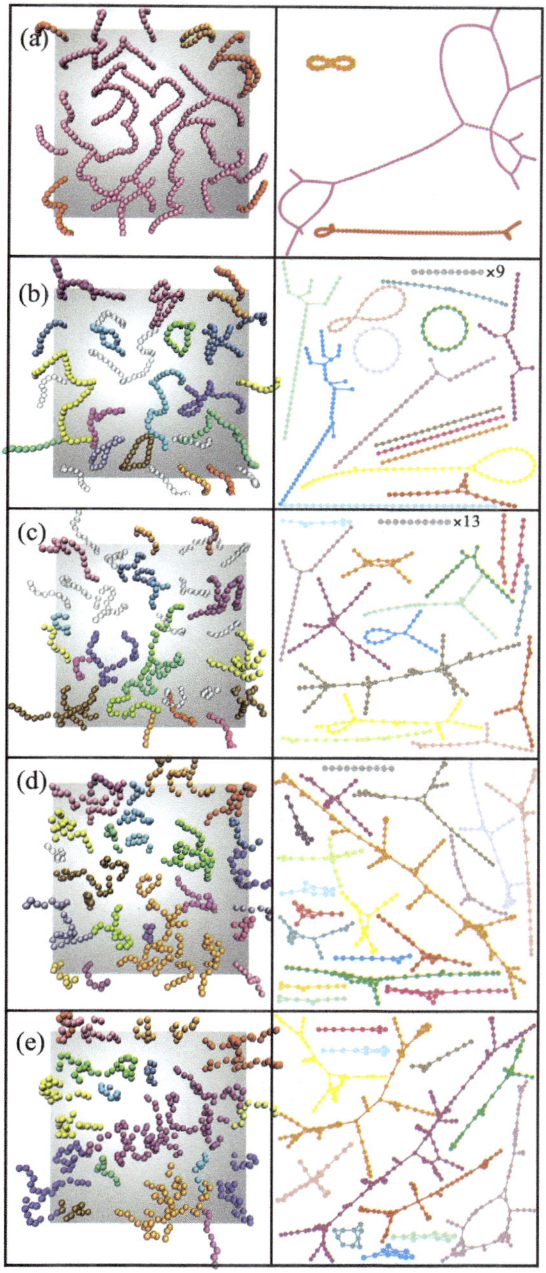

Fig. 3 Equilibrium configuration snapshots (left column, top view) and connectivity graphs (right column) at different temperatures for $\sigma = 0.040$, $H = 0$. Similar colours identify the same cluster in both the snapshot and the corresponding graph. The grey colour indicates filaments with no non-permanent connection. (a) $T = 0.35$. (b) $T = 0.75$. (c) $T = 1.5$. (d) $T = 2.5$. (e) $T = 3.75$.

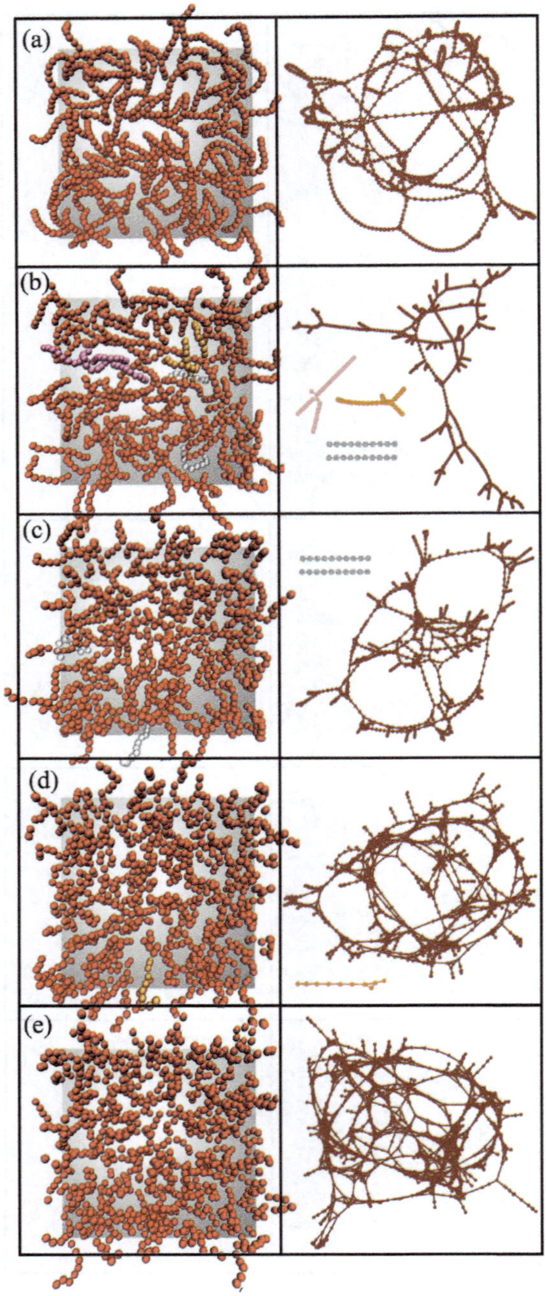

Fig. 4 Equilibrium configuration snapshots (left column, top view) and connectivity graphs (right column) at different temperatures for $\sigma = 0.111$, $H = 0$. Similar colours identify the same cluster in both the snapshot and the corresponding graph. The grey colour indicates filaments with no non-permanent connection. (a) $T = 0.35$. (b) $T = 0.75$. (c) $T = 1.5$. (d) $T = 2.5$. (e) $T = 3.75$.

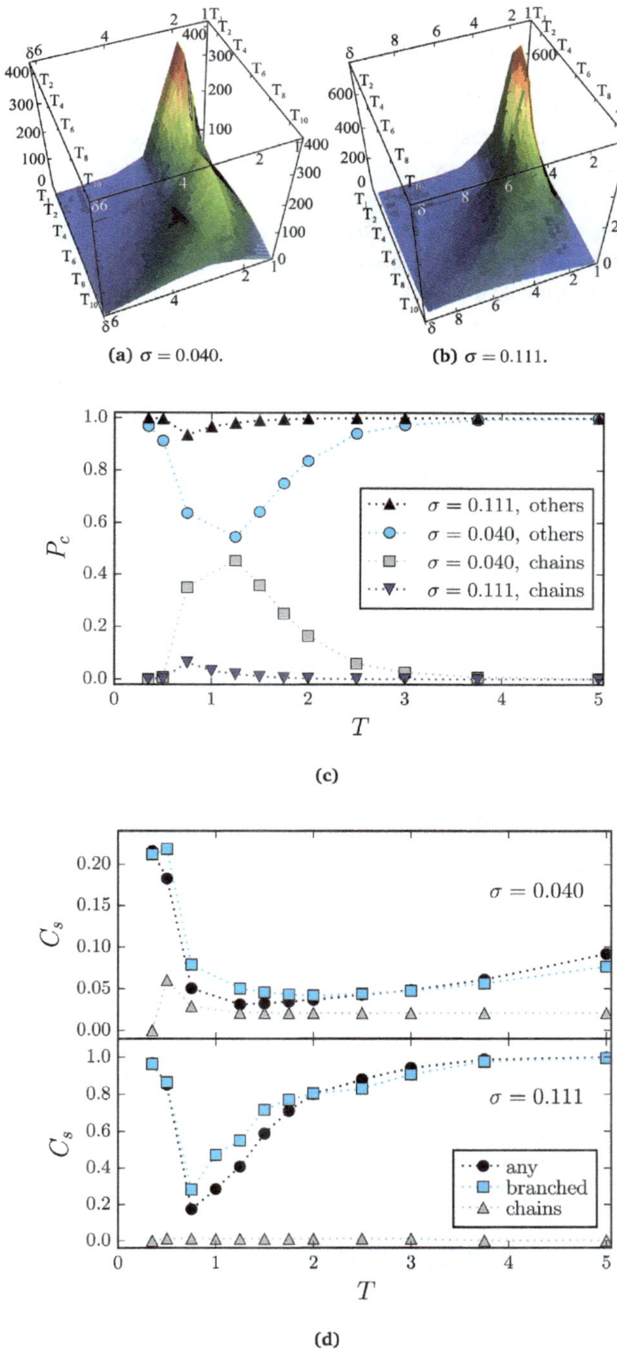

Fig. 5 Distributions of the diverse cluster analysis parameters as a function of the temperature obtained for different grafting densities and zero external field. (a) and (b) Distributions of the vertex degrees, δ. (c) The fraction of beads, P_c, belonging to chain-like or any other type of clusters. (d) Distributions of average normalised cluster sizes, C_s, corresponding to branched, chain-like and any type of cluster structures.

The second analysis approach is based on the fact that the peculiar behaviour that we observed in the aforementioned collective parameters for different values of both temperature and external magnetic field can only be explained by the existence of characteristic arrangements of the dipolar beads within the brush. In other words, the overall response of the filament brush to the external control parameters is determined by the self-assembly properties of the individual magnetic colloids under the geometrical constraints imposed by three factors: their belonging to a permanent chain; the presence of the neighbouring chains; the grafting surface. In order to characterise such self-assembly properties, we used an analysis approach based on graph theory.[41] In this approach, which we introduced in our preliminary work on magnetic filament brushes,[19] we study the arrangements of the beads by introducing the concept of generic connections between them. In particular, we consider two beads to be connected if their centre-to-centre distance does not exceed a certain threshold. Once all the connections in the system have been determined according to such simple criteria, the brush structure can be represented as a graph in which the dipolar beads are the vertices and the connections between them are the edges. A statistical analysis of the brush structure can then be performed by means of different graph parameters and their distributions for different conditions. The parameters analysed here are the distribution of edges, the vertex degree, the cluster size and the betweenness centrality. The vertex degree, δ, is the number of edges of a given vertex. To define the cluster size and the centrality, it is required to introduce the concepts of path, cluster and shortest path. A path between two given vertices, u and w, is any trajectory along the edges of the graph, p_{uw}, that connects such vertices, either directly or by crossing any number of other vertices. A cluster is a maximal set of vertices for which there exists at least one path that connects any pair of them, and its size is the number of vertices it has. Finally, a shortest path between vertices u and w, p^{s}_{uw}, is any path p_{uw} that crosses the minimum possible number of intermediate vertices. According to these definitions, the betweenness centrality of a given vertex v that belongs to a cluster V is defined as

$$C_{\mathrm{B}}(v) = \sum_{u,\, w \in V} \frac{p^{\mathrm{s}}_{uw}(v)}{p^{\mathrm{s}}_{uw}}, \tag{8}$$

where p^{s}_{uw} is the total number of shortest paths between each pair of vertices u and w that belong to the cluster V, and $p^{\mathrm{s}}_{uw}(v)$ is the number of those paths that pass through v. Therefore, $C_{\mathrm{B}}(v)$ is a measure of the importance in terms of optimal path construction of the vertex v in the cluster.

3 Results and discussion

For simplicity, here we measure all the physical parameters in reduced units, taking as reference the characteristic diameter of the colloids, $d = 1$, their mass, $m = 1$, and the prefactor of the soft-core steric potential (1) and (2), $\varepsilon_{\mathrm{S}} = 1$. The prefactor of the potential (5) is set to $K_{\mathrm{S}} = 30$, a value that, according to our previous studies with this model of magnetic filaments,[19,32,33] provides an average distance between bonded beads—or average bond length, $\langle b \rangle$—close to the reference bead soft core diameter, $\langle b \rangle \sim d = 1$. However, it is important to note

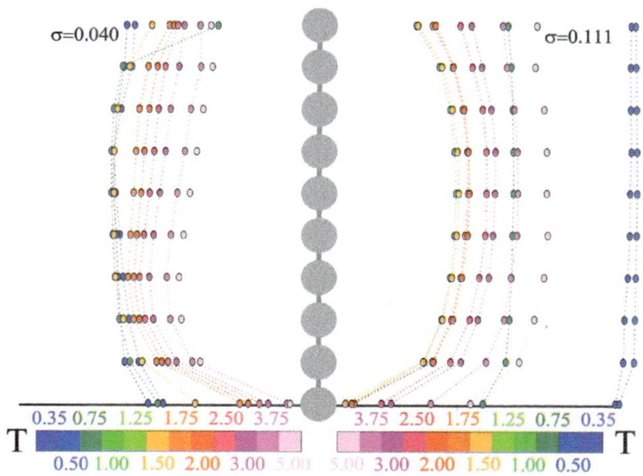

Fig. 6 Average betweenness in arbitrary units for the two sampled grafting densities (left and right horizontal scales) obtained for the different positions of the beads along the filaments (vertical scale). The grafted end corresponds to the lowest position, the free end to the upper one. The colours indicate the temperature corresponding to each curve.

that the bond length is not fixed: since the attractive part of the bonding interaction (5) is an unbonded harmonic potential, the bond length experiences thermal fluctuations by effect of the temperature. Therefore, its average value tends to grow as the temperature is increased.[33] For the parameters used in this study, the maximum distance between the surfaces of two bonded beads was approximately $d/2$. Finally, we have taken $\mu^2 = (\vec{\mu}\cdot\vec{\mu}) = 5$ for the squared dipole moment of the beads. We explored the equilibrium structures of the magnetic filament brushes for two sets of values of the control parameters. In one case, we analysed the effects of variations in temperature at zero field by taking a range of reduced temperatures within $T = 0.35$ and $T = 5$. In the other case, the effect of an external magnetic field perpendicular to the grafting surface at a reference reduced temperature $T = 1$ has been studied by varying its modulus from $H = 0.5$ to $H = 4$. In both cases, a single filament length of $N = 10$ and two grafting densities, $\sigma = 0.040$ and $\sigma = 0.111$, were chosen.

The set of parameters we described above can be associated with various experimental systems. Let $T = 1$ correspond to 300 K. As an example, we can choose magnetite with a bulk saturation magnetisation at this temperature equal to 450 kA m^{-1}. To reach the value $\mu^2 = 5$ at close contact, the diameter of the nanoparticle magnetic core should be approximately 30 nm.[7] With this magnetic core size, the thickness of the non-magnetic layer on the particle surface can be estimated to be on the order of 3 nm. To keep the distance between the centres of the crosslinked nanoparticles at approximately 50 nm, one needs the crosslinking polymers to be at least 7 nm in length, which corresponds to approximately 20 DNA base pairs.[42,43] Varying the temperature in experiments is not directly equivalent to that in simulations, as the absolute value of the saturation magnetisation in real ferromagnetic materials is temperature-dependent. In other words, the range of dimensionless temperatures chosen in our simulations is supposed to reflect the variation of the ratio between the magnetic and thermal

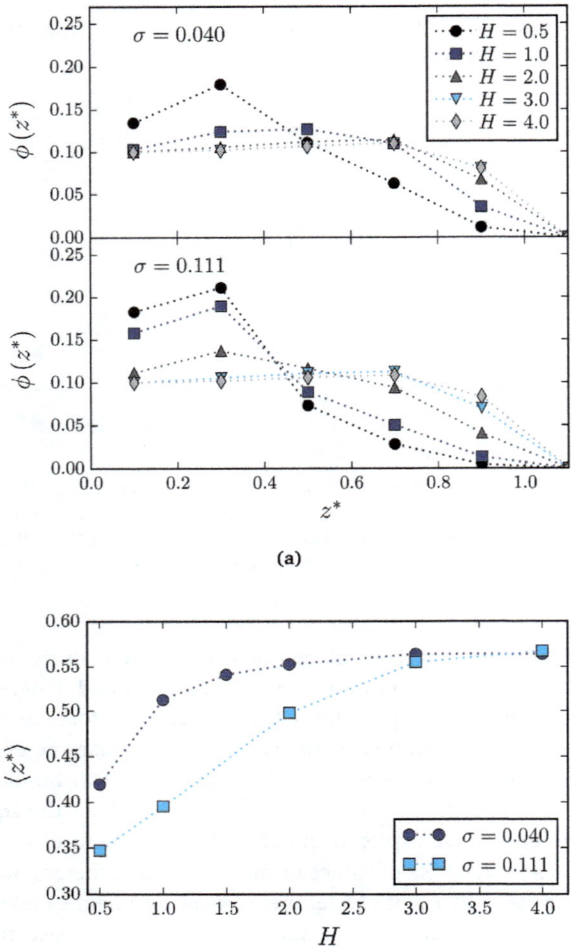

(a)

(b)

Fig. 7 Density profiles (a) and average brush heights (b) for different intensities of the external magnetic field and grafting densities, obtained at $T = 1$.

energies. Besides that, if the temperature becomes too low in experiments, the system might undergo undesirable phase transformations such as carrier liquid freezing or irreversible coagulation due to the loss of colloidal stability. However, it is experimentally possible to keep the suspension of magnetite nanoparticles stable down to 150 K,[44,45] which in our simulations corresponds to a value between $T = 0.35$ and $T = 0.75$.

3.1 Effects of temperature

It is known that the self-assembled structures of free magnetic colloids in suspension experience diverse structural transitions on cooling.[10,11] Computer simulations have also provided indications of structural changes in isolated

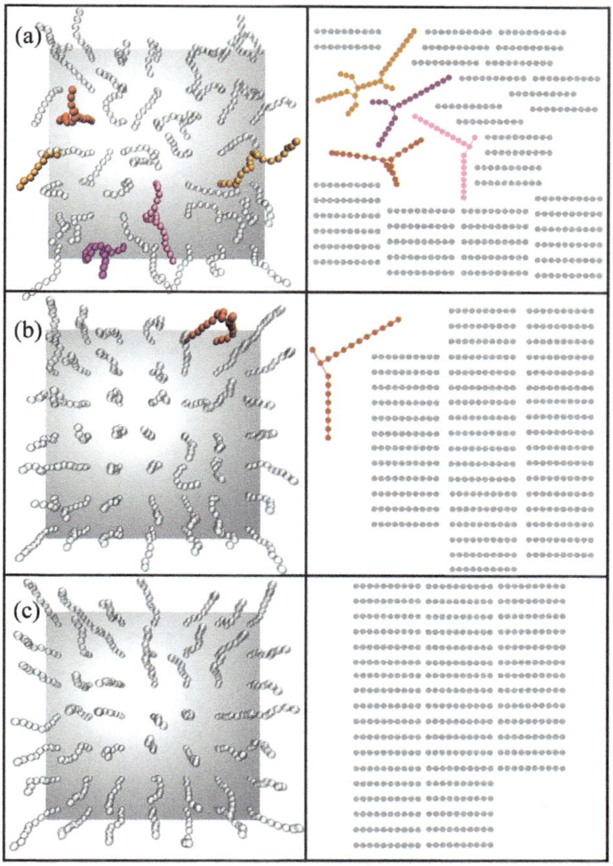

Fig. 8 Equilibrium configuration snapshots (left column) and connectivity graphs (right column) at different normal external fields for $\sigma = 0.040$, $T = 1$. (a) $H = 0.5$. (b) $H = 2$. (c) $H = 3$. For higher fields, no added connections were observed.

magnetic filaments as the importance of the thermal fluctuations decreases in front of the dipole–dipole interactions.[32,46] In magnetic filament brushes, the effects of the added geometrical restrictions and the crowded environment make the structural changes, which these systems may experience as a function of the temperature, hard to anticipate.

We begin the analysis of the effects of temperature on the equilibrium structure of the magnetic filament brush at zero field by examining the evolution of the density profiles, $\phi(z^*)$, and the average brush height, $\langle z^* \rangle$. Fig. 2 shows the results obtained for these parameters. The density profiles (Fig. 2a) evidence the highly compact overall structure of the magnetic brush, showing a rather abrupt decay of the density in the upper regions, as was discussed in our previous work for the case $T = 1$.[19] Here, we can see that this behaviour persists for both grafting densities within the whole range of explored temperatures, and becomes notably more pronounced as T decreases. At low temperatures, practically all the beads remain close to the grafting surface—within a region below 1/2 of the contour

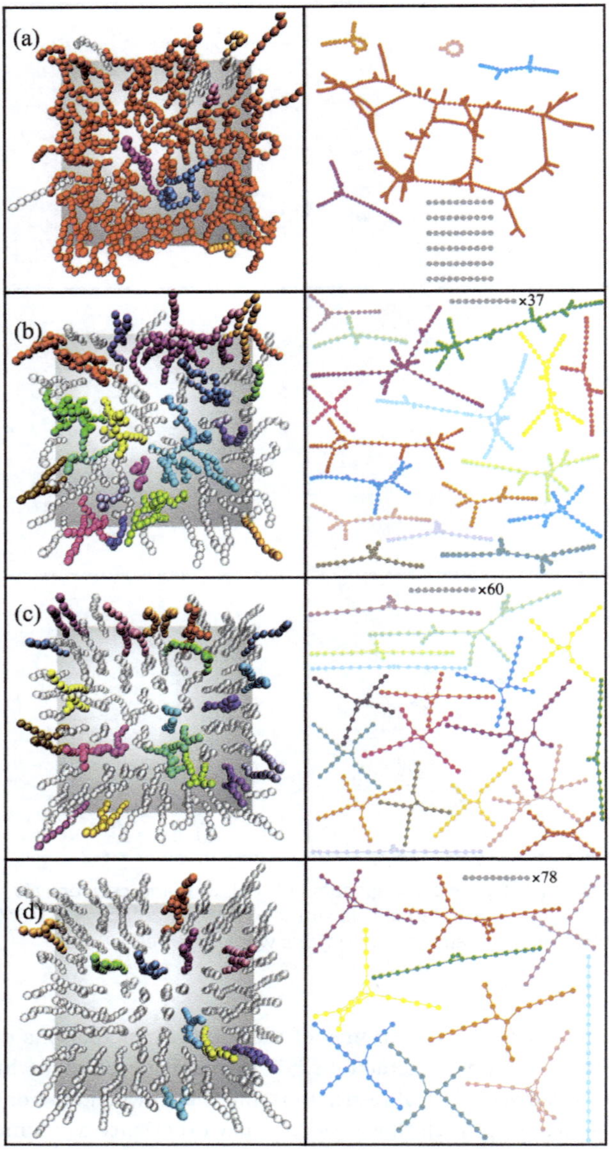

Fig. 9 Equilibrium configuration snapshots (left column) and connectivity graphs (right column) at different normal external fields for $\sigma = 0.111$, $T = 1$. (a) $H = 0.5$. (b) $H = 2$. (c) $H = 3$. (d) $H = 4$.

length of the filaments—whereas the decay of the profile above such a region is extremely abrupt. Interestingly, at low temperatures the profiles for both grafting densities are very similar. As the temperature increases, the profiles become wider and their decay gets smoother. At even higher temperatures, where the dipole–dipole interactions are suppressed by thermal fluctuations, the filament brush is expected to have a parabolic density profile, as predicted for a non-magnetic

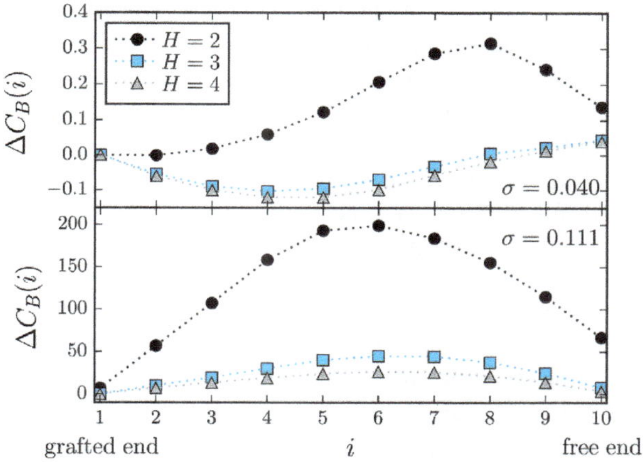

Fig. 10 Variation of the average betweenness centrality of the beads with respect to the case of an ideal chain, measured for two grafting densities as a function of the position of the bead along the filament.

polymer brush.[47,48] The average brush height, shown in Fig. 2b, reveals a non-trivial dependence of the filament brush thickness on the temperature and the grafting density. At a high temperature, the plateau of the equilibrium height is determined by the grafting density and, if the temperature starts decreasing, we can see that $\langle z^* \rangle$ slowly decreases from its saturation value. At approximately $T \sim 1.5$, the decrease of $\langle z^* \rangle$ with T changes significantly for both grafting densities, making it rather steep on further cooling. Importantly, as was observed for the density profiles, in the region $T \sim [0.5, 1.5]$, $\langle z^* \rangle$ assumes basically the same values for both grafting densities.

In order to understand the origin of the non-trivial dependence of the equilibrium overall brush structure on the temperature and the grafting density, we applied the connectivity graph analysis described above to these simulation results. The distance criterion used here to define the generic connections between the magnetic colloids in the brush—and, with them, the connectivity graphs that represent the detailed structure of the brush—takes into account two different origins of pair formation. Besides the connectivity imposed by the filaments' permanent links, in general we can expect magnetic beads to self-assemble, forming close contact pairs whose relative orientations correspond to the low energies of their dipole–dipole interaction. Due to its strongly directional nature, the role of the two dipoles' orientation phase space, corresponding to the attractive dipole–dipole pair interaction, is rather restricted. On the other hand, sufficiently strong thermal fluctuations may force a pair of beads to come into close contact or to separate regardless of their relative dipole orientations. All magnetically and entropically driven non-permanent connections, in addition to the permanent connections led by the crosslinkers along the filaments, can be taken into account by imposing a threshold centre-to-centre distance between connected beads equal to the maximum of the values $\max(r_{\text{cut}}, b_{\text{max}})$, where r_{cut} is the cutoff distance defined for the soft core repulsion potential (1), $r_{\text{cut}} = 2^{1/6}d$, and b_{max} is the maximum bond length observed for any pair of linked beads in the system at a given temperature. It should be noted that at high T one finds that $b_{max} > r_{cut}$, whereas at low T this relationship is reversed. With this criterion, the first step in the connectivity analysis approach is to define the graphs that represent the connections and, from them, the clusters of filaments connected through their beads.

Fig. 3 and 4 show a selection of snapshots of the brush equilibrium configurations obtained for different temperatures and the two sampled grafting densities. Next to each snapshot, the corresponding connectivity graph is also shown. In both snapshots and graphs, different clusters are identified by distinctive colours. For the lowest temperature, $T = 0.35$, and grafting density, $\sigma = 0.040$, one can see that the magnetic filament brush has a particular structure in which the permanent chains arch, connecting their free ends with the grafted or nearby end beads in neighbouring filaments. The topology of the corresponding connectivity graph also seems rather clear: all the beads are grouped into few clusters, in which the length of the linear segments—*i.e.* segments without beads that have a degree higher than 2—is large. In most cases, such linear segments form closed paths. Due to the high correlation between the orientation of the dipoles and the chain backbone of the magnetic filaments we use—particularly strong at low T—the closed paths in the connectivity graphs indicate the existence of structures that tend to minimise their net magnetic moment, forming a nearly closed trajectory for the magnetic flux of their dipoles. For the higher σ, these general traits are even more pronounced: the whole brush is connected into a single cluster with a topology of multiple closed loops. Therefore, the structure of the brush at this temperature is clearly determined by the self-assembly of the filaments driven by the dipole–dipole interactions. At $T = 0.75$ the dipolar interaction is still very strong, but the effects of the entropic contribution start to become noticeable with the appearance of free chain ends, even though ring-like structures are still visible for low σ. In this T-region, one can also observe the formation of branched structures, which is more pronounced if the grafting density is high. At this point, it is useful to distinguish between two types of cluster topologies: the *linear chains*, that are formed by beads with a degree no higher than two, and the *branched structures*, in which there is at least one particle with a degree higher than two and/or the total amount of edges is equal to or larger than the total amount of beads in the cluster. It is worth mentioning that, at $T \sim 1$, the branched structures contain mainly linear segments and do not have vertices with high degrees, not even at $\sigma = 0.111$. To illustrate this, we calculated the distribution of values of the vertex degrees averaged over all sampled configurations. Fig. 5a and b show 3D plots of such distributions for the two sampled grafting densities. One can see that up to $T_4 = 1$, the distributions are very narrow and have a clear maximum at $\delta = 2$. This corresponds to both the permanent connections between neighbours in the filaments and the connections between free and grafted ends in linear chain structures driven by the dipole–dipole interactions. The figures also evidence that, at higher temperatures, it becomes more probable to find branched structures for both values of the grafting density, as the most frequent vertex degree tends to $\delta = 4$, for low grafting densities, and $\delta = 6$, if the brush is denser. As mentioned above, the directional character of the dipole–dipole interaction makes such types of multiple junctions unlikely. Consequently, one can conclude that, for temperatures above $T > 1$, the internal structure of the brush is mainly determined by thermal fluctuations. The structural changeover, driven by the change of T, is also clearly quantified by Fig. 5c, where the fraction of particles in open chains and other structures as a function of the temperature is plotted. In this case, other structures include both branched ones and closed unbranched loops. For both systems, there is a clear minimum for the fraction of particles in these latter structures around $T \sim 1$. Below this temperature, the brush tends to be formed by closed loop structures with long linear segments. The fraction of particles in open chains is maximised at the temperature at which the entropic gain of the chains' free ends is no longer compensated for by the energetically favourable creation of an additional bond between them. If T increases further, the fraction of particles in branched structures begins to increase again due to the connections induced by the random thermal fluctuations. Finally, looking at the evolution with temperature of the sizes of the different cluster types, shown in Fig. 5d, it becomes clear that, once the entropy loss associated with forming large zero-magnetic flux clusters is no longer feasible for the system, the cluster size reaches its minimum and then grows fast again due to simple entanglement caused by thermal motion. This also explains why the high temperature cluster size increase is much more pronounced for the brush with a higher density.

The graph theory analysis used here allows us to not only characterise the overall structures of the clusters, but also to pinpoint how the bead position in a filament affects its part in the self-assembly. Here, in particular, we analyse the average betweenness of each particle in the brush at a given position along the filament for different temperatures, as shown in Fig. 6. In this representation, the vertical axis corresponds to the positions along the filament, from the grafted end (at the bottom) to the free end (at the top), whereas the values of the betweenness for the two sampled grafting densities are shown in the horizontal axes. The different values of the temperature are represented by the colours of the curves. As shown in our previous work,[19] for an individual chain, the average betweenness has a parabolic shape with the maximum corresponding to the chain middle point. The higher the value of the average betweenness for a certain bead position, the more active its participation in the path formation. In this way, Fig. 6 completes our understanding of the structural changeover induced by the temperature. First, at low T, the active participation of the grafted ends in self-assembly is confirmed: as we can see, their betweenness is only slightly lower than that of the free ends and middle particles for both values of σ. In general, a flat betweenness profile along the chain is the signature of loop formation. In this case, almost flat profiles are found for $\sigma = 0.111$ and $T \leq 0.75$. With growing temperature, the participation of the grafted ends drops drastically. However, for $\sigma = 0.040$ the decrease of the grafted ends' betweenness is monotonous with T, whereas for the denser brush a reversal is observed at $T \sim 1.5$. This reversal for $\sigma = 0.111$ is also accompanied by the shift of the maximum betweenness to the beads closer to the grafted surface on heating. This tendency is much less pronounced for the brush with $\sigma = 0.040$, due to a lower combinatoric probability of entanglement at high T. For both σ, the value of the betweenness of the free ends is rather high. For $\sigma = 0.111$, the betweenness of the free ends follows the reversal analogous to that of the grafted ends. Interestingly, this reversal, albeit on a smaller scale, is also observed for free ends for the brush with $\sigma = 0.040$. The minimum of the betweenness of the free ends for the latter coincides with the maximum of the particle fraction in the chains, as seen in Fig. 5c.

As a final conclusion, we underline that the interplay between entropy and energy in the constrained environment of a magnetic filament brush not only leads to a much broader range of temperature-induced self-assembly scenarios than those in non-crosslinked dipolar particles, but also opens one possibility to efficiently tune the height and the local structure of such a brush. In the next Section, we explore the use of external magnetic fields to also finely manipulate the brushes of magnetic filaments.

3.2 Effects of normal external magnetic fields

Following the strategy used above, we first calculate the density profiles and average brush heights for both grafting densities at various values of an applied magnetic field, H, perpendicular to the grafting surface. Fig. 7 shows the results for these parameters, obtained at $T = 1$. The most evident conclusion from the density profiles (Fig. 7a) is that the grafting density has a strong influence on the response of the system to the field. For $\sigma = 0.040$, even a low field can perturb the brush and straighten the chains, hindering the formation of non-permanent connections between the beads. This results in a flat rectangular density profile. For a denser brush, instead, the complex internal structure found at zero field at this temperature, discussed above, is found to be less sensitive to the presence of the external field: only starting from $H = 3$, the density profile approaches a rectangular shape. To the same conclusion leads Fig. 7b, in which we plot the average height of the magnetic filament brush as a function of H. Here, for $\sigma = 0.111$, $\langle z^* \rangle$ grows significantly slower with an external field than that for $\sigma = 0.040$. At high fields, neither entropy nor dipolar interactions can compete with the field-dipole interaction, which leads to the alignment of the filaments with the field and, as a result, makes the brush average height reach its natural saturation for both values of σ. In order to start the characterisation of the brush structure at the scale of individual beads, we present in Fig. 8 and 9 two sets of simulation snapshots and connectivity graphs analogous to the ones discussed in the Section above. In this case, the configurations correspond to selected values of the external field. A direct comparison of these plots with Fig. 3 and 4 evidence a drastic difference with respect to the evolution with temperature. For low σ, the self-assembly between chains is almost completely hindered, even by such a low field value as $H = 0.5$. One can observe only a few clusters, with a lot of linear segments, open ends and branching points with rather low degrees. Importantly, this topology does not require a significant misalignment of the dipoles with respect to the field direction. If the field increases, the clusters gradually disappear, preserving in any case such characteristic topology. For higher σ and $H = 0.5$, instead, the clusters are rather big, with a relatively large amount of long linear segments. The junctions are rather versatile and free chain ends are present in a significant amount. The topology of the clusters in Fig. 9a is very similar to those observed in Fig. 4b and c. In other words, the influence of this low external magnetic field is negligible in this range of parameters. Besides that, the chains keep forming junctions even at high fields. The topology of these branched structures is not versatile and clearly coincides with the branched structures for $\sigma = 0.040$ and lower fields: multiple junctions are basically absent, whereas the amount of linear segments rapidly grows. A direct inspection of these configurations suggests that there are two typical branched structures compatible with an external magnetic field perpendicular to the grafting surface: X-type junctions, in which two nearly parallel filaments establish a single branching connection between beads located in their middle regions (see, for example, the light blue clusters in Fig. 9d), or half-interlaced chains with free ends, in which connected filaments twist around each other (see the dark purple cluster in Fig. 9d, for example). Looking for confirmation of the latter observation, we decided to focus on the betweenness. This choice was driven by the fact that the distribution of degrees in this case provides very limited information, as it is very narrow and has a sharp peak at $\delta = 2$, due to the overwhelming dominance of linear segments. On the other hand, the average cluster size for both grafting densities is a monotonously decreasing function of the applied field, with a saturation value equal to the number of particles in a filament. If one plots the fractions of particles in chains and in branched structures, the chains contain almost all beads in the brushes starting from $H = 0.5$ for $\sigma = 0.040$ and from $H \sim 2.5$ for $\sigma = 0.111$. The betweenness, instead, can provide a more accurate measure after some manipulation. In our previous work, we showed that for a system of ideal N-particle chains—*i.e.*, chains with no other connections than those provided by the permanent links—the average betweenness as a function of the bead position along the filament, $j = \{1, \ldots, N\}$, can be written as:[19]

$$C_{\mathrm{B}}^{\mathrm{id}}(j) = -j^2 + j(N+1) - N. \tag{9}$$

Thus, we can subtract $C_{\mathrm{B}}^{\mathrm{id}}(j)$ from the averaged betweenness of the beads given by expression (8) to see even small deviations from the ideal system, $\Delta C_{\mathrm{B}}(j) = C_{\mathrm{B}}(j) - C_{\mathrm{B}}^{\mathrm{id}}(j)$. The results of this subtraction for high external fields are shown in Fig. 10. For $\sigma = 0.040$ the deviations are basically negligible, but the position of the maximum clearly points towards the middle of the chain. The full confirmation of our aforementioned assumption about the topology of junctions can be obtained by looking at Fig. 10. It becomes clear that the highest deviation from ideal chains is between particle numbers 5 and 6, with number 1 being the grafted end and number 10, the free chain end.

As a last comment, these kind of X-junctions have three clear advantages: they preserve the entropy of free chain ends, they add a negative dipolar connection, which lowers the system free energy, and basically do not perturb the dipoles aligned along the magnetic field. On the other hand, their presence lead to non-complete chain stretching at given fields.

4 Conclusions

In this work, we explored the control of the structural properties of a magnetic filament brush with the help of two external stimuli: temperature and an applied external magnetic field perpendicular to the grafting surface. Both parameters affect the self-assembly of magnetic colloids by changing the topology and the size distribution of formed clusters. The latter structural transformations manifest themselves in the brush density profiles and average brush height. In other words, we explored the possibility of manipulating the magnetic filament brush through fine tuning of the constituents' self-assembly. In order to carefully study the equilibrium properties of magnetic filament brushes, we employed the combination of molecular dynamics computer simulations and graph theory analysis methods. We investigated two grafting densities: $\sigma = 0.040$ and $\sigma = 0.111$. This allowed us to pinpoint the influence of chain concentration on the self-assembly. We showed that at zero field, low temperature and low grafting density, filament chains bend to connect their free ends to neighbouring grafted ends, creating a closed multi-ring-like cluster, whose magnetic moment is nearly zero. The same tendency, only with less regularity, is observed for a high grafting density in zero field at low-T regimes. Under these conditions, the brush is very compact, with a rather low height, and weakly magnetoresponsive. If the temperature increases slightly, the dipole–dipole interaction is still strong enough to drive the self-assembly in long linear structures that involve both free and grafted chain ends, but the entropy leads to the opening of the magnetically closed structures, gaining in such a way the entropic contributions of the chains' free ends. At this point the number of "chain-like" clusters reaches its maximum, and the number of branching points has its minimum. Further increase of T results in a completely random, highly branched network of bonds, mainly defined by the grafting density and thermal fluctuations. The analysis of the influence of the external field at $T = 1$ reveals a more drastic grafting density dependence. Thus, for a low grafting density, even a low field leads to an almost complete elimination of the self-assembly between neighbouring chains and results in the brush formed by extended chains, whose length is the result of the competition between thermal fluctuations and dipole–dipole interactions stimulated by the external field. For a high grafting density, instead, a weak external field cannot perturb the structures formed due to the dipolar interactions: the overall brush height remains rather small, and the brush continues to be rather dense up to moderate fields. For stronger fields, however, the same structural changeover, as observed for $\sigma = 0.040$ at low fields, takes place, and the filaments tend to straighten perpendicular to the grafting surface. We also observed that, even though the number of branching points decreases with the field, for $\sigma = 0.111$ there are two clear configurations of clusters that appear to "survive" even under strong external fields, namely, the X-shaped junction connecting two neighbouring chains by only one point of their middle regions, and the staggered configuration of two chains. Both configurations do not demand large deviations of the dipoles from the orientation of the field. Cluster size is a monotonously decreasing function of the applied field. In general, the field response of the denser brush suggests the potential existence of a hysteresis-like behaviour on low-high-low field cycles. However, this effect deserves special attention and will be the subject of future studies. This work, instead, clearly shows that magnetic filament brushes have a complex microstructure with versatile self-assembly in clusters of different size and topology which, importantly, can be uniquely predicted, based on the grafting density, temperature, dipole moment and intensity of externally applied magnetic fields.

Even though at this point there is no direct experimental realisation of a magnetic filament brush, modern soft matter physics actually offers all necessary building blocks. Thus, for example, a very good control has been achieved in the synthesis of monodisperse magnetite or cobalt permanently magnetised nanoparticles.[5,49] On the other hand, growing experience in DNA origami[50,51] opens up a new perspective in precisely tuneable crosslinking. Finally, the recent advances in the functionalisation of surfaces[12,52] provide various elaborated techniques to create an actual magnetic filament brush.

As a final comment, we briefly discuss the open questions that arose during the study of the field and temperature influence on the properties of the magnetic filament brush, which are far beyond the scope of this paper. First, it is important to understand the thermodynamics of these systems and analyse relevant parameters, like the specific heat, as a function of field and temperature. This can shed light on the differences between magnetic filament systems and self-assembling dipolar spheres resulting from the presence of the permanent links. On the other hand, it is known that there is a series of structural transitions in dipolar hard spheres on cooling, but the temperatures at which they occur is rather low.[10,11] For the filament brush system, the temperature range should be extended in order to perform a proper analysis and provide an extensive comparison. Another control parameter, which is known to be very efficient in tuning brush properties, is the length of the grafted chains. Here, we purposefully focused on only one chain length to avoid multiple effects and provide a clear description of the structural transitions under cooling and increasing field. Now, one can easily separate the influence of the number of particles in filaments. Finally, an open question, which has already been mentioned above, is related to the external magnetic field: whether there exists a hysteresis on the increasing-decreasing field cycle and, if that is the case, what is the influence on it of the grafting density and the rest of the brush parameters?

Acknowledgements

This research has been partially supported by the Austrian Research Fund (FWF): START-Projekt Y 627-N27. Authors are grateful to the Ural Federal University stimulating programme. S. S. K is supported by RFBR mol-a-ved 15-32-20549. E. V. N. acknowledges the support of President RF Grant No. MK-5216.2015.2. The authors are grateful to the Ministry of Education and Science of the Russian Federation (Project 3.12.2014/K) and EU-Project 642774 ETN-Colldense. J. J. C. and T. S. also acknowledge funding from a grant awarded by the Conselleria d'Innovació, Recerca i Turisme del Govern de les Illes Balears and the European Social Fund (ESF). Computer simulations were performed at the Vienna Scientific Cluster (Austrian universities consortium for High Performance Computing) and the Nuredduna high-throughput computing cluster (IFISC, UIB-CSIC). We also thank Michaela McCaffrey for language suggestions.

References

1 G. M. Whitesides and B. Grzybowski, *Science*, 2002, **295**, 2418–2421.

2 R. Thiruvengadathan, V. Korampally, A. Ghosh, N. Chanda, K. Gangopadhyay and S. Gangopadhyay, *Rep. Prog. Phys.*, 2013, **76**, 066501.

3 M. A. C. Stuart, W. T. S. Huck, J. Genzer, M. Muller, C. Ober, M. Stamm, G. B. Sukhorukov, I. Szleifer, V. V. Tsukruk, M. Urban, F. Winnik, S. Zauscher, I. Luzinov and S. Minko, *Nat. Mater.*, 2010, **9**, 101–113.

4 B. J. Park, F. F. Fang and H. J. Choi, *Soft Matter*, 2010, **6**, 5246–5253.

5 S. Singamaneni, V. N. Bliznyuk, C. Binek and E. Y. Tsymbal, *J. Mater. Chem.*, 2011, **21**, 16819–16845.

6 M. Colombo, S. Carregal-Romero, M. F. Casula, L. Gutierrez, M. P. Morales, I. B. Bohm, J. T. Heverhagen, D. Prosperi and W. J. Parak, *Chem. Soc. Rev.*, 2012, **41**, 4306–4334.

7 M. Klokkenburg, C. Vonk, E. M. Claesson, J. D. Meeldijk, B. H. Erné and A. P. Philipse, *J. Am. Chem. Soc.*, 2004, **126**, 16706–16707.

8 M. Klokkenburg, R. P. A. Dullens, W. K. Kegel, B. H. Erné and A. P. Philipse, *Phys. Rev. Lett.*, 2006, **96**, 037203.

9 E. M. Furst and A. P. Gast, *Phys. Rev. Lett.*, 1999, **82**, 4130–4133.

10 S. Kantorovich, A. O. Ivanov, L. Rovigatti, J. M. Tavares and F. Sciortino, *Phys. Rev. Lett.*, 2013, **110**, 148306.

11 S. S. Kantorovich, A. O. Ivanov, L. Rovigatti, J. M. Tavares and F. Sciortino, *Phys. Chem. Chem. Phys.*, 2015, **17**, 16601–16608.

12 A. C. Balazs, T. Emrick and T. P. Russell, *Science*, 2006, **314**, 1107–1110.

13 S. Behrens, *Nanoscale*, 2011, **3**, 877–892.

14 R. Dreyfus, J. Baudry, M. L. Roper, M. Fermigier, H. A. Stone and J. Bibette, *Nature*, 2005, **437**, 862–865.

15 J. J. Benkoski, S. E. Bowles, R. L. Jones, J. F. Douglas, J. Pyun and A. Karim, *J. Polym. Sci., Part B: Polym. Phys.*, 2008, **46**, 2267–2277.

16 K. Ērglis, D. Zhulenkovs, A. Sharipo and A. Cēbers, *J. Phys.: Condens. Matter*, 2008, **20**, 204107.

17 Z. Zhou, G. Liu and D. Han, *ACS Nano*, 2009, **3**, 165–172.

18 J. Byrom, P. Han, M. Savory and S. L. Biswal, *Langmuir*, 2014, **30**, 9045–9052.

19 P. A. Sánchez, E. S. Pyanzina, E. V. Novak, J. J. Cerdà, T. Sintes and S. S. Kantorovich, *Macromolecules*, 2015, **48**, 7658.

20 S. T. Milner, *Science*, 1991, **251**, 905–914.

21 A. Halperin, M. Tirrell and T. Lodge, in *Macromolecules: Synthesis, Order and Advanced Properties*, Springer Berlin Heidelberg, 1992, vol. 100/1, pp. 31–71.

22 F. Zhou and W. T. S. Huck, *Phys. Chem. Chem. Phys.*, 2006, **8**, 3815–3823.

23 W. S. Choi, H. Y. Koo, J. Y. Kim and W. T. S. Huck, *Adv. Mater.*, 2008, **20**, 4504–4508.

24 Y. Ye, Z. Pan, L. Zhang, L. He, A. Xia and H. Liang, *J. Polym. Sci., Part B: Polym. Phys.*, 2010, **48**, 1873–1881.

25 H. Singh, P. E. Laibinis and T. A. Hatton, *Nano Lett.*, 2005, **5**, 2149–2154.

26 A. Cēbers, *Curr. Opin. Colloid Interface Sci.*, 2005, **10**, 167–175.

27 M. Belovs and A. Cēbers, *Phys. Rev. E: Stat., Nonlinear, Soft Matter Phys.*, 2009, **79**, 051503.

28 F. Fahrni, M. W. J. Prins and L. J. van IJzendoorn, *Lab Chip*, 2009, **9**, 3413–3421.

29 A. Babataheri, M. Roper, M. Fermigier and O. D. Roure, *J. Fluid Mech.*, 2011, **678**, 5–13.

30 I. Javaitis and V. Zilgalve, *Adv. Mater. Res.*, 2011, **222**, 221–224.

31 Y. N. Kaznessis, D. A. Hill and E. J. Maginn, *Macromolecules*, 1998, **31**, 3116–3129.

32 J. J. Cerdà, P. A. Sánchez, T. Sintes and C. Holm, *Soft Matter*, 2013, **9**, 7185–7195.

33 P. A. Sánchez, J. J. Cerdà, T. Sintes, A. O. Ivanov and S. S. Kantorovich, *Soft Matter*, 2015, **11**, 2963–2972.

34 J. D. Weeks, D. Chandler and H. C. Andersen, *J. Chem. Phys.*, 1971, **54**, 5237–5247.

35 F. F. Abraham and Y. Singh, *J. Chem. Phys.*, 1977, **67**, 2384–2385.

36 M. P. Allen and D. J. Tildesley, *Computer Simulation of Liquids*, Clarendon Press, Oxford, 1st edn, 1987.
37 J. J. Cerdà, V. Ballenegger, O. Lenz and C. Holm, *J. Chem. Phys.*, 2008, **129**, 234104.
38 A. Bródka, *Chem. Phys. Lett.*, 2004, **400**, 62–67.
39 H. J. Limbach, A. Arnold, B. A. Mann and C. Holm, *Comput. Phys. Commun.*, 2006, **174**, 704–727.
40 A. Arnold, O. Lenz, S. Kesselheim, R. Weeber, F. Fahrenberger, D. Roehm, P. Košovan and C. Holm, in *Meshfree Methods for Partial Differential Equations VI*, ed. M. Griebel and M. A. Schweitzer, Springer Berlin Heidelberg, 2013, vol. 89, pp. 1–23.
41 C. Godsil and G. Royle, in *Algebraic Graph Theory*, Springer New York, 2001, vol. 207, pp. 163–192.
42 G. Zanchetta, F. Giavazzi, M. Nakata, M. Buscaglia, R. Cerbino, N. A. Clark and T. Bellini, *Proc. Natl. Acad. Sci. U. S. A.*, 2010, **107**, 17497–17502.
43 C. De Michele, L. Rovigatti, T. Bellini and F. Sciortino, *Soft Matter*, 2012, **8**, 8388–8398.
44 A. Lebedev, *Colloid J.*, 2010, **72**, 815–819.
45 A. Lebedev and S. Lysenko, *J. Magn. Magn. Mater.*, 2011, **323**, 1198–1202.
46 P. A. Sánchez, J. J. Cerdà, T. Sintes and C. Holm, *J. Chem. Phys.*, 2013, **139**, 044904.
47 M. Murat and G. S. Grest, *Macromolecules*, 1989, **22**, 4054–4059.
48 S. T. Milner, T. A. Witten and M. E. Cates, *Macromolecules*, 1988, **21**, 2610–2619.
49 J. Liu, *JOM*, 2010, **62**, 56–61.
50 S. Srivastava, D. Nykypanchuk, M. Fukuto, J. D. Halverson, A. V. Tkachenko, K. G. Yager and O. Gang, *J. Am. Chem. Soc.*, 2014, **136**, 8323–8332.
51 R. M. Zadegan and M. L. Norton, *Int. J. Mol. Sci.*, 2012, **13**, 7149–7162.
52 O. Azzaroni, *J. Polym. Sci., Part A: Polym. Chem.*, 2012, **50**, 3225–3258.

Faraday Discussions

PAPER

Membrane-mediated aggregation of anisotropically curved nanoparticles†

Alexander D. Olinger,[ab] Eric J. Spangler,[ab] P. B. Sunil Kumar*[c] and Mohamed Laradji*[d]

Received 12th October 2015, Accepted 3rd November 2015
DOI: 10.1039/c5fd00144g

Using systematic numerical simulations, we study the self-assembly of elongated curved nanoparticles on lipid vesicles. Our simulations are based on molecular dynamics of a coarse-grained implicit-solvent model of self-assembled lipid membranes with a Langevin thermostat. Here we consider only the case wherein the nanoparticle–nanoparticle interaction is repulsive, only the concave surface of the nanoparticle interacts attractively with the lipid head groups and only the outer surface of the vesicle is exposed to the nanoparticles. Upon their adhesion on the vesicle, the curved nanoparticles generate local curvature on the membrane. The resulting nanoparticle-generated membrane curvature leads in turn to nanoparticle self-assembly into two main types of aggregates corresponding to chain aggregates at low adhesion strengths and aster aggregates at high adhesion strength. The chain-like aggregates are due to the fact that at low values of adhesion strength, the nanoparticles prefer to lie parallel to each other. As the adhesion strength is increased, a splay angle between the nanoparticles is induced with a magnitude that increases with increasing adhesion strength. The origin of the splay angles between the nanoparticles is shown to be saddle-like membrane deformations induced by a tilt of the lipids around the nanoparticles. This phenomenon of membrane mediated self-assembly of anisotropically curved nanoparticles is explored for systems with varying nanoparticle number densities, adhesion strength, and nanoparticle intrinsic curvature.

I. Introduction

Matrix assisted assembly of nanomaterials can be achieved through their mixing with liquid crystals,[1] block copolymers,[2] or simply as a result of their aggregation

[a]Department of Biomedical Engineering, The University of Memphis, Memphis, TN 38152, USA
[b]Department of Physics and Materials Science, The University of Memphis, Memphis, TN 38152, USA
[c]Department of Physics, Indian Institute of Technology Madras, Chennai-600 036, India. E-mail: sunil@iitm.ac.in
[d]Department of Physics and Materials Science, The University of Memphis, Memphis, TN 38152, USA. E-mail: mlaradji@memphis.edu

† Submitted to Faraday Discussions on November 2, 2015.

at fluid–fluid interfaces.[3] The self-assembly of colloidal and nanoscale particles at fluid–fluid interfaces, in particular, is driven by the wetting properties of the particle at the interface, which is controlled in part by the interfacial tension of the coexisting fluids. Stable self-assemblies of small nanoparticles at fluid–fluid interfaces tend however to be dynamic and lack long-range order.[4] Lipid membranes are a different class of fluid surfaces which are, unlike other self-assembling media, highly flexible quasi-two-dimensional fluids, with their mechanics and thermodynamics being controlled by their bending energy since they are typically tensionless. They can readily deform to accommodate an adsorbed colloidal or nanoscale particle to a shape determined by the interplay between the adhesion of the particle on the lipid membrane and its curvature energy. Lipid membranes naturally serve as a medium for the assembly of a variety of biomolecules, including caveolin[5] and clathrin[6] and the Bin-Amphyphysin-Rvs (BAR) domain family of proteins.[7–9]

Nanoparticles have many promising biomedical applications in biosensing,[10] medical imaging,[11] gene therapy,[12] and the treatment of cancer.[13] Meanwhile, there is an increasing concern in regard to the potential toxic effects of some nanoparticles on living organisms.[14] As a result, many studies have been conducted to investigate the interactions between single nanoscale particles and lipid membranes.[15–20] The pioneering work of Koltover *et al.*[21] demonstrated that 300–900 nm spherical latex beads, which are bound on giant unilamellar vesicles, self-assemble into linear chains or close-packed hexagonal structures. Saric and Cacciuto showed, using coarse-grained simulations of triangulated membranes, that the induced membrane curvature due to adhesion of spherical nanoparticles leads to their self-assembly into linear or hexagonally-packed aggregates.[22] Studies of the self-assembly of anisotropic nanoparticles on lipid membranes are however lacking.

In this article, we present a computational investigation of the interaction between elongated and curved nanoparticles and lipid membranes and the ensuing self-assembly of the nanoparticles on the biomembrane. Specifically, we focus on elongated nanoparticles with the shape of a solid arc and square cross-section and with the inner face of the nanoparticle that interacts attractively with the lipid head groups. We will show that these nanoparticles create distinct self-assemblies that depend on the adhesion strength. At low values of adhesion strength, the nanoparticles arrange side-by-side creating linear chains. At high values of the adhesion strength, however, the nanoparticles self-assemble into aster clusters. These aster-shaped clusters can from a meshwork with triangular symmetry at large enough concentrations of nanoparticles.

II. Model

A mesoscale implicit-solvent model for lipid membranes is used here to study the self-assembly of anisotropically curved nanoparticles on lipid vesicles.[23] In this model, each lipid is coarse-grained into a semi-flexible linear chain composed of one head (h) bead and two (t) tail beads. Since the solvent in this model is implicit, the hydrophobic interactions that lead to the self-assembly of the lipids is effectively accounted for by making t–t interactions attractive.

As shown schematically in Fig. 1, each nanoparticle is modeled as a composite of two types of beads. The beads of the bottom layer (concave surface), labeled n,

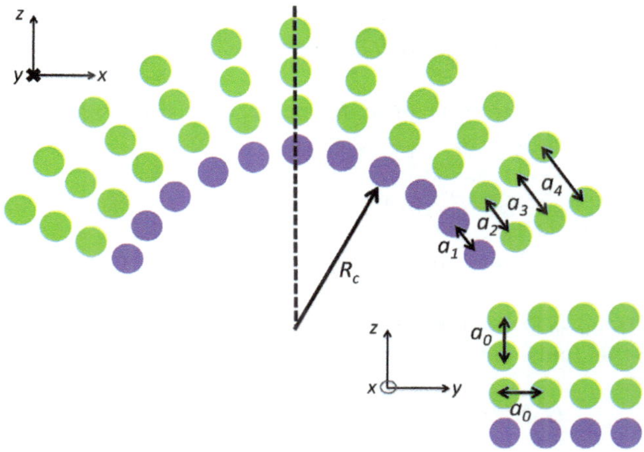

Fig. 1 Schematic representation of a longitudinal slice of an elongated nanoparticle along the xz-plane. Purple (n) beads interact attractively with the lipid head beads, while the green (p) beads interact repulsively with the lipid head beads. Both n and p beads interact repulsively with the lipid tail beads. A spontaneous curvature of the nanoparticle is induced by setting the preferred inter-particle separations to $a_1 < a_2 < a_3 < a_4$. A transverse slice of the nanoparticle along the yz-plane, shown by the dashed line, is displayed in the bottom right. Bond lengths in this plane are set to a_0.

interact attractively with the lipid head beads while the remaining beads in the nanoparticle, labeled p, interact repulsively with the lipid beads. This favours the adhesion of the bottom layer of the nanoparticle on the lipid bilayer, as shown in Fig. 1. Within a nanoparticle, the beads are arranged in a distorted simple cubic lattice and the nearest neighbour beads are linked by harmonic springs. The desired curvature of the nanoparticle is ensured by setting the rest length of the bonds a_1, a_2, a_3, and a_4 as shown in Fig. 1.

The potential energy of the system is given by,

$$U(\{\mathbf{r}_i\}) = \sum_{i,j} U_0^{\alpha_i\alpha_j}(r_{ij}) + \sum_i \sum_{j\in\Omega_i} U_{\text{bond}}^{\alpha_i\alpha_j}(r_{ij}) + \sum_i \sum_{k\in\Omega_i} \sum_{l\in\Omega_i} U_{\text{bend}}^{\alpha_i}(\mathbf{r}_{ik}, \ \mathbf{r}_{il}), \tag{1}$$

where $\mathbf{r}_i$ is the position of bead i and $r_{ij} = |\mathbf{r}_i - \mathbf{r}_j|$. The type of bead i, $\alpha_i = \text{h, t, n}$ and p for a lipid head bead, lipid tail bead, a nanoparticle bottom layer bead, or other beads in the nanoparticle, respectively. Ω_i is the set of beads bonded to bead i. In eqn (1), $U_0^{\alpha_i\alpha_j}$ is a soft two-body interaction between beads i and j and is given by,

$$U_0^{\alpha_i\alpha_j}(r) = \begin{cases} \left(U_{\max}^{\alpha_i\alpha_j} - U_{\min}^{\alpha_i\alpha_j}\right)\dfrac{(r_{\text{m}} - r)^2}{r_{\text{m}}^2} + U_{\min}^{\alpha_i\alpha_j} & \text{if} \quad r \le r_{\text{m}}, \\ -2U_{\min}^{\alpha_i\alpha_j}\dfrac{(r_{\text{c}} - r)^3}{(r_{\text{c}} - r_{\text{m}})^3} + 3U_{\min}^{\alpha_i\alpha_j}\dfrac{(r_{\text{c}} - r)^2}{(r_{\text{c}} - r_{\text{m}})^2} & \text{if} \quad r_{\text{m}} < r \le r_{\text{c}}, \\ 0 & \text{if} \quad r > r_{\text{c}}, \end{cases} \tag{2}$$

where $U_{\max}^{\alpha\beta} > 0$ for any pair (α, β). A negative value of $U_{\min}^{\alpha\beta}$ implies a short-range attraction between two beads of type α and β at intermediate distances. Except for $(\alpha, \beta) = (\text{t, t})$ or $(\alpha, \beta) = (\text{h, n})$, $U_{\min}^{\alpha\beta} = 0$ for any other interacting pair. A negative

U_{min}^{tt} allows for the self-assembly of the lipid molecules into thermodynamically stable bilayers,[23,24] and a negative U_{min}^{hn} allows the nanoparticles to adhere to the lipid bilayer.

Beads that belong to a single lipid molecule or a nanoparticle are linked through $U_{bond}^{\alpha\beta}$ given by

$$U_{bond}^{\alpha_i\alpha_j}(r_{ij}) = \frac{k_{bond}^{\alpha_i\alpha_j}}{2}\left(r_{ij} - a^{\alpha_i\alpha_j}\right)^2, \tag{3}$$

where $k_{bond}^{\alpha_i\alpha_j}$ and $a^{\alpha_i\alpha_j}$ are the stiffness and preferred length of bond α_i–α_j, respectively.

In eqn (1), U_{bend}^{α} is the three-body interaction potential energy which provides stiffness to the lipids and nanoparticles and is given by,

$$U_{bend}^{\alpha_i\alpha_k\alpha_l}(\mathbf{r}_{ik},\ \mathbf{r}_{kl}) = \frac{k_{bend}^{\alpha_i\alpha_k\alpha_l}}{2}\left(\cos\theta_{\alpha_i\alpha_k\alpha_l}^0 - \hat{\mathbf{r}}_{ik}\cdot\hat{\mathbf{r}}_{kl}\right)^2, \tag{4}$$

where $k_{bend}^{\alpha_i\alpha_k\alpha_l}$ and $\theta_{\alpha_i\alpha_k\alpha_l}^0$ are the bending stiffness of a triplet $(\alpha_j,\ \alpha_k,\ \alpha_l)$ and its preferred splay angle, respectively.

In the present study, we set the interaction parameters $U_{min}^{tt} = -6\epsilon$ and $U_{min}^{nh} = -\xi$. Otherwise, we set $U_{min}^{\alpha_i\alpha_l} = 0$. We also set $U_{max}^{tt} = U_{max}^{nh} = 200\epsilon$. Otherwise, $U_{max}^{\alpha_i\alpha_l} = 100\epsilon$. For pairs of beads belonging to the same nanoparticle, we set $U_{min}^{nn} = U_{min}^{pp} = U_{min}^{pp} = 0$, and $U_{max}^{nn} = U_{max}^{pp} = U_{max}^{pp} = 0$. In order to prevent the nanoparticles from aggregating in the absence of lipid membranes, interactions between two beads belonging to different nanoparticles are fully repulsive. For these pairs of beads, we set $U_{min}^{nn} = U_{min}^{pp} = U_{min}^{pp} = 0$ and $U_{max}^{nn} = U_{max}^{pp} = U_{max}^{pp} = 100\epsilon$.

The elastic coefficients of the two-body and three-body interactions for the lipids are set to $k_{bond}^{ht} = k_{bond}^{tt} = 100\epsilon/r_m^2$ and $k_{bend}^{htt} = 100\epsilon$, respectively. The rest bond length for the lipids is set to $a^{ht} = a^{tt} = 0.7r_m$ and the preferred splay angle of a lipid chain is set to $\theta_{htt}^0 = 180°$.

Each nanoparticle is made of 480 beads arranged in a distorted simple cubic lattice, having four layers each composed of 30 × 4 beads (see Fig. 1). Each bead is linked to its nearest neighbours by the bonding potential U_{bond}^{pp}, U_{bond}^{np} or U_{bond}^{nn} in eqn (3). The rigidity of the nanoparticle is maintained by the three-body potential U_{bend} with the angle $\theta_{\alpha_i\alpha_j\alpha_k}^0 = 90°$ if i and k in the triplet are next nearest neighbours and $\theta_{\alpha_i\alpha_j\alpha_k}^0 = 180°$ if i, j and k are collinear. In order to induce a radius of curvature R_c in the nanoparticle, the preferred bond lengths a along the longitudinal direction are set to values given in Table 1 (see Fig. 1). Along the two other directions, the bond length is set to $a_0 = 1.0r_m$ (see Fig. 1). The elastic coefficients of the two-body and three-body interactions for the lipids are set to $k_{bond}^{nn} = k_{bond}^{pp} = k_{bond}^{np} = 1000\epsilon/{r_m}^2$ and $k_{bend}^{ppp} = k_{bend}^{ppn} = k_{bend}^{nnp} = k_{bend}^{nun} = 1000\epsilon$,

Table 1 Tangential bond lengths a_1, a_2, a_3 and a_4 for the case of $R_c = 11r_m$ and $R_c = 7.5r_m$

	$R_c = 7.5r_m$	$R_c = 11r_m$
a_1/r_m	0.70	0.75
a_2/r_m	0.85	0.83
a_3/r_m	1.00	0.92
a_4/r_m	1.15	1.00

respectively. The desired radius of curvature, R_c, of the nanoparticle is set by the combined effects of the bending potential and the variation in bond length along the radial direction.

Beads are moved using a molecular dynamics scheme with a Langevin thermostat:[25]

$$\dot{\mathbf{r}}_i(t) = \mathbf{v}_i(t) \text{ and } m\dot{\mathbf{v}}_i(t) = -\nabla_i U(\{\mathbf{r}_i\}) - \Gamma \mathbf{v}_i(t) + \sigma \Xi_i(t), \tag{5}$$

where m is the mass of a each bead (same for all beads), Γ is a bead's friction coefficient, and $\sigma\Xi_i(t)$ is a random force originating from the heat bath. $\Xi_i(t)$ obeys

$$\langle \Xi_i(t) \rangle = 0, \tag{6}$$

$$\langle \Xi_i^{(\mu)}(t) \Xi_j^{(\nu)}(t') \rangle = \delta_{\mu\nu}\delta_{ij}\delta(t - t'), \tag{7}$$

with μ, $\nu = x$, y, or z. The dissipative and random forces are interrelated through the dissipation–fluctuation theorem, which leads to the relationship

$$\Gamma = \sigma^2/2k_BT. \tag{8}$$

The equations of motion are numerically integrated using the velocity-Verlet algorithm with $\Gamma = 6m/\tau$ where the timescale $\tau = r_m(m/\epsilon)^{1/2}$ with r_m and ϵ being used as scales for length and energy, respectively. To ensure that systems reach thermal equilibrium, the simulations were run for at least 10^7 Δt, with the time step $\Delta t = 0.01\tau$. The simulations are run at $k_BT = 3\epsilon$, a temperature at which the bilayer is in the fluid phase.[24] The vesicles in the simulations are composed of 180 000 lipids, corresponding to a diameter of $130r_m$.

III. Membrane-mediated interaction between nanoparticles

In order to understand the effect of lipid membranes on the effective interactions between anisotropically curved nanoparticles, we first performed simulations of two and three nanoparticles with an intrinsic radius of curvature $R_c = 11r_m$ on vesicles of diameter $130r_m$ and for adhesion strengths of $\xi = 1.0\epsilon$, 2ϵ, 3.0ϵ, 4.0ϵ, 5.0ϵ, and 6.0ϵ. In these simulations, to conserve time, the two or three nanoparticles are initially placed parallel to each other. Similar results are obtained when they are placed relatively far from each other. The resulting equilibrium configurations together with distributions of the splay angle, ϕ, between the nanoparticles are shown in Fig. 2(a) and (c). These figures show that for $\xi = 1.0\epsilon$, the two nanoparticles lie side-by-side and are very close to each other. For $\xi = 2.0$ and 3.0ϵ also, the nanoparticles are parallel to each other, but are not as close to each other as in the case of $\xi = 1.0\epsilon$. The angle distribution for the case of $\xi = 1.0\epsilon$ (purple line in Fig. 2(b)) is very narrow and is peaked at a very small splay angle, $\phi \approx 3°$. However, the peak position is higher for $\xi = 2\epsilon$ and 3ϵ, and the width of the distribution increases with increasing adhesion strength. This is evident in the corresponding distributions (black and red lines) in Fig. 2(b) and (d) which are peaked at a small angle around $\phi_{max} = 5°$. The width of the distribution for $\xi = 3.0\epsilon$ is however higher than that for 2.0ϵ, implying that there is a tendency for the nanoparticles to splay from each other with increasing ξ. For $\xi \geq 4.0\epsilon$, the

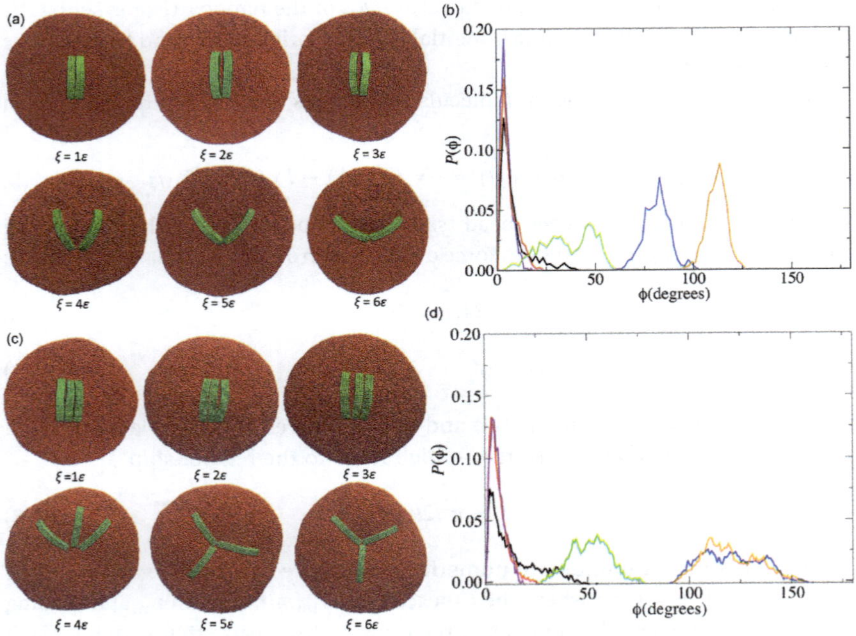

Fig. 2 (a) Equilibrium snapshots of lipid vesicles with two anisotropic nanoparticles having an intrinsic radius of curvature $R_c = 11r_m$ at adhesion strength ξ ranging between 1.0∊ and 6.0∊. Green beads represent the nanoparticle and the red beads indicate the lipid head groups. (b) Splay angle distributions between two nanoparticles. Purple curve corresponds to $\xi = 1.0\epsilon$, red curve corresponds to $\xi = 2.0\epsilon$, black curve corresponds to $\xi = 3.0\epsilon$, green curve corresponds to $\xi = 4.0\epsilon$, blue curve corresponds to $\xi = 5.0\epsilon$, and the orange curve corresponds to $\xi = 6.0\epsilon$. (c) Same as in (a) but for three nanoparticles. (d) Same as in (b) but for three nanoparticles with same colour coding.

nanoparticles are clearly splayed from each other as is evident from both configurations and splay angle distributions with the preferred splay angle that increases with increasing adhesion strength. It is noted that the splay angle between the nanoparticles is the result of membrane curvature mediated interactions and, for relatively high adhesion strength, it is influenced by the presence of other neighbouring nanoparticles as demonstrated by the configurations in Fig. 2(b) and the distributions of Fig. 2(d) which are broader and shifted for $\xi =$ 4.0 and 5.0∊. This implies that at high adhesion strength, effective membrane-induced many-body interactions between the nanoparticles play an important role in their self-assembly.

We now turn to the question of why a splay between the nanoparticles is induced at high adhesion strength. Since these conformations arise due to membrane mediated interactions, a better understanding of the nature of membrane deformation induced by nanoparticles is necessary. We therefore performed simulations of a single nanoparticle with $R_c = 11r_m$ on a planar lipid bilayer for values of adhesion strength varying between 1.0∊ and 5.0∊. For all values of adhesion strength considered here, the bilayer conforms to the curvature of the nanoparticle along its longitudinal direction. Profiles of the lipid bilayer at the plane transversally bisecting the nanoparticle at the midsection (dashed plane according to Fig. 1) are shown in Fig. 3(c) and (d) at $\xi = 2.0\epsilon$ and $\xi = 5.0\epsilon$, respectively. These figures clearly show that for $\xi = 2.0\epsilon$, the bilayer is not curved along the midsection plane of the nanoparticle. This implies that the nanoparticle locally induces positive mean curvature and zero Gaussian curvature of the bilayer. In contrast, for $\xi = 5\epsilon$, the bilayer is negatively curved in the midsection plane, while it is positively curved along the longitudinal direction. As a result, the nanoparticle locally induces a saddle profile of the lipid bilayer. This induced saddle region in the lipid bilayer must be due to the increased interaction between the lipid head groups and the bottom layer of the nanoparticle.

The induced negative curvature of the bilayer in the midsection is further quantified by the relationship between the tilt of the lipids, defined as the angle between the horizontal and the end-to-end vector of a lipid, and their distance from the nanoparticle in the proximal leaflet at the midsection plane, as illustrated in Fig. 3(a). Fig. 3(b) shows that for $\xi = 1.0\epsilon$ and 1.5ϵ, $|\cos\theta| \approx 0.17$ implying an angle $\theta \approx 80°$, independent of the distance between the lipids and the nanoparticle, which corresponds to the horizontal membrane profile at the midsection plane of the nanoparticle, as clearly indicated by Fig. 3(c). This favours a side-by-side configuration, of the two nanoparticles. The low but finite values of $|\cos\theta|$ at these low adhesion strengths are due to thermal fluctuations. As the adhesion strength is increased, the $|\cos\theta|$ distribution shows a clear peak with the peak position shifting to the right. This clearly shows that the lipid tilt increases with increasing adhesion strength and is maximum at the boundary of

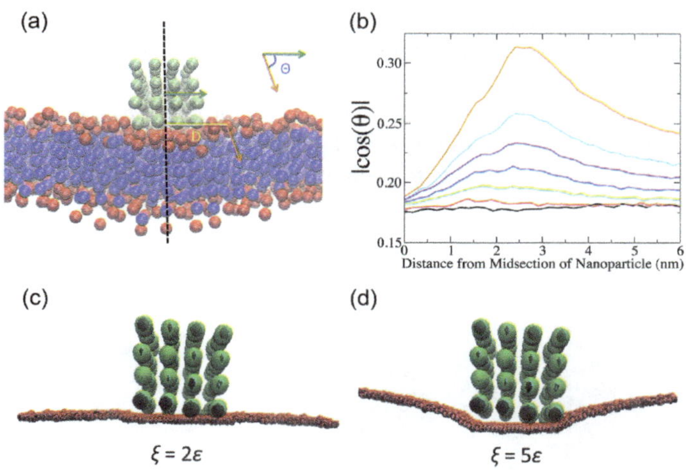

Fig. 3 (a) The cross section of a nanoparticle on a lipid bilayer. The yellow line represents the distance D of a lipid head particle from the median vertical plane bisecting the nanoparticle (shown by the black dashed line). The green vector is normal to the median plane and the orange vector is a typical head-to-tail vector for the lipid, with θ being the angle between them. (b) $|\cos(\theta)|$ *vs.* D in the case of a nanoparticle with $R_c = 11r_m$ bound to a $59r_m \times 59r_m$ lipid bilayer containing 10 000 lipids with constant lipid number density of $\rho = 3.20r_m^{-2}$. Only lipids in the top leaflet of the membrane and within a cutoff distance of $5r_m$ from the bottom layer of the nanoparticle are considered. In (b), curves from bottom to top, correspond to $\xi = 1.0\epsilon$, 1.5ϵ, 2.0ϵ, 2.5ϵ, 3.0ϵ, 4.0ϵ, and 5.0ϵ, respectively. (c and d) The height profile of the lipid bilayer, at the mid-section of the bound nanoparticle (yz-plane in Fig. 1) for the cases of $\xi = 2.0\epsilon$ and 5.0ϵ, respectively.

the nanoparticle. The resulting induced saddle deformation in the bilayer makes it difficult for another nanoparticle to be close and parallel to it.

IV. Self-assembly

In order to investigate the nature of aggregation of the nanoparticles on a vesicle, we performed an extensive amount of simulations with 100 and 200 nanoparticles of radius of curvature $R_c = 11r_m$, on vesicles with diameters of $130r_m$. In the initial configurations, the nanoparticles are positioned and oriented randomly very close to the spherical lipid vesicles. In Fig. 4, snapshots are shown for the case of 100 nanoparticles with $\xi = 2.0\epsilon$, 4.0ϵ and 6.0ϵ. This figure clearly shows that the adhesion strength strongly affects the nature of self-assembly of the nanoparticles. At low values of adhesion strength ($\xi = 2.0\epsilon$), the nanoparticles are self-assembled in linear chains while in the case of high values of adhesion strength ($\xi = 4.0\epsilon$ and 6.0ϵ), the nanoparticles are arranged into asters with the average number of nanoparticles per aster increasing with ξ. These results are in excellent agreement with those presented above for two and three nanoparticles. The clear bulges, between the nanoparticles, in the vesicle for the case of $\xi = 6.0\epsilon$ are due to the saddle regions in the membrane created by the nanoparticles at high adhesion strength (see Fig. 3).

Effect of nanoparticle number density on their self-assembly

We now turn to exploring the effect of the surface number density of the nanoparticles on their membrane-mediated self-assembly. Fig. 5 shows configurations of vesicles with $N = 100$ and 200 nanoparticles at $\xi = 2.0\epsilon$ and 6.0ϵ. In the case of $\xi = 2.0\epsilon$, chain-like aggregates are the dominant form of self-assembly. For $N = 200$, the chains intersect at triangular vertices and the average chain segment size is smaller than that for $N = 100$. These triangular clusters are defects that arise at high nanoparticle densities due to geometrical constraints induced by the finite spherical surface.

Fig. 4 Snapshots of 100 nanoparticles with intrinsic radius of curvature $R_c = 11r_m$ on a vesicle with diameter $130r_m$. Snapshots from left to right correspond to $\xi = 1.0\epsilon$, 4.0ϵ, and 6.0ϵ, respectively. In the case of $\xi = 1.0\epsilon$, a few nanoparticles are not bound to the vesicle or adhere sidewise to the membrane. This is due to the fact that the long chain clusters induce tension on the membrane which prevents the nanoparticles from generating local curvature and full binding. These nanoparticles will eventually fully bind after their diffusion to the ends of the chain clusters which are curvature friendly regions.

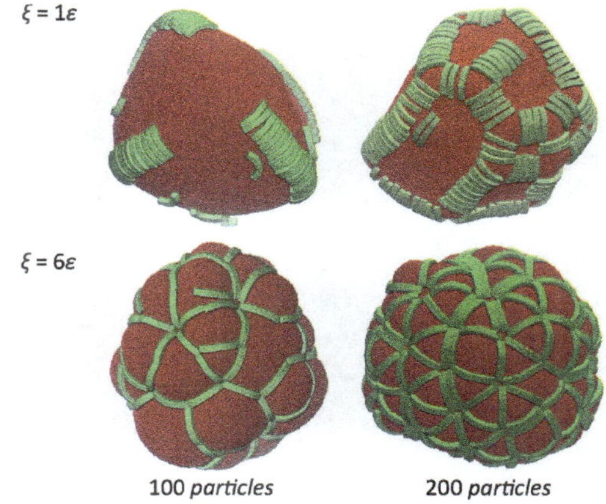

Fig. 5 Equilibrium snapshots of lipid vesicles with 100, and 200 nanoparticles (left to right) and intrinsic radius of curvature $R_c = 11r_m$, and at adhesion strengths $\xi = 1.0\epsilon$ (top row) and 6.0ϵ (bottom row).

For $\xi = 6.0\epsilon$, the splay angles between neighbouring nanoparticles for the case of $N = 100$ are clearly larger on average than those for the case of $N = 200$. In fact, for $N = 200$, most of the splay angles between neighbouring particles are about 60°, which is a result of the fact that a large number of nanoparticles cannot be self-assembled onto the vesicle with large splay angles as in the case of $N = 100$. The asters for the case of $N = 200$ have a somewhat hexagonal symmetry and form a meshwork spanning the whole vesicle. In this case, even though $\xi = 6.0\epsilon$ is high, crowding effects lead some nanoparticles to be arranged side-by-side, a situation that does not occur at low number density ($N = 100$).

Effect of nanoparticle curvature

We extended our study to also investigate the effect of the nanoparticle intrinsic radius of curvature, R_c. In Fig. 6, equilibrium configurations of vesicles with $N = 200$ nanoparticles for $R_c = 7.5r_m$ and $11r_m$ are shown for $\xi = 1.0\epsilon$, 2.0ϵ and 6.0ϵ. This figure shows that the weakly curved nanoparticles ($R_c = 11r_m$) are all bound to the membrane and self-assembled into chains at $\xi = 1.0\epsilon$. In contrast, none of the highly curved ($R_c = 7.5r_m$) nanoparticles are fully bound to the vesicle and only a few are adsorbed to the vesicle by their two ends. This is due to the fact for $R_c = 7.5r_m$ and at low values of ξ, the decrease in free energy through full adhesion of the nanoparticles to the vesicle is not enough to compensate for the increase in energy resulting from bending the lipid bilayer. At $\xi = 2.0\epsilon$, the weakly curved nanoparticles ($R_c = 11r_m$) self-assemble into chains, albeit with an average size that is shorter than that for $\xi = 1.0\epsilon$. The highly curved nanoparticles, at the same adhesion strength, however, are self-assembled into chains that are on average longer than for the case of $R_c = 11r_m$ at $\xi = 1.0\epsilon$. It is interesting to note that some of the highly curved nanoparticles at $\xi = 2.0\epsilon$ are merely adsorbed by their two ends. These monomeric nanoparticles cannot easily fully bind due to their inability to generate local curvature of the membrane because of the tension induced on the membrane by the long fully bound chains.

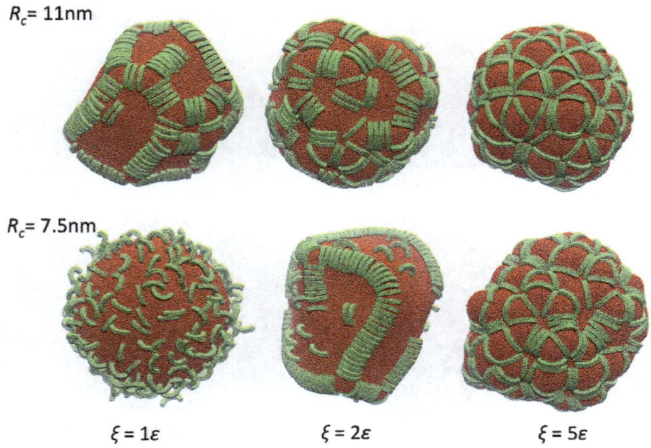

Fig. 6 Equilibrium snapshots of 200 nanoparticles with intrinsic radius of curvature $R_c = 11r_m$ (top row) or $R_c = 7.5r_m$ (bottom row) and adhesion strengths $\xi = 1.0\epsilon$ (left column), 2.0ϵ (center column) and 5.0ϵ (right column).

V. Conclusion

In conclusion, we presented in this article a large scale study of the self-assembly of anisotropically curved nanoparticles on lipid vesicles *via* coarse-grained molecular dynamics simulations of an implicit-solvent model. Our simulations show that the curved nanoparticles are able to generate a local curvature of the membrane which then mediates their self-assembly into either chains or asters. Chains, in which the nanoparticles are arranged side-by-side are formed at weak adhesion strength. As the adhesion strength increases, the nanoparticles induce saddle curvature deformations on the membrane which results in aster aggregates emerging. At high surface densities of the nanoparticles the effect of tension, induced by binding of the curvature-generating nanoparticles, also affects the binding affinity of the particles and plays a role in determining the type of self-assembly. In particular, the aster clusters tend to be isolated and have a small number of arms for small densities, while at high densities, the asters form a meshwork with hexagonal symmetry. We hope that the present study will incite future experiments on the self-assembly of anisotropically curved nanoparticles on lipid membranes. Crescent-shaped gold nanoparticles[26–28] with appropriate functionalization should be a good candidate to perform such experiments.

Acknowledgements

This work was supported by the National Science Foundation (DMR-0812470), the National Institute of General Medical Sciences of NIH (R15GM106326), the American Physical Society Travel Grant Program, and a Faculty Research Grant from the University of Memphis. The latter support does not necessarily imply endorsement by the University of research conclusions. All simulations were performed on computers of the High Performance Computing Facility of the University of Memphis.

References

1 X. Zeng, F. Liu, A. G. Fowler, G. Ungar, L. Cseh, G. H. Mehl and J. E. MacDonald, *Adv. Mater.*, 2009, **21**, 1746.
2 W. A. Lopes and H. M. Jaeger, *Nature*, 2001, **414**, 735.
3 Z. Niu, J. He, T. P. Russel and Q. Wang, *Angew. Chem., Int. Ed.*, 2010, **49**, 10052.
4 M. J. A. Hore and M. Laradji, *J. Chem. Phys.*, 2007, **126**, 244903.
5 I. Fernandez, Y. Ying, J. Albanesi and R. G. W. Anderson, *Proc. Natl. Acad. Sci. U. S. A.*, 2002, **99**, 11193.
6 R. Matthews and C. N. Likos, *Soft Matter*, 2013, **9**, 5794.
7 A. Frost, R. Perera, A. Roux, K. Spasov, O. Destaing, E. H. Egelman, P. De Camilli and V. M. Unger, *Cell*, 2008, **132**, 807.
8 B. J. Peter, H. M. Kent, I. G. Mills, Y. Vallis, P. J. G. Butler, *et al.*, *Science*, 2004, **303**, 495.
9 M. G. Ford, I. G. Mills, B. J. Peter, Y. Vallis, G. J. Praefcke, *et al.*, *Nature*, 2002, **419**, 361.
10 S. T. Selvan, T. T. Y. Tan, D. K. Yi and N. R. Jana, *Langmuir*, 2010, **26**, 11631–11641.
11 J. L. West and N. J. Hala, *Annu. Rev. Biomed. Eng.*, 2003, **5**, 285–292.
12 M. E. Davis, J. E. Zuckermann, C. H. J. Choi, D. Selingson, A. Tolcher, C. A. Alabi, Y. Yen, J. D. Heidel and A. Ribas, *Nature*, 2010, **464**, 1067.
13 K. K. Jain, *Technol. Cancer Res. Treat.*, 2005, **4**, 407–416.
14 N. Lewinski, V. Colvin and R. Drezek, *Small*, 2008, **4**, 26.
15 O. Le Bihan, P. Bonnafous, L. Marak, T. Bickel, S. Trépout, S. Mornet, F. de Haas, H. Talbot, J.-C. Taveau and O. Lambert, *J. Struct. Biol.*, 2009, **168**, 419–425.
16 M. Deserno and T. Bickel, *Europhys. Lett.*, 2003, **62**, 767–773.
17 T. Ruiz-Herrero, E. Velasco and M. F. Hagan, *J. Phys. Chem. B*, 2012, **116**, 9595.
18 X. Yi and H. Gao, *Soft Matter*, 2015, **11**, 1107–1115.
19 S. Dasgupta, T. Auth and G. Gompper, *Nano Lett.*, 2014, **14**, 687–693.
20 R. Vácha, F. J. Martinez-Veracoechea and D. Frenkel, *Nano Lett.*, 2011, **11**, 5391–5395.
21 I. Koltover, J. O. Rädler and C. R. Safinya, *Phys. Rev. Lett.*, 1999, **82**, 1991.
22 A. Saric and A. Cacciuto, *Phys. Rev. Lett.*, 2012, **108**, 118101.
23 J. D. Revalee, M. Laradji and P. B. Sunil Kumar, *J. Chem. Phys.*, 2008, **128**, 035102.
24 E. J. Spangler, P. B. Sunil Kumar and M. Laradji, *Soft Matter*, 2012, **8**, 10896.
25 G. S. Grest and K. Kremer, *Phys. Rev. A*, 1986, **33**, 3628.
26 J. S. Shumaker-Parrt, H. Rochholz and M. Kreiter, *Adv. Mater.*, 2005, **17**, 2131.
27 H. Rochholz, N. Bocchio and M. Kreiter, *New J. Phys.*, 2007, **9**, 53.
28 A. Unger, U. Rietzler, R. Berger and M. Kreiter, *Nano Lett.*, 2009, **9**, 2311.

Faraday Discussions

DISCUSSIONS

Nanocomposites: general discussion

Guruswamy Kumaraswamy, Sanat Kumar, Siddharth Kulkarni, Ajeet Srivastav, Priyadarshi Roy Chowdhury, G. V. Pavan Kumar, Erika Eiser, Alison Edwards, B. L. V. Prasad, Madivala G. Basavaraj, Nagaraj Shetti, Michael Bockstaller, Himani Medhi, Yogesh M. Joshi, Neena S. John, Charusita Chakravarty, Vandana Shinde, Rajdip Bandyopadhyaya, Nicholas Kotov, Ramanan Krishnamoorti, Marc Couty, Jacques Jestin and Daan Frenkel

DOI: 10.1039/C6FD90003H

Yogesh M. Joshi opened a discussion of the paper by Marc Couty: It is indeed a very novel way of manufacturing the composite: first dissolving and swelling Styrene butadiene copolymer in styrene and subsequently crosslinking the latter. Does such assembly have any special benefits in terms of mechanical or other properties over the usual Styrene butadiene copolymer?

While it has been claimed that post swelling occurs, it is possible to separate the filler into the small aggregates. The SEMs in Figure 3 in your manuscript do not show any appreciable change in aggregate size. As expected, it seems only the average inter-particle distance has increased.

Marc Couty responded: After polymerisation, the polystyrene content is so high that the sample is glassy at room temperature and of no use for tyre properties. The experiment should be repeated at a lower swelling ratio to test the mechanical properties of such a material.

Indeed, SEM does not show any reduction in size of the object. We must keep in mind that since SEM is revealing only 10 nm under the surface of the material, for such branched objects made of primary particles around 10 nm it gives only informations on cuts of the "branches". If one assimilates the aggregate with a human hand, SEM would be equivalent to image finger cuts.

B. L. V. Prasad enquired: 1. During the preparation process some reagents are added to the filler silica particles and it is seen that these reagents then adhere to the silica particles. What happens if these silica particles are pre-grafted with these reagents?

2. Is the chemical nature of the interactions between the grafted silica particles and then polymer well understood? How important is this for the final quality and function of the material?

Marc Couty responded: Some silica are pre-grafted with silanes. The main difficulty comes from adapting the melt mixing process to incorporate these fillers. I don't know any study of silica pre-grafted with polymers. It would be difficult to make because the polymer grafting step must be done before the silica aggregates are agglomerated. The chemical nature of the interactions is still debated. However, for silica dispersion what matters is tethering the chains to the surface whether it is chemical grafting or physical strong adsorption. For the final quality in usage like wear or tear resistance, it may make a difference.

Sanat Kumar asked: 1. If you fix the amount of functionalising chains, then, is the domain size fixed as an equilibrium quantity? Or, does this depend on the processing? How do I think about such data? Overall, how do I think about this problem? 2. You have previously said "I know I'm not at an optimum" presumably in terms of the dispersion state of the nanoparticles and hence the properties you achieve. How do you know that?

Marc Couty answered: I know, today, that I am not at an optimum, because I have glasses that allow me to see that the material structure is evolving with a strong coupling between processing conditions and raw materials used. My task is to master all these coupled mechanisms to gain in terms of material performance. We do have a dependence on the processing conditions at a fixed amount of functional chains because melt mixing is not just mechanics but chemistry. Depending on the temperature during the mixing cycle (and all the processing tools on the material transformation chain), one can activate or not the grafting at given stages of the process. The complexity goes one step further when one thinks that grafting can take place before fragmentation of agglomerates and thus generates "Janus" type aggregates *in situ* with one "face" grafted while the other, previously connected with the agglomerate, stays free of the grafts. I do not know how to characterize this type of object in the material but it would be very interesting to state on the existence and their probable impact on the material properties.

Yogesh M. Joshi commented: Polystyrene dissolves in styrene, but whether butadiene dissolves in styrene or not, I do not know. In case there is a different extent of swelling of the styrene part *vis-à-vis* the butadiene part in styrene, some of the parts of the chain will be more swollen than the other. It may lead to interesting scenarios and possibly residual stresses in the resultant crosslinked material. Any comments?

Marc Couty commented in reply: Unfortunately, I am not using block co-poly (styrene-butadiene) but a statistical polymer and I cannot give any hint on that interesting comment.

Rajdip Bandyopadhyaya asked: (a) In our paper on "Dynamics of chain aggregates of carbon nanoparticles in isolation and in polymer films: Implications for nanocomposite materials",[1] we had shown the role of a linear chain of nanoparticle-aggregate on enhancing mechanical properties of a polymer composite. This mechanism has direct implication in car tyres. Have you conducted any similar elongation and reinforcement type experiment for your silica-filled nanocomposite? Can you study mechanical properties by considering only laboratory-made individual nanoparticles, which will behave differently to the actual commercial silica aggregate?

(b) How do you define and distinguish between agglomerates *vs.* aggregates? In our case, individual carbon-filler nanoparticles were formed by laser ablation; it eventually generated a chain of carbon nanoparticles bonded at the particle-interface by C–C covalent bond, which were well-dispersed in the polymer matrix. Such a chain of carbon nanoparticles is called an aggregate, as it cannot be broken into individual nanoparticles, due to the strong C–C bond. On the other hand, agglomerates are reversible structures, which can be broken down by sonication. Are your definitions similar?

(c) How do you interpret the terms "dispersion" and "distribution" separately, while discussing your processing results at different steps of the nanocomposite study?

1 R. Bandyopadhyaya *et al.*, *Chem. Mater.*, 2004, **16**, 3147–3154.

Marc Couty replied: We have done many studies of mechanical reinforcement using shear or elongation since many of the material properties in tyre are linked with mechanics, and indeed the shape of the aggregates have a strong impact. In toy model systems of nanoparticles, we are mostly interested to extract parameters transferable to more complex systems and in general not looking for their direct mechanical properties that are in general very different for industrial commercial silica. In my "definition", agglomerates differ from aggregates essentially by their size. I suppose both are rigid objects and that the process of evolution from agglomerates to aggregates is irreversible. If there is a reversible process then I would associate it like you with an interaction between primary objects that I (implicitly) consider as rather weak compared to covalent bonding and not refer to these objects as agglomerates, that may not be rigid and evolve while applying stress.

As you point out, "dispersion" and "distribution" are extremely coupled since it is a matter of interaction strength between aggregates. The swelling process I have used allows to separate what is a direct (strong) interaction and a polymer mediated (weak) interaction.

Ajeet Srivastav questioned: What is the reason behind choosing nanoparticles rather than nanowires, though it has been established very well that the fibrous reinforcements are the most preferred way to enhance the mechanical properties of the composites?

Marc Couty answered: Again, melt mixing has some limitations: it is often difficult to separate wires at the nanoscale when they are assembled before mixing and it results in a poor dispersion with reinforcement properties that do not reach the ones with nanoparticles. When the wires diameter is in the order of 100 nm to 1 μm, they get fragile and one ends up in a dispersion of objects with no cylindrical character.

Sanat Kumar remarked: We, in academia, do research and work on problems we imagine that you, in industry, are interested in. For example, you say that it is not feasible to completely disperse nanoparticles into a polymer – but this is almost exactly what we (in academia) have been focusing on as our Holy Grail and end goal. What else are we missing/ignoring? What should we be thinking about in the big picture? Processing? Anything else?

Marc Couty responded: The tyre industry deals with huge quantities of materials (million tons per year) and investments have been made for robustness and efficiency. Using melt mixing techniques, it is probably difficult to achieve a dispersion of nanoparticles. Working on other processing techniques to achieve the ultimate dispersion of particles is itself very interesting for the industry to measure the distance to the asymptote or to access some parameters transferable to the industrial case of aggregated particles. Academia could address the industrial problems more and in the case of the present paper, focus on the driving mechanisms of agglomerate (or aggregate) fragmentation under shear. This requires accessing the mechanical stresses acting on "fractal" aggregates under shear and the material resistance of nanoscale silica to rupture that is today a field with limited research. I think that you, as representative of academia, are working on the right subject to generate a dispersion of particles and derive the thermodynamic mechanisms that drive the self assembly of the particles at the nanoscale.

Rajdip Bandyopadhyaya addressed Marc Couty, Jacques Jestin and Sanat Kumar: In order to study the reinforcement of polymers with filler nanoparticles, use of spherical, filler-nanoparticles is good as a toy system, as there is no isolated, spherical filler-nanoparticle in a real polymer nanocomposite sample.

The latter almost always has a short network of filler particles – like linear (or branched) chains of spherical, filler-nanoparticles. These chains are few hundred nm in length. They are also desirable, as the anisotropic, linear-chain of nanoparticles as they have more reinforcement, compared to spherical, filler-particles.[1]

It would be interesting to study dynamics of such linear chains, although polymer dynamics is well known for years. However, the dynamics of such linear chains (which constitute almost 30% of the polymer nanocomposite in the car tyre) is hardly studied or understood. Has there been any such development or observation in your work?

1 R. Bandyopadhyaya *et al.*, *Chem. Mater.*, 2004, **16**, 3147–3154.

Marc Couty responded: We haven't looked at the dynamics of the aggregates in the materials. There are a few studies of SAXS/SANS studies under deformation or XPCS under quiescent conditions but not much in terms of interpretation of the obtained signals.

Sanat Kumar replied: We have looked at the role of particle dispersion state on mechanical reinforcement.[1,2] What we find is that the optimal reinforcement occurs when the nanoparticles form percolated, system spanning structures. While this is a necessary condition it is not sufficient. In addition to percolating we find that chains need to be adsorbed on the particles, and the particle–particle overlap (necessary for percolation) be mediated by the presence of adsorbed polymer chains. It is the resulting "entropic spring" contribution to elasticity that we find increases the modulus and also toughness.

1 P. Akcora *et al.*, *Macromolecules*, 2010, **43**, 1003–1010.
2 J. Moll and S. K. Kumar, *Macromolecules*, 2012, **45**, 1131–1135.

Jacques Jestin answered: The dynamic of the chains in the presence of the particles in nanocomposites is indeed a critical point. Neutron scattering experiments offer the possibility to label the chains and to separate the contribution of the chains close to the particles from the ones in bulk. It allows also to probe a broad range of typical times by combining backscattering and spin echo experiments. However, such experiments are difficult to design properly because it can be difficult to find the right polymer with the convenient time scale while knowing and controlling the structural organization of the NPs inside the polymer. We can find some nice experiments of spin echo for high particle loading by Richter and of backscattering on silica/PMMA systems by Robertson and Ackora. Alternative approaches use dielectric spectroscopy (Sokolov) or NMR (Long/Lequeux). As far as I know, there are different scenarios which are discussed to explain the reinforcement: gradient of chain mobility, glassy layer or glassy bridges, increase of the entanglement density, entanglement between the grafted and the free chain.

Nicholas Kotov addressed Marc Couty and Jacques Jestin: I wonder if you tried the classical Halpin-Tsai model for description of these composites?

Marc Couty answered: There is no general model that can accommodate all the physics of these materials. The ideas behind composite models work well in general for macroscopic composites. When it comes to nanocomposites where dimensions of particles, interparticle distances and polymer chain characteristics are comparable, they tend to be crude. Several ideas are currently debated like the existence of a glassy polymer shell around particles and slowed down dynamics of polymers due to entanglements in a crowded environment, which are much more promising.

Jacques Jestin replied: We used the model of Takayanagi[1] which is a series parallel model combining the modulus of a reinforced phase (the filler) with the one of a soft phase (the matrix). The general idea of this model is very close to the Halpin-Tsai model but for isotropic fillers. The fillers are here spherical nanoparticles that are forming clusters (also more or less spherical). There is indeed no orientation parameter like in the Halpin-Tsai model more dedicated to anisotropic fillers than fibers.

1 M. Takayanagi *et al.*, *J. Polym. Sci., Part C: Polym. Symp.*, 1964, **5**, 113.

Guruswamy Kumaraswamy asked: I am curious about the effect of the molecular weight of the grafted polymer chain on the dispersion that is achieved. As the polymer molecular weight increases, there is a decrease in the concentration of the terminal groups that graft onto the nanoparticle surface. However, it is likely the stretching of high molecular weight chains (tethered to the nanoparticle surface) in the flow field that results in effective nanoparticle dispersion. Is there, therefore, an optimum molecular weight of the telechelic polymer chains to achieve dispersion?

Marc Couty responded: We have shown in previous work[1,2] that the grafting density is a key parameter impacting the dispersion. Indeed, when one decreases the polymer molecular weight, the grafting density is increasing which results in a reduction of material dissipation through a better dispersion. In the meantime, the number of free chain end increases and the dissipation due to the visco-elasticity of the matrix is higher and has to be balanced. As you mention, decreasing the polymer molecular weight decreases the viscosity and probably, the ability to rupture the agglomerate should decrease. Such an optimization has not yet been performed.

1 G. P. Baeza, A.-C. Genix, C. Degrandcourt, L. Petitjean, J. Gummel, R. Schweins, M. Couty and J. Oberdisse, *Macromolecules*, 2013, **46**, 6621–6633.
2 G. P. Baeza, A.-C. Genix, C. Degrandcourt, J. Gummel, A. Mujtaba, K. Saalwächter, T. Thurn-Albrecht, M. Couty and J. Oberdisse, *ACS Macro Lett.*, 2014, **3**, 448–452.

Priyadarshi Roy Chowdhury communicated: Is the dispersive process limited to SAXS?

Marc Couty communicated in reply: Swelling the material and separating the aggregates can also be used in TEM and SEM studies since one can image individual objects.

Priyadarshi Roy Chowdhury communicated: Which analytical technique do you think gives confirmation about particle-growth?

Marc Couty communicated in reply: Particle growth is one stage of the filler synthesis process and I don't have any access to this information.

Priyadarshi Roy Chowdhury communicated: Please discuss the low "Q" concentration associated with SAXS.

Marc Couty communicated in reply: We have not been able to understand clearly the origins of the strong intensity at low q. We suspect that we still have few large compact agglomerates that generate a strong diffusion.

Priyadarshi Roy Chowdhury further communicated: Do you think cross-linking or twisting is possible in the synthesis of the polymer nanocomposites, which you have mentioned in your paper?

Marc Couty communicated in reply: I don't think that cross-linking occurs during the synthesis of the nanocomposites since we don't have any chemicals that have an action apart from reacting on silica, and chains carry at most one functional group.

B. L. V. Prasad opened a discussion of the paper by Jacques Jestin: Studies similar to those carried out by Dr Marc Couty and Dr Jacques Jestin may have been reported in the literature. How useful are these for industrial applications? If they are not very useful what are the reasons? Are they too "ideal"? Are the academicians missing to include important stuff, that is relevant for industry, in their studies?

Marc Couty answered: I don't have much to add to the answer I made to Prof. Sanat Kumar's previous question. The fundamental knowledge gained on an "ideal" system is very important for industrial application. The industry is focused on technical processes to make materials at large scale based on that knowledge.

Jacques Jestin replied: The collaboration between academicians and industrials are supposed to permit avoiding to miss important stuff relevant for industry. Those studies are useful for industrial applications by providing quantitative directions to optimize the processes and to delineate boundaries of the relevant parameters.

Alison Edwards asked: Would you wish to comment on the need to actually undertake some experiments on systems of interest in order to verify that the expectation and the experiment actually match? In particular, I would value your views regarding how much extrapolation from one system to another can be relied upon when working with nanoscale materials where new and novel properties are sought.

Jacques Jestin replied: We can benefit from methodologies and tools (synthesis, sample processing and analytical or numerical analysis) given by such studies for extrapolation to other systems, either fundamental nanoscale or applied materials.

Alison Edwards questioned: It is not clear to me why you would do some of your SANS work at Saclay and some at ILL. Could you clarify for us whether the instruments were chosen to optimize particular aspects of your experiments, or whether there was insufficient beam-time available to complete the work at one or other location (if indeed this would have been preferred)?

Jacques Jestin answered: We have tried to optimize the beam time as function of the samples and the performances of the spectrometers between Saclay and ILL. The isotropic samples were measured at Saclay and the anisotropic stretched samples that required more flux were measured at ILL. We did not suffer from insufficient beam time at one or the other location.

Erika Eiser remarked: Did you also look at the mechanical properties of you polymer–nanoparticle composites in the presence of hydrogenated chains? Deuteration in polymer chains with high molecular weights can lead to phase separation even in apparently identical polymers, as shown by the group of Jacob Klein and others. Moreover, the replacement of H-atoms by deuterium along the polymer chain can alter the van der Waals attraction between the chains and the nano particles and therefore affect the nano-particle aggregation.

Jacques Jestin replied: The H/D mixtures are fully miscible due to low molecular masses (40k). We do not see any significant differences between the pure H matrix and the H/D mixtures for mechanical properties and nanoparticle dispersion.

Priyadarshi Roy Chowdhury addressed Jacques Jestin and Marc Couty by communicating: What are the effects that play a significant role in conformational changes associated with the polymeric chain?

Marc Couty communicated in reply: The grafting density has probably the major impact. The chain length of free and grafted polymer has also a great impact.

Jacques Jestin communicated in reply: The main conclusion of the study is that the mean chain conformation is not affected at rest by the presence of the particles whatever the nanoparticle loading and the polymer/filler interactions. Under deformation, for the crosslinked elastomers, the chain relaxation is reduced by the NPs that act as additional cross-linking junctions.

Priyadarshi Roy Chowdhury opened a discussion of the paper by Ramanan Krishnamoorti† by communicating: What will be the effect of small-angle neutron scattering on moderate dispersions of POEOMA-SiO_2?

Ramanan Krishnamoorti communicated in reply: The structural correlations between the hybrid nanoparticles will become increasingly important as the concentration of the particles is increased. For dilute solutions, as much of this study demonstrates, the scattering methods allow us to probe the single particle regime. However, as shown in the last figure of the manuscript at modest concentrations, a correlation hole scattering peak is observed and is quite well modeled using a Percus-Yevick type model.

Vandana Shinde communicated: Which type of information is received from SAXS and SANS, which are necessary for these nanocomposite poly(oligo(ethylene oxide)monomethyl methacrylate)-grafted silica nanoparticles?

Ramanan Krishnamoorti communicated in reply: Typically SAXS is sensitive to the electron density contrast and therefore is most sensitive to the presence of higher atomic number atoms. Therefore in polymer–silica hybrids (or in polymer–particle hybrids), SAXS is sensitive to the particle–particle correlations and the internal structure of the particles. On the other hand, SANS is sensitive to the neutron scattering length density differences. It turns out that one of the largest neutron SLD differences lies between that of a Hydrogen atom and a Deuterium atom. Therefore for polymer–silica hybrid particles, by varying the amount of H_2O to D_2O, the scattering length density of the solvent can be systematically varied, so that it can be made equivalent to that of the polymer or that of the silica particle. Therefore the overall structure of the polymer, the structure of the particle and the correlations between these hybrid particles can be systematically examined.

Priyadarshi Roy Chowdhury opened a discussion of the paper by Nicholas A. Kotov: How can one bridge the experimental and simulation results in a more comprehensive manner for chiral nanoparticles in atomic scale as well as in assemblies?

Nicholas Kotov replied: This is one of the biggest questions of the field. We made an attempt addressing it in a consistent manner in our recent paper about interparticle forces at nanoscale whose complexity seems to be under

† The paper by Ramanan Krishnamoorti was not presented at the meeting, but questions and responses have been provided afterwards.

appreciated.[1] One of the central problems, in my perception, is the development of a set of adequate force-fields for nanoparticles. They need to incorporate interaction characteristics of the inorganic and organic surface ligands. It is possible for the models combining DFT and molecular dynamics modules. These models are computationally heavy and the timescales needed for realistic description of nanoparticle assemblies are not available for them. Coarse grained models can account for characteristic times of nanoparticle self-organization (seconds to hours). The adequate transition from atomistic to coarse grained models is known to be very difficult for proteins despite decades of work put into it. It is still an art form for nanoparticles. Some approaches being used for protein interactions can be transitioned to nanoparticle self-assembly.

1 C. A. Batista, R. G. Larson and N. A. Kotov, *Science*, 2015, **350**, DOI: 10.1126/science.1242477.

Michael Bockstaller enquired: Thank you for the very interesting contribution. Could the author please elaborate further on the origin and implications of the partial oxidation. In particular, why does the formation of a silver oxide layer enhance the plasmic response? Is it related to the dielectric properties of the silver oxide shell or would other coatings result in a similar effect? Also, could you comment on the thickness of the oxide layer and what the threshold thickness would be to retain the reported effect.

Nicholas Kotov replied: The enhancement of polarization rotation upon oxidation of silver caps on ZnO nanorods is something counter-intuitive and unexpected. We attribute it, at least in part, to the increase in the volume of the plasmonic caps that improves electromagnetic coupling with incident photons. I need to mention also that not all of the silver is transformed into silver oxide in these caps. The core remains metallic.

Charusita Chakravarty said: How complex can self-assembled structures be? At some level of complexity in nanoparticle/colloidal aggregates, would you expect emergent properties that cannot be obtained by bottom–up integration?

Nicholas Kotov responded: In my estimates, the self-assembled structures from nanoparticles can approach in structural and functional complexity self-assembled structures from nanoscale biological components, for instance proteins.[1] A recent observation of an enantioselective assembly of left or right helices from simple water-soluble CdTe nanoparticles confirms this hypothesis.[2]

1 N. Kotov,*Science*, 2010, **330**, 188.
2 J. Yeom, B. Yeom, H. Chan, K. W. Smith, S. Dominguez-Medina, J. Hwan Bahng, G. Zhao, W.-S. Chang, S.-J. Chang, A. Chuvilin, D. Melnikau, A. L. Rogach, P. Zhang, S. Link, P. Král and N. A. Kotov, *Nature Mater.*, 2015, **14**, 66.

G. V. Pavan Kumar asked: Your work has literally added a twist to plasmons. I have two questions. Firstly, what is the nature of plasmons in your system? Are they localized or propagating plasmons or a combination of both? Secondly, chirality can be further utilized in generating light beams with orbital angular momentum. Can your structures be harnessed for the same?

Nicholas Kotov replied: When I think about the plasmons in, for instance, two gold rods in a "scissor" configuration (*e.g.* Ref. 1), I visualize them as localized plasmons. I might draw analogy here between the surface plasmons in such twisted assemblies and aromatic electrons in biphenyl compounds. Both of them are chiral and both of are delocalized within a particular space. Chiral inorganic nanostructures can be used to generate photons with circular polarization. This is one of the most exciting areas of research now. So far it was difficult to observe circularly polarized luminescence in chiral semiconductor nanostructures, but I am confident, it will be accomplished for multiple chiral nanoscale systems. Synthesis of assemblies with hybrid plasmon-exciton states is one of the routes that I can foresee to be fruitful.

1 W. Ma, H. Kuang, L. Wang, L. Xu, W.-S. Chang, H. Zhang, M. Sun, Y. Zhu, Y. Zhao, L. Liu, C. Xu, S. Link and N. A. Kotov, *Sci. Rep.*, 2013, **3**, 1934.

Ajeet Srivastav addressed Nicholas Kotov and Daan Frenkel: How should a mix of different shapes of nanoparticles in a crystallographic sense with different energies of the out-surfaces affect the kinetics of the assembly? If it favors different building blocks, as mentioned by Prof. Daan Frenkel, then how to choose those different shapes of the nanoparticles as building blocks?

Nicholas Kotov answered: The shape of the nanoparticles is not only a factor of the crystallography of the core albeit it is a big factor. The surface layer of the nanoparticles plays a big role in what effective shape the nanoparticle has. For instance a nanoparticle with a seemingly spherical core is known to behave anisotropically due to the uneven distribution of the surface ligands. Its effective shape could be quite a bit more complex than a sphere. Moreover, the shape is also a dynamic parameter. Although the dynamic reconfiguration of the crystalline particle core is not common, the reconfiguration of the particle shell is. The kinetics of the self-assembly is strongly affected by the particle shape. If we treat the particle shape as a sum of two components, static and dynamic, I can see how much faster the dynamic component of the particle shape speeds up and facilitates the self-assembly of nanoparticles.

The choice of the different shapes of nanoparticles as building blocks is governed by the symmetry relations characteristic of the assembled structures to be made. The symmetry of the static component of the nanoparticle shape and reconfigurability of its dynamic component are reflected in the resulting self-assembled structure. As a simple example, the nanoparticles with bipolar anisotropy (magnetic, electrical, *etc.*) will tend to form self-assembled systems with axial symmetry. The dynamic and entropic components can, however, change this trend dramatically and much more complex architectures with more diverse symmetry elements could be obtained. As an example, instead of the particle chain or ribbon a helix will form.

Daan Frenkel replied: The question of the shape of the building blocks is a very interesting one as it expands the degree to which interactions can be made specific. To obtain optimal interaction, one needs not only complementarity of the chemical moieties responsible for binding, but also complementarity of the shapes of the binding surfaces.

Nagaraj Shetti remarked: You have prepared a material of metallic silver on ZnO nanopillars, do you think it will act as an electrocatalyst? How will it be helpful to detect bio-active molecules in electrochemical sensors? As compared to other materials, silver has less of an electro-potential window.

Nicholas Kotov answered: I am confident that a ZnO–Ag–Ag_2O system can serve as an electrocatalyst. Chirality of the silver caps can be the key in improving the selectivity in biological compounds. Although it could be a problem to address later, I am not worried about the electropotential window of silver at the moment. Finding a "right" electrocatalytic reaction to perform on this potentially very active catalytic system would be my line of thinking for now. In case we want to expand the electrocatalytic window when it could become a problem, chiral nanocaps from gold or platinum can be created in the same way.

Priyadarshi Roy Chowdhury communicated: Please illustrate the complexity of the chiral plasmonic nanostructures, mentioned in your work.

Nicholas Kotov communicated in reply: Chiral plasmonic (or simply chiroplasmonic) nanostructures were made of great complexity using DNA bridges. For example, in collaboration with Prof. Chuanlai Xu we reported tetrahedrons made of four different nanoparticles[1] with some of them being made out of gold, silver, and semiconductor using this method. Importantly, they can be made of specific chirality not by accident but by design. It allowed us to see optical signatures of all constituent nanoparticles in mirror-image CD spectra. Most recently, propeller-like assemblies with geometries reminiscent of chiral metal complexes were made from gold nanorods and upconverting nanoparticles.

1 W. Yan, L. Xu, C. Xu, W. Ma, H. Kuang, L. Wang and N. A. Kotov, *J. Am. Chem. Soc.*, 2012, **134**, 15114.

Priyadarshi Roy Chowdhury communicated: What type of interactions do you think play a dominant role in chiral plasmonic nanostructures?

Nicholas Kotov communicated in reply: In my experience, dispersion interactions are quite strong for all plasmonic nanoparticles and need to be taken into account along with electrostatic interactions and hydrogen bonds. I see these three interactions as most universal and, to some degree, dominant for all plasmonic nanoparticles assemblies made in aqueous phase.

Himani Medhi communicated: The feasible utilization of Ag nanocaps on ZnO nanopillar arrays as an oxygen detector will really become helpful for the future as it has possibility to help people working in mines and also can help astronauts to detect the presence of oxygen in the atmosphere around them instantly. This is a nice work going to be published. I have a suggestion to try to develop a recycling method to recover the Ag nanocaps on ZnO nanopillar arrays from the oxidized one Ag_xO, which will make it more beneficial for sustainable development.

Nicholas Kotov communicated in reply: This is a very thoughtful suggestion. It is indeed necessary to think about the careful use of precious metals in chiroplasmonic materials. Recycling should be possible. I recall when I was a kid and did a lot of old school silver-based photography, the recycling of silver was possible by dissolving silver oxide with amine complexes.

Himani Medhi communicated: In your paper, it was mentioned that Ag nanocaps on ZnO nanopillar arrays with different thicknesses were prepared by spraying Ag vapour from different directions. I have some queries regarding this. Is there any chiroptical activity observed in pure ZnO nanopillars? Which directional coating did lead to better chiroptical activity and better oxygen sensitivity of Ag nanocaps on ZnO nanopillar arrays? What was the thickness of Ag nanocaps made on ZnO nanopillar arrays that led to a better oxygen sensitivity property to the material and how long can that be used as an oxygen detector?

Nicholas Kotov communicated in reply: The anisotropy of light matter interactions in thin films can come from both dichroism and birefrigence. In each case it can be linear and circular. Thin films are notorious for having all four of them at once which makes it difficult to assign CD spectra to a specific optical effect. It is difficult but not impossible. The use of Mueller matrix polarimetry (see Ref. 1 and the references therein) enables one to identify the contribution of each component. In the case of a ZnO nanowire forest, one could also expect all four components, but careful examination showed that under the same registration conditions, with incident angle and wavelength being the key ones, the ZnO nanowire forest did not have circular dichroism.

1 Y. Kim, B. Yeom, O. Arteaga, S. J. Yoo, S.-G. Lee, J.-G. Kim and N. A. Kotov, *Nature Mater.*, 2016, DOI: 10.1038/nmat4525.

Neena S. John communicated: Could you give more details regarding the "gas sensing" experiments of plasmonic silver nanocaps, particularly, what parameter one should monitor to show the sensing action? Is it the "g-factor" variation or CD intensity at a particular wavelength? In the paper, you have shown only variation with two exposure times to air. Have you done more experiments with respect to controlled oxygen flow and if so, what is the detection limit obtained using your material?

Nicholas Kotov communicated in reply: Unfortunately, Jun Li, a very diligent and hardworking student, had to go back home soon after completion of this work and I am still looking for a replacement for him. These experiments would certainly be essential for technological implementations of such gas sensors. When designing such sensors, I would measure polarization rotation at a maximum g-factor point. The optical set-up would be reminiscent of the CD spectrometer but a lot smaller in size.

Vandana Shinde communicated: 1. What is the gas sensing mechanism of metallic silver on ZnO nanopillar arrays? 2. What are the response and recovery time for this array? 3. Can you test the selectivity of this array for other gases?

Nicholas Kotov communicated in reply: Previously we showed that circular dichroism is a potent analytical tool when combined with plasmonic nanostructures.[1] The high sensitivity originates from the strong dependence of polarization rotation on the conformation of anisotropic plasmonic particles. The gas sensing mechanism for nanopillars is based on a similar effect. However, in addition to the very strong polarization rotation, we also have a wavelength shift. Both factors are synergistically combined in this system. Unfortunately, Jun Li had to go back home and we did not measure the recovery time or sensitivity to other gases. Testing hydrogen as a recovery agent and an example of a different gas would be promising.

1 M. Ma, H. Kuang, L. Xu, L. Ding, C. Xu, L. Wang and N. A. Kotov, *Nature Comm.*, 2013, **4**, 2689.

Priyadarshi Roy Chowdhury opened a discussion of the paper by Rajdip Bandyopadhyaya: In this paper, you have presented indeed a very good work on photocatalysis for degradation of a complex dye. If possible perform the Markov-Chain Monte Carlo (MCMC) simulations and co-relate the experimental and MCMC simulation results, as done by Prof. Zhao's group in his paper, published in PCCP and enlisted in Ref. 1.

1 B. Liu and X. Zhao, *Phys. Chem. Chem. Phys.*, 2014, **16**, 22343.

Rajdip Bandyopadhyaya answered: The present paper reports only the synthesis of hybrids of a porous host-nanoparticle and its optimization to maximize the photocatalytic dye degradation rate, without any model calculation. The work done in the PCCP paper referred to in the question is indeed interesting, wherein by using a Markov-Chain Monte Carlo method, the authors have calculated both photocurrent decay and first order rate constants of photocatalysis. Although we have not yet done any MC calculations for this system, it would be possible for us to perform MC simulations in the future to correlate our experimental data, since our group has extensively worked on predicting the nanoparticle size and its aggregation behaviour by Kinetic MC simulation. In this regard, we have already published an alternative deterministic model to calculate the dye degradation kinetics of SnO_2-SBA-15 hybrids in a previous paper.[1]

1 N. R. Srinivasan, P. Majumdar, N. K. R. Eswar and R. Bandyopadhyaya, *Appl. Catal. A*, 2015, **498**, 107–116.

Madivala G. Basavaraj commented: What is the typical size of pore compared to particle diameter? Can this be tuned by the synthesis approach you follow?

Rajdip Bandyopadhyaya responded: Cylindrical pores in our mesoporous silica (SBA-15) are arranged in an ordered, hexagonal manner. The mean pore diameter for different samples (depending on whether the pores had nanoparticles or not) were between 7.5 to 9.1 nm, whereas the nanoparticles varied from 3.3 to 6.7 nm (Table 1 in our manuscript). So, the ratio of pore to particle diameter typically ranged from being slightly greater than 1, to more than 2. The diameter of the pore can be easily tuned further – it can be increased or decreased by taking amphiphilic molecules of larger or smaller tail-length than that of P123 (used in this work) during synthesis, namely F127 or CTAB, respectively. Pore diameter can also be increased by adding hydrophobic molecules (decane, octane, mesitylene, *etc.*) as a pore expander, in the conventional sol–gel synthesis step of SBA-15. In essence, we can systematically control the ratio of pore to particle diameter over small incremental steps, which is useful to examine the relative roles of pore-diffusion, adsorption and reaction on the catalyst surface. In our group, this has been one of the goals of current research in this area.

Nagaraj Shetti said: You have prepared a $Sn_xTi_{(1-x)}O_2$ solid solution nanoparticle embedded mesoporous silica (SBA-15) hybrid as a photocatalyst and used to degrade a rhodamine B dye. But, you have not discussed the degradation products, or degradation mechanism. I think it will be better if you explain the degradation mechanism and the respective products.

Rajdip Bandyopadhyaya replied: The present work reports the synthesis of hybrids of a porous host–nanoparticle system, wherein each nanoparticle was successfully embedded inside the porous host (SBA-15). The approach followed in the present work to form solid-solution nanoparticles inside a porous host without causing pore blockage, and is a promising route towards increasing reaction rates in catalytic applications. To this end, Rhodamine-B was considered as a model dye and we investigated its degradation in the presence of the photocatalyst, with possible mechanisms *via* an energy band diagram. The goal of this work was to establish the efficacy of such a hybrid system to achieve an enhanced reaction rate, rather than exploring the reaction pathway and the kinetic scheme. However, currently we are in the process of measuring the degradation products using chromatographic techniques to precisely uncover these aspects. According to the literature, the possible degradation products are pthalic acid, malonic acid, benzoic acid, maleic acid, pthalic anhydride *etc.* which we also plan to measure.

Siddharth Kulkarni asked: Will the aggregation behaviour of the particles be different if particles are dispersed in complex anisotropic medium such as liquid crystal? Will the particles still showcase predominantly repulsive forces?

Liquid crystals consists of molecules with intermediate order and mobility between perfect solids and liquids? The liquid crystal medium mediates the forces between particles which are attractive in nature.

Rajdip Bandyopadhyaya responded: In the present work, pluronic block copolymer (P123) molecules form a hexagonal, liquid crystalline mesophase in solution. The hexagonal array of cylinders act as a template-structure, on which mesoporous silica (SBA-15) was formed, by sol–gel reaction and subsequent calcination to remove the P123 molecules. Finally, $Sn_xTi_{(1-x)}O_2$ photocatalyst nanoparticles were synthesized *in situ* within the pores of SBA-15 to obtain the hybrid nanomaterial. In our synthesis scheme, the liquid crystalline phase acted only as a template for nanoparticle formation, not as a medium for dispersion of nanoparticles. Therefore, being formed *in situ* within the pores, the particles were mostly separated from each other by the few nm thick, silica-walls of SBA-15 (intertwining between two pores), thereby preventing interparticle aggregation of the nanoparticles. On the other hand, if pre-formed nanoparticles are attempted to be impregnated into the pores, there is a much greater possibility of the particles becoming aggregated, either outside or inside the pores, which is not desirable. In general, dispersion of nanoparticles in liquid crystals can be understood in detail in the paper by Hegmann *et al.*[1]

1 T. Hegmann, H. Qi and M. V. Marx, *J. Inorganic and Organometallic Polymers and Mater.*, 2007, **17**, 483–508.

Ajeet Srivastav commented: How reliable is the localized elemental mapping with such a large length scale and poor statistical measures? The mapping should be done at smaller length scale with a reduced number of frames per second for a longer period of time.

Rajdip Bandyopadhyaya responded: Elemental mapping is a useful technique to get insight into both composition and spatial dispersion of different elements (Sn and Ti) in our sample. Area mapping was chosen instead of point mapping as the mesoporous silica (SBA-15) particle is relatively large, with a mean diameter of 630 nm and a standard deviation of 85 nm. Thus, elemental mapping using SEM for a large area (several tens of micron-square containing many SBA-15 particles, Fig. 4 in our paper) was done, to identify whether $Sn_xTi_{(1-x)}O_2$ solid solution nanoparticles were present with the correct ratio of Sn and Ti within SBA-15. Data of Sn/(Sn+Ti) in Table 1, in our paper, indeed validate the composition in the actual samples, in line with the intended stoichiometric ratios of reactants. For this, a high dwell time (40 s) was maintained to reduce any statistical errors. On the other hand, a well-dispersed state and small size of the nanoparticles (3–7 nm, Table 1 in our manuscript) in the porous hosts have both been ascertained from the bright field TEM images (Fig. 1 in our paper) of $Sn_xTi_{(1-x)}O_2$ solid solution nanoparticle-SBA-15. During the actual presentation of this paper in the Faraday Discussions meeting, elemental mapping using SEM of a single SBA-15 particle was additionally shown (with a much higher magnification than in the printed-version of the paper) to prove that in the local scale too, the hybrid consists of both Sn and Ti, throughout the solid solution nanoparticle.

Priyadarshi Roy Chowdhury communicated: From the photoluminescence analysis of the hybrid photocatalyst, what type of defect states could be evident?

Rajdip Bandyopadhyaya communicated in reply: Photoluminescence spectroscopy is an efficient technique to explore the defect state emission. The origin of defect state emission in the hybrid are oxygen vacancies and self-trapped charges. In the PL spectra (Fig. S2, supplementary information of our manuscript) the peaks around 410, 460 and 520 nm are ascribed to intrinsic state transitions of anatase phase TiO_2, trapped charges and oxygen vacancies, respectively. This information and its implication has already been elaborated in the supplementary information of this paper.

Priyadarshi Roy Chowdhury communicated: What is the thickness of the hybrid photocatalyst along the *c*-axis, mentioned in this work?

Rajdip Bandyopadhyaya communicated in reply: Both the photocatalyst ($Sn_xTi_{(1-x)}O_2$ nanoparticle) in the hybrid, as also the SBA-15 host particle (of the hybrid), is basically isotropic, mostly spherical/polygonal in shape (Fig. 4 in our paper). As a result there is no thickness to be reported. Typically, the estimated mean diameter of the nanoparticles is 3–7 nm (Table 1 in our paper).

Himani Medhi communicated: As I have gone through your paper, I have found that the material $Sn_{0.15}Ti_{0.85}O_2$-SBA-15 showed better characterization results like a comparatively higher specific surface area, pore size and volume and lower band gap compared to that of $Sn_{0.05}Ti_{0.95}O_2$-SBA-15, but $Sn_{0.05}Ti_{0.95}O_2$-SBA-15 showed better photocatalytic degradation. Will you please explain this anomaly?

Rajdip Bandyopadhyaya communicated in reply: The $Sn_{0.15}Ti_{0.85}O_2$-SBA-15 sample has a slightly higher pore volume and surface area (Table 1 in our paper), with little less band gap, compared to $Sn_{0.05}Ti_{0.95}O_2$-SBA-15. The reason for these differences is due to the introduction of strain in the $Sn_{0.15}Ti_{0.85}O_2$ sample during the replacement of Ti^{4+} with Sn^{4+}. The strain decreases the particle size (from 5.6 nm in $Sn_{0.05}Ti_{0.95}O_2$ to 4.3 nm in $Sn_{0.15}Ti_{0.85}O_2$, Table 1) and thereby increases pore volume and surface area. Now, to explain enhanced photocatalytic activity, it is important to look at various aspects – crystallinity of the nanoparticle, mobility of the charge carriers, pore surface area, nanoparticle size and band gap values, *etc.* It can be seen from the XRD curves (Fig. 2 in our paper) that the gradual loss of the anatase phase in TiO_2 (which is observed for nanoparticle samples with a high Sn fraction of 0.50 to 0.15) is not seen in $Sn_{0.05}Ti_{0.95}O_2$, as the 0.95 fraction of it is Ti, with very little (0.05 fraction) Sn. As a result, $Sn_{0.05}Ti_{0.95}O_2$-SBA-15 provides better photocatalytic activity than that of $Sn_{0.15}Ti_{0.85}O_2$-SBA-15. This has already been discussed in the paper in the Conclusions section.

Himani Medhi communicated: In the Table 1 on page no. 6 of your paper, it has been reported that the band gap of $Sn_{0.05}Ti_{0.95}O_2$-SBA-15 is 3.23 eV, $Sn_{0.15}Ti_{0.85}O_2$-SBA-15 is 3.19 eV and $Sn_{0.25}Ti_{0.75}O_2$-SBA-15 is 3.25 eV. But in the 3rd line of page no. 9, it has been mentioned, for $Sn_xTi_{(1-x)}O_2$-sphere like SBA-15, to be 3.23, 3.19 and 3.25 eV for $x = 0.05$, 0.10 and 0.15, respectively. Is it correct as it appears in the paper?

Rajdip Bandyopadhyaya communicated in reply: The band gap values reported in Table 1 (page no. 6 of the paper) are correct. However, there is a small mistake in page no. 9. It should be read as, $x = 0.05$, 0.15 and 0.25, respectively, instead of $x = 0.05$, 0.10 and 0.15. Otherwise, the numerical values of band gaps (given in eV) are correct in both pages, hence the ensuing conclusions remaining in the paper are valid.

Vandana Shinde communicated: On what basis did you fit the electrochemical circuit? Is the circuit really fitted properly to all impedance spectra? For SBA-SnO_2, the Nyquist plot shows a nearly straight line, while the others show a slightly arc type behavior. What is its meaning?

Rajdip Bandyopadhyaya communicated in reply: Separation and transport properties of charge carriers are crucial factors in determining photocatalytic activity. Electrochemical impedance spectroscopy (EIS) is a useful method to get physical insights into interfacial electron transfer at the electrode, during the photocatalytic process. The impedance spectra are plotted in terms of Zreal and Zimg, which comes from the resistance and capacitance elements of the electrochemical cell. The most common method of analyzing EIS spectra is the equivalent circuit model; the simplest model in the literature includes a solution resistance, a double layer capacitor and a charge transfer resistance. Moreover, this model is the starting point for more complex models, including the Warburg diffusion element. However, since our EIS spectra could be fitted properly with the simplest model itself, it was not required to add more elements in the equivalent circuit. The chi-square value for all samples is in the order of 10^{-3}.

In general, a semicircular arc in the EIS corresponds to effective separation of charge carriers and a faster interfacial charge transfer rate. In contrast, the straight line behaviour seen in SnO_2-SBA-15 (Fig. 6a in our paper) could be due to a diffusion-limited charge transport process, which resulted in a lower charge transfer rate. The smaller arc seen for the $Sn_xTi_{(1-x)}O_2$-SBA-15 hybrids (Fig. 6a), compared to SnO_2-SBA-15 indicated a faster interfacial charge transfer rate at the photocatalyst surface. These results show that replacing Ti^{4+} with Sn^{4+} ions enhances the separation and charge transfer efficiency of charge carriers, leading to a superior photocatalytic material. This was already discussed in the paper, under the Results and discussion section.

Faraday Discussions

PAPER

Revealing nanocomposite filler structures by swelling and small-angle X-ray scattering

Guilhem P. Baeza,†[ab] Anne-Caroline Genix,[a] Nathalie Paupy-Peyronnet,[b] Christophe Degrandcourt,[b] Marc Couty*[b] and Julian Oberdisse[a]

Received 27th July 2015, Accepted 1st September 2015
DOI: 10.1039/c5fd00117j

Polymer nanocomposites are used widely, mainly for the industrial application of car tyres. The rheological behavior of such nanocomposites depends in a crucial way on the dispersion of the hard filler particles – typically silica nanoparticles embedded in a soft polymer matrix. It is thus important to assess the filler structure, which may be quite difficult for aggregates of nanoparticles of high polydispersity, and with strong interactions at high loading. This has been achieved recently using a coupled TEM/SAXS structural model describing the filler microstructure of simplified industrial nanocomposites with grafted or ungrafted silica of high structural disorder. Here, we present an original method capable of reducing inter-aggregate interactions by swelling of nanocomposites, diluting the filler to low-volume fractions. Note that this is impossible to reach by solid mixing due to the large differences in viscoelasticity between the composite and the pure polymer. By combining matrix crosslinking, swelling in a good monomer solvent, and post-polymerization of these monomers, it is shown that it is possible to separate the filler into small aggregates. The latter have then been characterized by electron microscopy and small-angle X-ray scattering, confirming the conclusions of the above mentioned TEM-SAXS structural model applied directly to the highly loaded cases.

1. Introduction

Polymer nanocomposites (PNCs) are used widely in applications, in particular in the car tyre industry.[1] Their mechanical properties are governed by their microstructure and dynamics, which is why model systems where these properties can be easily evidenced are commonly investigated.[2–8] In particular the multiscale

[a]Laboratoire Charles Coulomb (L2C), UMR 5221 CNRS-Université de Montpellier, F-34095 Montpellier, France
[b]Manufacture Française des Pneumatiques MICHELIN, Site de Ladoux, 23 place des Carmes Déchaux, F-63 040 Clermont-Ferrand, France. E-mail: marc.couty@fr.michelin.com
† Current address: FORTH/IESL – Polymer group, 70013 Heraklion, Crete, Greece.

microstructure of filler nanoparticles (NPs), which form small aggregates, which themselves can be structured in space and build filler networks, has received considerable attention in the past.[9–11] Whatever the experimental approach, direct imaging[12] or scattering techniques,[13] two main difficulties arise: polydispersity in shape and size, and strong interactions. The latter point is obviously related to mechanical reinforcement, which can only be reached if the filler content is high enough. Considerable effort has therefore been undertaken to unravel the filler structure in PNCs, with quite some success in model systems. On the other hand, less attention has been devoted to complex multiscale filler of industrial origin, and most descriptions remain necessarily limited to a few global quantities such as average fractal dimensions.[14–20]

The standard procedure to synthesize PNCs with high viscosity polymers and precipitated silica is to use solid mixing. The typical size of the filler objects in the composite is commonly supposed to depend on the characteristics of the mixing process, such as temperature and torque *vs.* time, and thus on the raw materials used to formulate the composite. In a series of papers[21–24] on filler structure in styrene–butadiene (SB) nanocomposites, we have coupled small-angle X-ray scattering (SAXS) and transmission electron microscopy (TEM) on silica-filled PNCs of industrial relevance. A structural model extracting the average aggregation number N_{agg} (the number of primary particles in an aggregate) from combined TEM/SAXS data under the hypothesis of repulsive interactions between aggregates has been proposed.[23] Using a fully polydisperse description of aggregate structure and interactions deduced from Monte Carlo simulations, it has been shown that for systems with a coating agent and an equal mixture of non-functional and functional polymers, N_{agg} is independent of the silica volume fraction. Conversely, N_{agg} decreases for systems with an increasing amount of functional polymer at constant volume fraction. However, to validate this structural model, the best method would be to reduce interactions – instead of simulating them – in order to access the aggregate structure directly. This could be achieved by diluting the nanocomposite with polymer to low-volume fractions. Unfortunately, because of large differences in viscoelasticity between the composite and the pure polymer, solid mixing of both components – polymer and silica – at low filler volume fraction generates only a dispersion of the original high-volume-fraction material in the pure polymer matrix.

Swelling nanocomposites with solvents or monomers has been used in the field of intercalating nanoclays.[25–27] Timochenco *et al.* for example show that the interaction peak between clay platelets in X-ray diffraction disappears upon swelling with polystyrene.[25] Boukerrou *et al.* use swelling with aliphatic molecules in order to widen galleries between clay filler for polymerization of epoxy resins and nanocomposite formation.[26] Finally, swelling of polymer networks with organic solvents is a commonly used technique to characterize the entropic elasticity of the network, and thus crosslinking. The latter approach can be extended to the presence of filler nanoparticles and their influence on the network properties.[27] Concerning structural studies by small-angle scattering, Shinohara *et al.* have used swelling of SB nanocomposites with industrial filler NPs in organic solvents to draw conclusions qualitatively on the absence of interactions between aggregates.[18] Using imaging techniques with electron microscopy, Sarkawi *et al.* have swollen PNCs based on natural rubber filled with silica in styrene.[28] Their objective was two-fold. First, the degree of swelling could be used to analyze the apparent crosslink density. Secondly, the dilution of the filler structure reduced interactions between filler NPs, and thus geometrical overlap, allowing them to draw conclusions about the importance of filler–matrix interactions as a function of coupling agent.

In the present paper, we propose an original experimental approach using SAXS combined with matrix crosslinking of the PNC, swelling in a good polymer 'solvent' – styrene monomers – and post-polymerization of these monomers. We first compare the filler structures of the samples studied here to those of previously described PNCs in the unswollen state. Then, the structures after crosslinking and after swelling are investigated using electron microscopy and SAXS, and compared to the results of the structural model applied in the presence of interactions.

2. Materials and methods

Starting materials and nanocomposite preparation

The styrene–butadiene copolymer (with antioxidants 6PPD and AO2246) has been purpose-synthesized by Michelin, and the chain mass characterized by size-exclusion chromatography. The polymer matrix is made of two types of chains of molecular mass of *ca.* 190 kg mol^{-1} and PI below 1.1, see Table 1. Each chain is a statistical copolymer with styrene (27% wt) and butadiene (73% wt) units (42% of which are 1–2 and 58% 1–4). One type – termed D3 – carries a single graftable end function,[24] whereas the other one (NF) is not functionalized. The glass-transition temperature was measured by differential scanning calorimetry (DSC) (200F3 Maia from Netzsch) with a heating rate of 20 K min^{-1}.

Silica–SB nanocomposites have been formulated by stepwise introduction and mixing of SB chains with silica pellets (Zeosil 1165 MP from Solvay, with DPG, Vulcacit from Bayer, and covering agent octeo from Dynasylan) in an internal mixer (Haake) following the same method as before.[23] In addition, however, an organic crosslinking agent in a very small amount compared to standard crosslinking agents (dicumyl peroxide, 1–5 mg for 8 g of nanocomposite) was added during the external mixing phase (two roll mill) for samples intended for subsequent swelling experiments without dissolving the samples, as explained below. Note that any trace of carbon black, catalysing nanoparticles (ZnO), standard vulcanising or coupling agents was avoided, making these simplified industrial nanocomposites suitable for analysis by scattering. After external mixing, samples were crosslinked by thermal activation in a press at 160 °C for 40 minutes. Three series of nanocomposites have been studied. One series at a fixed matrix composition of 50% D3 as a function of silica volume fraction Φ_{Si}, and two series at silica volume fraction Φ_{Si} fixed to ~10%*V* and ~20%*V* as a function of matrix

Table 1 Characteristics of functionalized and non-functionalized polymer molecules

SB polymer	D3	NF
T_g (°C)	−36.0	−36.6
M_n (kg mol^{-1})	185	174
M_w (kg mol^{-1})	197	188
PI	1.07	1.08

composition. The real silica volume fractions in the nanocomposites reported here have been measured by thermogravimetric analysis (Mettler Toledo) using a first ramp at 30 K min^{-1} from 25 °C to 550 °C under nitrogen, followed by a second ramp at 20 K min^{-1} from 550 °C to 750 °C under air, and results are given below.

Swelling experiments

Crosslinked nanocomposite disks of thickness 0.6 mm were immersed for 65 h in styrene monomer mixed with benzoyl peroxide as an initiator of styrene polymerisation (polystyrene, PS) after swelling. While the maximum swelling ratio is usually reached after 24 h, most homogeneous polymerisation was found by waiting up to 65 h. Polymerisation of the styrene inside the nanocomposites was performed at 100 °C over 4 hours. The final samples were flat disks of approximate thickness 1.5 mm. The procedure is shown in Fig. 1, together with photographs of the initial nanocomposites and the final samples. Throughout this article, we will refer to the final samples as swollen samples (or "after swelling"), meaning that they have undergone the complete crosslinking/swelling/polymerization procedure.

The silica volume fraction in swollen samples was measured by TGA. The volume swelling ratio for the three nanocomposite series was determined from the silica volume fraction ratio of the initial nanocomposite with respect to the sample with polystyrene, and is reported in Tables 2 and 3. Note that excellent agreement with large-q SAXS normalization (intensity proportional to Φ_{Si}) was found. Concerning the swelling ratios, they are rather scattered around an average value of *ca.* 10, without any dominant tendency related to silica content and grafting density. Higher swelling seems to be obtained for the samples with least grafting.

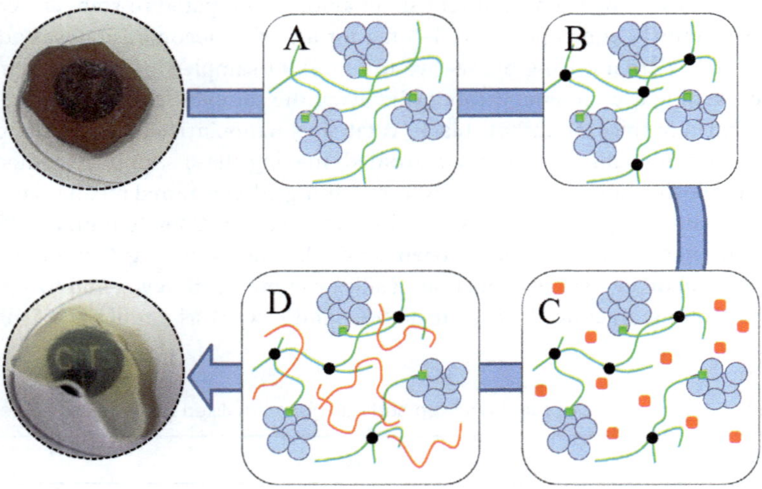

Fig. 1 Initial nanocomposites (upper photograph and A) with a schematic representation of the SB crosslinking (B; SB in green), the swelling with styrene monomers and initiator (in red, C) leading to polymerization (D), where PS is depicted in red. The swollen nanocomposite is shown to be transparent in the lower photograph.

Table 2 Filler volume fractions for the nanocomposite series at fixed matrix composition (50% D3). Φ_{Si} is reported before and after swelling, and the swelling ratio deduced

Initial Φ_{Si} (%V)	Final Φ_{Si} (%V)	Swelling ratio
8.2	0.8	10.5
13.3	1.8	7.2
16.8	2.0	8.6
21.3	1.1	19.8

Structural characterization

The silica microstructure of the nanocomposites has been studied by electron microscopy and SAXS. TEM pictures were obtained with samples prepared by ultracryomicrotomy at −80 °C on a LEICA FC-7 (Diatome ultra 35°, desired thickness set to 50–70 nm) with a Philips CM200 LaB6 (200 kV, bright-field mode). Scanning electron microscopy (SEM) pictures were obtained with samples cut from a macroscopic sample with a standard razor blade using a SEM Zeiss Ultra 55 (1 kV, secondary electron images with In Lens detector). SAXS experiments (beamline ID02, ESRF, Grenoble) were performed at a wavelength of 1.1 Å (12.46 keV), using two sample-to-detector distances (1 m and 10 m), yielding a total q range from 0.001 to 0.5 $Å^{-1}$. Even lower q data was measured on the Bonse–Hart setup on ID02 ($q_{min} = 10^{-4}$ $Å^{-1}$). Samples had an approximate thickness of 0.8 mm. The scattering cross section per unit sample volume $d\Sigma/d\Omega$ (in cm^{-1}) – which we term scattered intensity $I(q)$ – was obtained using standard procedures, including background subtraction and calibration, given by ESRF. In all cases, both PS-swollen and unswollen matrices were subtracted. The contrast of silica in SB polymer for SAXS experiments was calculated from the scattering length densities ($\rho_{SB} = 8.85 \times 10^{10}$ cm^{-2}, $\rho_{Si} = 1.97 \times 10^{11}$ cm^{-2}, $\Delta\rho = 1.09 \times 10^{11}$ cm^{-2}), which were themselves known from the chemical composition. Moreover, the contrast between polystyrene with silica (1.0×10^{11} cm^{-2}) is very close to the silica-SB contrast, whereas the styrene-SB contrast is a factor of ten weaker (0.9×10^{10} cm^{-2}). It is thus sufficient to rationalize the data in terms of polymer-silica contrast only, whatever the polymer.

A SAXS form factor analysis of the silica NPs was found to be compatible with a lognormal size distribution ($R_0 = 8.55$ nm, $\sigma = 27\%$).[23] The average bead volume is then ($V_{Si} = 3.6 \times 10^3$ nm^3), which will be used to estimate aggregation numbers of silica NPs in aggregates.

Table 3 Matrix compositions (grafting % D3) of the nanocomposite series at fixed initial silica volume fractions of *ca.* 8.2%V (left) and 16.7%V (right)

% D3	Initial Φ_{Si} (%V)	Final Φ_{Si} (%V)	Swelling ratio	Initial Φ_{Si} (%V)	Final Φ_{Si} (%V)	Swelling ratio
0	8.3	0.9	9.2	16.8	1.4	11.9
25	8.1	1.1	7.6	16.8	1.6	10.6
50	8.2	0.8	10.5	16.8	2.0	8.6
75	8.1	1.1	7.6	16.7	2.3	7.4
100	8.4	0.7	11.4	16.6	2.3	7.3

3. Results and discussion

The homogeneous distribution of filler nanoparticles throughout the polymer matrix has been checked by SEM in order to qualitatively analyze the material at large length scales. In Fig. 2, an example with a matrix composition of 50% D3, and $\Phi_{Si} = 16.8\%V$ (*i.e.*, native, unswollen PNCs), is shown at low magnification. The majority of the silica NPs is seen to be distributed homogeneously on the scale of microns and tens of microns, as can be seen from the background in Fig. 2, which actually has a fine structure of silica distributed homogeneously in the matrix, as well as some highly filled zones. In Fig. 3a, a SEM picture of the 'background' zones of the same PNC before swelling (*cf.* Fig. 2) is shown at high magnification. A dense assembly of small aggregates of typical radius 50 nm is visible. It is difficult to tell from the 2D picture if they are all connected, but the filler structure is very similar to the large-scale silica network of aggregates of nanoparticles, or locally connected aggregates with surrounding polymer channels observed by us before.[23,24] The micrograph on the right-hand side, Fig. 3b, finally, displays the filler structure of the swollen nanocomposites on a local scale. Clearly, the network-type structure is broken up on the scale of aggregates, which appear to be well-dispersed in space. If they are somewhat polydisperse in size and shape, it is still surprising that they stay in the same size range of typically below 100 nm. Thus electron microscopy confirms that the swelling approach separates the filler in homogeneously filled zones. Naturally, swelling is accompanied by a large reduction of the visible silica surface fraction (from Fig. 3a and b). TGA measurements given in Section 2 confirm the dilution of the original *ca.* 17% volume fraction PNCs to around 2%*V*.

In Fig. 4, TEM pictures of a swollen sample show more detail on the aggregate structure. The completely functionalized nanocomposite is seen to have a typical aggregate size in the sub-100 nm range, corresponding to aggregate radii of 40 or 50 nm on average. One can also count individual silica nanoparticles in order to estimate aggregation numbers, and one finds that there is considerable polydispersity around aggregation numbers of typically several tens. However, it is difficult to conclude statistically relevant information from these pictures.

Fig. 2 SEM of a silica–SB nanocomposite at low magnification showing the presence of highly filled zones ($\Phi_{Si} = 16.8\%V$, 50% D3).

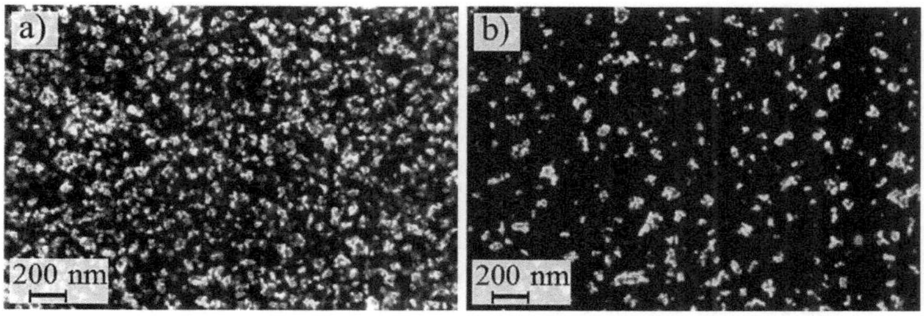

Fig. 3 SEM (a) of a silica–SB nanocomposite at high magnification showing the filler network structure (Φ_{Si} = 16.8%*V*, 50% D3), (b) of the same sample after swelling, as explained in the text (Φ_{Si} = 2.0%*V*).

Several highly filled zones are also observed as indicated by the arrow in Fig. 2. In Fig. 5a, the filler structure in such a zone is shown at high magnification. Electron microscopy demonstrates that these zones contain polymer, and can thus be swollen. Indeed in Fig. 5b, one can see denser regions corresponding to these zones in the swollen sample.

To summarize the electron microscopy results, the minority of the highly filled zones can be swollen, whereas the vast majority of the PNCs contain a homogeneous filler network, which is disintegrated into aggregates upon swelling.

In order to go beyond these qualitative results, a more quantitative and statistically significant analysis has been performed. Small-angle scattering is a powerful tool to investigate filler structure in samples at all stages, before and after the swelling procedure. Since swelling generates large inter-particle distances, the SAXS experiments should enable a direct measurement of the form factor of aggregates in the original material, and thus their average mass and shape. In Fig. 6, a newly formulated nanocomposite series is shown for different silica volume fractions and identical matrix composition (50% D3), as studied before.[23] In order to avoid the complete dissolution of nanocomposites by swelling with the solvent, the samples have been crosslinked following the procedure given in the Experimental section. As can be seen in Fig. 6, for the lowest and highest silica content, the effect of crosslinking on the silica microstructure is found to be weak. Scattering in both the high- and the low-*q* regimes superimpose rather well, and the minor difference around the shoulder indicates

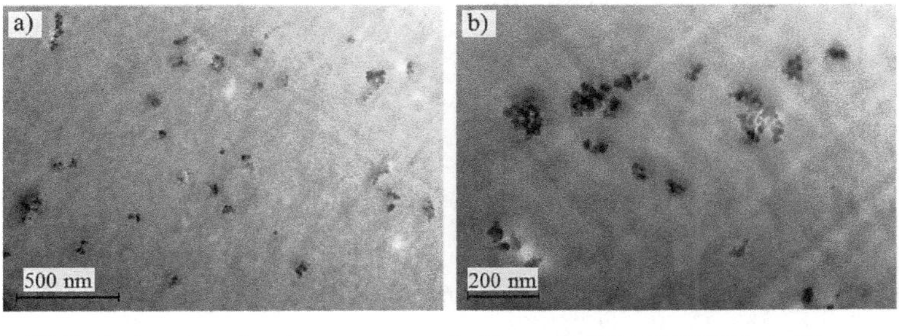

Fig. 4 TEM of a functionalized nanocomposite (Φ_{Si} = 16.6%*V*, 100% D3) at different scales (a) 500 nm, and (b) 200 nm.

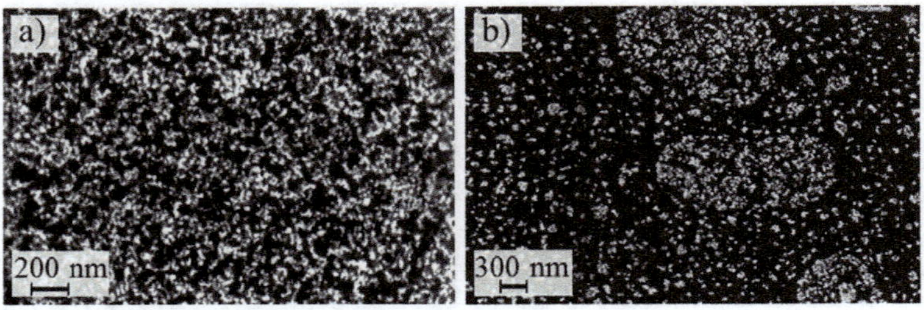

Fig. 5 SEM (a) of a silica–SB nanocomposite at high magnification showing the filler network structure inside a highly filled zone ($\Phi_{Si} = 16.8\%V$, 50% D3), (b) of the same sample highlighting the swelling of the highly filled zones ($\Phi_{Si} = 2.0\%V$).

that the size and shape of aggregates remains mostly unchanged upon crosslinking.

In order to check the similarity in aggregate structure, the previously published results[23] are compared to the present ones for two nanocomposites at low (8.2%V) and high (21.3%V) silica contents in the inset of Fig. 6. For both Φ_{Si} values, the reduced scattered intensities, $I(q)/\Phi_{Si}$, of the initial unswollen nanocomposites are found to be very close to the previously observed one. In particular, both the high-q and the low-q regimes coincide, which proves that both the local structure – NP and internal organization of filler aggregates – and the large-scale structure – fractal networks – are identical. The new nanocomposites display a slightly higher intensity of the intermediate-q shoulder, around 3×10^{-3} Å^{-1}, evidencing a marginally higher compacity, as

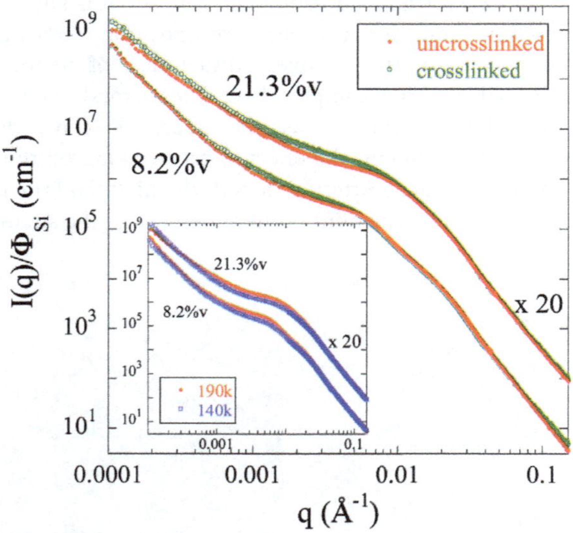

Fig. 6 Comparison of filler microstructure as seen by SAXS before (red) and after (green) crosslinking for the 8.2%V and 21.3%V nanocomposites. Curves have been shifted for clarity as indicated. In the inset, new 190k samples are compared to previously measured 140k nanocomposites.[23]

we will see from the application of the structural model below. It can thus be concluded from the comparisons in Fig. 6 that the nanocomposites produced for the present study have the same microstructure as before, and that it is unchanged upon crosslinking.

After crosslinking, nanocomposite samples were swollen with styrene and polymerized as illustrated in Fig. 1. In agreement with the low-silica fractions (Table 2), the resulting SAXS intensities are much weaker. However, by normalizing to the experimental Φ_{Si}, the latter can be brought to superposition in the high-q range, as shown in Fig. 7.

There are two families of curves shown in Fig. 7. First, the raw data of the unswollen series is shown in Fig. 7a (colored symbols): in the intermediate q range, the normalized intensities decrease with increasing filler fraction, which was successfully modelled in ref. 23 by the growing contribution of the repulsive inter-aggregate structure factor. Indeed, our structural model decorrelates in a self-consistent way the evolution of form factor describing aggregate size, and the structure factor, which is due to inter-aggregate interactions. The model shows that the form factor is mostly unchanged in nanocomposites of different volume fraction. In other words, aggregates were merely concentrated but not reorganized internally (besides a slight densification), contrary to what one could expect from the radically different mixing viscosities at various Φ_{Si} values. In Fig. 7b, the same data of the unswollen nanocomposites is plotted after subtraction of the matrix and the low-q network contribution, in order to apply the model in the intermediate q-range ($q = 3 \times 10^{-3}$ Å^{-1}). The results of the structural model are summarized in Table 4.

As compared to our previous study,[23] the average aggregation number of the silica aggregates remained in the same range, about 45, and the average aggregate radius decreased from about 40 nm to 32 nm. The combination of size and mass yields the compacity, which was found to increase slightly from 34 to 43%. As

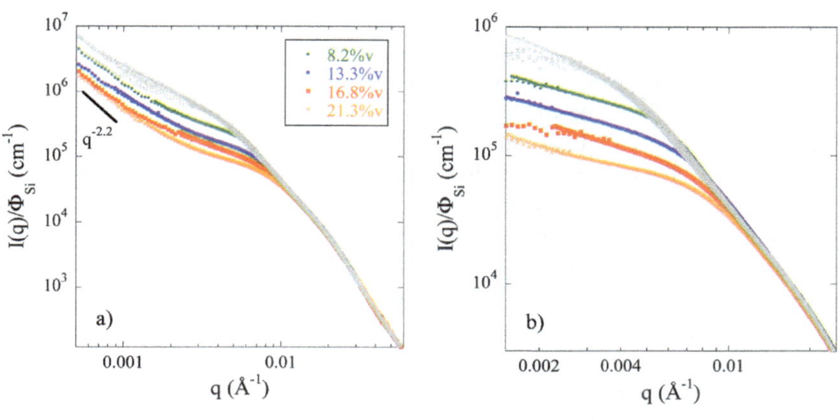

Fig. 7 (a) Comparison of microstructures of unswollen (plain, colored symbols) and swollen (empty, grey symbols) nanocomposites for the series with silica volume fractions of 50% D3. SAXS intensities are renormalized to the effective silica volume fraction (8%V to 21%V, as indicated in the legend) after subtraction of the corresponding matrix. (b) The same data of nanocomposites after subtraction of the fractal network contribution ($q^{-\alpha}$, with $\alpha = 2.2 \pm 0.1$ depending on samples).

Table 4 Average aggregate radius, average aggregation number, and compacity for a series of nanocomposites at different silica volume fractions (50% D3, no swelling)

Φ_{Si} (%V)	$\langle R_{agg} \rangle$ (nm)	$\langle N_{agg} \rangle$	κ
8.2	39.1	52	34%
13.3	35.4	46	40%
16.8	34.5	44	41%
21.3	32.4	37	42.5%

already noted in the comparison in the inset of Fig. 6, the new samples are thus structurally very similar to the previous ones.

The second series is shown in the same plots, see Fig. 7a and b. After swelling and styrene polymerization, the normalized intensities now overlap rather nicely over almost the entire q range. This constitutes the main result of the present study: diluting the sample *via* the swelling procedure amounts to reducing interactions and shows that aggregates are mostly identical in spite of the different original Φ_{Si} values. Ideally, swelling was expected to reduce the inter-aggregate structure factor to one, and to allow measuring the form factor. In Fig. 7b, the intensity data of swollen nanocomposites have been plotted after subtraction of the independently measured matrix contribution, and the low-q power law describing the large-scale network of aggregates. The intensities are seen to level off to a quasi-plateau at intermediate q, which is the origin of the shoulder in the scattered intensity in Fig. 7a. It turns out, however, that the very low-q scattering does not overlap perfectly in Fig. 7b. This can be understood from the swelling procedure, and the swelling ratio given in Table 2. Indeed, swelling is not equivalent to diluting the sample as one would dilute a colloidal solution, where lower concentrations imply a lower degree of order. Here, the pair correlation function is more probably simply stretched, corresponding to an affine displacement of all centers of mass of aggregates, without disturbing the correlation. The swelling ratio of about a factor of 10 leads to an extension by $10^{1/3}$ of the pair correlation function, and thus to a corresponding compression of $S(q)$. Unfortunately, this is not sufficient to obtain a structure factor of one in the q range of interest: swelling moves the low-q regime of $S(q)$ out of reach, without reaching the high-q regime where $S(q) = 1$. Another consequence is that the aggregate size determination is also hindered by the variations of $S(q)$ in the relevant q range. We must conclude that even if the qualitative result in Fig. 7 is convincing, our model is not appropriate for extracting compacity and aggregate characteristics from these scattering curves with high precision. For the sake of completeness, we have nonetheless applied our model to the data of swollen samples. From the intensity value, N_{agg} is found to be in the 20–30 range, which might even agree with the TEM pictures (for 100% D3) shown in Fig. 4. Secondly, the aggregate size determined from the Kratky plots increases slightly (*e.g.*, from 39 to 43 nm and from 32 to 33 nm for $\Phi_{Si} = 8.2$ and 21.3%V, respectively). This appears to be in line with the SEM/TEM pictures shown in Fig. 3b and 4, but one needs to keep in mind that this value is probably too high, and minor changes in R_{agg} affect the compacity strongly because $N_{agg} = \kappa(R_{agg}/R_{Si})^3$ The resulting compacity κ is far too low, about 15%, presumably due to the high value of R_{agg}, whereas the TEM pictures indicate values in the range $\kappa \approx$ 30–40%. The latter value also agrees with the model results of unswollen samples reported in Table 4. From the combined analysis of the TEM and SAXS data, it thus appears that aggregates are separated by swelling, sufficiently for the TEM to recognize individual aggregates having characteristics typically found in the unswollen state, but insufficiently to apply the structural model to the swollen state due to remaining interferences of the structure factor. Concerning the value of N_{agg} after swelling, which might suggest that aggregates have been partially destroyed by the swelling procedure, this should be seen with caution as the value is deduced from the model, which does not apply perfectly well to this case of intermediate dilution. Also, N_{agg} is too close to that of the original, unswollen case, making a clear claim impossible.

To summarize, the measurement of the microstructures of swollen nanocomposites validates the results of our original structural model.[23] Indeed, after swelling-induced dilution, the intensities superimpose for almost all q values in Fig. 7b, indicating that the aggregate form factors are identical, in spite of the different filler volume fractions of the initial formulation.

In previous articles, it has been established that the matrix composition (D3/NF) is the key parameter determining the filler structure due to the end-grafting of chains on the silica NPs.[22,24] As outlined in the introduction, the size and mass of aggregates was found to decrease with increasing D3 content. The analogous SAXS results for the system studied here (190 kg mol^{-1}) are shown in Fig. 8a, before (colored, plain symbols) and after swelling (grey, empty symbols). Before swelling, the large and low-q data superimpose, but a D3-dependent intermediate-q dependence is found. The higher the density of grafting D3 functions, the lower the intensity level of this shoulder. In the framework of our structural model,[23] this has been translated into a decrease in mass and size of filler aggregates.[24] This can be understood easily, as these nanocomposites have an identical filler content, and thus roughly identical inter-aggregate interactions described by $S(q)$.

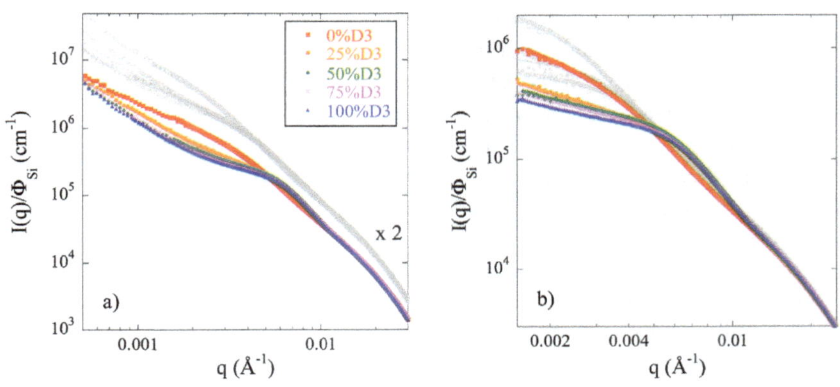

Fig. 8 (a) Comparison of microstructures of unswollen (plain, colored symbols) and swollen (empty, grey symbols) nanocomposites for the series, with matrix compositions (grafting) from 0 to 100% D3, at $\Phi_{Si} = 8.2\%V$, after subtraction of the corresponding matrix. (b) The same data for nanocomposites after subtraction of the fractal network contribution.

Accordingly, the decrease observed in Fig. 8a is due to the reduction in aggregate mass.

In analogy with Fig. 7b, the scattering of the same 8%V samples before and after swelling is shown in Fig. 8b, after subtraction of the matrix and the fractal power law. The observation is the same as in Fig. 8a: the differences below *ca.* 0.005 $Å^{-1}$ are systematic with the quantity of graftable end D3 function, and the decrease in intensity is directly related to that of the aggregate form factor, and thus mass. These measurements on swollen nanocomposites with a variation in grafting density again confirm the results of the structural model applied to the concentrated, unswollen case. Finally, it may be noted that a series analogous to the one shown in Fig. 8, but at higher silica volume fraction, confirms our findings (not shown).

4. Conclusion

The complex filler structure of silica nanoparticles in simplified industrial nanocomposites made of styrene–butadiene copolymers with parts of the chains – a fraction termed % D3 – carrying a single end function capable of grafting on the silica surface has been investigated by SAXS. Three series of nanocomposites have been formulated with a matrix chain mass of 190 kg mol^{-1}. The first series at a fixed matrix (grafting) composition of 50% D3 was a series with silica volume fractions of 8–21%V, for which a previously published structural model concluded the surprising result of mainly identical small aggregates containing some 40 or 50 nanoparticles, independent of Φ_{Si}. The model was based on a self-consistent description of aggregate interactions. In the present article, these results were confirmed by swelling such samples with styrene, thereby reducing aggregate interactions in parts. The dilution has first been illustrated by SEM and TEM images. Also, the different steps of our approach – crosslinking, swelling and polymerization – have been monitored carefully, both by electron microscopy and SAXS. The final SAXS analysis of swollen samples shown here confirms in a convincing manner the absence of the effect of filler content on aggregate size and shape. Secondly, two series at fixed silica volume fractions (*ca.* 8%V and 17%V) have been investigated, for matrix compositions ranging from 0% to 100% D3. In both cases, increasing the grafting fraction leads to smaller aggregates, as described by our structural model previously, and again this is confirmed nicely by the SAXS analysis of the swollen samples.

After having established the first structural model of 'non-model' polymer nanocomposites with polydisperse silica filler of complex multiscale structure using a quantitative combination of SAXS, TEM, and numerical simulations, the challenge was to cross-check these results with an independent method. The present article proposes such an approach, based on a physicochemical manipulation of such nanocomposites (swelling) together with SAXS, and the result gives unambiguous credit to our previous structural analysis. On the other hand, quantitative agreement is not found in terms of aggregation numbers, presumably because our model constructed for concentrated aggregates does not perform well in the diluted domain, in particular at intermediate dilution. Our approach can now be generalized to microstructures of increasing complexity, hopefully contributing to a deeper understanding of the mechanical properties of more realistic polymer nanocomposites.

Acknowledgements

We are thankful for a "Chercheur d'Avenir" grant by the Languedoc-Roussillon region (JO) and PhD funding "CIFRE" by Michelin (GPB). Parts of the swelling experiments were carried out by A. Garnero during her internship supervised by NPP and MC. SAXS experiments were performed on beamline ID02 at ESRF, with help from T. Narayanan and M. Sztucki. ACG and JO acknowledge support from ANR under Contract No. ANR-14-CE22-0001-01 (NANODYN).

References

1 J. E. Mark, B. Erman and F. R. Eirich, *Science and Technology of Rubber*, Academic Press, San Diego, 1994.

2 J. Jancar, J. F. Douglas, F. W. Starr, S. K. Kumar, P. Cassagnau, A. J. Lesser, S. S. Sternstein and M. J. Buehler, Current issues in research on structure–property relationships in polymer nanocomposites, *Polymer*, 2010, **51**, 3321–3343.

3 N. Jouault, P. Vallat, F. Dalmas, S. Said, J. Jestin and F. Boue, Well-Dispersed Fractal Aggregates as Filler in Polymer–Silica Nanocomposites: Long-Range Effects in Rheology, *Macromolecules*, 2009, **42**, 2031–2040.

4 C. Chevigny, J. Jestin, D. Gigmes, R. Schweins, E. Di-Cola, F. Dalmas, D. Bertin and F. Boue, Wet-to-Dry Conformational Transition of Polymer Layers Grafted to Nanoparticles in Nanocomposite, *Macromolecules*, 2010, **43**, 4833–4837.

5 M. Tatou, A. C. Genix, A. Imaz, J. Forcada, A. Banc, R. Schweins, I. Grillo and J. Oberdisse, Reinforcement and polymer mobility in silica-latex nanocomposites with controlled aggregation, *Macromolecules*, 2011, **44**, 9029–9039.

6 A. Banc, A. C. Genix, M. Chirat, C. Dupas, S. Caillol, M. Sztucki and J. Oberdisse, Tuning structure and rheology of silica-latex nanocomposites with the molecular weight of matrix chains: a coupled SAXS-TEM-simulation approach, *Macromolecules*, 2014, **47**, 3219–3230.

7 A. Papon, S. Merabia, L. Guy, F. Lequeux, H. Montes, P. Sotta and D. R. Long, Unique Nonlinear Behavior of Nano-Filled Elastomers: From the Onset of Strain Softening to Large Amplitude Shear Deformations, *Macromolecules*, 2012, **45**, 2891–2904.

8 S. K. Kumar, N. Jouault, B. Benicewicz and T. Neely, Nanocomposites with Polymer Grafted Nanoparticles, *Macromolecules*, 2013, **46**, 3199–3214.

9 N. Jouault, F. Dalmas, F. Boue and J. Jestin, Multiscale characterization of filler dispersion and origins of mechanical reinforcement in model nanocomposites, *Polymer*, 2012, **53**, 761–775.

10 J. Oberdisse, Aggregation of colloidal nanoparticles in polymer matrices, *Soft Matter*, 2006, **2**, 29–36.

11 G. Capuano, G. Filippone, G. Romeo and D. Acierno, Universal Features of the Melt Elasticity of Interacting Polymer Nanocomposites, *Langmuir*, 2012, **28**, 5458–5463.

12 L. Staniewicz, T. Vaudey, C. Degrandcourt, M. Couty, F. Gaboriaud and P. Midgley, Electron tomography provides a direct link between the Payne effect and the inter-particle spacing of rubber composites, *Sci. Rep.*, 2014, **4**, 7389.

13 J. Oberdisse, P. Hine and W. Pyckhout-Hintzen, Structure of interacting aggregates of silica particles for elastomer reinforcement, *Soft Matter*, 2007, **3**, 476–485.

14 D. W. Schaefer, T. Rieker, M. Agamalian, J. S. Lin, D. Fischer, S. Sukumaran, C. Y. Chen, G. Beaucage, C. Herd and J. Ivie, Multilevel structure of reinforcing silica and carbon, *J. Appl. Crystallogr.*, 2000, **33**, 587–591.

15 S. S. Choi, I. S. Kim, S. G. Lee and C. W. Joo, Filler–polymer interactions of styrene and butadiene units in silica-filled styrene–butadiene rubber compounds, *J. Polym. Sci., Part B: Polym. Phys.*, 2004, **42**, 577–584.

16 J. Ramier, C. Gauthier, L. Chazeau, L. Stelandre and L. Guy, Payne effect in silica-filled styrene–butadiene rubber: influence of surface treatment, *J. Polym. Sci., Part B: Polym. Phys.*, 2007, **45**, 286–298.

17 L. Conzatti, G. Costa, M. Castellano, A. Turturro, F. M. Negroni and J. F. Gerard, Morphology and viscoelastic behaviour of a silica filled styrene/butadiene random copolymer, *Macromol. Mater. Eng.*, 2008, **293**, 178–187.

18 Y. Shinohara, H. Kishimoto, N. Yagi and Y. Amemiya, Microscopic Observation of Aging of Silica Particles in Unvulcanized Rubber, *Macromolecules*, 2010, **43**, 9480–9487.

19 K. W. Stockelhuber, A. S. Svistkov, A. G. Pelevin and G. Heinrich, Impact of Filler Surface Modification on Large Scale Mechanics of Styrene Butadiene/Silica Rubber Composites, *Macromolecules*, 2011, **44**, 4366–4381.

20 A. Mujtaba, M. Keller, S. Ilisch, H. J. Radusch, T. Thurn-Albrecht, K. Saalwachter and M. Beiner, Mechanical Properties and Cross-Link Density of Styrene–Butadiene Model Composites Containing Fillers with Bimodal Particle Size Distribution, *Macromolecules*, 2012, **45**, 6504–6515.

21 G. P. Baeza, A. C. Genix, C. Degrandcourt, J. Gummel, M. Couty and J. Oberdisse, Mechanism of aggregate formation in simplified industrial silica styrene–butadiene nanocomposites: effect of chain mass and grafting on rheology and structure, *Soft Matter*, 2014, **10**, 6686–6695.

22 G. P. Baeza, A. C. Genix, C. Degrandcourt, J. Gummel, A. Mujtaba, K. Saalwachter, T. Thurn-Albrecht, M. Couty and J. Oberdisse, Studying twin-samples provides evidence for a unique structure-determining parameter in simplified industrial nanocomposites, *ACS Macro Lett.*, 2014, **3**, 448–452.

23 G. P. Baeza, A. C. Genix, C. Degrandcourt, L. Petitjean, J. Gummel, M. Couty and J. Oberdisse, Multiscale Filler Structure in Simplified Industrial Nanocomposite Silica/SBR Systems Studied by SAXS and TEM, *Macromolecules*, 2013, **46**, 317–329.

24 G. P. Baeza, A. C. Genix, C. Degrandcourt, L. Petitjean, J. Gummel, R. Schweins, M. Couty and J. Oberdisse, Effect of Grafting on Rheology and Structure of a Simplified Industrial Nanocomposite Silica/SBR, *Macromolecules*, 2013, **46**, 6388–6394.

25 L. Timochenco, V. G. Grassi, M. Dal Pizzol, J. M. Costa, L. G. Castellares, C. Sayer, R. A. F. Machado and P. H. H. Araujo, Swelling of organoclays in styrene. Effect on flammability in polystyrene nanocomposites, *eXPRESS Polym. Lett.*, 2010, **4**, 500–508.

26 A. Boukerrou, J. Duchet, S. Fellahi, M. Kaci and H. Sautereau, Morphology and mechanical and viscoelastic properties of rubbery epoxy/organoclay montmorillonite nanocomposites, *J. Appl. Polym. Sci.*, 2007, **103**, 3547–3552.

27 H. Acharya and S. K. Srivastava, Influence of nanodispersed organoclay on rheological and swelling properties of ethylene propylene diene terpolymer, *Macromol. Res.*, 2006, **14**, 132–139.

28 S. S. Sarkawi, W. K. Dierkes and J. W. M. Noordermeer, Elucidation of filler-to-filler and filler-to-rubber interactions in silica-reinforced natural rubber by TEM Network Visualization, *Eur. Polym. J.*, 2014, **54**, 118–127.

Faraday Discussions

PAPER

Structural characterization of aqueous solution poly(oligo(ethylene oxide) monomethyl methacrylate)-grafted silica nanoparticles†

Arnaldo T. Lorenzo, Ramakrishna Ponnapati, Tirtha Chatterjee and Ramanan Krishnamoorti*

Received 6th September 2015, Accepted 26th October 2015
DOI: 10.1039/c5fd00137d

The structure of aqueous dispersions of poly(oligo(ethylene oxide) monomethyl methacrylate)-grafted silica nanoparticles was characterized using contrast variation small-angle neutron scattering studies. Modeling the low hybrid concentration dispersion scattering data using a fuzzy sphere and a polydisperse core–shell model, demonstrated that the polymer chains are highly swollen in the dispersions as compared to the dimensions of the free polymer chains in dilute solution. At higher hybrid concentrations, the dispersions were well described using a Percus–Yevick approximation to describe the structure factor. These structural characterization tools are excellent starting points for effective molecular level descriptors of dewetting and macroscopic phase transitions for polymer tethered hybrid nanoparticle systems.

Introduction

Grafting polymers onto solid surfaces is attracting significant interest due to their important applications ranging from colloidal stabilizers to nanocomposite materials.[1,2] Polymer-grafted nanoparticles, with core–shell morphologies, can offer promising properties for applications in many varied fields.[3] Core–shell structured materials display a monomer density profile that decays from the particle center to its periphery and might provide superior properties useful for many applications, as for example, in controlled release or sensors,[4] or solution rheology modifiers.[5]

The "grafting-from" technique is commonly used to covalently attach vinyl polymers by conventional free radical polymerization in conjunction with surface-bound initiators.[6,7] However, this approach cannot provide a good control over the molecular weight, polydispersity, and terminal chain functionality of the

Department of Chemical and Biomolecular Engineering, University of Houston, Houston, TX 77204, USA. E-mail: ramanan@uh.edu

† Electronic supplementary information (ESI) available. See DOI: 10.1039/c5fd00137d

grafted polymers. The aforementioned problem can be reduced by the use of more controlled "living-like" polymerization methods where control over molecular weight, polydispersity, and structure of the resulting polymer can be attained.[8] Atom transfer radical polymerization (ATRP)[8,9] has proven to be a versatile approach for incorporating (grafting) organic (co)polymers of precise molar mass, composition, and functionality onto surfaces[6,10] and substrates.[4,11]

The synthesis of poly(oligo(ethylene oxide) monomethyl methacrylate) (POEOMA) polymers with various structures remains an active research area.[12] The self-assembly of brush-like POEOMA homopolymers were investigated in aqueous solutions by employing dynamic/static light scattering and transmission electron microscopy (TEM).[13] The thermoresponsive behavior of POEOMA has been explored extensively for possible temperature triggered encapsulation and release applications.[14] Moreover, POEOMA was copolymerized with other polymers such as methacrylic acid[15] and ethyl cellulose[16] to create materials with unique properties. Ion-sensitive responsive polymers were made to allow a phase transition to be triggered without a temperature change.[17] Copolymers with both thermoresponsive and photocrosslinkable properties were prepared using ATRP.[18] The copolymers can be crosslinked by photopolymerization through their multivinyl functional groups. Linear copolymers or 4-arm star-block copolymers have also been prepared with permanently hydrophilic poly(ethylene oxide) (PEO) inner blocks and thermoresponsive POEOMA outer blocks.[19]

In the present paper we present a comprehensive investigation by small-angle neutron scattering (SANS) of the structure of aqueous solutions of POEOMA-grafted silica nanoparticles.[20] Much interest has focused on understanding the structure of polymer chains in such hybrid materials[2,21–24] as the issues of wetting and de-wetting from such brush geometries have not been completely worked out.[1,25,26] A direct modeling expression for the scattering intensity profiles is presented and evaluated, which describes the experimental scattering intensity, $I(q)$ *vs.* wavevector, q, and data over a broad range of q, at different solvent isotopic ratios (different D_2O/H_2O mixture compositions). Using this model, the fuzzy sphere model,[27] the structure of the nanoparticles (core and shell sizes relationship) can be obtained and analyzed. These efforts parallel those performed on other grafted nanoparticle systems[28–31] and previous efforts to understand the chain conformations of multi-arm stars[30,32,33] and dendrimers[34] in solution. Moreover, for conditions where the polymer shell is poorly contrasted with the solvent, a simple core–shell model was employed to model the neutron scattering results. Understanding the structure of the nanoparticle–polymer hybrid dispersions is a crucial first step towards elucidating the structure and structural transitions in polymer matrices and the influence of thermodynamic interactions on these properties.[35] Two significant problems of interest include understanding the changes in brush structure as a result of a changing the thermodynamic quality of a matrix (or solvent)[35] and the impact of the brush structure on the effective interfacial activity of such hybrid nanoparticles in oil–water mixtures.[23]

Materials and methods

Materials

11-Bromo-1-undecanol (98%), 2-bromo-2-methyl propionyl bromide (98%), diethyl ether (99+%), magnesium sulfate ($MgSO_4$ anhydrous, 97+%), methanol

(MeOH, 99.8%), copper(I) bromide (CuBr, 98%), tetrahydrofuran (THF, 99+%) and cyclohexane (99+%) from Sigma-Aldrich, 2,2′-bipyridyl (98+%) and anisole (99+%) from Fluka, and trimethyl amine (33 wt% solution in EtOH) from Acros Organics, were used as received without further purification. Colloidal silica (SiO_2, 47–49 wt % in H_2O) was kindly supplied by Nissan Chemicals and was also used as received. Poly(ethylene glycol) methyl ether methacrylate (OEOMA475) was purchased from Sigma-Aldrich and passed through a column packed with alternating layers of inhibitor removers for hydroquinone (HQ), hydroquinone monomethyl ether (MEHQ, 4-methoxyphenol), 4-*tert*-butylcatechol (TBC) and with basic alumina to remove the inhibitor. Polyhydroxyethyl methacrylate (HEMA, ophthalmic grade 99+%) from Polysciences was also filtered through layers of HQ and MEHQ inhibitor removers and basic alumina before use.

Synthesis

POEOMA and SiO_2–POEOMA grafted nanoparticles were synthesized using the ATRP technique in the presence of CuBr as the catalyst and bipyridyl as the ligand. In the case of POEOMA, ethyl 2-bromo-2-methylpropionate was used as the ATRP initiator. SiO_2–POEOMA grafted nanoparticles were synthesized by a grafting from technique, where an ATRP initiator was grafted onto the Si particles prior to the polymerization. Complete details and synthesis procedure can be found in the literature published previously.[5] The chemical structure for the SiO_2–POEOMA nanocomposites is represented in Fig. 1.

The purification of the polymer-grafted nanoparticles followed a mixed solvent precipitation procedure to remove any remaining untethered or free chains. The dried sample was first redissolved in a minimum amount of THF and centrifuged for 10 minutes. Unfunctionalized silica nanoparticles, if any, remained in the bottom of the tube and were readily separated from the bulk of the sample. The supernatant was then transferred to another flask and placed in an oil bath at 50 °C for 30 minutes. This step ensured the homogenous dispersion of the hybrid nanoparticles as well as any unattached chains in the solvent. The flask was then cooled down to room temperature. A non-solvent (acetone) was added dropwise to the solution until the cloudiness of the system persisted. After which, the flask was reheated to 50 °C resulting in a clear dispersion. The mixture was transferred to a separatory funnel and was left undisturbed for 2 days. The formation of two layers was observed. The bottom layer which contained the pure hybrid nanoparticles dispersed in THF was collected and precipitated in an excess amount of acetone.

Determination of polymer molecular weight by size exclusion chromatography

In order to estimate the molecular weight of the synthesized polymers, size exclusion chromatography (SEC) measurements were carried out. Complete

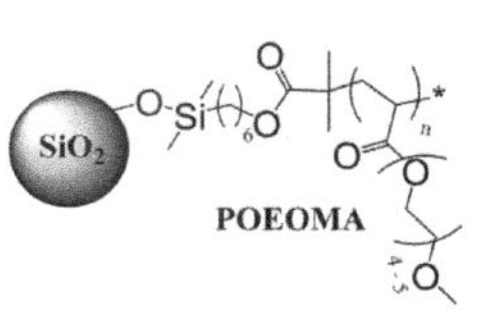

Fig. 1 SiO_2–POEOMA chemical structure.

details can be found elsewhere.[5] From the SEC analysis, the number average molecular weight (M_n) for the POEOMA homopolymer and for the grafted POEOMA chains (chains cleaved from the SiO_2) was found to be 76.1 and 62.7 kg mol^{-1}, respectively; with polydispersity indexes of 1.20 and 1.12, respectively.

Thermogravimetric analysis of SiO_2–POEOMA nanocomposites

Thermogravimetric analysis of the SiO_2–POEOMA composites was performed on a TA Instruments TGA Q500 Thermogravimetric Analyzer under pure Ar flow. During the TGA analysis, the sample was heated to 1000 °C at 10 °C min^{-1} after an initial 10 minute isotherm at 30 °C and the weight loss of the nanocomposites upon heating was monitored. The nanocomposite samples were heated in vacuum oven at 55 °C overnight prior to TGA analysis. The SiO_2 content of the nanocomposites was estimated to be 3.0 ± 0.2 wt%.

Small-angle X-ray scattering (SAXS)

Small angle X-ray scattering (SAXS) measurements were performed using a Rigaku S-MAX3000 X-ray scattering apparatus with a 1.2 kW microfocus source and equipped with a two-dimensional multiwire detector. The q-scale was calibrated using a silver behenide standard. Samples were loaded between mylar sheets and the intensity data corrected for detector dark current and empty cell scattering at 30 °C.

Small-angle neutron scattering (SANS)

SANS experiments were carried out on the Bio-SANS instrument at the High Flux Isotope Reactor of Oak Ridge National Laboratory. Measurements were carried out at room temperature in 1.0 mm path quartz cuvettes. Scattering data and the associated backgrounds were recorded using multiple detector distances, specifically 14.5 m, 6 m and 0.3 m. The wavelength was set to 6 Å with a wavelength spread, $\Delta\lambda/\lambda$, of 0.15. The detector settings provided scattering vectors (q) $0.003 < q < 0.3$ Å, which allowed sufficient angular coverage to enable effective estimation of the incoherent scattering and accurate background subtraction. Some measurements were also carried out on the NGB 30 m SANS beamline at the National Institute of Standards and Technology Center for Neutron Research (NCNR) in Gaithersburg, MD. The measurements were performed with a neutron wavelength of 6 Å and two sample-to-detector distances of 4 m and 1 m, thus accessing a q-range of 0.008 Å^{-1} to 0.5 Å^{-1}. Data reduction followed standard procedures to correct for dark current, detector response, transmission and sample incoherent scattering background prior to azimuthal averaging to produce the 1-dimensional scattering profile at 30 °C.[36]

Results and discussion

Before describing the results of the structural characterization of the POEOMA based silica hybrids, we present the dilute solution (in 100% D_2O) SANS characterization of POEOMA with a similar molecular weight (~76 kg mol^{-1}) to that present in the SiO_2–POEOMA nanocomposites samples, in Fig. 2. The excluded volume model (see ESI, Fig. S1†) was applied in order to calculate the radius of gyration (R_g) of the POEOMA chains in D_2O; in Fig. 2 the solid lines represent the fitting to the SANS experimental data for 1 and 2 wt% POEOMA solutions. From the excluded volume model an R_g of 80 ± 3 Å was found. Additionally, Table S1 (see the ESI†) shows the different R_g values obtained for the POEOMA homopolymer employing different models. It needs to be noted that this average R_g value will be used in order to obtain the stretching ratio between the homopolymer free-chains and the grafted-chains within the nanoparticles.

The melt-state scattering behavior of the pure SiO_2–POEOMA hybrid as measured by small angle X-ray scattering is shown in Fig. 3. The SAXS data were fitted using a hard sphere model with a log-normal distribution for the nanoparticle size, and indicated consistency with the structure that was a FCC lattice with a weak presence of both first and second order intensity peaks ($q_1^* : q_2^*$:1 : $\sqrt{3}$) (see Fig. 3).[24] The data described here are similar to that observed for other multi-arm star systems most notably the pioneering work of Vlassopoulos,[37,38] wherein they demonstrated similar scattering signatures from assemblies of multi-arm stars in the melt and in solutions. From this fitting, the mean diameter of the silica nanoparticle was found to be $D \sim 14.6 \pm 0.5$ nm ($R \sim 7.3 \pm 0.25$ nm), with the presence of an ordered structure with a mean distance between silica nanoparticles ~45.3 ± 0.5 nm (center-to-center distance). We note that the mean shell-to-shell distance is ~30.7 ± 1 nm, a value significantly larger than twice the unperturbed radius of gyration of comparable dilute free chains in water; this suggests that the grafted POEOMA chains are significantly stretched in such hybrids due to the grafting of such chains at high grafting density on the silica nanoparticles.

Fig. 4 shows the SANS data for low wt% SiO_2–POEOMA hybrid dispersions in water (low concentration regime: from 0.1 to 2 wt%). It is clearly visible that the data over these concentrations, we observe only the individual contributions from the hybrid material, *i.e.*, the inter-particle interactions (structure factor) are absent and the hybrid structure did not alter as a function of concentration. This can also be observed when the intensity is normalized by the volume fraction of scatterers and the data superposed over a broad concentration range as shown in Fig. 4.[39] Evaluating a SiO_2–POEOMA wt% equal to or lower than 2% will allow us to estimate only the neat structure (form factor) of individual scatters.

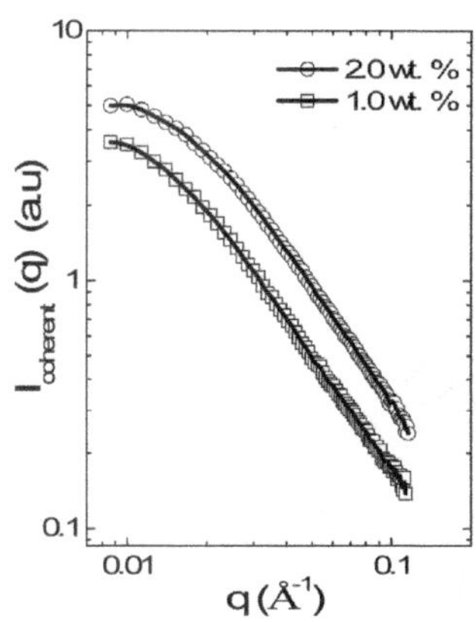

Fig. 2 SANS scattering data for the POEOMA homopolymer, in 100 vol% D_2O at a temperature of 30 °C. Solid lines represent the fitting to the excluded volume model.

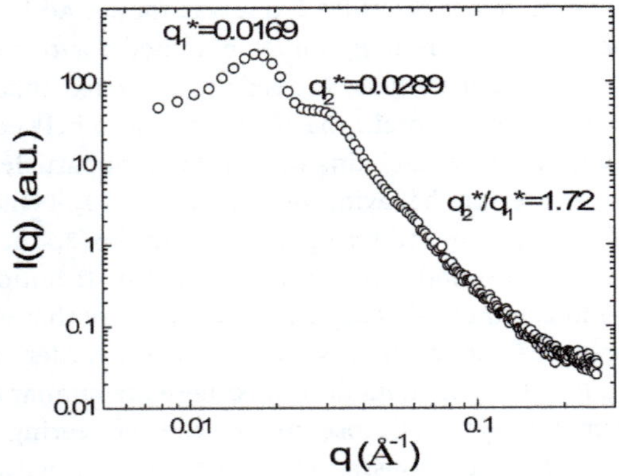

Fig. 3 SAXS data for a bulk sample of SiO_2–POEOMA nanoparticle hybrids at 30 °C, where the POEOMA chains are in their melt state.

In neutron scattering, the contrast variation method is used to isolate features of complex structures (*i.e.*, separate the scattering contributions of core from the shell in a core–shell like structure).[40,41] This can be achieved because hydrogen and deuterium have very different neutron scattering length densities (SLDs, denoted ρ_i for component i) characterizing the way they interact with neutrons (*i.e.*, for H_2O and D_2O the SLDs are -5.6×10^{-7} Å^{-2} and 6.3×10^{-6} Å^{-2}, respectively). In our SANS experiments, we consider that the SLD of a mixture (D_2O/H_2O) is a volume fraction weighted average of the SLDs of the individual components, allowing the SLD of a mixed solvent system to be determined. Table 1 summarizes the coherent neutron scattering length densities for all the materials employed in this work.

The results of a contrast variation SANS study of a 2 wt% SiO_2–POEOMA sample in mixtures of H_2O and D_2O are presented in Fig. 5. We note that the scattering intensity decreased with decreased D_2O content in the solvent mixture until about 25% D_2O. For the dispersions of the SiO_2–POEOMA hybrids in solvents, there are multiple contrasts (SiO_2 with solvent and POEOMA with

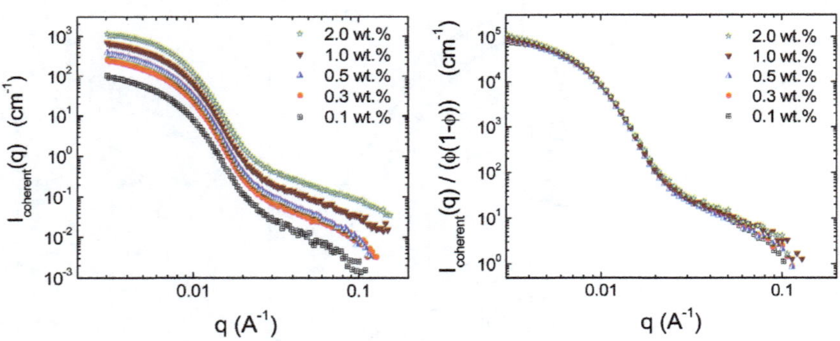

Fig. 4 SANS intensity scattering profiles (30 °C) in D_2O, for SiO_2–POEOMA samples in the low concentration regime: (left) original experimental data of scattered intensity as a function of wavevector; (right) volume-corrected profiles.

Table 1 Coherent neutron scattering length densities (ρ)

Material	ρ (Å^{-2})
H_2O	-5.6×10^{-7}
D_2O	6.3×10^{-6}
SiO_2	3.4×10^{-6}
POEOMA	8.3×10^{-7}

solvent) and therefore a single match point does not exist. In order to demonstrate the presence of multiple contrast match points (between the core and shell constituents with the solvent), the intensities at selected q values were evaluated as a function of D_2O fraction as seen in Fig. 6. At a lower q value ($q = 0.007$ Å) only a single contrast match point can be observed at 25 vol% D_2O, which can be assigned to the POEOMA chains match point (in close agreement with theoretical calculations shown in Fig. S2†). In this case, a scattering contribution originating from the SiO_2 core will be observed. On the other hand, at a higher q value ($q = 0.1$ Å), two contrast minima points are clearly visible: the first one at 25 vol% D_2O which has already been assigned to the POEOMA chains and a second one at 40 vol% D_2O that is assigned to the SiO_2 contrast match point. While the experimentally measured contrast match point for the polymer agrees closely with the theoretically calculated value (Fig. S2†), that is not the case for the contrast match point corresponding to the SiO_2 core. This presumably originates from a dense polymer chain layer close to the SiO_2 particles that essentially behaves as a dense hard sphere and thus reduces the core SLD (2.18×10^{-6} A^{-2} as opposed to the value for silica of $\sim 3.4 \times 10^{-6}$ A^{-2}). Fig. 5 allowed us therefore to assign contrast matching for the different portions (SiO_2 particles and POEOMA chains) and it also indicated that the total scattering intensity, in different q ranges, was dominated by different portions within the hybrid material.

In order to evaluate the structural characteristics of the SiO_2–POEOMA hybrids, the SANS scattering data were modeled using a fuzzy sphere model[27,42,43]

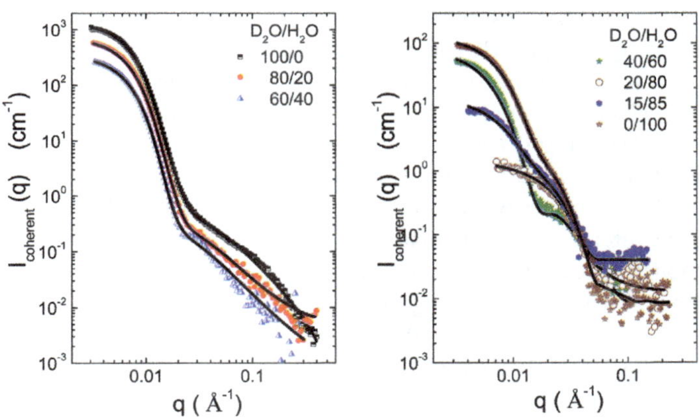

Fig. 5 SANS scattering curves (at 30 °C) for a 2 wt% SiO_2–POEOMA sample, in different D_2O/H_2O ratio solutions: (left) curves fitted with the fuzzy sphere model and, (right) curves fitted with the core–shell model.

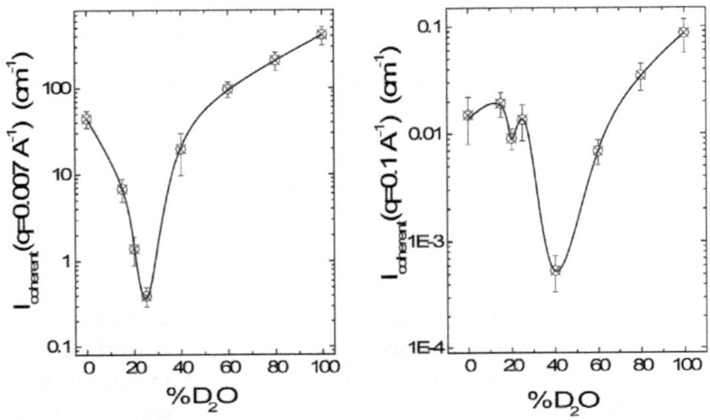

Fig. 6 Scattering intensities at two different q values as a function of the % D_2O content for a 2 wt% SiO_2–POEOMA sample.

and a spherical core–shell model.[40,42] The core–shell model considers spherical particles having a polydispersed core (and in this case with a fixed SLD) with a shell thickness of constant thickness. The form factor is normalized by the average particle volume fraction. This model appeared to be the most relevant when there was poor contrast between the outer polymer shell and the solvent, where effectively the exterior shell of the polymer was no longer clearly observed. On the other hand, the fuzzy sphere model was also used to model the data as it is able to calculate the scattering intensity from spherical particles with a "fuzzy" interface. In this model a Gaussian distribution of the particle radii is assumed and combined with a Lorentzian term that accounts for fluctuations in the shell (grafted chains) density, with the radial density profile decreased gradually from the sphere surface.[27,43] This model appeared to be most representative of the system wherein the polymer in the outer shell has a large contrast with the solvent and the silica particle (with the inner dense shell of the polymer) and the scattering measurements probed the entire structure of the hybrid nanoparticles.

For the hybrid samples considered here, in the swollen state the polymer segment density of a particle can be estimated to be not homogenously distributed but gradually decreases from the core through the shell surface, a schematic particle model can be seen in Fig. 7. This structure is consistent with the model proposed by Daoud and Cotton,[44] verified for block copolymer micelles[32] and subsequently for multi-arm stars.[38] Interestingly, the region close to the nanoparticles is largely impenetrable to the solvent, a reflection of the high density of chains at the interface. We anticipate that this region will change with changes in the thermodynamic character of the solvent and this has been studied for the case of a polymer matrix with changing phase behavior with grafted chains.[35]

As noted in Fig. 5, the fuzzy sphere model fits well when the D_2O content is equal to or higher than 60 vol% (high D_2O content systems). For lower D_2O concentrations, the fitting to the fuzzy sphere model deviates noticeably. In those cases, *i.e.*, for D_2O content equal to or lower than 40 vol%, the core–shell model was successfully applied (see Fig. 5, right). The contrast variation results shown in Fig. 5 provide an unequivocal methodology to understand quantitatively the

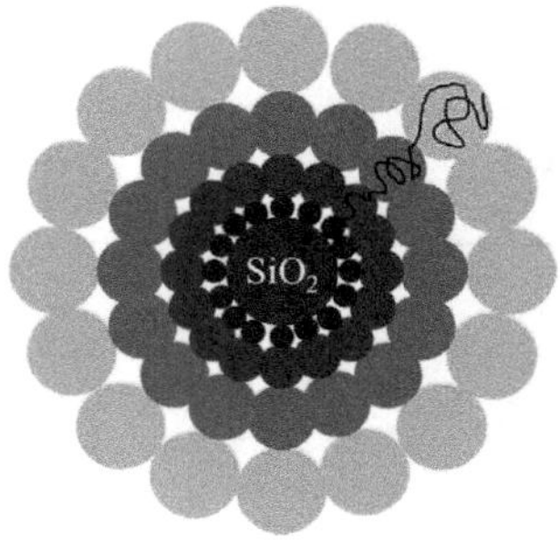

Fig. 7 Proposed particle model for the SiO_2–POEOMA nanoparticles. Due to the POEOMA grafting density at the core surface, a decreasing gradient density will be present from the SiO_2 core through the shell surface (darker to lighter color in the scheme).

structure of the polymer chains when tethered to the silica nanoparticles and lend support to previous reports on bottlebrush and grafted polymer structures.[22,26,28,45]

Using the average SiO_2 core radius (from the SAXS analysis) and the fitting parameter from the employed SANS model fit, the thickness of the POEOMA shell can be obtained. We identified two main components in the hybrid structure: (i) a dense hybrid core, which is formed by the SiO_2 core surrounded by a dense POEOMA shell, where the polymer chains are poorly solvated by the solvent, and; (ii) the nanoparticle shell, which is essentially formed by less dense POEOMA chains and strongly solvated by the solvent. The two most critical data sets to consider are those at 40% D_2O where the solvent SLD is close to the SLD of the silica-dense polymer hybrid core and the data set at 25% D_2O where the solvent SLD is close to that of the outer shell of the polymer chains. We note that the data for the 40% D_2O sample indicate that the chains attached to the silica nanoparticle are highly stretched compared to the solution scattering data described in Fig. 2 for the free polymer chains in water.

From the fitted values, a stretching ratio value can be calculated using the following equation:

$$R_{\text{stretching}} = \frac{(\text{hybrid core radius} + \text{shell thickness} - \text{SiO}_2\ \text{radius}_{(\text{SAXS})})}{R_{\text{g}}} \tag{1}$$

with the underlying assumption that there is no interpenetration of chains in the shell layers. Table 2 presents the POEOMA hybrid core radius, the shell thickness (t) and the stretching ratio for the 2 wt% SiO_2–POEOMA nanocomposite in the D_2O/H_2O mixture with higher D_2O content.

From the extracted fitting parameters presented in Table 2, we infer from the conditions where the solvent mixture SLD is different from the POEOMA contrast match point (15–25 vol% D_2O), that the POEOMA chains are stretched at least two times when grafted to the SiO_2 nanoparticles. For the conditions where the SLD of the solvent mixture is close to that of the POEOMA outer shell, the scattering models do not provide an accurate measure of the shell thickness and therefore a value of $R_{\text{stretching}}$ could not be obtained. Moreover, the extracted core radius obtained from SANS data fitting is slightly larger than the same obtained from SAXS measurements and consistent with the notion of a non-solvated polymer shell in proximity to the SiO_2 surface.

Table 2 Fitting values and the stretching ratios for the SiO_2–POEOMA nanocomposite in different D_2O/H_2O mixtures, at 30 °C

D_2O vol%	Hybrid core radius (Å)	Shell thickness (Å)	$R_{stretching}$	SLD shell (A^{-2})	SLD solvent (A^{-2})
Fuzzy sphere model					
100	181 ± 21	67 ± 5	2.2 ± 0.4	3.26 (±0.1) × 10^{-6}	6.30 × 10^{-6}
80	165 ± 18	78 ± 4	2.1 ± 0.2	2.54 (±0.3) × 10^{-6}	4.93 × 10^{-6}
60	178 ± 20	87 ± 6	2.4 ± 0.3	1.89 (±0.2) × 10^{-6}	3.56 × 10^{-6}
Core–shell model					
40	91 ± 11	133 ± 7	2.1 ± 0.3	1.25 (±0.2) × 10^{-6}	2.18 × 10^{-6}
25	88 ± 4	7 ± 3		1.14 (±0.2) × 10^{-6}	1.16 × 10^{-6}
20	84 ± 6	5 ± 4		1.08 (±0.6) × 10^{-6}	8.27 × 10^{-7}
15	88 ± 4	7 ± 3		1.85 (±0.5) × 10^{-6}	4.69 × 10^{-7}
0	94 ± 9	101 ± 13	1.6 ± 0.4	2.60 (±0.4) × 10^{-6}	−5.60 × 10^{-7}

Fig. 8 shows how the structure factor became more evident at higher nanoparticle concentrations. Using a spherical core–shell form factor ($P(q)$) combined with a hard sphere (HS) structure factor ($S(q)$) with a Percus–Yevick (PY) closure, it was possible to obtain reasonable fits for the scattering intensity ($I(q) = kP(q)S(q)$) from 3, 5 and 10 wt% SiO_2–POEOMA nanoparticles at the POEOMA match point (25/75 D_2O/H_2O). Working close to the POEOMA contrast match point greatly reduced the scattering contribution originating from the POEOMA chains. The applicability of a hard-sphere structure factor implies that the dense cores (of the SiO_2–polymer inner shell) in these dilute to semi-dilute solutions interact through volume exclusion. From the fitting data the average SiO_2 radius was found to be around 8.3 ± 0.5 nm, a value in close agreement with the result found using the

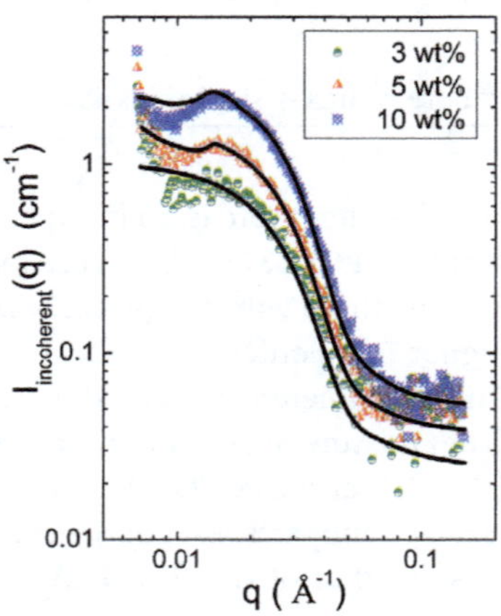

Fig. 8 SANS intensity scattering profiles, 30 °C, for SiO_2–POEOMA samples, in the high hybrid concentration regime, at 25/75 D_2O/H_2O. The Percus–Yevick approximation provides excellent agreement with the observed intensity data and indicates that under conditions where the polymer shell is contrast matched, the particles interact with each other through hard-sphere like interactions.

bulk sample SAXS scattering and the lower concentration core–shell modelling demonstrated in Table 2.

Conclusions

Contrast variation based small-angle neutron scattering of dilute dispersions of the POEOMA–SiO_2 nanoparticle hybrids is an excellent tool to understand the structure of the polymer chains when they are attached at high grafting densities to the nanoparticle. The tethered polymer chains are highly stretched compared to the dimensions of the free chains dissolved in the same solvent. Moreover, the scattering data from higher concentration dispersions of the nanoparticles, under conditions where the solvated polymer chain based outer shell is contrast matched by the solvent, demonstrate good agreement with a Percus–Yevick approximation to describe the interactions between the nanoparticles.

These characterization methods, and the presence of two distinct contrast minimum ratios of solvent isotopic mixtures corresponding to the matching of the contrast with the outer shell and the inner core with the solvent, indicate that these methods can be used to characterize the interactions between the brush chains with each other as well as with the solvent and therefore enable unique molecular level detail of the characterization of thermodynamic transitions. Specifically, we believe that the issues of dewetting of the polymer brush and the phase transition resulting from the incompatibility of the polymer brush and the matrix solvent or polymers would be ideally suited for such a study.[31,35]

Acknowledgements

SAXS data were obtained using an instrument obtained on a National Science Foundation (NSF DMR 1040446) funded instrument. A portion of this research at ORNL's High Flux Isotope Reactor was sponsored by the Scientific User Facilities Division, Office of Basic Energy Sciences, US Department of Energy. A portion of this work utilized the NIST Center for Neutron Research (NCNR) supported in part by the National Science Foundation under Grant #DMR-0944772.

References

1 V. Ganesan and A. Jayaraman, *Soft Matter*, 2014, **10**, 13–38; S. K. Kumar, N. Jouault, B. Benicewicz and T. Neely, *Macromolecules*, 2013, **46**, 3199–3214.

2 V. Goel, T. Chatterjee, L. Bombalski, K. Yurekli, K. Matyjaszewski and R. Krishnamoorti, *J. Polym. Sci., Part B: Polym. Phys.*, 2006, **44**, 2014–2023.

3 S. Nayak and L. A. Lyon, *Angew. Chem., Int. Ed.*, 2004, **43**, 6706–6709.

4 J. Pyun and K. Matyjaszewski, *Chem. Mater.*, 2001, **13**, 3436–3448.

5 R. Ponnapati, O. Karazincir, E. Dao, R. Ng, K. K. Mohanty and R. Krishnamoorti, *Ind. Eng. Chem. Res.*, 2011, **50**, 13030–13036.

6 Y.-P. Wang, X.-W. Pei, X.-Y. He and K. Yuan, *Eur. Polym. J.*, 2005, **41**, 1326–1332.

7 O. Prucker and J. Ruhe, *Macromolecules*, 1998, **31**, 602–613; O. Prucker and J. Ruhe, *Macromolecules*, 1998, **31**, 592–601; X. Fan, C. Xia and R. C. Advincula, *Langmuir*, 2003, **19**, 4381–4389; T. Stohr and J. Ruhe, *Macromolecules*, 2000, **33**, 4501–4511.

8 O. Webster, *Science*, 1991, **251**, 887–893.

9 J.-S. Wang and K. Matyjaszewski, *J. Am. Chem. Soc.*, 1995, **117**, 5614–5615; K. Matyjaszewski and J. Xia, *Chem. Rev.*, 2001, **101**, 2921–2990; M. Kamigaito, T. Ando and M. Sawamoto, *Chem. Rev.*, 2001, **101**, 3689–3746.

10 D. M. Jones, A. A. Brown and W. T. S. Huck, *Langmuir*, 2002, **18**, 1265–1269; W. Huang, J.-B. Kim, M. L. Bruening and G. L. Baker, *Macromolecules*, 2002, **35**, 1175–1179; W. Huang, G. L. Baker and M. L. Bruening, *Angew. Chem., Int. Ed.*, 2001, **40**, 1510–1512.

11 X. Chen, D. P. Randall, C. Perruchot, J. F. Watts, T. E. Patten, T. von Werne and S. P. Armes, *J. Colloid Interface Sci.*, 2003, **257**, 56–64; J. N. Kizhakkedathu and D. E. Brooks, *Macromolecules*, 2003, **36**, 591–598; P. von Natzmer, D. Bontempo and N. Tirelli, *Chem. Commun.*, 2003, 1600–1601; G. Zheng and H. D. H. Stöver, *Macromolecules*, 2003, **36**, 1808–1814; T. von Werne and T. E. Patten, *J. Am. Chem. Soc.*, 1999, **121**, 7409–7410; J. Pyun, S. Jia, T. Kowalewski, G. D. Patterson and K. Matyjaszewski, *Macromolecules*, 2003, **36**, 5094–5104; P. Liu, W. Liu and Q. Xue, *J. Macromol. Sci., Part A: Pure Appl. Chem.*, 2004, **41**, 1001–1010.

12 T. Ishizone, A. Seki, M. Hagiwara, S. Han, H. Yokoyama, A. Oyane, A. Deffieux and S. Carlotti, *Macromolecules*, 2008, **41**, 2963–2967.

13 H. Hussain, K. Y. Mya and C. He, *Langmuir*, 2008, **24**, 13279–13286.

14 J. F. Lutz, *J. Polym. Sci., Part A: Polym. Chem.*, 2008, **46**, 3459–3470; J. F. Lutz and A. Hoth, *Macromolecules*, 2006, **39**, 893–896.

15 C. R. Becer, S. Hahn, M. W. M. Fijten, H. M. L. Thijs, R. Hoogenboom and U. S. Schubert, *J. Polym. Sci., Part A: Polym. Chem.*, 2008, **46**, 7138–7147.

16 Y. Li, R. Liu, W. Liu, H. Kang, M. Wu and Y. Huang, *J. Polym. Sci., Part A: Polym. Chem.*, 2008, **46**, 6907–6915.

17 J. P. Magnusson, A. Khan, G. Pasparakis, A. O. Saeed, W. Wang and C. Alexander, *J. Am. Chem. Soc.*, 2008, **130**, 10852–10853.

18 H. Tai, W. Wang, T. Vermonden, F. Heath, W. E. Hennink, C. Alexander, K. M. Shakesheff and S. M. Howdle, *Biomacromolecules*, 2009, **10**, 822–828.

19 N. Fechler, N. Badi, K. Schade, S. Pfeifer and J.-F. O. Lutz, *Macromolecules*, 2008, **42**, 33–36; N. Badi and J.-F. Lutz, *J. Controlled Release*, 2009, **140**, 224–229.

20 S. Rose, A. Marcellan, T. Narita, F. Boue, F. Cousin and D. Hourdet, *Soft Matter*, 2015, **11**, 5905–5917.

21 C. J. Kim, M. Kraska, M. Mazurowski, K. Sondergeld, M. Gallei, M. Rehahn and B. Stuhn, *Soft Mater.*, 2014, **12**, S41–S48; V. Lebedev, G. Torok and L. Vinogradova, *J. Macromol. Sci., Part B: Phys.*, 2013, **52**, 1736–1755; C. J. Kim, K. Sondergeld, M. Mazurowski, M. Gallei, M. Rehahn, T. Spehr, H. Frielinghaus and B. Stuhn, *Colloid Polym. Sci.*, 2013, **291**, 2087–2099; M. J. A. Hore, J. Ford, K. Ohno, R. J. Composto and B. Hammouda, *Macromolecules*, 2013, **46**, 9341–9348; F. Gal, H. Perez, V. Noel and G. Carrot, *J. Polym. Sci., Part A: Polym. Chem.*, 2012, **50**, 289–296; V. Goel, J. Pietrasik, K. Matyjaszewski and R. Krishnamoorti, *Ind. Eng. Chem. Res.*, 2010, **49**, 11985–11990.

22 C. Chevigny, D. Gigmes, D. Bertin, J. Jestin and F. Boue, *Soft Matter*, 2009, **5**, 3741–3753.

23 D. Kim and R. Krishnamoorti, *Ind. Eng. Chem. Res.*, 2015, **54**, 3648–3656.

24 V. Goel, J. Pietrasik, H. C. Dong, J. Sharma, K. Matyjaszewski and R. Krishnamoorti, *Macromolecules*, 2011, **44**, 8129–8135.

25 C. E. Estridge and A. Jayaraman, *J. Polym. Sci., Part B: Polym. Phys.*, 2015, **53**, 76–88; M. Asai, A. Cacciuto and S. K. Kumar, *Soft Matter*, 2015, **11**, 793–797; T. B. Martin and A. Jayaraman, *J. Polym. Sci., Part B: Polym. Phys.*, 2014, **52**, 1661–1668; B. Lin, T. B. Martin and A. Jayaraman, *ACS Macro Lett.*, 2014, **3**, 628–632; J. Moll, S. K. Kumar, F. Snijkers, D. Vlassopoulos, A. Rungta, B. C. Benicewicz, E. Gomez, J. Ilavsky and R. H. Colby, *ACS Macro Lett.*, 2013, **2**, 1051–1055; T. B. Martin and A. Jayaraman, *Soft Matter*, 2013, **9**, 6876–6889; T. B. Martin and A. Jayaraman, *Macromolecules*, 2013, **46**, 9144–9150; A. Jayaraman, *J. Polym. Sci., Part B: Polym. Phys.*, 2013, **51**, 524–534; P. M. Dodd and A. Jayaraman, *J. Polym. Sci., Part B: Polym. Phys.*, 2012, **50**, 694–705; D. Maillard, S. K. Kumar, A. Rungta, B. C. Benicewicz and R. E. Prud'homme, *Nano Lett.*, 2011, **11**, 4569–4573; L. M. Hall, A. Jayaraman and K. S. Schweizer, *Curr. Opin. Solid State Mater. Sci.*, 2010, **14**, 38–48; P. Akcora, H. Liu, S. K. Kumar, J. Moll, Y. Li, B. C. Benicewicz, L. S. Schadler, D. Acehan, A. Z. Panagiotopoulos, V. Pryamitsyn, V. Ganesan, J. Ilavsky, P. Thiyagarajan, R. H. Colby and J. F. Douglas, *Nat. Mater.*, 2009, **8**, 354.

26 J. Kalb, D. Dukes, S. K. Kumar, R. S. Hoy and G. S. Grest, *Soft Matter*, 2011, **7**, 1418–1425.

27 M. Stieger, W. Richtering, J. S. Pedersen and P. Lindner, *J. Chem. Phys.*, 2004, **120**, 4750–4752.

28 M. J. A. Hore, X. C. Ye, J. Ford, Y. Z. Gao, J. Y. Fei, Q. Wu, S. J. Rowan, R. J. Composto, C. B. Murray and B. Hammouda, *Nano Lett.*, 2015, **15**, 5730–5738; C. Picot, F. Audouin and C. Mathis, *Macromolecules*, 2007, **40**, 1643–1656.

29 Y. Jiao and P. Akcora, *Phys. Rev. E: Stat., Nonlinear, Soft Matter Phys.*, 2014, **90**, 9; G. G. Vogiatzis and D. N. Theodorou, *Macromolecules*, 2013, **46**, 4670–4683; H. Lee, D. H. Kim, K. N. Witte, K. Ohn, J. Choi, B. Akgun, S. Satija and Y. Y. Won, *J. Phys. Chem. B*, 2012, **116**, 7367–7378; S. Lecommandoux, F. Checot, R. Borsali, M. Schappacher, A. Deffieux, A. Brulet and J. P. Cotton, *Macromolecules*, 2002, **35**, 8878–8881.

30 M. Mazurowski, K. Sondergeld, J. Elbert, C. J. Kim, J. Y. Li, H. Frielinghaus, M. Gallei, B. Stuhn and M. Rehahn, *Macromol. Chem. Phys.*, 2013, **214**, 1094–1106.

31 M. Zackrisson, A. Stradner, P. Schurtenberger and J. Bergenholtz, *Langmuir*, 2005, **21**, 10835–10845.

32 R. Krishnamoorti, S. Rai and M. B. Shek, *Soft Mater.*, 2003, **1**, 263–275.

33 M. Moffitt, Y. S. Yu, D. Nguyen, V. Graziano, D. K. Schneider and A. Eisenberg, *Macromolecules*, 1998, **31**, 2190–2197; L. Willner, O. Jucknischke, D. Richter, B. Farago, L. J. Fetters and J. S. Huang, *Europhys. Lett.*, 1992, **19**, 297–303.

34 M. A. J. Gillissen, T. Terashima, E. W. Meijer, A. R. A. Palmans and I. K. Voets, *Macromolecules*, 2013, **46**, 4120–4125; O. V. Borisov, E. B. Zhulina and T. M. Birshtein, *ACS Macro Lett.*, 2012, **1**, 1166–1169; D. Potschke, M. Ballauff, P. Lindner, M. Fischer and F. Vogtle, *Macromol. Chem. Phys.*, 2000, **201**, 330–339.

35 T. B. Martin, K. I. S. Mongcopa, R. Ashkar, P. Butler, R. Krishnamoorti and A. Jayaraman, *J. Am. Chem. Soc.*, 2015, **137**, 10624–10631.

36 S. R. Kline, *J. Appl. Crystallogr.*, 2006, **39**, 895–900; C. J. Glinka, J. M. Nicol, G. D. Stucky, E. Ramli, D. Margolese, Q. Huo, J. B. Higgins and M. E. Leonowicz, *J. Porous Mater.*, 1996, **3**, 93–98.

37 D. Vlassopoulos, G. Fytas, J. Roovers, T. Pakula and G. Fleischer, *Faraday Discuss.*, 1999, **112**, 225–235; R. Seghrouchni, G. Petekidis, D. Vlassopoulos, G. Fytas, A. N. Semenov, J. Roovers and G. Fleischer, *Europhys. Lett.*, 1998, **42**, 271–276; D. Vlassopoulos, T. Pakula, G. Fytas, J. Roovers, K. Karatasos and N. Hadjichristidis, *Europhys. Lett.*, 1997, **39**, 617–622.

38 T. Pakula, D. Vlassopoulos, G. Fytas and J. Roovers, *Macromolecules*, 1998, **31**, 8931–8940.

39 A. P. Gast, *Langmuir*, 1996, **12**, 4060–4067; C. M. Stancik, A. R. Lavoie, P. A. Achurra, R. M. Waymouth and A. P. Gast, *Langmuir*, 2004, **20**, 8975–8987; C. M. Stancik, A. R. Lavoie, J. Schütz, P. A. Achurra, P. Lindner, A. P. Gast and R. M. Waymouth, *Langmuir*, 2003, **20**, 596–605.

40 A. Ramzi, J. F. Cristianne, F. J. Theo, D. Schwahn, W. E. Hennink and C. F. van Nostrum, *J. Phys. Chem. B*, 2008, **112**, 784–792.

41 *Polymers and Neutrons Scattering*, ed. J. S. B. Higgins and H. C. Benoit, Clarendon Press, Oxford, U.K., 1994; M. Zackrisson, A. Stradner, P. Schurtenberger and J. Bergenholtz, *Langmuir*, 2005, **21**, 10835–10845; C. D. Jones and L. A. Lyon, *Macromolecules*, 2000, **33**, 8301–8306.

42 I. Berndt, J. S. Pedersen and W. Richtering, *Angew. Chem., Int. Ed.*, 2006, **45**, 1737–1741.

43 M. Stieger, J. S. Pedersen, P. Lindner and W. Richtering, *Langmuir*, 2004, **20**, 7283–7292.

44 M. Daoud and J. P. Cotton, *J. Phys.*, 1982, **43**, 531–538.

45 C. Chevigny, D. Gigmes, D. Bertin, R. Schweins, J. Jestin and F. Boue, *Polym. Chem.*, 2011, **2**, 567–571.

Faraday Discussions

PAPER

Interplay between polymer chain conformation and nanoparticle assembly in model industrial silica/rubber nanocomposites

Adrien Bouty,[ab] Laurent Petitjean,[b] Julien Chatard,[b] Rachid Matmour,[b] Christophe Degrandcourt,[b] Ralf Schweins,[c] Florian Meneau,[d] Paweł Kwaśniewski,[e] François Boué,[a] Marc Couty[b] and Jacques Jestin*[a]

Received 24th August 2015, Accepted 10th September 2015
DOI: 10.1039/c5fd00130g

The question of the influence of nanoparticles (NPs) on chain dimensions in polymer nanocomposites (PNCs) has been treated mainly through the fundamental way using theoretical or simulation tools and experiments on well-defined model PNCs. Here we present the first experimental study on the influence of NPs on the polymer chain conformation for PNCs designed to be as close as possible to industrial systems employed in the tire industry. PNCs are silica nanoparticles dispersed in a styrene-butadiene-rubber (SBR) matrix whose NP dispersion can be managed by NP loading with interfacial coatings or coupling additives usually employed in the manufacturing mixing process. We associated specific chain (d) labeling, and the so-called zero average contrast (ZAC) method, with SANS, *in situ* SANS and SAXS/TEM experiments to extract the polymer chain scattering signal at rest for non-cross linked and under stretching for cross-linked PNCs. NP loading, individual clusters or connected networks, as well as the influence of the type, the quantity of interfacial agent and the influence of the elongation rate have been evaluated on the chain conformation and on its related deformation. We clearly distinguish the situations where the silica is perfectly matched from those with unperfected matching by direct comparison of SANS and SAXS structure factors. Whatever the silica matching situation, the additive type and quantity and the filler content, there is no significant change in the polymer dimension for NP loading up to 15% v/v within a range of 5%. One can see an extra scattering contribution at low *Q*, as often encountered, enhanced for non-perfect silica matching but also visible for perfect filler matching. This contribution can be qualitatively attributed to specific h or d chain adsorption on the NP surface inside the NP cluster

[a]*Laboratoire Léon Brillouin, CEA Saclay, 91191 Gif-sur-Yvette Cedex, France. E-mail: jacques.jestin@cea.fr*
[b]*Manufacture Française des Pneumatiques MICHELIN, Site de Ladoux, 23 Place des Carmes Déchaux, F-63 040 Clermont-Ferrand, Cedex 9, France*
[c]*Institut Laue-Langevin, BP 156, 6, Rue Jules Horowitz, 38042 Grenoble Cedex 9, France*
[d]*Synchrotron SOLEIL, L'Orme des Merisiers, PO Box 48, Saint-Aubin, 91192 Gif/Yvette, France*
[e]*European Synchrotron Radiation Facility, 6 Rue Jules Horowitz, 38043 Grenoble, France*

that modifies the average scattering neutron contrast of the silica cluster. Under elongation, NPs act as additional cross-linking junctions preventing chain relaxation and giving a deformation of the chain with the NP closer to a theoretical phantom network prediction than a pure matrix.

I. Introduction

Progress in the understanding of the mechanisms that govern the reinforcement of polymer matrices by NPs is of strategic interest to improve the design and industrial processes of the conception of new materials, specifically in tire industries. One of the critical breakthroughs is to be able to understand and describe the local structure of the filler (individual dispersion, connected network, aggregates or agglomerates…) as well as that of the polymer chain (unmodified, stretched or compressed…) and to correlate them with the macroscopic mechanical behaviors of the materials like reinforcement, flow behavior or non-linear deformation.[1,2] Despite the lack of a general model to describe the filler contribution (NP or assembly of NP), many progresses have been made recently especially to dissociate the filler–filler interactions from the filler–matrix ones in reinforcement on both well-defined models[3–8] and simplified industrial systems.[9–15] On the contrary, the question of the matrix chain conformation and its modification due to the presence of NPs has been, up to now, only addressed from a fundamental point of view. Theories have been developed using mainly simulations[16–22] considering the polymer radius of gyration R_g and the NP radius R_p ratio (R_g/R_p). They lead to contradictory conclusions, either swelling or decreasing of R_g as a function of the NP volume fraction, mostly when the polymer radius of gyration R_g is equal or superior to the NP radius R_p. For example, Termonia *et al.* and Sung *et al.*[16,17] reported a significant swelling of the chains for $R_g/R_p > 1$ while Vacatello *et al.*[18,19] reported a contraction of the chains. Sharaf *et al.*[20] have reported a swelling of the chain but for $R_g/R_p < 1$. The situation has also been probed from an experimental point of view on well-defined model systems[23–31] using SANS and conventional polymers for which the labeled (d) forms can be conventionally accessible. These studies also yielded controversial behaviors whose origins can be discussed according to different issues. A first possible origin is the difficulty to extract the experimental signal of the chain without ambiguous matching or undefined extra scattering. Nakatani *et al.*,[23] Mackay *et al.*[24] (for the C60/PS system) and also Botti *et al.*[25] extracted the conformation of the polymer chain from a complex scattering signal in which the interactions between labeled (d) and non-labeled (h) chains have to be taken into account. Sen *et al.*,[26] Nusser *et al.*[27] and Genix *et al.*[28] (on latex/silica systems) used the ZAC method, which cancels inter- and intra-chain correlations, under unperfected matching conditions[26,27] or in the presence of an extra scattering signal coming from chain diffusion.[28] Jouault *et al.*[29] have demonstrated with a random (h/d) matrix which matched the silica NP the unambiguous validity of the ZAC method while also observing an unresolved extra scattering contribution. An unambiguous contrast matched method that could be used is also that of Tuteja *et al.*[30] with cross-linked PS particles as fillers but concerns a very specific case of PNCs filled with soft fillers. Another contribution comes from the role played by the NP dispersion into the polymer matrix. Despite the general prediction proposed by

Mackay *et al.*[24] that suggested that the particle dispersion is only governed by the R_g/R_p ratio, it has been also demonstrated that the particle dispersion strongly depends on the processing conditions.[6,31] Indeed, some of the previous studies suffer from a clear NP dispersion characterization that can render a confusing reading of the results if the NPs are aggregated or form connected networks. When reducing the discussion to the specific case for which the polymer radius of gyration is close to the NP radius, the situation is however relatively clear when most of the experiments have concluded that the chain conformation seems to be unaffected by the NPs.[24–26,28–30] Only two previous studies[23,27] show some effects that arise only at high NP loading, up to 40% v/v. The absence of NP influence has also been confirmed recently by Crawford *et al.*[32] for PNCs with uniformly dispersed NPs. The previous examples examine the possible pre-strain of the chain induced by the NPs. Another relevant contribution is whether the chain deformation is also affected by the NP under non-linear (large) deformations. This point has been addressed with simulations[33] or with indirect experiments like NMR.[34] A few experiments have probed the chain deformation scattering signal directly in the presence of NPs[35–37] on model PNCs. Homogeneous overstrain has been established in triblock copolymer micelles of PI–PS–PI in a cross-linked polyisoprene (PI) matrix where no overstrain effects have been measured in uncross-linked model silica/polystyrene PNCs.[37]

Beyond the fundamental interest of the previous studies mainly performed on model systems, the question of the chain conformation inside a composite is also a main preoccupation of industrial researchers. In the current context of petroleum resource depletion, a new challenge of the tire industry is to develop new materials that can reduce oil consumption as well as CO_2 emissions while offering high mechanical performances (reinforcement, adhesion, tensile strength…). One possible way is to act on the NP, for example, by replacing the classical carbon black filler with silica nanoparticles. Another possible way is to act on the rubber matrix by modifying the chain characteristics (molecular mass, polydispersity index, chemical functions…) to modify the polymer–NP interactions. Consequently, matrix chain conformation/deformation modification in PNCs and its correlation with the mechanical performances of the material becomes a central question that we have addressed in the present paper.

We present here a complete study of model industrial silica/SBR systems using complementary SAXS/SANS/TEM experiments. We have previously demonstrated[15] that the silica NP dispersion can be controlled by using coating (octyltriethoxysilane – OCTEO) and coupling (sulfur functional organosilane bis(triethoxysilyl)propyl tetrasulfide – Si69) additives in SBR for non-cross-linked PNCs similarly whatever the sample preparation method, solvent casting or internal mixing. Thanks to covalent bonds with the chains of the matrix, the coupling agent gives rise to smaller and denser aggregates than can form an interpenetrated percolating network when increasing the NP loading. The coating agent has a more repulsive action that reduces the network density. In other words, it is demonstrated that interfacial additives can tune the matrix chain confinement and thus, they appear to be a relevant way, as well as NP loading, to evaluate the effect on the chain conformation. The synthesis of (d) labelled chains, restricted to low molecular mass, limits the capacity to probe a broad range of NP dispersion leads to treat the case of NPs organized as clusters. We varied the type and the quantity of interfacial additives for different NP loadings from 1 to 15% vol. We applied the principle of ZAC by mixing labelled (d) and non-labelled (h) SBR chains to extract the single chain scattering signal and see whether or not the NPs are potentially inducing some pre-straining effect on the chains. The clustering of the NPs is evaluated from SAXS/TEM analysis. To evaluate the effect of NPs on the chain deformation at a large non-linear elongation rate, we completed the study with *in situ* SANS measurements on cross-linked silica NP/SBR PNCs for different elongation rates.

II. Material and methods

1. Synthesis of the (d) SBR chains

d8-Styrene and d6-butadiene deuterated monomers were supplied by Cortecnet (purity of 99%) and Eurisotop (purity of 99%), respectively. Before their use for copolymerization, the monomers were first dried over BuLi for d6-butadiene and over calcium hydride and dibutyl magnesium for d8-styrene and then distilled to obtain purified monomers. Similarly to hydrogenated SBR copolymers in classical runs, the deuterated SBR copolymer samples were synthesized by anionic polymerization by the Michelin Company. The copolymerization was initiated by BuLi in cyclohexane at 50 °C. Deuterated monomers were mixed in appropriate conditions to reach a microstructure as close as possible to that of the hydrogenated chains used here (26% wt of styrene, 74% wt of butadiene; $M_n = 40.000$ g mol^{-1}, $M_w = 41.700$ g mol^{-1}, PDI = 1.05). Two samples of deuterated SBR copolymer were obtained with the same microstructure as the hydrogenated ones (26% wt of styrene, 74% wt of butadiene) and sufficiently satisfactory chain mass distribution: Batch 1: $M_n = 49\ 700$ g mol^{-1}, $M_w = 100\ 500$ g mol^{-1}, PDI = 2.0; Batch 2: $M_n = 45\ 000$ g mol^{-1}, $M_w = 96.270$ g mol^{-1}, PDI = 2.1. The large polydispersities (PDI ≥ 2.0) obtained for the deuterated copolymers are due to the presence of butene derivative impurities present in non-negligible quantities in the supplied deuterated butadiene.

2. Preparation of the PNCs

The matrix consists systematically of a mixture of chains from the hydrogenated and the deuterated batches, in the appropriate volume proportions to match the silica. Considering the neutron scattering length densities of the three components ($\rho_{SiO_2} = 3.41 \times 10^{10}$ cm^{-2}, $\rho_{h\text{-}SBR} = 0.59 \times 10^{10}$ cm^{-2} and $\rho_{d\text{-}SBR} = 6.67 \times 10^{10}$ cm^{-2}), the matrix composition is fixed to 54% v h-SBR + 46% v d-SBR as it theoretically matches silica. All samples were prepared by a solvent casting route with dimethylacetamide (DMAc) as the solvent. The filler was Ludox type colloidal silica supplied by Aldrich (LS-30); the size distribution of these nanoparticles was finely characterized by SANS and gives a mean particle radius $R_0 = 7.8$ nm with a log-normal distribution of 0.15.

The uncross-linked samples were synthesized with deuterated chains from Batch 1. The steps and conditions of this synthesis – Ludox transfer from water to DMAc, dissolution of the matrix, addition of the Ludox/DMAc suspension at 130 °C, solvent casting under nitrogen flow at 130 °C – were precisely explained in a previous paper[15] and rigorously reproduced here for the uncross-linked samples. The dispersing agents, coating (OCTEO) and coupling (Si69) additives are in a liquid form at room temperature. They are diluted at 7% v/v in methanol

(MeOH), and this solution is then introduced in the Ludox–DMAc suspension in the nominal quantity of 1 or 4 triethoxysilane groups per nm^2 of silica for OCTEO and 1/2 or 1 triethoxysilane groups per nm^2 of silica for Si69. Quantities were only adapted to both the small amount of Batch 1 and the following SANS experiment conditions (beam size). Uncross-linked samples consist of malleable bulks of about 50 mg, which can be fixed in a calibrated cell.

Cross-linked samples were synthesized with deuterated chains from Batch 2. The modifications relative to uncross-linked samples are the following: five minutes before starting the casting phase, a cross-linking mixture is added to the polymer/silica/DMAc mixture at 130 °C; soluble sulfur as the cross-linker (2% wt with matrix) and CBS as the vulcanization accelerator (2% wt with matrix). These cross-linking agents are fine powders under normal conditions, and their solubility was previously checked in pure DMAc at 130 °C. In the following phase, mixtures are poured into round Teflon molds (diameter = 50 mm, thickness = 10 mm) and let cast in an oven at a constant temperature T_{cast} = 130 °C under nitrogen flow for 4 hours. During this phase both solvent casting and crosslinking are taking place. This yields dry films of about 1 mm thickness, which do not creep at rest and under moderate traction.

2.1. Small angle neutron scattering (SANS). The composition of the matrix in all samples (54% v h-SBR + 46% v d-SBR) should reach the zero average contrast condition in the SANS experiments. In this case, the scattering is directly proportional to the form factor of a single chain. SANS measurements in ZAC conditions were performed on two spectrometers. Uncross-linked systems were measured at rest at the Laboratoire Léon Brillouin (LLB Saclay) on the spectrometer PAXE, by varying the wavelength λ and the sample-to-detector distance D. The three configurations defined by $\lambda = 4$ Å/$D = 1$ m, $\lambda = 6$ Å/$D = 5$ m and $\lambda = 17$ Å/$D = 5$ m cover a total Q-range from 3×10^{-3} to 5×10^{-1} Å^{-1}. Samples were measured in calibrated quartz cells. The 2D patterns were reduced to 1D spectra $I(Q)$ after radial averaging around the center of scattering. Standard corrections by sample thickness, neutron beam transmission, empty cell signal subtraction, detector efficiency and subtraction of incoherent scattering were applied to get the scattered intensities on an absolute scale. Cross-linked systems were measured at the Institut Laue Langevin (ILL Grenoble) on the spectrometer D11. The three configurations defined by $\lambda = 8$ Å/$D = 1.2$ m, $\lambda = 8$ Å/$D = 5$ m and $\lambda = 8$ Å/$D = 20$ m cover a total Q-range from 3×10^{-3} to 4×10^{-1} Å^{-1}. Samples were fixed in a home-made traction device allowing us to perform measurements at a fixed traction rate $\zeta = L/L_0$, ranging from 1 (sample at rest) to 1.8. The resulting 2D SANS patterns were anisotropic and radially averaged in 30° sectors along the parallel and the perpendicular direction of stretching.

2.2. Small angle X-ray scattering (SAXS) and ultra small angle X-ray scattering (USAXS). In SAXS experiments, we measure the silica scattering in filled systems due to the electronic contrast existing between SBR and silica: the scattering length densities are respectively $\rho_{SBR} = 8.41 \times 10^{10}$ cm^{-2} and $\rho_{silica} = 17.36 \times 10^{10}$ cm^{-2}, calculated from the chemical composition, giving the electronic contrast $\Delta\rho = 8.95 \times 10^{10}$ cm^{-2}. Uncross-linked systems were measured in their glass cell, at SOLEIL (Saclay) on the beam-line SWING, with a single configuration defined by a wavelength of $\lambda = 1.77$ Å ($E = 7$ keV) and a sample-to-detector distance of 6.5 m. As the center of scattering did not coincide with that of the detector, this single configuration yields a remarkably large total Q-range from 1×10^{-3} to 1.6×10^{-1}

Å^{-1}, by merging two measurements with two different beam-stops (a small one and a large one). Standard data treatments include merging, subtraction and normalization. Cross-linked systems were measured at ESRF on the beam-line ID02, with a wavelength of 1 Å (E = 12.4 keV) and two sample-to-detector distances of 1 and 10 m, yielding a total Q range from 1×10^{-3} to 5×10^{-1} Å^{-1}. On the same beam-line, the Bonse–Hart setup allowed us to reach the lower Q value of 3×10^{-4} Å^{-1}. SAXS measurements were performed exclusively at rest, after accurate measurement of the thicknesses of the samples. Standard treatments of data include isotropic averaging of the 2D isotropic scattering patterns, merging and normalization. After measuring the scattering of the unfilled system, the pure silica scattering in nanocomposites of silica volume fraction φ_{SiO_2} is obtained by subtracting the matrix contribution for each sample, according to the following operation: $I_{SiO_2} = I_{sample} - (1 - \varphi_{SiO_2}) \times I_{matrix}$.

2.3. Transmission electronic microscopy (TEM). For TEM characterization, samples were cut by ultracryomicrotomy (mechanical cutting) at −80 °C (lower than the glass transition temperature of SBR) at a desired thickness of about 50 nm, then laid on a TEM grid before observation. Observations were made using a Philips CM200 LaB6 (200 kV) microscope in bright field mode.

III. Results and discussion

1. Chain conformation

The scattering intensity of a single chain in a nanocomposite can be obtained directly by SANS using the zero average contrast (ZAC) method. This method was first used for a mixture of normal, *i.e.* non deuterated (h), and deuterated (d) chains in a solvent; the total scattering intensity $I(Q)$ can be expressed by the relation:[38]

$$I(Q) = (\rho_d - \rho_h)^2 \phi_h \phi_d \nu \Phi N P(Q) + (\phi_d \rho_d + \phi_h \rho_h - \rho_0)^2 [\nu \Phi N P(Q) + V \Phi^2 S(Q)] \quad (1)$$

where ρ_d, ρ_h and ρ_0 are respectively the scattering length density of the deuterated chains, hydrogenated chains and solvent, ϕ_h, ϕ_d are the volume fractions of the hydrogenated and deuterated chains, ν is the volume of the monomer, Φ is the volume fraction of the polymer, V is the global volume, $P(Q)$ is the form factor of the chain, and $S(Q)$ is the inter-chain structure factor. When the average contrast between polymer mixture and solvent is adjusted to zero, *i.e.*:

$$\phi_d \rho_d + \phi_h \rho_h - \rho_0 = 0 \quad (2)$$

the scattering intensity is only related to the form factor of one chain:

$$I(Q) = (\rho_d - \rho_h)^2 \phi_h \phi_d \nu \Phi N P(Q) \quad (3)$$

In our case, the role of the solvent is played by the silica particles. The scattering length density has been determined by previous contrast variation experiments[39] in solvent and found to be equal to 3.41×10^{10} cm^{-2}. For our system, the matching composition to satisfy eqn (2) is achieved by mixing 54% v/v of hydrogenated SBR chains (SLD = 0.59×10^{10} cm^{-2}) and 46% v/v of deuterated SBR chains (SLD = 6.74×10^{10} cm^{-2}). When both hydrogenated and deuterated chains have equivalent molecular weights and there is no interaction between the monomer, for ideal mixtures, the form factor of a single chain corresponds to a Gaussian chain in a melt whose scattering function can be described by the Debye function:[40]

$$P(Q) = 2\Big/\left(Q^2 R_g^{\,2}\right)^2 \left(e^{-Q^2 R_g^{\,2}} - 1 + Q^2 R_g^{\,2}\right) \tag{4}$$

where R_g is the radius of gyration of the chain. However mixtures of hydrogenated and deuterated chains are usually not ideal and one needs to introduce a monomeric interaction parameter (Flory–Huggins) χ. Eqn (3) is extended to the random phase approximation (RPA):[41]

$$\frac{1}{I(Q)} = \frac{1}{\Delta\rho^2 \phi_h N_h \nu_h P_h(Q)} + \frac{1}{\Delta\rho^2 \phi_d N_d \nu_d P_d(Q)} - \frac{2\chi}{\Delta\rho^2 V} \tag{5}$$

where $\phi_{d,h}$, $N_{d,h}$, $\nu_{d,h}$, $P_{d,h}(Q)$ are the volume fraction, the number of monomeric units per chain, the volume of monomeric units and the form factor of the chains respectively for the hydrogenated and deuterated SBR polymers. V is the average monomer volume $= \phi_h\nu_h + \phi_d\nu_d$. The molecular weight determination using the SEC experiments permits the extraction of the $N_{d,h}\nu_{d,h}$ product as $M_{wd,h} = dN_A N_{d,h}\nu_{d,h}$ where d is the density in g cm^{-3} and N_A is the Avogadro number. As the two monomers have the same structure, the number of parameters can be simplified according to $R_{gd} = R_{gh}(M_{wd}/M_{wh})^{1/2}$.

1.1. Uncross-linked systems. Fig. 1 presents the normalized SANS curve obtained for the pure h/d SBR uncross-linked mixture (Batch 1) under the ZAC condition, without filler or additives. The curve is perfectly fitted with the RPA function that enables the extraction of the following parameters with an uncertainty equal to 5%: $R_{gh} = 7.00$ nm and $\chi = 6 \times 10^{-4}$. The radius of gyration of the d chains can be deduced from the molecular mass ratio and is equal to $R_{gd} = 8.30$ nm. We now consider the situation where we added increasing amounts of filler to the h/d polymer matrix first without dispersing agents. The SANS curves are presented in Fig. 2 for 1, 10 and 15% vol of silica. We can observe that down to a Q value of 0.01 Å^{-1}, all the curves superimpose perfectly with the pure h/d matrix.

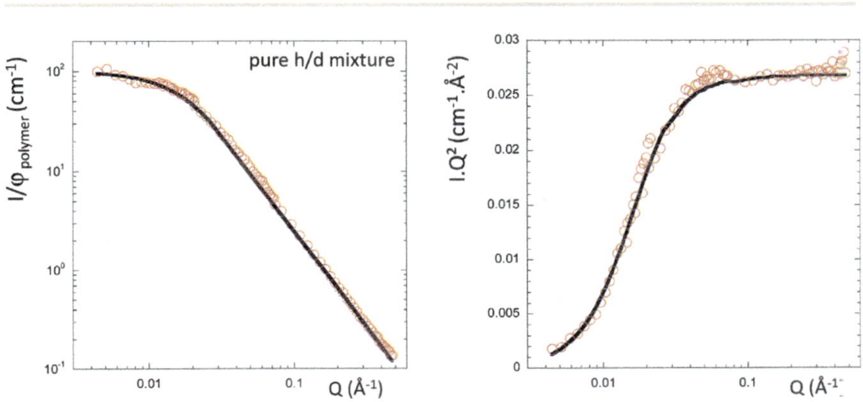

Fig. 1 SANS curves for the pure h/d SBR chain mixtures at the ZAC composition in log–log representation (left) and IQ^2 *versus* Q Kratky representation (right). The full line is the best fit obtained using the RPA function with the following parameters $R_{gh} = 7.00$ nm and $\chi = 6 \times 10^{-4}$. The error bar on the R_g determination is ±2%.

Below this Q value, one can see an upturn of the low Q scattering that will be discussed in detail below. We performed the SANS analysis with the RPA function (full black lines) by fixing the χ parameter to the h/d blend value and fitted the R_{gh} to the height of the Kratky plateau (only one fitting parameter). The resulting fitting R_{gh} values are reported in Table 1.

We now consider the influence of the dispersing additive using the coating agent OCTEO that can covalently bind to the surface of the silica and is known to improve the compatibility with the polymer. The different series containing respectively 1 and 4 equivalent (1 and 4 TEOS groups per nm^2 of silica) of coating agent have been probed with SANS for 1, 10 and 15% vol of silica filler (Fig. 3). We previously double-checked that the additives do not contribute to the SANS scattering by comparing the pure h/d blend with the same blend with the same amount of OCTEO or Si69. No specific scattering coming from the additive can be detected with SANS.

As for the case of silica without additives, the scatterings follow the pure blend up to 0.01 $Å^{-1}$ and also reveal an upturn for the low Q region. This effect is especially pronounced when increasing the amount of coating agent (O_4) for high filler contents of 10 and 15% vol of silica. This is illustrated by a decrease of the agreement between the experimental data and the RPA calculation, which is particularly visible in the Kratky representation.

A similar trend is observed with 1/2 and 1 equivalents of the coupling agent Si69 (Fig. 4) that can be covalently bonded to the particles and the polymer chains. Up to 0.01 $Å^{-1}$, the scattering again follows the blend signal but an upturn appears at low Q that disturbs the agreement between the experimental data and the RPA calculation. We performed the RPA analysis by fixing the Flory parameter to 6×10^{-4} and fitted the R_g of the h chains. Results are reported in Table 1 for the different experimental conditions.

1.2. Cross-linked systems. Fig. 5 presents the SANS curves obtained with the cross-linked samples (Batch 2). We notice that crosslinking modified the h/d blend with the appearance of a peak visible in the Kratky representation. This

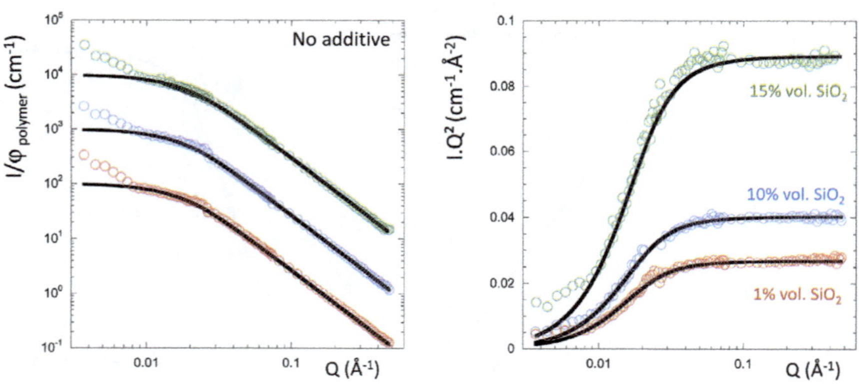

Fig. 2 SANS curves for the silica filled h/d SBR chain mixtures at the ZAC composition without additives for 1% vol (red dots), 10% vol (blue dots) and 15% vol of silica (green dots) in log–log (left) and Kratky (right) representation. The full black lines are the best fit obtained with the RPA function (see text). The curves have been scaled for clarity.

Table 1 R_{gh} values deduced from the RPA analysis by fixing χ to the initial blend value 6×10^{-4}. The uncertainty is ±5%

	Pure h/d blend R_{gh} (nm)	1% vol SiO_2 R_{gh} (nm)	10% vol SiO_2 R_{gh} (nm)	15% vol SiO_2 R_{gh} (nm)
No additive	7.00	7.00	7.00	6.65
Coating agent O_1	7.00	7.05	6.60	6.80
Coating agent O_4	7.00	7.25	7.00	6.75
Coupling agent $Si_{1/2}$	7.00	6.95	6.85	6.90
Coupling agent Si_1	7.00	6.75	6.70	6.65

peak is related to an intra-chain correlation induced by the crosslinking process. The mean distance between two junctions in the Gaussian pathway is reduced with a resulting modification of the chain form factor. Such a form factor has already been computed by Bastide *et al.*[42] We choose not to use this formula for the present study as it is beyond the scope of the paper. In addition, RPA is still

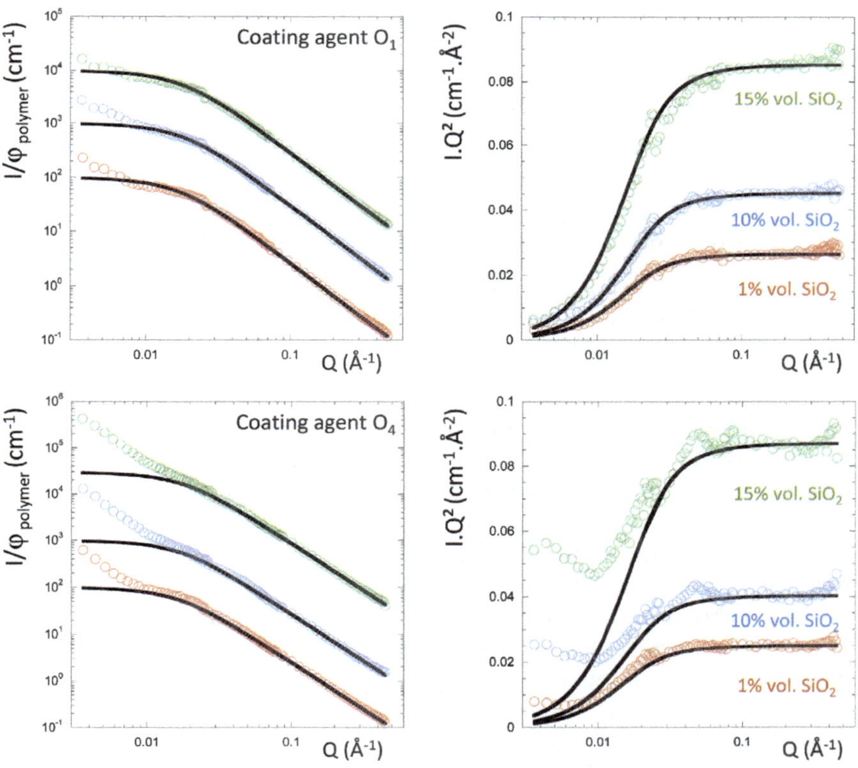

Fig. 3 SANS curves for the silica filled h/d SBR chain mixtures at the ZAC composition with (top) 1 equivalent of coating agent (O_1) and 1% vol (red dots), 10% vol (blue dots) and 15% vol of silica (green dots), and (bottom) with 4 equivalent of coating agent (O_4) and 1% vol (red dots), 10% vol (blue dots) and 15% vol of silica (green dots) in log–log (left) and Kratky (right) representations. The black full lines are the best fit obtained with the RPA function. The curves have been scaled for clarity.

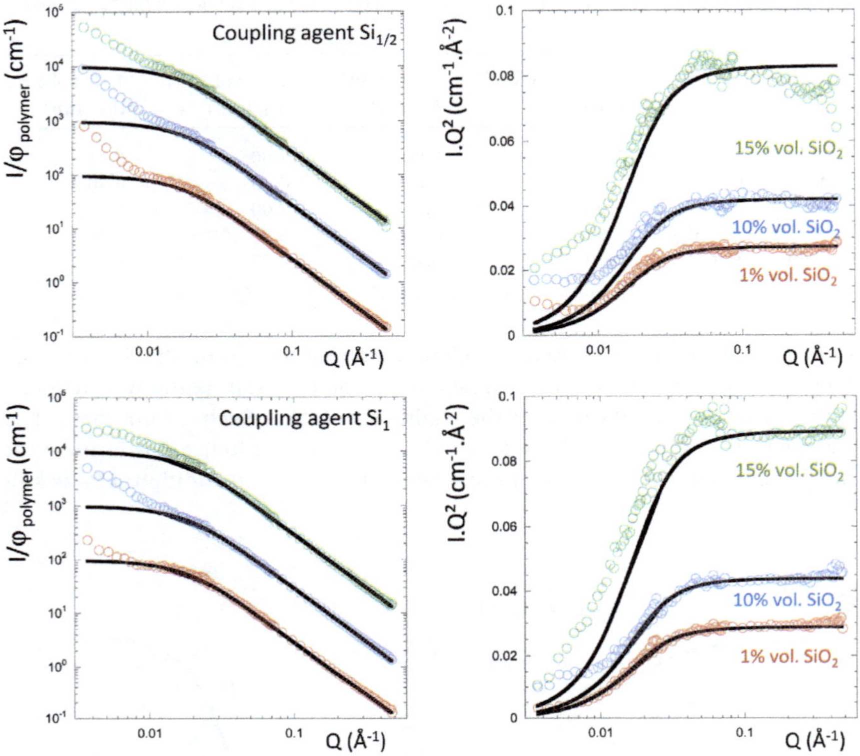

Fig. 4 SANS curves for the silica filled h/d SBR chain mixtures at the ZAC composition with (top) 1/2 equivalent of coupling agent ($Si_{1/2}$) and 1% vol (red dots), 10% vol (blue dots) and 15% vol of silica (green dots), and with (bottom) 1 equivalent of coupling agent (Si_1) and 1% vol (red dots), 10% vol (blue dots) and 15% vol of silica (green dots) in log–log (left) and Kratky (right) representations. The full line is the best fit obtained with the RPA function. The curves have been shifted for clarity.

valuable in the Guinier regime to capture the R_g values and their variations as functions of the filler content, additive and elongation rates. We obtain the following parameters with a larger uncertainty (10%) than for uncross-linked systems (2%): $R_{gh} = 6.80$ nm, $R_{gd} = 10.78$ nm and $\chi = 0$. The value for the h chains is a little bit smaller than that obtained for uncross-linked chains as expected from the intra-chain correlation effects. For the same reason, we can see a drop in the Flory parameter. One gets a higher value for the d chains due to the larger dispersion of the molecular masses for Batch 2. We applied the same analysis in the presence of coating and coupling agents for 1% vol and 5% vol of silica. The best fitted values of R_{gh} are reported in Table 2.

Beyond the range of the error bars, one can see that at rest for $\zeta = 1$, the R_g values of the h chain in the presence of the filler and additives do not seem to be different to the uncross-linked situation. This suggests that cross-linking does not significantly modify the polymer chain conformation in the nanocomposites. We stretched the samples for different elongation rates $\zeta = 1.3$, 1.5 and 1.8 and compared the pure chain deformation with that in the presence of silica (1% and 5% vol) and 1 equivalent of coating and coupling agents. We show only two cases,

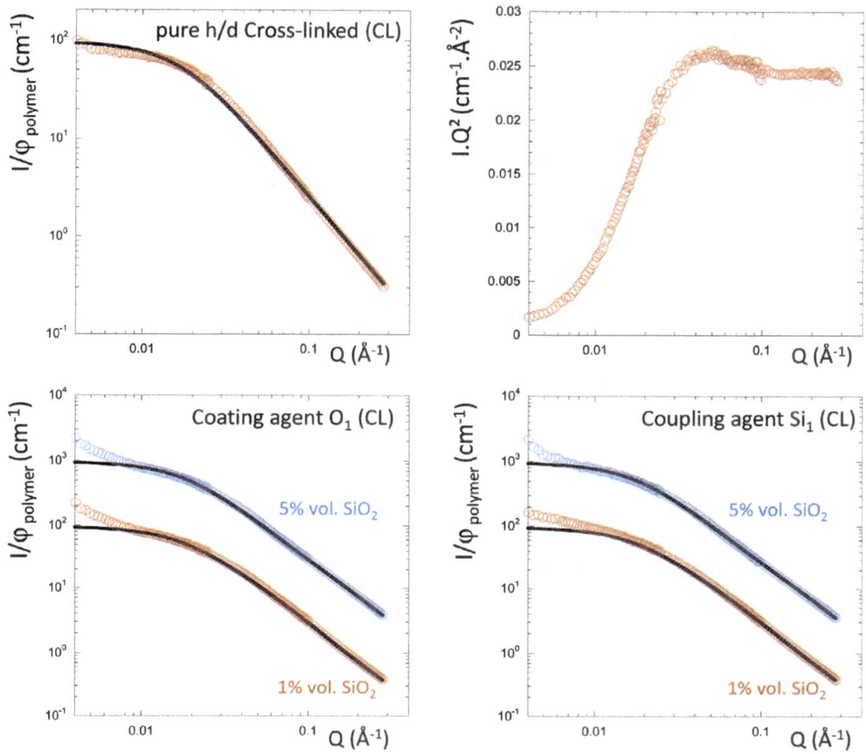

Fig. 5 (top) SANS curves for the pure h/d SBR chain cross-linked (CL) mixtures at the ZAC composition in log–log (left) and Kratky (right) representation. (bottom) SANS curves for silica filled h/d SBR chain cross-linked (CL) composites with (left) 1 equivalent of coating agent (O_1) and 1% vol (red dots) and 5% vol (blue dots) of silica, and (right) 1 equivalent of coupling agent (Si_1) and 1% vol (red dots) and 5% vol (blue dotes) of silica. The full line is the best fit obtained with the RPA function. The curves have been shifted for clarity.

the pure h/d and that with coating O_1 at 5% vol of silica. The results are reported in Fig. 6 for two specific averaging directions, parallel (left) and perpendicular (right) to the stretching direction in log–log and Kratky representations.

We applied the RPA analysis to the elongated samples in the same way as the non-elongated ones. The variations of R_{gh} as a function of the elongation rate are reported in Table 2. Under elongation, the deformation was lost at a large angle (loss of affinity), and the chains typically adopted their initial configuration for distances smaller than the mean distance between two cross-linking junctions.

2. Nanoparticle assembly

We performed SAXS experiments to evaluate the NP dispersion inside the polymer blend. For X-ray radiation the contrast is mainly between silica particles and the polymer matrix (there is no contribution of the labeled h/d chains). The experiments have been made on the same samples prepared for SANS. The SAXS curves are presented in Fig. 7 for the different conditions (no additive, coating and coupling agent) as a function of the silica loading. For comparison, we reported

Table 2 R_{gh} values deduced from the RPA analysis by fixing χ to the initial blend value equal to 0. The uncertainty is ±(nm) 10%

Sample	$\zeta = 1$, $R_{gh\ isotropic}$, (nm)	$\zeta = 1.3$, $R_{gh\parallel} - R_{gh\perp}$, (nm)	$\zeta = 1.5$, $R_{gh\parallel} - R_{gh\perp}$, (nm)	$\zeta = 1.8$, $R_{gh\parallel} - R_{gh\perp}$, (nm)
Pure h/d blend	7.00	7.10–5.90	7.50–5.80	8.20–5.60
Coating O_1 1% vol SiO_2	6.95	7.80–6.10	8.60–5.80	10.00–5.60
Coating O_1 5% vol SiO_2	6.70	7.50–5.70	8.10–5.40	9.50–5.20
Coupling Si_1 1% vol SiO_2	6.60	7.10–5.70	7.90–5.70	—
Coupling Si_1 5% vol SiO_2	7.10	7.70–6.10	8.80–6.00	9.50–5.70

the spherical form factors of the primary particles (full black line) previously determined.[15]

We observe a systematic increase of the low Q scattering suggesting that the primary particles are aggregated inside the polymer matrix and form clusters of

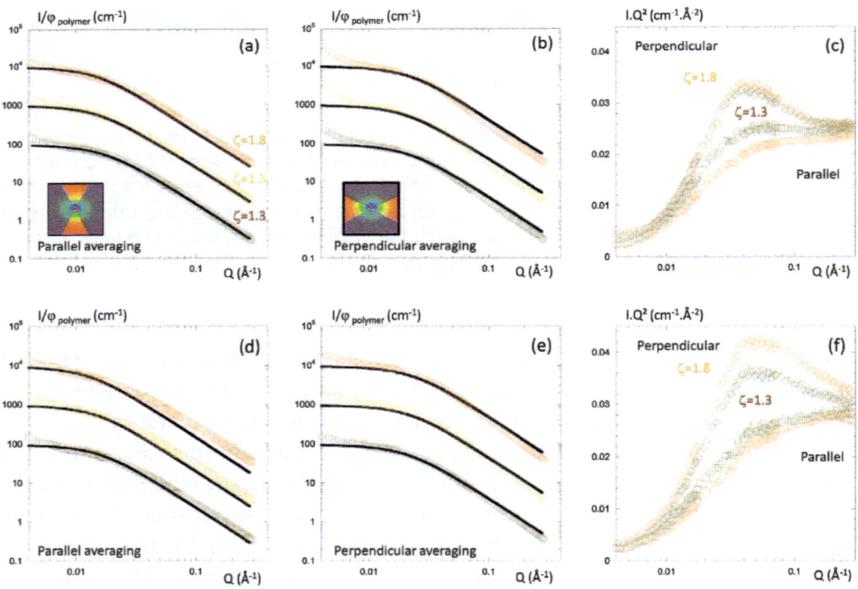

Fig. 6 SANS curves for the cross-linked stretched samples at different elongation ratios $\zeta = 1.3$ (brown), 1.5 (yellow) and 1.8 (orange), (a) pure matrix averaged in the parallel direction, (b) pure matrix averaged in the perpendicular direction, (c) pure matrix averaged in the parallel and perpendicular directions in Kratky representation, (d) composite filled with coating agent O_1 at 5% vol of silica averaged in the parallel direction, (e) composite filled with coating agent O_1 at 5% vol of silica averaged in the perpendicular direction, (f) composite filled with coating agent O_1 at 5% vol of silica averaged in the parallel and perpendicular directions in Kratky representation. The inset pictures illustrate the parallel and perpendicular averaging of the scattering intensity. The full black lines are RPA calculations. The curves in log–log representation have been scaled for clarity.

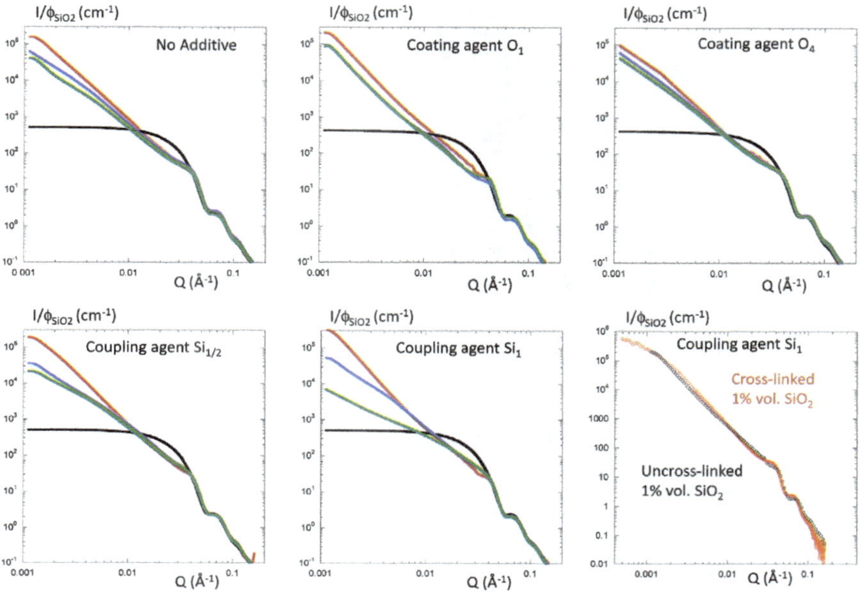

Fig. 7 SAXS curves for the different conditions: (top) no additive and coating agent (O_1 and O_4), (bottom) coupling agent ($Si_{1/2}$ and Si_1) as a function of the silica volume fraction, 1% vol (red line), 10% vol (blue line) and 15% vol (green line). The black line is the form factor of the primary particles calculated using a spherical function convoluted with a log-normal distribution with $R_0 = 7.8$ nm and $\sigma = 0.15$. The bottom right plot illustrates the effect of cross-linking on the NP dispersion. The curves have been normalized by the silica volume fraction.

NPs. The slope of the decrease of the intensity in the intermediate Q range, close to 3, indicates the formation of compact aggregates. We also observe a diminution of the low Q scattering level when increasing the particle volume fraction resulting from an increase of the interaction between the clusters; this increase arises when the clusters form a continuous connected network as a result of a percolation process. The absence of a plateau prevents the determination of the characteristic size of the clusters. The formation of clusters seems to be systematic and not controlled by the nature and quantity of the additives. This can be due to the low molecular weight of the matrix used (40k) that gives rise to a very low viscosity that cannot balance the cluster formation during the solvent casting process. To evaluate the influence of crosslinking on the NP assembly, we compared the SAXS curves for all the samples, uncross-linked and cross-linked. One example is given in Fig. 7 (bottom right). We can see that the scattering superimposed perfectly illustrating that cross-linking does not modify the NP assembly inside the polymer matrix. We completed the analysis with the TEM images presented in Fig. 8. The TEM images confirmed the formation of dense and compact aggregates that are interacting at shorter distances when increasing the silica volume fraction. We can clearly see the transition from individual clusters to the formation of a continuous network visible at 10% vol of silica. To evaluate the size of the clusters, we calculated the radially averaged autocorrelation function $C(r)$ using the image processing software ImageJ.[43] $C(r)$ basically

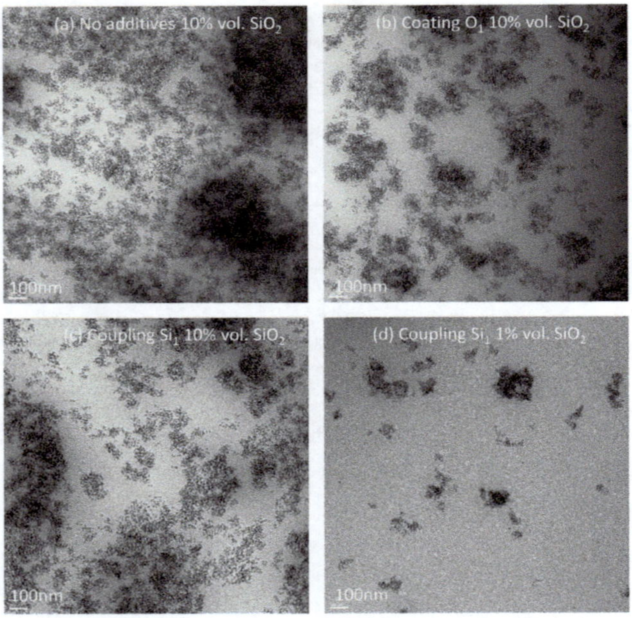

Fig. 8 TEM images for h/d blends filled with (a) no additive at 10% vol of silica, (b) coating agent at 10% vol of silica, (c) coupling agent at 10% vol of silica and (d) coupling agent at 1% vol of silica.

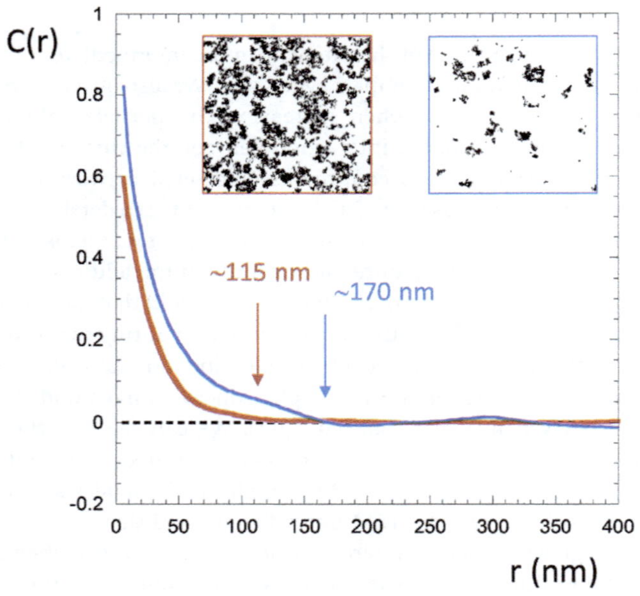

Fig. 9 *C*(*r*) in the TEM images for (red) coating agent O_1 at 10% vol of silica and (blue) coupling agent Si_1 at 1% vol of silica.

calculates the two-point correlation of the pixels of an image, after applying a binary filter, as a function of distance, and gives information about the typical feature size in an image. The intercepted abscissa line with zero gives the cluster size as illustrated in Fig. 9 which are respectively 170 nm and 115 nm for 1 and 10% vol of silica, consistent with the filler network formation.

IV. Discussion

As shown through the different results, we have access to the radius of gyration of the SBR chains for uncross-linked and cross-linked filled nanocomposites at rest and under uniaxial elongation. We can control NP assembly from individual clusters to a continuous filler network with NP loading, coating and coupling additives. We successfully used the RPA function to analyze the SANS data with almost consistent agreement between the model and the experimental data for both uncross-linked and cross-linked systems. However, one can observe for some situations a discrepancy, especially in Fig. 3 and 4. The most probable explanation of this discrepancy is imperfect matching of the silica.

This can be illustrated with the following idea: for imperfect matching, the SANS curve will be modified by a contribution of the NP structure factor. We thus divide the SANS scattering intensity by the chain form factor, calculated with the RPA function, to extract an "apparent" structure factor. We compared this apparent structure factor deduced from SANS with the "effective" NP structure factor that can be calculated by dividing the SAXS intensity by the well-known NP form factor. The structure factors have been determined from the intensities in absolute units and normalized by the X-ray and neutron contrast terms. We report two examples of such a comparison in Fig. 10 for the two series containing 1 (top) and 4 equivalents (bottom) of coating agent. For complete silica matching, the SANS signals remain below the SAXS signals as can be seen for the O_1 series. However, for uncompleted silica matching, the SANS signals become equal to or even larger than the SAXS scattering signals. In addition, we can see the appearance of a peak (in the insets) that corresponds exactly to the NP–NP interactions: the NP in close contact inside the cluster that is visible on both apparent and effective structure factors. There is always a low Q contribution to SANS scattering, beyond the size of the polymer chain, which is enhanced for imperfect matching of the filler. The typical size of this contribution can be evaluated approximately using a simple Guinier function that gives $R_g = 50$ nm and 40 nm for 1% vol and 10% vol of silica. One can evaluate the corresponding sphere equivalent size using the relation ${R_g}^2 = 3/5R^2$ and we obtain typical diameters of 130 nm and 105 nm for respectively 1% and 10% vol of silica. These values are similar whatever the additive used. The interesting feature here is that these values are close to the silica cluster size determined from the TEM analysis in Fig. 9. This means that we can see the large cluster with SANS even if the contrast matching condition is reached. This effect can be due to specific random adsorption of h or d chains at the surface of the particles during sample processing. A recent study has estimated an average grafting density of 0.01 chains per nm^2 corresponding to the adsorbed bound layer in similar NP–polymer systems.[44] This amount, around 1–2 nm adsorbed monolayer, is sufficient to modify the mean scattering neutron contrast of the cluster that makes them visible even under the ZAC condition. An alternative explanation could be that the

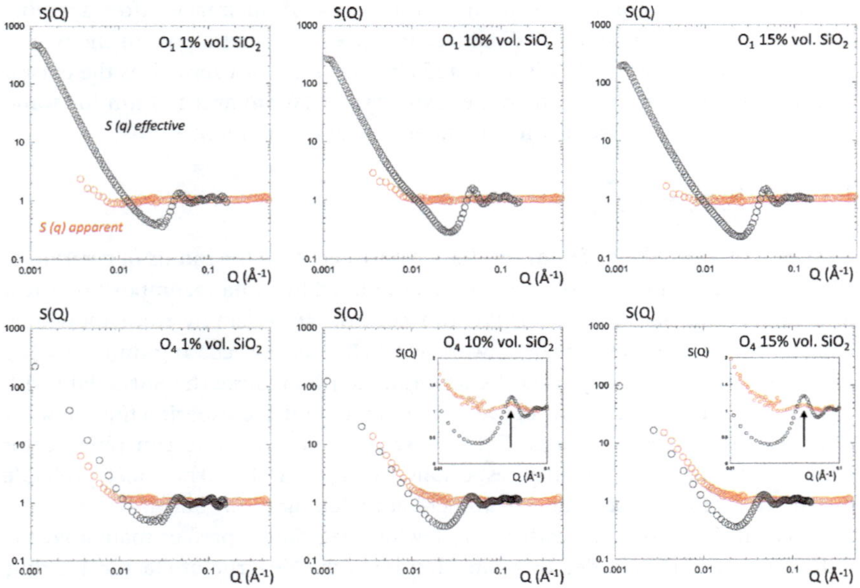

Fig. 10 Comparison between the effective structure factor $S(Q)$ deduced from SAXS and the apparent structure factor deduced from SANS for (top) the coating agent O_1 series and (bottom) the coating agent O_4 series. The insets are magnifications to highlight the NP–NP peak interactions.

NPs induced the formation of h/d phase separation domains of typical sizes close to the silica cluster size ($R_g \sim 30$–50 nm). A rapid simulation shows us that even a very low amount of the domains, typically 0.01% in volume fraction, is enough to create the low Q additional scattering. However, due to the limited amount of data in the low Q regime (5–6 points), it is difficult to proceed further in the quantitative analysis and to clearly conclude between the different explanations. One way to check this idea deeply could be to anneal the sample for a longer time at a high temperature to see whether the low Q contribution is decreasing due to chain desorption or not.

The question of the low Q contribution and its influence on the determination of the polymer chain conformation is always a critical point when considering the chain conformation in the presence of a filler. Here we demonstrate that whatever the matching condition (perfect or not), the low Q contribution, certainly related to the silica cluster, can be treated separately from the determination of the radius of gyration of the polymer chains. As a result, we can follow the evolution of the normalized radius of gyration of the uncross-linked h chains from the values reported in Table 1. The normalized R_{gh} values, the ratios between the filled and unfilled values, are reported in Fig. 11a. This representation shows clearly that the polymer chain conformation at rest is not modified within a range of $\pm 5\%$ whatever the degree of confinement by the filler (managed by the NP loading) and whatever the degree of interaction between the NP and the matrix (the nature and amount of additives). This is the first conclusion of the present study which is in line with several previous conclusions on model systems.[24–26,28–30] The cross-linking of the samples does not modify the NP assembly and the chain

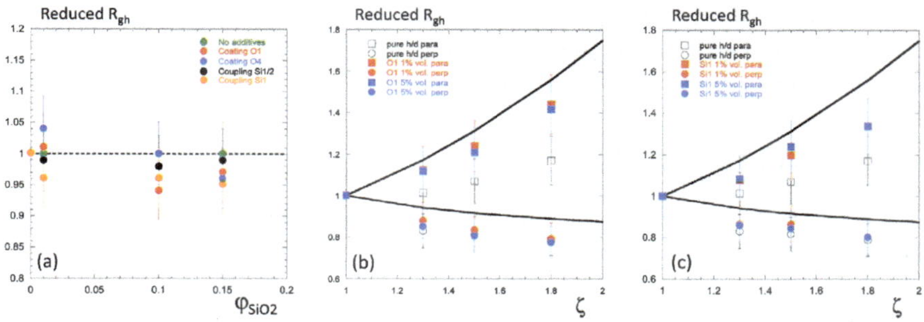

Fig. 11 Reduced radius of gyration of h chains (a) for uncross-linked samples as a function of the silica volume fraction, (b) for cross-linked stretched samples containing 1 equivalent of coating agent for 0, 1 and 5% vol of silica as a function of the elongation rate and (c) for cross-linked stretched samples containing 1 equivalent of coupling agent for 0, 1 and 5% vol of silica as a function of the elongation rate. The full black line is the phantom network deformation model.

conformation. Then one can directly compare the pure chain deformation with that in the presence of filler, coating (Fig. 11b) and coupling agent (Fig. 11c). We also computed the theoretical phantom network model deformation[45] for a functionality $f = 4$ for comparison. On can see that the pure polymer is less deformed compared to the theoretical prediction especially in the parallel direction. This can be due to the chain relaxation during the deformation that can occur for low cross-linking, namely only 3 cross-linking junctions per chain. Surprisingly this effect is reduced in the presence of a filler for which the chain deformation is closer to the theoretical prediction in the parallel direction. The silica clusters can then act as additional cross-linking junctions reducing the chain relaxation and then enhancing the chain deformation. Such conclusions are of great interest for further understanding of the macroscopic mechanical properties of these materials.

V. Summary and conclusions

We report a complete study of the interplay between polymer chain conformations and NP assemblies in model industrial SBR/silica nanocomposites. By mixing d and h chains, we applied the ZAC method with SANS that enables the extraction of the polymer chain form factor while matching the silica filler contribution. SAXS/TEM permits characterization of the NP assembly as a function of NP loading and dispersing agent, coating and coupling additives. NPs form individual clusters at low loading that can connect to form a continuous network at high silica loading. Using RPA analysis, we show that the radius of gyration of the polymer chain remains unchanged – compared to the pure polymer – within a range of ±5% whatever the crosslinking, the degree of confinement induced by NP loading and the degree of interaction between NP and polymer balanced by the nature (coating or coupling) and the amount of the dispersing additives. The extra scattering at low Q, that does not influenced the R_g determination, can be attributed to NP clusters whose neutron contrast is enhanced from ZAC conditions by trapped h/d chains due to random chain adsorption at the surface of the NPs. Under stretching, the NP acts as an additional cross-linked junction preventing the chain relaxations that are deformed at a larger amplitude than the pure polymer.

Acknowledgements

We thank G. Baeza, A.-C. Genix and J. Oberdisse (LCC, Montpellier) for fruitful discussions.

References

1 J. Jancar, J. F. Douglas, F. W. Starr, S. K. Kumar, P. Cassagnau, A. J. Lesser, S. S. Sternstein and M. J. Buehler, *Polymer*, 2010, **51**(15), 3321–3343.

2 S. C. Tjong, *Mater. Sci. Eng., R*, 2006, **53**, 73–197.

3 P. Akcora, H. Liu, S. K. Kumar, J. Moll, Y. Li, B. C. Benicewicz, L. S. Schadler, D. Acehan, A. Z. Panagiotopoulos, V. Pryamitsyn, V. Ganesan, J. Ilavsky, P. Thiyagarajan, R. H. Colby and J. Douglas, *Nat. Mater.*, 2009, **8**, 354–359.

4 C. Chevigny, N. Jouault, F. Dalmas, F. Boué and J. Jestin, *J. Polym. Sci., Part B: Polym. Phys.*, 2011, **49**(11), 781–791.

5 D. Sunday, J. Ilavsky and D. L. Green, *Macromolecules*, 2012, **45**, 4007–4011.

6 N. Jouault, P. Vallat, F. Dalmas, S. Said, J. Jestin and F. Boué, *Macromolecules*, 2009, **42**(6), 2031–2040.

7 M. Tatou, A.-C. Genix, A. Imaz, J. Forcada, A. Banc, R. Schweins, I. Grillo and J. Oberdisse, *Macromolecules*, 2011, **44**(22), 9029–9039.

8 A.-S. Robbes, F. Cousin, F. Meneau, F. Dalmas, F. Boué and J. Jestin, *Macromolecules*, 2011, **44**(22), 8858–8865.

9 M. Klüppel and G. Heinrich, *Rubber Chem. Technol.*, 1995, **68**, 623–651.

10 J. Ramier, C. Gauthier, L. Chazeau, L. Stelandre and L. Guy, *J. Polym. Sci., Part B: Polym. Phys.*, 2007, **45**(3), 286–298.

11 L. Conzatti, G. Costa, M. Castellano, A. Turturro, F. M. Negroni and J. F. Gerard, *Macromol. Mater. Eng.*, 2008, **293**(3), 178–187.

12 P. Mele, S. Marceau, D. Brown, Y. de Puydt and N. D. Alberola, *Polymer*, 2002, **43**(20), 5577–5586.

13 K. W. Stockelhuber, A. S. Svistkov, A. G. Pelevin and G. Heinrich, *Macromolecules*, 2011, **44**(11), 4366–4381.

14 G. P. Baeza, A.-C. Genix, C. Degrandcourt, L. Petitjean, J. Gummel, M. Couty and J. Oberdisse, *Macromolecules*, 2013, **46**(1), 317–329.

15 A. Bouty, L. Petitjean, C. Degrandcourt, J. Gummel, P. Kwasniewski, F. Meneau, F. Boué, M. Couty and J. Jestin, *Macromolecules*, 2014, **47**(15), 5365–5378.

16 Y. Termonia, *Polymer*, 2009, **50**, 1062–1066.

17 B. J. Sung, R. Chang and A. Yethiraj, *J. Chem. Phys.*, 2009, **130**, 124908.

18 M. Vacatello, *Macromol. Theory Simul.*, 2002, **11**, 757–765.

19 M. Vacatello, *Macromolecules*, 2002, **35**, 8191–8193.

20 M. A. Sharaf, A. Kloczkowski, T. Z. Sen, K. I. Jacob and J. E. Mark, *Eur. Polym. J.*, 2006, **42**, 796–806.

21 R. C. Picu and M. S. Ozmusul, *J. Chem. Phys.*, 2003, **118**, 11239–11248.

22 M. S. Ozmusul, R. C. Picu, S. S. Sternstein and S. K. Kumar, *Macromolecules*, 2005, **38**, 4495.

23 A. I. Nakatani, W. Chen, R. G. Schmidt, R. V. Gordon and G. C. Han, *Polymer*, 2001, **42**, 3713–3722.

24 M. E. Mackay, A. Tuteja, P. M. Duxbury, C. J. Hawker, B. van Horn, Z. Guan, G. Chen and R. S. Krishnan, *Science*, 2006, **311**, 1740–1743.

25 A. Botti, W. Pyckhout-Hintzen, D. Richter, V. Urban, E. Straube and J. Kohlbrecher, *Polymer*, 2003, **44**, 7505–7512.

26 S. Sen, Y. Xie, S. K. Kumar, H. Yang, A. Bansal, D. L. Ho, L. Hall, J. B. Hooper and K. S. Schweizer, *Phys. Rev. Lett.*, 2007, **98**, 128302.

27 K. Nusser, S. Neueder, G. J. Schneider, M. Meyer, W. Pyckhout-Hintzen, L. Willner, A. Radulescu and D. Richter, *Macromolecules*, 2010, **43**, 98379847.

28 A.-C. Genix, M. Tatou, A. Imaz, J. Forcada, R. Schweins, I. Grillo and J. Oberdisse, *Macromolecules*, 2012, **45**(3), 1663–1675.

29 N. Jouault, F. Dalmas, S. Said, E. Di-Cola, R. Schweins, J. Jestin and F. Boué, *Macromolecules*, 2010, **43**(23), 9881–9891.

30 A. Tuteja, P. M. Duxbury and M. E. Mackay, *Phys. Rev. Lett.*, 2008, **100**, 077801.

31 S. Sen, Y. Xie, A. Bansal, H. Yang, K. Cho, L. S. Schadler and S. K. Kumar, *Eur. Phys. J.: Spec. Top.*, 2007, **141**, 161–165.

32 M. K. Crawford, R. J. Smalley, G. Cohen, B. Hogan, B. Wood, S. K. Kumar, Y. B. Melnichenko, L. He, W. Guise and B. Hammouda, *Phys. Rev. Lett.*, 2013, **110**, 196001.

33 D. Long and P. Sotta, *Macromolecules*, 2006, **39**, 6282.

34 S. Dupres, D. Long, P. A. Albouy and P. Sotta, *Macromolecules*, 2009, **42**, 2634.

35 S. Westermann, M. Kreitschmann, W. Pyckhout-Hintzen, D. Richter, E. Straube, B. Farago and G. Goerigk, *Macromolecules*, 1999, **32**, 5793.

36 E. Straube, V. Urban, W. Pyckhout-Hintzen, D. Richter and C. J. Glinka, *Phys. Rev. Lett.*, 1995, **74**, 4464.

37 N. Jouault, F. Dalmas, S. Said, E. Di Cola, R. Schweins, J. Jestin and F. Boué, *Phys. Rev. E: Stat., Nonlinear, Soft Matter Phys.*, 2010, **82**, 031801.

38 M. Nierlich, C. E. Williams, F. Boue, J.-P. Cotton, M. Daoud, B. Farnoux, G. Jannink, C. Picot, M. Noan, C. Wolff, M. Rinaudo and P. G. de Gennes, *J. Phys.*, 1979, **40**, 701–704.

39 C. Chevigny, D. Gigmes, D. Bertin, J. Jestin and F. Boué, *Soft Matter*, 2009, **5**(19), 3741–3753.

40 J. S. Higgins and H. Benoit, *Polymers and neutron scattering*, Oxford University Press, New York, 1994.

41 P. G. de Gennes, *Scaling, Concepts in Polymer Physics*, Cornell University Press, Ithaca, NY, 1979.

42 J. Bastide, J. Herz and F. Boué, *J. Phys. France*, 1985, **46**(11), 1967–1979.

43 N. Jouault, J. F. Moll, D. Meng, K. Windsor, S. Ramcharan, C. Kearney and S. K. Kumar, *ACS Macro Lett.*, 2013, **2**, 371–374.

44 N. Jouault, D. Lee, D. Zhao and S. K. Kumar, *Adv. Mater.*, 2014, **26**(24), 4031–4036.

45 H. M. James and E. Guth, *J. Chem. Phys.*, 1943, **11**, 455–481.

Faraday Discussions

PAPER

Circular extinction of plasmonic silver nanocaps and gas sensing†

Jun Li[abc] and Nicholas. A. Kotov*[abde]

Received 7th September 2015, Accepted 6th November 2015
DOI: 10.1039/c5fd00138b

Chiral plasmonic nanostructures exhibit strong rotatory optical activity and are expected to enrich the field of metaoptical materials. Potential applications of chiroplasmonic nanostructures include circular polarizers, optical polarization detectors, asymmetric catalysts, and sensors. However, chiral plasmonic materials require subwavelength structural control and involve laborious chemical or lithographic procedures for their manufacturing. Moreover, strong rotatory activity of subwavelength structures whose chirality was imparted by microfabrication, has been obtained for the red and infrared parts of the spectrum but faces new challenges for the blue and violet spectral ranges even with plasmonic materials with plasmonic bands in the 200–400 nm window. In this study, we address this problem by preparing chiral subwavelength nanostructures by glancing angle sputtering of metallic silver on ZnO nanopillar arrays. Silver deposition in two different planes is a convenient method for preparation of silver chiroplasmonic nanocaps (Ag CPNCs) with controlled asymmetry. Circular dichroism spectroscopy was used to examine the circular extinction for the left-handed nanocaps (L-CPNCs) with understanding that not only circular dichroism but also many other optical effects contribute to the amplitude of these bands. The pillared silver films exhibit circular extinction in the violet area of the electromagnetic spectrum. Partial oxidation of Ag to Ag_xO causes the absorption and corresponding circular extinction band obtained using a conventional CD spectrometer at 400–525 nm to increase and shift. This optical material may be used to detect oxygen and extends the spectrum of application of chiroplasmonic materials to gas sensing.

Chiral plasmonic nanostructures exhibit strong optical circular dichroism (CD) and other modes of light-matter interaction activity over a broad range of the

[a]Department of Chemical Engineering, University of Michigan, Ann Arbor, MI 48109, USA. E-mail: kotov@umich.edu
[b]Biointerfaces Institute, University of Michigan, Ann Arbor, MI 48109, USA
[c]School of Chemical Engineering and Technology, State Key Laboratory of Urban Water Resource and Environment, Harbin Institute of Technology, Harbin 150001, P. R. China
[d]Department of Materials Science and Engineering, University of Michigan, Ann Arbor, MI 48109, USA
[e]Department of Biomedical Engineering, University of Michigan, Ann Arbor, MI 48109, USA

† Electronic supplementary information (ESI) available. See DOI: 10.1039/c5fd00138b

electromagnetic spectrum.[1] Chiroplasmonic materials are expected to enrich the field of optical materials because of their ability to display unusual optical effects.[2–4] Potential applications of chiral plasmonic nanostructures include circular polarizers,[5] detectors for circularly polarized light,[6] asymmetric catalysts,[7] and sensors.[1a,8] These applications will benefit from materials with an anisotropy g factor ($g = \Delta\varepsilon/\varepsilon$, where $\Delta\varepsilon$ and ε represent the molar circular dichroism and molar extinction, respectively) larger by 2–3 orders of magnitude than those of chiral biomolecules.[9] Therefore, chiral materials based on individual nanoparticles (NPs),[10] or their assemblies,[11] have attracted much attention because of the tunability of their absorption bands and strong rotatory activity. Plasmonic NP systems are especially promising in this respect because the extinction of plasmons can yield chiral structures with exceptionally high g values and exhibit exceptionally high polarization rotation power.[12]

In our previous study, we described the preparation of chiral gold (Au) nanoscale caps on zinc oxide (ZnO) nanopillars without the use of chiral templates, biological molecules subject to environmental decomposition, or multistep lithography. The method utilizing metal deposition on nanopillars is simple and applicable to a variety of different sized substrates.[13] We were able to obtain chiroplasmonic films with exceptionally high rotatory activity and mirror-image circular dichroism (CD) spectra when the chirality of the Au nanocaps was reversed.

Evaluation of the contribution of linear artefacts to the "true" circular optical activity demonstrated that it was small. Nevertheless, a good deal of caution needs to be taken when evaluating apparent CD spectra in anisotropic optical media. In this paper, silver (Ag), which shows greater plasmonic activity and chemical activity than gold, was used to construct chiral silver nanocaps that show a strong chiral response in the violet area. Since we already demonstrated the possibility of both positive and negative rotatory power when changing the glancing angle of metal deposition, this study will be devoted to better understanding how different the optical and chemical properties of chiral plasmonic films can be when one sputtered material is replaced with another using only left handed nanocaps. Distinct resonance between surface plasmonic electrons of the metal and oscillating charge carriers in this structure was determined, and this chiral plasmonic Ag nanostructure appeared promising for use as a detector of trace oxygen in special environments.

To obtain the chiral Ag nanostructures, we initially grew ZnO hexagonal nanopillars onto glass substrates *via* hydrothermal synthesis[14] (Fig. 1a). Then, the first Ag layer was evaporated on the ZnO nanopillars. Note that the flux of Ag atoms was inclined with respect to the direction of the ZnO nanopillars, 315° of the X axis, in order to have one-sided deposition (Fig. 1b). The asymmetric deposition of an approximately 5 nm thick Ag layer covered the entire sample at the same time. After the first deposition from the X axis, the flux of Ag atoms was then rotated to 45° of the Y axis in order to create nanocaps with chiral asymmetry (Fig. 1c). The thickness of the second Ag layer was controlled to be 10 nm. This process produced arrays of nanopillars coated with left-handed chiral plasmonic nanocaps (L-CPNCs), shown as Fig. 1d and e. The high resolution SEM images of single Ag L-CPNCs on the ZnO nanopillars are shown in Fig. 1f and g.

Generally, for high aspect ratios of anisotropic CPNCs, the longitudinal mode of surface plasmon dominates over the transverse mode, which could be expected given better coupling of incident light with the protruding 3D features, and thereby resulting in stronger optical rotation.[15] However, ZnO nanopillars also resulted in excessive light reflection and scattering, which made the chiral testing

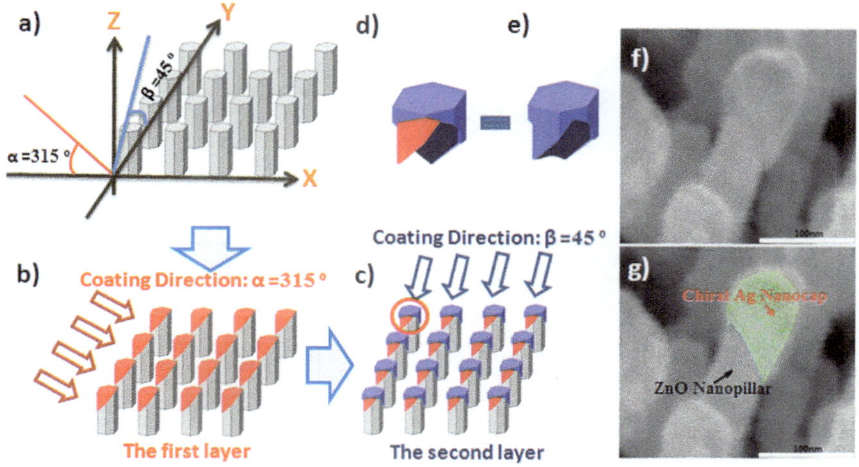

Fig. 1 Schematics of the fabrication process (a–e), SEM images of the obtained chiral Ag nanocaps (f and g).

of the Ag nanocaps difficult. Therefore, in this study, ZnO nanopillars with relatively small aspect ratio of ~4.2 were used. The diameters of the ZnO nanopillars varied from 30 nm to 75 nm, and their average lengths were about 200 nm (Fig. 2).

The optical activity of the Ag L-CPNCs was examined by CD spectroscopy (Fig. 3). Note that in the context of this study and in Fig. 3 we use notation CD as a reference to the method and not to the optical process. Since the nanopillar must exibit multiple optical effects we actually measure here circular extinction.

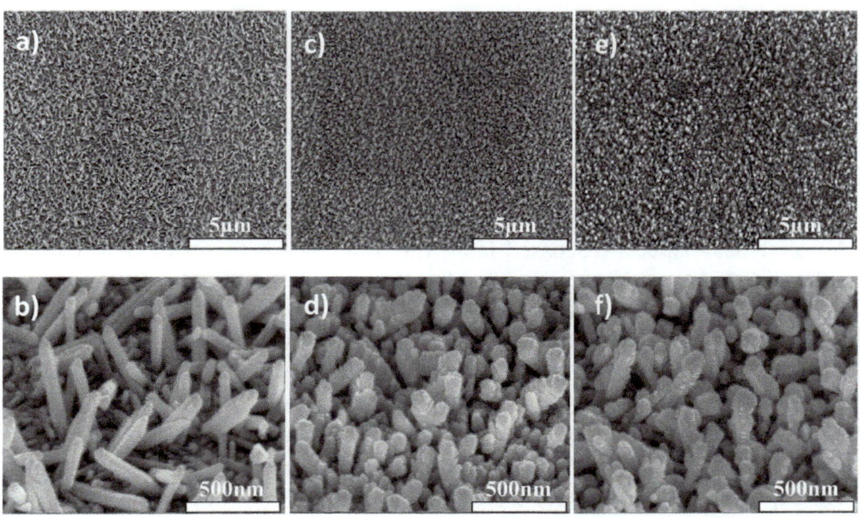

Fig. 2 SEM images of ZnO nanopillars (a and b), ZnO nanopillars coated with Ag L-CPNCs (c and d) and ZnO nanopillars coated with Ag L-CPNCs after exposure to air for 96 h (e and f).

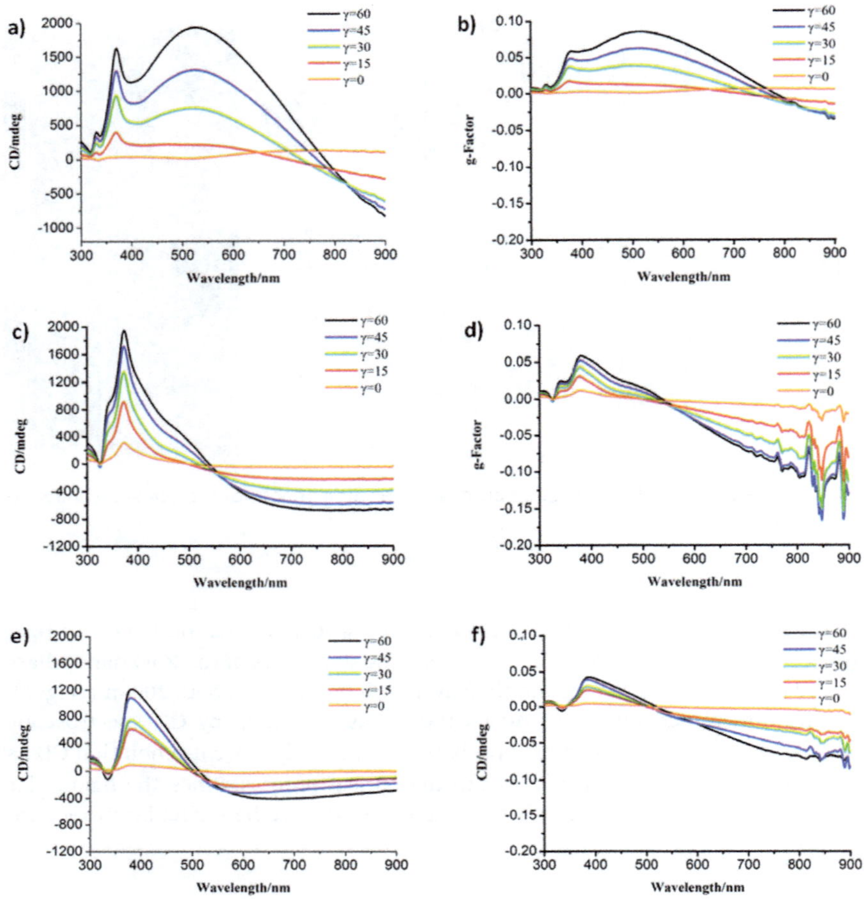

Fig. 3 Circular extinction spectra obtained on a CD spectrometer and the corresponding apparent g-factor curves of ZnO nanopillars coated with Ag: (a and b) as made, (c and d) exposed to air at room temperature for 24 h, (e and f) exposed to air at room temperature for 96 h obtained for different incident angles given in the legends. Note that besides circular effects there are multiple optical processes contributing to the amplitude of the registered peak registered by the conventional CD spectrometer.

For analytical aspects discussed later, the difference between CD and circular extinction is less relevant than for (meta)material optics. Circular extinction spectra obtained for the ZnO nanopillars with different aspect ratios demonstrate that it is possible to manipulate the chiroptical activity of nanopillar-based Ag L-CPNCs in terms of the spectral position and amplitude of the plasmonic band. Since CPNCs are vertically aligned to the substrates and highly anisotropic, one might expect that their orientation with respect to the incident light would affect the chiroptical properties.[16] The optical activities of the Ag L-CPNCs were investigated, while substrates were tilted to an angle of γ as presented in Fig. 3 and Fig. S1.† A change of tilting angle from 0° to 60°strongly enhanced the circular extinction peak maximum of the Ag L-CPNCs. The amplitude of the circular extinction band for the Ag L-CPNCs increased from +20 mdeg to +1980 mdeg; the

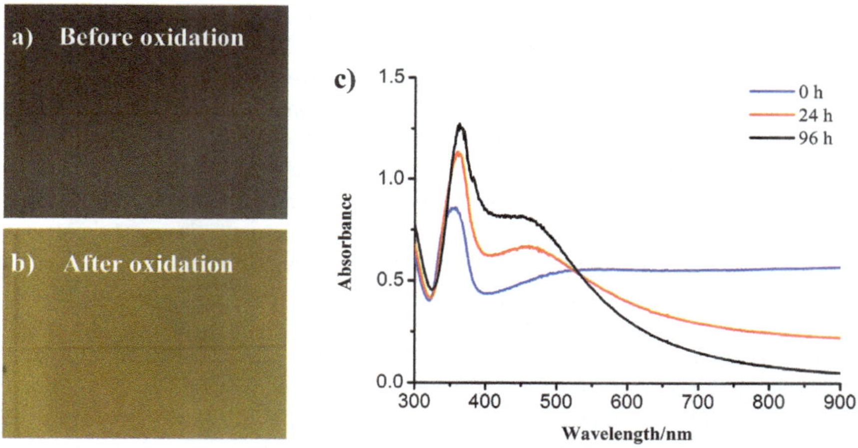

Fig. 4 Photographs of the ZnO nanopillars coated with Ag L-CPNCs as made and after exposure to air for 96 h, respectively (a and b). (c) UV spectra of the samples exposed to air for different lengths of time.

amplitude was nearly 100 times higher as γ changed from 0° to 60°. Meanwhile, the position of the peaks shifted to higher wavelengths compared to the incidence light. An important comment needs to be made here. These spectral effects cannot be attributed completely to "real" circular dichroism. The nanopillar arrays represent complex optical media especially for non-zero angles of light incidence and both circular and linear effects need to be taken into account here. Ultimately, the distinction between linear and circular effect, polarization rotation and retardation must be resolved using Mueller matrix polarimetry or a similar method.

As expected, the chiral Ag nanocaps have the ability to impart chiroptical activity to the achiral ZnO nanopillars, as indicated by the appearance of chiroptical peaks at 375 nm corresponding to optical extinction of the ZnO nanopillars (Fig. 3). ZnO nanopillars grown by hydrothermal methods do not display any optical activity on their own. However, when the ZnO nanopillars are covered with chiral Ag nanocaps, the asymmetric electrostatic field generated by the chiral plasmons in the Ag nanocaps couples with oscillation of the charge carriers in the ZnO nanopillars, resulting in chiral absorption of the achiral ZnO nanopillars. Similar phenomena have been observed for achiral nanoparticles when chiral organic molecules were adsorbed onto their surfaces.[17] The difference in the origin of the induced chirality here is that the dissymmetric field originates from chiral Ag nanocaps rather than from natural amino acids or other components. Note that due to limitations of the hydrothermal method, the obtained ZnO arrays were not all parallel and some rods were tilted at different angles to the glass substrate, which could have affected the optical properties by broadening the CD bands. However, the presence of tilted nanorods does not change the interpretation of the circular extinction activity of these films.

The CD instrument response at ~525 nm was attributed to the complex circular extinction of the chiral Ag nanocaps. The band intensity and apparent *g* factor of the Ag nanocaps on the ZnO nanopillars are about 100 times larger than those of the nanocaps dispersed in water, where the distribution is isotropic and is likely (but not necessarily) free from the linear artefacts.

We found that partial oxidation of the chiral Ag nanocaps greatly affected the chiroptical properties of the ZnO nanopillars coated with Ag L-CPNCs. Oxidation was completed by exposing the Ag nanocaps on the ZnO nanopillars to air, at ambient atmospheric pressure, and at room temperature. Fig. 4a and b show photographs of the initial sample and doped sample, respectively. The color of the initial sample changed from dark gray to yellow-gray after 48 h of oxidation. As can be seen from the energy dispersive spectroscopy (EDS) results (Fig. S2 and Table S1†), the content of oxygen in the surface layer of the initial sample increased greatly with time, which indicated that the partial oxidation of the Ag nanocaps had occurred, and part of the Ag had been oxidized to Ag_xO. Note that the oxidation of Ag could have been facilitated by the photocatalytic generation of holes in ZnO. Despite the change of appearance and optical absorption bands, no shape deformation of the ZnO nanopillar arrays and Ag nanocaps was found after the partial oxidation (Fig. 3).

Fig. 4c presents the UV spectra of the ZnO nanopillar arrays coated with Ag L-CPNCs after oxidation for different lengths of time. The absorption at 400–525 nm increased with increasing time indicating the presence of Ag_xO. We also note that the sample displayed small red-shifts of UV-Vis absorption bands at ~360 nm with respect to the initial structure. The red-shift is consistent with changes in the plasmonic bands of the Ag NPs because of the Ag_xO layer, which influences the electromagnetic field of the Ag NPs.

In addition, the formation of Ag_xO causes spectroscopic changes in circular extinction spectra of Ag L-CNCs, as shown in Fig. 4. The circular extinction band at ~525 nm decreased with oxidation time and eventually disappeared while the circular extinction band at ~375 nm did not change much during the oxidation, which means it is possible to manipulate the chiroptical activity of the Ag L-CPNCs in terms of the spectral position and amplitude of the plasmonic band. In our understanding, the circular extinction band at ~525 nm, which is caused by the chiral Ag nanocaps, disappeared because of the generation of Ag_xO. It is well known that Ag_xO has a broad absorbance of 1.2–3.4 eV,[19] or 364–1044 nm. The Ag_xO layer shows no chirality and absorbed most of the incident light before the light interacted with the chiral Ag nanocaps and generated polarization rotation. Therefore, the circular extinction between 364–1044 nm decreased gradually. However, due to coupling of the asymmetric electromagnetic field from the Ag nanocaps with the charge oscillation in the ZnO nanopillars, the apparent CD peak at ~375 nm did not change much. This property of the chiral Ag nanocaps makes it feasible to use them as an oxygen detector, which extends the field of applications of chiroplasmonic materials previously used primarily for biosensing.[20]

In conclusion, we have prepared chiral Ag nanocaps on ZnO nanopillar arrays by using evaporation of Ag with different thicknesses and from different directions. Partial oxidation of the Ag nanocaps caused great changes in the UV and circular extinction spectra registered on a conventional CD spectroscopy instrument of the Ag L-CPNCs, which is suggested to be related to the formation of a Ag_xO structure. This chiroplasmonic surface shows the ability to detect oxygen and extends the means to obtain chiral materials for gas sensing applications. A future extension of this work could be the elucidation of other gas sensing applications, where the gas to be detected changes the shape of the nanocaps, which should change the position of the circular extinction bands.

Note added after first publication

This article replaces the version published on 8th March 2016, which contained errors in the title and the text. Clarifications were made regarding the terms "chiroptical activity", "CD" and "circular extinction" leading to changes in the title and text of the manuscript. Corrections to incorrect figure citations were also made.

Acknowledgements

The central part of this work was supported by the Center for Photonic and Multiscale Nanomaterials (C-PHOM) funded by the National Science Foundation (NSF) Materials Research Science and Engineering Center program DMR 1120923. Partial support of this work was also made by NSF projects 1403777; 1411014; 463474. The authors thank the University of Michigan's EMAL for its assistance with electron microscopy, and for the NSF grant #DMR-9871177 for funding of the JEOL 2010F analytical electron microscope used in this work.

References

1 (*a*) A. Guerrero-Martínez, J. L. Alonso-Gómez, B. Auguié, M. M. Cid and L. M. Liz-Marzán, *Nano Today*, 2011, **6**, 381–400; (*b*) V. K. Valev, J. J. Baumberg, C. Sibilia and T. Verbiest, *Adv. Mater.*, 2013, **25**, 2517–2534.

2 (*a*) S. Tretyakov, I. Nefedov, A. Sihvola, S. Maslovski and C. Simovski, *J. Electromagn. Waves Appl.*, 2003, **17**, 695–706; (*b*) J. B. Pendry, *Science*, 2004, **306**, 1353–1355; (*c*) Z. Li, M. Mutlu and E. Ozbay, *J. Opt.*, 2013, **15**, 023001.

3 U. Gubler and C. Bosshard, *Nat. Mater.*, 2002, **1**, 209–210.

4 R. Zhao, J. Zhou, T. Koschny, E. N. Economou and C. M. Soukoulis, *Phys. Rev. Lett.*, 2009, **103**, 103602.

5 J. K. Gansel, M. Thiel, M. S. Rill, M. Decker, K. Bade, V. Saile, G. Von Freymann, S. Linden and M. Wegener, *Science*, 2009, **325**, 1513–1515.

6 I. Hodgkinson and Q. H. Wu, *Adv. Mater.*, 2001, **13**, 889–897.

7 Y. Tang and A. E. Cohen, *Science*, 2011, **332**, 333–336.

8 E. Hendry, T. Carpy, J. Johnston, M. Popland, R. V. Mikhaylovskiy, A. J. Lapthorn, S. M. Kelly, L. D. Barron, N. Gadegaard and M. Kadodwala, *Nat. Nanotechnol.*, 2010, **5**, 783–787.

9 Y. H. Chen, J. T. Yang and H. M. Martinez, *Biochemistry*, 1972, **11**, 4120–4131.

10 (*a*) T. G. Schaaff and R. L. Whetten, *J. Phys. Chem. B*, 2000, **104**, 2630–2641; (*b*) Y. Zhou, M. Yang, K. Sun, Z. Tang and N. A. Kotov, *J. Am. Chem. Soc.*, 2010, **132**, 6006–6013.

11 W. Chen, A. Bian, A. Agarwal, L. Liu, H. Shen, L. Wang, C. Xu and N. A. Kotov, *Nano Lett.*, 2009, **9**, 2153–2159.

12 (*a*) A. Kuzyk, R. Schreiber, Z. Fan, G. Pardatscher, E. M. Roller, A. Hogele, F. C. Simmel, A. O. Govorov and T. Liedl, *Nature*, 2012, **483**, 311–314; (*b*) Y. Tang and A. E. Cohen, *Phys. Rev. Lett.*, 2010, **104**, 163901; (*c*) M. Kuwata-

Gonokami, N. Saito, Y. Ino, M. Kauranen, K. Jefimovs, T. Vallius, J. Turunen and Y. Svirko, *Phys. Rev. Lett.*, 2005, **95**, 227401.

13 B. Yeom, H. N. Zhang, H. Zhang, J. Il Park, K. Kim, A. O. Govorov and N. A. Kotov, *Nano Lett.*, 2013, **13**, 5277–5283.

14 L. Greene, M. Law, J. Goldberger, F. Kim, J. Johnson, Y. Zhang, R. Saykally and P. Yang, *Angew. Chem., Int. Ed.*, 2003, **42**, 3031–3034.

15 (*a*) B. Auguié, J. L. Alonso-Gómez, A. S. Guerrero-Martínez and L. M. Liz-Marzán, *J. Phys. Chem. Lett.*, 2011, **2**, 846–851; (*b*) P. Mulvaney, *Langmuir*, 1996, **12**, 788–800; (*c*) L. S. Slaughter, Y. Wu, B. A. Willingham, P. Nordlander and S. Link, *ACS Nano*, 2010, **4**, 4657–4666.

16 (*a*) E. Plum, V. A. Fedotov and N. I. Zheludev, *J. Opt. A: Pure Appl. Opt.*, 2009, **11**, 074009; (*b*) A. S. Schwanecke, V. A. Fedotov, V. V. Khardikov, S. L. Prosvirnin, Y. Chen and N. I. Zheludev, *Nano Lett.*, 2008, **8**, 2940–2943.

17 (*a*) W. J. Yan, L. G. Xu, C. L. Xu, W. Ma, H. Kuang, L. B. Wang and N. A. Kotov, *J. Am. Chem. Soc.*, 2012, **134**, 15114; (*b*) H. Yao, K. Miki, N. Nishida, A. Sasaki and K. Kimura, *J. Am. Chem. Soc.*, 2005, **127**, 15536–15543; (*c*) S. D. Elliott, M. P. Moloney and Y. K. Gun, *Nano Lett.*, 2008, **8**, 2452–2457.

18 Z. Fan, H. Zhang and A. O. Govorov, *J. Phys. Chem. C*, 2013, **117**, 14770–14777.

19 (*a*) Y. Gao, D. Shan, F. Cao, J. Gong, X. Li, H. Ma, Z. Su and L. Qu, *J. Phys. Chem. C*, 2009, **113**, 15175; (*b*) H. Y. Chuang and D. H. Chen, *Nanotechnology*, 2009, **20**, 105704.

20 (*a*) W. Chen, A. Bian, A. Agarwal, L. Liu, H. Shen, L. Wang, C. Xu and N. A. Kotov, *Nano Lett.*, 2009, **5**, 2153–2159; (*b*) X. Wu, L. Xu, L. Liu, W. Ma, H. Yin, H. Kuang, L. Wang, C. Xu and N. A. Kotov, *J. Am. Chem. Soc.*, 2014, **49**, 18629–18636; (*c*) M. Ma, H. Kuang, L. Xu, L. Ding, C. Xu, L. Wang and N. A. Kotov, *Nat. Commun.*, 2013, **4**, 2689.

Faraday Discussions

PAPER

$Sn_xTi_{1-x}O_2$ solid-solution-nanoparticle embedded mesoporous silica (SBA-15) hybrid as an engineered photocatalyst with enhanced activity†

N. R. Srinivasan and Rajdip Bandyopadhyaya*

Received 24th August 2015, Accepted 12th October 2015
DOI: 10.1039/c5fd00126a

Synthesis of hybrids of a porous host-material (with well-dispersed embedded nanoparticles inside the pore), wherein each nanoparticle has precisely controlled properties (size and composition) poses a generic challenge. To this end, a new strategy is proposed to form $Sn_xTi_{1-x}O_2$ solid-solution-nanoparticles inside the pores of sphere-like mesoporous silica (SBA-15), with different percentages of Sn in the nanoparticle (varying from 5 to 50 at%), for enhanced photocatalysis. X-ray diffraction confirms the formation of solid-solution nanoparticles in the porous silica hybrid, while the location of nanoparticles and elemental composition are identified using electron microscopy. The hybrid with 5 at% of Sn ($Sn_{0.05}Ti_{0.95}O_2$-sphere-like SBA-15) shows the maximum photocatalytic activity for degradation of rhodamine-B dye (first order rate constant for degradation, $k = 1.86$ h^{-1}), compared to both pure TiO_2-sphere-like SBA-15 ($k = 1.38$ h^{-1}) or pure SnO_2-sphere-like SBA-15 ($k = 0.14$ h^{-1}) or other hybrids in this series. XPS and PL spectra suggest the formation of more oxygen vacancies during the replacement of Ti^{4+} with Sn^{4+}. Electrochemical studies reveal that there is a reduction of charge transfer resistance from 910 kΩ cm^{-2} for TiO_2-sphere-like SBA-15, to 332 kΩ cm^{-2} for $Sn_{0.05}Ti_{0.95}O_2$-sphere-like SBA-15. These results imply that the enhancement in photocatalytic performance is as a result of delay in recombination of charge carriers. Therefore, the approach followed in the present work to form solid-solution nanoparticles inside a porous host without causing pore blockage, would be a promising route towards increasing reaction rates in catalytic applications of hybrid materials.

Introduction

Semiconductor mediated photocatalysts have been extensively investigated to degrade organic pollutants such as toxic dyes, pesticides, pharmaceutical

Department of Chemical Engineering, Indian Institute of Technology Bombay, Powai, Mumbai 400076, India. E-mail: rajdip@che.iitb.ac.in; Fax: +91 22 2572 6895; Tel: +91 22 2576 7209

† Electronic supplementary information (ESI) available: XPS and PL spectra. See DOI: 10.1039/c5fd00126a

compounds *etc.*, present in wastewater.[1–4] This technique is a suitable alternative to treat wastewater without releasing harmful products, as other technologies like adsorption by activated carbon, solvent extraction, ozonation *etc.* produce secondary pollutants that are hazardous.[5,6] Photocatalysis over a semiconductor particle surface is initiated by absorption of a photon having energy equal to or greater than the semiconductor band gap, thereby producing charge carriers such as electrons and holes.[7] Once these charge carriers reach the semiconductor surface, they undergo a series of reactions with adsorbed molecules to degrade the latter completely.[8]

Among semiconductor photocatalysts, TiO_2 is the most promising due to its high photocatalytic efficiency, non-toxicity, and low cost.[9,10] However, the major drawback of TiO_2 is the very fast recombination of its charge carriers (~10 to 100 ns). As a result, only a fraction of them are responsible for reaction with the adsorbed molecules (~100 ns to 10 ms). The remaining charge carriers recombine and release energy in the form of heat or luminescence.[8] Hence, this leads to a significant reduction in photocatalytic efficiency. Thus, TiO_2 is not suitable for industrial applications because of recombination of charge carriers, interparticle aggregation, and reusability issues.

Over the past few years, coupled semiconductor photocatalysts have been in focus, primarily to exploit the synergetic effect arising from the presence of two different electronic structures of two semiconductor materials, compared to only one semiconductor material. To achieve this, the following strategies such as deposition of noble metals (*e.g.*, Ag, Au, Pt, Pd, *etc.*) on the semiconductor particle;[11–13] doping by metal ion (Fe, Cr, Mn, V, *etc.*);[14–16] or non-metal doping (C, N, S, B, F *etc.*) into semiconductor;[1,17–20] and coupled semiconductors (SnO_2/TiO_2, CdS/TiO_2, Bi_2S_3/Bi_2WO_6 *etc.*)[21–23] have been explored so far. For example, coupling of SnO_2 (band gap of bulk $SnO_2 \sim 3.8$ eV) with TiO_2 (band gap of bulk $TiO_2 \sim 3.2$ eV) shows an enhancement in photocatalytic activity, because of a good match in the energy levels of the conduction and valence bands (ECB for $SnO_2 = 0$ V, *vs.* NHE at pH 7; ECB for $TiO_2 = -0.5$ V *vs.* NHE at pH 7).[23] More importantly, photoexcited electrons shuttle from the conduction band of TiO_2 to the conduction band of SnO_2 due to this potential difference, whereas photoexcited holes shuttle from the valence band of SnO_2 to the valence band of TiO_2. This shuttling effect increases the charge separation, and thereby decreases the recombination of charge carriers.[10]

Coupled semiconductor photocatalysts are mostly made either in the form of individual core shell nanoparticles or thin films containing such core–shell nanoparticles to increase their accessibility against diffusing molecules.[24–26] In this regard, the thin film form will increase reusability of the photocatalyst, but results in low photocatalytic activity, because of its lower specific surface area than individual nanoparticles.[26] Similarly, the main challenge in the core–shell structure lies in achieving a uniform layer of shell-coating on the core nanoparticle, and also prevention of particle aggregation. Hence, coupled semiconductor nanoparticles in the form of a solid-solution do not have the issue of a uniform shell-coating, as the two metal ions are dispersed in the same lattice structure.[27–31] Moreover, it also provides the possibility of fine tuning of the band gap energy, by altering composition of the two metal components in the solid-solution. Therefore, it is necessary to focus on pathways to form nanostructured solid-solution photocatalysts.

The other challenge in prevention of interparticle aggregation can be addressed through embedding of the solid-solution nanoparticulate photocatalyst in porous hosts (zeolite, mesoporous silica *etc.*).[32–37] Among these hosts, mesoporous silica (SBA-15) has a uniform, but tunable, mesopore diameter (5–10 nm), high specific surface area (800–1000 m^2 g^{-1}) and high specific pore volume (0.8–1.2 cm^3 g^{-1}),[38] which are ideal features for a catalyst-host. These impart the following advantages to SBA-15 as a host, if used for photocatalysis: (i) the presence of mesopores of large diameter in the SBA-15 particle provides easy passage for efficient transport of reactants and products during photocatalytic reactions; (ii) the hydrophilic nature of SBA-15 ensures that wastewater with contaminants (like dyes) in it can access the surface of the embedded photocatalyst without much diffusional resistance; (iii) the stability of silica against reactive oxygen species during UV light irradiation. In many reports on semiconductor particle (TiO_2, SnO_2, ZnO *etc.*) embedded porous hosts (SBA-15)[39] as a photocatalyst, the pores are either blocked by semiconductor particles or the latter grows on the external surface of SBA-15.[40–44] Therefore, embedding of a nanostructured solid-solution photocatalyst within the porous host without pore blockage is still not achieved, which would translate into more photocatalytic sites for a possible higher rate of dye degradation.

With this motivation, a new approach has been pursued to form $Sn_xTi_{1-x}O_2$ solid-solution-nanoparticles inside the mesopores of submicron sphere-like SBA-15 particles, without causing pore blockage. The influence of Sn content (5 to 50 at %) in this hybrid has been demonstrated by relating the photocurrent response with the photocatalytic activity towards degradation of rhodamine B under UV irradiation. The mechanism of photocatalytic activity has also been explained in detail for the first time, *via* their energy band diagram. The strategy used in this work would therefore provide a methodology to prepare an assembly of solid-solution nanoparticles in a porous, host-material for catalysis.

Experimental

Synthesis of $Sn_xTi_{1-x}O_2$-sphere-like SBA-15

The synthesis protocol for submicron sphere-like SBA-15 was according to our earlier work,[33] with inclusion of mesitylene in the reactant, for controlling the growth of the SBA-15 particle. Typically, 3.6 g of P123 was added in a polypropylene beaker containing 140 ml of 2.5 M HCl solution and stirred at 35 °C until P123 was completely dissolved. Then, 7.74 g of TEOS was added drop wise with vigorous stirring (500 rpm) for 2 min and was kept under static condition at 35 °C. After 30 min, mesitylene (1.8 g) was added into the beaker containing silica particles (white precipitate), and stirred at 200 rpm for 2 min. Then, the mixture was kept (without stirring) at 35 °C for 24 h, and subsequently transferred into a Teflon coated closed vessel, which was kept at 100 °C for 24 h in an air oven. Thereafter, the resultant solid product was filtered, washed with water, and dried at 100 °C. Finally, the sample was calcined at 550 °C (temperature ramp of 1 °C min^{-1}) in the presence of air for 4 h.

The protocol for embedding SnO_2 nanoparticles inside the pores of SBA-15 was as follows. Typically, 100 mg of sphere-like SBA-15 particles were mixed with a solution containing 50 μl of $SnCl_4$ and 0.6 ml of isoproponal. Then, the mixture was put in a vial and kept under vacuum at 80 °C for 24 h to diffuse the Sn precursor inside the mesopores of the sphere-like SBA-15. The samples were subsequently heated in the presence of air at 550 °C for 6 h to obtain SnO_2-SBA-15. For the preparation of TiO_2-SBA-15, 100 mg of sphere-like SBA-15 particles were mixed with a solution containing 0.225 ml of TTIP and 0.5 ml of isoproponal. Afterwards, a similar procedure was followed, as above for the SnO_2-SBA-15 protocol. For the $Sn_xTi_{1-x}O_2$ synthesis, different concentrations of $SnCl_4$ and TTIP were mixed to vary x in the preparation of the solid-solution. Subsequently, similar steps were followed to form solid-solution-nanoparticles inside the SBA-15 particles.

Characterization techniques

Particle morphology and composition were analyzed using a FE-SEM (Model: JSM-7600F), attached with an EDX spectrometer, and also using a FEG-TEM (Model: JEM-2100F). XRD was done using a Philips X'Pert Pro diffractometer, with CuKα radiation. Nitrogen and hydrogen sorption studies were performed using a Micromeritics ASAP 2020 system. UV-Vis diffuse reflectance spectra for all samples were recorded using a lambda 35 (Perkin-Elmer) spectrophotometer. X-ray photoelectron spectroscopy (XPS) experiments were carried out with a VG Scientific ESCALAB 220IXL instrument, using Al Kα X-ray radiation. Photoluminescence (PL) spectra were recorded using a Perkin-Elmer LS 55 spectrofluorometer. Electrochemical impedance spectroscopy (EIS) was performed on a CHI660 electrochemical workstation (CH Instruments).

Electrochemical analysis

10 mg of the photocatalyst and 20 μl of nafion solution (5 wt%) were dispersed in 2 ml of ethanol (sonication for 2 min), after which, 100 μl of the above dispersion was added onto FTO conductive glass and spin coated to form a working electrode. Electrochemical studies were carried out under UV irradiation using a conventional three-electrode system made from a quartz cell. Platinum foil and saturated calomel electrodes were used as counter and reference electrodes, respectively. 50 ml of 0.1 M KOH aqueous solution was degassed with nitrogen and used as an electrolyte.

Photodegradation of rhodamine B dye

For photocatalytic degradation experiments, 100 mg of the hybrid powder sample was put in a quartz tube containing 100 ml of 50 μM rhodamine B solution and kept under stirring. Above this dispersion, a 250 W high-pressure mercury lamp was placed, housed within a jacketed tube. Water was circulated through the annulus of the tube to avoid heating of the dispersion. After every 30 min, samples were withdrawn from the reactor, and centrifuged for 10 min at 15 000 rpm, to avoid any scattering from suspended solids during absorbance measurements in the UV-Vis spectrophotometer. The concentration of rhodamine B dye was measured by the characteristic absorption peak of rhodamine B at 554 nm.

Results and discussion

Morphological and structural analysis

Both the morphology and pore diameter of sphere-like SBA-15, as well as the location of the SnO_2, TiO_2 and $Sn_{0.25}Ti_{0.75}O_2$ solid-solution-nanoparticles inside the mesopores of sphere-like SBA-15 were investigated by SEM and TEM images, as shown in Fig. 1. The most commonly obtained SBA-15 morphology in the literature is fiber-like,[45] which has a long aspect ratio (length to diameter of SBA-15 particles) of 10 or more, leading to additional diffusional-resistance of reactants and products inside pores of the porous host. Therefore, it is necessary to reduce the diffusion path-length of dye molecules for its efficient uptake and release of products. In order to do so, a hydrophobic substance (mesitylene) was added during synthesis in the present work, after formation of the SBA-15 particle. It is believed that mesitylene eventually surrounds the silica particles and restricts their growth.

Consequently, SBA-15 particles are found to be of sphere-like morphology (with an aspect ratio near unity) in low magnification SEM images (Fig. 1a and b), having a mean diameter of 630 nm and a standard deviation of 85 nm. The TEM image of the sphere-like SBA-15 (Fig. 1c) particles show well-ordered cylindrical pores, which are orientated parallel to each other. The mean pore diameter calculated from many such TEM images is 7.1 nm, with a standard deviation of 0.4 nm.

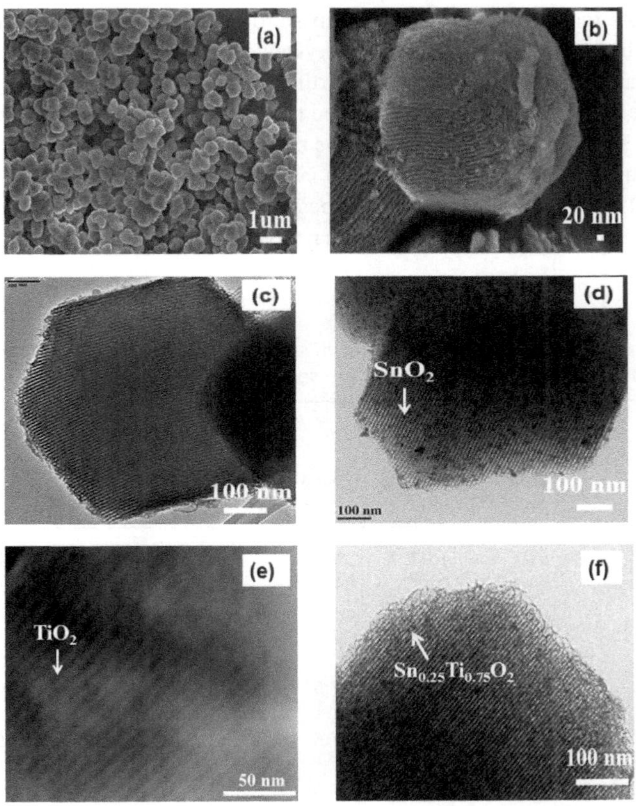

Fig. 1 (a) and (b) SEM images of sphere-like SBA-15 (c) HRTEM image of sphere-like SBA-15 (d) HRTEM image of SnO_2-sphere like SBA-15 (e) HRTEM image of TiO_2-sphere-like SBA-15 (f) HRTEM image of $Sn_{0.25}Ti_{0.75}O_2$-sphere-like SBA-15.

Subsequently, metal oxide precursors ($SnCl_4$ and TTIP) were allowed to impregnate under vacuum and to diffuse inside the pores of the sphere-like SBA-15 in order to form the corresponding metal oxide particles (SnO_2 and TiO_2 respectively). For the $Sn_xTi_{1-x}O_2$ solid-solution, different amounts of $SnCl_4$ and TTIP precursors were added together, which would diffuse inside the pores. This alters the atomic ratio of Sn and Ti, and provides the optimum composition with respect to the maximum photocatalytic activity against rhodamine B dye. To gain physical insight into the photocatalysis, pure SnO_2 and pure TiO_2 nanoparticles were also embedded inside the pores of sphere-like SBA-15, as a control sample (Fig. 1d and e). A representative HRTEM image of SnO_2-sphere-like SBA-15 is shown in Fig. 1d. It can be seen that SnO_2 nanoparticles are well dispersed inside the pores, without any pore blockage. The measured mean diameter of the SnO_2 particle from the HRTEM images of SnO_2-sphere-like SBA-15 is found to be 5.6 nm with a standard deviation of 0.9 nm.

Fig. 1e shows the dark field HRTEM image of TiO_2-sphere-like SBA-15. It is difficult to locate the TiO_2 nanoparticles inside the pores, in the bright field imaging mode, due to the minimal difference in their atomic numbers. As shown in Fig. 1e, the measured mean diameter of TiO_2 nanoparticle is 5.8 nm. In the case of the $Sn_xTi_{1-x}O_2$-sphere-like SBA-15 samples (Fig. 1f), with an increase in the atomic percentage of Sn, the nanoparticle size first decreases and then increases. The possible reason for the reduction in solid-solution particle size (Table 1) is due to the introduction of strain in the crystal, on replacement of Ti^{4+} with Sn^{4+}.[28]

The peaks in the wide angle X-ray diffraction pattern (Fig. 2) of the SnO_2-sphere-like SBA-15 sample can be indexed to the cassiterite tetragonal structure of crystalline SnO_2 (JCPDS no. 41-1445). The mean SnO_2 particle size from Scherrer's formula was found to be 6.7 nm, which is close to the value (5.6 nm), measured from the TEM images. Similarly, the peaks in TiO_2-sphere-like SBA-15 can be assigned to the anatase phase [JCPDS card no. 21-1272]. Moreover, there are no

Table 1 Textural and band gap properties of sphere-like SBA-15, SnO_2-sphere like SBA-15, TiO_2-sphere like SBA-15 and $Sn_xTi_{1-x}O_2$-sphere-like SBA-15

Samples	Specific surface area ($m^2\ g^{-1}$)	Pore volume ($cm^3\ g^{-1}$)	Mean pore diameter	Mean nanoparticle size by XRD (nm)	Sn/(Sn + Ti)		Band gap (eV)
					Mean	Std dev.	
Sphere-like SBA-15	640	1.05	9.1	NA	NA	NA	NA
SnO_2-sphere-like SBA-15	448	0.71	7.7	6.7	NA	NA	3.90
TiO_2-sphere-like SBA-15	409	0.62	7.7	6.2	NA	NA	3.39
$Sn_{0.05}Ti_{0.95}O_2$-sphere-like SBA-15	420	0.65	7.5	5.6	0.05	0.02	3.23
$Sn_{0.15}Ti_{0.85}O_2$-sphere-like SBA-15	428	0.68	7.6	4.3	0.14	0.01	3.19
$Sn_{0.25}Ti_{0.75}O_2$-sphere-like SBA-15	411	0.64	7.6	3.3	0.24	0.01	3.25
$Sn_{0.50}Ti_{0.50}O_2$-sphere-like SBA-15	424	0.67	7.7	6.0	0.51	0.01	3.52

peaks of other phases (rutile) in the sample, which confirms the presence of only pure anatase. The mean TiO_2 particle diameter of TiO_2 in sphere-like SBA-15 is 6.2 nm, which is close to the particle diameter (5.8 nm), measured from the TEM images.

From the diffraction patterns in Fig. 2(b–e), we note that diffraction peaks of solid-solution samples are located in-between pure SnO_2 and pure TiO_2 peaks. It is clearly seen that the anatase phase of titania is retained even after the addition of 5 at% of Sn. This is evidence of solid solution nanoparticle formation inside the porous SBA-15 matrix. With further addition of Sn (15 and 25 at%) in $Sn_xTi_{1-x}O_2$-sphere-like SBA-15, the X-ray diffraction pattern exhibits both titania (anatase) and tin oxide phases, which are in good agreement with the literature.[28] Therefore, it is believed that only 5 at% of Sn forms solid solution completely, when embedded in the lattice sites of Ti. Simultaneously, with increasing Sn amount in the $Sn_xTi_{1-x}O_2$-sphere-like SBA-15, the mean particle size is found to be 5.6, 4.3, and 3.3 nm for the solid-solution samples with 5, 15 and 25 at% Sn, respectively. This shows a decreasing trend in particle size with increases in Sn amount from 5–25 at% Sn, which could be due to the strain arising from the replacement of Ti^{4+} with Sn^{4+}. Furthermore, it can be seen that there is an increase in peak broadening upon increased SnO_2 amount, which indicates a reduction in particle size.

It can be seen from Fig. 3a that the sphere-like SBA-15 particles have a type IV adsorption isotherm with a hysteresis loop, which confirms the presence of mesopores. Furthermore, capillary condensation is observed in the relative pressure range of 0.45–0.9, suggesting that the mesostructure has been retained in the sample. The features of sphere-like SBA-15 in terms of porous nature do not change after loading with SnO_2, TiO_2 or $Sn_xTi_{1-x}O_2$ particles. Moreover, there is a shift in lower relative pressure in the nanoparticle loaded samples, as compared to pure sphere-like SBA-15, which indicates the reduction in mean pore size, due

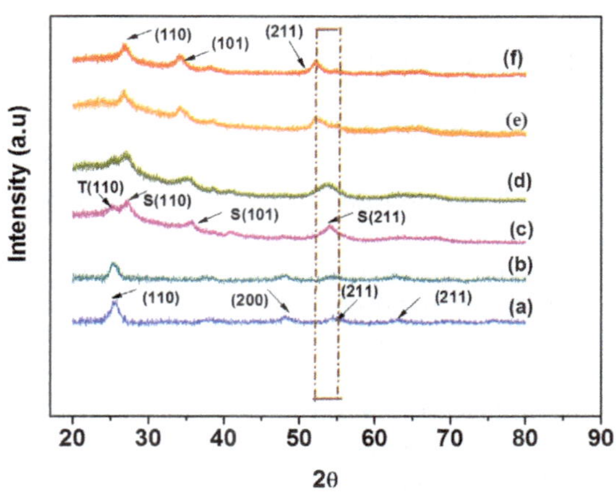

Fig. 2 XRD spectra of (a) TiO_2-sphere-like SBA-15, (b) $Sn_{0.05}Ti_{0.95}O_2$-sphere-like SBA-15, (c) $Sn_{0.15}Ti_{0.85}O_2$-sphere-like SBA-15, (d) $Sn_{0.25}Ti_{0.75}O_2$-sphere-like SBA-15, (e) $Sn_{0.50}Ti_{0.50}O_2$-sphere-like SBA-15, and (f) SnO_2-sphere-like SBA-15. Rectangular bar shows changes in peak position when different percentages of Sn^{4+} are embedded in the lattice sites of Ti^{4+}.

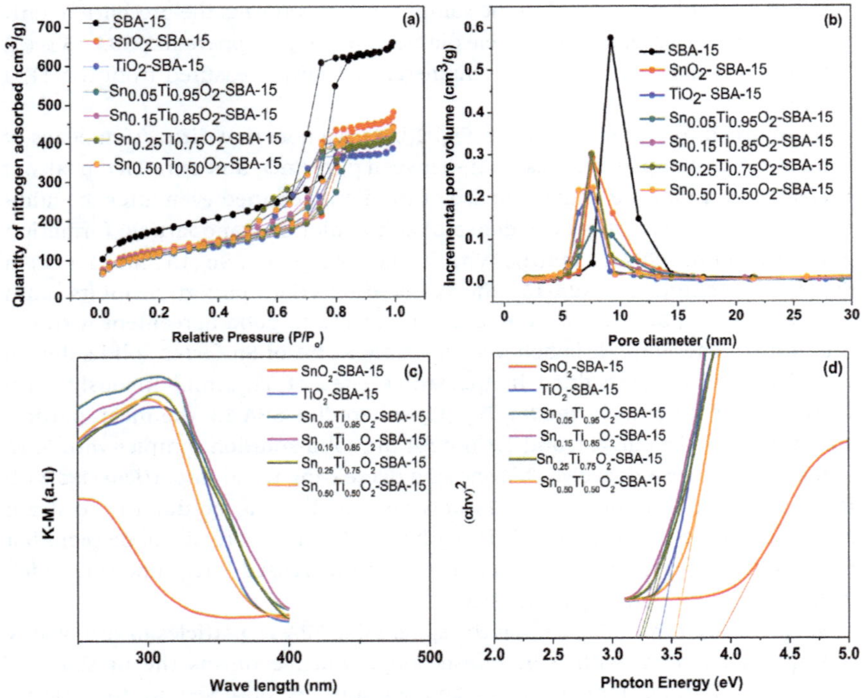

Fig. 3 (a) Nitrogen adsorption and desorption isotherms and (b) pore size distributions of sphere-like SBA-15, SnO_2-sphere-like SBA-15, TiO_2-sphere-like SBA-15, and $Sn_xTi_{1-x}O_2$-sphere-like SBA-15. (c) Diffuse reflectance spectra and (d) plots of the $(\alpha h\nu)^2$ *vs.* the photon energy of SnO_2-sphere-like SBA-15, TiO_2-sphere-like SBA-15 and $Sn_xTi_{1-x}O_2$-sphere-like SBA-15.

to the presence of nanoparticles. This observation is consistent with the pore size distribution plot as shown in Fig. 3b. The mean pore diameter, specific surface area and specific pore volume results are listed for all samples in Table 1. As shown in Table 2, the specific surface area and mean pore diameter of the nanoparticles embedded in sphere-like SBA-15 are in the range of 409–448 $m^2\ g^{-1}$ and 7.5–7.7 nm, respectively, which is less than that of the pure sphere-like SBA-15 particles. This indirectly confirms the presence of nanoparticles inside the pores of the hybrid.

Diffuse reflectance UV-Vis spectra were recorded for all hybrid samples (Fig. 3c and d) to calculate band gap values (Table 1) and understand their influence on the photocatalytic performance. The band gap value is obtained by extrapolating the linear portion of $(\alpha h\nu)^2$ *versus* photon energy (eV). Moreover, it is observed that there is a red shift of the absorption edge in the case of solid-solution particle embedded sphere-like SBA-15 in comparison to SnO_2-sphere-like SBA-15, which shows the electronic structure of the solid-solution particle has been altered. The reduction in band gap values (Table 1) after addition of Sn shows that the hybrids are not a simple mixture of SnO_2 and TiO_2, but they are solid-solution. Furthermore, the band gap value of TiO_2-sphere-like SBA-15 (3.39 eV) is higher than that of $Sn_xTi_{1-x}O_2$-sphere-like SBA-

15 (3.23, 3.19 and 3.25 eV for $x = 0.05$, 0.10 and 0.15, respectively), which indicates that the solid-solution sample has a narrow band gap. This phenomenon will eventually result in promoting more charge carriers from the surface of the photocatalyst. The band gap value increases further from 3.25 to 3.52 eV when the value of x in $Sn_xTi_{1-x}O_2$-sphere-like SBA-15 increases from 0.25 to 0.50. Therefore, there could be a drastic reduction of photocatalytic activity in comparison with only TiO_2-sphere-like SBA-15. The percentage of elements present in the hybrid is measured by EDX analysis and shown in Table 1. The EDX spectra of all hybrids show the presence of both Sn and Ti in the $Sn_xTi_{1-x}O_2$-sphere-like SBA-15 samples. Fig. 4 and Table 1 show that the atomic ratios from EDX spectra for all samples are very close to the atomic ratio added stoichiometrically during the synthesis. This confirms complete conversion of both the Sn and Ti precursors into their respective oxides.

The percentage of photocatalyst in the final samples for all the hybrids was kept the same at 17.7 wt%.

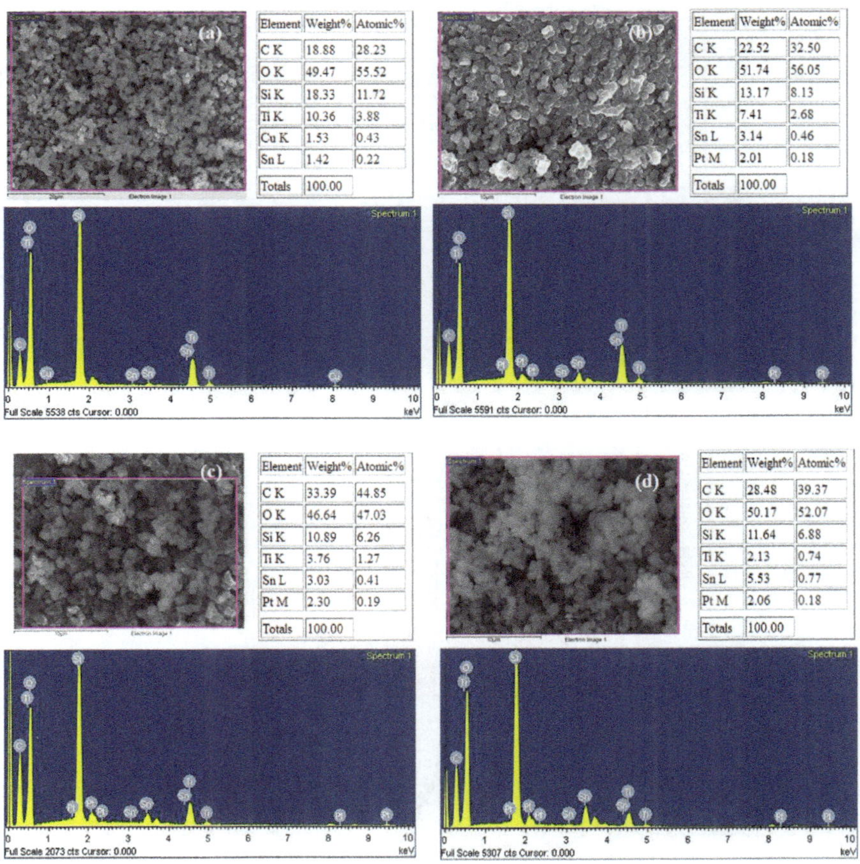

Fig. 4 EDX spectra of (a) $Sn_{0.05}Ti_{0.95}O_2$-sphere-like SBA-15, (b) $Sn_{0.15}Ti_{0.85}O_2$-sphere-like SBA-15, (c) $Sn_{0.25}Ti_{0.75}O_2$-sphere-like SBA-15, and (d) $Sn_{0.50}Ti_{0.50}O_2$-sphere-like SBA-15. Rectangular bar in SEM images are the areas selected for elemental analysis.

Photocatalytic activity of $Sn_xTi_{1-x}O_2$-sphere-like SBA-15

To investigate the photocatalytic activity, rhodamine B dye was allowed to react with $Sn_xTi_{1-x}O_2$-sphere-like SBA-15 under UV light. The change in concentration of rhodamine B as a function of irradiation time in the presence of different atomic percentages (5, 15, 25, and 50%) of Sn in $Sn_xTi_{1-x}O_2$-sphere-like SBA-15, SnO_2-SBA-15 and TiO_2-SBA-15 is shown in Fig. 5. A pseudo-first order dye degradation equation, $-\ln(C/C_0) = kt$, was used for calculating the first order rate constant (k) for all hybrid samples, where C and C_0 are dye concentrations at any later time t, and at initial time, $t = 0$, respectively. As seen in Fig. 5, the photocatalytic activity of SnO_2-sphere-like SBA-15 ($k = 0.14\ h^{-1}$) is much less compared to TiO_2-sphere-like SBA-15 ($k = 1.38\ h^{-1}$).

This is because of the wide band gap of SnO_2-sphere-like SBA-15 (3.9 eV, from Table 1), leading to a limited absorption of photons. On the other hand, TiO_2-sphere-like SBA-15 has a narrower band gap (3.39 eV, from Table 1), which results in a much better photocatalytic activity. When the amount of Sn was slowly increased, it was found that the sample with 5 at% of Sn ($Sn_{0.05}Ti_{0.95}O_2$-sphere-like SBA-15) showed even better photocatalytic activity

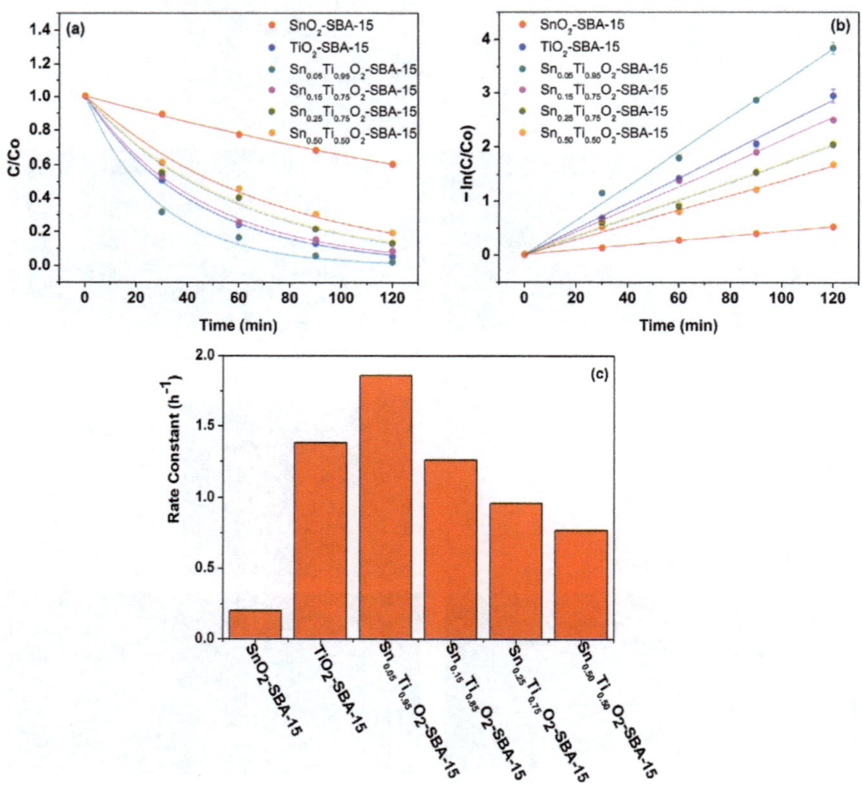

Fig. 5 (a) Normalized concentration profile of rhodamine B dye with UV irradiation time for different photocatalysts; (b) comparison of experiments with fitted kinetics of rhodamine B dye for different photocatalysts; (c) effect of Sn content in $Sn_xTi_{1-x}O_2$-sphere-like SBA-15 on degradation rate constant.

(1.86 h^{-1}) than that of pure TiO_2-sphere-like SBA-15 ($k = 1.38\ h^{-1}$). However, the photocatalytic activity decreased slowly with further increases in atomic percentage of Sn in the hybrid. The photocatalytic activity of a semiconductor is also dependent on charge carrier transfer and its separation efficiency. To this end, electrochemical impedance spectroscopy (EIS) was measured in order to understand these aspects at the interface of a semiconductor particle.[46] The Nyquist plot of the EIS measurement for all samples is shown in Fig. 6a. The equivalent-circuit model is also fitted for all EIS spectra (Fig. 6a), where R_s is the solution resistance, R_{ct} is the charge transfer resistance, and C_{dl} is the double layer capacitance. To gain further insight on the charge transport and recombination kinetics, model parameters were calculated using the Zsimp win software and are listed in Table 2.

In general, the electron-transfer behaviour of the samples is reflected by the arc of the spectrum. A smaller arc in the EIS spectra indicates an effective separation of charge carriers and a faster interfacial charge transfer. It can be seen from Fig. 6 that the arc of $Sn_{0.05}Ti_{0.95}O_2$-sphere-like SBA-15 is much smaller than SnO_2-sphere-like SBA-15 or TiO_2-sphere-like SBA-15, suggesting lower electron transfer resistance and a faster interfacial charge transfer rate in the former. The arc slowly vanishes as the loading of Sn increases, which suggests that the charge transfer resistance increases.

From Table 2 it can be seen that the solution resistance (R_s) value for all hybrids is almost the same. The charge transfer resistance, (R_{ct}) however, decreases from 910 kΩ cm^{-2} to 332 kΩ cm^{-2}, on addition of 5 at% of Sn in the TiO_2 matrix, which could be due to the increase in faradic current, as a result of a reduction in recombination of the charge carriers overall, the solid-solution in sphere-like SBA-15 reduces the recombination of charge carriers, resulting in superior photocatalytic activity.

Photocurrent responses (Fig. 6b) were recorded for a few on–off cycles of UV irradiation. The value of the photocurrent quickly goes down to zero once the UV light is switched off, whereas it will return to a steady value, when the light is again turned on. The above trend is reproducible and shows stability over a few cycles. The $Sn_{0.05}Ti_{0.95}O_2$-sphere-like SBA-15 shows much higher photocurrent

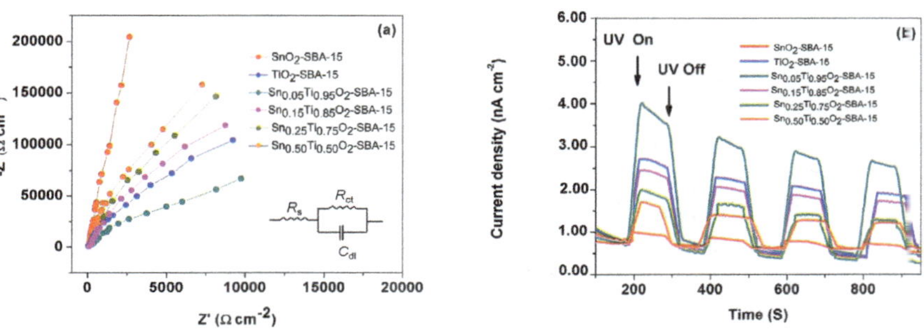

Fig. 6 (a) Nyquist plots and (b) transient photocurrent responses of SnO_2-sphere-like SBA-15, TiO_2-sphere-like SBA-15, $Sn_{0.05}Ti_{0.95}O_2$-sphere-like SBA-15, $Sn_{0.15}Ti_{0.85}O_2$-sphere-like SBA-15, $Sn_{0.25}Ti_{0.75}O_2$-sphere-like SBA-15 and $Sn_{0.50}Ti_{0.50}O_2$-sphere-like SBA-15.

Table 2 Model parameters of all hybrids based on EIS results

Sample	Solution resistance R_s (Ω cm^{-2})	Charge transfer resistance R_{ct} (kΩ cm^{-2})	Double layer capacitance C_{dl} (μF cm^{-2})
SnO_2-SBA-15	68.1	8920	3.32
TiO_2-SBA-15	67.4	910	5.71
$Sn_{0.05}Ti_{0.95}O_2$-SBA-15	67.7	332	6.10
$Sn_{0.15}Ti_{0.85}O_2$-SBA-15	65.4	1236	5.28
$Sn_{0.25}Ti_{0.75}O_2$-SBA-15	66.2	1798	4.97
$Sn_{0.50}Ti_{0.50}O_2$-SBA-15	67.1	2300	4.68

intensity than other hybrids, indicating that the sample can reduce charge carrier recombination, by improving interfacial charge transfer under UV irradiation.

To gain further insight into the maximum enhancement of photocatalytic activity in the $Sn_{0.05}Ti_{0.95}O_2$-sphere-like SBA-15 sample, compared to TiO_2-sphere-like SBA-15 – so as to emphasize the role of surface states – XPS spectra of both samples are compared in Fig. S1 (ESI†). This confirms the presence of elements and chemical states. Fig. S1† shows spectra from XPS of the complete survey of elements (Fig. S1a, ESI†), Ti2p (Fig. S1b, ESI†) and O1s (Fig. S1c and d, ESI†). The complete survey (Fig. S1a, ESI†) shows the presence of Sn, Ti, C, Si and O elements in $Sn_{0.05}Ti_{0.95}O_2$-sphere-like SBA-15, while Ti, C, Si and O elements occur in TiO_2-sphere-like SBA-15, respectively. The presence of elemental C in both samples could be due to atmospheric contamination during sample exposure in air.[47] Fig S1b (ESI†) shows the high resolution spectra of the Ti2p region, for both samples at about 458.8 and 464.6 eV corresponding to Ti2p3/2 and Ti2p1/2. The two peaks of the Ti2p for $Sn_{0.05}Ti_{0.95}O_2$-sphere-like SBA-15 show a slight shift towards higher binding energy levels, in comparison to TiO_2-sphere-like SBA-15.[48,49] This is mainly due to the variation of electronegativities of the Ti and Sn elements (Ti – 1.54 and Sn – 1.96) and confirms the bonding of Ti–O–Sn in $Sn_{0.05}Ti_{0.95}O_2$-sphere-like SBA-15. Furthermore, the position of the Ti2p doublet peaks and the splitting value of 5.8 eV indicate the presence of the Ti^{4+} oxidation state. In order to quantify oxygen vacancies in both samples, the O1s peak (Fig. S1c, ESI†) was deconvoluted into three peaks at about 530.1, 531.5 and 532.5 eV, which are associated with lattice oxygen species (O_L), oxygen vacancies (O_V) and hydroxyl oxygen species (O_H), respectively. The percentages of O_V (10.85%) and O_H (63.53%) were increased for $Sn_{0.05}Ti_{0.95}O_2$-sphere-like SBA-15, compared to TiO_2-sphere-like SBA-15 O_V (8.35%) and O_H (61.57%). The increase in oxygen vacancy is the result of substitution of Sn^{4+} into the Ti^{4+} lattice due to the weakening of the Ti–O bonds.

To explore the possible mechanism of the higher photodegradation by the $Sn_{0.05}Ti_{0.95}O_2$-sphere-like SBA-15 hybrid compared to other samples, their band positions were calculated from the Mulliken electronegativity, using the following empirical formulae.[50,51]

$$E_{VB} = X - E_e + 0.5E_g \quad (1)$$

$$E_{CB} = E_{VB} - E_g \quad (2)$$

where, E_{VB} represents the valence band edge potential, E_{CB} represents the conduction band edge potential, E_g represents the band gap energy of the semiconductor, E_e is the energy of free electrons on the hydrogen scale (4.5 eV), and X is the electronegativity of the semiconductor, which is calculated based on the arithmetic mean of the atomic electron affinity and the ionization potential. The calculated values of X, E_{CB}, and E_{VB} are given in Table 3.

Based on the energy band structure calculation, the energy level diagram for all hybrids is shown in Fig 7. This band energy structure calculation is necessary as the redox ability of the photogenerated electron–hole pairs depends on the band position of the semiconductor. The conduction band and valence band potentials of the $Sn_xTi_{1-x}O_2$ sphere-like SBA-15 hybrid are positioned between the SnO_2-sphere-like SBA-15 and TiO_2-sphere-like SBA-15. It can be clearly seen that the potentials can be adjusted by changing the ratio of Sn and Ti in the hybrid. It has been found that the values of the redox couples (OH^-/OH and $O_2/O_2^{\bullet-}$) display the potential (*vs.* SHE) of 2.38 and −0.33 V, respectively.[52] These potential values are considered as the conduction and valence band positions, if radicals are present in the system.

From Fig. 7, it can be seen that the valence band of $Sn_xTi_{1-x}O_2$-sphere-like SBA-15 is more positive than that of E° ($^{\bullet}OH/OH^-$) (2.38 V *vs.* SHE). As a result, the most prominent way of degradation could be the oxidation of OH^- to OH by photogenerated holes, rather than the reduction of O_2 to $O_2^{\bullet-}$ by photogenerated electrons. Furthermore, the addition of Sn into the TiO_2 matrix ($Sn_{0.05}Ti_{0.95}O_2$-sphere-like SBA-15) decreases the conduction band and valence band potential, which are less than that of pure TiO_2-sphere-like SBA-15. This leads to a narrow band gap and reduction in recombination of charge carriers, thereby increasing the photocatalytic activity of the sample. However, addition of 15, 25 and 50 at% Sn in the hybrid leads to reduction in photocatalytic activity. The possible reason for this could be due to the appearance of more crystalline features of SnO_2 and the prevention or delay of crystallization of the TiO_2 phase in the solid-solution. The other reasons may be the formation of SnO_2 arising from the non-substituted Sn^{4+} in the crystal lattice of solid-solution-nanoparticles and an increase in band gap values. Thus, based on the photocatalytic activity in this work, the maximum percentage of Sn to be used in $Sn_xTi_{1-x}O_2$ is 5 at%. The narrow band gap due to band structure modification is also supported by photoluminescence (Fig. S2, ESI†), which is mainly the result of recombination of charge carriers. The sample ($Sn_{0.05}Ti_{0.95}O_2$-sphere-like SBA-15) is less luminescent in comparison to TiO_2-sphere-like SBA-15. A possible reason is the presence of oxygen vacancies and consequent reduction of recombination of charge carriers.[53] Therefore, the

Table 3 Calculated values of X, E_{VB}, and E_{CB} for $Sn_xTi_{1-x}O_2$-sphere-like SBA-15 hybrids

Sample	X (eV)	E_{VB} (eV)	E_{CB} (eV)
TiO_2-SBA-15	5.83	3.03	−0.36
$Sn_{0.05}Ti_{0.95}O_2$-SBA-15	5.85	2.96	−0.27
$Sn_{0.15}Ti_{0.85}O_2$-SBA-15	5.89	2.98	−0.21
$Sn_{0.25}Ti_{0.75}O_2$-SBA-15	5.93	3.06	−0.19
$Sn_{0.50}Ti_{0.50}O_2$-SBA-15	6.03	3.29	−0.23
SnO_2-SBA-15	6.23	3.68	−0.22

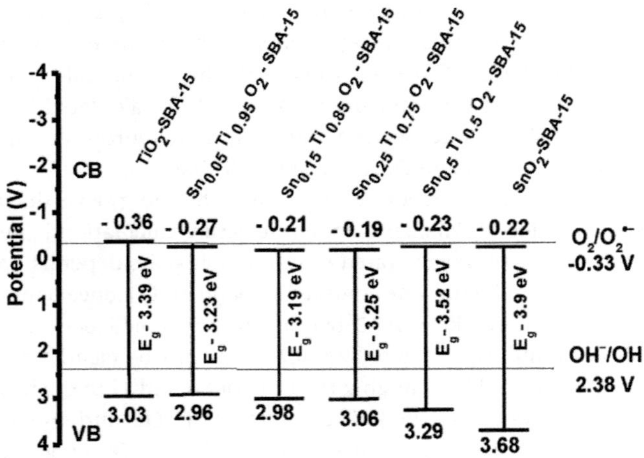

Fig. 7 Schematic energy level diagram of $Sn_xTi_{1-x}O_2$-sphere-like SBA-15 hybrids.

photocatalytic mechanism of $Sn_xTi_{1-x}O_2$-sphere-like SBA-15 is explained by surface states, electronic structure, luminescence and charge carrier transfer at the interface by EIS.

Table 4 shows a comparison of recent studies on the photodegradation of the dye by $Sn_xTi_{1-x}O_2$-sphere-like SBA-15, with different forms of SnO_2–TiO_2 hybrids. The goal is to identify important parameters such as catalyst loading, degradation time, *etc.* Other than the effect of nanoparticle composition (shown in this work), factors such as pore blockage and the SBA-15 morphology (geometrical features) of the porous host also play an important role in enhancing photodegradation. For example, removal of pore blockage (by reducing the embedded SnO_2 nanoparticle size from 6.8 to 3.5 nm) in SnO_2-fiber-like SBA-15 increased the first order degradation rate constant from 0.067 to 0.14 h^{-1} (Table 4, sl no. 2) – about 2.1 times – emphasizing the need for smaller embedded nanoparticles. Similarly,

Table 4 Comparison of photocatalytic performance of $Sn_xTi_{1-x}O_2$-SBA-15 hybrid with similar reports based on SnO_2 and TiO_2

Sl no.	Photocatalyst	Rhodamine B dye concentration (μM)	Catalyst (%)	Degradation time (min)	Rate constant (h^{-1})	Reference no.
1	SnO_2	10	100	150	0.20	4
2	SnO_2-SBA-15	50	5	600	0.067, 0.14	33
3	SnO_2-SBA-15	50	5	360	0.54	34
4	TiO_2	10	100	180	0.43	28
5	TiO_2-SBA-16	10	5	120	1.63	39
6	TIO_2-SBA-15	50	17.7	120	1.38	Present work
7	SnO_2–TiO_2	20	100	60	0.91	48
8	SnO_2–TiO_2	10	100	180	0.73	28
9	$Sn_{0.05}Ti_{0.95}O_2$-SBA-15	50	17.7	120	1.86	Present work

reducing the aspect ratio by changing the SBA-15 particle's external morphology from fiber-like (aspect ratio ~ 11) to sphere-like (aspect ratio ~ 1), further improved the rate constant by a factor of 3.8 from 0.14 to 0.54 h^{-1} (Table 4, sl no. 3). Incorporating these improvements and pushing further in the present work, the optimum composition (of $x = 0.05$) in $Sn_xTi_{1-x}O_2$ achieved a very significant increase in the rate constant, from 0.54 to 1.86 h^{-1}, *i.e.* by an additional factor of 3.4. Therefore, embedding solid-solution-nanoparticles inside a porous host (as shown in this work), together with ensuring no pore blockage (by using 5.6 nm diameter catalytic particles in 7 nm diameter SBA-15 host) and a shorter pore-channel length (using sphere-like, short aspect ratio SBA-15 host) helped in achieving an overall very impressive rate improvement (from 0.067 to 1.86 h^{-1}, a combined factor of about 27.8 times) in photocatalytic degradation.

It can be further seen from Table 4 (sl no. 1 and 4) that only TiO_2 without any porous host ($k = 0.43$ h^{-1}) is a better photocatalyst in comparison to only SnO_2 without any porous host ($k = 0.20$ h^{-1}), obviously because of the smaller band gap of the former. However, these materials perform far better when they are embedded in the porous SBA-15 matrix. The porous matrix increases the accessible specific surface area of the photocatalyst particles, by preventing nanoparticle inter-aggregation due to attractive surface forces. Therefore, a higher value of $k = 1.38$ h^{-1} is obtained (sl no. 6 in Table 4) for TiO_2-SBA-15, as compared to pure TiO_2 ($k = 0.43$ h^{-1}, in sl no. 4 in Table 4). A further improvement is achieved when the two oxides are combined into a heterostructure, with rate constants ($k = 0.73$ and 0.91 h^{-1}, sl no. 8 and 7 in Table 4, respectively), being better than that of the individual oxide particles ($k = 0.43$ h^{-1}, sl no. 4 in Table 4). The decrease in recombination of charge carriers in the heterostructure could be the reason for such an improvement. Therefore, the enhancements obtained in the current work are from the following aspects: (i) using a porous host particle of nearly spherical shape (*i.e.* shorter pore-length to increase the accessibility of the nanoparticle surface for the dye molecule) and to prevent pore blockage and (ii) using a combination of two oxides (in the form of a solid-solution in each nanoparticle). As a result, $Sn_{0.05}Ti_{0.95}O_2$-sphere-like SBA-15 ($k = 1.86$ h^{-1}, sl no. 9 in Table 4) provides the fastest rate of dye degradation of any of the hybrids in Table 4. Furthermore, these changes in the hybrid will make an impact in the application of dye-sensitized solar cells (DSSCs), where recombination of charge carriers is one of the loss factors. Another impact would be in application areas such as lithium-ion batteries and gas sensors, where transport of electrons in a short path length is important to enhance their performances. The presence of semiconductor nanoparticles along with the shorter length of a porous matrix results in a faster response, whenever it is exposed to diffusing molecules in a gas sensor or ions in a lithium-ion battery. Therefore, these functional differences can make an impact not only in the area of photocatalysis, but also in other areas where recombination of charge carriers and transport of electrons are important.

Conclusions

In summary, this work demonstrates the formation of controlled $Sn_xTi_{1-x}O_2$ solid-solution-nanoparticles inside sphere-like SBA-15 particles. This ensures confinement of the small oxide nanoparticles within the mesopores of SBA-15, without interparticle aggregation, thereby eliminating pore blockage. On varying

x between 0 (pure TiO_2) to 1.0 (pure SnO_2) with other intermediate values, it is observed that the hybrid with 5 at% of Sn, *i.e.*, $Sn_{0.05}Ti_{0.95}O_2$-SBA-15 shows the maximum photocatalytic activity for degradation of rhodamine-B dye (first order rate constant, $k = 1.86\ h^{-1}$), compared to both pure TiO_2-sphere-like SBA-15 ($k = 1.38\ h^{-1}$) or pure SnO_2-sphere-like SBA-15 ($k = 0.14\ h^{-1}$). Electrochemical impedance spectroscopy shows an increase in charge carrier lifetime to the maximum extent in $Sn_{0.05}Ti_{0.95}O_2$-sphere-like SBA-15, leading to enhanced photocatalytic activity for this hybrid. This enhancement is well supported by XPS and PL spectra, showing the effect of the increase in oxygen vacancies in $Sn_{0.05}Ti_{0.95}O_2$-SBA-15, compared to TiO_2-sphere-like SBA-15. However, a further increase in Sn amount from 5 to 50 at%, led to a reduction in photocatalytic activity, due to gradual loss of the anatase phase in TiO_2. Therefore, $Sn_{0.05}Ti_{0.95}O_2$-sphere-like SBA-15 displays a higher dye degradation rate constant because of its structural features (crystallinity, particle size, specific surface area) and electronic properties (band structure and mobility of charge carriers) than that of previous reports based on only SnO_2, only TiO_2 or other hybrids. This work not only demonstrates a methodology to form solid-solution nanoparticles of SnO_2 and TiO_2, well-dispersed and embedded inside sphere-like SBA-15 particles without pore blockage, but also provides new insight into understanding the optimum atomic percentages of Sn and Ti in terms of their photocatalytic performance. Thus, as shown above in Table 4, optimization of geometrical structure and composition in any hybrid is crucial to achieve maximum rate enhancement in any heterogeneous catalytic reaction. Therefore, these findings can be translated to alter the electronic structure of semiconductor nanoparticles inside porous hosts (with maximum accessibility of nanoparticle surface), for other different applications of hybrid materials, as in dye-sensitized solar cells, hydrogen production, *etc.*

Acknowledgements

Authors acknowledge the National Centre for Nanoscience and Nanotechnology, University of Madras, Chennai, and Department of Chemistry, Indian Institute of Technology Madras for XPS and PL spectra, respectively.

Notes and references

1 R. Asahi, T. Morikawa, T. Ohwaki, K. Aoki and Y. Taga, *Science*, 2001, **293**, 269–271.
2 A. Hagfeldt and M. Graetzel, *Chem. Rev.*, 1995, **95**, 49–68.
3 Z. Zou, J. Ye, K. Sayama and H. Arakawa, *Nature*, 2001, **414**, 625–627.
4 S. Wu, H. Cao, S. Yin, X. Liu and X. Zhang, *J. Phys. Chem. C*, 2009, **113**, 17893–17898.
5 G. Ciardelli and N. Ranieri, *Water Res.*, 2001, **35**, 567–572.
6 P. R. Gogate and A. B. Pandit, *Adv. Environ. Res.*, 2004, **8**, 501–551.
7 A. Fujishima and K. Honda, *Nature*, 1972, **238**, 37–38.
8 M. R. Hoffmann, S. T. Martin, W. Choi and D. W. Bahnemannt, *Chem. Rev.*, 1995, **95**, 69–96.
9 Y. Qu and X. Duan, *Chem. Soc. Rev.*, 2013, **42**, 2568–2580.
10 H. Zhou, Y. Qu, T. Zeid and X. Duan, *Energy Environ. Sci.*, 2012, **5**, 6732.

11 B. Sun, A. V. Vorontsov and P. G. Smirniotis, *Langmuir*, 2003, **19**, 3151–3156.
12 S. Sakthivel, M. V Shankar, M. Palanichamy, B. Arabindoo, D. W. Bahnemann and V. Murugesan, *Water Res.*, 2004, **38**, 3001–3008.
13 Z. Bian, T. Tachikawa, W. Kim, W. Choi and T. Majima, *J. Phys. Chem. C*, 2012, **116**, 25444–25453.
14 W. Choi, A. Termin and M. R. Hoffmann, *J. Phys. Chem.*, 1994, **98**, 13669–13679.
15 S. G. Kumar and L. G. Devi, *J. Phys. Chem. A*, 2011, **115**, 13211–13241.
16 H. Yamashita, M. Harada, J. Misaka, M. Takeuchi, K. Ikeue and M. Anpo, *J. Photochem. Photobiol., A*, 2002, **148**, 257–261.
17 Q. Xiang, J. Yu, W. Wang and M. Jaroniec, *Chem. Commun.*, 2011, **47**, 6906–6908.
18 J. C. Yu, J. Yu, W. Ho, Z. Jiang and L. Zhang, *Chem. Mater.*, 2002, **14**, 3808–3816.
19 M. Zhou and J. Yu, *J. Hazard. Mater.*, 2008, **152**, 1229–1236.
20 S. Sakthivel and H. Kisch, *Angew. Chem., Int. Ed.*, 2003, **42**, 4908–4911.
21 D. Kannaiyan, E. Kim, N. Won, K. W. Kim, Y. H. Jang, M.-A. Cha, D. Y. Ryu, S. Kim and D. H. Kim, *J. Mater. Chem.*, 2010, **20**, 677.
22 Z. Zhang, W. Wang, L. Wang and S. Sun, *ACS Appl. Mater. Interfaces*, 2012, **4**, 593–597.
23 K. Vinodgopal and P. V. Kamat, *Environ. Sci. Technol.*, 1995, **29**, 841–845.
24 A. Fontes Garcia, M. Fernandes and P. Coutinho, *Nanoscale Res. Lett.*, 2011, **6**, 426.
25 L. Lin, Y. Yang, L. Men, X. Wang, D. He, Y. Chai, B. Zhao, S. Ghoshroy and Q. Tang, *Nanoscale*, 2013, **5**, 588–593.
26 E. M. El-Maghraby, *Phys. B*, 2010, **405**, 2385–2389.
27 Y. Wang, J. Wu, J. Zheng, R. Jiang and R. Xu, *Catal. Sci. Technol.*, 2012, **2**, 581–588.
28 M. F. Abdel-Messih, M. A. Ahmed and A. S. El-Sayed, *J. Photochem. Photobiol., A*, 2013, **260**, 1–8.
29 Y. Huang, Z. Zheng, Z. Ai, L. Zhang, X. Fan and Z. Zou, *J. Phys. Chem. B*, 2006, **110**, 19323–19328.
30 S. Ponja, S. Sathasivam, N. Chadwick, A. Kafizas, S. M. Bawaked, A. Y. Obaid, S. Al-Thabaiti, S. N. Basahel, I. P. Parkin and C. J. Carmalt, *J. Mater. Chem. A*, 2013, **1**, 6271–6278.
31 M. Liu, L. Wang, G. (Max) Lu, X. Yao and L. Guo, *Energy Environ. Sci.*, 2011, **4**, 1372–1378.
32 M. V. Phanikrishna Sharma, V. Durga Kumari and M. Subrahmanyam, *Chemosphere*, 2008, **73**, 1562–1569.
33 N. R. Srinivasan and R. Bandyopadhyaya, *Microporous Mesoporous Mater.*, 2012, **149**, 166–171.
34 N. R. Srinivasan, P. Majumdar, N. K. R. Eswar and R. Bandyopadhyaya, *Appl. Catal., A*, 2015, **498**, 107–116.
35 T. Kamegawa, R. Kido, D. Yamahana and H. Yamashita, *Microporous Mesoporous Mater.*, 2013, **165**, 142–147.
36 Z. Liu, Z. Liu, T. Cui, J. Li, J. Zhang, T. Chen, X. Wang and X. Liang, *Chem. Eng. J.*, 2014, **235**, 257–263.
37 S. Liu, E. Guo and L. Yin, *J. Mater. Chem.*, 2012, **22**, 5031–5041.
38 D. Zhao, J. Feng, Q. Huo, N. Melosh, G. Fredrickson, B. Chmelka and G. Stucky, *Science*, 1998, **279**, 548–552.

39 J. Ma, L. Qiang, X. Tang and H. Li, *Catal. Lett.*, 2010, **138**, 88–95.

40 X. Wang, F. Li, Y. Hao, S. Liu and M. Yang, *Mater. Lett.*, 2013, **99**, 38–41.

41 X. F. Qian, T. Kamegawa, K. Mori, H. X. Li and H. Yamashita, *J. Phys. Chem. C*, 2013, **117**, 19544–19551.

42 S.-Y. Chen, T. Mochizuki, Y. Abe, M. Toba and Y. Yoshimura, *Appl. Catal., B*, 2014, **148–149**, 344–356.

43 Y. J. Acosta-Silva, R. Nava, V. Hernández-Morales, S. A. Macías-Sánchez, M. L. Gómez-Herrera and B. Pawelec, *Appl. Catal., B*, 2011, **110**, 108–117.

44 K. de Witte, A. M. Busuioc, V. Meynen, M. Mertens, N. Bilba, G. van Tendeloo, P. Cool and E. F. Vansant, *Microporous Mesoporous Mater.*, 2008, **110**, 100–110.

45 H. Zhang, J. Sun, D. Ma, G. Weinberg, D. S. Su and X. Bao, *J. Phys. Chem. B*, 2006, **110**, 25908–25915.

46 L. Huang, H. Xu, Y. Li, H. Li, X. Cheng, J. Xia, Y. Xu and G. Cai, *Dalton Trans.*, 2013, **42**, 8606–8616.

47 J.-M. Wu, *J. Phys. Chem. C*, 2008, **112**, 13192–13199.

48 C. Wang, C. Shao, X. Zhang and Y. Liu, *Inorg. Chem.*, 2009, **48**, 7261–7268.

49 Y. Duan, N. Fu, Q. Liu, Y. Fang, X. Zhou, J. Zhang and Y. Lin, *J. Phys. Chem. C*, 2012, **116**, 8888–8893.

50 M. Long, W. Cai, J. Cai, B. Zhou, X. Chai and Y. Wu, *J. Phys. Chem. B*, 2006, **110**, 20211–20216.

51 C. Xing, Y. Zhang, W. Yan and L. Guo, *Int. J. Hydrogen Energy*, 2006, **31**, 2018–2024.

52 P. Wardman, *J. Phys. Chem. Ref. Data*, 1989, **18**, 1637–1755.

53 A. Leelavathi, G. Madras and N. Ravishankar, *Phys. Chem. Chem. Phys.*, 2013, **15**, 10795–10802.

Faraday Discussions

DISCUSSIONS

Modeling and Theory: general discussion

Guruswamy Kumaraswamy, Vinothan N. Manoharan, Jayant Singh, Nirmalya Bachhar, Siddharth Kulkarni, Ranjini Bandyopadhyay, Sudeep Punnathanam, Ajeet Srivastav, Priyadarshi Roy Chowdhury, Daan Frenkel, G V Pavan Kumar, Vimala Sridurai, Debdas Dhabal, Erika Eiser, Mukta Tripathy, Alison Edwards, B. L. V Prasad, Madivala G. Basavaraj, Hari Om Sharanam Yadav, Lynn Walker, Sanat Kumar, Charusita Chakravarty, Vandana Shinde, Rajdip Bandyopadhyaya, Jacques Jestin, Mohamed Laradji, Sofia Kantorovich, Nicholas Kotov, Oleg Gang, Sunil Kumar, Robert Botet, Hima Bindu Kolli, Yogesh M. Joshi and Pedro A. Sánchez

DOI: 10.1039/C6FD90002J

Sanat Kumar began the discussion of the paper by Jayant K. Singh: You talked about gold and copper; it is apparent that you are missing the shape transition of gold. Is copper different? Is behaviour "universal" among this class of metals? Can we look at classes of metals and make a reduced model ? Or, is the behaviour chemically specific?

Jayant Singh responded: The behaviour of copper is similar to that seen for gold, as far as the dewetting mechanism on graphene is concerned. We do observe a liquid to solid phase transition of gold nanoparticles within our simulation time frame. However, we could not capture the formation of triangular-shape gold nanoparticles. We are not sure if this is due to imperfections in the model, if there are any. Considering the similar behaviour seen for some metals on graphene, it seems plausible to make a reduced model for metals on a weakly interacting substrate such as graphene. Certainly this would be a worthwhile attempt.

Guruswamy Kumaraswamy remarked: Your images of the break-up of thin (0.5 nm) films suggest that a specific lengthscale characterizes the initial structure, immediately after the film breaks up to form holes. What physics governs this lengthscale? You have suggested that the Navier–Stokes equations could be used to describe the film break-up process. Would bulk constitutive equations such as the Navier–Stokes work in this regime of such thin films?

Jayant Singh responded: It is well known that thin film rupture leading to the formation of holes is mainly a spinodal process.[1,2] Usually, a thin film having a thickness lower than a critical thickness breaks up to form holes; however in our case the holes do form initially (most likely due to spinodal instability) but merge to make bigger holes, and later the rim contracts leading to drop formation. This could be attributed to competitive inertial and viscous forces. However at higher thicknesses, such as 10 Å or higher which turns out to be greater than the critical thickness (for crossover behaviour), the viscous forces dominate which is seen in our work. The dewetting behaviour for thicknesses of <0.5 nm, as observed in our work, is also seen for other metallic films,[3,4] which is in good agreement with the experimental results.[5] The Navier–Stokes equations may be used to address the dewetting mechanism of the metallic thin film as shown recently by Nguyen *et al.*[6] The authors used a continuum model (CM) based on a long wave limit of the Navier–Stokes equation and molecular dynamics (MD) simulations to study the breakup of nanometer sized thin Cu rings. The results of the CM and MD were found to be in good agreement.

1 T. D. Nguyen, M. Fuentes-Cabrera, J. D. Fowlkes and P. D. Rack, *Phys. Rev. E*, 2014, **89**, 032403.
2 A. Atena and M. Khenner, *Phys. Rev. B*, 2009, **80**, 075402.
3 X. Li, Y. He, Y. Wang, J. Dong, and H. Li, *Sci. Rep.*, 2014, **4**, 3938.
4 Y. Li, C. Tang, J. Zhong and L. Meng, *J. Appl. Phys.*, 2015, **117**, 064304.
5 C. Favazza, R. Kalyanaraman and R. Sureshkumar, *Nanotechnology*, 2006, **17**, 4229.
6 T. D. Nguyen, M. Fuentes-Cabrera, J. D. Fowlkes, J. A. Diez, A. G. González, L. Kondic and P. D. Rack, *Langmuir*, 2012, **28**, 13960.

Debdas Dhabal addressed Jayant Singh: Upon heating at 1073 K, what is the reason behind the apparent lack of hole formation in the case of the 10 Å film? Whereas, hole formation is observed in the 5 angstrom film. Fig. 5 in your paper shows that in the case of the 10 Å film, the centre of mass (CM) velocity is going towards a negative *z*-direction up to 30 ps. Does this mean that it is trying to initiate hole formation but somehow it couldn't? If so, why?

Jayant Singh replied: As mentioned in response to the above question, the dewetting behaviour is clearly due to the competing forces. At higher thickness viscous forces dominate over inertial forces leading to the contraction behaviour without any hole formation.[1] This type of dewetting behaviour for thicknesses of <0.5 nm is also observed for other metallic films.[2,3]

The initial structures for the simulations are generated by placing the gold nanofilm on the surface of the graphene at a vertical distance of 0.3 nm. During the first 30 ps the film adjusts itself by moving towards the surface, which results in the negative velocity of the CM in the *z*-direction. Similar behaviour is seen for all film thicknesses. Hence, the movement of the CM in the negative *z*-direction does not indicate hole formation.

1 N. Péron, F. Brochard-Wyart and H. Duval, *Langmuir*, 2012, **28**, 15844.
2 X. Li, Y. He, Y. Wang, J. Dong, and H. Li, *Sci. Rep.*, 2014, **4**, 3938.
3 Y. Li, C. Tang, J. Zhong and L. Meng, *J. Appl. Phys.*, 2015, **117**, 064304.

Daan Frenkel commented: I notice that the position of the maximum of the free energy barrier for cluster nucleation (Fig. 8 in your manuscript) is almost independent of temperature, even though the temperature changes by more than 30%.

This is not the behaviour that one would expect on the basis of classical nucleation theory, and I wonder if there is a simple explanation for the observed behaviour.

Jayant Singh replied: The individual size of the cluster used in the study is 2 nm. The cluster size in Fig. 8 of the manuscript is the number of such particles in the cluster. The position of the maximum of the free energy barrier is not constant, and is found to vary slightly. For example, the critical cluster size is approximately 7, 8, 9, 7, and 7 at 673 K, 773 K, 873 K, 973 K, and 1073 K, respectively. Certainly, there appears to be some competing effects contributing to the unexpected behaviour of the critical cluster size. Unfortunately, with the current data from our simulation we cannot compare the nucleation behaviour observed in our case with classical nucleation theory. However, it will be investigated rigorously in our next work.

Mohamed Laradji remarked: Graphene sheets are characterized by a high bending elastic modulus. How are the dewetting kinetics affected by the bending modulus of the substrate?

Jayant Singh answered: It has already been shown in previous studies of bilayer systems that the elastic strain energy required to deform the top layer decreases when the bottom layer is changed from a rigid to viscoelastic layer.[1–4] Hence, we believe that a bending elastic modulus would certainly affect the dewetting behaviour of the gold film, which needs to be explored in future work.

1 S. Kumar and O. K. Matar, *J. Colloid Interface Sci.*, 2004, **273**, 581.
2 A. E. Hosoi and L. Mahadevan, *Phys. Rev. Lett.*, 2004, **93**, 137802.
3 D. Bandyopadhyay, A. Sharma and V. Shankar, *Europhys. Lett.*,2010, **89**,36002.
4 B. Ray, G. Biswas and A. Sharma, *Phys. Fluids*, 2012, **24**, 082108.

Hari Om Sharanam Yadav asked: What would be the effect of any additive such as some ligands or lattice defect on dewetting?

Jayant Singh responded: There is no simple answer as it depends on the order of roughness created by the lattice or additive. Lattice defects or the roughness of a substrate can change the dewetting behaviour significantly. For example, creating roughness such as pillar height and gaps on the surface may increase the dewettability of the substrate. Thus, detachment of the metallic droplet from the substrate can be tuned by patterning the substrate. We haven't yet looked into the role of such patterns in the current work. However, it has been explored for a Cu film on graphene.[1]

1 X. Li , Y. He , Y. Wang , J. Dong and H. Li, *Scientific Reports*, 2014, **4**, 3938.

Priyadarshi Roy Chowdhury communicated: What is the burn-in period of the simulations associated with your work mentioned in this paper?

Jayant Singh communicated in reply: Assuming the heating of the film is the 'burning period' for the dewetting study, we carried out the simulations for 2 ns for all the thicknesses considered in our study.

Daan Frenkel opened the discussion on the paper by Achille Giacometti:† As you correctly point out in your paper, it is usually impossible to make the pitch of

† Achille Giacometti's paper was presented by Hima Bindu Kolli, University of Oslo, Oslo, Norway.

a cholesteric phase compatible with the periodic boundary conditions used in a simulation. Hence, simulations tend to probe cholesterics in a twist-free (*i.e.* stressed) state.

Can you estimate (*e.g.* using density functional theory (DFT)) how much this torsional stress would affect the location of the phase boundaries that you have computed?

Hima Bindu Kolli answered: Results on the cholesteric phase reported in the presentation stem from DFT calculations (see page 5, lines 20-25 of the provided paper). Depending on the shape of the particles (as given by *r* and *p*), as well as on the packing fraction, estimates of the obtained pitches range from a few hundred to several thousands of units of length (Fig. 5). Hence, in most cases, these values are clearly significantly larger than the computational box and the torsional effect provided by the cholesteric twist cannot be probed by periodic boundary conditions, as remarked. The above estimates were obtained by using a DFT hinging upon a second-virial approximation. Increasing precision could be achieved by extending the calculation to account for the third virial coefficient, along the lines suggested in our previous work (Ref. 24 in the manuscript). We plan to do this in a future contribution.

The presented phase diagrams reported in Fig. 6 and 9 only deal with the screw-like phases. Here, unlike the cholesteric phase, the pitch of the phase is of the order of the pitch of the particle *p*, that is, a few units. In this case, the periodic boundary conditions are perfectly compatible with the phase, so we expect the estimates for the location of the phase boundaries reported in Fig. 6 and 9 to be very reliable. Of course these phase diagrams do not include the cholesteric phase explicitly. However, we have made progress in the direct simulation of the cholesteric phase by embedding the system between two parallel plane hard walls, as alluded to on page 13, lines 5–10. These preliminary results[1] indicate that i) the screw-like and the cholesteric phase are mutually exclusive; ii) the whole regions indicated by N^* in the phase diagram reported in Fig. 6 and 9, are cholesteric phases. There is also a recent preprint by Dussi and Dijkstra[2] that reports similar simulations obtaining a cholesteric phase for a different, but similar in spirit, particle model.

1 part of an oral communication entitled "Self-assembly of hard helices: cholesteric and screw-like, nematic and smectic, phases" given at the XII International Conference on Materials Chemistry, York, England, 20-23 July 2015.
2 S. Dussi and Marjolein Dijkstra, 2015, http://arxiv.org/abs/1511.09318.

Nicholas Kotov asked: What would be the difference in structure of the cholesteric phase as well as the phase diagram between hard and soft helices? Why and how is it different?

Hima Bindu Kolli responded: We would not expect significant qualitative differences between the structure of the cholesteric phase as formed by hard or soft helices. This should hold true in particular if one considers soft helices still interacting between themselves with short-range repulsive interactions. These expectations are based on past experience with soft short-range repulsive rod-like particles as compared to hard rod-like particles,[1] as well as on numerical simulations we are currently carrying out in which the bead–bead interaction is changed from hard-sphere-like to soft repulsive (specifically, Weeks-Chandler-Andersen-like, *i.e.* the repulsive part of the properly shifted Lennard-Jones potential energy function). The addition of attractive interactions is again expected not to alter the main qualitative features of the structure of the cholesteric phase but they will have a role in modifying the value of the pitch as well as the location of the isotropic–cholesteric phase transition.

1 *e.g.* G. Cinacchi, L. De Gaetani and A. Tani, *Phys. Rev. E*, 2005, **71**, 031703.

Vimala Sridurai said: In liquid crystal systems with high chirality, Blue phases with microscopic cubic symmetry which are optically isotropic but show birefringence are obtained between the isotropic and the cholesteric phases.

So, were any such phases encountered in your studies? And one humble comment. While presenting the phase diagram, you mentioned various liquid crystalline phases like isotropic, nematic, smectic and so on. Typically in liquid crystal physics, the isotropic phase is not considered as a liquid crystalline phase. As liquid crystalline phases are mesophases between the isotropic and crystal phases.

Hima Bindu Kolli answered: In our simulations, performed using Periodic Boundary Conditions (PBCs) and not-large-enough systems, we could not find phases with long-scale periodicity, like the cholesteric phase and Blue Phases. In such simulations the isotropic phase was followed, at higher density, by a nematic phase with a uniform director. However, both second-virial Density Functional Theory (reported in the paper) and recent simulations where the PBC constraint was relaxed along one direction (Ref. 36 in the manuscript), showed that the liquid crystal phase following the isotropic phase is indeed cholesteric (see my above reply to Daan Frenkel's question). Likewise, we cannot exclude that, at least for some of our systems, Blue Phases can exist between the isotropic and the cholesteric phase. However, their identification may pose significant problems because of their 3-dimensional periodicity and especially their stability range, which is usually very narrow. We concur that this would be a very interesting study that is left for further investigation.

Daan Frenkel asked: The 'screw phase' of helical particles is fascinating and may have exciting dynamical properties. In particular, the translation–rotation coupling will be extremely large.

Hence, one could imagine that a helical 'tracer' particle with a dipole moment perpendicular to its axis could be 'propelled' by (rotating) micro-wave radiation. Are you aware of any studies of such effects and, if not, is this something that you could study by simulation?

Hima Bindu Kolli replied: We very much appreciate this comment. While we did not consider this very interesting possibility, nor are we aware of any studies of such helical tracer particles, an analysis of the dynamical properties would indeed be very exciting and is very high on our agenda. Imagine having a well equilibrated system of hard helices in a screw-nematic phase. Then imagine randomly selecting one of the helices and start spinning it about its own axis. This could be achieved, for instance, using the trick that you have suggested. According to our predictions on the translational–rotational coupling, we could expect the helix to move like a "nut in a bolt". How and how fast this motion would be achieved is inherently interesting.

Charusita Chakravarty communicated: 1. The papers show phase diagrams for hard helices. Soft spheres are well-studied. Can the authors comment on a way to add a simple softness parameter and the effect on the phase diagram? 2. While biological peptides may not fit the models here, well-developed methods for synthesising peptides and peptido-mimetics exist. Could these be adapted to fit the models studied here, especially to generate the smectic phases?

Hima Bindu Kolli communicated in reply: 1. Indeed, there are several issues related to this question. We did consider soft repulsive potentials in a few particular representative cases and found almost identical results. We expect this to be a general feature with possibly only minor changes in the exact locations of the phase boundaries. In addition, we also implemented successive umbrella sampling grand-canonical calculations, patterned after a similar calculation on sphero-cylinders,[1] with finite repulsive energy. This clearly introduces a dependence on temperature that can eventually be eliminated by sending this energy to infinity to recover the hard-core potential. This procedure is necessary to account for the extremely low insertion probability of hard helices at high density. A related interesting case, that we plan to tackle soon, hinges on the addition of a charge to hard helices within a DLVO description, that introduces a softness related to the Debye screening length of the DLVO potential.

2. Helical peptides can form liquid crystal phases; a well studied system is poly(gamma-benzyl-L-glutamate), which exhibits nematic and, if monodisperse, also smectic phases. This and other helical peptides, as well as other types of helical polymers, are possible candidates for the formation of the phases observed for our model helices. However, the existence of azimuthal correlations with the same periodicity of the peptide helix (less than one nm), which would be a signature of such phases, may be difficult to prove.

1 R. Vink, S. Wolfsheimer and T. Schilling, *J. Chem. Phys.*, 2005, **123**, 074901.

Priyadarshi Roy Chowdhury communicated: What is the iteration period of the simulation associated with your paper?

Hima Bindu Kolli communicated in reply: As mentioned in the paper (section II, page 5, top), a typical simulation has about 3×10^6 MC steps of equilibration time, followed by an additional 2×10^6 MC steps.

Guruswamy Kumaraswamy opened the discussion on the paper by Sudeep Punnathanam: What is the size of the critical nucleus compared to the size of the trough on the rough surface? Does this size ratio play a role in the enhancement of nucleation?

Sudeep Punnathanam answered: The size of the critical nucleus (defined as the number of particles in the nucleus) varies between 50 and 80. Assuming a spherical shape, this corresponds to a diameter of nearly 5 s. The surface is modeled as a sinusoidal wave with an amplitude of 15 s and a wavelength of 30 s.

Daan Frenkel posed two questions: 1. In your simulations, you find that the effect of nano-wrinkles on crystal nucleation only becomes significant if the range of the attractive particle–wall interaction is larger than approximately 20% of the protein diameter. Can you indicate if this regime is likely to be accessible in experiments? 2. In the simulations of van Meel *et al.* (Ref. 11 of your manuscript) capillary condensation played a key role in enhancing crystal nucleation. In your simulation, you observe enhancement of the particle protein concentration in the troughs of the wrinkles. It is not clear to me if this accumulation can be viewed as capillary condensation.

Sudeep Punnathanam replied: One of the conclusions from our simulations is that enhancement of protein nucleation due to nano-wrinkles can occur if there is sufficient attraction between the wall and the protein molecules. In this work this effect was simulated by increasing the range of the square well interaction. The effect of nano-wrinkles was felt only when the interaction range was 20% of the protein diameter. However, as you are rightly hinting at, this is too large a range for any realistic wall–protein interaction. Reflecting upon your question, another way to simulate an increase in the wall–protein interaction would have been to increase the well depth of the square well interaction while keeping the range at 0.076 times the protein diameter. Still, we felt that our conclusion is experimentally relevant since the range of attraction is limited to a single monolayer of protein molecules at the surface. This formation of a dense monolayer can also be caused by increasing the well depth of the square well. Whether this increase is greater at the troughs of the nano-wrinkles will need further study.

Our comparison of this work with that of van Meel *et al.* is based on the following similarity. van Meel observed capillary condensation, which is the formation of a liquid (a high density fluid region) inside pores, as the cause for enhanced nucleation. We observed formation of a dense fluid layer at the troughs which in turn enhances nucleation.

Sanat Kumar commented: It is possible that the critical nucleation size depends on the state that you come out of, namely gas *vs.* liquid. So, at a flat surface you might have nucleation from the gas phase nucleating out; on the other hand, in a pocket there might be liquid and the starting state is quite different. Does the critical nucleus size depend on where you start from? Is the pathway different and therefore is the critical nucleus different in the two cases?

Sudeep Punnathanam responded: The critical nucleus size as well as the nucleation barrier do depend on the initial state. A pocket of liquid will definitely lower the nucleation barrier. Such a situation was observed by van Meel *et al.*[1] where a pocket of liquid was formed inside nano-pores due to capillary condensation. In our simulations, there is a dense layer of proteins near the surface. This density is higher near the troughs of the wrinkled surface compared to the regions with zero curvature. This higher density is the reason for the enhanced nucleation rate on the nano-wrinkled surfaces.

1 J. A. van Meel, R. P. Sear and D. Frenkel, *Phys. Rev. Lett.*, 2010, **105**, 205501.

Sofia Kantorovich remarked: Is there a non-monotonic dependence of the nucleation rate/size of nuclei on the size of the wrinkles?

Sudeep Punnathanam answered: In this study, we haven't investigated the dependence of nucleation rate on the size of the wrinkles. A non-monotonic dependence can be expected when the amplitude and the wavelength of the wrinkles become comparable to the size of the protein molecules. At such conditions, the nucleation rate would be much higher when there is a match between the lattice spacing of the growing crystal nuclei and the wavelength of the wrinkles.

Daan Frenkel asked: As was shown by Page and Sear[1] there is an optimal pore size for enhanced nucleation, namely under conditions where the nucleation barrier for crystal formation inside a pore is comparable to the barrier for 'break-out' of a crystal into the bulk.

1 A. J. Page and R. P. Sear, *Phys. Rev. Lett.*, 2006, **97**, 065701.

Ajeet Srivastav said: How do the nano-wrinkled surfaces facilitate the supersaturation and affect the activation barrier and critical size of the nucleus?

Sudeep Punnathanam responded: There is an enhancement in the density of the protein molecules near the troughs of the nano-wrinkled surfaces. This enhancement in density is strongest near the metastable vapor–liquid critical point. The higher density at the troughs leads to a lowering of the activation barrier for the critical nucleus as well as the number of particles required to form the critical nucleus.

Jayant Singh remarked: Does the wetting state affect the nucleation behaviour? In other words, how does the nucleation behaviour change if the proteins sit on the top of the rough substrate (a metastable state typically) as opposed to the results presented in the meeting.

Sudeep Punnathanam answered: The lengthscale of the roughness studied is much larger than the size of the proteins. Hence the critical nucleus is formed at the locations depicted in Fig. 5. If the roughness is of the order of the size of the protein molecules, compatibility between the periodicity of the roughness and the structure of the nucleating protein crystal would strongly affect the nucleation barrier. This aspect has been explored by van Meel *et al.*.[1]

1 J. A. van Meel, R. P. Sear and D. Frenkel, *Phys. Rev. Lett.*, 2010, **105**, 205501.

Priyadarshi Roy Chowdhury communicated: Have you used other simulation models besides Monte Carlo? Or is the Monte Carlo model the best amongst other simulation models with respect to your presented paper?

Sudeep Punnathanam communicated in reply: We only used Monte Carlo simulation techniques. One can use NPT molecular dynamics to generate trajectories however each trajectory needs to be accepted or rejected using Metropolis criteria.

Priyadarshi Roy Chowdhury communicated: What is the iteration period of the Monte Carlo simulation results associated with your paper?

Sudeep Punnathanam communicated in reply: As mentioned in section 3.3, our simulations consisted of 100000 trajectories for equilibration and 4000000 trajectories for production. We defined a trajectory as a series of Monte Carlo moves consisting of on average 20 displacement, 1 insertion and 1 deletion move per particle.

Sanat Kumar began the discussion on the paper by Yogesh M. Joshi: Some of your data suggest that a gel forms in the time scale of 3 h (or some such small time). The charge conduction changes on the other hand are taking hundreds of hours. Are you suggesting the structure is changing on these slower time scales? What is the change? Is it big? What is this slower process? If it is ion bridging related then you can prove the hypothesis by using a higher valence countering – maybe replace sodium with calcium. Do you expect the changes to take much longer? Maybe you can also vary the monovalent salt concentration to address this issue.

Yogesh M. Joshi answered: As rightly mentioned the gel formation takes place over a timescale of hours, while an increase in ionic conductivity continues over a timescale of hundreds of hours. It is proposed that even after gel formation, the stacks of Laponite plates continue to delaminate, and these delaminated plates participate in the network structure. Consequently, the network becomes progressively dense as a function of time, which also gets reflected in the time evolution of G' and G'' over the mentioned timescales. To best of our knowledge, ion bridging has not been observed among Laponite particles in aqueous media. Multivalent cations, on the other hand, either counterions or those externally added, are known to have a profound effect on the gelation kinetics of clay suspensions in general and Laponite suspensions in particular. Typically the higher the valency of the cations, the more pronounced the shielding of the electrostatic repulsive interactions. Consequently, gel formation accelerates in the presence of multivalent cations. In our previous work we studied an effect of a monovalent salt on the aging behaviour of Laponite in great detail.[1]

1 A. Shahin and Y. M. Joshi, *Langmuir*, 2012, **28**, 15674–15686.

B. L. V Prasad asked: Is it possible to remove the sodium ions faster by increasing the temperature or following any other means? Does this lead to faster gelation?

Yogesh M. Joshi replied: An increase in temperature indeed enhances the ionic conductivity suggesting dissociation of the sodium ions, and therefore the delamination process, is faster at higher temperatures as observed by Dhavale *et al.*[1] An increase in temperature also leads to faster gelation.[2] To the best of our knowledge, we have not come across any other means by which the delamination process could be accelerated.

1 T. P. Dhavale, S. Jatava and Y. M. Joshi, *Soft Matter*, 2013, **9**, 7751–7756.
2 A. Shahin and Y. M. Joshi, *Langmuir*, 2012, **28**, 15674–15686.

Jacques Jestin questioned: What is the effect of the particle concentration? According to the orientation of the platelets, one should expect to see different percolation thresholds?

Yogesh M. Joshi responded: Our unpublished results show that an increase in the concentration of Laponite in the suspension reduces the gelation time, while it increases the fractal dimension associated with the critical gel state.

Sanat Kumar remarked: If you apply steady shear do you get gelation? I presume you realign the Laponite particles?

Yogesh M. Joshi responded: Laponite will eventually gel no matter whether its suspension is subjected to shear or not.[1,2] However, as rightly pointed out, the particles will align and therefore the system will not show a critical gel state as reported by Jatav and Joshi.[3]

1 A. Shukla and Y. M. Joshi, *Chemical Engineering Science*, 2009, **64**, 4668–4674.
2 Y. M. Joshi, A. Shahin and M. E. Cates, *Faraday Discuss.*, 2012, **158**, 313–324.
3 S. Jatav and Y. M. Joshi, *J. Rheol.*, 2014, **58**, 1535–1554.

Erika Eiser commented: I worked with Daniel Bonn on Laponite; one way to decouple timescales would be to strongly shear the samples and filter them, which also induces strong shear. This procedure leads to reproducible results. Have you tried such a procedure? In this way one could first let the sample reach its final delamination state and then only shear it. A second question is about the magnitude of the interactions holding the platelets edge-on-face together. Are these simply van der Waals interactions? In this context, does it matter whether you work in a nitrogen atmosphere or at ambient conditions?

Yogesh M. Joshi responded: Either applying strong shear or filtering the sample using a micron sized mesh indeed leads to reproducible results only if the time difference between the various experiments is very short compared to the time elapsed since preparation of a suspension. It has been reported by Shahin and Joshi[1] that a Laponite suspension undergoes irreversible aging wherein the application of strong shear cannot break the network to the individual particle scale such that the average size of the cluster that can be obtained after shear melting increases with time elapsed since preparation of the same.

The fact that the faces of the Laponite disks are negatively charged while the edges of the same are positively charged suggests the interactions holding the edges and faces together are electrostatic in nature.

Our experiments suggest that if the concentration of Laponite in the suspension is sufficiently large, whether one prepares and preserves the suspension under nitrogen or air does not matter. However for small concentrations of Laponite, a nitrogen atmosphere ensures the magnesium does not leach from the Laponite particles. The chemical stability limits for different Laponite concentrations after various days elapsed since preparation of the suspension have been reported by Joshi and coworkers.[2,3]

1 A. Shahin and Y. M. Joshi, *Langmuir*, 2010, **26**, 4219–4225.
2 S. Jatav and Y. M. Joshi, *App. Clay Sci.*, 2014, 97–98, 72–77.

3 R. Mohanty and Y. M. Joshi, *App. Clay Sci.*, 2016, **119**, 243–248.

Sanat Kumar remarked: Have you tried sonication as a means of separating the clay (Laponite) sheets?

Yogesh M. Joshi answered: We often perform sonication on a Laponite suspension. Sonication is observed to only marginally increase the rate of time evolution of ionic conductivity of a Laponite suspension. Consequently sonication does not seem to significantly influence the delamination process in a Laponite suspension.

Erika Eiser asked: The problem is the slow delamination process of clumps in the Laponite powder, which becomes increasingly more difficult for increasing concentrations. At the same time going to low concentrations, say 1 wt% or lower increases the gelation time making measurements more challenging. However, one possibility to accelerate the ageing process in these solutions is to add small amounts of salt. In this way better delamination and rejuvenation of the sample due to shear could be reached.

Yogesh M. Joshi responded: We agree that in a low concentration Laponite suspension, complete delamination may occur before a critical gel state is reached. We also agree that by adding salt, the time at which a critical gel state is reached could be altered. However, as mentioned before, shear cannot break all the inter-particle bonds among the Laponite particles.

Ranjini Bandyopadhyay asked: Could you discuss the mechanism of the clay delamination process under the freezing–thawing cycle?

Guruswamy Kumaraswamy added: I am also curious about how the freeze–thaw cycle results in delamination of the Laponite stacks. I would have imagined that expulsion of the Laponite particles to the ice domain boundaries would result in (re)stacking of the plate like particles. Could you please comment on how your experimental protocol results in enhancing the delamination of Laponite stacks? Also, I would like to comment that it should be possible to use SAXS to characterize the Laponite aggregate structure through the freeze–thaw process. This might provide direct structural evidence for the delamination that you propose.

Yogesh M. Joshi answered: The delamination of Laponite stacks is proposed to occur as a function of time over the timescale of days. The newly delaminated clay plates participate in the gel/network structure. What we believe the freeze–thaw cycle does is to break the gel/network structure into primary components. We propose that upon increasing the temperature of the frozen Laponite suspension, in the vicinity of the melting point, ice surrounding the Laponite particles undergoes a phase transformation from solid to liquid. Since the relative permittivity of liquid water at the freezing point is 88 while that of ice is 3.2, the presence of crystallites along with water obstructs the electrostatic interactions, which is the very backbone of a structure. Consequently, after defrosting, the suspension acquires a liquid state of low viscosity. Consequently carrying out a freeze–thaw cycle at a later date since preparation of the suspension, as done in the present experimental protocol, leads to a greater extent of delaminated Laponite particles in a suspension. We agree that SAXS can be used to provide direct structural evidence to confirm the proposed delamination mechanism.

Guruswamy Kumaraswamy commented: One of the reasons that the research community started working on Laponite was that it appeared to be an excellent model system for plate-like particles. It is commercially available, synthetic (and therefore, has controlled concentration), has moderate size polydispersity and a large aspect ratio. However, your work clearly indicates the challenges inherent in this system. Could you comment on the suitability of using Laponite as a model system to understand the phase behaviour of plate-like particles?

Yogesh M. Joshi replied: Certainly Laponite is not a model system to study the phase behaviour of plate-like particles due to their highly charged nature (under certain circumstances the particles even possess dissimilar charges). However, an aqueous Laponite suspension can certainly be considered as a model soft glassy material that shows spectacular aging and rejuvenation behaviour. Moreover, even though a large volume of literature exists on the Laponite suspension, there are several questions that are still puzzling. On the other hand, smectite clays in general and Laponite clay in particular are very popular additives used to modify the rheological behaviour in a number of industries that deal with water based paste-like products. Therefore it is important that we understand this system well.

Vinothan N. Manoharan commented: In Prof. Joshi's paper the results are interpreted using DLVO theory. But DLVO theory has several limitations: it is a mean-field theory; it assumes constant charge density; and to understand structure formation one has to assume pairwise additivity. Prof. Joshi, is there any reason to believe, *a priori*, that DLVO theory should give sensible results for Laponite suspensions? A broader question for all delegates: Has DLVO theory been "proven innocent" for modeling interactions in nanoparticle dispersions? In other words, is it more helpful than harmful to our understanding of structure formation?

Jayant Singh answered: It is evident from our recent work on silica deposition on a soft surface that DLVO cannot capture many aspects of nanoparticle structural formation. Was the use of DLVO harmful? I guess not as long as one is aware of the danger of approximation. This may be true for any mean-field theory based prediction.

Nicholas Kotov responded: The application of DLVO theory to nanoparticles reminds me of the story about Cinderella's sisters. They tried very hard to put on the shoe, but it did not fit any way they tried it. We can conclusively state that DLVO theory is not applicable to nanoparticles below 20 nm in diameter because nearly all key assumptions of DLVO theory are violated at this scale.[1] One of the central fundamental problems is the separation of interactions into electrostatic, van der Waals, hydrophobic and others. For nanoparticles at small separations (20 nm and less) these interactions are non-additive and become interdependent on each other. Atomistic molecular dynamic modeling can help to resolve this problem. Potentially other methods are possible, too.

1 C. A. Batista, R. G. Larson, N. A. Kotov, *Science*, 2015, **350**, 1242477.

Yogesh M. Joshi answered: We agree with the mentioned criticism that DLVO theory makes a number of assumptions, and therefore before employing DLVO theory to analyze any experimental data it is important to assess to what extent the assumptions are valid. Firstly regarding the Laponite particle, it is safe to assume that negative and positive charges are distributed uniformly across the faces and the edge, respectively. DLVO considers the addition of van-der Waals interactions and electrostatic interactions. The former is dominant only over inter-particle lengthscales of less than 1 nm. In this work we used DLVO theory to essentially make a point. We solved the theory only for a case when two plates approach each other in a parallel fashion. DLVO theory predicts that with an increase in counterion concentration the height of the repulsive barrier increases while its width decreases. Consequently, when two particles approach each other in a non-parallel fashion, with the edges being positive and the faces being negative, DLVO theory implicitly suggests that edge to face bond formation is faster with an increase in ionic conductivity as observed experimentally. Regarding the broader question on DLVO theory, we still feel that DLVO theory gives good qualitative information about the interparticle interactions. However it is important to judiciously examine the assumptions involved in DLVO theory with respect to the experimental system before interpreting the results.

Nicholas Kotov addressed Vinothan N. Manoharan and Yogesh M. Joshi: Thank you Vinothan Manoharan for bringing this issue up. The validity of DLVO at the nanoscale is the core of this meeting. Referring to my earlier analogy to the story of Cinderella, maybe it is a good time to turn this story around and say that DLVO is indeed the wrong shoe for this foot. The foundational assumptions of DLVO are not valid for many if not all nanoparticles especially when they have such an anisotropic shape as clay nanosheets.

Vinothan N. Manoharan answered: I agree. As Prof. Kotov notes in his recent review paper,[1] atomistic simulation is a tractable alternative to DLVO for small nanoparticles. For larger (10–100 nm) particles, it will be important to develop experimental tools that allow direct or indirect measurement of the effective potential between particles. We recently demonstrated an optical technique that allows us to measure the interaction between two 50 nm nanoparticles at separations close to contact.[2] Other techniques, such as fluid-cell electron microscopy, offer even higher resolution.

1 C. A. S. Batista, R. G. Larson and N. A. Kotov, *Science*, 2015, **350** (6527), 1242477.
2 S. Faez, Y. Lahini, S. Weidlich, R. F. Garmann, K. Wondraczek, M. Zeisberger, M. A. Schmidt, M. Orrit and V. N. Manoharan, *ACS Nano*, 2015, **9**, 12349.

Yogesh M. Joshi replied: Kotov and coworkers[1] described the following four fundamental assumptions of DLVO theory that are not valid at the nanometric lengthscale: 1. Contrary to assumption the sizes of the solvated ions and solvent molecules are not negligibly small over the lengthscales of 1 to 20 nm; 2. The additivity assumption of van der Waals and electrostatic interactions fails over the considered nanometric lengthscales; 3. The thickness of an interfacial layer on a nanoparticle could be as large as the nanoparticle size itself raising a question over the assumption of uniform continua outside a particle; and 4. The assumption of nanoparticles having a simple shape. In the present work we consider a special case of two flat plates approaching each other in a parallel fashion over lengthscales of typically up to 7 nm with a Debye screening length of around 3 nm. On the other hand the sizes of the water molecule as well as Na^+ ions are around 0.3 and 0.1 nm respectively, which clearly suggests that the first assumption of DLVO theory is not too bad if not good. Regarding the failure of the additivity assumption of van der Waals and electrostatic interactions, it has been observed that for the Laponite system van der Waals interactions are negligible beyond 1 nm of inter-surface distance of the plates.[2] Therefore for the lengthscales from 1 to 7 nm we are primarily accounting for only the electrostatic interactions between the faces of the particles. Consequently violation of the non-additivity assumption does not affect the present system over most of the considered region. The third issue is about a layer on the nanoparticle surface. Since the present system contains only water, salt and the Laponite particles, the layer on the face of the Laponite particle is essentially a Stern layer of the order of 0.1 nm thickness (which is followed by a diffuse layer in the solvent). Finally regarding the shape of a particle, it is well known that clays are plate-like particles (in the case of Laponite the diameter is around 30 nm). As mentioned before we consider only a case when two plates are parallel to each other. The electrostatic interaction between the two charged plates parallel to each other is a classic problem having an analytical solution, and examples of clays are often given as textbook problems. We understand that in all the above cases various assumptions may not be perfectly validated owing to the finite size of ions involved as mentioned by Kotov and coworkers. However isn't it the case that DLVO theory tends to show significant interactions for many systems only below 10 nm of inter-surface distance irrespective of the size of a particle.[3] Therefore we believe that there is nothing specific about Laponite that would violate the DLVO assumptions more than in the case of larger particles as the distances over which significant interactions are observed are similar in both. Finally, it is important to note that the present work uses DLVO theory for a specific configuration just to make a qualitative point. To conclude metaphorically, in the whole kingdom only Cinderella had the precisely same sized feet for the shoes. All the other girls, let alone her sisters, can still make use of oversized or undersized shoes as no perfect sized shoes are available with an understanding that there is indeed a mismatch, in some cases greater while in other cases smaller.

1 C. A. S. Batista, R. G. Larson and N. A. Kotov, *Science*, 2015, **350** (6527), 1242477.
2 D. Saha, R. Bandyopadhyay and Y. M. Joshi, *Langmuir*, 2015, **31**, 3012–3020.
3 J. N. Israelachvili, *Intermolecular and Surface Forces*, Academic Press, London, UK, 3rd edn, 2010.

Madivala G. Basavaraj asked: The analysis of interactions in a system of clay particles in terms of classical DLVO interactions seems an over simplification, especially because of the fact that the clay particles are "patchy" – that is – the rim and the face are differently charged. In such cases one has to worry about orientation dependent interactions in asymmetrically charged anisotropic colloids. How does one visualize a double layer in such cases? Would the screening of the double layer be different in different directions? Do you include the contribution of Na^+ ions that dissociate from clay particles into the calculation of the Debye screening length?

Yogesh M. Joshi responded: Firstly even though the rim and the faces have different charges, it is safe to assume the charge distribution to be uniform as well as continuous. Secondly as mentioned before we solve the DLVO theory only for a case when two Laponite particles approach each other in a face to face parallel fashion. We therefore do not solve DLVO theory for interactions between an edge and a face, when two particles approach each other in a nonparallel fashion. We only implicitly infer about such interactions. Similar to an electric double layer on the negatively charged faces of Laponite particles, there exists a double layer on the positively charged edges as well. However as mentioned before we did not analyze the interaction between the edges and faces using DLVO theory.

We indeed considered the concentration of sodium ions while computing the Debye screening length. On the other hand, the concentration of dissociated sodium ions also leads to surface charge density on the faces of the Laponite particles.

Lynn Walker commented: I worry that DLVO theory is dangerous in a system as complex as this one. My concern is that it might be leading to some erroneous interpretation of the gelation behaviour. I do recognize that it is the obvious zeroth order place to start, but some of the observations might suggest the need for more complex theory. This is more of a comment than a question.

Is it possible that some of the behaviour in these systems is governed by aggregate interactions rather than individual particle interactions? In a number of colloidal systems, there are observations of critical aggregates forming that have very different inter-aggregate interactions than the individual particles.

Yogesh M. Joshi responded: We agree that the DLVO theory makes a number of assumptions and it is important to carefully scrutinize the assumptions with respect to the experimental system. It is important to note that in this work we use DLVO theory essentially to make a point. We solve the theory only for a case when two plates approach each other in a parallel fashion. The DLVO theory predicts that with an increase in counterion concentration the height of the repulsive energy barrier increases while its width decreases. Consequently when two particles approach each other in a non-parallel fashion, with the edges being positive and the faces being negative, DLVO theory implicitly suggests edge to face bond formation is faster with an increase in ionic conductivity as observed experimentally. The very fact that gelation precedes with complete delamination of the stacks with the difference in timescale being at least one order of magnitude suggests that, to begin with, the stacks of Laponite particles undergo gelation. As time progresses, more plates delaminate and the size of a stack reduces. Furthermore, while the surface of the stacks or aggregates is formed by the faces and edges of the Laponite particles themselves, as rightly mentioned the inter-aggregate interactions could be different from those of the interparticle interactions.

Siddharth Kulkarni remarked: Why does the crossover between G' and G'' get delayed with a decrease in angular frequency (Fig. 4)? Is this indicative of liquid-like behaviour?

Yogesh M. Joshi responded: In gel forming systems sampling a critical gel state, such as the aqueous Laponite suspension, the frequency dependence of time at which G' and G'' cross each other depends on the fractal dimension of the critical gel state (or the unique value where all the iso-frequency tan δ curves meet). If the fractal dimension associated with a critical gel state is above 2, as is the case with the Laponite suspension studied in the present work, crossover between G' and G'' will get delayed with a decrease in frequency. On the other hand if the fractal dimension of a critical gel state is below 2, crossover between G' and G'' will get delayed with an increase in frequency.

Priyadarshi Roy Chowdhury communicated: In Fig. 2, the behaviour of free energy interactions obtained from DLVO theory, of the Laponite particles, having face to face interactions is shown. What would be the behaviour of these Laponite particles when they are moderately apart and far apart from each other from a theoretical point of view?

Yogesh M. Joshi communicated in reply: DLVO theory assumes electrostatic interactions and van-der Waals interactions to be additive. The latter are dominant only at very short length-scales (<1 nm). The interaction associated with the former on the other hand depends on the surface potential as well as the Debye screening length. In the present work the DLVO theory has been solved to obtain the interaction between two Laponite particles approaching each other in a parallel fashion. For this configuration, over moderate length-scales (less than 10 nm but more than 1 nm), a decrease in distance between the same will lead to an increase in repulsion as shown in Fig. 2. However at large distances between the plates (>10 nm), interaction among them will be negligibly small.

Guruswamy Kumaraswamy opened the discussion of the paper by Daan Frenkel: If we consider pathways to the ordered state, could we derive inspiration from protein folding where chaperones play an important role in the folding process? Are there any examples that you are aware of where such biomimetic strategies have been employed to form ordered structures from complex mixtures? Is there a way to obtain insight from simulations on how to design such chaperones for ordering?

Daan Frenkel answered: In the design of complex, self-assembling structures, the building blocks must obviously encode for the target structure, but as Dr Kumaraswamy points out: this is not enough. It is crucial that the self-assembly pathway does not terminate in a metastable 'off-target' structure, of which there are very many. Protein folding is an excellent example: many proteins would misfold if, in their nascent state, they were not be protected by a variety of chaperones. Similarly, in synthetic chemistry, protecting groups are commonly used to inhibit undesired ('off-pathway') reactions. Clearly, in complex self-assembly, the use of protecting groups or chaperone-like units is (or will be) of crucial importance. In existing DNA-brick structures, the DNA bricks on the surface are commonly made non-reactive by terminating them with poly-T chains. I am not aware of the use of DNA-brick based chaperones, but the potential is obvious.

G V Pavan Kumar asked: In your simulations on aggregates, the coordinate number will influence the lattice phonon modes. Can you simulate them?

Daan Frenkel replied: You are absolutely right that different coordination numbers will affect the phonon density-of-states of the self-assembled structures. At this stage, not much is known (at least not to me) about the phonon spectrum of DNA-brick structures. Of course, as the structures form in aqueous solution, most lattice vibrations are likely to be overdamped. But even so, the low-frequency phonon spectrum is of crucial importance for the elastic properties of the self-assembled structures. It would be interesting to investigate this, but I fear that the highly coarse-grained model is not sufficiently realistic to capture the phonon spectrum.

Priyadarshi Roy Chowdhury asked: Is it possible to achieve high co-ordination numbers in many component self assembly in spite of having high structural barriers and complexities?

Daan Frenkel answered: We were indeed surprised to find that, in simulations, we could self-assemble structures with high coordination numbers because at low super-saturations the nucleation barriers for such clusters are high. We feared that at higher super-saturation, many defects would be incorporated into the growing structures. The simulations showed that the situation was not as we feared: we found that even structures with a coordination number of 12 could self-assemble in a reasonable time and with some, but not very many, point defects. Whether this can be achieved in experiments remains to be seen.

Sofia Kantorovich said: Is it possible to introduce temperature activated patches, so that not only the combination of active sites per particle determines the final self assembled structure, but the latter can also be altered afterwards by changing *T*?

Daan Frenkel replied: Thus far, the main tool that has been used to change functionality is the composition of the mixture from which the structures self-assemble.

However, there are techniques by which the functionality of patches can be made temperature-dependent. Examples can be found in the literature.[1,2]

But, of course, there is huge scope for new approaches to add active 'control' to complex self-assembly.

1 S. Angioletti-Uberti, B. M. Mognetti and D. Frenkel, *Nat. Mater.* 2012, **11**, 518–522.
2 W. B. Rogers and V. N. Manoharan, *Science*, 2015, **347**(6222), 639–642.

Vinothan N. Manoharan asked: In response to the question from Dr Kantorovich: Arvind Murugan and colleagues[1] conducted a theoretical study of how to make nanoparticle dispersions that can self-assemble into multiple different prescribed structures. Different structures can be assembled from the same mixture of components by changing, for example, the temperature. It should be possible to make such mixtures using DNA-labeled particles and strand displacement reactions. Michael Brenner's group[2] has also used simulations to examine the assembly of maximally specific colloidal particles with high coordination number. Just as in Prof. Frenkel's work, the self-assembled structures contained defects, but those defects did not prevent the target structure from forming.

Experimentally, colloids with 'addressable complexity' can be made using grafted DNA strands, but it is difficult to make multiple species all having the same melting temperature. The melting temperature depends on the sequence of the grafted DNA strands as well as the grafting density, which is difficult to control precisely. An alternative approach is to use 'blunt-end stacking' interactions between grafted DNA sequences.[3] These interactions are specific to the shape of the DNA constructs, and they can have a different temperature dependence than base-pairing interactions. Prof. Frenkel, have you considered studying the self-assembly of particles that have both sequence-specific and shape-specific interactions?

1 A. Murugan, Z. Zeravcic, M. P. Brenner and S. Leibler, *PNAS*, 2015, **112**, 54.
2 Z. Zeravcic, V. N. Manoharan and M. P. Brenner, *PNAS*, 2014, **111**, 15918.
3 S. Woo and P. W. K. Rothemund, *Nature Chem.*, 2011, **3**, 620.

Daan Frenkel replied: In my talk, I did not really touch upon the 'multifarious' structures considered by Murugan *et al.*, but such systems that use the same building blocks to encode for a (large) number of different target structures are clearly extremely important. They are used in biology and there seems to be little doubt that they can be used in DNA brick assembly. However, in the work reported here, we did not consider this option.

The second suggestion (blunt-end stacking) is indeed a technique to expand the 'alphabet' of interactions. It is therefore likely to become increasingly important as we move to larger structures. Blunt-end stacking exploits shape complementarity and that is a good way to add selectivity to the interaction between different building blocks. I agree completely that the potential of blunt-end stacking deserves to be explored numerically in a generalised coarse-grained model.

Alison Edwards asked: It is refreshing to see published the initial hypothesis on which a study has been undertaken – especially when the results are contrary to the hypothesis. In the context of A. L. MacKay's classic[1] on a dense non-crystallographic packing of spheres based on 12-fold coordination, I wonder whether you have pondered the energetic consequences that such a 12-fold 'non-crystallographic' regime might facilitate in terms of optimizing entropy during nucleation which could allow a particle to expand out prior to collapse (or not) to an ordered CCP packing which would correspond to the tetrahedral preference you held at the beginning? Might your work demonstrate that nucleation as icosahedra in the first instance is a worthwhile hypothesis even for the particles which are ultimately found to be tetrahedral?

1 A. L. MacKay, *Acta Cryst.*, 1962, **15**, 916.

Daan Frenkel responded: In crystal nucleation, the Mackay solids play the role of the widely ignored elephant in the room. Yes, it is true that Ostwald's 'rule of stages' argues that the crystal phase that nucleates need not be the most stable phase, but the phase that is closest in stability to the parent phase (a statement that, in its generality is also not quite correct). However, Mackay realised that finite structures might nucleate ('Mackay solids') that have a symmetry that is incompatible with three-dimensional periodicity (the paper by Mackay also contains a tantalising reference to a comment made by J. D. Bernal about extended, non-periodic structures). Could DNA brick structures also nucleate *via* intermediate ordered structures that do not have the symmetry of the target structure? It seems more difficult than for systems that have only one or two distinct building blocks. However, it should be possible to design bricks that could nucleate *via* two (or more) competing routes. I am not aware of work in this direction.

Mukta Tripathy said: The manuscript shows that at high coordination numbers the free energy barrier to forming self-assembled structures becomes quite high (20–25 kT). In spite of such high barriers, you report few defects. Could this be because the excess free energy of a defect is large compared to the thermal energy (perhaps because of the specificity of the interactions)?

More generally, are there good simulation techniques to estimate the excess free energy of a defect?

Daan Frenkel replied: To answer the second question first, there are good numerical techniques to compute the free energy (and hence the equilibrium concentration) of vacancies and interstitials in 'atomic' crystals.[1,2] These techniques can be extended to molecular crystals and structures with addressable complexity. In that case, we should also consider another type of (double) point defect: an exchange between two bricks. However, insufficient equilibration during growth is likely to be the main source of point defects in DNA brick structures. The free energy cost of all these defects is likely to be high – but out of equilibrium they may still form.

1 C. H. Bennett and B. J. Alder, *J. Chem. Phys.*, 1970, **54**, 4796.
2 S. Pronk and D. Frenkel, *J Phys. Chem. B*, 1970, **105**, 6722–6727.

Jayant Singh asked: Considering the complexity of the system ,it would be useful if a comment is given on difficulties in the simulation, and in particular the necessity of a biasing method in such simulations. In other words, is there scope or future direction in the development of better methodologies for such systems.

Daan Frenkel answered: The biasing methods that we use are relatively simple and can certainly be improved. Hence there is huge scope for methodological improvement.

Priyadarshi Roy Chowdhury communicated: Could you elaborate on the 'narrow temperature window' associated with the many component self assembly with special emphasis to the nucleation behaviour.

Daan Frenkel communicated in reply: The temperature window for successful nucleation is bounded at high temperatures by the regime where the target structures are unstable: small structures can only form fleetingly and will then decompose. At lower temperatures, non-selective interactions (*e.g.* due to not fully complementary DNA strands) will become strong enough to bind bricks together. As a result such bricks will bind in the wrong place. The phenomenon is similar to self-poisoning during crystal growth. The temperature where this happens depends on the strength (and the number) of non-specific interactions. For DNA bricks, there is a window where self-assembly favours the target structure. However, one can easily envisage linker molecules other than DNA for which there is no window for successful aggregation.

Priyadarshi Roy Chowdhury communicated: In order to co-relate the experimental results with those of theoretical interpretations, in a very realistic way, which model do you think is the best at present for studying the nucleation behaviour?

Daan Frenkel communicated in reply: To get a quick impression of the typical height and location of the nucleation barrier, cheap, coarse-grained models are arguably best. However, any model that aims at quantitative predictions needs a more realistic description of the DNA structure and dynamics. Possibly, the OxDNA model (dna.physics.ox.ac.uk) could be used as a compromise between cost and accuracy.

Oleg Gang remarked: How do you see the relationship between thermodynamic entropy and Shannon entropy?

Can you imagine an algorithmic approach for the position of the particles, *i.e.* instead of the *x, y, z* of particles, to have a function or recursive relationship? Would it be more concise in terms of the amount of information for self-assembly in comparison with conventional methods (interaction, shapes) or not? In this respect, more generally: how one can define a minimum amount of information that is required to make a structure from particles? And the next question: how to pass that information to the system efficiently in order to prescribe the self-assembly process?

Nicholas Kotov said: Bulding on Oleg Gang's original question about the information needed to be encoded in each block to allow for specific self-assembly to occur, it seems to me that the seemingly simple nanostructures with relatively low contents of coded information, that is molecular and nanoscale engineering, can produce structures of high complexity. Their geometry can be competitive with those observed in biology. Examples can be multiple, but tobacco–mosaic–virus-like assemblies of inorganic nanoparticles with a core and helical shell with specific symmetry come to my mind first. The information that is needed to be encoded into CdTe nanoparticles was quite minimal compared to what we would expect from a self-assembled viral particle. This brings me to the fact that we do not really have good tools to assess and predict the assembly patterns of nanoparticles in solutions. Besides computational hardware that is constantly improving, we possibly need to plug new math into the description of self-organization phenomena.

Vinothan N. Manoharan addressed Daan Frenkel and Oleg Gang: In response to Dr Gang's question about the connection between information and DNA-mediated self-assembly and to Prof. Frenkel's comment about the connection to neural networks: the paper by Murugan and coworkers[1] discusses the relation between the ability of a nanoparticle suspension to self-assemble into multiple structures and the ability of associative neural networks to store memories, within the context of the Hopfield model.

1 A. Murugan, Z. Zeravcic, M. P. Brenner and S. Leibler, *PNAS*, 2015, **112**, 54.

Priyadarshi Roy Chowdhury opened a general discussion on the paper by Robert Botet by communicating:§ How can simulation be done in a better way so that the co-relation between experimental and theoretical results becomes much more realistic?

Robert Botet communicated in reply: The major improvement to do is to include the liquid phase in the Monte-Carlo simulations. Indeed, the liquid phase is expected to be a kind of reservoir of particles unfit for the expected crystalline phases. This idea is only an assumption at this stage and should be checked. The issue is to compute the correct free energy of the liquid phase for a polydisperse system at a given volume fraction.

Priyadarshi Roy Chowdhury communicated: Which simulation model do you consider to give a reliable and comprehensive explanation of molecular reaction dynamics?

Robert Botet communicated in reply: I probably do not grasp the question. The simulations presented in paper 8522 are about regular Monte-Carlo (*i.e.* performing random exchanges of system states according to the thermodynamic rules) and not reaction dynamics. For the latter, one may refer to the book by R. Levine.

Vandana Shinde communicated: 1. Can we find the % amount of FCC and BCC crystalline structures in the system? 2. Is this theory applicable for all types of binary nanostructure crystalline systems?

Robert Botet communicated in reply: 1. Yes, since the proportion of FCC and BCC is allowed to vary during a numerical simulation, the final state provides a value for the respective volume proportions. 2. This approach is quite general, providing the final system is, in some way, an equilibrated system (in the thermodynamic sense). Actually, the theory is very close to the original Monte-Carlo idea: one follows the biased random walk of the representative point of the system in the configuration space. The main issue is that one has to 'guess', before starting a simulation, which crystalline phases could be present. If one forgets a phase, only a part of the configuration space is investigated, which may lead to the wrong final state.

Nicholas Kotov opened a general discussion on the paper by Pedro A. Sánchez: One can, in principle, manipulate magnetic particles inside the body using external magnetic fields.

§ The paper by Robert Botet was not presented at the meeting, but questions and responses have been provided afterwards.

Pedro A. Sánchez answered: Yes, that is, for instance, the idea behind the use of magnetic nanoparticles for hyperthermia treatments against cancer tumors, in which the particles are forced to quickly fluctuate under the action of oscillating external fields, heating the surrounding environment and destroying the cancer cells. Magnetic microswimmers have also been proposed as propellers for drug delivery microrobots driven by external magnetic fields.

B. L. V Prasad asked: A few nanoparticles assemble into a brush like structure, how do the brushes interact with each other? Does this influence the final structures being formed?

Pedro A. Sánchez answered: The interaction between two brushes is expected to be really complex due to the interplay between the bonding constraints and the steric and magnetic forces. Therefore, it is really hard to foresee the resulting structures without further studies. This is an interesting topic that we will surely address in the future.

Ajeet Srivastav commented: The magnetic anisotropies and the dipole–dipole interactions should be considered when designing the magnetic field interactions.

Pedro A. Sánchez responded: The design complexity of the external field depends on the scale at which one wants to control the structure of the brush. If one is interested only in determining the overall height of the brush, then it is enough to apply a homogeneous field perpendicular to the grafting surface with a proper intensity. The control of the structure at the scale of individual particles is much more complicated and it is even possible that some arbitrary configurations will never be accessible.

Daan Frenkel asked: In your presentation you discuss one possible application of the field-induced stretching of filaments of magnetic colloids, namely the design of switchable filters. Have you considered using such filaments in conjunction with a periodically modulated, rotating field to make magnetic 'cilia' that could be used to pump liquids?

Pedro A. Sánchez answered: That is actually one of the most largely studied applications of chains of magnetic colloids and nanoparticles to date, even though the amount of research groups around the world working on the topic is still relatively small. Both simply self-assembled chains and polymer crosslinked chains of magnetic particles, which are what we call magnetic filaments, have been considered for this functionality, either grafted to other small bodies to form a magnetically propelled microswimmer or grafted to the wall of a channel to pump the background fluid. In the latter case, that is closer to the brush structure discussed in our paper, there should be a minimum distance between the grafted chains in order to obtain an effective pumping. Therefore, the related studies consider only grafting densities of filaments that are significantly lower than the ones analyzed in our work.

Nirmalya Bachhar remarked: What happens if you change/tilt the direction of the magnetic field? Would you expect a different kind of shape?

Pedro A. Sánchez responded: We have not studied these conditions yet, but we can guess what will be the overall response of the brush structure. In any case, in the limit of high field intensities the filaments experience a straightening and tend to remain parallel to the field direction. When the field is perpendicular to the grafting surface, the brush reaches its maximum height, given by the contour length of the filaments. If the field is not perpendicular, the filaments will still tend to remain parallel to the field, resulting in a lower maximum vertical height of the brush. A more difficult to predict question is what will happen to the structure at the scale of individual particles. The tilting of the filaments, following the field direction, will reduce the distance between neighboring chains, maybe favoring the appearance of certain interchain connections, but probably also increasing the repulsions. Therefore, it is hard to predict the small scale structure without further analysis and eventual simulations.

Priyadarshi Roy Chowdhury communicated: What limitations arise if the temperature is kept in the low range rather than using an extended range with respect to the filament brush system?

Pedro A. Sánchez communicated in reply: At low temperatures, the structure of the magnetic brush experiences larger changes by switching the external magnetic field on and off. Without the field, the filaments bend towards the grafting surface to form a fully connected network and the average brush height reaches its minimum. In this situation, the structure of the filament network is mainly determined by the dipole–dipole interactions. When a field perpendicular to the grafting surface is switched on, the filaments tend to align with the field, remaining in a straight conformation. This hinders the connections between neighboring filaments. Therefore, at low temperatures the overall impact of the field is higher and the network of connected filaments tends to disappear for strong fields.

Oleg Gang opened a general discussion on the paper by P. B. Sunil Kumar:‡ What is the main origin of the different phases that you observed?

Mohamed Laradji answered: The clustering of the nanoparticles is the result of effective membrane-mediated interactions between them. In particular, at high adhesion strengths, the nanoparticles locally induce a negative Gaussian curvature of the membrane, preventing them from clustering in chains, leading to star-shaped clusters.

Sudeep Punnathanam asked: Did you compute the surface diffusivity of the nanoparticles on the lipid surface? What is its role in the formation of ordered structures both in the low adhesion case as well as the high adhesion case?

Mohamed Laradji replied: We calculated the diffusion coefficient of a single nanoparticle on the lipid membrane and found that the diffusivity slightly decreases with increasing adhesion strength. We did not however compute the

‡ The paper by P. B. Sunil Kumar was presented by Mohamed Laradji, University of Memphis, Memphis, USA.

diffusivity of an ensemble of nanoparticles on the membrane. We anticipate that it will decrease with increasing the number density of the nanoparticles. However, the relationship between the diffusivity and the type of assembly is not known at the moment. We will compute it in the near future.

Rajdip Bandyopadhyaya asked two questions: 1. Have you conducted simulations where the membrane surface fluctuates or modulates spontaneously, creating zones of positive and negative curvature? Will this lead to transient nanoparticle assemblies? 2. What is the motivation for starting with anisotropic nanoparticles? Rather, if you start with spherical particles one can maybe compare the resulting simulated structures with experimental work in drug delivery, where spherical nanoparticles adsorb on lipid membranes (on the outer cell-boundary) before internalization into the cells.

Mohamed Laradji answered: 1. The membrane undergoes spontaneous thermally induced fluctuations with different local curvatures. We have performed simulations of planar membranes with different values of lateral tension. We found that for high values of tension, the adhesion of the anisotropic nanoparticles is hindered by the fact that the membrane is mainly flat. Small local fluctuations in curvature allow for the adhesion of the nanoparticles and their self-assembly. For low tension, the amount of thermally induced fluctuations increases, leading to an increase in the amount of self-assembled clusters. Once formed, the clusters tend to be very robust and stable. 2. We have indeed performed some simulations of spherical nanoparticles. In particular, we recently looked at the adhesion, wrapping and endocytosis of spherical nanoparticles.[1] We are also currently looking in detail at the self-assembly of spherical nanoparticles in bilayers. We already know that there exist spherical nanoparticles with a wide range of sizes and adhesion strengths which self-assemble into linear chains. They may also self-assemble into compact clusters with triangular symmetry. The stability of these structures is under investigation.

1 E. J. Spangler, S. Upreti and M. Laradji, *J. Chem. Phys.*, 2016, **144**, 044901.

Vinothan N. Manoharan commented: Prof. Bandyopadhyaya asked about the assembly of spherical nanoparticles at a membrane interface. My group has done experiments on a much simpler system consisting of spherical colloidal particles on the surface of a spherical droplet.[1] We found that the particles form quasi-one-dimensional ribbons (see Fig. 1), similar to the assemblies seen in Prof. Kumar's system of anisotropic particles on the surface of a vesicle. The difference is that in our system, the particles do not alter the curvature, and there are isotropic attractive interactions between the particles. The reason for the quasi-1D assembly was first proposed by Schneider and Gompper in a theoretical paper from 2005:[2] the formation of ribbons minimizes the elastic energy associated with stretching the assemblies onto a surface with positive Gaussian curvature. So in addition to the mechanism shown by Prof. Kumar for the formation of chain-like assemblies from anisotropic particles, there is another mechanism for the assembly of anisotropic assemblies from isotropic particles.

1 G. Meng, J. Paulose, D. R. Nelson and V. N. Manoharan, *Science*, 2014, **343**, 634.
2 S. Schneider and G. Gompper, *Europhys. Lett.*, 2005, **70**, 136.

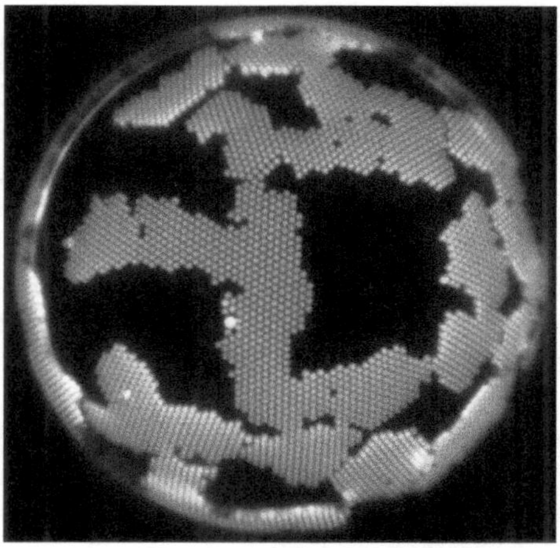

Fig. 1 Confocal microscope image of 1-micrometer, attractive colloidal particles assembling into quasi-one-dimensional crystals ('ribbons') on the surface of a spherical water droplet. See G. Meng, J. Paulose, D. R. Nelson and V. N. Manoharan, *Science*, 2014, **343**, 634. Image courtesy of G. Meng.

Sunil Kumar responded: I agree that there could be multiple mechanisms acting to form the one dimensional assembly. In our simulations, there is not explicit attractive interactions between the particles. Particle assembly is through membrane mediated interactions. We see sidewise attraction between pairs of particles and then these pairs attracting sidewise form bigger clusters *etc.* We also show that, depending on the parameters of the simulation, there is a preferred angle between a pair of particles, indicating that pairwise interactions are the dominant reason for the shape of the assemblies formed.

Rajdip Bandyopadhyaya replied: It is nice to learn of contrasting examples of assembly of nanoparticles on a membrane surface – in one case it is repulsive anisotropic particles (Sunil Kumar and Mohammed Laradji's paper in this Faraday Discussion volume) influencing local membrane curvature to bring about chain-like or splay aggregation of nanoparticles, while in the example of this comment[1] it is that of attractive, spherical nanoparticles forming much larger global-scale branched-ribbon-like patterns. It will be interesting to explore other possible assembly-patterns theoretically and validate them through experiments, a concerted experimental plus simulation study, which will be a logical next step from these two papers.

1 G. Meng, J. Paulose, D. R. Nelson and V. N. Manoharan, *Science*, 2014, **343**, 634.

Madivala G. Basavaraj addressed Mohamed Laradji and Sunil Kumar: One way to realize the simulation results you have presented *via* experiments is by using a system of cationic vesicles and oppositely charged rods through electrostatic attraction. Would there be any shape transitions in the vesicles due to the fact that particles are charged, which would accompany the release of counterions? Have you considered the effect of such counterions? Would the presence of particles in the interior of the vesicles change the phase behaviour?

Mohamed Laradji answered: The proposal seems very reasonable. The nanoparticles used in the simulations are crescent-shaped. The aggregation of the anisotropically curved nanoparticles in our simulations is different from that of rods. For example, aster-shaped aggregates do not form when rods adhere to vesicles. Counterions are not explicitly accounted for in our simulations. As alluded, we do expect that aggregation of the anisotropically curved nanoparticles on the interior side of the vesicle to be different from the case where they are on the outer side of the vesicle. In particular, the binding of the nanoparticles to the vesicle inner side will occur at a higher binding threshold than that to the outer side. This is currently being investigated by us.

Sunil Kumar responded: Yes, electrostatic attraction would change the shape of the vesicles. Please see our recent work "Curvature inducing macroion condensation driven shape changes of fluid vesicles" on this.[1] In the current manuscript we do not consider any explicit electrostatic interactions. The shape of the vesicles can depend on the presence of particles inside the vesicles as this will restrict the volume change. We are currently working on this problem.

1 K. K. Sreeja, J. H. Ipsen, P. B. Sunil Kumar, *J. Chem. Phys.*, 2015, **143**, 194902.

Daan Frenkel addressed Sunil Kumar and Mohamed Laradji: Your curved nanoparticles reproduce many of the functions of BAR proteins.

As you know there are also proteins that stabilise phospholipid tubes between vesicles. Can this behaviour be captured by your coarse-grained model?

Mohamed Laradji answered: As stated in the introduction of our article, our work is in part motivated by BAR proteins. It is very possible that the anisotropic nanoparticles considered by us will indeed stabilize nanoscale tubes between vesicles or a single vesicle to a planar membrane. This is a very interesting point which we will certainly investigate in the near future.

Sunil Kumar replied: Yes, BAR proteins are believed to be involved in the formation of lips tubes and the ramified shapes of organelles. Though the exact mechanism by which they achieve this is still unclear, curvature induction and curvature sensing are supposed to play important roles in the ability of BAR proteins to deform the membrane. In our simulations, we do see stabilisation of highly curved regions, like ridges, on the membrane. However, to see tube formation we need to go into much larger system sizes. Please note that earlier Monte Carlo Simulations, which treat these inclusions as an anisotropic curvature inducing field, have seen tube formation.[1]

1 N. Ramakrishnan, P. B. Sunil Kumar and J. H. Ipsen, *Biophys. J.*, 2013, **104**,1018.

Rajdip Bandyopadhyaya addressed Daan Frenkel, Mohamed Laradji and Pedro A. Sánchez: Steve Granick mentioned the Science paper on constructing a periodic table of different protein complexes.[1]

Similarly, what is the possibility of nanoparticle-aggregate motifs? How many parameters influence this problem? What is the role of the individual nanoparticle and where does the effect of particle–particle interaction end? What diverse topological structures are possible? For example, maybe the number of amino-acid units and pH define protein aggregates. Likewise what are the minimum number of parameters required to capture nanoparticle aggregates, in the absence of any external field? Is it possible to develop some finite number of measures of resulting nanoparticle aggregate-structures, so as to capture all structures and classify them systematically? Otherwise a large set of different aggregates is possible.

1 S. E. Ahnert, J. A. Marsh, H. Hernández, C. V. Robinson and S. A. Teichmann, *Science*, 2015, **350**, DOI: 10.1126/science.aaa2245.

Daan Frenkel replied: I should stress that our models are, at present, highly simplified and that therefore we cannot really assess the effect on the aggregates of the shape of the DNA bricks (or the constructs with a higher coordination number that we discuss in our paper). Apart from steric factors, the interaction between DNA bricks is determined exclusively by the complementary DNA sequences that link neighbouring particles together. As was shown in Ref. 4, DNA bricks can self-assemble into a variety of topologically distinct structures. These can be characterised by their topological genus. However, the topological genus is but one parameter to characterise DNA-brick structures.

Mohamed Laradji responded: These are excellent questions. I think that others should interject in answering them. I would guess that the shape of the nanoparticles, and the way they are decorated should affect very strongly the details of their aggregation.

Pedro A. Sánchez answered: In the simplest (but still reasonably representative) model of magnetic nanoparticles self-assembly, corresponding to monodisperse hard spheres with point dipoles located at their centers, the parameters that determine the self-assembly motifs are the ratio between the dipolar moment and the temperature, and the particle concentration. The latter determines the appearance of frustrated configurations in the aggregates. Even considering just the ground states of the system, the possibilities change strongly with any added parameter that we may want to consider, like the polydispersity of the particles or any shape anisotropy.

In an exercise of maximum abstraction, I would say that the minimum general factors that should determine the motifs in aggregates of nanoparticles are:

(1) The range of particle pair interactions (steric, patchy: short ranged; electrostatic: long ranged but usually screened; magnetic: long ranged, not screened ...); (2) The relative strength of the particle pair interactions with respect to the thermal fluctuations; (3) The anisotropies of the particle pair interactions; (4) The constraints (walls, density ...). Of course this exercise is just a starting point without any predictive power

Oleg Gang questioned: To what degree are the local membrane deformations important?

Mohamed Laradji replied: Clustering of the nanoparticles is the result of membrane deformations induced by the nanoparticles.

Sanat Kumar addressed Mohamed Laradji and Sunil Kumar: You use coarse-grained surfactants in your simulations to form the membrane. Cacciuto's work uses an abstraction for the membrane – in that he treats an elastic sheet with bending rigidity incorporated. In terms of the minimal information needed, what do you need to put in in terms of the membrane? Do you need surfactants or is a sheet-like description enough? In the spirit of discussion, can I coarse grain the surfactant away? What is the minimum model to do what you're doing?

Mohamed Laradji responded: Indeed, Saric and Cacciuto's work[1] treats the membrane as an elastic sheet through a triangulated surface model. The model to be used depends on the ratio between the lengthscales of the nanoparticles of interest and the thickness of the lipid bilayer. In our work, we focus on elongated and curved nanoparticles with an arclength only a few times that of the bilayer thickness. The formation of aster-shape clusters at high adhesion strength, which we observe in our simulations is the result of the nanoparticles inducing regions with negative Gaussian curvature which is the result of an induced inhomogeneity in the properties of the lipid bilayer, including an increased lipid density of the vicinal leaflet and increased chain order parameter of the lipids close to the nanoparticles. Meanwhile the lipids in the distal leaflet are not as affected by the nanoparticle. These effects cannot properly be accounted for by an elastic model. We recently investigated the interaction and endocytosis of small spherical nanoparticles with tensionless lipid bilayers[2] using the same lipid model and found that the nanoparticles can be partially wrapped by the tensionless bilayer. This is in contrast to results based on the treatment of the bilayer as an elastic sheet with zero thickness.[3] We showed in this work that the finite range of the interaction (while being shorter than the thickness of the bilayer) and inhomogeneities of the mechanical properties of the bilayer, induced by the nanoparticle lead to partial wrapping. When the size of the nanoparticle is much larger than the bilayer thickness, we retrieve the results of the elastic theory, which treats the bilayer as an infinitesimally elastic sheet.[3] Therefore, if the interest is on small nanoparticles, then a model similar to ours is needed. For studies of large nanoparticles, an elastic sheet model for the bilayer might be sufficient.

1 A. Saric and A. Cacciuto, *Phys. Rev. Lett.*, 2012, **108**, 118101.
2 E. J. Spangler, S. Upreti and M. Laradji, *J. Chem. Phys.*, 2016, **144**, 044901.
3 M. Deserno and T. Bickel, *Europhys. Lett.*, **62**, 767.

Sunil Kumar responded: This will depend on the scale of deformation involved. The thickness of the bilayer lipid membrane is of the order of 5 nm. If one is considering particles which are much bigger than this length (the radius of curvature of the induced deformation being much higher than the smallest length of the anisotropic particle) and there is no lipid flip flop, then an elastic sheet approximation is valid. Also please note that the elastic sheet model is much harder to use for dynamics.

Faraday Discussions

PAPER

Dynamics of dissipative self-assembly of particles interacting through oscillatory forces†

M. Tagliazucchi[ab] and I. Szleifer*[a]

Received 10th July 2015, Accepted 4th August 2015
DOI: 10.1039/c5fd00115c

Dissipative self-assembly is the formation of ordered structures far from equilibrium, which continuously uptake energy and dissipate it into the environment. Due to its dynamical nature, dissipative self-assembly can lead to new phenomena and possibilities of self-organization that are unavailable to equilibrium systems. Understanding the dynamics of dissipative self-assembly is required in order to direct the assembly to structures of interest. In the present work, Brownian dynamics simulations and analytical theory were used to study the dynamics of self-assembly of a mixture of particles coated with weak acids and bases under continuous oscillations of the pH. The pH of the system modulates the charge of the particles and, therefore, the interparticle forces oscillate in time. This system produces a variety of self-assembled structures, including colloidal molecules, fibers and different types of crystalline lattices. The most important conclusions of our study are: (i) in the limit of fast oscillations, the whole dynamics (and not only those at the non-equilibrium steady state) of a system of particles interacting through time-oscillating interparticle forces can be described by an effective potential that is the time average of the time-dependent potential over one oscillation period; (ii) the oscillation period is critical to determine the order of the system. In some cases the order is favored by very fast oscillations while in others small oscillation frequencies increase the order. In the latter case, it is shown that slow oscillations remove kinetic traps and, thus, allow the system to evolve towards the most stable non-equilibrium steady state.

Introduction

Colloidal self-assembly is a topic of both fundamental interest and practical relevance. Colloids are unique and versatile tools to study important problems in self-assembly, such as the effect of shape and directional interactions,[1–7]

[a]Department of Biomedical Engineering, Department of Chemistry and Chemistry of Life Processes Institute, Northwestern University, Evanston, IL 60208, USA. E-mail: igalsz@northwestern.edu
[b]INQUIMAE-CONICET, Ciudad Universitaria, Pabellón 2, Ciudad Autónoma de Buenos Aires C1428EHA, Argentina

† Electronic supplementary information (ESI) available. See DOI: 10.1039/c5fd00115c

symmetry breaking,[8] kinetic *versus* thermodynamic control,[9,10] the role of entropy[1,7,11–14] and far-from-equilibrium self-organization.[8,15–21] Furthermore, developments in colloidal self-assembly enable new avenues for the fabrication of nanoparticle-based devices for applications in electronics,[22,23] batteries,[24] photodetectors,[25] photonics[3,26] and bioanalysis.[27] The great majority of experimental and theoretical work in the area of colloidal self-assembly is focused on equilibrium systems. However, there is a recent and widespread interest in self-organization far from thermodynamic equilibrium. Far-from-equilibrium self-organization, also known as dissipative self-assembly,[16] requires a continuous influx of energy into the system and energy dissipation by the system into its surroundings. While equilibrium colloidal self-assembly is relatively well understood, we still know very little about the rules that govern dissipative self-assembly as well as its potential advantages and limitations.[28] Here, we build on our previous work on dissipative self-assembly *via* the oscillation of interparticle interactions[20] and provide new insights about the dynamics of this process.

In dissipative self-assembly, the particles in the system couple to a source of energy that keeps the system far from equilibrium. This coupling can be either invariant in time or periodically modulated. Examples of systems in which the coupling remain invariant in time are colloidal spinners driven by a constant torque,[19,29] charged particles in constant electric fields[8,18] and particles that self-propel with a constant force.[15,21,30] On the other hand, when the coupling between the system and the energy source is periodically modulated, some external or internal variable must oscillate in time, such as the strength of the interactions between particles[17,20,31–33] or the magnitude and direction of an external field.[5,7,16,34–37] In these cases, the system evolves in time toward a non-equilibrium steady state (NESS) that is oscillatory in nature: the properties of the system in the NESS are equivalent at times t and $t + \tau$ (where τ is the oscillation period).

Dissipative systems may have more than one NESS,[38] so an important and open question is how to control the evolution of these systems toward a specific NESS. This problem is, of course, analogous to the problem of finding the conditions that direct the evolution of equilibrium self-assembly toward a determined local minimum of the free energy. However, dissipative systems involve an additional level of complexity due to the lack of variational principles, such as the minimization of free energy that rules equilibrium systems.

In the present work, we explore the dynamics of a dissipative system of particles interacting through oscillatory forces as they evolve toward a NESS. Our model dissipative system, introduced in our previous work,[20] is a mixture of colloids under periodic pH oscillations, in which half of the particles are coated by weak acids and the other half by weak bases. We analytically prove here that the whole dynamics of dissipative self-assembly in the limit of fast oscillations is described by an effective potential that is the time average of the oscillatory potential. We numerically confirm this result showing that two systems that have different oscillatory potentials, but the same effective potential, exhibit the same dynamics in the limit of fast oscillations. Finally, we examine the effect of the oscillation period on the dynamics and outcome of dissipative self-assembly and show that in some cases the order is favored by fast oscillations, but in others slow oscillations increase the homogeneity of the system by allowing particles to escape from kinetic traps.

Theoretical methods

We studied the dissipative self-assembly of particles interacting through oscillatory potentials in a mixture of colloids coated by weak acidic functionalities (*a*-type particles) and colloids coated by weak basic functionalities (*b*-type particles) under continuous oscillations of the solution pH. The charges of the *a*-type and *b*-type particles change with pH according to the acid–base equilibrium of the charged groups on their surfaces. Neglecting charge regulation effects,[39–41] the charges of the particles are determined by the expressions for the acid–base equilibria, *i.e.*

$$z_a = -z_0 \frac{1}{1 + 10^{-(\mathrm{pH} - \mathrm{p}K_\mathrm{a})}} \tag{1}$$

for *a*-type particles and

$$z_b = z_0 \frac{1}{1 + 10^{(\mathrm{pH} - \mathrm{p}K_\mathrm{a})}} \tag{2}$$

for *b*-type particles. For simplicity, we use here the same maximum absolute charge (z_0) and pK_a (p$K_\mathrm{a} = 5$) for both types of particles. Fig. 1A shows a plot of the charge of the particles as the pH is varied from 3 to 7 and back to 3. For pH = 3, *a*-type particles are almost neutral and *b*-type particles have a charge of z_0. For pH = p$K_\mathrm{a} = 5$, *a*-type particles have a charge of $-1/2z_0$ and *b*-type particles have a charge of $1/2z_0$. Finally, for pH = 7, *a*-type have a charge of $-z_0$ and *b*-type particles are almost neutral. We oscillate the pH of the system with a period τ using either a triangular (Fig. 1B) or sawtooth (Fig. 1C) pH–time program. We also explore in this work a situation in which the pH is oscillated between 3 and 5 with a triangular program, Fig. 1D.

We model the evolution of the system using Brownian dynamics (BD) in a 2D box with periodic boundary conditions. The equation of motion in BD is:

$$-\frac{\mathrm{d}\mathbf{r}_i}{\mathrm{d}t}\xi + \sqrt{2\xi k_\mathrm{B}T}\mathbf{R}_i(t) - \nabla u_i(\mathbf{r}_i) = 0 \tag{3}$$

where r_i and u_i are the position and potential energy of particle *i*, respectively, ξ is the drag coefficient and $\mathbf{R}_i(t)$ is a Gaussian random force acting on particle *i* with components $R_{ij}(t)$ that have zero mean, $\langle R_{ij}(t)\rangle = 0$, and delta-function correlation, $\langle R_{ij}(t)R_{ij}(t')\rangle = \delta(t - t')$. The drag coefficient is related to the particle diffusion coefficient and diameter (σ) and the solvent viscosity (η) by

$$\xi = \frac{k_\mathrm{B}T}{D} = 3\pi\eta\sigma \tag{4}$$

Eqn (3) is similar to the Langevin equation,[42] but neglects the inertial term as the characteristic inertial timescale is much smaller than the diffusional timescale. The first two terms in eqn (3) model the friction of the solvent and the random collisions of its molecules with the particles, respectively. The last term in eqn (3) models the conservative forces. For simplicity, we neglect the hydrodynamic forces between the particles and between the particles and the substrate.

The interparticle potential is pairwise-additive and results from the combination of a screened electrostatic potential and a short-range repulsive potential (we neglect many body effects[43]). Since the ions are considered implicitly in our

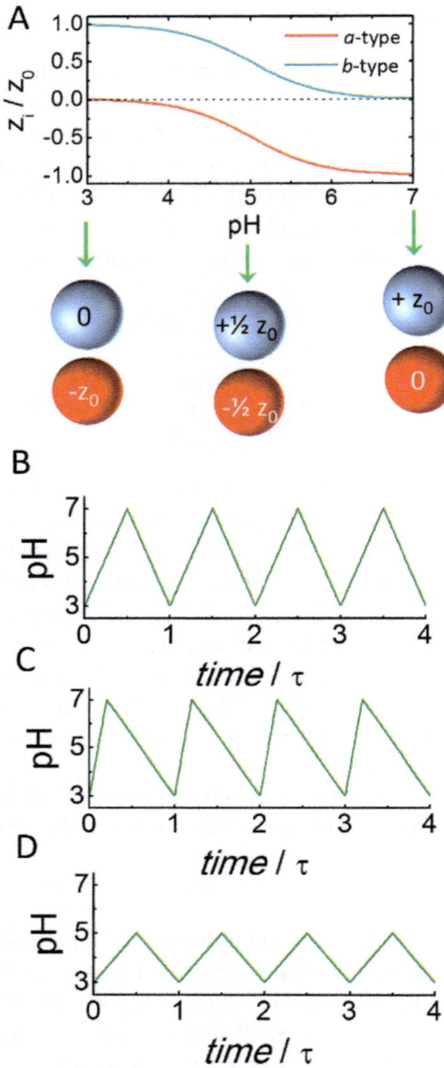

Fig. 1 (A) Charge of a-type and b-type particles as a function of the solution pH (determined with eqn (1) and (2)). In this work, the pH was oscillated in time between pH 3 and pH 7 following a linear triangular (B) or sawtooth (C) program, or between pH 3 and pH 5 with a triangular program (D).

system, the interaction between particles is modeled with a Yukawa potential, which describes the electrostatic interactions screened by mobile ions,

$$u_{ij}^{\mathrm{Yuk}}(r) = \frac{z_i}{z_0}\frac{z_j}{z_0}\frac{C}{r}\mathrm{e}^{-\frac{r}{\lambda_\mathrm{D}}} \tag{5}$$

In eqn (5), λ_D is the solution Debye length (we used $\lambda_\mathrm{D} = 1.5\sigma$ in all calculations) and C is a constant that determines the strength of the electrostatic potential in the system, given by

$$C = \frac{4{z_0}^2 e^{\frac{\sigma}{\lambda_D}} \lambda_B}{(2 + \kappa\sigma)^2} \tag{6}$$

where λ_B is the Bjerrum length. The short-range repulsions are modeled using the repulsive component of the Lennard-Jones potential,

$$u_{ij}^{\text{rep}}(r) = \left(\frac{\sigma}{r}\right)^{12} \tag{7}$$

Finally, we cut off and shift the total interparticle as:

$$u_{ij}(r) = \begin{cases} u_{ij}^{\text{rep}}(r) + u_{ij}^{\text{Yuk}}(r) - \left[u_{ij}^{\text{rep}}(r_{\text{cut-off}}) + u_{ij}^{\text{Yuk}}(r_{\text{cut-off}})\right] & \text{for } r < r_{\text{cut-off}} \\ 0 & \text{for } r \geq r_{\text{cut-off}} \end{cases} \tag{8}$$

We use a long cut-off radius $r_{\text{cut-off}}$ of 8σ in order to ensure that $u_{ij}^{\text{rep}}(r_{\text{cut-off}}) + u_{ij}^{\text{Yuk}}(r_{\text{cut-off}}) < 0.05k_BT$ for all conditions studied.

We performed BD simulations in two dimensions with a home-developed parallel code that allows modification of the interparticle potential in each time step. In all cases, we simulated 2500 particles and started from a disordered system of particles interacting only through repulsive LJ potentials. The simulation time step was $10^{-6}t_d$.

We use dimensionless variables: lengths are normalized by the diameters of the colloids, σ, and times by the characteristic diffusion timescale, $t_d = \sigma^2/D$ (where D is the particle diffusion coefficient). In the ESI of our previous work,[20] we discussed the conversion from dimensionless units to real units and showed that dissipative self-assembly of particles with diameters of 10 nm to 1 μm would require oscillation periods of microseconds to seconds.

Results and discussion

Dissipative self-assembled structures

In our previous work,[20] we performed a systematic investigation of the self-assembled morphologies that form upon pH oscillation in a system with an equal number of *a*-type and *b*-type particles. Fig. 2 shows the morphologies reported in our previous work and additional morphologies that result from changing the ratio of *a*-type to *b*-type particles. All these morphologies were obtained in the limit of very fast oscillations ($\tau \to 0$).

Dimers are the predominant structures for a 1 : 1 composition at density $\rho = 0.11\,part\,\sigma^{-2}$. However, under these conditions we also observe the formation of a few trimers (both a_2b and ab_2) as well as isolated a and b particles. Increasing the fraction of *a*-type particles to 2 : 1 produces a_2b trimers (again with some a, b and ab impurities). Interestingly, increasing the $a : b$ ratio beyond 2 : 1 does not produce colloidal molecules with more than two *a*-type particles per *b*-type particle (*i.e.* we do not obtain a_3b, a_4b, *etc.*). We will show that such colloidal molecules are unstable when the pH is oscillated between 3 and 7, but they can form by oscillating the pH between 3 and 5.

For $\rho = 0.39\,part\,\sigma^{-2}$ and $a : b = 1 : 1$, we observe fibers formed by the head-to-tail association of dimers. For $\rho = 1.0\,part\,\sigma^{-2}$, we observe crystalline lattices for

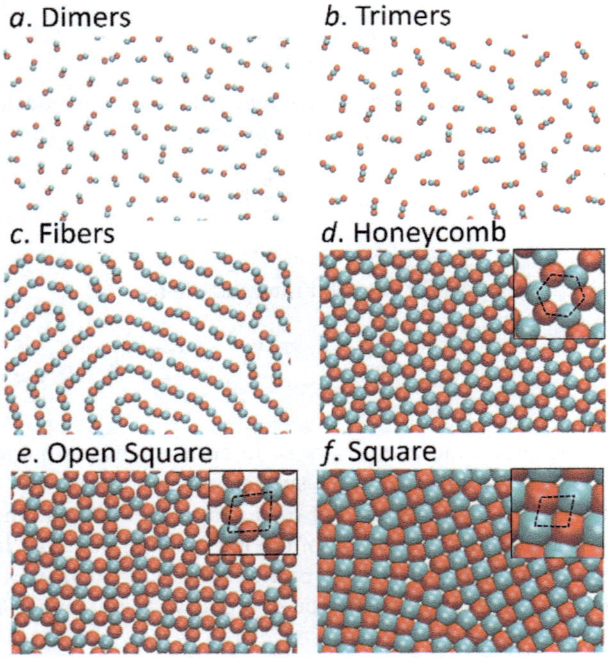

Fig. 2 Morphologies of the dissipative self-assembled systems. The snapshots show the morphologies obtained in the limit of very fast oscillations ($\tau \to 0$, obtained using an effective potential), $C = 1000k_BT\sigma$, a triangular pH–time program (Fig. 1B) and the following densities and compositions: dimers, $\rho = 0.11$ *part* σ^{-2} and $a:b = 1:1$; trimers, $\rho = 0.11$ *part* σ^{-2} and $a:b = 2:1$; fibers, $\rho = 0.39$ *part* σ^{-2} and $a:b = 1:1$; honeycomb, $\rho = 1.0$ *part* σ^{-2} and $a:b = 1:1$; open squares, $\rho = 1.0$ *part* σ^{-2} and $a:b = 2:1$; square, $\rho = 1.56$ *part* σ^{-2} and $a:b = 1:1$. Only part of the simulation box is shown. The diameter of the colloids is σ in all cases; red and blue particles represent a-type and b-type particles, respectively. In the cases where the system forms a periodic lattice, we show the unit cell of the lattice as an inset in the upper right corner of the panel.

$a:b = 1:1$ and $2:1$. Even though these crystals have a large number of defects, it is possible to identify their unit cell, whose composition is equal to the global composition of the system. For $a:b = 1:1$, we obtain a honeycomb structure with a hexagonal unit cell (already reported in our previous work[20]). For $a:b = 2:1$, the system forms an open square lattice.

Finally, for $\rho = 1.56$ *part* σ^{-2}, the system is compressed beyond the density required to close pack hard disks with diameter σ (1.15 *part* σ^{-2}). In these conditions (which are not experimentally achievable but are discussed here for completeness) we observe the formation of a square lattice (with a small hexagonal distortion) for $a:b = 1:1$, as was reported in our previous work.[20]

Analytical analysis of the dynamical evolution of particles interacting through oscillatory potentials in the limit of fast oscillations

In our previous work,[20] we demonstrated that the non-equilibrium steady state (NESS) of a system of particles interacting through oscillatory forces in the limit of very fast oscillations is equivalent to the equilibrium state of a system of particles interacting *via* a non-oscillatory effective potential. In that demonstration, we used a power series expansion of the Fokker–Planck equation in terms of τ, the period of the oscillation. We now use a different approach in order to prove a more general result: we show here that the effective potential describes the full kinetics of the non-equilibrium system for very fast oscillations (not only for the NESS), *i.e.* the effective potential describes the whole dynamics of the system as it approaches its NESS.

We start by writing down the Fokker–Planck equation for the system,[42]

$$\frac{\partial p(\mathbf{r},t)}{\partial t} = \hat{L}p(\mathbf{r},t) \tag{9}$$

where $p(\mathbf{r}, t)$ is the probability density at time t, *i.e.* the probability of finding the system at $\mathbf{r}$ (where $\mathbf{r}$ is the vector of the positions of all particles in the system) at time t. The Fokker–Planck operator $\hat{L}$ is defined as:

$$\hat{L}p(\mathbf{r}, t) = D\nabla(\nabla p(\mathbf{r}, t) + p(\mathbf{r}, t)\nabla U(\mathbf{r}, t)) \tag{10}$$

where $U(\mathbf{r}, t)$ is the potential energy of the system. We will consider the evolution of the system during the period of an oscillation, τ. Let us first assume that there exists a time-independent effective potential, U_{eff}, that describes the evolution of the system. In that case, U (and therefore $\hat{L}_{\text{eff}}$) are time-independent and thus there is a formal solution for eqn (9), which is given by[32]

$$p(\mathbf{r}, t + \tau) = \exp(\tau\hat{L}_{\text{eff}})p(\mathbf{r}, t) \tag{11}$$

here, $\hat{L}_{\text{eff}}$ is the time-independent effective Fokker–Planck operator. Expanding the exponential as a power series of $\tau\hat{L}_{\text{eff}}$ results in:

$$p(\mathbf{r},t+\tau) = \left(1 + \tau\hat{L}_{\text{eff}} + \frac{1}{2!}\tau^2\hat{L}_{\text{eff}}{}^2 + \ldots\right)p(\mathbf{r},t) \tag{12}$$

Let us now focus on the system of particles interacting through oscillatory forces. In that case, U and $\hat{L}$ depend on time and thus eqn (11) is no longer a solution of the Fokker–Planck equation. However, we can divide the oscillation period τ into infinitesimally small time steps $\mathrm{d}t = \tau/N$, such that U and $\hat{L}$ are time-independent within each time step. We then propagate the probability density from t to $t + \tau$ using these small time steps:

$$\begin{aligned} p(\mathbf{r},t+\tau) &= \exp\left(\mathrm{d}t\hat{L}(t+N\mathrm{d}t)\right)\ldots\exp\left(\mathrm{d}t\hat{L}(t+\mathrm{d}t)\right)p(\mathbf{r},t) \\ &= \prod_{j=1,N}\exp\left((\tau/N)\hat{L}(t+j\tau/N)\right)p(\mathbf{r},t) \end{aligned} \tag{13}$$

Expanding the exponential as a power series yield results in:

$$p(\mathbf{r},t+\tau) = \prod_{j=1,N}\left(1 + (\tau/N)\hat{L}(t+j\tau/N) + \frac{1}{2!}(\tau/N)^2\hat{L}(t+j\tau/N)^2 + \ldots\right)p(\mathbf{r},t) \tag{14}$$

We expand the multiplications and group terms of the same power in τ, to result in:

$$p(\mathbf{r},t+\tau)=\left(1+\tau\sum_{j=1,N}\frac{1}{N}\hat{L}(t+j\tau/N)+\tau^2\left[\sum_{j=1,N}\left(\frac{\hat{L}(t+j\tau/N)}{N}\sum_{i=j,N}\frac{\hat{L}(t+j\tau/N)}{N}\right)\right.\right.$$
$$\left.\left.-\frac{1}{2}\sum_{j=1,N}\frac{\hat{L}(t+j\tau/N)^2}{N^2}+\ldots\right]\right)p(\mathbf{r},t) \tag{15}$$

The term of order τ^2 in eqn (15) cannot be factorized as the square of a summation due to the fact that $\hat{L}(t)$ and $\hat{L}(t')$ do not commute for $t \neq t'$ because U depends on t.

We define $h = j/N$, $k = i/N$, $\mathrm{d}h = \mathrm{d}k = 1/N$ and take the limit $\mathrm{d}t \to 0$ and $N \to \infty$, to result in:

$$p(\mathbf{r},t+\tau)=\left(1+\tau\int_0^1\mathrm{d}h\hat{L}(t+\tau h)+\tau^2\left[\int_0^1\mathrm{d}h\hat{L}(t+\tau h)\int_h^1\mathrm{d}k\hat{L}(t+\tau k)\right]+\ldots\right)p(\mathbf{r},t) \tag{16}$$

We note that

$$\left(\int_0^1\mathrm{d}h\hat{L}(t+\tau h)\right)p(\mathbf{r},t)=D\nabla\left(\nabla p(\mathbf{r},t)+p(\mathbf{r},t)\nabla\left(\int_0^1U(\mathbf{r},t+\tau h)\mathrm{d}h\right)\right)$$
$$=D\nabla\left(\nabla p(\mathbf{r},t)+p(\mathbf{r},t)\nabla U_{\mathrm{eff}}(\mathbf{r})\right) \tag{17}$$

where $U_{\mathrm{eff}}(\mathbf{r})$ is the average of the potential energies of the system over one oscillation period, *i.e.*

$$U_{\mathrm{eff}}(\mathbf{r})=\int_0^1U(\mathbf{r},t+\tau h)\mathrm{d}h \tag{18}$$

Note that since $U(\mathbf{r}, t)$ is a periodic function with period τ, $U_{\mathrm{eff}}(\mathbf{r})$ does not depend on t. Eqn (17) suggests the convenience of defining the operator,

$$\hat{L}_{\mathrm{eff}}p(\mathbf{r},t) = D\nabla(\nabla p(\mathbf{r},t) + p(\mathbf{r},t)\nabla U_{\mathrm{eff}}(\mathbf{r})) \tag{19}$$

and we rewrite eqn (16) in terms of this operator:

$$p(\mathbf{r},t+\tau)=\left(1+\tau\hat{L}_{\mathrm{eff}}+\tau^2\left[\int_0^1\mathrm{d}h\hat{L}(t+\tau h)\int_h^1\mathrm{d}k\hat{L}(t+k\tau)\right]+\ldots\right)p(\mathbf{r},t) \tag{20}$$

A comparison between eqn (12) and (20) shows that these two equations agree up to first order in τ. In other words, the whole evolution of the system of particles interacting *via* oscillatory potentials can be described in terms of a system of particles interacting through an effective non-oscillatory potential U_{eff} only when the terms of second order and greater in τ can be neglected in eqn (20). This is the limit of very fast oscillations, $\tau \to 0$. Note that we did not require non-equilibrium steady-state (NESS) conditions in this derivation (unlike the derivation in our previous work[20]), therefore, we proved that the effective non-oscillatory potential can be used to describe the whole dynamics of the oscillatory system (and not only the NESS) in the limit of fast oscillations.

If only pairwise interactions are present, the effective potential is given by

$$U_{\text{eff}} = \frac{1}{2}\sum_{i=1}^{N}\sum_{j=1}^{N} u_{ij}^{\text{eff}} \tag{21}$$

where u_{ij}^{eff} is the effective potential between particles i and j. In the present example, the interparticle potential is a combination of a short-range LJ repulsion and a screened electrostatic interaction (Yukawa potential). The pH oscillations affect the charge of the particle and, therefore, the screened electrostatic potential changes in time, but the repulsive part of the potential is time-independent. The effective potential, therefore, is

$$u_{ij}^{\text{eff}}(r) = \langle u_{ij}^{\text{rep}}(r)\rangle + \left\langle u_{ij}^{\text{Yuk}}(r)\right\rangle = u_{ij}^{\text{rep}}(r) + \frac{\langle z_i z_j\rangle}{{z_0}^2}\frac{C}{r}\text{e}^{-\frac{r}{\lambda_{\text{D}}}} \tag{22}$$

where $\langle\ldots\rangle = \frac{1}{\tau}\int_t^{t+\tau}\ldots\text{d}t$ and we omitted the cut-off scheme for simplicity. The effective potential is, hence, dictated by the factor $\langle z_i z_j\rangle$, whose exact value depends on the shape of the pH–time program. The potentials in Fig. 1B (linear triangular potential) and Fig. 1C (linear sawtooth potential) have the same values of $\langle z_i z_j\rangle$ because the value of $\langle z_i z_j\rangle$ when the pH is linearly increased from 3 to 7 is equal to that when the pH linearly decreases from 7 to 3 and it does not depend on the rate at which the pH changes. Therefore, according to our analytical results, the two potential programs in Fig. 1B and C should produce the same dynamics and non-equilibrium steady state in the limit $\tau \to 0$. We will show subsequent simulation results that confirm this prediction.

Simulation of the dynamical evolution of particles interacting through oscillatory potentials in the limit of fast oscillations

Fig. 3A shows the time evolution of the average bond-order parameter of order 2, $\bar{\psi}_2$, as a function of the simulation time for the formation of fibers ($a:b = 1:1$, $\rho = 0.39$ *part* σ^{-2}). We define the average bond-order parameter of order n as

$$\overline{\psi}_n(t) = \frac{1}{N_{\text{part}}}\sum_{j=1}^{N_{\text{part}}}\psi_n(j,t) \tag{23}$$

where N_{part} is the total number of particles and $\psi_n(j, t)$ is the bond-order parameter of particle j at time t, which we define as[44–46]

$$\psi_n(j,t) = \begin{cases} \left|\dfrac{1}{NN(j)}\displaystyle\sum_{k=1}^{NN(j)}\exp\left(n\theta_{jk}(t)i\right)\right| & \text{for } NN(j)\geq n \\ 0 & \text{for } NN(j)<n \end{cases} \tag{24}$$

where $i = \sqrt{-1}$, $NN(j)$ is the number of nearest neighbors of the particle j within a distance of 1.5σ and θ_{jk} is the angle between the vector formed by the centers of particles j and k and a fixed (arbitrary) axis. The order parameter $\psi_n(j, t)$ has a maximum value of one when particle j has n neighbours separated by angles of $2\pi/n$. Therefore, the fiber structure is well characterized by the order-parameter of order two: the larger $\bar{\psi}_2$ is, the closer the system is to a perfect straight fiber.

The black and red lines in Fig. 3A show the time-evolution of $\bar{\psi}_2$ for a simulation in which the pH was oscillated with the triangular linear program (Fig. 1B) and the

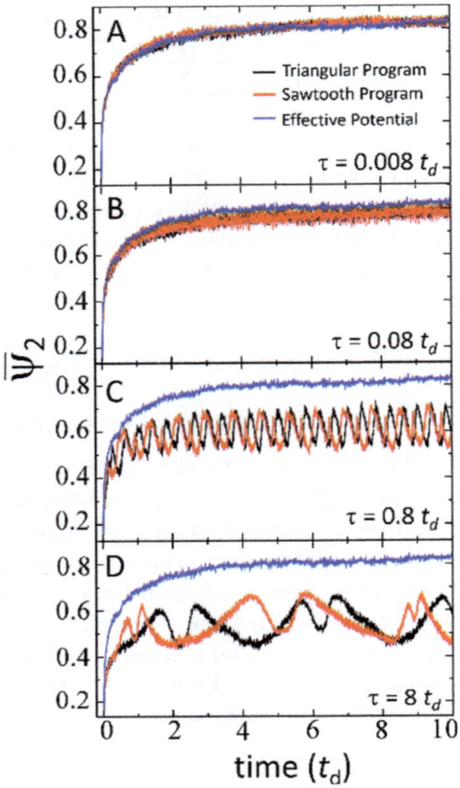

Fig. 3 Bond-order parameter of order 2 ($\bar{\psi}_2$) as a function of simulation time for $a : b = 1 : 1$, $\rho = 0.39$ *part* σ^{-2}, $C = 1000k_BT\sigma$ and an oscillation period τ of (A) $0.008t_d$, (B) $0.08t_d$, (C) $0.8t_d$ and (D) $8t_d$. Values of $\bar{\psi}_2$ close to one indicate the formation of a well-ordered fiber morphology. Black and red lines show the results of simulations in which the pH was varied with the triangular program (shown in Fig. 1B) and the sawtooth program (shown in Fig. 1C), respectively. Blue lines show the result of a simulation using the effective potential given by eqn (22) (this result is independent of τ and, thus, the same curve is shown in all panels).

sawtooth linear program (Fig. 1C), respectively. In this panel, the oscillation period ($\tau = 0.008t_d$) was much smaller than the characteristic diffusional time scale of the system (t_d). The blue line shows the result of a simulation performed with the time-independent effective potential given by eqn (22). As we already mentioned, both the triangular and the sawtooth linear programs have exactly the same effective potentials. We observe that, in the limit of fast oscillations, the $\bar{\psi}_2$ *vs.* time curves for the triangular program, the sawtooth program and the effective potential overlap, as predicted by our analytical results. Note that the degree of agreement of the $\bar{\psi}_2$ *vs.* time curves decreases as the frequency of the oscillation decreases (Fig. 3B–D). Fig. 4 shows snapshots of the system for the triangular and sawtooth potentials at a simulation time of $4t_d$ for $\tau = 0.008t_d$ and $8t_d$. For $\tau = 8t_d$, the morphologies produced by the triangular and the sawtooth potentials are very different. However, both pH–time programs yield well-ordered fibers for $\tau = 0.008t_d$, in agreement with the results in Fig. 3 and our analytical predictions.

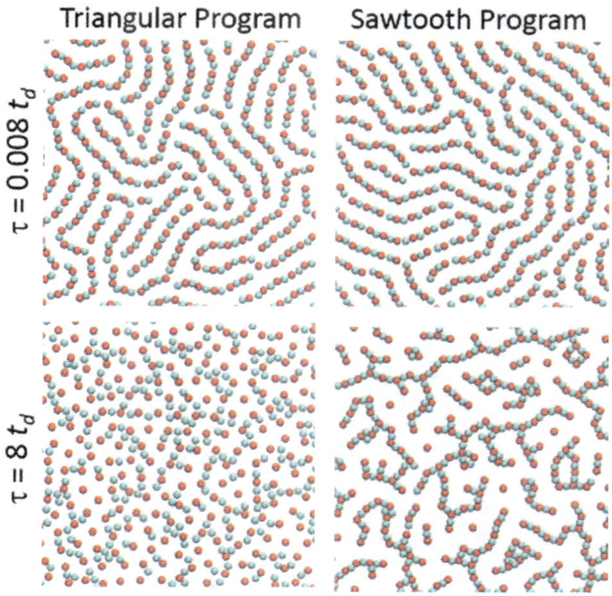

Fig. 4 Simulation snapshots at a simulation time of $4t_d$ for the same conditions as in Fig. 3.

Effect of the frequency of pH oscillations on the dissipative self-assembly of colloidal molecules

The values of $\bar{\psi}_2$ in Fig. 3 are maximal in the limit of very fast oscillations, $\tau \rightarrow 0$, *i.e.* fast oscillations favor the formation of well-ordered fibers. However, we show in this section that this conclusion is not general and that small oscillation frequencies can lead to more homogeneous mixtures of colloidal molecules than those obtained in the limit of very large frequencies.

As we mentioned before, when the pH is oscillated between 3 and 7, we only observe a, b, ab and a_2b particles (and ab_2 for systems rich in b-type particles). The effective potential allows us to assess the relative stability of each molecule in the system by calculating its potential energy in the limit of fast oscillations. Fig. 5 shows the energy gain (or loss) upon the formation of different colloidal molecules based on the effective potential. The predictions show that when the pH is oscillated from 3 to 7, the only stable a_ib colloidal molecules are ab and a_2b, in agreement with the simulation results. The rest of the colloidal molecules are unstable because in the effective potential, the a–a repulsions are stronger than the a–b attractions.[20] This situation changes when oscillating the pH from 3 to 5 (see Fig. 1D), since in those conditions the charge of the a-type particles is (on average) smaller than that of the b-type particles. Therefore, a–a repulsions are weaker than a–b attractions and the latter are weaker than b–b repulsions. Fig. 5 shows that when the pH is oscillated between 3 and 5, colloidal molecules with stoichiometries ab, a_2b, a_3b and a_4b can be obtained. Colloidal molecules with stoichiometries a_5b and a_6b have $U < 0$, but they are unstable with respect to the loss of satellite particles: breaking down a_5b particles into a_4b and a particles is energetically favourable.

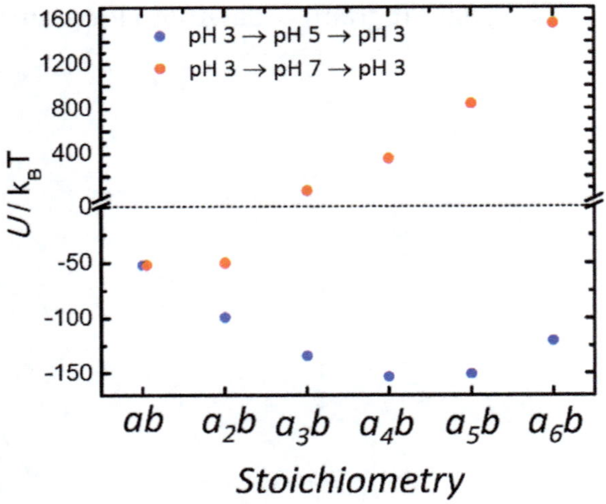

Fig. 5 Potential energy of a_ib colloidal molecules determined in the limit $\tau \rightarrow 0$ with the effective potential, when the pH was oscillated with a triangular program between 3 and 7 (red points, pH program in Fig. 1B) or between 3 and 5 (blue points, pH program in Fig. 1D).

We estimated the population of each colloidal molecule in the NESS as a function of the density ρ and composition $a : b$ using the energies in Fig. 5 (see details in the ESI†). The calculation was done assuming equilibrium statistical mechanics with the effective potentials, since in the fast oscillation limit we have proven that to be the proper description.[20] Table 1 shows the predicted fractions, $f^{\mathrm{P}}(a_ib)$, which we define as the number of particles in the system that belong to a colloidal molecule with composition a_ib. We observe that for compositions 1 : 1, 2 : 1 and 3 : 1, the predicted stoichiometry of the predominant molecules is the same as the global composition of the system, while for composition 9 : 1, our calculations predict a mixture of isolated a particles and a_4b colloidal molecules because a_5b and a_6b particles are unstable.

Fig. 6 shows snapshots for the simulations in which the pH was oscillated between 3 and 5 in the limit of high frequencies, for ρ = 0.11 *part* σ^{-2} and different compositions $a : b$. In all cases, we observe mixtures of different colloidal molecules. Table 2 shows the fractions of the different colloidal molecules in the systems of Fig. 6. The populations of colloidal molecules predicted using the potential energies of the molecules (Table 1) and those actually

Table 1 Fraction of particles in the system that belong to colloidal molecules of different stoichiometries, $f^{\mathrm{P}}(a_ib)$ for ρ = 0.11 *part* σ^{-2} and different $a : b$ compositions calculated from the potential energy of each molecule (see the ESI). The fractions of a_5b and a_6b are zero in all cases

$a : b$	a	b	ab	a_2b	a_3b	a_4b
1 : 1	0.00	0.03	0.87	0.10	0.00	0.00
2 : 1	0.00	0.00	0.00	0.99	0.00	0.00
3 : 1	0.00	0.00	0.00	0.00	1.00	0.00
9 : 1	0.50	0.00	0.00	0.00	0.00	0.50

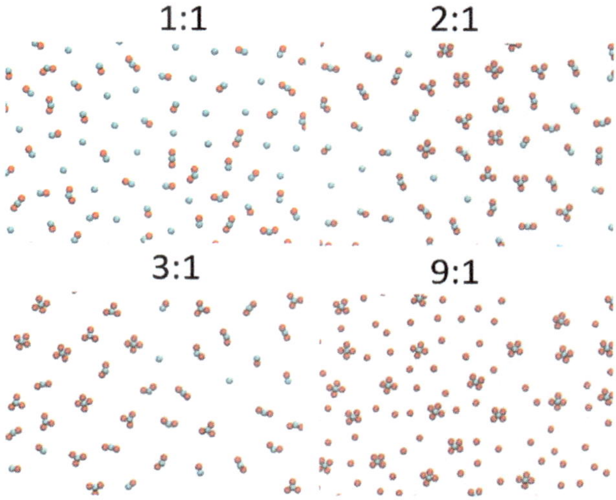

Fig. 6 Snapshots showing the colloidal molecules formed in simulations in which the pH was oscillated between 3 and 5 (program in Fig. 1D), for $\tau \rightarrow 0$, $\rho = 0.11$ *part* σ^{-2} and different $a:b$ compositions. The fractions of each type of colloidal molecule for each different composition are summarized in Table 2.

observed in the simulations (Table 2) are very different. In the simulations, we observe a broad distribution of stoichiometries, while energetic arguments predict that almost all colloidal molecules should have a stoichiometry equal to the global composition of the system. This discrepancy arises because the colloidal molecules form within the first 1–3t_d of simulation time (we start from a disordered system) and, after that time, particles cannot be exchanged between colloidal molecules. The fraction of each colloidal molecule in the system is, therefore, dictated by its formation kinetics during the initial stage of the simulation. This result means that for fast pH oscillations the system gets trapped in a non-equilibrium steady state rather than the most stable one, exactly in the same way as an equilibrium system can get trapped in a local (rather than the global) minimum of free energy.

So far, we have shown that the effective potential allows us to determine the conditions required to form dissipative self-assembled colloidal molecules, but it does not provide a way to avoid the kinetic traps that prevent the system from reaching its most stable NESS. We show next that these kinetic traps can be

Table 2 Fraction of particles in the system that belong to colloidal molecules of different stoichiometries, $f^{p}(a_i b)$ obtained from simulations for $\rho = 0.11$ *part* σ^{-2} and different $a:b$ compositions in the limit $\tau \rightarrow 0$ (see Fig. 6). The fractions of a_5b and a_6b are zero in all cases

$a:b$	a	b	ab	a_2b	a_3b	a_4b
1 : 1	0.00	0.17	0.40	0.32	0.11	0.00
2 : 1	0.00	0.03	0.18	0.30	0.33	0.16
3 : 1	0.02	0.01	0.05	0.16	0.29	0.47
9 : 1	0.47	0.00	0.03	0.01	0.00	0.47

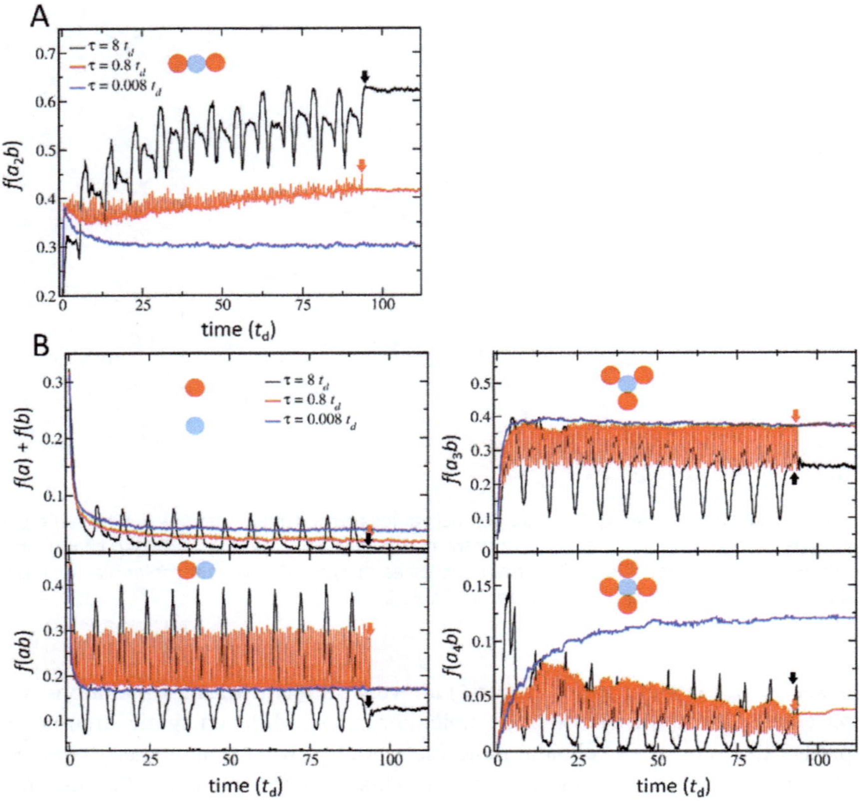

Fig. 7 Fractions of the particles in the system that belong to a_ib colloidal molecules, $f^p(a_ib)$, as a function of the simulation time for different values of the oscillation period, τ and pH oscillations between 3 and 5 (pH program in Fig. 1D). At the simulation times indicated with arrows, the period of the pH oscillations was switched to $\tau = 0.008t_d$. Other simulation conditions: $a : b = 2 : 1$, $C = 1000k_BT\sigma$ and $\rho = 0.11$ *part* σ^{-2}.

partially removed by an appropriate choice of the pH-oscillation period. In Fig. 7A we plot the time course of the fraction of particles that belong to a_2b molecules in a system with $a : b = 2 : 1$ and $\rho = 0.11$ *part* σ^{-2} for oscillation periods of $0.008t_d$, $0.8t_d$ and $8t_d$. For $\tau = 0.8t_d$ and $8t_d$, $f(a_2b)$ oscillates due to the oscillations of pH, but its value averaged over one period in the NESS (0.43 for $\tau = 0.8t_d$ and 0.57 for $\tau = 8t_d$) is clearly larger than the value observed for very fast oscillations (0.30 for $\tau = 0.008t_d$). Inspection of the simulations shows that decreasing the oscillation period allows the exchange of *a*-type particles between colloidal molecules. Particle exchange occurs because *a*-type particles are almost neutral for pH = 3 and therefore at that pH there is no net *a*–*b* attraction. For very fast pH oscillations, *a*-type particles at pH 3 cannot diffuse fast enough to travel from one colloidal molecule to the other before the pH changes. Note that the electrostatic interactions in our system are rather strong (we used $C = 1000k_BT\sigma$ in eqn (5)), therefore breaking the electrostatic bonds is very difficult if both *a*-type and *b*-type particles are charged. As the oscillation frequency decreases, the particles become able to escape from their original colloidal molecule, diffuse and attach to a different colloidal assembly. In the case of a 2 : 1 mixture, this process leads to a

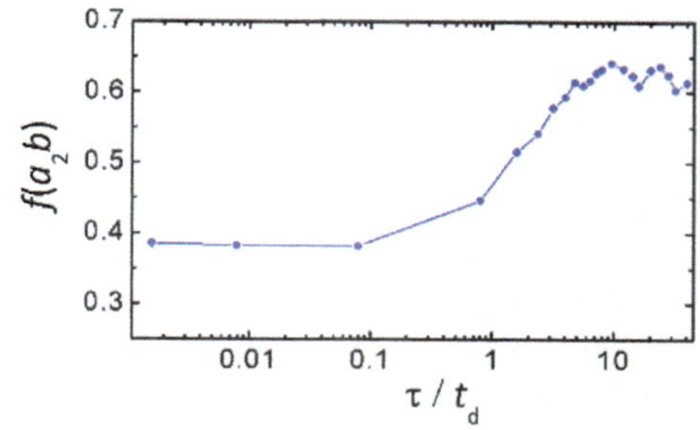

Fig. 8 Fraction of a_2b colloidal molecules at the end of the simulation as a function of the oscillation period.

neat formation of a_2b molecules, which is the expected product according the energetics of the system, see Table 1.

For $\tau = 8t_d$, the fraction of a_2b particles displays large oscillations in time. In an attempt to maximize $f(a_2b)$, we switched the oscillation period from $\tau = 0.8t_d$ or $8t_d$ to $\tau = 0.008t_d$ when $f^p(a_2b)$ passed through a maximum. Fig. 7A shows that $f^p(a_2b)$ remains stable after switching to fast oscillations (the switching times are shown with arrows). Fig. 7B shows the time evolution of the fractions of particles for the colloidal molecules different from a_2b. The increase of $f^p(a_2b)$ as the oscillation period increases is accompanied by a decrease in the fraction of the other colloidal molecules, especially free a-type, free b-type and a_4b particles, which drop to almost zero for $\tau = 8t_d$.

Fig. 8 shows a plot of the final fraction of a_2b particles as a function of the period of the oscillations. We observe that $f^p(a_2b)$ increases up to $\tau \sim 8t_d$ and then

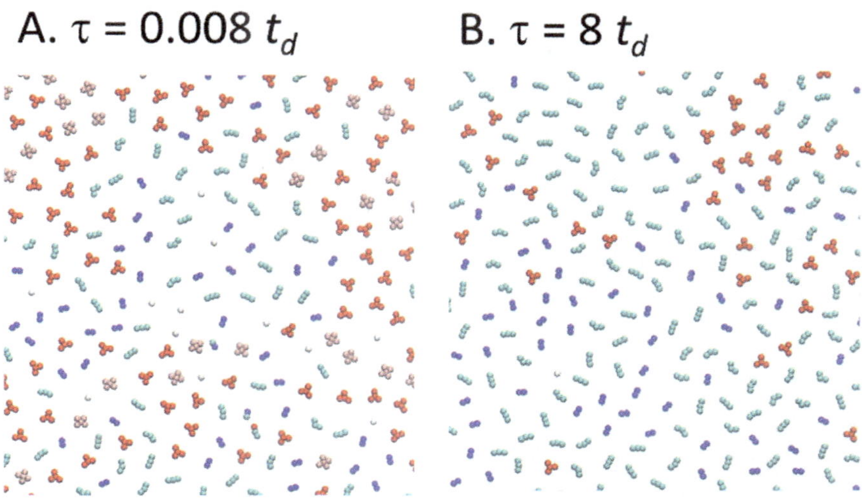

Fig. 9 Simulation snapshots at $t = 112t_d$ for the simulation of Fig. 7. A quarter of the simulation box (~625 particles) is shown in each panel. The particles are colored according to their stoichiometry: white for isolated a and b particles, blue for ab, cyan for a_2b, red for a_3b and pink for a_4b.

remains constant (our data are noisy for large τ, for which the pH only oscillates a few times during the simulation). In the critical time scale of $t \sim 8t_d$, a particle can diffuse at an average distance of $(4Dt)^{1/2} = 5.7\sigma$. This distance agrees well with the average distance between colloidal molecules (estimated from the number of colloidal molecules and the area of the simulation box) of 5.2σ, which confirms that the rate of a_2b formation is limited by the ability of a-type particles to exchange between colloidal molecules. It is also interesting to note that $f(a_2b)$ saturates around 0.63. In order to understand why larger $f(a_2b)$ cannot be obtained, we show in Fig. 9 snapshots of the system at the end of the simulation ($112t_d$) for $\tau = 0.008t_d$ and $8t_d$. Molecules of different stoichiometries are shown in different colors. For $\tau = 8t_d$, the system is mainly composed of ab, a_2b and a_3b molecules. These three types of molecules are not homogeneously distributed; instead we observe regions rich in ab assemblies and other regions rich in a_3b colloids. Regions that have different local compositions, which form due to random fluctuations during the initial assembly process, prevent complete conversion to a_2b in the time scale of the simulation: a low oscillation frequency allows particles to exchange within colloidal molecules in a region, but it cannot eliminate longer-range heterogeneities of local composition.

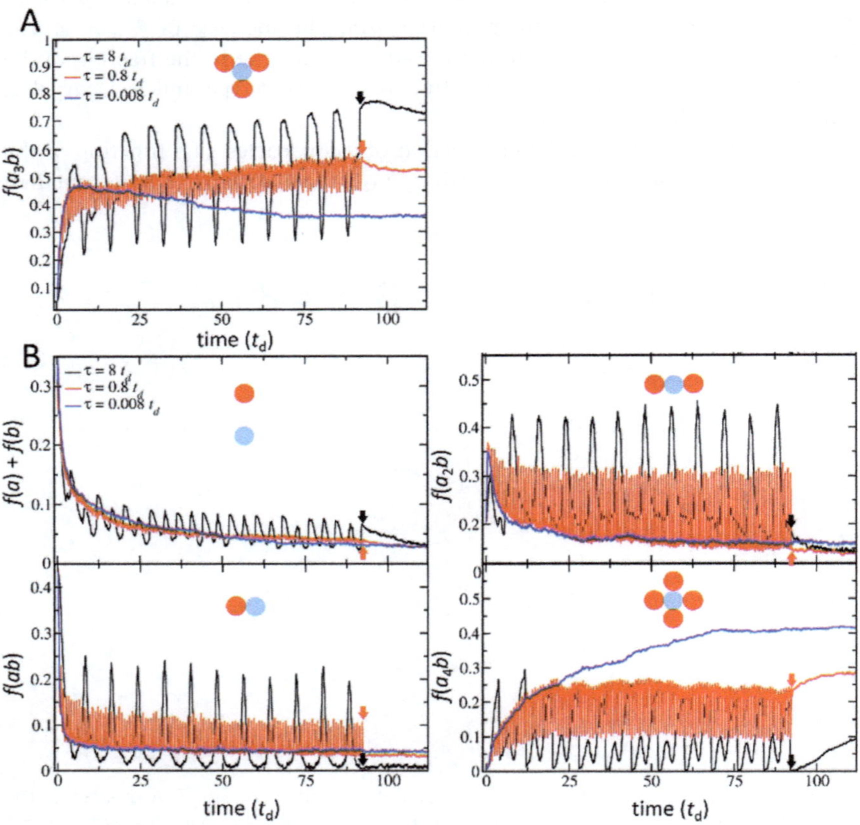

Fig. 10 Fractions of the particles in the system that belong to a_ib colloidal molecules (details given in Fig. 7) for $a : b = 3 : 1$.

The results presented in this section show that slow pH oscillations can produce more ordered systems than fast oscillations. This result is applicable to the formation of colloidal molecules in systems with compositions different from 2 : 1, for example Fig. 10 shows results for a system with a 3 : 1 composition. In this case, slow frequencies favor the formation of a_3b assemblies, which is the expected stoichiometry considering the global composition of the system. Moreover, our results are consistent with work in the literature on the self-assembly of particles interacting through oscillatory forces. For example, Swan *et al.* studied the self-assembly of a magnetorheological fluid in the present of a toggling magnetic field, whose oscillations allow the particles in the system to escape kinetic traps and form ordered structures.[34,35] Simulations by Jha *et al.*[31] and Risbud *et al.*[32] investigated the crystallization of particles whose interactions were turned on and off periodically and showed that there is an optimal and finite switching period that maximizes long-range order in the system. A common strategy that allows systems to jump over potential barriers, used in equilibrium systems, is to increase the temperature. Periodically modulating the potential of the system also allows systems to jump over potential barriers, but it grants greater control over the final product than increasing the global temperature. A temperature jump increases the kinetic energy of the whole system, while changing the frequency of the oscillations provides for control of the length that a particle can diffuse and the initial and final states that the changing potentials can provide. Thus, as a general conclusion, it is worthwhile to stress the role of the oscillation period as a handle to control the outcome of a dissipative self-assembly process. This handle has no equilibrium counterpart, which reinforces the idea that future synthetic dissipative systems may exhibit abilities for dynamical response and error correction well beyond the potential of equilibrium self-assembly.

Conclusions

In this work, we examined the dynamics of a dissipative system as it evolved toward a non-equilibrium steady state. We demonstrated two important results. First, we showed using analytical theory and exemplified with numerical simulations that the whole dynamics of systems with time-oscillating interparticle potentials can be described, in the limit of fast oscillations, by an effective time-independent potential. This effective potential is the time average of the oscillatory potential. Dissipative systems may exhibit more than one non-equilibrium steady state, for example magnetic spinners on the air–water interface may arrange in different patterns, each one corresponding to a different steady state.[16,38] It is unknown which of these steady states is the most stable one or if it is even correct to think of one of them as being more stable than the others. In the particular case of particles interacting through fast oscillatory forces, we solved this problem by demonstrating the existence of an effective potential, which maps the non-equilibrium problem into an equilibrium one. This result allows the application of the tools of equilibrium thermodynamics to the study of dissipative systems. For example, we used the effective potential to calculate the energy of colloidal molecules of different stoichiometries and then applied the tools of equilibrium statistical mechanics to calculate the fraction of each of these molecules in the (most stable) non-equilibrium steady state of the system.

The second important result of the present paper is related to the effect of the oscillation period on the formation of order. We observed that in some cases (*i.e.* formation of fibers), ordered structures are favored by fast oscillations, while in others (formation of colloidal molecules) slow oscillations increase the order of the system by allowing it to escape from kinetic traps. Previous works have shown that oscillating interparticle forces can assist in the crystallization of colloids.[31,32,34] Our results on colloidal molecules are related to these effects previously reported in the literature, with the difference that in previous works the colloidal crystals were an equilibrium product of the system (and the oscillations were introduced as perturbations), while in our case the colloidal molecules are dissipative structures that cannot be obtained in the absence of pH oscillations. The efficiency of trap removal by oscillating interparticle forces will, of course, depend on the details of the interparticle potential and should probably be analyzed on a case-per-case basis.

We arrived at the conclusions mentioned in the previous paragraphs by analyzing a system of particles coated by weak acids and bases under pH oscillations. This system is appealing to study dissipative self-assembly due to the richness of structures that it produces as well as its chemical relevance. However, our conclusion about the existence of a non-oscillatory potential that describes the dynamics of particles interacting through oscillatory forces is valid for any oscillatory potential. We believe that this conclusion also applies to systems under oscillating external fields and we will analyze this case in the future. Our work exemplifies two interesting aspects of dissipative self-assembly that are unavailable to equilibrium self-assembly: (i) the effective potential depends on the way in which the interparticle potential is oscillated in time (*e.g.* the shape of the pH–time program). This dependence provides a dynamic handle to engineer the interactions among particles that is unique to dissipative systems; (ii) the oscillations play the dual role of controlling the effective potential among particles and removing kinetic traps, which results in a unique ability of dissipative systems to avoid kinetically arrested states and provides a potential avenue for self-healing and error correction.

Acknowledgements

This work was supported as part of the Center for Bio-Inspired Energy Science, an Energy Frontier Research Center funded by the U.S. Department of Energy, Office of Science, Basic Energy Sciences, under Award no. DE-SC0000989. This research was supported in part through the computational resources and staff contributions provided by the Quest high performance computing facility at Northwestern University which is jointly supported by the Office of the Provost, the Office for Research, and Northwestern University Information Technology. MT is a fellow of CONICET (Argentina).

Notes and references

1 G. van Anders, D. Klotsa, N. K. Ahmed, M. Engel and S. C. Glotzer, *Proc. Natl. Acad. Sci. U. S. A.*, 2014, **111**, E4812–E4821.

2 N. R. Jana, *Angew. Chem., Int. Ed.*, 2004, **43**, 1536–1540.

3 X. Ye, J. E. Collins, Y. Kang, J. Chen, D. T. N. Chen, A. G. Yodh and C. B. Murray, *Proc. Natl. Acad. Sci. U. S. A.*, 2010, **107**, 22430–22435.

4 Y. Wang, Y. Wang, D. R. Breed, V. N. Manoharan, L. Feng, A. D. Hollingsworth, M. Weck and D. J. Pine, *Nature*, 2012, **491**, 51–55.

5 Q. Chen, S. C. Bae and S. Granick, *Nature*, 2011, **469**, 381–384.

6 Q. Chen, E. Diesel, J. K. Whitmer, S. C. Bae, E. Luijten and S. Granick, *J. Am. Chem. Soc.*, 2011, **133**, 7725–7727.

7 P. F. Damasceno, M. Engel and S. C. Glotzer, *Science*, 2012, **337**, 453–457.

8 A. Bricard, J.-B. Caussin, N. Desreumaux, O. Dauchot and D. Bartolo, *Nature*, 2013, **503**, 95–98.

9 M. Tagliazucchi, F. Zou and E. A. Weiss, *J. Phys. Chem. Lett.*, 2014, **5**, 2775–2780.

10 J. A. Balmer, O. O. Mykhaylyk, J. Patrick A Falrclough, A. J. Ryan, S. P. Armes, M. W. Murray, K. A. Murray and N. S. J. Williams, *J. Am. Chem. Soc.*, 2010, **132**, 2166–2168.

11 X. Mao, Q. Chen and S. Granick, *Nat. Mater.*, 2013, **12**, 217–222.

12 M. D. Eldridge, P. A. Madden and D. Frenkel, *Nature*, 1993, **365**, 35–XI.

13 W. M. Jacobs, A. Reinhardt and D. Frenkel, *Proc. Natl. Acad. Sci. U. S. A.*, 2015, **112**, 6313–6318.

14 B. de Nijs, S. Dussi, F. Smallenburg, J. D. Meeldijk, D. J. Groenendijk, L. Filion, A. Imhof, A. van Blaaderen and M. Dijkstra, *Nat. Mater.*, 2015, **14**, 56–60.

15 D. Helbing, I. J. Farkas and T. Vicsek, *Phys. Rev. Lett.*, 2000, **84**, 1240–1243.

16 B. A. Grzybowski, H. A. Stone and G. M. Whitesides, *Nature*, 2000, **405**, 1033–1036.

17 R. Klajn, K. J. M. Bishop and B. A. Grzybowski, *Proc. Natl. Acad. Sci. U. S. A.*, 2007, **104**, 10305–10309.

18 M. E. Leunissen, C. G. Christova, A. P. Hynninen, C. P. Royall, A. I. Campbell, A. Imhof, M. Dijkstra, R. van Roij and A. van Blaaderen, *Nature*, 2005, **437**, 235–240.

19 N. H. P. Nguyen, D. Klotsa, M. Engel and S. C. Glotzer, *Phys. Rev. Lett.*, 2014, **112**, 075701.

20 M. Tagliazucchi, E. A. Weiss and I. Szleifer, *Proc. Natl. Acad. Sci. U. S. A.*, 2014, **111**, 9751–9756.

21 R. Ni, M. A. C. Stuart and M. Dijkstra, *Nat. Commun.*, 2013, **4**, 2704.

22 D. K. Kim, Y. Lai, B. T. Diroll, C. B. Murray and C. R. Kagan, *Nat. Commun.*, 2012, **3**, 1216.

23 W.-k. Koh, S. R. Saudari, A. T. Fafarman, C. R. Kagan and C. B. Murray, *Nano Lett.*, 2011, **11**, 4764–4767.

24 H. Zhang, X. Yu and P. V. Braun, *Nat. Nanotechnol.*, 2011, **6**, 277–281.

25 S. A. McDonald, G. Konstantatos, S. Zhang, P. W. Cyr, E. J. D. Klem, L. Levina and E. H. Sargent, *Nat. Mater.*, 2005, **4**, 138–142.

26 D. J. Park, C. Zhang, J. C. Ku, Y. Zhou, G. C. Schatz and C. A. Mirkin, *Proc. Natl. Acad. Sci. U. S. A.*, 2015, **112**, 977–981.

27 R. A. Alvarez-Puebla, A. Agarwal, P. Manna, B. P. Khanal, P. Aldeanueva-Potel, E. Carbó-Argibay, N. Pazos-Pérez, L. Vigderman, E. R. Zubarev, N. A. Kotov and L. M. Liz-Marzán, *Proc. Natl. Acad. Sci. U. S. A.*, 2011, **108**, 8157–8161.

28 M. Fialkowski, K. J. M. Bishop, R. Klajn, S. K. Smoukov, C. J. Campbell and B. A. Grzybowski, *J. Phys. Chem. B*, 2006, **110**, 2482–2496.

29 K. Yeo, E. Lushi and P. M. Vlahovska, *Phys. Rev. Lett.*, 2015, **114**, 188301.

30 N. H. P. Nguyen, E. Jankowski and S. C. Glotzer, *Phys. Rev. E: Stat., Nonlinear, Soft Matter Phys.*, 2012, **86**, 011136.
31 P. K. Jha, V. Kuzovkov, B. A. Grzybowski and M. Olvera de la Cruz, *Soft Matter*, 2012, **8**, 227–234.
32 S. R. Risbud and J. W. Swan, *Soft Matter*, 2015, **11**, 3232–3240.
33 J. R. Green, A. B. Costa, B. A. Grzybowski and I. Szleifer, *Proc. Natl. Acad. Sci. U. S. A.*, 2013, **110**, 16339–16343.
34 J. W. Swan, P. A. Vasquez, P. A. Whitson, E. M. Fincke, K. Wakata, S. H. Magnus, F. de Winne, M. R. Barratt, J. H. Agui, R. D. Green, N. R. Hall, D. Y. Bohman, C. T. Bunnell, A. P. Gast and E. M. Furst, *Proc. Natl. Acad. Sci. U. S. A.*, 2012, **109**, 16023–16028.
35 J. W. Swan, J. L. Bauer, Y. Liu and E. M. Furst, *Soft Matter*, 2014, **10**, 1102–1109.
36 J. Yan, M. Bloom, S. C. Bae, E. Luijten and S. Granick, *Nature*, 2012, **491**, 578–581.
37 J. E. Martin and A. Snezhko, *Rep. Prog. Phys.*, 2013, **76**, 126601.
38 K. V. Tretiakov, I. Szleifer and B. A. Grzybowski, *Angew. Chem., Int. Ed.*, 2013, **52**, 10304–10308.
39 D. W. Wang, R. J. Nap, I. Lagzi, B. Kowalczyk, S. B. Han, B. A. Grzybowski and I. Szleifer, *J. Am. Chem. Soc.*, 2011, **133**, 2192–2197.
40 M. Tagliazucchi and I. Szleifer, *Soft Matter*, 2012, **8**, 7292–7305.
41 B. W. Ninham and A. Parsegian, *J. Theor. Biol.*, 1971, **31**, 405–428.
42 W. Coffey, *The Langevin equation: with applications to stochastic problems in physics, chemistry, and electrical engineering*, World Scientific, River Edge, NJ, 3rd edn, 2012.
43 J. Dobnikar, Y. Chen, R. Rzehak and H. H. von Grünberg, *J. Phys.: Condens. Matter*, 2003, **15**, S263–S268.
44 W. Lechner and C. Dellago, *J. Chem. Phys.*, 2008, **129**, 114707.
45 R. Zhang, P. K. Jha and M. Olvera de La Cruz, *Soft Matter*, 2013, **9**, 5042–5051.
46 D. R. Nelson and B. I. Halperin, *Phys. Rev. B: Condens. Matter Mater. Phys.*, 1979, **19**, 2457–2484.

Faraday Discussions

PAPER

Nano ellipsoids at the fluid–fluid interface: effect of surface charge on adsorption, buckling and emulsification†

Venkateshwar Rao Dugyala, Thiriveni G. Anjali, Siliveru Upendar, Ethayaraja Mani and Madivala G. Basavaraj*

Received 4th September 2015, Accepted 15th October 2015

DOI: 10.1039/c5fd00136f

In this contribution, we discuss the role of surface charge on the adsorption of shape anisotropic particles to fluid–fluid interfaces in the context of their application in particle-stabilized emulsions. Starting with a pendent aqueous drop containing nano-ellipsoids of known surface charge density suspended in an oil medium, we study the kinetics of adsorption of the ellipsoids to the water–decane interface using pendant drop tensiometry. The interfacial tension of the drop is recorded as a function of time by analyzing the shape of the drop. We show that the particles that are weakly charged readily adsorb to the water–decane interface and the adsorption behavior is influenced by the particle surface charge density. Furthermore, as the area available for the particles deposited at the interface is reduced, the interface populated with self-assembled ellipsoids shows wrinkles indicating buckling of the particle-laden interface under compression. However, the buckling is not observed if nano-ellipsoids are highly charged confirming that the particles do not adsorb to the interface when they are highly charged. This suggests that in several examples where the particles at interfaces concept is exploited, the repulsive energy barrier due to the particle surface charge plays a key role in the adsorption of particles to the interfaces. However, once the particles are adsorbed, the interfacial properties of the monolayer depend on the particle–particle interactions. Thus a combination of these interactions determines the concentration of particles at the interface, their microstructure and interfacial properties. The effect of these interactions on the quantity and size of the emulsion drops stabilized by ellipsoidal particles is also explored.

Introduction

The formation and stability of emulsions has received considerable attention in recent years due to their relevance in several different fields such as cosmetics,

Polymer Engineering and Colloid Science Lab (PECS Lab), Department of Chemical Engineering, Indian Institute of Technology Madras, Chennai-600036, India. E-mail: basa@iitm.ac.in

† Electronic supplementary information (ESI) available: Contains image charge interaction calculations and compression videos at different pH. See DOI: 10.1039/c5fd00136f

food and the oil recovery industry.[1–4] In the emulsification process, two or more immiscible liquids are stabilized by the adsorption of a third component at the interface. These adsorbed components stabilize the emulsions by kinetically arresting the coalescence of the liquid drops thus formed. In general, the third components that are typically used are surfactants, polymers, hard/rigid particles and more recently, soft microgel particles.[5–16] Therefore there is considerable interest in the study of the kinetics of adsorption of various species to interfaces and their response to compression, shear or a combination of the two. In particular, due to the additional advantages of "solid or particle stabilized" Pickering emulsions – such as longer term stability than the emulsion stabilized by the surfactants and polymers,[17,18] the coalescence and buckling phenomena in particle coated droplets has been investigated.[19–21]

Over the last decade, there has been a growing interest in using anisotropic particles for emulsion preparation due to the important role of particle shape in modifying the characteristics of a particle-laden interface.[16,22–26] This, in most studies, is attributed to the shape induced long-range capillary attraction between anisotropic particles trapped at fluid–fluid interfaces.[27–30] This arises due to non-uniform deformation of the interface due to particle curvature. The surface rheology and the formation of emulsions with different types of ellipsoidal particles show a remarkable dependence of the particle aspect ratio on the mechanical properties of the interfaces as well as on emulsion stabilization.[22,23] In order to effect emulsion stabilization, particles in the first place have to adsorb to the interface created during the emulsification process. To this end, a common strategy used in several studies is to screen surface charge by the addition of a salt or oppositely charged surfactant.[5,31–33] Thus emulsification is an effect of the reduced repulsion energy barrier for the adsorption of particles to the interface created during emulsification. Wang *et al.*, showed that the energy barrier for the adsorption of particles to the interface is due to image charge repulsion.[34] However, with the application of sufficient energy, it is possible to overcome this energy barrier and hence effect emulsion stabilization even if the particles used for emulsification are highly charged. While several studies have harnessed shape effects in emulsion and foam stabilization, the adsorption of model anisotropic particles has not been studied.

During the formation of particle stabilized emulsions, coarsening of the particle-coated drops occurs until an interface with the jammed particle monolayer forms. Beyond the jammed state, the interface shows buckling phenomena upon further reduction in the interfacial area available for the particles.[19,27,35] To visualize and quantify the buckling phenomena, typically compression experiments are carried out *via* the pendant drop method or with a Langmuir trough – where a planar monolayer is compressed with the help of movable barriers. In the pendant drop method, the interface of a particle coated drop suspended in an immiscible fluid is compressed by reducing the droplet volume and thereby decreasing the effective interfacial area per particle. Compression beyond a certain volume leads to the maximum possible surface coverage of particles at the interface and further compression results in buckling and wrinkle formation.[19,36–38] During compression, the increase in the Laplace pressure will push the adsorbed particles away from the interface.[35] So, the buckling of the film will depend on the detachment energy of the particles from the interface and the surface pressure. Desorption of gold nano particles from the oil–water interface was observed during withdrawal of the liquid, owing to the higher surface pressure developed during compression.[39] Sun *et al.* observed the buckling of an oil/water interface with graphene oxide nano sheets and a block copolymer, which were dispersed in water and toluene respectively.[40] In contrast to several studies on the compression behavior of spherical particle laden interfaces, buckling and wrinkling of interfaces containing anisotropic particles is far less investigated.[19,38,41] Two parameters that can possibly influence the buckling behavior of interfaces containing anisotropic particles are the higher possible packing density of ellipsoids in two-dimensions and the larger elastic modulus of ellipsoid-laden interfaces than those with spherical particles at the same coverage.

In the present work, we investigate the effect of particle shape on the adsorption and eventual buckling of a particle coated interface by using pendant drop tensiometry. The adsorption, buckling and wrinkling of particle monolayers at the water–decane interface is experimentally probed by using mono-dispersed hematite ellipsoids of 4.7 ± 0.6 aspect ratio whose zeta-potential can be tuned. The dynamic surface tension measurements recorded with drops of aqueous dispersion of ellipsoids suspended in decane medium revealed a decrease in surface tension with time when the suspension pH is 6.5, where particles are weakly charged. When the particles are highly charged, the surface tension remained almost constant indicating the presence of an energy barrier for adsorption. Consequently, under conditions where particle adsorption is favored – that is – when ellipsoids are weakly charged, pendent drops show buckling under compression. This buckling process is reversible and detachment of particles is not observed during the compression–expansion cycles. We also studied the effect of particle charge, concentration and mechanical energy on the formation of the emulsion. For weakly charged particles, the average droplet size decreases with an increase in particle concentration and energy input. For highly charged particles, emulsions are observed only at high homogenization speeds. The results presented in this manuscript suggest that in several examples where the particles at interfaces concept is exploited, the repulsive energy barrier due to the particle surface charge plays a key role in the adsorption of particles to the interfaces. While these interactions determine the concentration of particles at the interface, their microstructure and interfacial properties depend on particle–particle interactions at the interface, which in the case of anisotropic particles are dominated by shape induced capillary forces.

Materials and methods

Synthesis and characterization of hematite ellipsoids

The forced hydrolysis of iron(III) perchlorate in the presence of urea and a small amount of phosphate anion gives uniform α-Fe_2O_3 ellipsoidal particles.[42] In a typical synthesis, 18.488 g of $Fe(ClO_4)_3 \cdot 6H_2O$ (Sigma-Aldrich), 2.4024 g of NH_2CONH_2 (Fisher Scientific) and 0.2639 g of NaH_2PO_4 (Merck) are added to 400 ml of MilliQ water in a cleaned Pyrex bottle. The tightly stoppered bottle is then placed in a laboratory hot air oven at 100 °C for 24 h. At the end of the reaction time, the Pyrex bottle is taken out of the oven and allowed to cool to room temperature. The resulting reddish-brown suspension is centrifuged at 6000 rpm for 30 min to separate the hematite particles. The particles are washed with MilliQ water three more times and finally the ellipsoidal hematite particles are dispersed into MilliQ water.

The size distribution of the particles was analyzed from the scanning electron microscopy (Hitachi S-4800, Japan) images of the hematite particles. A representative electron microscopy image of the hematite ellipsoids synthesized is shown in Fig. 1, the size and aspect ratio of the particles resulting from the data of 200 particles is tabulated in Table 1. The zeta potential of the hematite suspensions at different pH (pH = 2, 4 and 6.5) was measured using an electrophoretic dynamic light scattering technique (nano-partica SZ100, Horiba, Japan) at 0.001 M NaCl concentration. The pH values of the hematite particle suspensions were adjusted by using aqueous HNO_3. The hematite particles were dispersed into an aqueous solution at a particular pH and centrifuged multiple times to reach the desired pH value. The zeta potential values are shown in Table 2 for different pH hematite particle suspensions.

Pendent drop tensiometry

To prepare the aqueous suspension of hematite ellipsoids, deionized (DI) water from a Millipore system with a specific resistivity of 18.2 MΩ was used. The oil phase used was *n*-decane (anhydrous ≥99% purity from Sigma-Aldrich) and was pretreated with alumina (neutral, activity grade super 1 from Sigma-Aldrich) to remove all the surface active water soluble impurities in it.

Dynamic interfacial tension (DST) measurements of the *n*-decane–water system were performed by the pendant-drop method and the surface tension was obtained by analysis of the drop shape. The experiments were conducted in goniometer (GBX Digidrop, France) equipment at a temperature of 25 ± 2 °C. The water–decane interface is created by forming a pendant drop of 25 μl volume using a surgical needle with a flat tip that is suspended in a decane medium in a rectangular cuvette. The pendant drop image is recorded by using a high speed CCD camera programmed to capture the droplet images at a specified rate for 3000 seconds. The recorded images (one image each 10 seconds) are analyzed using software based on the axisymmetric pendant droplet profile. The shape of the liquid droplet pending from the needle is described by the Young–Laplace

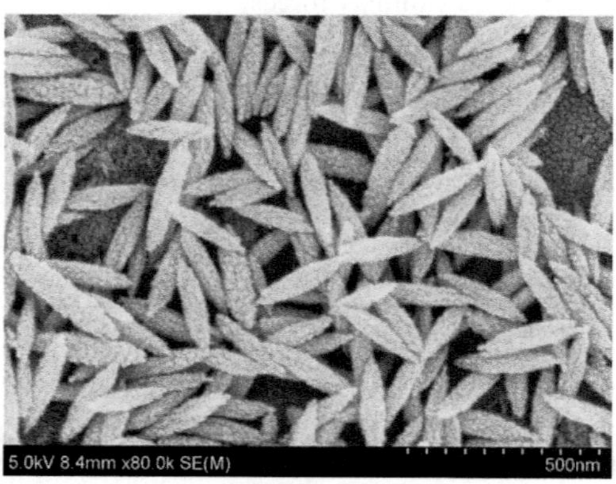

Fig. 1 Scanning electron microscopy (SEM) image of hematite particles of 4.6 ± 0.6 aspect ratio.

Table 1 Physical dimensions of hematite ellipsoids – size distribution of length, width and aspect ratio (AR)

	Mean size	Standard deviation
Length of major axis (nm)	254	50
Length of minor axis (nm)	55	12
Aspect ratio (AR)	4.6	0.6

equation and its numerical solution gives the surface tension of the interface under consideration.

Prior to experiments with the aqueous suspension of ellipsoids, the interfacial tension measurement of pristine decane–water is measured and a near constant value of surface tension indicated the absence of any surface active water soluble impurities in the decane phase. To study the adsorption behavior of the hematite ellipsoidal particles at the decane–water interface, 0.3 wt% of the hematite particle suspension is used to create the pendant drop. The dispersions at three different pH values – 2, 4 and 6.5 were studied as the hematite particles at these pH conditions bear different surface charge densities. The dynamic surface tension of at least three (or more in some cases) droplets of each case was measured to confirm the reproducibility of experimental data.

Pendent drop compression

The objective of these experiments was to study the behavior of the particle laden film formed at the surface of the pendant drop by subjecting it to expansion (or compression) by injecting (or withdrawing) the liquid into (or from) the droplet. Thus the drop volume and hence interface area can be systematically changed so that the response of particle laden interface to expansion or compression can be recorded. We first created a pendant drop of the aqueous hematite suspension hanging at the tip of a flat tipped needle in the decane (oil phase) in a rectangular cuvette. The droplet was left undisturbed for 1 hour to equilibrate. The interface was then compressed by slowly withdrawing the liquid at an average rate of $\sim$0.8 μl s^{-1} from the droplet. The droplet behavior during liquid withdrawal was recorded using a CCD camera. Experiments were conducted by considering pendent drops of hematite ellipsoids at pH = 2 and pH = 6.5. In some cases, the pendant drop during the final stages of compression exhibited buckling. To visualize the self-assembly of ellipsoids in these films, the compressed pendent drop was physically detached from the syringe tip and prepared for scanning

Table 2 Zeta potential of ellipsoidal hematite particles and decane–water interface as a function of pH

pH	2	4	6.5
Zeta potential of interface (mV)[a]	−5.0	−20.0	−60.0
Zeta potential of particles (mV)	+36.4	+50	+6.4

[a] Zeta potential values of water–decane interface as a function of pH is extrapolated from literature data.[34]

electron microscopy imaging. The detached film from a pendant droplet of the hematite suspension collected for SEM may contain the particles that are inside the droplet. Therefore, we also studied the compression behavior by creating an oil droplet inside an aqueous phase of hematite ellipsoids. Analogous to previous experiments, the drop was undisturbed for 1 hour to equilibrate and further compressed by slowly withdrawing the oil from the droplet. One of the advantages of working with this configuration is that when the compressed droplet is detached from the needle, it rises to the top and floats on the surface of the cuvette and therefore it can easily be collected on a substrate for visualization of particle self-assembly.

Preparation of emulsions

To prepare the particle stabilized emulsions, 1 ml of the hematite particle suspension of a known concentration (w/v%) and pH was emulsified with 1 ml of decane in a 5 ml glass vial. All samples were homogenized (with digital ULTRA-TURRAX®) for 2 minutes. The effect of particle concentration, homogenization speed and suspension pH on emulsion formation and size distribution of the resulting Pickering emulsion droplets was studied.

Results and discussion

Influence of surface charge on adsorption of ellipsoids to oil–water interface

Pendant drop tensiometry was used to measure the adsorption kinetics of the hematite ellipsoids at the water–decane interface. The zeta potential and hence the surface charge on the hematite particles in the pendent drops is controlled by varying the pH of the suspension. From Table 2, it can be seen that the hematite particles are highly charged and more stable in acid medium. The hematite particles are weakly charged at pH 6.5, which is close to the iso-electric point of around 8–9 pH.[41] We used three different pH suspensions – 2, 4 and 6.5 – for adsorption studies, where the zeta potentials respectively are +36.5 mV, +50 mV and +6.5 mV. The dynamic surface tension (DST) data for the hematite suspensions at different pH and pure water–decane are shown in Fig. 2. The DST for pristine decane–water is nearly constant, which is expected since surface active components either in water or in decane are not present. It is known that the adsorption of any such species to an interface will reduce the surface tension. In the case where water is replaced with the aqueous hematite suspension at pH 2, the DST data are similar to the case of pristine decane–water showing an average surface tension of 52 $\pm$ 0.5 mN m^{-1}. This observation indicates that there is no adsorption of particles to the interface. For the pendant drop of hematite suspensions at pH 4 and 6.5, the DST data show a decreasing trend with time indicating possible adsorption of ellipsoids to the water–decane interface. Since the particle concentration and the aspect ratio of the ellipsoids is the same in all the cases, the observed difference in the adsorption at different pH is due to the effect of electrostatic interaction energy as the particles approach the interface due to inherent Brownian motion. For particles in the vicinity of the interface, there will be a contribution of van der Waals attraction and electrostatic double layer interaction between the hematite ellipsoids and the

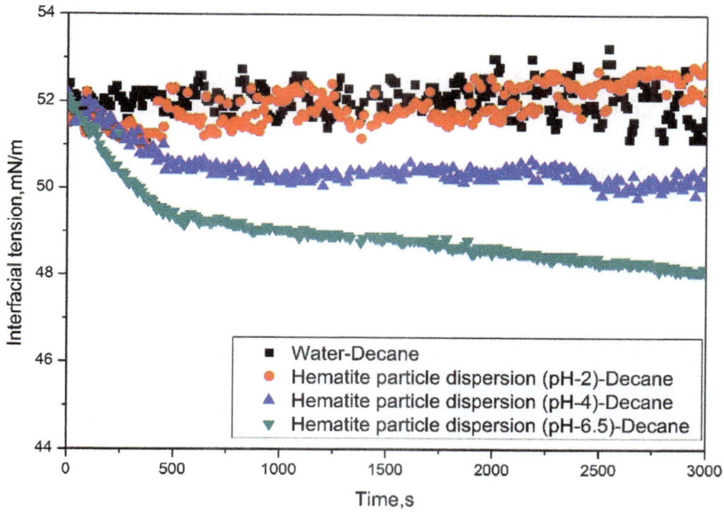

Fig. 2 The dynamic surface tension of the hematite suspension and decane interface as a function of suspension pH. The particle concentration is 0.3 w/v%. The surface tension is almost constant for the pH 2 suspension, where the image charge repulsion is stronger. A decrease in surface tension is observed at pH 4 and 6.5. The decrease in surface tension is more significant for the pH 6.5 suspension due to weak image charge repulsion.

decane–water interface. As the hematite-ellipsoids under the pH conditions studied are always positively charged and the interface is negatively charged, the overall potential is always attractive. Thus if particle–interface interactions alone are important, adsorption will have to occur for all pH cases. A clear difference in the DST data shown in Fig. 2 at different pH, indicates the presence of an energy barrier for the adsorption of the ellipsoids. In recent studies, this energy barrier has been attributed to image charge repulsion.[34] When a charged particle is approaching a low dielectric medium (oil or air) through a high dielectric medium (water), the original particle experiences repulsion from an image charge in the oil medium. The sign of the image charge is the same as that of the original particle when the particles dispersed in a high dielectric medium approach the interface and will be of opposite signs if the particles dispersed in a low dielectric medium approach the interface. Therefore the interaction between the original particle and the image charge particle plays an important role in the adsorption of charged particles to fluid interfaces. We have calculated the image charge interactions and particle–interface interactions for different pH suspensions (see ESI†). The calculated overall interactions between ellipsoidal particles qualitatively support our experimental findings. At pH 2, the net interaction is repulsive because the particle–interface interactions are weakly attractive and image charge repulsion is stronger. The net repulsion at the interface results in an energy barrier for the particles to reach the interface. As the suspension pH increases (pH 6.5), the image charge interactions are weakly repulsive because of the low zeta potential and the particle–interface interactions are more attractive because of the high interface charge (interface charge will increase with increasing pH, Table 2). So in the case of pH 6.5, the net attractive interaction favors the adsorption of particles at the interface. In the case of pH 4, the image charge repulsion and particle–interface interactions are comparable and therefore a very small change in the interfacial tension is observed.

From Fig. 2, the DST data for pH 4 and 6.5 show different slopes due to the different adsorption mechanisms. At initial times, a steep drop in the interfacial tension is observed due to a higher adsorption rate because of the pristine interface and because the entire interface is available for adsorption. Upon an increase in particle concentration at the interface, the adsorption rate drops due to repulsion from the already adsorbed particles. That is – the adsorbed particles at the interface create an electrostatic repulsion for the new particles to reach the interface.[43,44] Similarly, the probability to find a particle free interface will decrease as the particles coverage increases.[45] Therefore, the initial slope in the DST data is due to adsorption of the particles in the absence of any repulsion and a different slope in the second regime is due to partial coverage of the interface and the resistance from the adsorbed particles. From the DST experiments, we conclude that the adsorption of particles at the interface strongly depends on the overall interactions at the interface and the image charge interactions contribute to the major energy barrier for particle adsorption to interfaces.

Buckling and wrinkling of ellipsoid-laden interface under compression

While the DST data presented in the previous section confirm the adsorption of particles, we could not directly visualize their adsorption *in situ* due to the small particle size. Therefore we performed compression experiments, as droplets coated with particles are known to be viscoelastic and thus exhibit interesting shape changes during compression. In the case of the pendant drop method, compression is achieved by reducing the drop volume at a slow rate ($\sim$0.8 μl s^{-1}). The shape transitions during the compression of the pendant suspension drops containing hematite ellipsoids at pH 2 and 6.5 that were observed are shown in Fig. 3. In the pH 2 case, the droplet shapes are similar and are axi-symmetric at different compression ratios (V/V_0, where V_0 is the initial drop volume and V is the instantaneous drop volume during compression), however there is a decrease in the drop size. But in the case of the weakly charged particles (pH 6.5), with an increase in compression, the droplet shape becomes non-symmetric and the droplet starts to buckle and eventually wrinkles appear. Thus, compression studies also confirm that there is no adsorption of particles at the interface when the image charge repulsion dominates (pH 2) (Fig. 3A). As the image charge repulsion reduces (pH 6.5), the compression experiments clearly demonstrate that there is adsorption of particles at the interface (Fig. 3B). In Fig. 3B, it can be seen that initial buckling appeared at a volume ratio of 0.16, this stage possibly indicates the maximum coverage of the drop surface with hematite ellipsoids. Initially, the droplet interface behaves like a fluid, because of the minimal coverage of the particles (see ESI Video†). As the compression is continued, the coverage increases, so the droplet interface changes from fluid to solid-like. This fluid–solid transition has been observed in planar monolayers of ellipsoids. The

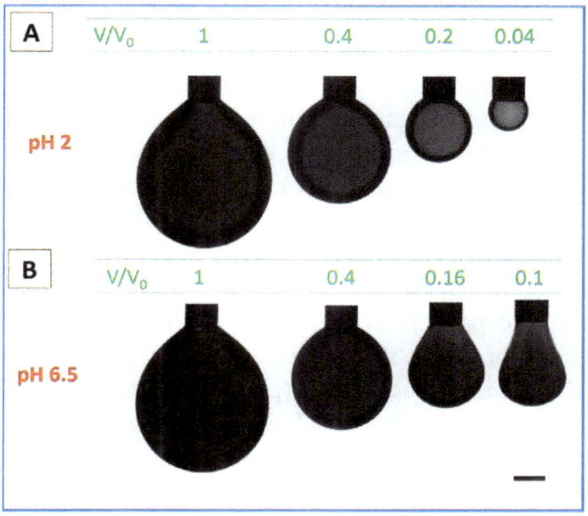

Fig. 3 Compression of pH 2 and 6.5 hematite droplet suspended in decane medium. The concentration of the particles is 0.3 w/v%. Withdrawing the liquid reduces the interfacial area available for the adsorbed particles, which results in wrinkling and buckling of the particle laden interface. (A) Buckling is not observed for pH 2 suspensions at different compression ratios. The shape of the droplet is identical at all compression ratios. (B) A clear buckling is observed in suspensions of pH 6.5 at a compression ratio smaller than 0.16. The droplet is highly deformed at higher compression (or low compression ratios). Scale bar is ~0.81 mm.

behavior of the interface with micron and sub-micron ellipsoidal particles at low particle coverage is viscous or fluid-like.[27] With further increases in particle coverage at the interface, the interface becomes viscoelastic. The compression studies of nano sized ellipsoidal particles on the drop surface that are presented here show a similar fluid to viscoelastic transition during the compression.

The buckling and wrinkles in the pH 6.5 pendant drop are shown in Fig. 4. At higher compression (smaller compression ratio), the buckled particulate

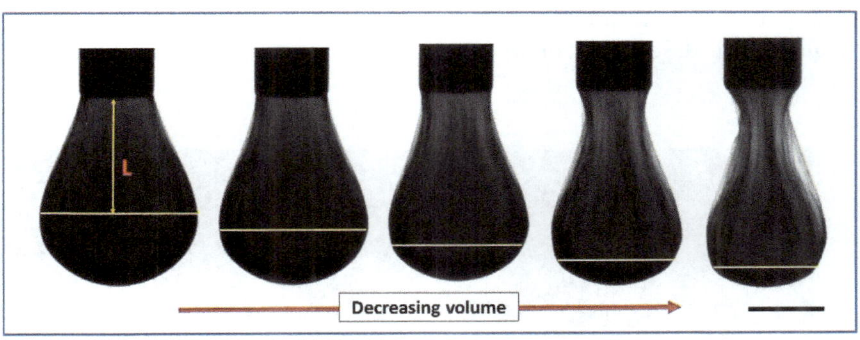

Fig. 4 Buckling behavior of the pendant drop at different compressions. Under compression, the wrinkle length (L) and amplitude (λ) changes. At higher compression, deep wrinkles are prominent. Scale bar ~0.81 mm.

film starts to show wrinkles. During high compression, surface tension reduction is more in the parallel direction than the meridional direction which supports the pull of gravity in the vertical direction. These anisotropic tensions and related strains cause wrinkles to form and align in the meridional direction.[38] In the compression experiments, wrinkles with smaller amplitude and short wavelength are observed to coexist with deep wrinkles that are of large wavelength. At high compression, the deep wrinkles are more numerous and irregular. The wrinkle length and wavelength increases with the compression.

To understand the particle arrangement at the interface during buckling, compression was performed in the reverse configuration by creating a suspended decane drop in the hematite suspension (pH = 6.5). After 1 h, we slowly withdrew the decane until the decane droplet buckled (see ESI Video†). The buckling of the droplet under compression is shown in Fig. 5. This behavior is similar to the case of the suspension drop compression shown in Fig. 3B and 4. To conclude, as a result of low particle surface charge (pH = 6.5) and their subsequent adsorption to the water–decane interface, buckling is observed in both cases – (a) compressed drops of the aqueous dispersion of ellipsoids suspended in the decane medium and (b) compressed decane drops suspended in the dispersion of ellipsoids (pH = 6.5). The shape changes observed depicted in Fig. 3B–5 are due to a transition from a fluid-like particle film to a solid-like particle film that is highly viscoelastic. When the buckled drops are un-compressed by increasing their volume, wrinkles are observed to disappear. Soon after the disappearance of all the wrinkles, the drops are observed to expand isotropically. Upon re-compression, the wrinkles reappear (Fig. 6). That is – the buckling and accompanying fluid–solid transition is reversible. This implies that the ellipsoidal particles adsorb irreversibly to the interface and that these particles are not expelled from the interface during compression/expansion cycles.

There have been excellent theoretical and experimental developments on the quantitative understanding of buckling and wrinkling of compressed planar monolayers of spherical particles.[35,46–49] Vella *et al.* studied the buckling behavior of spherical particle coated interfaces under uniaxial compression.[46] They estimated that the average wavelength of wrinkles depends on the particle size, coverage and Young's modulus. Further, the buckling phenomena in compression of pendant drops coated with particle–polymer[40] and particle–surfactant[50] mixtures have been studied. In these studies, a qualitative description of wrinkle formation as a result of compression

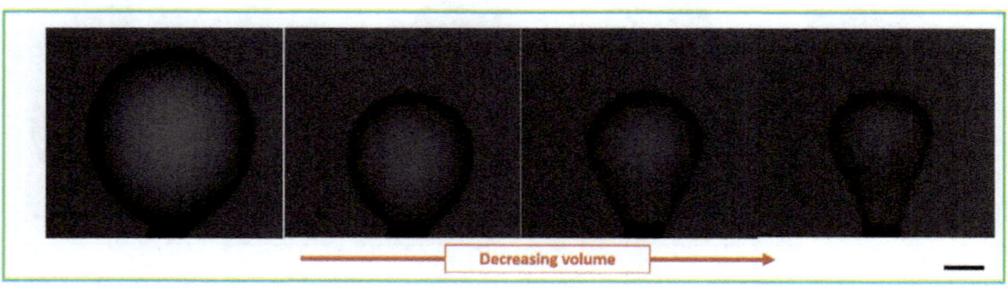

Fig. 5 Compression of decane droplet suspended in hematite suspension at pH of 6.5. The droplet buckles and shows wrinkles upon compression. Scale bar is 0.55 mm.

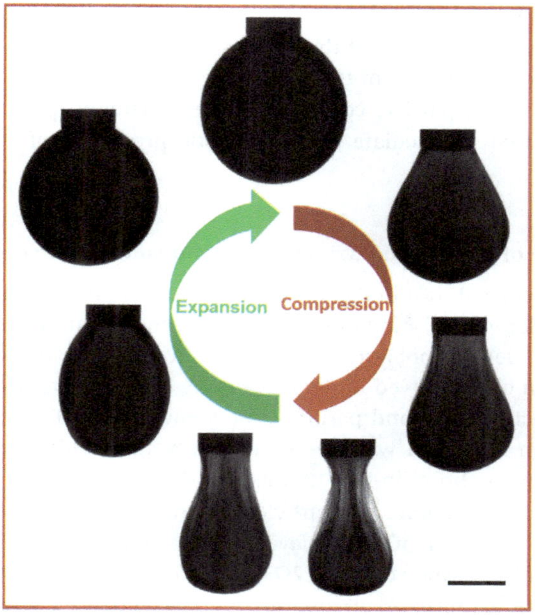

Fig. 6 Reversible buckling phenomena of pH 6.5 suspension suspended in decane medium. Scale bar is 0.81 mm.

driven jamming of the particle network is discussed. There are several reports on the pendant drop compression of interfaces containing polymers,[37] bacteria,[38] and proteins[36] where features that are similar to those observed in the compression of particle monolayers are observed. A detailed theoretical description of buckling of thin films such as those that occur during compression of pendant drops has been developed.[36] A comparison of the buckling behavior of monolayers of ellipsoids with the monolayers of spherical particles will require additional experiments and development of suitable theories that consider particle shape effects. However, the theoretical studies that consider the behavior of elastic shells to compression can be used to calculate the interface properties of ellipsoid particle coated interfaces.[36,37] Assuming that a viscoelastic sheet of ellipsoidal particles is formed prior to the onset of buckling, the elastic modulus of the interface is calculated using eqn (1)[37]

$$G' = \frac{\lambda^4(1-\nu)T}{32(\pi L h)^2} \tag{1}$$

where, λ is the wavelength of the wrinkle, ν is the Poisson ratio, L is the wrinkle length, h is the film thickness and T is the meridional stress. From the images of the pendant drops under compression, we obtained an average wrinkle wavelength of 115 μm and an average wrinkle length of 800 μm. These values were obtained at a compression ratio of 0.16. Using the values of wrinkle wavelength, wrinkle length, film thickness (equal to the length of the minor axis (50 nm)), Poisson ratio (0.5, suggested previously for rigid interfacial adsorption layers)[37] and meridional tension (5 mN m^{-1}, obtained from similar pendant drop experiments),[38] the elastic modulus of the film was calculated to be ~0.6 N m^{-1}. The estimated elastic modulus is of the same order of magnitude as that previously reported for ellipsoid particle coated interfaces.[23] Thus in principle, this technique can be used to calculate the viscoelastic properties of a particle laden interface.

Micro-structure of the buckled/wrinkled ellipsoid-laden interface

The buckled droplet shown in Fig. 5 was detached from the needle and collected on a substrate for visualization of the particle arrangement in the buckled/wrinkled ellipsoid-laden monolayer. After evaporation of any residual water and decane, the film was observed under a scanning electron microscope for microstructure characterization and particle arrangement. The SEM images in Fig. 7 show a uniform film with wrinkles that are several hundred micrometers in length. This film, as shown in Fig. 7B, was found to be a single particle thick and this confirms that the ellipsoidal particles adsorb at the water–decane interface resulting in the formation of a monolayer of ellipsoids. The particle alignment in the wrinkled region as shown in Fig. 7C is such that on average the major axis of ellipsoids is parallel to the wrinkling direction. This particle alignment is most likely a result of particle reorientation during compression. It must be noted that such an alignment is not seen in other regions (away from wrinkled or buckled regions) of the monolayer, wherein only local ordering of a few ellipsoids attached side-by-side similar to those experimentally observed for micron size ellipsoid particles[27] is present. Since the monolayer of ellipsoids in the buckled state

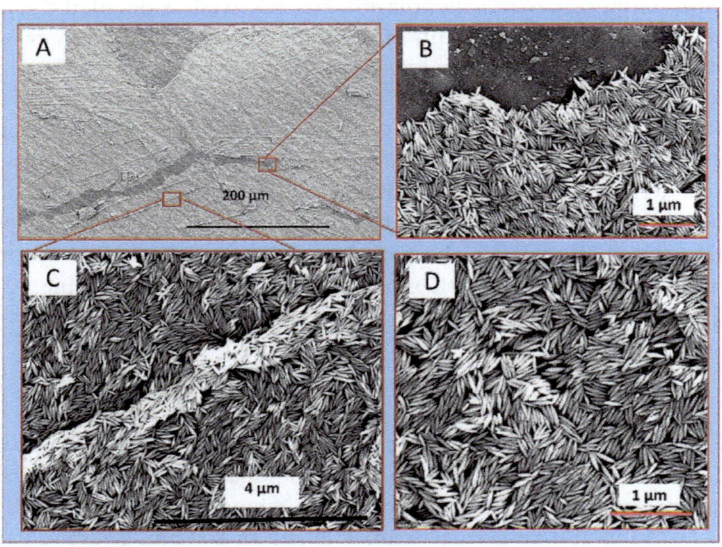

Fig. 7 SEM images of particulate film collected from the decane droplet suspended in hematite suspension. (A) Macroscopic view of the particulate film with wrinkles. (B) Microscopic view confirming that the film is a monolayer of ellipsoids. (C and D) Local ordering of the ellipsoidal particles near the wrinkled region.

is close packed and in a jammed state, the evaporation of residual fluid is not likely to influence any microstructural changes in the compressed film of ellipsoids.

Use of ellipsoids for emulsion stabilization – effect of surface charge

The adsorption as well as elastic nature of the interface is important when particles are used for emulsion stabilization. To study the effect of the charge of ellipsoids on emulsification, emulsions were prepared by taking a 2 w/v% concentration of an aqueous suspension of hematite ellipsoids (4.7 ± 0.6 aspect ratio) at pH values 2, 4 and 6.5 at a fixed speed of 10 000 rpm. Emulsification was observed in all samples and the emulsion was found to be of the oil-in-water type as confirmed by the drop test. The volume of the emulsion phase obtained was observed to increase and the size of emulsion droplet was observed to decrease with increasing pH. Since the particle concentration, aspect ratio, energy input, and volume of oil/water was the same, this effect is due to the adsorption of more particles to the interface with increasing pH or a decrease of surface charge. We also prepared a series of emulsions with (1) a hematite particle suspension of 2 w/v% concentration and pH = 2 and (2) a hematite particle suspension of 2 w/v% concentration and pH = 6.5 and equal volume of decane, however, the homogenization speed was varied from 1000 to 15 000 rpm. For pH = 2, the emulsions formed at 1000 and 5000 rpm showed no stability, but the solid stabilized emulsion droplets formed at 10 000 and 15 000 rpm were observed to be stable for several months. At pH = 6.5, stable emulsions were formed at speeds of 5000 rpm onwards. The emulsion droplet images are shown in Fig. 8. At pH 6.5, emulsion formation can be observed even at 1000 rpm by increasing the homogenization time. We also studied the effect of particle concentration on the volume of emulsion formed and droplet size distribution (at a fixed oil water ratio and homogenization speed), since the results show a similar trend as earlier work, they are not reported here.[22]

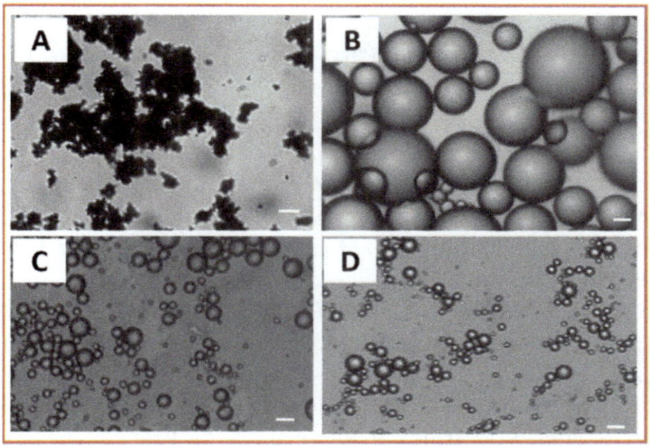

Fig. 8 Microscopy images of pH 6.5 emulsion droplets at (a) 1000 rpm, (b) 5000 rpm, (c) 10 000 rpm, (d) 15 000 rpm for 2 min. Particle concentration is the same in all samples (2 w/v%). Images are captured after 1 day of emulsification. Scale bar is 50 μm.

Conclusions

From several experiments discussed in the previous sections, we conclude that the surface charge of ellipsoids and the different interactions in the vicinity of the interface play an important role in the adsorption of ellipsoids to the water–decane interface. This is evident from both dynamic surface tension and pendant drop compression experiments. Under compression, the particle-laden interface containing ellipsoids (formed when the suspensions at pH = 6.5 are used) is observed to buckle and wrinkle due to a transition from a fluid-like to a solid-like particle film that is highly viscoelastic. The buckling and accompanying fluid–solid transition is reversible. The elastic modulus calculated from the analysis of the wrinkle pattern was found to be of the same order of magnitude as that experimentally reported and is considerably higher than the modulus of the monolayer of spherical particles.[23] In agreement with literature reports, ellipsoidal particles are observed to effectively stabilize oil-in-water emulsions. The mechanical energy required for emulsification as well as the volume of emulsion formed were observed to depend on the surface charge of the ellipsoids.

Acknowledgements

We would like to acknowledge the Goniometer facility at nano mechanics & nano materials lab (Prof. Pijush Ghosh, Department of Applied Mechanics, IIT Madras) and Scanning Electron Microscopy facility at the Department of Chemical Engineering, IIT Madras (procured through DST-FIST grant).

References

1 M. Nitschke and S. Costa, *Trends Food Sci. Technol.*, 2007, **18**, 252–259.
2 D. J. Miller, T. Henning and W. Grünbein, *Colloids Surf., A*, 2001, **183**, 681–688.
3 E. Dickinson, *Curr. Opin. Colloid Interface Sci.*, 2010, **15**, 40–49.
4 A. J. Worthen, H. G. Bagaria, Y. Chen, S. L. Bryant, C. Huh and K. P. Johnston, *J. Colloid Interface Sci.*, 2013, **391**, 142–151.
5 B. P. Binks and S. O. Lumsdon, *Langmuir*, 2001, **17**, 4540–4547.
6 B. P. Binks and S. O. Lumsdon, *Phys. Chem. Chem. Phys.*, 1999, **1**, 3007–3016.
7 R. Aveyard, B. P. Binks and J. H. Clint, *Adv. Colloid Interface Sci.*, 2003, **100**, 503–546.
8 O. B. Ho, *J. Colloid Interface Sci.*, 1998, **198**, 249–260.
9 C. Holtze, A. Rowat, J. Agresti, J. Hutchison, F. Angile, C. Schmitz, S. Köster, H. Duan, K. Humphry and R. Scanga, *Lab Chip*, 2008, **8**, 1632–1639.
10 J.-H. Park, C. Oh, S.-I. Shin, S.-K. Moon and S.-G. Oh, *J. Colloid Interface Sci.*, 2003, **266**, 107–114.
11 A. Benichou, A. Aserin and N. Garti, *Adv. Colloid Interface Sci.*, 2004, **108**, 29–41.
12 M. Destribats, V. Lapeyre, M. Wolfs, E. Sellier, F. Leal-Calderon, V. Ravaine and V. Schmitt, *Soft Matter*, 2011, **7**, 7689–7698.
13 Z. Li, T. Ming, J. Wang and T. Ngai, *Angew. Chem., Int. Ed.*, 2009, **48**, 8490–8493.
14 D. Dupin, S. Fujii, S. P. Armes, P. Reeve and S. M. Baxter, *Langmuir*, 2006, **22**, 3381–3387.
15 Z. Li, D. Harbottle, E. Pensini, T. Ngai, W. Richtering and Z. Xu, *Langmuir*, 2015, **31**, 6282–6288.

16 R. G. Alargova, D. S. Warhadpande, V. N. Paunov and O. D. Velev, *Langmuir*, 2004, **20**, 10371–10374.
17 B. P. Binks and C. P. Whitby, *Colloids Surf., A*, 2005, **253**, 105–115.
18 S. U. Pickering, *J. Chem. Soc., Trans.*, 1907, **91**, 2001–2021.
19 H. Xu, S. Melle, K. Golemanov and G. Fuller, *Langmuir*, 2005, **21**, 10016–10020.
20 D. E. Tambe and M. M. Sharma, *J. Colloid Interface Sci.*, 1993, **157**, 244–253.
21 A. B. Pawar, M. Caggioni, R. Ergun, R. W. Hartel and P. T. Spicer, *Soft Matter*, 2011, **7**, 7710–7716.
22 B. Madivala, S. Vandebril, J. Fransaer and J. Vermant, *Soft Matter*, 2009, **5**, 1717–1727.
23 B. Madivala, J. Fransaer and J. Vermant, *Langmuir*, 2009, **25**, 2718–2728.
24 J. W. J. de Folter, E. M. Hutter, S. I. R. Castillo, K. E. Klop, A. P. Philipse and W. K. Kegel, *Langmuir*, 2014, **30**, 955–964.
25 V. R. Dugyala, S. V. Daware and M. G. Basavaraj, *Soft Matter*, 2013, **9**, 6711–6725.
26 P. F. Noble, O. J. Cayre, R. G. Alargova, O. D. Velev and V. N. Paunov, *J. Am. Chem. Soc.*, 2004, **126**, 8092–8093.
27 M. G. Basavaraj, G. G. Fuller, J. Fransaer and J. Vermant, *Langmuir*, 2006, **22**, 6605–6612.
28 L. Botto, E. P. Lewandowski, M. Cavallaro and K. J. Stebe, *Soft Matter*, 2012, **8**, 9957–9971.
29 M. Oettel and S. Dietrich, *Langmuir*, 2008, **24**, 1425–1441.
30 J. C. Loudet, A. M. Alsayed, J. Zhang and A. G. Yodh, *Phys. Rev. Lett.*, 2005, **94**, 018301.
31 F. Gautier, M. Destribats, R. Perrier-Cornet, J.-F. Dechézelles, J. Giermanska, V. Héroguez, S. Ravaine, F. Leal-Calderon and V. Schmitt, *Phys. Chem. Chem. Phys.*, 2007, **9**, 6455–6462.
32 B. P. Binks and J. A. Rodrigues, *Langmuir*, 2007, **23**, 7436–7439.
33 N. J. Alvarez, S. L. Anna, T. Saigal, R. D. Tilton and L. M. Walker, *Langmuir*, 2012, **28**, 8052–8063.
34 H. Wang, V. Singh and S. H. Behrens, *J. Phys. Chem. Lett.*, 2012, **3**, 2986–2990.
35 S. Razavi, K. D. Cao, B. Lin, K. Y. C. Lee, R. S. Tu and I. Kretzschmar, *Langmuir*, 2015, **31**, 7764–7775.
36 S. Knoche, D. Vella, E. Aumaitre, P. Degen, H. Rehage, P. Cicuta and J. Kierfeld, *Langmuir*, 2013, **29**, 12463–12471.
37 P. Erni, H. A. Jerri, K. Wong and A. Parker, *Soft Matter*, 2012, **8**, 6958–6967.
38 L. Vaccari, D. Allan, N. Sharifi-Mood, A. Singh, R. Leheny and K. Stebe, 2015, arXiv preprint arXiv:1503.07255.
39 V. Garbin, J. C. Crocker and K. J. Stebe, *Langmuir*, 2011, **28**, 1663–1667.
40 Z. Sun, T. Feng and T. P. Russell, *Langmuir*, 2013, **29**, 13407–13413.
41 V. R. Dugyala and M. G. Basavaraj, *Langmuir*, 2014, **30**, 8680–8686.
42 M. Ocaña, M. P. Morales and C. J. Serna, *J. Colloid Interface Sci.*, 1999, **212**, 317–323.
43 N. Bizmark, M. A. Ioannidis and D. E. Henneke, *Langmuir*, 2014, **30**, 710–717.
44 F. Reincke, W. K. Kegel, H. Zhang, M. Nolte, D. Wang, D. Vanmaekelbergh and H. Möhwald, *Phys. Chem. Chem. Phys.*, 2006, **8**, 3828–3835.
45 S. Kutuzov, J. He, R. Tangirala, T. Emrick, T. Russell and A. Böker, *Phys. Chem. Chem. Phys.*, 2007, **9**, 6351–6358.
46 D. Vella, P. Aussillous and L. Mahadevan, *Europhys. Lett.*, 2004, **68**, 212.

47 Y.-L. Lee, Z.-C. Du, W.-X. Lin and Y.-M. Yang, *J. Colloid Interface Sci.*, 2006, **296**, 233–241.
48 T. S. Horozov, B. P. Binks, R. Aveyard and J. H. Clint, *Colloids Surf., A*, 2006, **282–283**, 377–386.
49 C. Monteux, E. Jung and G. G. Fuller, *Langmuir*, 2007, **23**, 3975–3980.
50 C. P. Whitby, D. Fornasiero, J. Ralston, L. Liggieri and F. Ravera, *J. Phys. Chem. C*, 2012, **116**, 3050–3058.

Faraday Discussions

PAPER

Transport of nanoparticulate material in self-assembled block copolymer micelle solutions and crystals

Vicki A. Cheng and Lynn M. Walker*

Received 17th August 2015, Accepted 5th November 2015
DOI: 10.1039/c5fd00122f

Water soluble poly(ethylene oxide)–poly(propylene oxide)–poly(ethylene oxide) [PEO–PPO–PEO] triblock copolymers self-assemble into thermoreversible micellar crystals comprised of periodically spaced micelles. The micelles have PPO cores surrounded by hydrated PEO coronas and the dimensions of the unit cell of the organized micelles is on the order of several to tens of nanometers. Fluorescence recovery after photobleaching (FRAP) is used to quantify nanoparticle transport in these nanostructured polymer micelle systems. Diffusivity of bovine serum albumin (BSA, $D_h \sim 7$ nm) is quantified across a wide range of polymer, or micelle, concentrations covering both the disordered fluid as well as the structured micellar crystal to understand the effects of nanoscale structure on particle transport. Measured particle diffusivity in these micellar systems is reduced by as much as four orders of magnitude when compared to diffusivity in free solution. Diffusivity in the disordered micellar fluid is best understood in terms of diffusion through a polymeric solution, while transport in the structured micellar phase is possibly due to hopping between interstitial sites. These results not only show that the nanoscale structures of the micelles have a measureable impact on particle diffusivity, but also demonstrate the ability to tune nanoscale transport in self-assembled materials.

Introduction

Block copolymers exhibit complex phase behavior, particularly when dispersed in a solvent that selectively solubilizes specific polymer blocks. At low polymer concentration, block copolymers in selective solvents form micelles, with aggregation numbers in the tens to hundreds and dimensions of nanometers. As the polymer concentration increases, these micelles self-assemble into packed nanocrystalline structures with geometries including cubic arrangements of spherical micelles and hexagonal ordering of cylindrical micelles.[1–5] These

Center for Complex Fluids Engineering, Department of Chemical Engineering, Carnegie Mellon University, Pittsburgh, PA, USA. E-mail: lwalker@andrew.cmu.edu

micellar crystals have short range and long range crystalline order and mechanical behavior that is best described as a soft solid.

Poly(ethylene oxide)–poly(propylene oxide)–poly(ethylene oxide) block copolymers (PEO_n–PPO_m–PEO_n), commercially known under the trade name Pluronic®, or Poloxamer®, are a class of water-soluble block copolymers that undergo a reversible temperature-induced transition from a disordered micellar fluid to a crystalline nanostructured system.[5–8] While the PEO and PPO blocks are both water soluble at temperatures below about 10 °C, the block copolymers micellize at higher temperatures, with the PPO dehydrating first and self-assembling to form the micellar core. At sufficiently high temperatures and concentrations, the block copolymer micelles will pack into crystalline structures. This self-assembling behavior has been characterized in phase diagrams for several types of Pluronic systems using a combination of small-angle scattering techniques; these same techniques provide details about the block copolymer packing structure, micelle size and shape.[7–14]

There has been recent interest in generating hydrophilic nanocomposites based on interspersing nanoparticulate material in these nanostructured block copolymer micelles.[15–20] Sequestering hydrophilic nanoparticulate material into the continuous phase of these packed block copolymer systems has the potential to control the spatial arrangement of the nanoparticulate material using the self-assembled structure of the block copolymer micellar solution. This technique has been demonstrated in several different architectures of Pluronic®; F127 (PEO_{106}–PPO_{70}–PEO_{106}), F87 (PEO_{75}–PPO_{46}–PEO_{75}) and P123 (PEO_{20}–PPO_{70}–PEO_{20}) for a variety of nanoparticulate materials with hydrodynamic diameters ranging from 4–7 nm, including silica, lysozyme, and bovine serum albumin (BSA).[15–19,21–23] This structure has been verified by contrast matching small-angle neutron scattering (SANS).[15,16,18] In developing this templating process, the nanoparticle systems have been selected such that they are comparable in size to the interstitial domains of the structured block copolymer.

Previous studies on these nanoparticle–block copolymer composites have characterized the static structure and mechanics of these systems, but little is known about the transport behavior of nanoparticles within the ordered block copolymer. The structure and composition of these composites presents several possibilities for the dynamic behavior of the nanoparticles. Because the nanoparticle dimensions are closely matched to that of the block copolymer interstitial spacing, it is possible that the nanoparticles are trapped within the structured matrix formed by the block copolymer, and no significant particle transport occurs. However, these block copolymer composites are also 'soft' systems largely comprised of water, so nanoparticles also diffuse through the solvent-filled free volume available in the block copolymer. Furthermore, these composites form an intercalated crystal structure, and so it is plausible for particle transport to resemble diffusive behavior in solid crystalline materials. Clearly, several possible transport mechanisms exist in these nanostructured block copolymers, and it is critical to understand the dynamic behavior of nanoparticles in these systems, especially when considering their potential applications as nanocomposites,[24,25] systems for storing ordered nanoparticle arrays,[17,19] protein protection,[26] as media for DNA separation,[27,28] or as nanostructured templates for mineralization.

Here, we quantify nanoparticle diffusion through micellar block copolymer systems. Considerable work has been performed to understand and model particle diffusion in polymer solutions and melts.[3,29–34] Studies on diffusion through polymer gels and hydrogels have typically been performed on materials whose structures have no long-range order, and so the nanoscale structure of these block copolymers provide a unique medium for studying particle transport, especially when the interstitial voids are on similar lengthscales as the particles. Fluorescence recovery after photobleaching (FRAP) is used to determine the diffusive properties of small molecules and macromolecules through biological environments and polymeric gels,[35–37] and is suitable for studying nanoparticle transport through these self-assembled block copolymers. For this study, fluorescently-tagged BSA, a globular protein, is used as a model monodisperse tracer particle to complement the prior structural studies done on the nanocomposite system. Protein (probe) diffusivity is studied across a range of polymer concentrations covering both the disordered fluid phase as well as the structured micellar phase to understand how micellar ordering affects particle transport. The effects of protein loading, shear alignment, particle type, and block copolymer composition are also examined in this study.

Materials and methods

Materials

Pluronic® block copolymers F127 (PEO_{106}–PPO_{70}–PEO_{106}), F87 (PEO_{75}–PPO_{46}–PEO_{75}), and P123 (PEO_{20}–PPO_{70}–PEO_{20}) (BASF; Mount Olive, NJ) were used as received. These are commercial materials with both molecular weight and structural polydispersity, but they have been studied extensively in the literature and applied broadly, so the materials are used as received here. Properties of the three block copolymers used are reported in Table 1; the reported overall molecular weights, degree of polymerization of the blocks and ratio of number of PEO to PPO units are provided. The CMC is reported from the literature. 5-Hexadecanoylamino-fluorescein (HDFL) was purchased from Sigma Aldrich (St. Louis, MO) and used as received. Fluorescein isothiocyanate (FITC) was used as received for labeling the protein bovine serum albumin (BSA) (Sigma Aldrich; St. Louis, MO).

Protein labeling

Bovine serum albumin is fluorescently labeled with fluorescein isothiocyanate (FITC) at a 1 : 1 fluorescence to protein (F/P) molar ratio using a 0.1 M sodium carbonate-bicarbonate buffer (pH = 9.0), following standard protocols described

Table 1 Physical properties of different block copolymers, with molecular formula PEO_x–PPO_y–PEO_x. Molecular properties are reported by the manufacturer, the critical micelle concentration (CMC) at 25 °C is reported in the literature. The transition concentration, C_{gl}, is determined here

Polymer	Molecular weight (g mol^{-1})	PEO units (x)	PPO units (y)	PEO : PPO ratio	CMC[60,61]	C_{gl} (wt%)
P123	5750	20	70	0.57	0.03	28
F127	12 600	106	70	3.0	1	18
F87	7700	75	46	3.3	2	34

elsewhere.[38,39] The labeled protein is separated from unreacted FITC using a Sephadex G-25M column (Sigma Aldrich; St. Louis, MO), followed by dialysis. The concentrations of BSA and FITC are determined spectrophotometrically using extinction coefficients previously determined in the buffer: ${}^{495}_{\mathrm{FITC}}\varepsilon = 70\,000\ \mathrm{M^{-1}\ cm^{-1}}$,[40] ${}^{280}_{\mathrm{FITC}}\varepsilon = 10\,950\ \mathrm{M^{-1}\ cm^{-1}}$,[41] or measured ${}^{280}_{\mathrm{BSA}}\varepsilon = 34\,512 \pm 820\ \mathrm{M^{-1}\ cm^{-1}}$. The addition of the FITC tag (389 g mol^{-1}) to BSA (69 324 g mol^{-1}) at this low level of tagging does not significantly alter the hydrodynamic size of the BSA probe and is not expected to change the interaction of BSA with the block copolymer matrix.

Block copolymer–protein composite preparation

The FITC–BSA is combined with unlabeled BSA to reduce the average F/P ratio, maintaining the total FITC concentration at 47 μM. This FITC concentration was verified to be within the linear region of fluorescence intensity as a function of fluorophore concentration. The BSA is then incorporated into three Pluronic® block copolymer systems (F127, F87, and P123), following procedures outlined previously.[15,16,23] All dispersions were prepared in 0.1 M sodium carbonate-sodium bicarbonate buffer (pH = 9.0) to maximize the fluorescence of the FITC. The BSA concentration in the block copolymer–protein composites ranged from 0.25 to 5.4 wt%, while the block copolymer concentration ranged from 7 to 43 wt% of the total composition.

Rheological characterization

Viscosity measurements are performed for the fluid solutions using a stress-controlled rheometer (Bohlin Gemini, Malvern Instruments; West Borough, MA) with a Couette geometry (cup and bob, $O_{\mathrm{D}} = 27.5$ mm, $I_{\mathrm{D}} = 25$ mm, gap = 1.25 mm). The temperature is controlled with a Peltier unit, and measurements are made at 25 °C. For the micellar crystals, measurements are made with a parallel plate geometry ($D = 25$ mm) and a gap of 800 μm. The concentration at which the block copolymer first resembles a solid, C_{gl}, is defined here as the concentration at which G' first exceeds G'' at 25 °C and is identified by performing temperature sweeps (1 °C min^{-1}, 1 rad s^{-1}) on block copolymer samples, and is given in Table 1. Measurements were performed on samples in increments of 1 wt% to find the critical concentration. This is a somewhat arbitrary definition; however, it is used consistently for all block copolymer samples to provide a convenient comparison. The block copolymer systems are allowed to cool at 5 °C for at least 5 minutes before proceeding with experiments. A thin layer of mineral oil is placed around the sample geometry to prevent evaporation during the experiment. First, the linear viscoelastic region is determined by increasing the stress amplitude at fixed frequency (1 rad s^{-1}) and identifying the region where the moduli are invariant to stress amplitude. The magnitude of the complex modulus ($|G^*|$, or for simplicity, G^*) is then measured in the linear viscoelastic region; for the systems here, at an applied stress of 50 Pa at a frequency of 1 rad s^{-1}.

Fluorescence recovery after photobleaching (FRAP)

Block copolymer–protein samples are loaded into rectangular capillaries (0.4 × 4 mm cross sectional area, Vitrocom; Mountain Lakes, NJ), and the ends are sealed with wax. The capillaries are mounted on a microscope slide and imaged on a fluorescence microscope (Nikon Eclipse Ti-U; Melville, NY) at the 10× objective (NA = 0.3) using a mercury arc lamp (XCite 120 Q, Lumen Dynamics; Ontario CA). An ND4 filter is used to reduce exposure of the sample to the light source. An image is captured prior to the bleaching, and the sample is then bleached with a 488 nm laser (Sapphire™, Coherent Inc; Santa Clara, CA) for 5 seconds. Subsequent images are taken using the mercury lamp for illumination; the shutter remains closed in between image acquisition to reduce the effects of sample bleaching from the imaging process.

Images are normalized based on a double normalization procedure.[42] Briefly, the intensity of the bleached spot at each time point $I_s(t)$ is first normalized with respect to the pre-bleached image, $I_{s,i}$. Then, the intensity is corrected for bleaching during the imaging process using the ratio of the pre-bleach intensity of an unbleached portion of the sample, $I_{ub,i}$ to its intensity at a particular time point, $I_{ub}(t)$, as a correction factor, yielding a final normalized intensity of $I_N(t) = \frac{I_s(t)}{I_{s,i}}\frac{I_{ub,i}}{I_{ub}(t)}$. The normalized recovery curves are obtained by plotting the normalized intensity within the central region of the bleached spot over time. The subsequent recovery curves $I_N(t)$ are then fitted to the recovery function of the form $I_N(t) = f_m\left[\sum_{n=0}^{\infty}\frac{-K^n}{n!}[1+n(1+2t/\tau)]^{-1}\right] + (1-f_m)\frac{I_N(0+)}{I_{N,i}}$, where f_m is the mobile fraction of the probe, K is a bleaching parameter describing the bleach depth, τ is the time constant for fluorescence recovery, and $I_N(0+)$ is the sample intensity immediately following bleaching.[43] Fits to the recovery function indicate that the probe species is at essentially full mobility in all samples, with mobile fractions f_m above 0.9 for all systems studied. Diffusion coefficients are determined from the relaxation time using $D = \frac{w^2}{4\tau}$ where w is the initial radius of the bleached spot and the typical spot size is (48 ± 3) μm. Average values of D are found by taking measurements at a minimum of three different positions in each sample.

To verify that the long range alignment of the samples does not significantly impact the diffusivity, FRAP experiments on shear aligned samples are performed on a fluorescence microscope equipped with an *XY* microscope stage controller (Applied Scientific Instrumentation; Eugene, OR). A block copolymer sample is loaded onto a glass microscope slide and placed in the slide holder. A second glass slide is positioned over the first slide such that the gap between the two slides is 0.5 mm, and is held stationary for the duration of the experiment. An initial FRAP experiment is performed on the unaligned soft solid sample. The *XY* stage is then programmed to oscillate in a simple linear manner at 1 Hz and 500% strain for 10 minutes. This is shown to be a large enough deformation to align the crystal for these materials.[2,17,22] On cessation of shear, FRAP experiments are performed on the sample. No asymmetry in the recovery is observed. The fluorescence recovery behavior for both sheared and unsheared F127 gels (25 wt% and 30 wt%) containing 1 wt% BSA was quantified. The results indicate that the recovery of sheared samples is similar to that of the unsheared samples, suggesting that adjusting the alignment of the micellar crystals does not affect the overall transport behavior. This also suggests that nanoparticles are not sequestered at the grain boundaries of the polycrystalline material as has been seen in other studies.[20]

Results

Fig. 1 shows the mechanical behavior of Pluronic® F127, F87, and P123 both below and above C_{gl}, values are given in Table 1. Below C_{gl}, the rheology is characterized by zero shear viscosity while above C_{gl}, the rheology is characterized by the magnitude of the complex modulus $G^*(\omega)$. Again, this transition point is defined as the point that the in and out of phase components of the moduli are equivalent at 25 °C and 1 rad s^{-1}. The samples defined as "fluid" flow in reasonable timescales while those defined as "micellar crystal" do not. In many publications, this is referred to as a gel and, in other cases and more correctly, a glassy system.[44] Here, we avoid that terminology as the rheology of these soft solids, or micellar crystals, is much more complicated than that of a traditional polymer gel and lacks structural connectivity.[45] The disordered micellar fluids show an exponential increase in viscosity as block copolymer concentration increases, which is consistent with results seen by others and attributed to the increasing number of micelles present in the system which are interacting hydrodynamically.[46] Above C_{gl}, the block copolymer micelles arrange into packed configurations and have moduli on the order of 100 kPa, with $G^*(\omega)$ exhibiting moderate increases with increasing concentration. We would expect the modulus to scale with $\sim kT/R^3$ and the moduli to follow the correct dependence with micelle size, but the range of R sampled is not large enough to verify this scaling. The mechanical behavior is tunable by changes in block copolymer concentration, and the transition between fluid and solid behavior in the bulk can be monitored rheologically.

Fig. 2 shows the resulting FRAP measurements of the two types of probes in ordered block copolymer systems. Hydrophilic BSA tagged with fluorescein

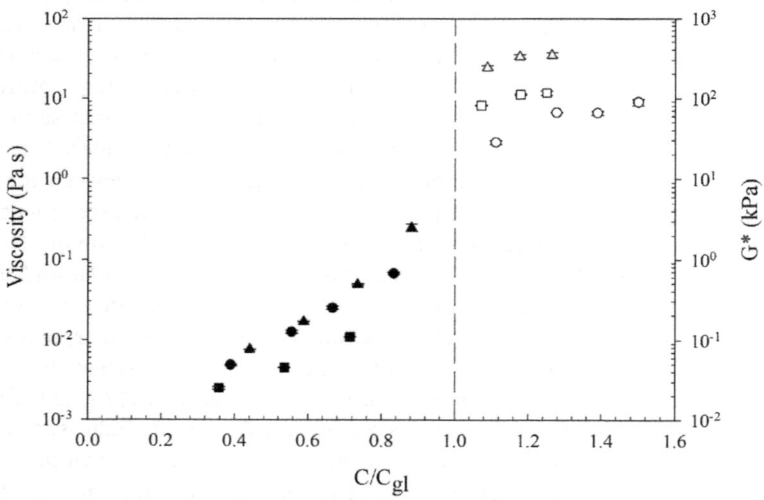

Fig. 1 Zero shear rate viscosity and complex modulus (small amplitude, 1 rad s^{-1}) for block copolymer samples in the fluid (filled symbols) and micellar crystal (open symbols) phases, measured at $T = 25$ °C. Block copolymer concentration has been normalized by C_{gl}. The symbols correspond to F127 (●), F87 (▲), and P123 (■).

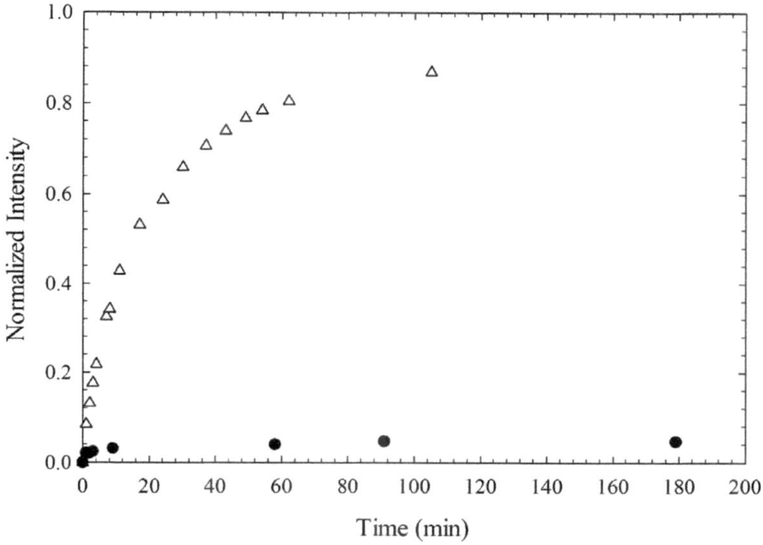

Fig. 2 Fluorescence intensity normalized to the value immediately after photobleaching, as a function of time for FITC–BSA (open symbols) and hydrophobic HDFL dye (filled symbols) in 25 wt% F127 ($C/C_{gl} = 1.4$) at ambient temperature.

isothiocyanate (FITC–BSA), and a hydrophobic dye, 5-hexadecanoylamino-fluorescein (HDFL), a C_{16} alkane derivative of fluorescein. The data show the normalized intensity as a function of time for each of the two probes in 25 wt% F127 ($C/C_{gl} = 1.4$). In this system, FITC–BSA exhibits fluorescence recovery, a measure of the probe diffusing over lengthscales of the spot in the timescales of the experiment. The fluorescence recovery of FITC–BSA in the sample occurs on the order of hours, and a diffusion coefficient of $0.32 \pm 0.01\ \mu m^2\ s^{-1}$ is extracted from this data. The diffusivity of free FITC is much faster compared to that expected for diffusion of solvent molecules and too rapid for analysis using our FRAP experiment.

In contrast, the hydrophobic dye shows little recovered intensity over even longer time scales. Because of the hydrophobic nature of HDFL, the dye resides closer to the hydrophobic interior of the micelle. The fact that so little fluorescence recovery is observed with the hydrophobic dye indicates that movement of the ordered micelles is much slower compared to that of FITC–BSA. This result suggests that the packed structures in the micellar crystal phase restrict the movement of the micelles and is consistent with other measurements.[47] The hydrophobic dye is also tested in the disordered micellar fluid phase (7 wt% F127), and fluorescence recovery is observed, with a diffusion coefficient of $3.2 \pm 0.1\ \mu m^2\ s^{-1}$, consistent with diffusion of an object of the size of a micelle in water.

The diffusion coefficient of FITC–BSA is measured in three different block copolymer systems with concentration ranges spanning across the fluid and micellar crystal phases. The protein concentration (1 wt%, or 10 mg mL^{-1}, BSA) is the same for all experiments and kept low to avoid protein–protein interactions. Fig. 3 shows the normalized FITC–BSA diffusivity as a function of block copolymer concentration. Protein diffusivity in the block copolymer solution has been

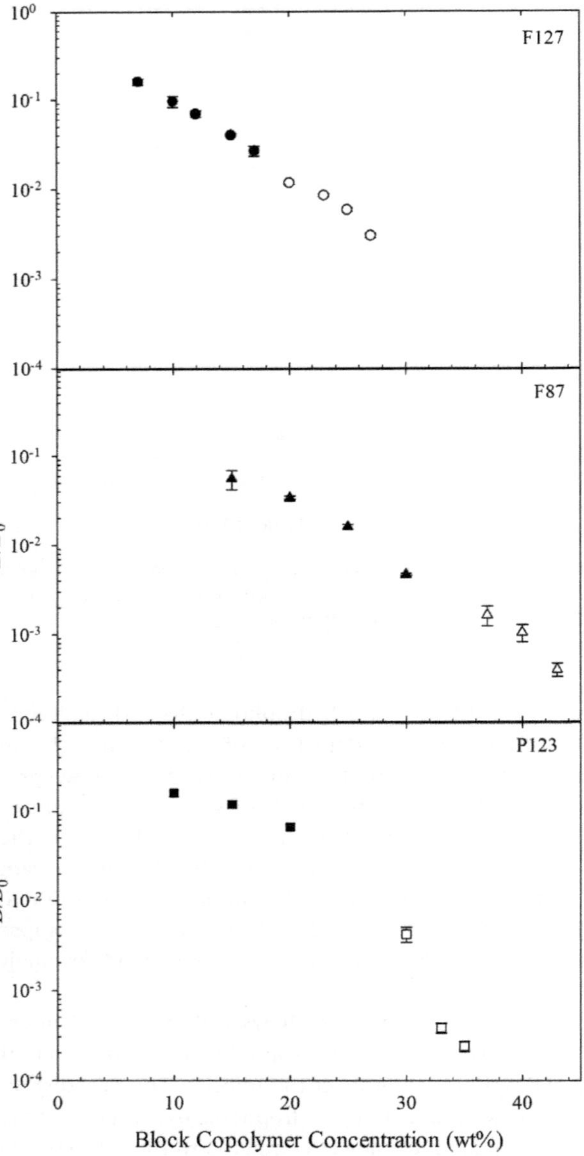

Fig. 3 Normalized FITC–BSA diffusivity (D_o is the diffusivity of BSA in water) as a function of block copolymer concentration for both disordered fluid (filled symbols, $C < C_{gl}$) and ordered micelle (open symbols, $C > C_{gl}$) concentrations and ambient temperature.

normalized to the protein diffusivity in free solution, D_o, which is $55.1 \pm 0.5\ \mu m^2\ s^{-1}$. The filled symbols correspond to fluid samples while the open symbols correspond to soft solids. The results indicate that the protein diffusivity decreases with the addition of the block copolymer. Even at the lowest block copolymer concentrations micelles are present but disordered, and the BSA diffusivity in this case is already one order of magnitude slower than for BSA in free solution. In all three systems, the normalized diffusivity decreases smoothly as the block copolymer concentration increases in the fluid phase. As the concentration increases above C_{gl}, protein diffusivity decreases by at least two orders of magnitude. For the F127 and F87 systems, the dependence of diffusivity on concentration is smooth through the transition from a fluid to a soft solid. For the P123 samples, the decrease in protein diffusivity when the polymer is above C_{gl} is more significant showing a significant drop at C_{gl} and then an even steeper drop as concentration increases. Nanoparticles of size similar to BSA are expected to be most significantly hindered in mobility in P123 at $C/C_{gl} > 1.2$. Note from Table 1 that P123 has the shortest PEO chains relative to PPO, suggesting micelles with large cores and thinner corona are best for hindering particle mobility.

To demonstrate the impact of the block copolymer structure on the diffusion of BSA, Fig. 4 shows the same data as Fig. 3, with the concentration scaled with the concentration at which the sample transitions to a soft solid, C_{gl}. The diffusion coefficient is again scaled with the free diffusivity of BSA in water, so it is the same for all solutions. The solid symbols are for block copolymer solutions that are fluid while the open symbols are soft solids; hence the sharp division at $C/C_{gl} = 1$. In this figure, two of the block copolymers, F127 and P123 show similar behavior in the fluid state but then separate quickly in the soft solid state as the diffusivity of P123 drops rapidly with increasing concentration. The F87 sample shows slower diffusion than the other two except for the highest concentrations of P123. It is also clear that none of these samples follow a simple power law dependence with overall block copolymer concentration.

Significant increases in particle loading have been shown to change the composition, rheology, and structure of the composite systems.[15,21] Fig. 5 demonstrates the impact of protein diffusivity on various F127 micellar crystals as the BSA loading is increased. Analysis of the micelle crystal structure by SANS[21] has shown that the block copolymer micelle structure is not affected in studies where the BSA loading has been kept at 1 wt%, which is a concentration where

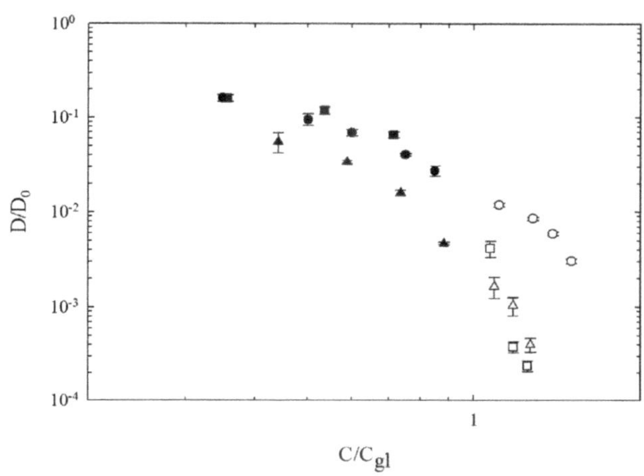

Fig. 4 The normalized FITC–BSA diffusivity (D_o is the diffusivity of BSA in water) as a function of block copolymer concentration for both disordered fluid (filled symbols, $C < C_{gl}$) and ordered micelle (open symbols, $C > C_{gl}$) concentrations and ambient temperature.

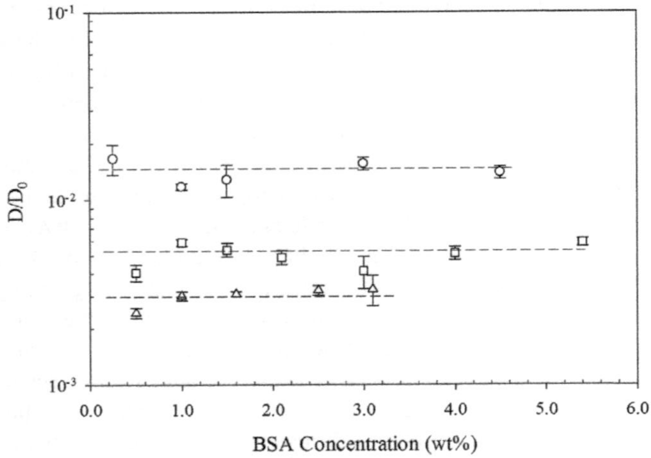

Fig. 5 Reduced diffusivity as a function of total BSA particle loading. Three different block copolymer concentrations are examined: 20 wt% F127 (○), 25 wt% F127 (□) and 27 wt% F127 (△).

there are fewer than one BSA per micelle. Mechanical characterization of the block copolymer nanocomposite has also shown as much as a 40 kPa decrease in complex modulus with the addition of nanoparticles at higher loadings, indicating a significant weakening of the soft solid.[15,16,18] Despite the significant structural and mechanical changes in the block copolymer nanocomposites at higher nanoparticle concentrations, the BSA diffusivity exhibits essentially no change with increased particle loading.

BSA is a globular protein and it is often treated as a spherical probe. This has been successful in studying the transport of BSA in other polymer matrices. Stokes–Einstein–Sutherland diffusion of a sphere in a solvent provides an established relationship between probe size, solvent viscosity and diffusivity. Assuming that the medium is homogeneous, the product of the particle diffusivity and system viscosity $D\eta$ is related to the inverse of the hydrodynamic size of the probe R in the medium and a group of constants as shown in eqn (1), where k is the Boltzmann constant and T is the temperature,

$$D\eta = \frac{kT}{6\pi R}. \tag{1}$$

When normalized with D_o, the particle diffusivity through the pure solvent, and η_o, the viscosity of the pure solvent, the quantity $D\eta/D_o\eta_o$ should be a constant of value unity for a probe experiencing simple transport through the solvent. The value of $D\eta/D_o\eta_o$ is used to evaluate whether a system conforms to this behavior, and can also provide some measure of the local behavior of the solution.[32] The quantity $D\eta/D_o\eta_o$ for F87, F127, and P123 is plotted against block copolymer concentration in Fig. 6 using measured values for the viscosities of the solutions (see Fig. 1). Particle diffusivities deviate from simple behavior for nearly all concentrations, indicating that protein transport is influenced by local solution properties other than the macroscopic viscosity, but this is not unexpected

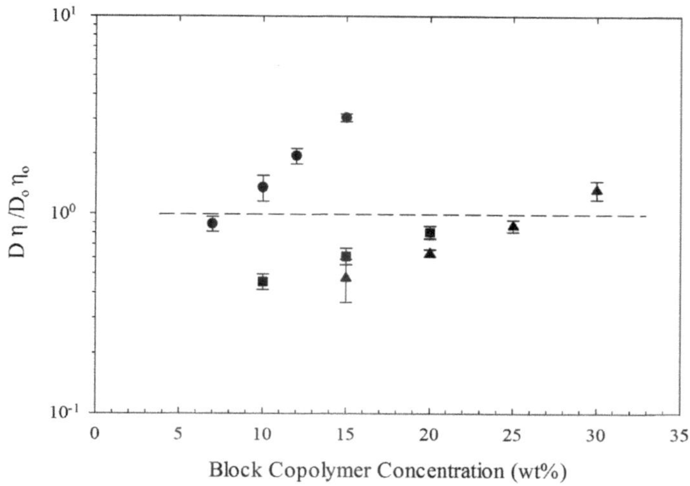

Fig. 6 $D\eta/D_o\eta_o$, normalized relative to free solution values for FITC–BSA in block copolymer solutions F127 (●), F87 (▲), and P123 (■) at concentrations below C_{gl}. The dashed line shows the expected behavior for a probe undergoing Stokes–Einstein–Sutherland diffusion.

given the relatively high concentration of micelles present. Similar analysis performed with BSA dispersed in aqueous solutions of 6000 g mol^{-1} PEO showed reasonable agreement with Stokes–Einstein–Sutherland for concentrations below 10 wt% PEO.

For cross-linked and entangled polymer systems, the crosslink density is proportional to the elastic modulus of the material,[48] and impacts particle transport, in that an increase in crosslink density is directly related to a measureable reduction in particle diffusivity.[49,50] In this system, there are no entanglements and the modulus arises from the glassy-like behavior of the packed micelles. Although these block copolymers form micellar crystals by physical rather than chemical association, they still possess a measurable modulus which may be related to transport of particles. In Fig. 7, the reduced diffusivity of BSA is shown *versus* the complex modulus G^* for a series of micellar crystal block copolymer compositions. Here, we report the complex modulus, G^*, noting that for these soft solids, the phase angle is close to zero and the complex modulus is dominated by the elastic modulus, G'. Protein diffusivity decreases with increasing values of the modulus for each block copolymer system. However, the results indicate that macroscopic properties such as G^* are not the sole predictor for nanoparticle transport, as the F87 systems exhibit the highest values of G^* but intermediate diffusivity values.

It is interesting to note that two systems (F127 and P123) seem to follow a behavior of a similar decrease in diffusivity with increase in modulus. While it would be too much to argue a universal behavior from the overlap of the limits, these are the two systems that exhibit an FCC unit cell. The F87 system forms a BCC unit cell and has significantly higher modulus for similar transport of BSA through the soft solid.

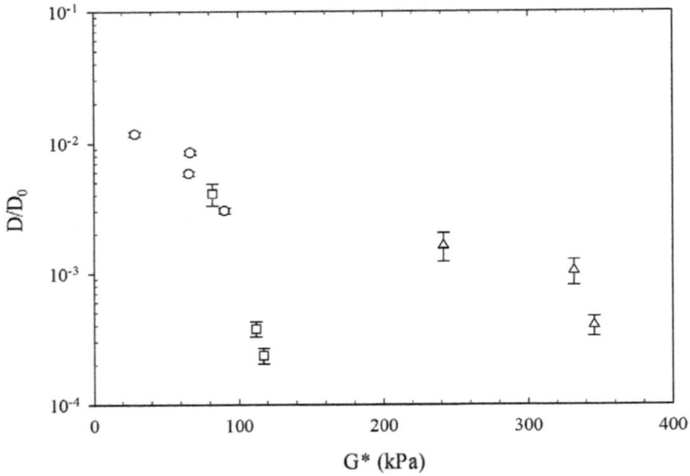

Fig. 7 Reduced diffusivity of FITC–BSA as a function of complex modulus of the composite for the block copolymer solutions at $C > C_{gl}$ and room temperature. Symbols correspond to F127 (○), F87 (△) and P123 (□).

Discussion

Diffusivity of nanoparticulate material in these self-assembled block copolymer systems is complicated by the different lengthscales involved. In the fluid phase, the block copolymers are in a combination of unimers and micelles on the order of nanometers to tens of nanometers and are comparable to the size of the BSA probe particle ($D_h \sim 7$ nm). As the block copolymer micelles pack into ordered crystalline structures, similarities between the lengthscales of the polymer and particle remain. The dimensions of the block copolymer micelles and interstitial regions can be determined from SANS measurements on the ordered block copolymer; for this study, the interstitial regions in between packed micelles range from 1.8 to 4.5 nm, which corresponds closely to the hydrodynamic radius of BSA.

Diffusion in the micellar fluid

The lack of agreement with Stokes–Einstein–Sutherland using the solvent or bulk viscosity indicates that, rather than a homogeneous bulk, the probes are likely sampling the PEO coronas of the micelles, better modeled as a polymer solution.[30–32] BSA diffusion probes the nanostructure of a block copolymer solution. Since this work is performed on isothermal samples, the block copolymers persist as micelles in solution at all concentrations. The micelles partition the PPO into cores so diffusion is through the solvent and the PEO coronas. As concentration increases and the system forms a soft solid, the BSA will sense a solution that is more and more reminiscent of a PEO solution in water. The PEO blocks in these materials (see Table 1) are short, below 5000 g mol^{-1}. If free in solution, the radius of gyration of these PEO chains would be less than the radius of the probe ($R_o = 3.5$ nm). To compare the measured diffusion coefficients to those expected in PEO solutions, the volume of the cores is calculated using the bulk density of PPO and assuming that the cores are fully dehydrated. Assuming that the BSA does not penetrate these cores, the remaining volume contains the PEO chains. These are structured into micelle coronas, but for an initial comparison, the effective volume fraction of PEO, ϕ_{PEO}, is calculated as the concentration of PEO chains (two per block copolymer chain) in the solvent omitted from the PPO cores. The PPO cores make up less than 10% of the volume for most of the solutions studied. The radius of gyration of the PEO chains in water has been shown to be $R_g = 0.02M_w^{0.58}$, giving a R_g of less than 2–3 nm.[51] The value of ϕ^*, or the overlap concentration, is calculated using $\phi^* = 3M_w/(4\pi\rho N_A R_g^3)$ and found to correspond to concentrations in the fluid state for F127 and F87, but is above the studied concentrations for all of the P123 solutions. This is due to the shorter PEO blocks of the P123 system. Above ϕ^*, the relevant lengthscale is the correlation length, $\xi \approx R_g(\phi/\phi^*)^{-0.76}$, which is concentration dependent. For the three block copolymers studied here, $\xi < R_o$, for most of the solutions. For F127 and P123, only the lowest concentrations have correlation lengths larger than the probe size, but those samples have effective PEO concentrations below ϕ^*. For F87, all of the concentrations have correlation lengths shorter than the probe size.

Fig. 8 shows the diffusivity scaled for comparison to the stretched exponential dependence reported for the PEG solutions. Here, the scaled diffusivity is plotted with concentration expressed in terms of relevant lengthscales, the R_g of the PEO block and the correlation length which is inversely proportional to the effective PEO volume fraction. Based on observed scaling in the literature for PEG solutions, we expect the diffusivity to follow a stretched exponential, $D_0/D = \exp(\beta(R_g/\xi)^\delta)$.[33] This functional form is simply convenient for comparison. Again, filled

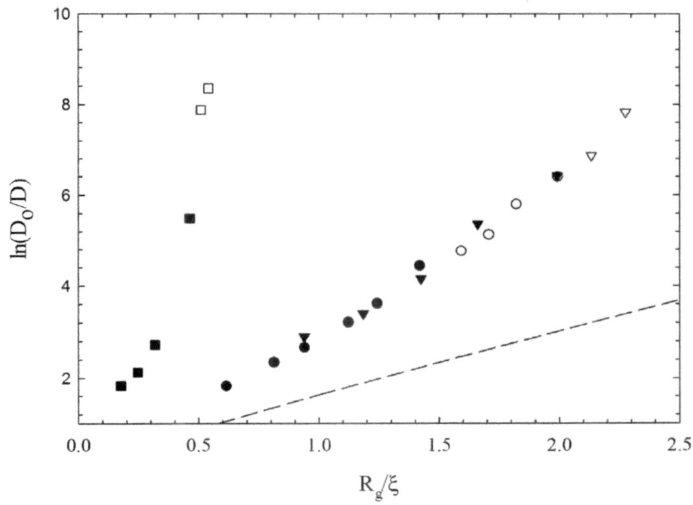

Fig. 8 Inverse of the reduced diffusivity of FITC–BSA as a function of polymer radius of gyration over correlation length for each of the block copolymer systems below (filled) and above (open) the C_{gl}. Symbol shapes are the same as the previous figures. The dashed line represents the measured relationship presented by Kohli *et al.*[33] for gold nanoparticles diffusing through PEO solutions.

symbols are for samples that are macroscopic fluids and the open symbols are for soft solids. Two of the samples, F127 and F87, collapse in this representation. Since these have similar molecular weights of the PEO blocks, this might not be surprising, but the collapse is extremely good with a large region of overlap and through C_{gl}. The P123 samples are significantly different with much slower diffusivity even in the fluid state.

The dotted line shows the expected behavior of the PEG solutions, with $\beta = 1.63$ and $\delta = 0.89$ which underestimates the relationship;[33] this is not unexpected as the block copolymer micelle environment is diffusion through packed coronas rather than free PEO chains. The data for F127 and F87 does agree with a stretched exponential with $\beta \sim 3$ and $\delta \sim 1.1$ but this imparts no physical insight into the diffusion mechanism although it might prove a useful tool to predict transport behavior. The collapse of the F127 and F87 samples suggests that probe diffusion in these materials is similar to diffusion through a semidilute polymer solution, with hindrances due to the structure of the corona or probe–polymer interactions.

A recent approach to modeling diffusion of non-interacting (*i.e.*, non-sticky) probes in unentangled and entangled polymer solutions and melts is also considered.[29,52] A scaling analysis has been presented in which particle diffusivity is shown to be affected by the size of the particle, $d = 2R_o$, relative to the polymer correlation length, ξ.[29] For a polymer in a good solvent, the volume fraction of the polymer at which ξ is on the same order of the particle size scales as

$$\phi_{d,\xi} \cong (d/b)^{-1.32}, \tag{2}$$

where d is particle diameter, and b is the Kuhn length of the monomer. Below this concentration, diffusivity is expected to follow Stokes–Einstein–Sutherland behavior with respect to the solvent viscosity, while above this concentration diffusivity is expected to have a power law dependence on the polymer concentration,

$$D(\phi)/D_0 = \alpha(\phi/\phi_{d,\xi})^{-1.52}, \tag{3}$$

where α is a scaling factor determined from fitting of the data, D_0 is the diffusivity of the probe in pure solvent, ϕ is polymer concentration, and $\phi_{d,\xi}$ is the concentration at which ξ is the same size as the probe particle diameter.

For the block copolymer micelles, the PPO blocks are sequestered in the micelle cores, while the PEO coronas are exposed to water. The systems here are modeled as PEO solutions, in which the polymer concentration, c, is determined by calculating an effective PEO concentration which excludes the volume occupied by fully dehydrated PPO cores. Water is a good solvent for PEO and all of the polymer concentrations in this study are above the "crossover" concentration $\phi_{d,\xi}$ ($\phi_{d,\xi} \sim 7\%$ for a BSA and PEO probe–polymer system),[53,54] so the diffusivity data is examined with eqn (3). The fitting parameter α is obtained for each block copolymer system by vertically shifting the lowest concentration (For F127 and F87, $\alpha \sim 0.13$ while the value is slightly lower for P123, $\alpha \sim 0.1$). The reduced diffusivity is plotted against normalized concentration, $\phi/\phi_{d,\xi}$ in Fig. 9, and the systems in the fluid phase are differentiated from the micelle crystal phase with filled and unfilled symbols, respectively. The power law dependence predicted by

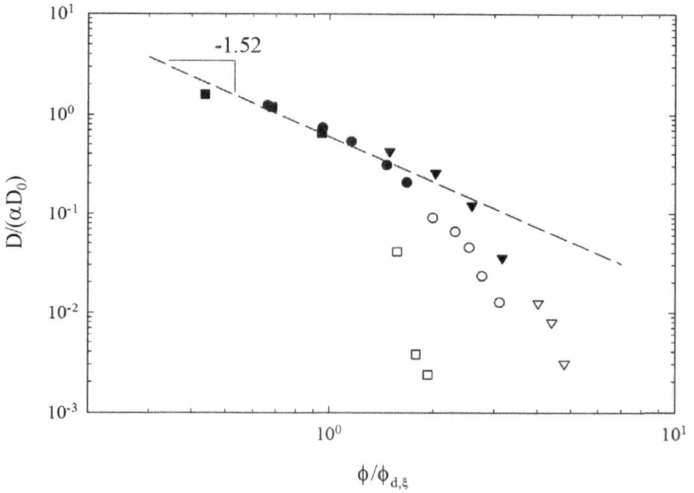

Fig. 9 Reduced diffusivity of FITC–BSA as a function of normalized polymer concentration for each of the block copolymer systems below (filled) and above (open) the C_{gl}. Symbol shapes are the same as the previous figures. The dashed line represents the scaling relationship presented by Cai *et al.*[29] for non-sticky nanoparticles diffusing through polymer solutions.

the scaling relationship is denoted with the dashed line, and the data for the lower concentrations in each polymer system fall onto this curve. The data deviate from this scaling at higher concentrations, suggesting a change in the diffusion mechanism as the local environment becomes increasingly crowded with micelles. In all cases, the scaling appears to hold for concentrations below about $\phi/\phi_{d,\xi} \sim 2$.

Diffusion in the ordered block copolymers

As the block copolymer concentration increases, the system reaches a point when the micelles pack into ordered configurations. Neutron scattering measurements on these ordered materials have produced structural information on the specific packing configurations, as well as relevant lengthscales, such as the crystal lattice parameters and hard sphere dimensions.[18,19,21,23] Table 2 lists these parameters for some specific block copolymer compositions; for the diffusion studies, the

Table 2 Corresponding micellar core dimensions and volume fraction for different block copolymer compositions obtained from SANS measurements and the literature ($T = 25$ °C)

Polymer	Conc. (wt%)	Crystal lattice	Unit cell lattice (nm)	Hard sphere radius (nm)	R_{core} (nm)	ϕ_{core}	R_g PEO[62] (nm)
F127	25	FCC	28.7	10.2	3.8	0.04	2.7
F87	40	BCC	15.6	6.8	3.7	0.11	2.3
P123	35	FCC	25.3	9.0	5.1	0.13	1.3

block copolymer soft solids are cubic (FCC or BCC), and the crystal type remained consistent for each block copolymer system.

Fig. 10 shows a schematic of the nanoscale structure of the three block copolymer systems used in this work. Dimensions for a specific concentration and temperature are determined using small angle scattering (SAXS and SANS) assuming that the PPO core is fully dehydrated. Important dimensions of the micellar crystal are the unit cell size, a, measured directly from scattering, the center-to-center distance of the micelles, D_{cc}, determined from the unit cell size and knowledge of the crystal structure, the size of the PPO core, R_{core}, determined from the copolymer molecular properties assuming that the core is fully dehydrated. The interstitial region around and between the micelles is a solvent and corona (PEO) filled region. This is a simple schematic and the details of the interstitial regions should be interpreted with caution; but one goal of this work is an improved understanding of this region.

Physical obstructions impact diffusion in gels, with diffusivity decreasing proportionally to the volume fraction of obstruction. Here, the impact of PPO cores as obstructions in the ordered micellar system is considered. The reduced diffusivity can be expressed as $D/D_0 = (1 - \alpha\phi)$ where α is a shape factor that is 3/2 for spheres, and ϕ is the volume fraction of obstructions.[55] For this system, the PPO micellar cores create the obstructions and their volume fraction is estimated by assuming fully dehydrated PPO cores and calculating the volume fraction of the cores for a given unit cell.[11] Table 2 shows the values for R_{core} and ϕ_c, showing that the PPO core occupies a small fraction of the sample volume. The predicted reduction in diffusivity from the PPO obstructions is no more than 20%, which is not likely to be the cause of the drop in D observed in Fig. 3. However, in the case of P123, the probe size approaches that of the distance between the PPO cores.

SANS measurements suggest that BSA and other similarly sized particles template within the interstitial regions of ordered block copolymer micelles. Because the block copolymer–nanoparticle composite structure resembles that of an intercalated crystal, the probe particle diffusivity in the ordered micellar phases is modeled here with an interstitial diffusion mechanism similar to atom migration in solid crystalline materials.[56] In interstitial diffusion, the diffusion coefficient is related to the crystal dimension by $D = \Gamma r^2/6$, where Γ is the total

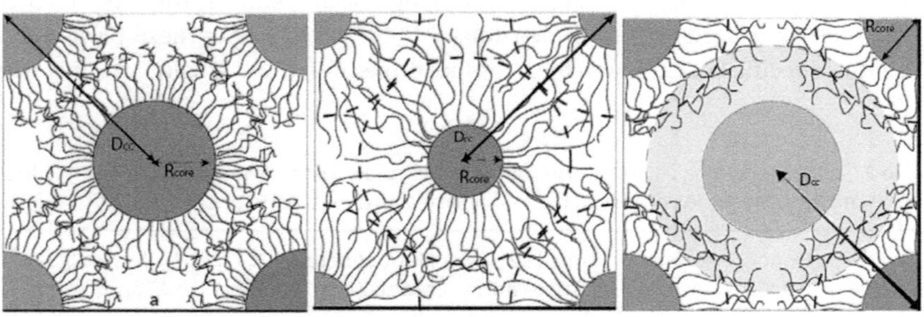

Fig. 10 Unit cell depictions of P123 (left), F127 (center), and F87 (right), with micelle cores depicted as dark grey regions. Unit cell size and structure are determined from small angle scattering. The PPO core is estimated by assuming a fully dehydrated core. The dotted line is micelle size determined by assuming hard spheres packed in a crystal. P123 and F127 form FCC crystals, while F87 forms a BCC crystal. Dimensions are provided in Table 2.

jump rate of the particle, and r^2 is the jump distance. The jump distance can be written in terms of the lattice parameter a, which is dependent on the specific crystalline structure; for a BCC structure, $r_{BCC} = a/2$, while for an FCC crystal $r_{FCC} = 2^{1/2}a/2$. Thus, $D_{BCC} = \Gamma a^2/24$ and $D_{FCC} = \Gamma a^2/12$. For the block copolymers in this study, the lattice parameters of the ordered micellar structures are obtained through SANS measurements, and the lattice parameter scales inversely with polymer concentration in the concentration range studied, $a \sim 1/c$.[21,23] Other experimental observations on these block copolymers also note an inverse relationship between the lattice parameter and polymer concentration.[57] Then,

$$D \sim \Gamma c^{-2}, \tag{4}$$

where c is polymer concentration. From the open symbols in Fig. 4, this underestimates the dependence of D on concentration if Γ is not a function of concentration. So, an attempt is made to provide an expression for the jump rate. The total jump rate of the particle in the material is $\Gamma = Z\nu \exp(-\Delta E/kT)$, where Z is the number of nearest interstitial sites ($Z_{BCC} = 4$, $Z_{FCC} = 12$), ν is the attempt frequency of the particle, ΔE is the energy barrier that the particle must overcome in order to successfully move from one interstice to another, k is the Boltzmann constant, and T is temperature. The product $\nu \exp(-\Delta E/kT)$ is the jump rate of one particle from one site to a neighboring site and describes the frequency at which the particle is able to successfully complete the jump. Since BSA is diffusing through the PEO coronas of the micelles, we treat the environment as one of a polymer solution so the energy barrier is regarded as the elastic energy of deformation that the polymer network experiences to accommodate passage of the probe particle through the micellar structures. In this case, $\Delta E \sim kT\gamma(d/\xi)^2$, where γ is related to the number of polymer meshes that undergo deformation, d is the probe particle diameter, and ξ is the characteristic size of the unperturbed polymer network. In a good solvent, ξ is related to polymer concentration, c, as $\xi \sim c^{-3/4}$.[45,58,59] So diffusion from an activated hopping process in these micellar crystals will have a complex dependence on concentration,

$$D(c) \sim \beta' c^{-2} \exp(-\gamma d^2 c^{3/2}) \tag{5}$$

where β' is a prefactor related to the crystal geometry and jump attempt frequency, and γ is a prefactor related to the number of polymer chains involved in the stretching of the network.

The open symbols in Fig. 4 show the BSA diffusivity in soft solids with enough nanoscale structure to be studied with this simple model. While an exponential dependence of reduced diffusion coefficient on concentration is observed, there is not enough data to study the detailed $D(c)$ suggested by eqn (5) for samples in the ordered micellar phase. In all three cases, the value of D in the soft solid follows an exponential dependence on concentration. With limited data, all that should be extracted is that the slope of this dependence is similar for F87 and F127, but higher for P123. Since this slope and γ, are proportional to the number of polymer chains that are deformed during the jumping process, then, the overall energy required for jumping to occur will increase for a system with more densely packed coronas. Of the three systems, P123 has the highest aggregation number, followed by F87 and then F127,[7,9] which is consistent with the physical description suggested by the model. While both F127 and P123 form FCC configurations in the ordered micellar phases, the PEO block length is nearly four times shorter in P123 than in F127. As a result, the P123 corona size, which ranges from approximately 1–2 nm, is very nearly the same as the BSA radius ($R_o \sim 3.5$ nm) which would be more restrictive to BSA diffusivity than the F127 corona size of ~2–3 nm.

Conclusions

The complexity that self-assembling block copolymers exhibit in their structural and mechanical properties clearly extends to nanoparticle transport through these systems, with results indicating that nanoparticle diffusivity is sensitive to block copolymer type, concentration, and structure. The resulting range of protein diffusivity spans across many orders of magnitude for the concentrations studied, and demonstrates the versatility of these self-assembled systems for controlling nanoparticle transport. Macroscopic properties are not enough to predict the level of hindrance to transport observed, the architecture of the block copolymer and micelle structure need to be considered. In a solution of concentrated micelles (near C_{gl}, defined mechanically), probes of similar size to the micelles exhibit scaling expected for diffusion through a polymer solution with probe size similar to the correlation length of the polymer solution. The transition of the block copolymer micelle solution to a soft solid does not result in a discontinuous drop in probe diffusivity. In the soft solid the more relevant lengthscales will be those of the unit cell.

Acknowledgements

This project is funded by the NSF grant CBET 0931025. Also, thanks to Michael Rubinstein and Liheng Cai for helpful discussions regarding the activated hopping mechanisms.

References

1 F. S. Bates and G. H. Fredrickson, *Annu. Rev. Phys. Chem.*, 1990, **41**, 525–557.
2 I. W. Hamley, *J. Phys.: Condens. Matter*, 2001, **13**, R643–R671.
3 T. P. Lodge, *Macromol. Chem. Phys.*, 2003, **204**, 265–273.
4 K. Mortensen, W. Brown and B. Norden, *Phys. Rev. Lett.*, 1992, **68**, 2340–2343.
5 P. Alexandridis, J. F. Holzwarth and T. A. Hatton, *Macromolecules*, 1994, **27**, 2414–2425.
6 P. Alexandridis and T. A. Hatton, *Colloids Surf., A*, 1995, **96**, 1–46.
7 G. Wanka, H. Hoffmann and W. Ulbricht, *Macromolecules*, 1994, **27**, 4145–4159.
8 K. Mortensen, *J. Phys.: Condens. Matter*, 1996, **8**, A103–A124.
9 K. Mortensen and W. Brown, *Macromolecules*, 1993, **26**, 4128–4135.
10 P. Linse, *Macromolecules*, 1993, **26**, 4437–4449.
11 R. K. Prudhomme, G. W. Wu and D. K. Schneider, *Langmuir*, 1996, **12**, 4651–4659.
12 I. Goldmints, G. E. Yu, C. Booth, K. A. Smith and T. A. Hatton, *Langmuir*, 1999, **15**, 1651–1656.

13 I. W. Hamley and V. Castelletto, *Prog. Polym. Sci.*, 2004, **29**, 909–948.

14 H. Yardimci, B. Chung, J. L. Harden and R. L. Leheny, *J. Chem. Phys.*, 2005, **123**, 244908.

15 D. C. Pozzo, K. R. Hollabaugh and L. M. Walker, *J. Rheol.*, 2005, **49**, 759–782.

16 D. C. Pozzo and L. M. Walker, *Macromol. Symp.*, 2005, **227**, 203–210.

17 D. C. Pozzo and L. M. Walker, *Macromolecules*, 2007, **40**, 5801–5811.

18 D. C. Pozzo and L. M. Walker, *Colloids Surf., A*, 2007, **294**, 117–129.

19 D. C. Pozzo and L. M. Walker, *Eur. Phys. J. E: Soft Matter Biol. Phys.*, 2008, **26**, 183–189.

20 A. Louhichi, E. Tamborini, J. Oberdisse, L. Cipelletti and L. Ramos, *Phys. Rev. E: Stat., Nonlinear, Soft Matter Phys.*, 2015, **92**, 32307.

21 T. A. LaFollette, Thesis (PhD), Carnegie Mellon University, 2010.

22 T. A. LaFollette and L. M. Walker, *Polymers*, 2011, **3**, 281–298.

23 D. C. Pozzo, Thesis (PhD), Carnegie Mellon University, 2006.

24 M. R. Bockstaller, R. A. Mickiewicz and E. L. Thomas, *Adv. Mater.*, 2005, **17**, 1331–1349.

25 B. Sarkar and P. Alexandridis, *Prog. Polym. Sci.*, 2015, **40**, 33–62.

26 M. M. Domach and L. M. Walker, *JALA*, 2010, **15**, 136–144.

27 A. W. Kirkeminde, T. Torres, T. Ito and D. A. Higgins, *J. Phys. Chem. B*, 2011, **115**, 12736–12743.

28 R. Svingen, P. Alexandridis and B. Akerman, *Langmuir*, 2002, **18**, 8616–8619.

29 L.-H. Cai, S. Panyukov and M. Rubinstein, *Macromolecules*, 2011, **44**, 7853–7863.

30 Y. Cheng, R. K. Prud'homme and J. L. Thomas, *Macromolecules*, 2002, **35**, 8111–8121.

31 G. D. J. Phillies, G. S. Ullmann, K. Ullmann and T. H. Lin, *J. Chem. Phys.*, 1985, **82**, 5242–5246.

32 J. Won, C. Onyenemezu, W. G. Miller and T. P. Lodge, *Macromolecules*, 1994, **27**, 7389–7396.

33 I. Kohli and A. Mukhopadhyay, *Macromolecules*, 2012, **45**, 6143–6149.

34 R. Holyst, A. Bielejewska, J. Szymanski, A. Wilk, A. Patkowski, J. Gapinski, A. Zywocinski, T. Kalwarczyk, E. Kalwarczyk, M. Tabaka, N. Ziebacz and S. A. Wieczorek, *Phys. Chem. Chem. Phys.*, 2009, **11**, 9025–9032.

35 T. K. L. Meyvis, S. C. de Smedt, P. van Oostveldt and J. Demeester, *Pharm. Res.*, 1999, **16**, 1153–1162.

36 E. M. Johnson, D. A. Berk, R. K. Jain and W. M. Deen, *Biophys. J.*, 1996, **70**, 1017–1023.

37 E. E. Holly, S. K. Venkataraman, F. Chambon and H. H. Winter, *J. Non-Newtonian Fluid Mech.*, 1988, **27**, 17–26.

38 T. H. The and T. E. Feltkamp, *Immunology*, 1970, **18**, 865.

39 T. H. The and T. E. Feltkamp, *Immunology*, 1970, **18**, 875.

40 Pierce Biotechnology: FITC and TRITC, http://www.piercenet.com/instructions/2162081.pdf.

41 M. Malmsten, *Biopolymers at interfaces*, ed. M. Malmsten, Marcel Dekker, New York, 2nd edn, 2003.

42 R. D. Phair, S. A. Gorski and T. Misteli, *Methods Enzymol.*, 2004, **375**, 393–414.

43 J. Tong and J. L. Anderson, *Biophys. J.*, 1996, **70**, 1505–1513.

44 S. Gupta, J. Stellbrink, E. Zaccarelli, C. N. Likos, M. Camargo, P. Holmqvist, J. Allgaier, L. Willner and D. Richter, *Phys. Rev. Lett.*, 2015, **115**, 128302.

45 M. Rubinstein and R. H. Colby, *Polymer physics*, Oxford University Press, Oxford, New York, 2003.
46 L. C. P. Trong, M. Djabourov and A. Ponton, *J. Colloid Interface Sci.*, 2008, **328**, 278–287.
47 V. Castelletto, I. W. Hamley and T. A. Waigh, *J. Chem. Phys.*, 2004, **121**, 111474.
48 L. R. G. Treloar, *The physics of rubber elasticity*, Clarendon Press, Oxford University Press, Oxford, New York, 3rd edn, 2005.
49 C. T. Reinhart and N. A. Peppas, *J. Membr. Sci.*, 1984, **18**, 227–239.
50 M. S. Green and A. V. Tobolsky, *J. Chem. Phys.*, 1946, **14**, 80–92.
51 K. Devanand and J. C. Selser, *Macromolecules*, 1991, **24**, 5943–5947.
52 L.-H. Cai, S. Panyukov and M. Rubinstein, *Macromolecules*, 2015, **48**, 847–862.
53 K. Devanand and J. C. Selser, *Nature*, 1990, **343**, 739–741.
54 S. Kawaguchi, G. Imai, J. Suzuki, A. Miyahara and T. Kitano, *Polymer*, 1997, **38**, 2885–2891.
55 M. A. Lauffer, *Biophys. J.*, 1961, **1**, 205–213.
56 R. W. Balluffi, S. M. Allen, W. C. Carter and R. A. Kemper, *Kinetics of materials*, J. Wiley & Sons, Hoboken, N.J., 2005.
57 P. Alexandridis, D. L. Zhou and A. Khan, *Langmuir*, 1996, **12**, 2690–2700.
58 A. Halperin, M. Tirrell and T. P. Lodge, in *Macromolecules: Synthesis, Order and Advanced Properties*, Springer, Berlin, Heidelberg, 1992, vol. 100/1, ch. 3, pp. 31–71.
59 P. G. Degennes, *J. Chem. Phys.*, 1982, **76**, 3316–3321.
60 B. Lindman and P. Alexandridis, *Amphiphilic block copolymers: self-assembly and applications*, Elsevier, Amsterdam, New York, 1st edn, 2000.
61 J. R. Lopes and W. Loh, *Langmuir*, 1998, **14**, 750–756.
62 R. Bhat and S. N. Timasheff, *Protein Sci.*, 1992, **1**, 1133–1143.

Faraday Discussions

PAPER

Aggregation and stability of anisotropic charged clay colloids in aqueous medium in the presence of salt†

Samim Ali and Ranjini Bandyopadhyay*

Received 20th August 2015, Accepted 9th November 2015
DOI: 10.1039/c5fd00124b

Na-montmorillonite nanoclay is a colloid of layered mineral silicate. When dispersed in water, this mineral swells on absorption of water and exfoliates into platelets with electric double layers on their surfaces. Even at low particle concentration, the aqueous dispersion can exhibit a spontaneous ergodicity breaking phase transition from a free flowing liquid to nonequilibrium, kinetically arrested and disordered states such as gels and glasses. In an earlier publication [*Applied Clay Science*, 2015, **114**, 8592], we showed that the stability of clay gels can be enhanced by adding a salt later to the clay dispersion prepared in deionized water, rather than by adding the clay mineral to a previously mixed salt solution. Here, we directly track the collapsing interface of sedimenting clay gels using an optical method and show that adding salt after dispersing the clay mineral does indeed result in more stable gels even in very dilute dispersions. These weak gels are seen to exhibit a transient collapse after a finite delay time, a phenomenon observed previously in depletion gels. The velocity of the collapse oscillates with the age of the sample. However, the average velocity of collapse increases with sample age up to a peak value before decreasing at higher ages. With increasing salt concentration, the delay time for transient collapse decreases, while the peak value of the collapsing velocity increases. Using ultrasound attenuation spectroscopy, rheometry and cryogenic scanning electron microscopy, we confirm that morphological changes of the gel network assembly, facilitated by thermal fluctuations, lead to the observed collapse phenomenon. Since clay minerals are used extensively in polymer nanocomposites, as rheological modifiers, stabilizers and gas absorbents, we believe that the results reported in this work are extremely useful for several practical applications and also for understanding geophysical phenomena such as the formation and stability of quicksand and river deltas.

Raman Research Institute, C. V. Raman Avenue, Sadashivanagar, Bangalore 560080, India. E-mail: samim@rri.res.in; ranjini@rri.res.in

† Electronic supplementary information (ESI) available: Theory used for PSD measurement using ultrasound attenuation spectroscopy (Section I), supplied parameter values to estimate PSDs (Section II), the electroacoustic setup used in this work (Section III and Fig. S1), plot of I_{CVI} *versus* t_s for varying clay concentrations (Fig. S2) and plot of I_{TVI} *versus* t_s for varying salt concentrations (Fig. S3). See DOI: 10.1039/c5fd00124b

I. Introduction

Na-montmorillonite is a naturally occurring nanoclay mineral belonging to the smectite group and is available in abundance in nature. The unit particles, known as platelets, of this clay mineral are generally polydisperse in their size and shape, with negative charges on their surfaces and pH dependent positive charges on the edges.[1] The aqueous dispersions of such platelets show spontaneous phase transitions from liquid-like states to nonequilibrium kinetically arrested phases such as glasses and gels.[2–6] The glassy phase has a disordered arrangement of clay platelets stabilized by repulsive interactions, while in the gel phase, the clay platelets assemble into a disordered and volume spanning network structure through attractive bond formation.[7] The formation of such phases, and their rheological and stability behaviors, play a crucial role in many geophysical phenomena such as river delta and quick-sand formation.[8,9] Besides its fundamental importance in many nonequilibrium phenomena, the understanding of the aggregation and stability of dispersions of colloidal clay particles is also relevant to several technological applications.[10–12]

In dry form, the minerals exist in the form of tactoids, which are stacks of clay platelets having thicknesses of the order of nanometers.[13] When these tactoids are dispersed in water, they exfoliate into single platelets in the dilute concentration regime. The degree of exfoliation of the tactoids in an aqueous medium depends on the clay concentration and the ionic conditions of the dispersing medium. In a recent study using a model smectite LAPONITE®, we showed that the exfoliation rate of the tactoids in aqueous dispersion slows down significantly as the dispersion approaches the glass transition. This was attributed to intertactoid repulsions.[14] Our investigation of Na-montmorillonite dispersions showed that the tactoids exfoliate incompletely even below the glass transition concentration due to the reverse osmotic pressure of the hydrated Na^+ counterions in the bulk medium.[15] The presence of added salt in the dispersing medium also inhibits the tactoid exfoliation process. This has an important influence on the stability and the rheological behavior of the resultant dispersions. Above a critical volume fraction of clay particles, the addition of salt in the clay dispersions leads to gel formation. The stability and rheological behavior of the gels with varying salt concentrations depend on the nature of platelet association for attractive bond formation and the strength of these bonds between the platelets. The association of the platelets in the form of overlapping coins (OC) and house of cards (HoC) configurations contribute significantly to the increasing yield stress of the clay gels up to a critical salt concentration.[16] Above the critical salt concentration, face–face aggregation of clay platelets due to van der Waals attractions is dominant due to considerable charge screening. This increases the thicknesses, and therefore the weights, of the gel strands, thereby making the gels susceptible to gravitational collapse. The concentration of salt in the dispersing medium is therefore a very critical factor in determining the stability of the gels formed from charged colloidal clay particles.

In this paper, we report a detailed study of the stability behavior of Na-montmorillonite clay platelets and their aggregates in dilute aqueous dispersions at different ionic conditions. It is observed that dilute salt-free dispersions, with clay concentrations below the glass transition concentration, are not stable under gravity. By directly monitoring the position of the interface of a sedimentating clay gel, we find that a gel that is formed by adding clay powder to a brine solution is also highly unstable under gravity. We demonstrate here that the stability of the dispersions can be enhanced if attractive interactions are induced between the platelets by adding salt *after* dispersing the clay in deionized water. The weak gels that are formed remain stable for a duration that depends on the salt concentration in the medium and eventually exhibit transient collapse. Furthermore, the velocity of the collapsing gel is seen to oscillate with increasing gel age, with the average magnitude of the velocity increasing up to a peak value before decreasing subsequently.

The phenomenon of transient collapse was first observed by Poon *et al.* in 1993,[17] and has been a subject of intense research since then. In the last two decades, the transient collapse phenomenon of colloidal gels has been studied in depletion gels[18–21] and in colloidal gels flocculated at their secondary minima.[22,23] However, a complete understanding of the microscopic origin of transient collapse remains elusive. Our study, using a combination of rheological measurements, ultrasound attenuation spectroscopy and cryogenic scanning electronic microscopy, reveals that the continuous restructuring and association of the platelets in overlapping coin configurations lead to increases in the widths and lengths of the gel network strands. We argue that competition between the gravitational stresses on these network strands and the local yield stresses plays an important role in determining the stability of gels formed from charged colloidal clay particles.

II. Experimental section

A. Material structure

Na-montmorillonite is a natural smectite clay mineral having a 2 : 1 layered phyllosilicate structure.[1] A unit layer, known as a platelet, consists of an aluminum octahedral sheet sandwiched between two tetrahedral silica sheets. The flat surfaces (the basal planes) of the platelet acquire negative charges due to the isomorphic substitution of Al^{3+} ions in the octahedral sheet by Mg^{2+} and Fe^{2+} ions. The negative charges thus arising on the basal surfaces are compensated by exchangeable intercalated Na^+ counterions. The edges of the platelet have pH dependent positive charges.[24] The thickness of a single-layer platelet is around 1 nm, while its lateral sizes can vary from tens of nanometers to a few micrometers. In dry form, when the negative charges on the basal surfaces are neutralized by the exchangeable Na^+ counterions, several platelets assemble in a stack, known as a tactoid, under the influence of van der Waals attraction. In an aqueous medium, the Na^+ counterions get hydrated due to the absorption of water molecules in the intratactoid spaces. As a result, tactoids slowly swell and exfoliate, producing laminar flexible platelets with electric double layers (EDLs) on their surfaces.[25] The highly polydisperse nature of the sizes and shapes of the Na-montmorillonite platelets can be seen in the representative SEM micrograph in the inset of Fig. 1. The lateral sizes of these platelets, calculated from the SEM micrographs, vary from tens of nanometers to a few micrometers with an average lateral size of 420 nm (Fig. 1).

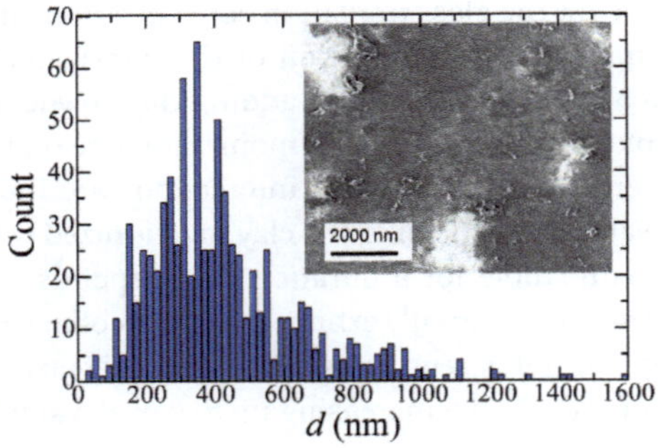

Fig. 1 Lateral size distribution of Na-montmorillonite platelets measured using SEM. Calculation of the platelet size distribution was done with more than 700 single layers identified in SEM micrographs. The average lateral size calculated from the distribution is 420 nm. Inset shows a representative SEM micrograph containing Na-montmorillonite platelets.

B. Sample preparation

The Na-montmorillonite powder used in this work was purchased from Nanocor, Inc. The powder is baked for 24 h in an oven at 120 °C to remove any absorbed moisture. The powder is next dispersed in water using a magnetic stirrer under vigorous stirring conditions. Three batches of samples were prepared at different ionic conditions as noted below:

1. **Samples without salt.** Clay powder is dispersed in highly deionized Milli-Q water of resistivity 18.2 MΩ cm. The dispersion is stirred for seven days before using it for experiments.

2. **SBC sample.** Clay powder is dispersed in brine solution. The brine solution is prepared by adding a predefined amount of NaCl in Milli-Q water. The mixture is kept under stirring conditions for seven days before experiments. We designate such a sample as SBC (Salt added Before Clay).

3. **SAC sample.** A predefined amount of salt is added after seven days to a clay dispersion prepared in a salt free condition. The mixture is further stirred for three hours before experiments. We designate such a sample as SAC (Salt added After Clay).

C. Experimental arrangements

1. **Light based experimental setup.** A homemade light based experimental setup, shown in Fig. 2(a), is used to monitor the sedimentation of the clay dispersions. A schematic diagram of this setup is shown in Fig. 2(b). A linear light source made from red LEDs (midband wavelength = 633 nm) is housed within a rectangular box (L′) containing two layers of light shaping diffusers procured from Luminit™. This arrangement produces a homogeneous and diffused light source which is used to illuminate the samples. The intensity of the diffused light is controlled by changing the current to the LEDs. Rectangular glass cuvettes (S) of inner dimensions 10 mm (w) × 9 mm (d) × 48 mm (h) are used to hold the clay dispersions. The cuvettes are then housed in a cuvette-holder (H) fixed to the goniometer (G) arm. Cuvettes containing the samples of height h_0 = 45 mm are sealed using a silicone sealant to prevent evaporation of water. The temperature

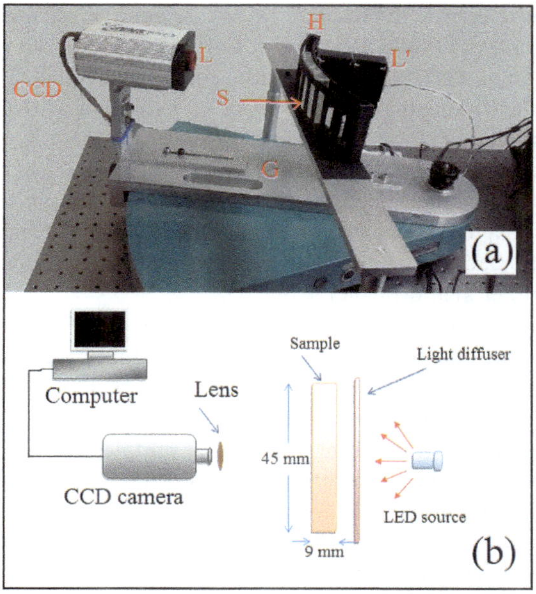

Fig. 2 (a) Snapshot of the light based experimental setup used for the sedimentation study. The main components of this setup are – a light source (L′), sample cuvettes (S), cuvette holder (H), goniometer (G), CCD camera and lens assembly (L). (b) Schematic diagram of the experimental setup.

of the sample is kept constant at 25 ± 0.2 °C. The diffused light is incident perpendicularly to the cuvette wall. The light transmitted through the sample is collected using a CCD camera (Photometrics CoolSnap ES, Roper Scientific Inc.) and a lens (L). A regular sequence of images is captured at different ages t_s of the sample. $t_s = 0$ is defined as the time at which the stirring of the sample is stopped inside the sample cuvette. The intensity of the incident light is adjusted so that the transmitted intensity detected by the CCD camera remains below the saturation level of the CCD sensor. The CCD images are finally analyzed using Winview and Matlab softwares. The transmitted intensity I_{tr} at each pixel of the CCD image is measured in terms of the grey value. To eliminate the spatial inhomogeneity of the incident light, the value of I_{tr} at each pixel of the CCD image is normalized by the background intensity of light that passes through clear water.

2. **Rheological measurements.** Rheological measurements are performed at 25 ± 0.1 °C using a stress controlled rheometer (Anton Paar MCR501) with a concentric cylindrical measuring head of length 25 mm and diameter 16.6 mm. The sample is kept in a cylindrical container of inner diameter 27.6 mm and length 42 mm. The measuring head is inserted into the sample up to 25 mm from the top surface of the sample. After loading the sample, an oscillatory strain of amplitude 1000% and angular frequency 10 rad s^{-1} is applied for two minutes to shear melt the sample. A small oscillatory strain of amplitude 0.1% and angular frequency 6 rad s^{-1} is subsequently applied to the sample to monitor the evolution of the elastic modulus G' and the viscous modulus G'' with sample age t_s. Here, $t_s = 0$ is defined as the time at which the shear melting of the sample is stopped in the measuring cell.

3. Scanning electron microscopy (SEM). A field effect scanning electron microscope (ULTRA-PLUS FESEM) from Carl Zeiss is used to scan the clay platelets dried on an ITO (indium tin oxide) coated glass plate. The ITO surface, made hydrophilic by keeping the glass plate in a solution of H_2O_2, NH_3OH and water in the ratio 1 : 3 : 7 at a temperature of 80 °C for 5 h, is washed with deionized water and dried. A 10 μL drop of 0.5% w/v Na-montmorillonite dispersion is spread on the hydrophilized ITO–glass plate and dried very fast (within 10 seconds) in an oven at a temperature of 200 °C to avoid coagulation of the clay tactoids. Data acquisition and size measurements are carried out using SmartSEM (Carl Zeiss) and ImageJ softwares.

4. Cryogenic scanning electron microscopy. For cryo-SEM characterization, Na-montmorillonite dispersions with different salt concentrations, C_s, are loaded in glass capillary tubes (procured from Hampton Research, USA) of bore size 1 mm. The ends of the capillaries are then sealed. Samples are then kept in an undisturbed condition for different ages t_s. The samples are then vitrified using a liquid nitrogen slush of temperature −200 °C. The vitrified samples are fractured and sublimated for 30 min at a temperature −90 °C and then coated with a thin layer of platinum (thickness approximately 1 nm) at a temperature of −150 °C in a vacuum using a cryotransfer system (PP3000T from Quorum Technologies). The imaging of these samples is then performed using an ULTRA-PLUS FESEM from Carl Zeiss at an electron beam strength of 2 keV.

5. Size measurement using ultrasound attenuation spectroscopy. Clay particle size distributions at different ionic conditions are estimated by measuring the ultrasound attenuation spectra (attenuation coefficient α_v *vs.* ultrasound frequency ω) and ultrasound speeds using a DT-1200 acoustic spectrometer (procured from Dispersion Technology Inc.). The details of this technique can be found in the literature.[14,26] For each attenuation measurement, 130 mL of the clay dispersion is loaded in a cell containing two identical ultrasound transducers facing each other. The attenuation spectra are measured at different ionic conditions of the colloidal dispersions by varying the ultrasound frequency in the range 3 to 99.5 MHz and for 12 different sample widths ranging from 5 mm to 20 mm. A sample temperature of 25 ± 0.2 °C is maintained for each measurement. The unimodal particle size distribution (PSD) of the dispersed clay is estimated from the attenuation spectrum using an analysis algorithm developed by the manufacturer. The theory of viscous attenuation of ultrasound, which was used to fit the attenuation data, is described in Section I of the ESI.† The parameter values that were used to calculate the PSDs are supplied in Section II of the ESI.†

6. Electroacoustic measurements. Electroacoustic measurements are used to indirectly track the positions of the collapsing interfaces at higher clay concentrations. The technique is explained in detail in Section III of the ESI.† The results obtained from the electroacoustic measurements are used for comparison with the optical tracking data of the collapsing interfaces reported here for dilute clay concentrations. A photograph of the electroacoustic setup is provided in Fig. S1 of the ESI.†

III. Results and discussion

We first monitor the stability of the clay platelets in dilute aqueous salt-free dispersions using the light transmission method. Fig. 3 shows the variation of the

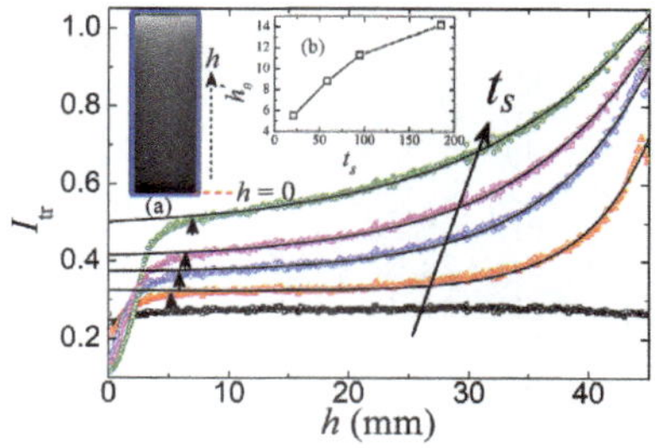

Fig. 3 Variation of the transmitted intensity (I_{tr}) along the height h of the cuvette for a 0.8% w/v Na-montmorillonite dispersion at ages t_s = 0, 21, 59, 95 and 185 h (bottom to top). Here, $h = 0$ is at the bottom of the sample column and $t_s = 0$ is defined as the time when the stirring of the sample in the cuvette is stopped. The arrow heads indicate the positions of the top surfaces of the sediment at different t_s. The solid lines are the exponential fits $[I_{tr}(h) = I_0 + I' \exp(h/h'_0)]$ to the data. Here, h'_0 is a length scale characterizing the particle concentration gradient. The inset (a) shows an image of the sample acquired at t_s = 59 h using the CCD camera. The variation of h'_0 with t_s is shown in the inset (b).

transmitted intensity (I_{tr}) along the height, h, of the cuvette for a 0.8% w/v aqueous dispersion of Na-montmorillonite clay at several t_s values. Here, t_s is the time since stirring of the sample is stopped and $h = 0$ is at the bottom of the sample cuvette. At age $t_s = 0$, I_{tr} remains almost constant with varying h. This indicates that the clay platelets are distributed uniformly in the dispersion. For $t_s > 0$, I_{tr} increases with h. A representative CCD image of the sample acquired at t_s = 59 h is shown in the inset (a) of Fig. 3. The I_{tr} *vs.* h data can be fit to $I_{tr}(h) = I_0 + I' \exp(h/h'_0)$, where I_0, I' and h'_0 are the fitting parameters. h'_0 is a length scale characterizing the intensity gradient, and therefore is a measure of the particle concentration gradient, along the sample height. At low h, a sharp decrease in I_{tr} can be observed due to the deposition of the clay platelets at the bottom of the sample cuvette. The positions where the I_{tr} data deviate from the exponential fits correspond approximately to the clay sediment heights and are indicated by arrow heads in Fig. 3. At larger h, the exponential variation of I_{tr} with h indicates a continuously varying concentration of platelets along the height of the sample. As a result of the wide size distribution of the clay platelets (Fig. 1), the bigger platelets settle faster than the smaller ones. The smaller particles diffuse out over larger length scales with increasing t_s, giving rise to a gradual increase in h'_0 with age as observed in the inset (b) of Fig. 3. The absence of any distinct interface between the supernatant and the clay rich region [inset (a) of Fig. 3] confirms the absence of a gel phase.[27,28]

Due to the high concentration of negative charges on the basal surfaces and the wide lateral sizes of the Na-montmorillonite platelets, there is a small spill-over of negative potential onto the positive edges of the platelets in dispersions of pH > 7.[29–31] As a result, the effective potential on the edges are also negative. Hence, the Na-montmorillonite platelets in dilute dispersions, such as the ones reported in Fig. 3, do not show any aggregation and sedimentation under gravity. However, salt-free aqueous dispersions of Na-montmorillonite clay platelets exhibit strong stability above the glass transition concentration (4% w/v) of the sample. In this higher concentration regime, the clay dispersions are stabilized by long range repulsive interactions and form kinetically constrained, structurally

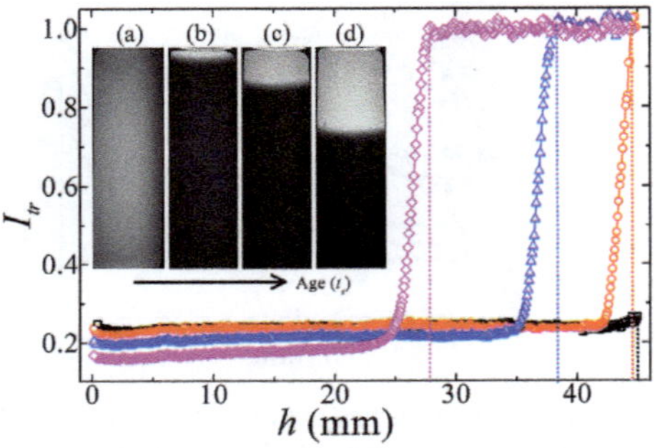

Fig. 4 Variation of the transmitted intensity (I_{tr}) with sample column height, h, of a 0.8% w/v Na-montmorillonite dispersion (SAC) with 60 mM salt at ages $t_s = 0.5$ h (□), 3 h (○), 10 h (△) and 95 h (◇). The corresponding CCD images of the gel at different t_s are shown in the insets (a)–(d). The vertical dotted lines indicate the positions of the interface between the supernatant and the collapsing gel at different t_s.

disordered soft glasses.[14,15] This has been probed using the electroacoustic method (Fig. S1 of the ESI†) and the results are plotted in Fig. S2 of the ESI.†

We next investigate the assembly and stability behaviors of the dilute clay dispersions in the presence of added salt. Attractive interactions between the clay platelets in a 0.8% w/v Na-montmorillonite dispersion are induced by the addition of salt to the medium which results in the reduction of the Debye screening lengths characterizing the Na-montmorillonite platelets in the dispersions. CCD images captured at different ages (t_s) of a 0.8% w/v Na-montmorillonite dispersion (SAC) with 60 mM salt, prepared by adding salt to the dispersions after dispersing clay powder in deionized water, are shown in the insets of Fig. 4. At $t_s =$ 0.5 h, the transmitted intensity, I_{tr}, remains constant with varying h (□ in Fig. 4), indicating that the clay particles are uniformly distributed along the sample height. At higher t_s, a distinct interface between the clear supernatant and the clay rich phase appears in the dispersion, with the transmitted intensity through the supernatant reaching 100% at decreasing values of h with increasing sample age. The positions of the interface at different ages are shown by dotted lines in Fig. 4.

Representative time evolution data of the normalized positions h/h_0 of the interfaces in 0.8% clay dispersions with 40 mM salt, prepared at two different ionic conditions (SBC and SAC), are shown in Fig. 5. Here, h_0 is the total sample height. It is seen in this figure that the sample designated as SBC (□), prepared by dispersing clay powder at a concentration of 0.8% w/v in a 40 mM salt solution, is highly unstable under gravity, with the normalized interface height (h/h_0) decreasing continuously with t_s. On the other hand, the sample designated as SAC (○ in Fig. 5), prepared by adding 40 mM salt to the 0.8% w/v clay dispersion, remains stable under gravity for a finite delay time, $\tau_d = 9$ h, during which time the interface remains in a quiescent state. At $t_s > \tau_d$, the interface height of the SAC sample decreases rapidly with t_s due to a transient collapse phenomenon and eventually saturates at longer times.

To understand the microstructures of these dispersions, we perform frequency sweep rheological experiments for the samples at $t_s = 30$ min by varying the imposed oscillatory frequency in the range 0.1–50 rad s^{-1} at a strain amplitude of

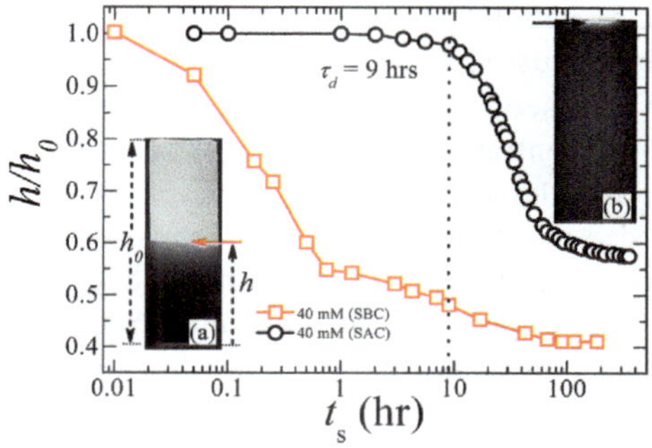

Fig. 5 Variation of the normalized height, h/h_0, of the interface between the clear supernatant and the collapsing gel *versus* age, t_s, in a 0.8% w/v Na-montmorillonite dispersion (SBC), prepared by adding clay powder to 40 mM brine (□), and a 0.8% w/v Na-montmorillonite dispersion (SAC), prepared by adding 40 mM salt after dispersing the clay powder (○). Here, $h = 0$ is the bottom of the sample column and $t_s = 0$ is defined as the time when stirring of the sample is stopped in the cuvette. The delay time, $\tau_d = 9$ h, for the onset of collapse of the SAC sample is indicated by the vertical dotted line. The insets (a) and (b) show the CCD images acquired at $t_s = 10$ h of the SBC and SAC gels, respectively.

0.1%. The frequency sweep measurement could not be extended to very low frequencies for the SBC gel due to the unstable nature of this sample. The SAC gel, in comparison, is stable for a very long duration, allowing data collection even for very small oscillatory frequencies. It can be seen from Fig. 6 that the that elastic modulus G' is much higher than the viscous modulus G'' for both the samples. Furthermore, G' is almost independent of the applied oscillatory frequency, whereas G'' is weakly dependent on frequency for values of less than 10 rad s^{-1}. These are typical rheological signatures exhibited by kinetically arrested colloidal particles in dispersion.[32] Since the clay concentration (0.8% w/v) investigated here is well below the glass transition concentration (4% w/v),[15] the observed kinetic arrest clearly arises from the formation of gel network structures that are facilitated by attractions between the clay platelets induced by the added salt. Indeed,

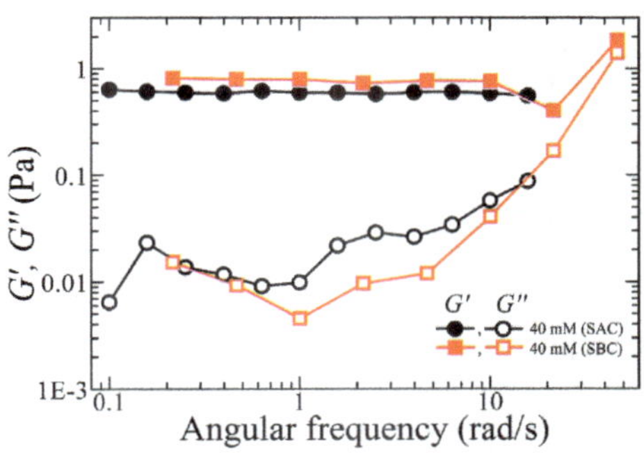

Fig. 6 Frequency dependence of storage modulus G' (solid symbols) and loss modulus G'' (empty symbols) of the SBC sample (□), prepared by dispersing 0.8% w/v Na-montmorillonite clay powder in 40 mM salt solution, and the SAC sample (○), prepared by adding 40 mM salt to a 0.8% w/v Na-montmorillonite dispersion. An oscillatory shear strain of amplitude 0.1% was applied in these experiments.

the formation of clay gels has been verified by us using direct visualization methods and will be discussed in detail later.

To understand the observed stability behavior of these gels prepared at identical clay and salt concentrations, we estimate the particle size distributions (PSDs) of the clay particles in their dispersions using ultrasound attenuation spectroscopy. The size measurements are performed while stirring the dispersions using a magnetic stirrer to avoid sedimentation of the particles. Fig. 7 shows the measured attenuation spectra in the ultrasound frequency range 3–99.5 MHz for 0.8% w/v Na-montmorillonite dispersions with 40 mM salt concentration prepared for the two different ionic conditions (SBC and SAC) described earlier. The attenuation spectra are fitted to the theoretical estimates for the attenuation spectra (solid lines in Fig. 7) considering a unimodal size distribution [eqn (S1)–(S6) (Section I) of the ESI†]. The particle size distributions thus estimated from these spectra are plotted in the inset of Fig. 7. It can be seen that the equivalent spherical median sizes are $d_{SBC} = 54$ nm for the SBC sample (□ in Fig. 7) and $d_{SAC} = 41$ nm for the SAC sample (○ in Fig. 7). The average thickness of the tactoids in the SBC sample is therefore higher than that in the SAC sample. The lower degree of exfoliation of the tactoids in the SBC sample arises due to the presence of salt in the dispersing medium during the preparation stage. The salt present in the water reduces the osmotic pressure difference of the hydrated Na^+ counterions between the intratactoid spaces and the bulk medium. This inhibits the diffusion of the native Na^+ counterions from the intertactoid spaces into the bulk water, thereby reducing the rate of exfoliation of the clay tactoids and leading to larger particle sizes in the SBC sample.[15] Owing to the large d_{SBC} values, the gravitational stress on the SBC gel network dominates over the local yield stress and eventually leads to the rapid gravitational collapse of the gel immediately after stirring is stopped. In contrast, the degree of exfoliation in the SAC sample (○ in Fig. 7) is higher, resulting in a larger number of exfoliated particles in this dispersion compared to

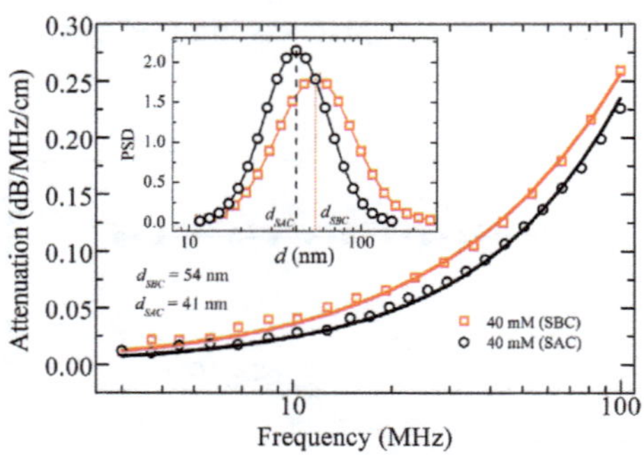

Fig. 7 Ultrasound attenuation spectra measured in the frequency range of 3–99.5 MHz at $t_s = 0$ for a 0.8% w/v Na-montmorillonite dispersion (SBC sample), prepared by adding the clay powder in 40 mM brine (□), and for a 0.8% w/v Na-montmorillonite dispersion (SAC sample), prepared by adding 40 mM salt after dispersing the clay powder in water (○). Solid lines are theoretical fits to the attenuation spectrum using eqn (S1)–(S6) (Section I) of the ESI† for a unimodal size distribution of the clay particles. The inset shows the corresponding particles size distribution of the two samples. It can be seen that $d_{SBC} = 54$ nm and $d_{SAC} = 41$ nm are the median values for the equivalent spherical sizes of the particles in the dispersions prepared in SBC (□) and SAC conditions (○), respectively.

the SBC sample. Hence, the network connectivity in the SAC sample is expected to be higher. At the same time, the gravitational stress on the gel network in the SAC sample is comparatively weaker due to the smaller sizes of the constituent particles. In contrast to the SBC gel, a quiescent or steady state of the SAC sample can therefore be observed up to a delay time $\tau_d = 9$ h (Fig. 5).

The sedimentation data for the 0.8% Na-montmorillonite dispersions prepared under SAC conditions with varying salt concentrations C_s are shown in Fig. 8. All samples here exhibit quiescent regimes followed by collapse regimes. When C_s is increased, the delay time τ_d decreases (inset of Fig. 8), indicating faster collapse of the gels. To better describe the collapse process, we calculate the

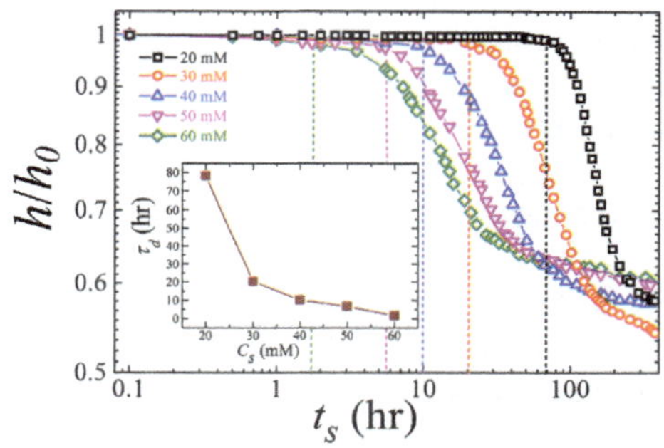

Fig. 8 Evolution of the normalized heights, h/h_0, of the interface between the clear supernatant and the collapsing gel *versus* age, t_s, of 0.8% w/v Na-montmorillonite dispersions (SAC) with salt concentrations of 20 mM (□), 30 mM (○), 40 mM (△), 50 mM (▽) and 60 mM (◇). Gelation is induced in these samples by dispersing clay powder in deionized water and then adding salt. The inset shows the variation of the delay time, τ_d, with varying salt concentrations. The vertical dashed lines indicate the positions of the delay time, τ_d, for the different salt concentrations.

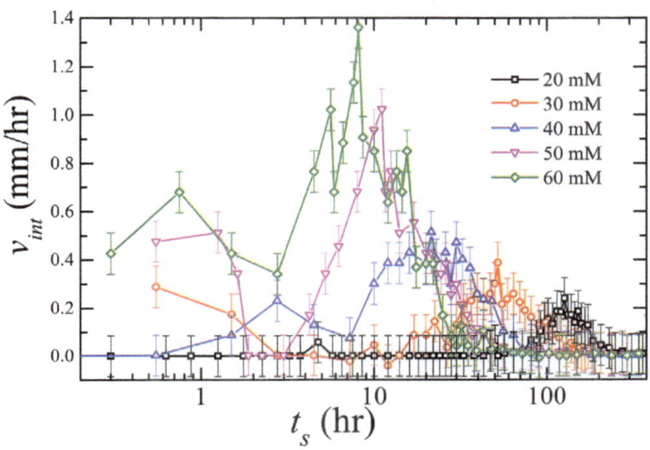

Fig. 9 Plot of the collapse velocity, v_{int}, of the interface with age t_s for 0.8% w/v Na-montmorillonite dispersions (SAC) at salt concentrations 20 mM (□), 30 mM (○), 40 mM (△), 50 mM (▽) and 60 mM (◇). The error bars quantify the systematic errors that arise from the detection of the interface using the optical method described earlier.

velocity, $v_{\rm int}$, of the collapsing interface (Fig. 9) from the $I_{\rm tr}$ *versus* $t_{\rm s}$ data (Fig. 8) for different $C_{\rm s}$. Irregular oscillations in the values of $v_{\rm int}$ with $t_{\rm s}$ can be seen during collapse. It can be seen that the average magnitude of $v_{\rm int}$ increases with $t_{\rm s}$ and shows a peak before decreasing at higher $t_{\rm s}$. It is also observed that the peak shifts to lower $t_{\rm s}$ for higher salt concentrations. This observation will be discussed later.

The variation of the elastic modulus G' with $t_{\rm s}$ is next measured by applying a small oscillatory strain of amplitude 0.1% and angular frequency 10 rad s^{-1} to samples prepared under SAC conditions with salt concentrations 20 mM, 40 mM and 60 mM. These data are shown in Fig. 10. It can be seen that G' increases monotonically up to age $t_{\rm s} = t_{\rm r}$ (indicated by arrows in Fig. 10 for different salt concentrations). This is followed by a slight decrease in G', with the data showing oscillations as $t_{\rm s}$ is increased further. However, the average magnitude of G' increases with $t_{\rm s}$. The variations of $t_{\rm r}$ (time at which the first oscillation in G' is observed) and the delay time, $\tau_{\rm d}$, measured from the interface collapse profile (Fig. 8) are plotted in the inset of Fig. 10. These two characteristic times show very good agreement with each other for all $C_{\rm s}$ values. This suggests that during the quiescent period ($t_{\rm s} < \tau_{\rm d}$), G', and hence the yield stress of the gel, both increase. However, a mechanical failure of the gel network occurs at $t_{\rm s} = t_{\rm r} \sim \tau_{\rm d}$, leading to transient collapse.

When $C_{\rm s}$ is increased, the platelets constituting the gels are expected to aggregate faster through the formation of stronger attractive bonds.[16] The weights of the gel strands increase substantially as a result, accelerating the transient collapse of the gels (inset of Fig. 10). Once transient collapse is initiated, the gel strands approach each other. This gives rise to even faster platelet aggregation. This, in turn, leads to an increase in the gravitational stress on the gel network, with the velocity of the collapsing interface increasing sharply with $t_{\rm s}$ as observed in Fig. 9. However, due to gel compaction and the formation of progressively stronger bonds between the particles in the network, the yield stress also increases simultaneously during the sedimentation process and eventually overcomes the gravitational stress. This results in a decreasing interface velocity at large $t_{\rm s}$ values. The competition between the gravitational and yield stress on

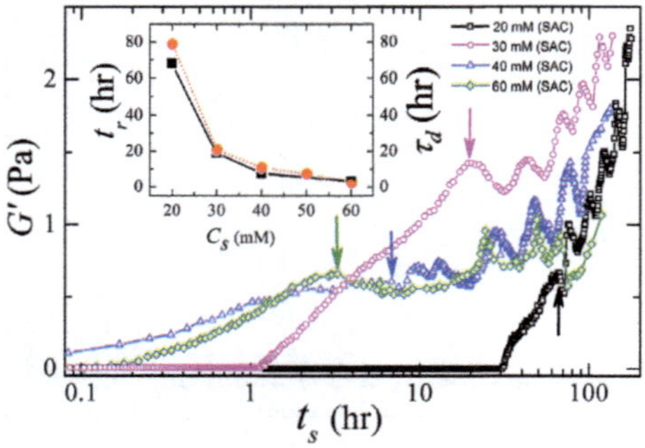

Fig. 10 Evolution of the elastic modulus, G', with age, $t_{\rm s}$, of 0.8% w/v Na-montmorillonite (SAC) dispersions at salt concentrations 20 mM (□), 30 mM (○), 40 mM (△) and 60 mM (◇). The moduli are measured by applying an oscillatory strain of amplitude of 0.1% and angular frequency of 10 rad s^{-1}. The arrows indicate the positions of the delay times, $t_{\rm r}$, at which G's start exhibiting oscillations for all four salt concentrations. The inset shows the plot of $t_{\rm r}$ (■) and $\tau_{\rm d}$ (○) with varying salt concentration.

the gel gives rise to a repetitive process of mechanical failure and build up of the gel structure in the collapse regime and results in the observed oscillations of v_{int} and G' at $t_s > t_r \sim \tau_d$.

To understand the mechanical failure of the gel network with age, we directly study the variation of the microscopic structures of the aging gels using cryogenic scanning electron microscopy (cryo-SEM). Fig. 11 shows the cryo-SEM micrographs of the weak SAC gels formed with $C_s = 20$ mM and 40 mM at $t_s = 11$ h, 48 h and 138 h. It can be seen from these images that the clay platelets form highly disordered network structures. It is clearly observed that the average lengths and widths of the network strands are higher than the average lateral size of each clay platelet (420 nm reported in Fig. 1). This indicates that the platelets in the network structures are connected in overlapping coin (OC) configurations.[33] In an OC configuration, the positive edge of a platelet is attached to the negative basal surface near the edge of another platelet in a parallel fashion (Fig. 4 in ref. 16). It is also clearly visible from the images in Fig. 11 that the widths and lengths of the network strands grow with age as an increasing number of platelets participate in network formation through OC configurations. This suggests that the gels restructure continuously due to thermal fluctuations and reorient in a way that

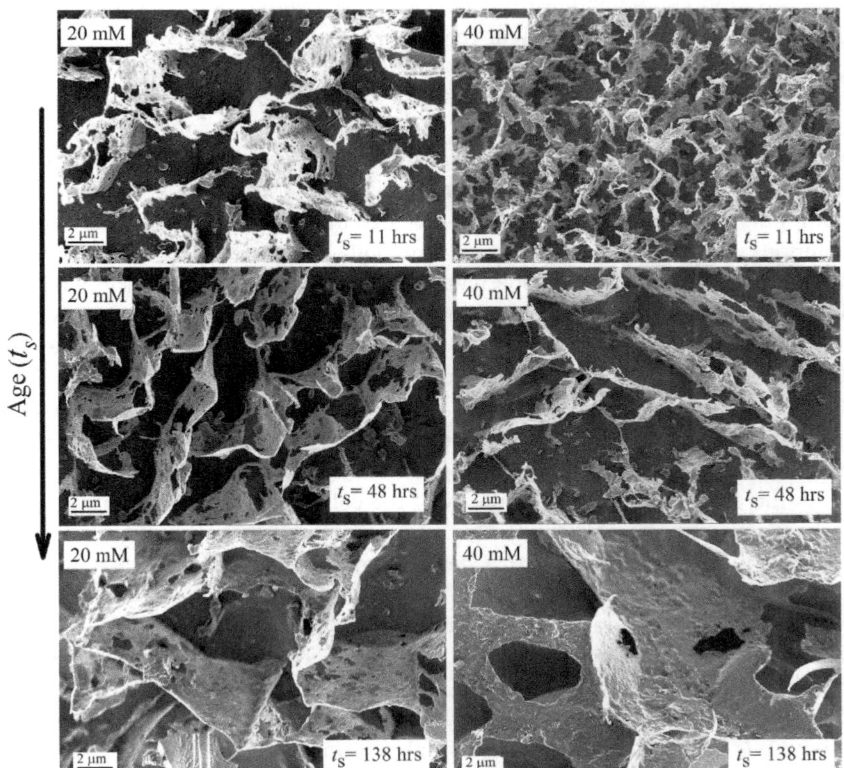

Fig. 11 Representative micrographs obtained using cryo-SEM for 0.8% w/v Na-montmorillonite dispersions (SAC) with $C_s = 20$ mM (left panel) and 40 mM (right panel) at ages $t_s = 11$ h, 48 h and 138 h (top to bottom). The scale bars represent 2 µm.

eventually leads to the formation of stronger attractive bonds.[34] As a result, the elastic modulus, and hence the yield stress, increase continuously with age up to a time $t_r \sim \tau_d$ as observed in the rheological measurements earlier (Fig. 10). On the other hand, due to the increase in the dimensions of the network strands as observed in Fig. 11, the gravitational stresses on the network strands increase simultaneously with t_s. At $t_s = \tau_d$, the gravitational stresses overcome the local yield stresses of the gels leading to mechanical failure and the subsequent collapse of the gels as presented in Fig. 8 and 10. In the SBC gel samples, on the other hand, the network strands are thicker and longer, and the gravitational stress typically always exceeds the local yield stress, resulting in accelerated gel collapse (Fig. 5).

IV. Conclusions

In this work, we have studied the aggregation and stability behaviors of charged anisotropic clay particles of Na-montmorillonite in aqueous dispersions at a concentration (0.8% w/v) well below that required to achieve the glass transition. Microscopic imaging using scanning electron microscopy shows that the clay platelets are highly polydisperse in their shapes and sizes (Fig. 1). We show that these platelets are not stable in salt-free aqueous dispersions in the dilute concentration regime and sediment under gravity with a particle concentration gradient along the sample height that gets more pronounced with sample age (Fig. 3).

We next investigate the stability of gels formed by inducing attractive interactions between the clay platelets in a dilute concentration regime by directly monitoring the gel interface in optical transmission experiments. Attraction is induced by screening the negative charges on the basal surfaces of the platelets by adding salt into the medium. The gels prepared by dispersing clay powder in NaCl solutions are not stable under gravity and collapse immediately with a distinct interface between the clear supernatant and a clay-rich sediment (the SBC gel indicated by □ in Fig. 5). In contrast, we report a remarkable enhancement of the stability of the dispersions if identical amounts of salt are added to the clay dispersions after adding the clay powder to water (the SAC gel indicated by ○ in Fig. 5). We further observe that the SAC gel is stable for a finite delay time and eventually exhibits transient collapse under gravity with a sharp interface height that decreases with increasing age (Fig. 5).

In the collapse regime, the velocity of the collapsing interface exhibits oscillations with sample age (Fig. 9). The average velocity of the interface increases with age up to a peak value before decreasing at higher ages. With a systematic increase in the salt concentration, the delay time for transient collapse decreases (Fig. 8), while the average collapse velocity increases (Fig. 9). The microscopic mechanism involved in the observed transient collapse is investigated using rheological measurements and with direct visualization of the gel microstructures using cryogenic scanning electron microscopy. It can be seen that the elastic modulus G' of the gel increases with sample age up to a time t_r (Fig. 10), indicating the growing attractive strength of the gel network. The G' data shows oscillations at ages higher than t_r for all salt concentrations investigated here. The times, t_r, at which the oscillations in G' start are found to be in good agreement with the delay times τ_d for transient collapse (inset of Fig. 10).

The direct visualization of the microscopic structures at different salt concentrations and ages of the gels confirms that the gels undergo microscopic restructuring with time (Fig. 11), thereby attaining higher elastic moduli and hence yield stresses with age since preparation. During the restructuring process, an increasing number of platelets is seen to participate in network formation through overlapping coin (OC) configurations (Fig. 12(a)). This leads to an increase in the lengths and widths of the network strands (Fig. 12(b)). The gravitational stress on the gel strands increases as a result and eventually overcomes the local yield stress after a delay period. This leads to the mechanical failure of the network strands and the subsequent transient collapse of the gels (Fig. 12(c)). During the collapse, the gravitational stresses on the gel network increase further due to the coarsening of the gel strands as they approach each other. This results in a further increase in the velocity of collapse. The processes of gradual gel compaction during sedimentation and the continuous restructuring lead to a substantial increase in the local yield stresses of the gel network. A condition eventually arises in which the yield stresses characterizing the network strands overcome the gravitational stresses, resulting in a decrease in collapse velocity at higher ages (Fig. 9). Such competition between the local yield stresses and the gravitational stresses on the gel network strands during gel collapse leads to oscillations in the interface collapse velocity (Fig. 9), and in the elastic modulus (Fig. 10).

Further insight into the transient gel collapse phenomenon reported here can be obtained by theoretically modelling the experimental data acquired in this work. To model the observed delayed gel collapse and the subsequent densification, two important microscopic quantities, the association rate K_A and the dissociation rate K_D of the clay platelets, have to be measured and combined in a poroelastic model of the collapsing gel. However, the dependences of the quantities K_A and K_D on the sample age, its exfoliation state, the salt concentration and the local volume fractions are extremely non-trivial and need to be first quantified accurately.

It is important to conclude by noting that our study of the stability behavior of clay dispersions at a higher clay concentration (3% w/v) in the presence of added salt using electroacoustic methods (Section III and Fig. S1 of the ESI†) shows similar stability behaviors of the SBC and SAC gels (Fig. S3 of the ESI†) as observed in the dilute concentration regime studied here. However, it can be seen for the

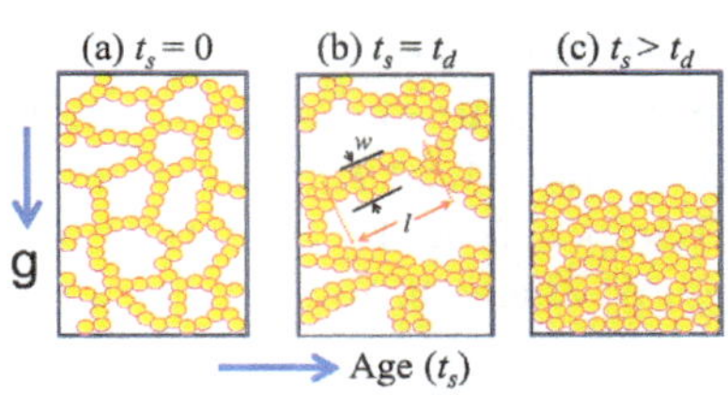

Fig. 12 Schematic of a clay gel under gravity (the direction is shown by an arrow pointing downward) illustrating the evolution of network morphology and subsequent gel collapse with increase in age, t_s. (a) A gel network dominated by overlapping coin (OC) configurations of clay platelets at a time immediately after preparation, (b) increase in the widths (w) and lengths (l) of network strands at higher ages due to the rearrangement of platelets, and (c) the collapsed state of the gel network at age $t_s > t_d$.

3% w/v gel that its yield stress dominates over the gravitational stress for relatively longer time durations due to the higher network connectivity and the slower rearrangement processes at higher clay concentrations.

Finally, we would like to point out here that the Na-montmorillonite used for this study belongs to the naturally occurring phyllosilicate group of clay minerals and is an important component of soil. We therefore believe that the results presented here on the tactoid exfoliation, aggregation and stability behaviors of aqueous dispersions of these minerals at different ionic conditions will be very useful in understanding many geophysical phenomena such as the formation of river deltas and the collapse of quicksand.[8,9] On the other hand, our observations of the collapse behavior in salt-stabilized colloidal clay particle dispersions also demonstrate the ubiquity of the transient collapse phenomenon previously only observed in depletion gels[18–21] and in colloidal gels flocculated at their secondary minima.[22,23]

V. Acknowledgements

We thank Mr A. Dhasan and Mr K. M. Yatheendran for their help with cryo-SEM imaging and Prof. V. A. Raghunathan for loaning us the goniometer used in this work. We also thank Mr T. R. Venketesh for his help during sample preparation and data collection.

References

1 F. Bergaya and G. Lagaly, in *Handbook of Clay Science*, ed. F. Bergaya and G. Lagaly, Elsevier, 2013, vol. 5, pp. 1–19.

2 S. Abend and G. Lagaly, *Appl. Clay Sci.*, 2000, **16**, 201–227.

3 R. Bandyopadhyay, D. Liang, H. Yardimci, D. A. Sessoms, M. A. Borthwick, S. G. J. Mochrie, J. L. Harden and R. L. Leheny, *Phys. Rev. Lett.*, 2004, **93**, 228302.

4 B. Ruzicka and E. Zaccarelli, *Soft Matter*, 2011, **7**, 1268–1286.

5 D. Saha, Y. M. Joshi and R. Bandyopadhyay, *Soft Matter*, 2014, **10**, 3292–3300.

6 E. Paineau, L. J. Michot, I. Bihannic and C. Baravian, *Langmuir*, 2011, **27**, 7806–7819.

7 E. Zaccarelli and W. C. K. Poon, *Proc. Natl. Acad. Sci. U. S. A.*, 2009, **106**, 15203–15208.

8 A. Thill, S. Moustier, J.-M. Garnier, C. Estournel, J.-J. Naudin and J.-Y. Bottero, *Cont. Shelf Res.*, 2001, **21**, 2127–2140.

9 A. Khaldoun, E. Eiser, G. Wegdam and D. Bonn, *Nature*, 2005, **437**, 635.

10 F. Uddin, *Metall. Mater. Trans. A*, 2008, **39**, 2804–2814.

11 H. H. Murray, *Appl. Clay Sci.*, 2000, **17**, 207–221.

12 Y. Liu, M. Zhu, X. Liu, W. Zhang, B. Sun, Y. Chen and H.-J. P. Adler, *Polymer*, 2006, **47**, 1–5.

13 G. Lagaly and S. Ziesmer, *Adv. Colloid Interface Sci.*, 2003, **100–102**, 105–128.

14 S. Ali and R. Bandyopadhyay, *Langmuir*, 2013, **29**, 12663–12669.

15 S. Ali and R. Bandyopadhyay, *Appl. Clay Sci.*, 2015, **114**, 85–92.

16 S. Ali and R. Bandyopadhyay, *Soft Matter*, 2016, **12**, 414.

17 W. Poon, J. Selfe, M. Robertson, S. Ilett, A. Pirie and P. Pusey, *J. Phys. II*, 1993, **3**, 1075–1086.

18 S. Manley, J. M. Skotheim, L. Mahadevan and D. A. Weitz, *Phys. Rev. Lett.*, 2005, **94**, 218302.

19 R. Buscall, T. H. Choudhury, M. A. Faers, J. W. Goodwin, P. A. Luckham and S. J. Partridge, *Soft Matter*, 2009, **5**, 1345–1349.

20 P. Bartlett, L. J. Teece and M. A. Faers, *Phys. Rev. E: Stat., Nonlinear, Soft Matter Phys.*, 2012, **85**, 021404.

21 E. Secchi, S. Buzzaccaro and R. Piazza, *Soft Matter*, 2014, **10**, 5296–5310.

22 L. Bergstrom, *J. Chem. Soc., Faraday Trans.*, 1992, **88**, 3201–3211.

23 A. A. Potanin and W. B. Russel, *Phys. Rev. E: Stat. Phys., Plasmas, Fluids, Relat. Interdiscip. Top.*, 1996, **53**, 3702–3709.

24 H. van Olphen, *An Introduction to Clay Colloid Chemistry*, John Wiley and Sons Inc., New York, 1977, vol. 53, pp. 230–230.

25 J. D. F. Ramsay, S. W. Swanton and J. Bunce, *J. Chem. Soc., Faraday Trans.*, 1990, **86**, 3919–3926.

26 A. S. Dukhin and P. J. Goetz, *Characterization of Liquids, Nano- and Microparticulates, and Porous Bodies using Ultrasound*, Elsevier, New York, 2010.

27 P. Snabre and B. Pouligny, *Langmuir*, 2008, **24**, 13338–13347.

28 É. Guazzelli and J. Hinch, *Annu. Rev. Fluid Mech.*, 2011, **43**, 97–116.

29 R. Secor and C. Radke, *J. Colloid Interface Sci.*, 1985, **103**, 237–244.

30 E. C. Y. Yan and K. B. Eisenthal, *J. Phys. Chem. B*, 1999, **103**, 6056–6060.

31 D. Zhou, A. I. Abdel-Fattah and A. A. Keller, *Environ. Sci. Technol.*, 2012, **46**, 7520–7526.

32 P. Sollich, F. Lequeux, P. Hébraud and M. E. Cates, *Phys. Rev. Lett.*, 1997, **78**, 2020–2023.

33 M. Delhorme, B. Jonsson and C. Labbez, *Soft Matter*, 2012, **8**, 9691–9704.

34 E. Zaccarelli, *J. Phys.: Condens. Matter*, 2007, **19**, 323101.

Faraday Discussions

PAPER

Designing disordered materials using DNA-coated colloids of bacteriophage fd and gold†

Z. Ruff,‡[a] S. H. Nathan,‡[a] R. R. Unwin,‡[a] M. Zupkauskas,[a] D. Joshi,[a] G. P. C. Salmond,[b] C. P. Grey[c] and E. Eiser*[ad]

Received 17th August 2015, Accepted 1st October 2015
DOI: 10.1039/c5fd00120j

DNA has emerged as an exciting binding agent for programmable colloidal self-assembly. Its popularity derives from its unique properties: it provides highly specific short-ranged interactions and at the same time it acts as a steric stabilizer against non-specific van der Waals and Coulomb interactions. Because complementary DNA strands are linked only *via* hydrogen bonds, DNA-mediated binding is thermally reversible: it provides an effective attraction that can be switched off by raising the temperature only by a few degrees. In this article we introduce a new binary system made of DNA-functionalized filamentous fd viruses of ~880 nm length with an aspect ratio of ~100, and 50 nm gold nanoparticles (gold NPs) coated with the complementary DNA strands. When quenching mixtures below the melt temperature T_m, at which the attraction is switched on, we observe aggregation. Conversely, above T_m the system melts into a homogenous particulate 'gas'. We present the aggregation behavior of three different gold NP to virus ratios and compare them to a gel made solely of gold NPs. In particular, we have investigated the aggregate structures as a function of cooling rate and determine how they evolve as function of time for given quench depths, employing fluorescence microscopy. Structural information was extracted in the form of an effective structure factor and chord length distributions. Rapid cooling rates lead to open aggregates, while slower controlled cooling rates closer to equilibrium DNA hybridization lead to more fine-stranded gels. Despite the different structures we find that for both cooling rates the quench into the two-phase region leads to initial spinodal decomposition, which becomes arrested. Surprisingly, although the fine-stranded gel is disordered, the

[a]Cavendish Laboratory, University of Cambridge, JJ Thomson Ave, Cambridge, CB3 0HE, UK. E-mail: ee247@cam.ac.uk
[b]Department of Biochemistry, University of Cambridge, Tennis Court Rd, Cambridge, CB2 1QW, UK
[c]Department of Chemistry, University of Cambridge, Lensfield Road, CB2 1EW, Cambridge, UK
[d]BP Institute, Bullard Laboratories, Madingley Rd, CB3 0EZ Cambridge, UK

† Electronic supplementary information (ESI) available: Video SV1, SV2, SV3 and SV4. See DOI: 10.1039/c5fd00120j

‡ Equal contributions.

overall structure and the corresponding length scale distributions in the system are remarkably reproducible. Such highly porous systems can be developed into new functional materials.

Introduction

Since the introduction of grafting dense coatings of short single-stranded (ss) DNA on to nanoparticles,[1,2] it has become possible to direct colloid–colloid self-aggregation. By barcoding different colloids, specific binding rules could be introduced to enable hierarchical colloidal aggregation, which was not possible with standard non-specific interactions. The main focus was on building colloidal crystals[3–6] with prescribed order for the development of photonic applications. The sharp melt transition of DNA-coated colloids, typically only 1–2 °C wide, made forming colloidal crystals difficult. Instead these systems would form dynamically arrested gels when quenched into the two-phase region.[7,8] Furthermore, the narrow melt region made annealing aggregates into equilibrium ordered solids almost impossible.[6]

However, many colloidal systems in nature, such as photonic colours in birds and insects[9,10] and applications such as paints, random lasers,[11] ceramics[12] and battery materials,[13,14] are inherently disordered. This makes the structural and dynamic study of colloidal gels and glasses particularly relevant. Rich literature on the dynamic behaviour of very dense hard sphere suspensions in the glassy or super-cooled phase has led to a good understanding of these systems.[15–18] In contrast, the aggregation and gelation dynamics of low-density suspensions of colloids interacting *via* van der Waals or short-range depletion interactions is less established. In recent years, the depletion-driven aggregation in low-density systems of spherical colloids has been studied both in experiments and in simulations.[19–23] In all these systems, the overall range and depth of attraction are fixed during a quench into the two-phase region.[24] Here we show that, similar to colloidal aggregation induced by depleting polymers, single and mixed DNA-coated colloidal suspension mixtures initially exhibit spinodal decomposition when quenched, followed by kinetic arrest. However, very different disordered structures can be achieved depending on the quench rate.

Previously we have shown that we can exploit the selectivity and strength of the short-ranged interactions to form bigels composed of two independent porous inter-percolating gels made of spherical sub-micron particles.[25–27] In order to develop new, high surface area, functional materials with tuneable porosity made with a variety of colloidal materials, we introduce a new system, namely a binary mixture of semi-flexible rods and nanoparticles. In our model system, monodisperse filamentous fd viruses were used as the semi-flexible rods. Short ssDNA (S) was attached to the side-coat proteins of the virus.[28] The complementary ssDNA (S′) was grafted to gold nanoparticles (gold NPs), as shown in the schematic in Fig. 1, and the DNA sequence is given in the Experimental section. Fluorescence microscopy was used to monitor the time evolution of the gel structures for different quench rates and depths.

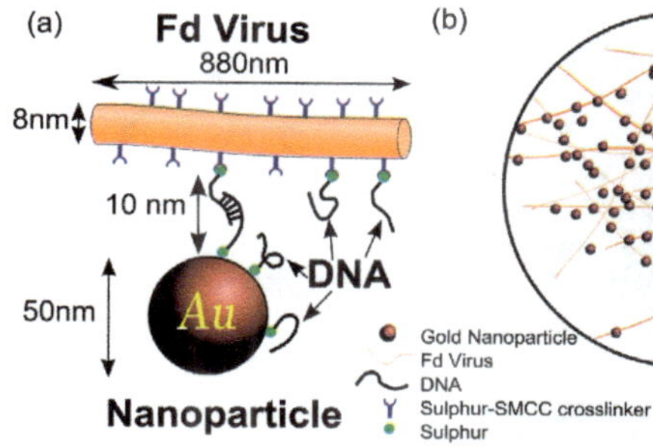

Fig. 1 Schematic of the system. (a) Gold nanoparticles are attached reversibly to fd viruses (filamentous ssDNA bacteriophages) *via* pairs of complementary DNA strands. (b) Cartoon depicting the expected local network structure of viruses and nanoparticles in a kinetically arrested state.

Results and discussion

Virus and Au nanoparticle gels – fast *versus* slow quench

We studied gold NP mixtures with three different number ratios (R_{GV}) of gold NPs to virus particles. The volume fraction of gold colloids, ϕ_{gold}, was 0.1% in all samples. Systems with $R_{GV} = 2.5$, 5 and 10 were studied, along with a control sample comprising only gold NPs, for which half of the spherical particles were coated with S DNA and the other with S′. For details see the Experimental section. Fluorescence microscope images were then obtained from these samples before heating, after heating, and after a controlled quench to room temperature (Fig. 2). The aqueous solvent is fluorescently labeled and hence the regions containing a high concentration of the gold colloids are visible as dark patches in the microscope images. Although the fd virions are almost one micrometer long they are too thin to be visible in optical microscopy. Individual gold NPs are also too small and highly diffusive to be visible. Images acquired of the samples in the melted or gas phase showed that the viruses and the gold NPs were fully dispersed and no sedimentation was observed (Fig. 2(c)). In the right-hand pan of Fig. 2, we show the structure factors corresponding to the real-space images in the left-hand pane. Very different structures were observed before (Fig. 2(a)) and after (Fig. 2(b)) the melting process. These structures will be referred to as '*open*' and '*homogenised*', respectively. As discussed in the next paragraphs, the differences in morphology arise from the different cooling protocols from the uniform, high-temperature phase.

The open structures were obtained for all three sample ratios when the samples were quenched from 60 °C (the temperature used during the preparation of the cuvettes containing the samples, outside the microscope) to room temperature (RT) (the initial temperature inside the microscope). Since RT is well below the DNA melt temperature, T_m ($\approx$42 °C), the rods and spheres start to aggregate and the suspension can undergo macroscopic phase separation. The rapid loading of the sample cells in the microscope resulted in a fast cooling process, as illustrated in Fig. 3. By subsequently heating the sample inside the microscope to 50 °C for a sufficiently long time, we return the colloids to the non-aggregated 'gas' phase. We can quench the sample back to RT to obtain a more

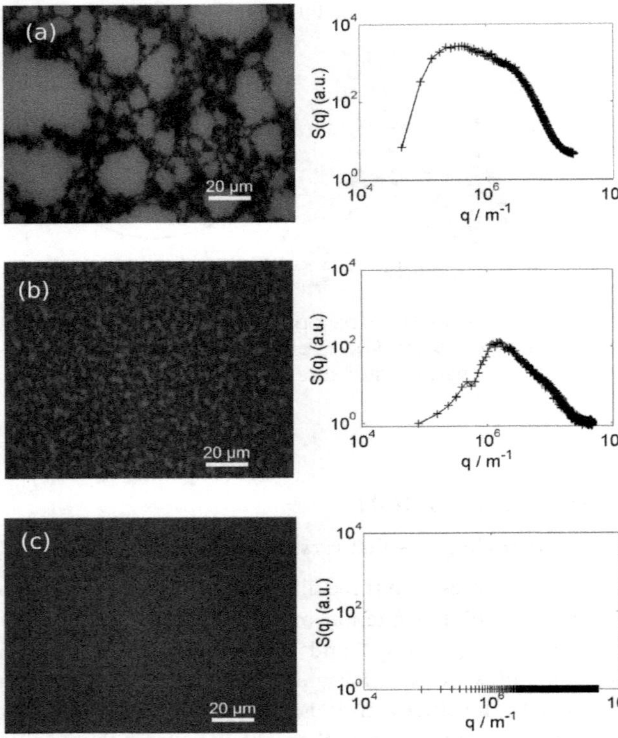

Fig. 2 Epifluorescence microscope images and the corresponding effective structure factor showing the difference in structure for (a) rapidly cooled (open), (b) slowly cooled (homogenised) and (c) melted virus–colloid structures, with their associated $S(q)$. A gold NP to virus ratio $R_{GV} = 5$ and gold NP volume fraction, $\phi_{gold} = 0.1\%$, were used. The images shown in (a) and (b) were obtained at RT while (c) was obtained at 50 °C. The light regions show the fluorescently labelled aqueous phase.

homogenous, fine-stranded gel structure (Fig. 2(b)). Gelled aggregates settle to the lower part of the sample chamber, leaving an aggregate-free top layer. Using a Peltier stage for controlled heating and cooling, we could reproduce this characteristic fine-stranded gel structure for many melting and cooling cycles around the melt temperature.

By comparing the results for the quench rates, it becomes apparent that the fast quench resulted in an approximately exponential time dependence of the temperature, while during the slower controlled cooling rates the T–t relation was nearly linear for temperatures between T_m and RT (Fig. 3(a)). In order to understand the difference in the resulting open and more homogenised gel structures we need to consider the rather slow process of DNA hybridisation that requires the correct zipping into the desired duplex, which is observed in DNA melt curve measurements. Only for sufficiently slow cooling or heating rates is no hysteresis observed in the hybridisation curves. A schematic cooling curve is drawn in Fig. 3(b). On rapid cooling, the fraction of faulty base pairs is high. This effect leads to a lower apparent melt temperature, reflecting that the structure that has formed is, on average, less stable than a structure that is formed under

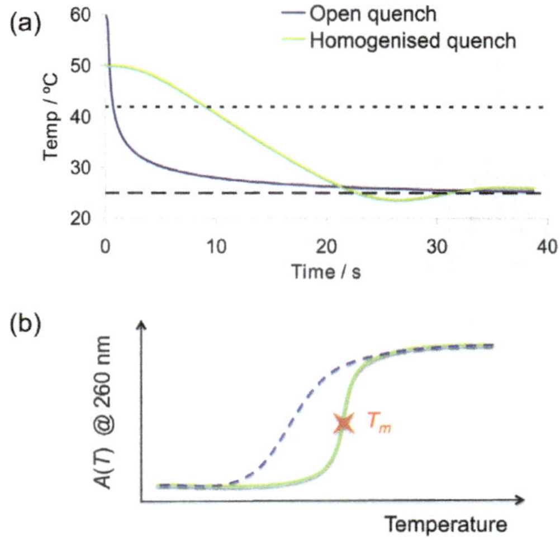

Fig. 3 (a) Curves showing the difference in quench profiles between the homogeneous and open structures. When the temperature goes through T_m (dashed line), the open structure is obtained for the fast cooling rate of -12 °C s^{-1}, and the homogenized one is obtained for the slower cooling rate of -1.5 °C s^{-1}. (b) Schematic melt curves. The solid line corresponds to equilibrium cooling and heating curves, while the dashed line indicates a hypothetical out-of-equilibrium cooling curve.

equilibrium. Rapid cooling therefore results in a shifted (and broadened) melting curve (see Fig. 3(b)). At first sight, the structures reported in Fig. 2(a) and (b) seem incompatible with the assumption of spinodal decomposition, as a shallow quench would allow the evolution of longer wavelength instabilities than a deep quench. However, one should bear in mind that what we observe is the 'aged' structure. We hypothesize that in the rapid quench, many small clusters form. These clusters can still diffuse and will aggregate into the morphology shown in Fig. 2(a). In fact, the thickness of the (dark) filaments in Fig. 2(a) is compatible with the assumption that it consists of smaller clusters. Additional evidence for this scenario comes from the chord analysis to be discussed in the context of Fig. 4. In contrast, the slow quench will result in much larger clusters that diffuse much more slowly and, eventually, get structurally arrested.

Further evidence for the above scenario comes from an analysis of the structure factors $S(q)$ plotted in Fig. 2. As expected, $S(q)$ is featureless when the sample is in the gas phase. However, for the slow quench (Fig. 2(a) and (b)) a well-defined peak in $S(q)$ appears, as one would expect for (arrested) spinodal decomposition. In contrast, for the fast quench (Fig. 2(a)), the peak in $S(q)$ is unusually broad, compatible with the assumption that spinodal decomposition is followed by aggregation of the primary clusters.

From our results for $S(q)$ we can estimate an average length scale L_o defined through $\sum qS(q)/\sum S(q) \equiv q_{ava} \equiv 2\pi/L_o$, where q_{ava} denotes the average scattering vector. Using this definition, we can estimate an average length scale $L_o = 3.4$ μm for the open, and $L_h = 1.3$ μm for the homogenised structure, of a gold NPs to virus ratio $R_{GV} = 5$ and a gold colloid volume fraction $\phi_{gold} = 0.1\%$. We also analysed the structure factor of the open structures of samples with the same ϕ_{gold}

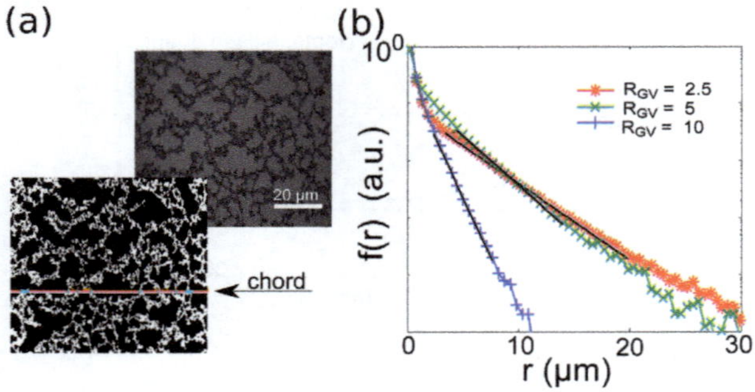

Fig. 4 Chord distribution analysis of open-structured gels. (a) A fluorescence image (back) of a sample with a colloid to virus ratio of 2.5 was binarized (front). Horizontal and vertical chords are then drawn through the image. (b) The resulting histograms of the distribution of lengths *f*(*r*) passing through the gel phase (light) and the 'empty liquid' (dark – not shown in the histogram) are shown for all three gold colloid to virus ratios. The fitted solid lines followed an exponential decay with characteristic length λ_{oa} – respective values are plotted in Table 1.

but different R_{GV} (see Table 1). We observe that L_o becomes smaller as the number of colloids per fd virion increased. This makes sense when interpreting the gold NPs as linkers that can bind virions together. The larger the number of gold NPs, the higher the maximum effective coordination number of the virions, in analogy to spherical colloids with a varying number of sticky patches.[29] Similarly, our spheres with isotropic binding potential can only bind maximally 12 other spheres, but because of the stick conditions provided by the DNA the effective coordination number will be much smaller, reflecting the larger L_o and λ_{oa}.

Using UV-vis spectrometry we estimated the S DNA grafting density on the virus to be roughly 160 strands per virion.[28] In our previous work,[28] we estimate the DNA-coated virus's diameter to be 12 nm compared to the diameter of 6–8 nm for the uncoated virus at our experimental solvent ionic strength. The gold NPs carry a much denser S′ DNA coating. Hence the 50 nm large gold NP can bind several fd viruses (if available). In all our experiments on NP–virus mixtures, the number of NPs per virus is larger than one (allowing each virion to bind at least two others, but most likely many more) and hence, below T_m the fd virions should

Table 1 Length scales for the open-structured gels with different rod–sphere ratios, as well as a sphere–sphere control sample following a fast quench. The characteristic length scales of the open structure, defined as $L_o = 2\pi/q_{ava}$, were obtained from $S(q)$. From chord analysis (Fig. 4) the decay lengths $\lambda_{oa,op}$ of the aggregates and the pores are given respectively. The same gold NP volume fraction $\phi_{gold} = 0.1\%$ was used in all samples

Ratio of gold NPs to virus R_{GV}	$L_o(S(q))/\mu m$	$\lambda_{oa}/\mu m$ Aggregates	$\lambda_{op}/\mu m$ Pores
2.5	7.4	4.36	11.01
5	5.4	2.98	14.57
10	3.5	1.77	23.70
Control	10.5	7.41	9.82

always be able to form percolating structures. However, as the volume fraction of virions is quite low, the percolating structures cannot be uniform. The initial non-uniformity emerges during spinodal decomposition. As the driving force for spinodal decomposition is larger for systems with large R_{GV} values, we expect that the wavelength of the maximally unstable density fluctuations, and hence L_o, is shortest for large R_{GV} values. Further coarsening is then arrested by percolation, and this again will happen earlier for large R_{GV} values.

In conjunction with studying the structure factor, we performed a chord distribution analysis (see Computational methods section) on the open structures. The resulting histograms (Fig. 4) can be fitted using an exponential with characteristic lengths of $\lambda_{0a} = 4.36$ µm, 2.98 µm and 1.77 µm for $R_{GV} = 2.5$, 5 and 10 respectively. The values obtained from chord analysis are consistently smaller than L_o, which is not surprising, as the chord analysis and $S(q)$ do not measure the same quantity. Nevertheless both measurements show the same tendency. We note that for a heterogeneous system consisting of two uniform phases, the ratio of the chord lengths is equal to the volume fractions occupied by the two phases. If we interpret the data in Table 1 in terms of a two-phase system, we must conclude that, at constant height (our microscopy images probe a cross section of the system), the system with $R_{GV} = 10$ has an effective gel volume fraction of 7%, whereas the system with $R_{GV} = 2.5$ has a volume fraction of 28%. This suggests that the system with $R_{GV} = 2.5$ can sediment much more than the system with $R_{GV} = 10$. We will come back to this point later.

The length scales present in the $S(q)$ peak were obtained by averaging over the entire sample, hence including aggregates and pores. The chord analysis technique allows us to separate these two phases. This difference will be further discussed for the homogenized samples.

We also measured a characteristic chord length for the control sample made of the same 50 nm large gold NPs, where half were coated with S and the other half with S′ DNA (details are given in the Experimental section). Keeping the gold NP volume fraction at $\phi_0 \approx 0.1\%$, we obtained a $\lambda_{0a} \approx 7.4$ µm for the open structure, which is considerably larger than that of the virus–colloid structure not only for the fast quench but also for the homogeneous one, as we will see later. Similarly, we obtain a larger $L_o = 10.5$ µm from the $S(q)$ peak maximum. Note that due to the different geometry of the contact region between two spheres compared to that between a sphere and a thin rod, the melt temperature of the pure gold NP system is increased to $T_m \approx 55$ °C.

Structural evolution as a function of quench depth

Using the controlled temperature quench from the gas phase to below T_m, we also studied the time evolution of the emerging 'homogenised' phase-separated samples as a function of quench depth. For the time evolution study, we cooled the well-homogenized samples to a given temperature below T_m, following the intensity of the emerging peak of the system's structure factor, which we define as $S(q_{max},t)$. For each quench temperature we first homogenized the sample by heating it to the gas phase. $S(q)$ was calculated from fluorescence microscope images recorded at 20 s intervals after the quench. In Fig. 5 such a time evolution is presented for a sample with $R_{GV} = 5$. ESI videos and corresponding $S(q,t)$ values for the NP–virus system and the control sample are provided as ESI videos SV1 and SV2.†

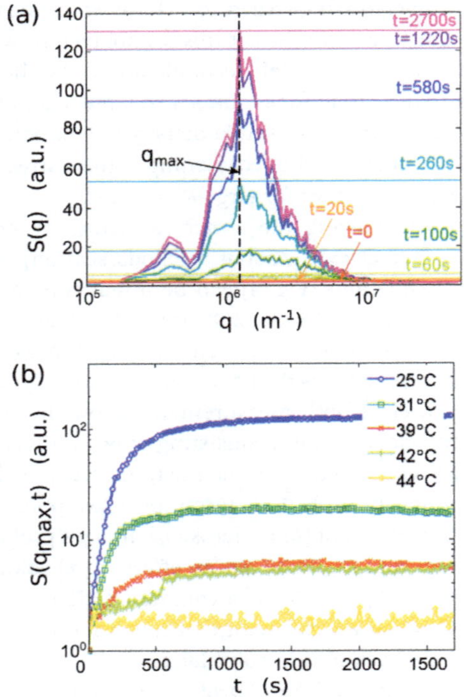

Fig. 5 (a) Time evolution of the structure factor $S(q)$ for a gold NP–virus sample quenched to 25 °C leading to a homogenized structure; $R_{GV} = 5$. The wavevector $q_{max} = 1.27 \times 10^6$ m^{-1} of the main peak is marked by the dashed vertical line. The times are measured from the start of the cooling ramp. (b) The height of the maximum of the structure factor over time for various quench depths for the same rod–sphere gel as in (a) – the $S(q_{max},t)$ was extracted from fluorescence images taken in the middle of the sample every 20 seconds.

Two important observations become apparent; firstly, in $S(q)$ the growing structural peak only becomes visible after roughly 20 s, which means that during the growth phase, the temperature of the sample is already at the target temperature. Secondly, the intensity of the emerging structural peak increases until reaching a plateau, while remaining centred about a characteristic wavevector $q_{max} = 1.27 \times 10^6$ m, corresponding to a characteristic length scale of about 5 μm. The rather constant position of q_{max} reflects the fact that the sample undergoes a spinodal phase separation, in which the longest characteristic wavelength grows fastest, but no subsequent coarsening occurs: the density fluctuation with the characteristic wavevector q_{max} grows until gelation arrests further coarsening of the aggregates, which is marked by the plateau.

In addition we observe the transition from a completely fluid sample with no structural peak at 44 °C to the onset of gelation, which we identified to be at $T_m \approx$ 42 °C. As can be seen in Fig. 5(b), when dropping just to this critical point the peak in the structure factor already shows evidence of demixing, and indeed, a video recording (see ESI videos SV1 and SV2†) clearly shows clusters forming, even though the sample as a whole still seems to be fluid-like. We speculate that the incipient phase separation at 42 °C is due to the heterogeneity in the NP–virion interaction (SV1†).

It is surprising that the wavelength of the maximum instability at this temperature is the same as that at lower quench temperatures, because a Cahn–Hilliard-style analysis would suggest that the wavelength of the dominant instabilities should grow as the temperature approaches the spinodal from below. At 39 °C most of the images show larger clusters and a clearly arrested phase in the form of the peak in $S(q)$ and the plateau. Reheating back into the gas phase and then cooling even further to 39 °C shows that more of the gel phase of a characteristic length scale is formed (SV2†). Comparing the slopes of the increase in $S(q_{max},t)$ also suggests that the phase separation process becomes faster as we cool deeper into the two-phase region, but the overall structure of the fine-stranded gel does not change much. This may be interpreted in terms of the finite attractive interactions provided by the DNA hybridization, which will be of the order of tens of k_BT, k_B being the Boltzmann constant. As the melt transition region is rather narrow (only a few degrees) the full attractive potential is reached already a few degrees below T_m. Hence it is only the entropic term that provides lowering of the system's Gibbs free energy upon further cooling.

Following the evolution of $S(q_{max},t)$ for different quench depths for the control sample of a pure gold NP gel with $\phi_0 \approx 0.1\%$, we observe a very similar behaviour: while q_{max} remains constant, the corresponding intensity plateaus after a few minutes (Fig. 6). ESI videos SV3 and SV4† show $S(q,t)$ for quench temperatures from 60 °C to about the melt temperature $T_m \approx 55$ °C and to 48 °C.

Finally, we also performed a chord analysis on all of the homogenized samples for the different quench depths, both after 400 s when $S(q_{max},t)$ was still growing, and after 1200 s when the plateau region was reached (Fig. 7). Similar to the gelled open structure, the homogenized samples also show an exponential length distribution with characteristic decay lengths λ_{ha}. However, unlike the open and homogenized structures of the control sample, the gelled gold NP virus structures seem to display two different decay length scales. Both are listed in Table 2. In Fig. 7 we present typical chord analysis results for a sample with $R_{GV} = 5$, showing the different regimes in the aggregates. However, we stress that, owing to the delicate measurements, large variations in the chord analysis from one set of samples to another need to be considered. Nevertheless, these short length scales,

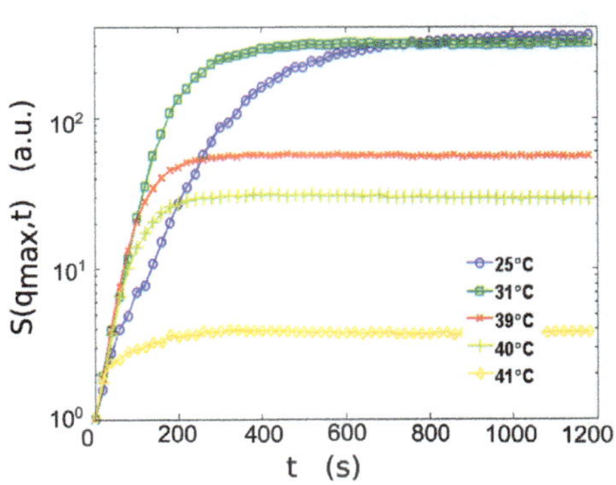

Fig. 6 Time evolution of the structure factor maximum $S(q_{max},t)$ for the control sample made of only gold NPs; $\phi_0 \approx 0.1\%$. The melt temperature of this system was at about 44 °C.

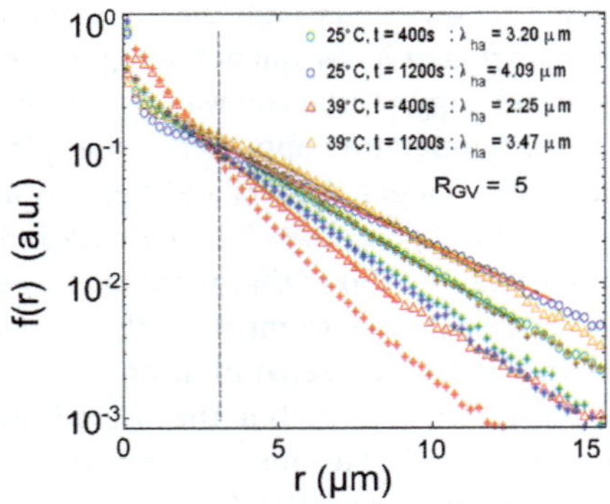

Fig. 7 Chord distribution analysis of a homogenous fine-stranded gel made of fd rods and gold NPs. The distributions for the aggregates are the lines with the solid fitted curves, while the non-fitted data represent the corresponding pore distributions. The vertical dashed line marks the divide between the long decay lengths λ_{ha} (right) and the short ones (left).

which are roughly a third of the length of the semi-flexible fd rods, do suggest that the stiffness of the rods combined with the tendency for the gold NPs to bind as many rods as possible influence the overall gel structure of the system. In contrast, the gold NP gels only made of spherical 50 nm large particles always show a single decay length λ_{ha}, similar to gels made of DNA-functionalized 500 nm polystyrene spheres, which we studied earlier.[26]

Further, we note a significant difference between the open and closed aggregate structures for the different R_{GV} values. While the typical long length scales for the homogeneous aggregates are all very similar, at 4–5 μm for all three gold NP to virus ratios, they do show significant changes in the open structure. This supports our earlier hypothesis that the very fast quench results in open structures due to the out-of-equilibrium hybridization of DNA, which leads to initially weaker

Table 2 Characteristic chord lengths extracted from fluorescence microscope images of samples with different gold NP to virus ratios and control samples, all quenched from 50 °C to RT. The given length sales are obtained from averages of 4–5 images from different regions in the 1 cm × 1 cm large sample, but the same height above the bottom surface

Homogenized samples	Time after quenching/s	λ_{ha}/μm		λ_{hp}/μm
		Aggregates		Pores
R_{GV}		Short lengths	Long lengths	
2	400	0.37	4.03	4.09
5	400	0.42	3.20	3.75
10	400	0.36	3.76	2.68
Control	400	1.89	1.95	3.19
2	1200	0.41	4.41	4.74
5	1200	0.39	4.09	3.02
10	1200	0.27	5.25	2.53
Control	1200	—	5.18	3.89

bonding, enabling loosely bound particles to detach and reattach onto larger aggregates. This is further supported by the same tendencies for the chord distribution of the pores.

Summarizing our results we demonstrate that the quench rate into the two-phase region governs the aggregation dynamics of DNA-functionalized colloids, as equilibrium DNA hybridization requires finite times. This is very different to colloid–polymer systems, in which the depth (strength) of the short-ranged interactions are fixed throughout the quench for a given depleting polymer concentration.[18] The strength of the short-ranged attraction between particles with complementary ssDNA will be initially reduced due to faulty binding events, allowing particles to unbind and rearrange, leading to an open gel structure. However, when cooled down sufficiently slowly, the maximum attractive well is attained fast enough, leading to the more homogenised fine-stranded gel. Using both time-dependent structure factor and chord distribution analysis on 2D fluorescence images, we were able to study these open and homogenized gel structures. Both techniques allowed us to determine that the segregation mechanism is initially due to spinodal decomposition in all cases, which is subsequently dynamically arrested. It is remarkable that the controlled slower cooling reproducibly led to a self-similar homogenized structure with what we call 'designed disorder'. Furthermore, while the $S(q,t)$ measurements provide an excellent tool to follow the dynamics of the aggregation into an arrested state, chord analysis provided us with further information on the aggregate and 'empty liquid' phase separately, giving insight into the dependence on the system's composition for both the rod–sphere and the pure sphere–sphere systems.

Finally we note that, unlike sphere–sphere colloidal gels that aggregate in the same way for all sizes between 10 nm and a few micrometers, the aggregation behaviour of the rod–sphere system does depend on the size and concentration ratio of the two. In our case we used 50 nm large particles and 880 nm long and 6–8 nm thick semi-flexible rods. When using 500 nm or larger DNA-functionalized spherical colloids we no longer observed significant aggregation in the presence of many more rods per colloid (data not shown here). This is most likely due to the semi-flexible nature of the virions; the gain of hybridization free energy per SS′ pair is of the order of 10 k_BT. Hence, the energy to bend the fd virion will be smaller than the gain from this binding energy, provided we have enough ssDNA grafted to the virion. Therefore the viruses will 'wrap' around the colloids, inverting the initial ssDNA coating into a complementary coating. The resulting fd-wrapped colloids cannot bind to each other, or only weakly if a virus bridges two colloids.

Experimental

Materials and methods

DNA strands were purchased from Integrated DNA Technologies. The sequences were modified from previously published work and had the following structure: 5′(15T)-(binding sequence)-(thiol)3′. The 15 thymine (T) bases are used to increase the entropy of the system by increasing the volume the sticky end can explore around the colloids. Two complementary binding sequences were designed based on minor modifications of a previously published sequence by the Gang group:[4] they are denoted S for 5′CCA ATC CAA T3′ and S′ for 5′ATT GGA TTG G3′, with S DNA being attached to the fd viruses and S′ DNA to the gold nanoparticles.

Virus functionalization

The filamentous fd virion is about 880 nm long and has a diameter of 6–8 nm. Its capsid coat is made of approximately 2700 copies of the gp8 coat proteins whose NH_2 terminus is pointing outwards.[30] DNA was covalently linked to these gp8 major-coat proteins using a slightly modified protocol developed by Unwin *et al.*,[28] which also describes the growth and purification of fd viruses. An amount of concentrated fd stock solution was diluted into 10 mM 4-(2-hydroxyethyl)-1-piperazineethanesulfonic acid buffer (HEPES buffer, Sigma-Aldrich) at pH 7.0 to make 0.5 mL of fd solution at a concentration of 0.25 mg mL^{-1}. To this solution, 0.5 mg of sulfosuccinimidyl-4-(*N*-maleimidomethyl) cyclohexane-1-carboxylate (sulfo-SMCC, Pierce) was added and the solution was mixed thoroughly; the mixture was then left to sit at room temperature (RT) for 1 hour in order to functionalize the fd major-coat proteins with thiol-reactive maleimide groups (Fig. 1). At the end of the hour, the fd–SMCC solution was twice desalted using desalting columns (40 kDa MWCO Zebra, Thermo Scientific), in order to remove spent reactants and any remaining crosslinker. Then 20 nM of thiolated S DNA was suspended in 100 μL of phosphate buffer and 100 μL of a 100 mM DTT stock solution, and was left to react for one hour. The activated thiolated DNA was purified using size elution chromatography (Illustra NAP-25, GE Lifesciences), equilibrated with phosphate buffer and eluted.

To the S DNA solution, 0.25 mL of the activated fd solution was added and the total mixture brought up to a salinity of 200 mM using 2 M NaCl phosphate buffer (Sigma-Aldrich). Due to changes in volume incurred by the desalting process, the final mixture of activated fd and thiolated DNA was 1.02 mL with an fd molarity of 3.74 nM. The mixture was left to sit at room temperature for 2 hours and then dialysed (100 kDa MWCO Spectrum Labs Float-A-Lyzer) against 10 mM Tris–EDTA buffer solution (TE, Sigma-Aldrich) in a refrigerator overnight. The resulting dialysed sample was then collected and stored at 4 °C.

Gold nanoparticle functionalization

Citrate-capped gold nanoparticles were purchased (Nanocomposix) at a concentration of 0.05 mg mL^{-1}. Sodium dodecyl sulphate (0.01%, Sigma-Aldrich) was added to 10 mL of gold nanoparticle solution and the suspension was concentrated to 1.0 mg mL^{-1} before adding 25 nM of activated thiol-functionalized DNA in 10 mM phosphate buffer (Sigma-Aldrich). Subsequently, 2 M NaCl was added in a stepwise manner to bring the final salt concentration of the solution to 0.5 M.[31] After incubating overnight, the DNA-functionalized gold nanoparticle suspension was then washed three times with phosphate buffer, before transferring the nanoparticles to 10 mM TE buffer.

Mixing process

188 μg of gold colloids coated with S′ were concentrated from the stock solution (ϕ = 0.025%) by centrifuging (17 000g for 30 minutes). After each spin, as much supernatant as possible was removed with a minimum loss of colloid. This process was repeated until less than 5 μL of solution remained (ϕ > 0.2%).

To the DNA–nanoparticle solution, a DNA-functionalised fd suspension was added to achieve the required number ratio of viruses to colloids, and the solution was mixed gently. 0.3 μg of fluorescein sodium salt (Sigma-Aldrich), a fluorescent dye (494/521 nm), was added for inverse staining of the samples, and further TE buffer and NaCl were added to bring the sample up to 10 μL with 50 mM salt to screen the negative charge of the phosphate backbones on complementary strands of DNA. The sample was then heated to 60 °C for 3 minutes, and finally gently mixed with a vortex.

Sample chambers

Sample chambers were prepared by forming a wedge between a cover slip (18 × 18 × 0.15 mm) and a standard glass slide using another, identical, cover slip as a spacer. One edge of the cover slip was glued to the glass slide using Norland Optical Adhesive, while the opposite edge rested on the spacer, which was placed directly on the glass slide. After the glue was allowed to spread *via* capillary action across the whole of the joint between the glass slide and the cover slip, the epoxy was set using UV light.

To prevent colloids from sticking to the chamber walls, the hydrophobicity of the sample chamber surfaces was increased using a previously published protocol.[25] Before chamber construction, both the glass slide and coverslips were washed with Hellmanex and 3 M NaOH. The assembled chamber also was treated with oxygen plasma immediately before loading the sample. To avoid aggregation of the suspension during sample loading, both the sample chamber and sample were heated to 60 °C. After sample loading, the chamber was sealed with a two-part epoxy (Araldite) and then immediately quenched by placing the chamber on a room temperature surface to obtain the open structure.

Imaging

The samples were imaged using a Nikon Eclipse Ti-E inverted microscope equipped with a digital camera (Grasshopper3, Point Grey Research Inc., Sony IMX174 CMOS sensor) using either a PLAN-APO 20/0.75 dry objective or a PLAN-APO 40/0.75 dry objective. The chambers were attached to a homemade heating/cooling stage consisting of a Peltier element and a copper block, which provided homogenous temperature control across the entire sample.

Before heating, the sample was imaged to obtain information on the open structure. Subsequently, the sample was heated to 50 °C and left for 20 minutes to achieve a homogenous gas phase of the viruses and gold particles. Afterwards, the samples were quenched to a given temperature below the melt temperature, and time evolution imaging was used to follow the ageing process. Exposure times were kept to a minimum and images were taken 20 s apart to minimise local heating due gold NP optical absorption.

Computational methods

Cord analysis. The porous structure of the samples was analysed using the chord distribution analysis method.[32] This analysis involves tracing lines (chords) through a binarized image of a biphasic medium, identifying the points at which these lines intersect the boundaries between the gel and the fluid phases. Here we distinguish between pores (p) and aggregate matrix (a).[33] The distances between neighbouring points represent the length through which the chord passes through a particular domain of one of the two phases. These lengths are tallied in histograms for their respective phases and then normalised to yield probability density functions for the pores ($f_p(r)$) and the aggregate matrix ($f_a(r)$). The probability density functions, which we call the chord distributions, represent the likelihood of finding a chord length r between r and $r + dr$ for a particular phase.[34]

Fluorescence microscope images of the aggregates were binarized using ImageJ to find the most appropriate pixel threshold values and then de-noised using standard methods.[25] The resulting images were then analysed by taking chords along horizontal and vertical rows of image pixels, tabulating the respective phase lengths into two histograms.

By inspecting the logarithmic plots of chord distributions against chord length the characteristic length scales of the system can be found. In our porous gels we found a negative exponential relationship[33,34] of the form $f_{a,p}(r) \propto \exp(-r/\lambda_{a,p})$, where $\lambda_{a,p}$ represents the exponential folding length of the chord distribution for the respective phases.

Structure factor. The structure of the gas phase and the ageing colloidal gel were quantitatively assessed by extracting an effective structure factor $S(q)$ from our epifluorescence microscope images. For a colloidal concentration $c(x, y)$, the structure factor is given by

$$S(q) = \frac{\langle |\tilde{C}(q_x, q_y)|^2 \rangle}{|f(q)|^2},$$

where $\tilde{C}$ denotes the 2D Fourier transform of $c(x, y)$, and $\langle \cdots \rangle$ denotes the radial average. The form factor $f(q)$ is the Fourier transform of the shape of a single colloid. We assumed that the observed intensity of the image $i(x, y)$ was linearly related to $c(x, y)$, and therefore $S(q)$ could be measured in arbitrary units by replacing $\tilde{C}$ with $\tilde{I}$.[26] We estimated $f(q)$ by taking the Fourier transform of the image of the sample in the gaseous phase such that the structure factor could be expressed as

$$S(q) = \frac{\langle |\tilde{I}(q_x, q_y, t)|^2 \rangle}{\langle |\tilde{I}(q_x, q_y, t = 0)|^2 \rangle}.$$

Conclusions

We introduced a new binary system of long filamentous viruses and gold nanoparticles, in which both were functionalized with complementary single-stranded DNA, providing temperature-reversible aggregation. The binding rules we established were such that viruses could only bind to gold NPs, while binding between like particles was forbidden. We studied the aggregation of these mixtures as a function of fast and slow cooling rates using fluorescence microscopy. The results were analysed in terms of 2D structure factors and chord length analysis. Our findings suggest that the structure formation in these NP–virion systems is somewhat different from that in systems of spherical particles. In particular, we find that the wavelength of the dominant density fluctuations hardly depends on the quench depth – in contrast to what the Cahn–Hilliard theory would predict. In addition, we find that the systems that are most deeply quenched have the largest density inhomogeneities. Whilst this seems logical at first, one should bear in mind that we work in a regime where NP–virion binding is very strong. Hence, at low temperatures the growth of inhomogeneities would seem to be difficult. It seems plausible that the initial instability in the system occurs early in the quench and has the same wavelength for all systems.

As the chord analysis suggests, systems with a low gold NP to virus ratio, R_{GV}, can sediment more than systems with high R_{GV}. A possible explanation is that the low R_{GV} samples 'consume' all NPs to bind virions together at an early stage in the phase separation. As a result, the resulting clusters interact only weakly and can sediment well. In contrast, the high R_{GV} clusters will have exposed nanoparticles on their surface. These particles will therefore tend to stick to each other and will sediment to form a more open structure. This two-stage phase separation scenario suggests a new method to design the pore structure in disordered nanoparticle systems. This behaviour can also be interpreted in terms of an effective coordination number: due to the large aspect ratio the virions can bind many nanoparticles. Hence, while keeping the gold NP volume fraction constant but changing the number of available virions to bind to, the effective coordination number of the system changes, leading to the different aggregation characteristics.

Acknowledgements

This work would not have been possible without the financial support of various agencies: EE, CPG and ZR thank the Winton Program for the Physics of Sustainability. CPG and ZR acknowledge support from the EU ERC FP7 programme *via* an advanced fellowship for CPG. RU and SHN received support from the Engineering and Physical Sciences Research Council (EPSRC) for financial support. Work in the GPCS Lab is funded by the BBSRC, UK. DJ acknowledges the financial support from Udayan Care, NTCU and Schlumberger Foundation's FFTF program. Finally, EE would like to thank Francesco Sciortino and Daan Frenkel for many valuable discussions.

References

1 C. A. Mirkin, R. L. Letsinger, R. C. Mucic and J. J. Storhoff, *Nature*, 1996, **382**, 607–609.
2 A. P. Alivisatos, K. P. Johnsson, X. Peng, T. E. Wilson, C. J. Loweth, M. P. Bruchez and P. G. Schultz, *Nature*, 1996, **382**, 609–611.
3 S. Y. Park, A. K. R. Lytton-Jean, B. Lee, S. Weigand, G. C. Schatz and C. A. Mirkin, *Nature*, 2008, **451**, 553–556.
4 D. Nykypanchuk, M. M. Maye, D. van der Lelie and O. Gang, *Nature*, 2008, **451**, 549–552.
5 E. Auyeung, T. I. Li, A. J. Senesi, A. L. Schmucker, B. C. Pals, M. O. de la Cruz and C. A. Mirkin, *Nature*, 2014, **505**, 73–77.
6 Y. Wang, Y. Wang, X. Zheng, E. Ducrot, J. S. Yodh, M. Weck and D. J. Pine, *Nat. Commun.*, 2015, **6**, 7253.
7 N. Geerts and E. Eiser, *Soft Matter*, 2010, **6**, 4647.
8 D. Lukatsky and D. Frenkel, *Phys. Rev. Lett.*, 2004, **92**, 068302.
9 H. Yin, B. Dong, X. Liu, T. Zhan, L. Shi, J. Zi and E. Yablonovitch, *Proc. Natl. Acad. Sci. U. S. A.*, 2012, **109**, 10798–10801.

10 E. R. Dufresne, H. Noh, V. Saranathan, S. G. J. Mochrie, H. Cao and R. O. Prum, *Soft Matter*, 2009, **5**, 1792–1795.
11 D. S. Wiersma, *Nat. Photonics*, 2013, **7**, 188–196.
12 J. A. Lewis, *J. Am. Ceram. Soc.*, 2000, **83**, 2341–2359.
13 A. Paolella, G. Bertoni, S. Marras, E. Dilena, M. Colombo, M. Prato, A. Riedinger, M. Povia, A. Ansaldo, K. Zaghib, L. Manna and C. George, *Nano Lett.*, 2014, **14**, 6828–6835.
14 H. Zhang, X. Yu and P. V. Braun, *Nat. Nanotechnol.*, 2011, **6**, 277–281.
15 A. M. Puertas and M. Fuchs, Glasses in Colloidal Systems. Attractive Interactions and Gelation, arXiv:0810.0681.
16 L. Cipelletti, L. Ramos, S. Manley, E. Pitard, D. A. Weitz, E. E. Pashkovski and M. Johansson, *Faraday Discuss.*, 2003, **123**, 237–251.
17 *Soft and Fragile Matter: Nonequilibrium Dynamics, Metastability and Flow (PBK)*, ed. M. E. Cates and M. R. Evans, Institute of Physics Publishing, Bristol and Philadelphia, 2000.
18 W. C. K. Poon, A. D. Pirie and P. N. Pusey, *Faraday Discuss.*, 1995, **101**, 65–76.
19 V. Prasad, V. Trappe, A. D. Dinsmore, P. N. Segre, L. Cipelletti and D. A. Weitz, *Faraday Discuss.*, 2003, **123**, 1–12.
20 S. Manley, H. M. Wyss, K. Miyazaki, J. C. Conrad, V. Trappe, L. J. Kaufman, D. R. Reichman and D. A. Weitz, *Phys. Rev. Lett.*, 2005, **95**, 238302.
21 A. Fierro, E. D. Gado, A. D. Candia and A. Coniglio, *J. Stat. Mech.: Theory Exp.*, 2008, **2008**, L04002.
22 V. Testard, L. Berthier and W. Kob, *J. Chem. Phys.*, 2014, **140**, 164502.
23 G. Foffi, C. D. Michele, F. Sciortino and P. Tartaglia, *Phys. Rev. Lett.*, 2005, **94**, 078301.
24 P. J. Lu, E. Zaccarelli, F. Ciulla, A. B. Schofield, F. Sciortino and D. A. Weitz, *Nature*, 2008, **453**, 499–503.
25 F. Varrato, L. Di Michele, M. Belushkin, N. Dorsaz, S. H. Nathan, E. Eiser and G. Foffi, *Proc. Natl. Acad. Sci. U. S. A.*, 2012, **109**, 19155, DOI: 10.1073/pnas.1214971109.
26 L. Di Michele, F. Varrato, J. Kotar, S. H. Nathan, G. Foffi and E. Eiser, *Nat. Commun.*, 2013, **4**, 2007.
27 L. Di Michele, D. Fiocco, F. Varrato, S. Sastry, E. Eiser and G. Foffi, *Soft Matter*, 2014, **10**, 3633–3648.
28 R. R. Unwin, R. A. Cabanas, T. Yanagishima, T. R. Blower, H. Takahashi, G. P. C. Salmond, J. M. Edwardson, S. Fraden and E. Eiser, *Phys. Chem. Chem. Phys.*, 2015, **17**, 8194–8202.
29 F. Sciortino and E. Zaccarelli, *Curr. Opin. Solid State Mater. Sci.*, 2011, **15**, 246–253.
30 D. A. Marvin, *J. Mol. Biol.*, 1966, **15**, 8–IN4.
31 S. J. Hurst, A. K. R. Lytton-Jean and C. A. Mirkin, *Anal. Chem.*, 2006, **78**, 8313–8318.
32 J. Méring and D. Tchoubar, *J. Appl. Crystallogr.*, 1968, **1**, 153–165.
33 P. Levitz, *Adv. Colloid Interface Sci.*, 1998, **76–77**, 71–106.
34 P. Levitz and D. Tchoubar, *J. Phys. I*, 1992, **2**, 771–790.

Faraday Discussions

PAPER

Modular, polymer-directed nanoparticle assembly for fabricating metamaterials†

Gurunatha K. Laxminarayana, Matthew Rozin, Stephanie Smith and Andrea R. Tao*

Received 1st September 2015, Accepted 14th September 2015
DOI: 10.1039/c5fd00134j

We achieve the fabrication of plasmonic meta-atoms by utilizing a novel, modular approach to nanoparticle self-assembly that utilizes polymer templating to control meta-atom size and geometry. Ag nanocubes are deposited and embedded into a polymer thin-film, where the polymer embedding depth is used to dictate which nanocube faces are available for further nanocrystal binding. Horizontal and vertical nanocube dimers were successfully fabricated with remarkably high yield using a bifunctional molecular linker to bind a second nanocube. Surface plasmon coupling can be readily tuned by varying the size, shape, and orientation of the second nanoparticle. We show that meta-atoms can be fabricated to exhibit angle- and polarization-dependent optical properties. This scalable technique for meta-atom assembly can be used to fabricate large-area metasurfaces for polarization- and phase-sensitive applications, such as optical sensing.

Introduction

Optical metamaterials pose major challenges in their fabrication over large areas, which can be prohibitively expensive and time-consuming using conventional nanolithography techniques. In optical metamaterials, metal nanostructures called "meta-atoms" behave as individual optical resonators that must be organized into macroscale arrays to produce the optical response of an effective medium.[1–3] Metamaterial parameters such as effective permittivity and permeability are dictated not only by the size and shape of the meta-atom, but are also strongly dependent on the spacing and arrangement of meta-atoms within the ordered array. This geometric complexity enables tuning of the metamaterial without changing the dielectric environment.[4,5] On the other hand, the ability to precisely fabricate meta-atom arrays becomes critical for optical performance.

NanoEngineering Department, University of California, 9500 Gilman Drive MC 0448, La Jolla, San Diego, CA 92093-0448, USA. E-mail: atao@ucsd.edu; Fax: +1 858 534 9533; Tel: +1 858 822 4237

† Electronic supplementary information (ESI) available: Extinction spectra of Ag nanocubes with various embedding depths and corresponding AFM analysis; Raman spectra of surface-functionalized Ag nanocubes. See DOI: 10.1039/c5fd00134j

Many optical applications such as subwavelength imaging or electromagnetic cloaking also require metamaterials to possess three-dimensional, bulk-like structures.[6–8] As a result, metamaterials are inherently challenging to fabricate due to this multiscale architecture.

Traditional top-down methods for fabricating micrometer-sized arrays of meta-atoms rely on photolithography methods to produce features with a lateral resolution of 20 nm, or direct-write electron-beam lithography methods for sub-10 nm resolution.[9–11] While lithographic techniques are able to define the nanostructure location with a high degree of precision, it is difficult to implement fabrication for manufacturing metamaterials or scalable metamaterials processing. In addition, the metal deposition associated with these techniques typically forms nanostructures with rough surfaces or small coalescing grains.[12] As a result, nanolithography produces surface irregularities that present a challenge when fabricating electromagnetic junctions (*e.g.* the gap in a split-ring resonator, or the spacing between wire-pairs) where metal nanostructures are separated by only a few nanometers. This has led to long-standing difficulties in reproducing nanojunctions that have been demonstrated to achieve high quality optical resonances. Well-defined, single-crystalline metal nanostructures are highly desirable for improving the quality and consistency of metamaterial performance.

The only nanomaterials that meet these requirements are chemically synthesized nanocrystals (NCs). Metal nanostructures composed of Au and Ag behave like optical antennae by supporting the excitation of localized surface plasmon resonances (LSPRs), where conduction electrons of the metal oscillate in resonance with the incident light wave to produce intense electromagnetic fields localized at the metal surface. Colloidal metal NCs with controlled shapes are ideal building blocks because they can be synthesized in batches using surfactants or polymers that can selectively stabilize low-energy crystal facets[13–16] and are typically single-crystalline. More importantly, these shaped NCs possess LSPRs that can be tuned through chemical synthesis.[17]

Self-assembly provides an ideal approach for metamaterial fabrication that overcomes limitations in scalability, tunability, and design. Self-assembly methods, in contrast with serial methods or lithography, can be carried out on a massively parallel scale using colloidal metal NC building blocks.[18,19] We and others have successfully demonstrated metal NCs for the assembly of large-area plasmonic materials.[20–22] The NCs have the potential to serve as optical meta-atoms by supporting the excitation of localized surface plasmon resonances (LSPRs) where light impinging on a metal nanostructure couples to the free electrons in the metal.[23,24] A significant challenge in the self-assembly of colloidal metamaterials is the ability to exhibit precise control over both short-range and long-range NC order to achieve architectures with targeted electromagnetic properties. Several strategies have been developed for achieving the programmable assembly of metal NCs, including the use of DNA linkers,[25,26] grafted polymers,[5,27,28] or patchy particles.[29–32] However, these methods are typically capable of assembling only small NC clusters. While NCs have been demonstrated to assemble into larger hierarchical structures through the use of capillary or convective assembly,[33,34] these methods often produce close-packed or diffusion-limited NC clusters that do not provide optimal geometries for optical performance.

To utilize NCs as building blocks for meta-atoms, the ability to assemble NCs with controlled densities and locations is necessary. For example, periodic arrays of metal NCs deposited at an absorber interface can serve as plasmonic concentrators in a photovoltaic device[35,36] and two-dimensional arrays of nanoscale split-ring resonators[37] and paired metal stripes[38,39] have been demonstrated as negative-index metamaterials that have the potential to enable optical elements operating beyond the diffraction limit (*e.g.* a superlens). We previously demonstrated that polymer-directed assembly could be used to direct the assembly of nanoparticle homojunctions (clusters composed of NC building blocks with the same size and shape)[40] and heterojunctions (clusters composed of different NC building blocks).[41] NCs grafted with different polymer ligands could be assembled into orientations that facilitate maximum electromagnetic field localization. While this polymer-directed assembly method exhibits excellent control over interparticle orientation, it presents a major difficulty in controlling NC cluster size and shape. Self-assembly is dictated by NC diffusion through the polymer matrix, and a broad distribution of cluster sizes ranging from populations of single NCs to long NC multimers are produced. We also observed the formation of both linear and branched NC chains, indicating no preferential location for NC attachment during the assembly process. As a result, the optical response of these self-assembled NC clusters is broadband and does not give rise to high quality optical resonances. In order to fabricate meta-atoms based on colloidal NCs, it is imperative to develop self-assembly methods that enable precise control over NC attachment and NC cluster size.

To address the aforementioned obstacles in self-assembled metamaterials, we move towards a modular, layer-by-layer approach for assembling and positioning colloidal NCs over large areas (Fig. 1). In this approach, NCs will first be assembled into an ordered array at an air–water interface and then deposited onto a polymer support. These NCs serve as templates for sequential NC conjugation to

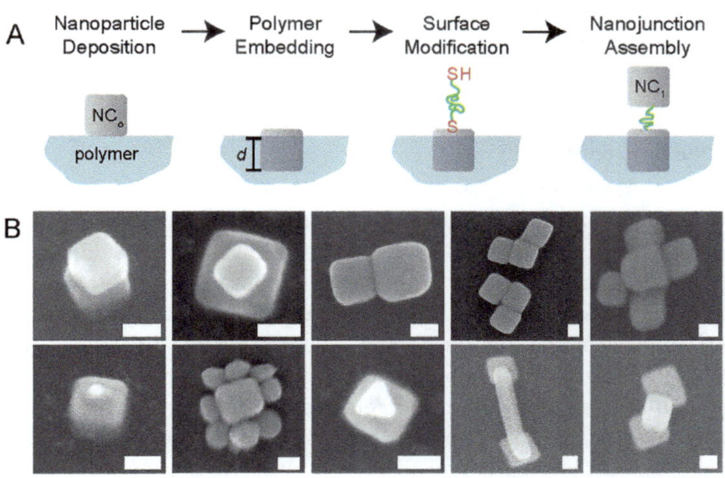

Fig. 1 (A) Schematic of the stepwise, modular NC assembly using Ag nanocubes to form dimer and cross-like meta-atoms. (B) Meta-atom assemblies for various combinations of NC_o and NC_1. Scale bar = 50 nm.

form more complex meta-atom structures. Both lateral (in-plane) and vertical (out-of-plane) NC arrangements can be precisely controlled to modulate the performance of the resulting colloidal metamaterial. Our assembly experiments target the formation of NC isomers that resemble meta-atoms fabricated using electron-beam and photolithography, such as rods and wires, chains, split-rings, and cross-bars. This stepwise approach to NC assembly bypasses the difficulties associated with diffusion-limited polymer-directed assembly, while at the same time employs techniques that are compatible with scalable nanomanufacturing such as spin-coating and dip-coating.

Experimental methods

Chemicals

Silver nitrate ($AgNO_3$), copper chloride ($CuCl_2$), polyvinylpyrrolidone (PVP, M_w = 55 000), citric acid monohydrate ($C_6H_8O_7 \cdot H_2O$) were purchased from Sigma Aldrich and used as received. 1,5-Pentanediol was purchased from Fluka. Polystyrene (M_w = 11 900) was purchased from Polymer source and poly(ethylene glycol) dithiol (PEG dithiol, M_w = 1000) was purchased from Nanocs chemicals. Analytical grade solvents were used throughout the reaction.

Ag nanocube synthesis and purification

Ag nanocubes were synthesized by a polyol method reported elsewhere.[40] Briefly, $AgNO_3$ was reduced in a solution of pentanediol, $CuCl_2$ and polyvinylpyrrolidone. A change in the solution color to an opaque yellow green indicates the formation of nanocubes. PVP plays an active role as a capping agent that controls nanocube nucleation and growth. The as synthesized nanocrystals were then characterized by UV-vis spectroscopy (Agilent Chemstation 8453). Purification of the nanocrystals was carried out by diluting the original solution to 50 mL using absolute ethanol and centrifugation using a Thermo Scientific CL2 Centrifuge (2500 rpm for 10 min). The supernatant was isolated and subjected to vacuum filtration with decreasing membrane pore sizes (Millipore Durapore membranes 0.65, 0.40 and 0.22 μm). The final solution was centrifuged at 3400 rpm for 60 min and re-dispersed with ethanol for the desired concentration.

Nanocomposite fabrication

Piranha cleaned soda-lime glass substrates (University wafers) were treated with hexamethyldisilazane vapor under vacuum at room temperature for 3 hours to obtain hydrophobic surfaces. 3.5 wt% of 11k polystyrene was dissolved in a toluene solution and filtered through a filter membrane with a pore size of 0.02 μm. A 270 nm thick polystyrene thin-film on the glass substrate was obtained by two-step spinning at 800 rpm for 3 s and 1700 rpm for 30 s (WS-400A-6NPP/LITE Laurell Technology Corp). To prepare the Ag nanocube films, 2 mL of the nanocrystal solution was washed 3 times in ethanol at 3400 rpm for 15 min and then the residue was re-suspended in 200 μL chloroform ($CHCl_3$). A clean glass Petri dish was washed thoroughly and filled with water to a convex meniscus. The nanocrystal dispersion was then carefully spread over the air–water interface and the resulting monolayer was left for 15 min to allow for evaporation of organic solvent and monolayer equilibration. Polystyrene coated glass wafers were then dip-coated to make the nanocube composite film. The area fraction or surface coverage of the nanocube was determined by ImageJ software and SEM images.

Nanocube embedding

Solvent annealing was carried out by exposing the substrates to $CHCl_3$ vapor in a closed container and samples were taken out and monitored optically and by AFM at different time intervals. Assembly experiments were then carried out.

Surface modification and assembly of Ag nanocubes

For composite films (NC_o), samples were exposed to 2 mL of 12 mM citric acid in ethanol for 15 hours. This was followed by incubation in a 7 mM PEG dithiol solution for 18 hours. The nanocube–polymer substrates were then washed with ethanol to remove excess dithiol from the surface. The removal of excess PVP from the Ag nanocube dispersions (NC_1) was carried out by adding 3 mL of 12 mM citric acid in ethanol to 6 mL of the as-synthesized colloidal dispersion, which was stirred vigorously for 3 hours. This dispersion solution was then washed several times by centrifugation and redispersion in neat ethanol. The final precipitate was redispersed in 10 mL of neat ethanol. The nanocube–polymer substrates (NC_o) were then exposed to the washed nanocube dispersions (NC_1) for 15 hours, followed by chemical, optical, and electron microscopy characterization.

Characterization

Raman measurement of the samples was performed on a microRaman spectrometer (Renishaw inVia) in the range of 200–1800 cm^{-1}. Samples were illuminated with a 514 nm, 50 mW argon ion laser. The back-scattered signal was collected on a Peltier-cooled (−70 °C) CCD detector with 30 s acquisition times. Scanning electron microscopy (SEM) characterization was carried out using an ultra-high resolution field emission electron microscope (FEI XL30-SFEG) with an accelerating voltage between 5 and 15 kV. Optical transmission and extinction spectra were measured using an Agilent Chemstation 8453 in the wavelength range of 200–1100 nm and optical reflectance spectra were obtained using a Perkin-Elmer Lambda 1050 UV-Vis-NIR Spectrometer. Height measurements of the Ag nanocubes were carried out upon annealing at different time intervals by an atomic force microscope (BRUKER Innova) operated in the tapping mode. The raw data was processed with Nanoscope IV software by third-order plane fitting and second-order flattening. Measurements were obtained using commercial silicon cantilevers (OLTESPA-R3) with resonance frequencies of 50–90 kHz and a spring constant near 3.5 N m^{-1}.

Finite-difference time-domain (FDTD) simulations

We use two-dimensional FDTD methods to simulate arrays of both horizontally- and vertically-assembled nanocube dimer structures. The horizontal dimer was modeled with NC_o with an edge length = 100 nm and an edge/corner radius of curvature = 8 nm, and partially embedded into the polymer matrix with ~40% of the nanocubes below the polymer surface. NC_1 (edge length = 90 nm, radius of curvature = 10 nm) is situated on top of the polymer matrix, its face flush with the polymer surface. The dimer nanojunction is created between the parallel surfaces of the two adjacent nanocube facets, spanning approximately 60 nm high with a gap width of 3 nm. A plane wave is injected normal to the polymer surface, with polarization parallel to the dimer axis. In order to simulate an array of like dimers supported on a continuous substrate, we use periodic (Bloch) boundary conditions in the plane of the supporting substrate, and perfectly-matched layer (PML) boundary conditions normal to the continuous plane. Scattered light is collected behind the light source in the far-field. The vertical dimer utilizes the same nanocube dimensions and boundary conditions as the horizontal dimer, but with an incident source tuned to excite the dimer's coupled dipole–dipole resonance. The plane wave source is injected with an angle of $\theta = 15°$ from normal to the plane of the substrate, with polarization parallel to the plane of incidence. In the vertical dimer geometry, NC_o is 90% embedded in the polymer matrix, and NC_1 is situated directly on top, with a 3 nm gap between the parallel nanocube facets. All FDTD simulations use a 0.5×0.5 nm maximum mesh size, and simulated reflectance spectra span a wavelength range of 200–1000 nm.

Results and discussion

Modular nanocrystal assembly

To build meta-atoms with complex architectures, we use a self-assembly approach where colloidal NCs are assembled onto a polymer or polymer-coated support in a stepwise manner. Dimers, trimers, and other NC isomers were fabricated by engineering the NCs to undergo sequential covalent binding events at the polymer interface (Fig. 1). First, colloidal NCs are deposited onto the polymer support. These initial NCs (which we term NC_o) serve as the initial binding platform for all subsequent assembly steps. Second, we carry out solvent or thermal annealing of the polymer support to immobilize and embed the deposited NCs into the polymer. Third, the exposed surfaces of the immobilized NC_o are chemically modified with a covalent molecular linker such as a dithiol. Finally, NC meta-atoms are formed by introducing a second particle (NC_1) that reacts with the molecular linker at the surface of NC_o. This highly modular protocol enables the formation of NC oligomers with controlled geometries and sizes since self-assembly is dictated by the accessibility of the molecular linker. In principle, these assembly steps can be cycled to produce hierarchical or dendritic NC motifs, or can be applied for the fabrication of hetero-NC oligomers composed of mismatched NCs. In this manner, the combinatorial formation of NC meta-atoms can be highly controlled by using different types of plasmonic nanoparticle for NC_o and NC_1. Fig. 1B shows an example of our results when Ag nanocubes (edge length = 70 nm), Au nanospheres (diameter = 25 nm), Ag nanospheres (diameter = 44 nm), Ag nanoprisms, or Ag nanorods are employed as NC_1.

To demonstrate the feasibility of generating meta-atom arrays using this modular approach, we focused on the formation of meta-atoms where Ag nanocubes are used as both NC_o and NC_1. Colloidal Ag nanocubes were synthesized *via* a polyol method,[40,42] where $AgNO_3$ is reduced in a solution of pentanediol, $CuCl_2$, and polyvinylpyrrolidone (PVP) ($M_w = 55\ 000$). PVP serves as a selective capping agent that controls nanocube nucleation and growth and also serves to passivate the surface of the Ag NCs during the assembly process. The as-synthesized Ag nanocube dispersion was concentrated by centrifuging the dispersion (3400 rpm for 60 min) and reconstituting the precipitate in chloroform. This dispersion was then deposited drop-wise onto the air–water interface of a glass Petri dish filled with deionized water (18 MΩ) and allowed to equilibrate for 15 min. NC arrays were generated by dip-coating polystyrene (PS)-coated glass slides through the air–water interface. These nanocubes serve as NC_0 during the assembly process. Large interparticle spacings (>100 nm) within the NC array are preserved by the presence of long PVP grafts that remain chemisorbed to the Ag surface of the colloidal nanocubes.

The deposited arrays were then subjected to solvent annealing with chloroform vapor to embed the nanocubes in the underlying polystyrene thin-film. The polymer embedding depth of NC_0 constrains the meta-atom assembly by exposing or protecting the metal surface of NC_0 from chemical modification with a molecular linker. The embedding depth can be precisely controlled by the solvent annealing time. Analogous to a two-dimensional melt processing step,[43] solvent absorption leads to a free volume increase in the polymer and reduced adhesion between the polymer chains. An increase in the mobility of the polymer chains allows vertical diffusion of the NC into the polymer film. Fig. 2 shows scanning electron microscopy (SEM) images of 100 nm Ag nanocubes that have been deposited onto a PS thin-film that is 270 nm thick, and then solvent annealed for 33, 53, and 63 minutes. The extinction spectra (Fig. S1†) and atomic force microscopy (AFM) topology maps of these arrays (Fig. S2†) clearly show the vertical mobility of the Ag NCs. This is further confirmed by a continuous red shift of the dipolar and quadrupolar LSPRs of the Ag NCs with increasing solvent annealing time, which is attributed to an increase in the refractive index of the NC environment as it diffuses into the PS film (Fig. S1†). The corresponding AFM height measurements of the PS-embedded nanocubes show that embedding depth can be controlled down to 10 nm increments.

Surface modification and self-assembly

The partially embedded Ag NCs were then surface functionalized with a bifunctional PEG dithiol linker in order to facilitate binding of a second NC. To carry out surface functionalization, removal of chemisorbed PVP from the Ag surface of NC_0 was necessary. This was carried out by treating the exposed surfaces of NC_0 with citric acid (see the Experimental section for the detailed procedure), which has been shown to displace PVP chains by mild oxidative etching.[44] Raman

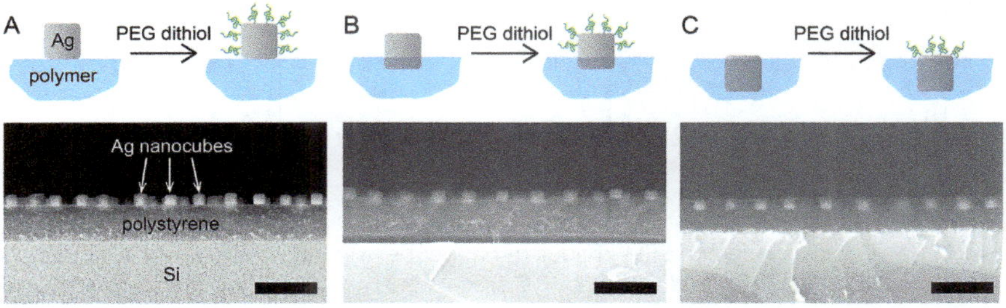

Fig. 2 Schematic of nanocube embedding into a polymer thin-film and surface modification with a bifunctional polymer graft. Tilted SEM pictures of Nanocube on PS thin-film embedded at (A) 15 nm, (B) 42 nm and (C) 61 nm. Scale bar = 500 nm.

spectroscopy of the treated NC_0 nanocubes (Fig. S3†) confirms the partial removal of PVP and the presence of citric acid on the nanocube surface. Next, the treated NC_0 surfaces were incubated with a solution of PEG dithiol, which serves as the molecular linker. Raman spectroscopy shows complete displacement of both PVP and citrate by the dithiol. Raman spectra also indicate the presence of the terminal –SH group (Fig. S3†), which is available for binding upon exposure to NC_1 nanocubes.

The dithiol-modified nanocubes were then exposed to dispersions of as-made Ag nanocubes with edge lengths of 70 nm, which serve as NC_1. Fig. 3 shows the results of nanocube assembly for three different NC_0 embedding depths: 15 nm, 42 nm and 61 nm. Inspection of the SEM images over an area of 25 μm^2 revealed that with an embedding depth of 15 nm (Fig. 3A), the resulting nanocube assemblies consist of only 16% dimers (Fig. 3D). Additionally, 39% of the meta-atoms are trimers and 36% are larger multimers with NC_1 binding horizontally on the NC_0 sidewalls as well as vertically on the top NC_0 surface. With an increased embedding depth (Fig. 3B and C), we observe a reduction in the population of

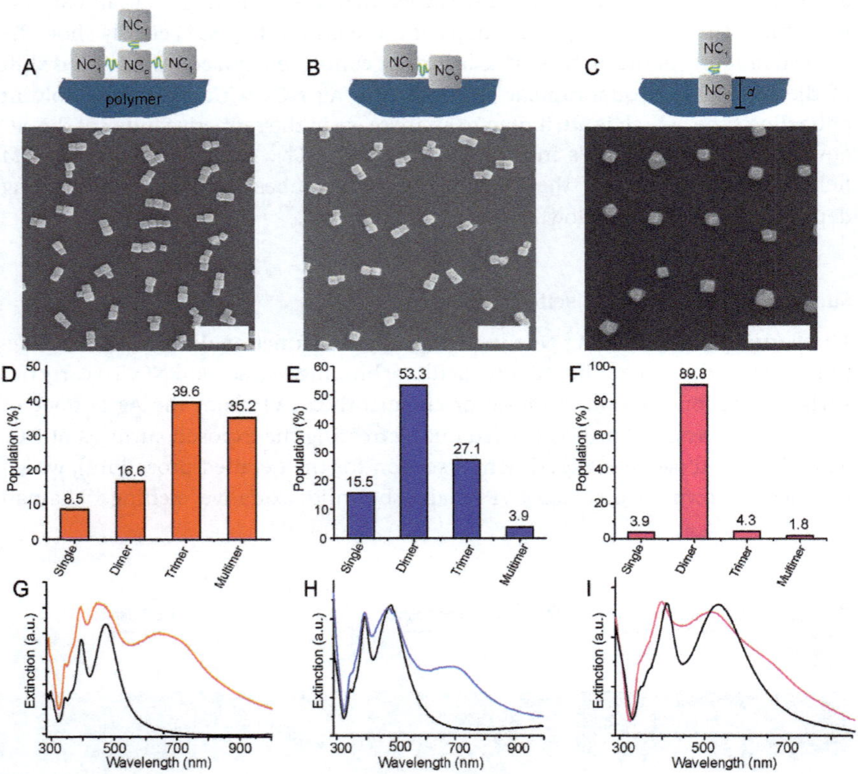

Fig. 3 (A–C) Schematic and SEM images of nanocube assembly observed at embedding depths of 15 nm, 42 nm and 61 nm of NC_0. Scale bar = 500 nm. (D–F) Corresponding statistical analyses showing the population distributions of monomers, dimers, trimers, and multimers in each assembly. (G–I) Corresponding optical extinction spectra taken at incident illumination with a broadband white light source. The extinction spectra of the as-deposited NC_0 nanocubes prior to modular assemblyare shown for reference (black lines).

trimers and multimers and a large increase in the population of nanocube dimers. This is attributed to the decreased surface area available for NC_1 binding. For an embedding depth of 60 nm (Fig. 3C) where nearly all of the NC_o side surfaces are protected by PS, a dimer population of 89% was achieved.

Optical extinction spectra were collected for the meta-atom assemblies at normal incidence and reveal the appearance of new LSPR bands that arise from surface plasmon coupling. This LSPR band appears at wavelengths between 650 and 700 nm for the assembled structures. In addition, we observe a blue-shift of the dipolar and quadrupolar LSPR bands attributed to isolated Ag NC_o nanocubes (black curves). This LSPR shift is correlated to both the embedding depth of NC_o and to plasmon coupling. For an embedding depth of 60 nm, the LSPR peak attributed to a coupled mode at 651 nm (Fig. 3I) is very weak despite the dominant dimer population. This is attributed to the dimer geometry, which is oriented orthogonal to the polymer–air interface of the substrate. This suggests the meta-atom assemblies are highly polarized, depending on the assembly orientation and geometry.

Optical characterization

Using FDTD methods, we modeled the optical response from horizontal (dimer axis oriented parallel to the substrate) and vertical (dimer axis oriented orthogonal to the substrate) dimers. Fig. 4 shows the simulated optical reflectance spectra and representative resonant electric field distributions for horizontal and vertical Ag nanocube dimers where $NC_o = 100$ nm and $NC_1 = 90$ nm. We find that spectral features below 500 nm are correlated with LSPR modes where the electric

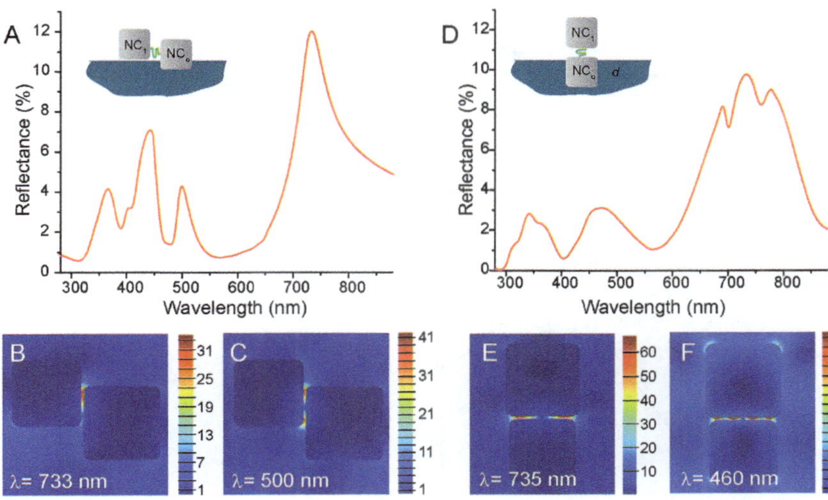

Fig. 4 Simulated optical reflectance and near-field distributions for (A–C) horizontally-assembled and (D–F) vertically-assembled Ag nanocube dimer arrays. In all simulations, $NC_O = 100$ nm Ag nanocubes and $NC_1 = 90$ nm Ag nanocubes. (B and C) Horizontal dimer cross-sections showing the electric-field distributions for resonances at (B) $\lambda = 733$ nm and (C) $\lambda = 500$ nm. (E and F) Vertical dimer cross-sections showing the electric-field distributions for resonances at (E) $\lambda = 735$ and (F) $\lambda = 460$ nm, respectively.

fields are localized on individual nanocubes, whereas features above 500 nm possess electric fields that are highly localized in the interstitial gap between the nanocubes. These gap modes are the result of strong electromagnetic coupling. In the reflectance spectra for both dimer orientations, the most intense and lowest energy reflectance peak correlates with excitation of the fundamental, or lowest-order dipolar gap mode. For the horizontal dimer, this fundamental gap mode is excited at $\lambda = 733$ nm and the optical field is highly concentrated between the corner of one cube and the face of the other cube, as shown in the field distribution map in Fig. 4B. For the vertical dimer, the fundamental gap mode is excited at $\lambda = 735$ nm and the field is highly symmetric within the interstitial gap (Fig. 4E). Reflectance peaks at shorter wavelengths resulting from higher-order gap modes, in some cases, overlap significantly with other LSPR modes. A quadrupolar gap mode is observed for the horizontal dimer at $\lambda = 500$ (Fig. 4C) and for the vertical dimer at $\lambda = 460$ nm (Fig. 4E). In the case of the vertical dimer, interference between multiple reflections within the nanojunction results in peak splitting and broadening of the reflectance peak associated with the fundamental mode.

To confirm these simulation results, we carried out angle- and polarization-dependent reflectance measurements from horizontal dimers with $NC_o = 100$ nm Ag nanocubes at an embedding depth of 42 nm, and $NC_1 = 90$ nm Ag nanocubes, as shown in Fig. 5. Reflectance spectra were recorded at two different incident angles of $\theta = 4°$ (black) and $\theta = 33°$ (red) with both s- and p-polarized light. At incident angle $\theta = 4°$, the LSPR modes resulting from coupling appear as intense reflectance bands at $\lambda = 732$ nm (from the dipolar mode) and at $\lambda = 504$ nm (from the quadrupolar mode) irrespective of s or p (Fig. 5C and D, black curves). This is because the field components of both s- and p-polarized light match with the

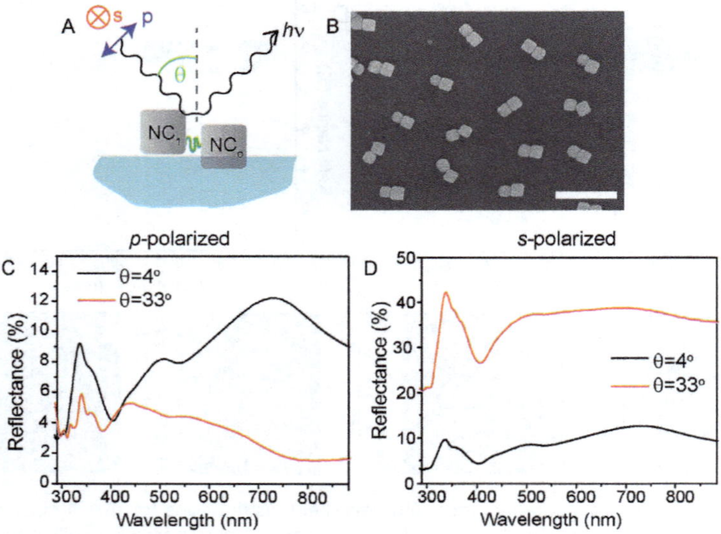

Fig. 5 (A) Schematic of the optical reflectance measurements showing the directions of s- and p-polarized light for a horizontal nanocube dimer as shown in the SEM image in (B). (C and D) Corresponding optical reflectance spectra taken at incident angles of $\theta = 4°$ (black) and $\theta = 33°$ (red) with p- and s-polarized light, respectively.

geometric axes of the horizontal dimers, exciting the fundamental gap mode of the nanocube pair. At incident angle $\theta = 33°$, illumination with p-polarized light results in the disappearance of this coupled LSPR mode, since the direction of polarization is no longer aligned with the nanocube dimer axes (Fig. 5C, red curve). For the same angle of illumination with s-polarized light, we observe a large increase in overall reflectance, since the polymer thin-film behaves as a standard dielectric interface. The dipolar coupled LSPR mode is still not excited due to the mismatch between polarization and dimer orientation (Fig. 5D, red curve).

Fig. 6 shows the angle- and polarization-dependent reflectance spectra for vertical nanocube dimers with NC_o = 100 nm Ag nanocubes at an embedding depth of 60 nm, and NC_1 = 90 nm. At near-incidence illumination angles of $\theta =$ 4°, both s- and p-polarization show limited evidence of plasmon coupling and only the LSPR modes associated with individual nanocubes are excited. For these dimers, illumination with p-polarized light at an incident angle of $\theta = 33°$ is the only configuration that can give rise to the coupled dipole–dipole LSPR mode. This is observed in Fig. 6C (red curve) at $\lambda = 735$ nm, where the intensity of the nanocube gap mode dominates almost the entire reflectance spectrum. For s-polarized light and an incident angle of $\theta = 33°$, the reflectance spectrum gives rise to a broadband feature in the longer wavelengths, similar to what is observed for horizontal dimers.

Finally, we demonstrate that the LSPR wavelength and strength of the coupled mode can be tuned by varying the NC_1 size. Fig. 7 shows the optical extinction spectra for horizontal nanocube dimers composed of NC_o = 100 nm Ag nanocubes at an embedding depth of 42 nm, and NC_1 = 64 nm, 70 nm, and 90 nm. For NC_1 = 64 nm, the coupled LSPR mode appears at $\lambda = 677$ nm (Fig. 7A). As the NC_1

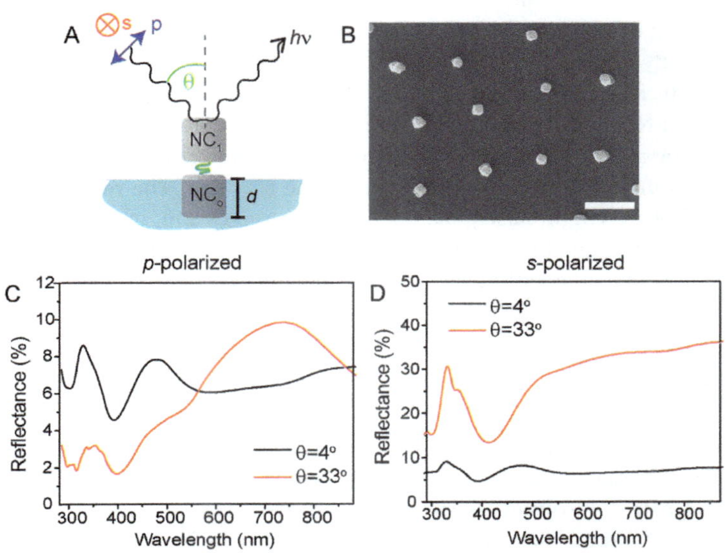

Fig. 6 (A) Schematic of the optical reflectance measurements showing the directions of s- and p-polarized light for a vertical nanocube dimer as shown in the SEM image in (B). (C and D) Corresponding optical reflectance spectra taken at incident angles of $\theta = 4°$ (black) and $\theta = 33°$ (red) with p- and s-polarized light, respectively.

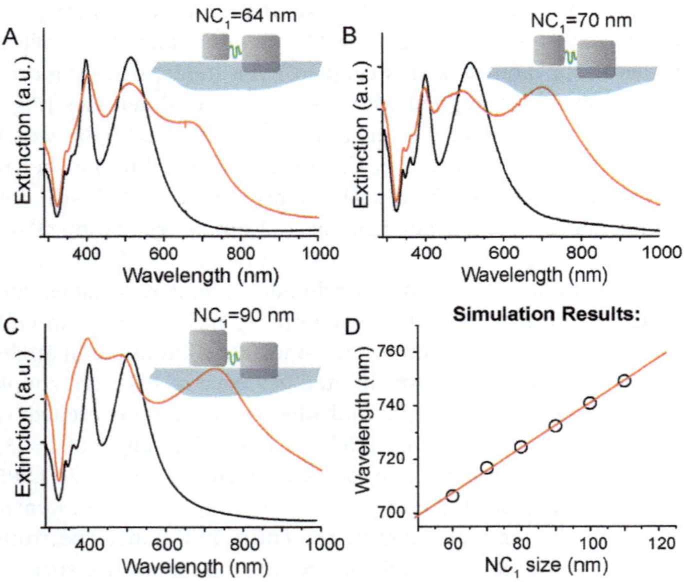

Fig. 7 (A–C) Extinction spectra of horizontal dimers with varying NC_1 nanocube sizes of (A) 64 nm, (B) 70 nm, and (C) 90 nm. The extinction spectra of the as-deposited NC_o nanocubes prior to modular assembly are shown in black for each sample. (D) Plot of the shift in dipole–dipole coupling mode with respect to the NC_1 size, obtained from simulated reflectance spectra.

size is increased, the wavelength of this fundamental mode red-shifts and increases in extinction intensity. The LPSR mode was red shifted to $\lambda = 700$ for $NC_1 = 70$ nm cubes (Fig. 7B) and to $\lambda = 726$ nm for $NC_1 = 90$ nm cubes (Fig. 7C). The intensity of this coupled mode is correlated to the strength of light confinement in the dimer gap. Fig. 7D shows simulation results for horizontal dimers with $NC_o = 100$ nm Ag nanocubes and $NC_1 = 60$–110 nm Ag nanocubes separated by a gap of 3 nm. The wavelength position of the lowest-order fundamental gap mode is plotted with respect to the NC_1 size. The redshift in reflectance peak wavelength closely matches a linear regression, which is consistent with the electromagnetic response for dipole–dipole coupling. We also find that the strongest optical coupling occurs for nanocube pairs where size mismatch is minimal.

Conclusions

We demonstrate that meta-atoms can be fabricated using a modular, polymer-templated approach that can be readily scaled over large surface areas. In particular, nanocube dimer arrays can be fabricated over a few square centimeters with yields of almost 90%. This is a promising proof-of-concept demonstration that self-assembly is a viable technique for bottom-up, polymer-based metamaterials. Furthermore, our method has the potential to achieve the assembly of more complex meta-atom structures by assembling NCs in a combinatorial fashion, or by carrying out multiple layer-by-layer assembly.

Acknowledgements

This work is funded by the Office of Naval Research (Award No. N000141210574) and a grant from the National Science Foundation (CMMI, Award No. 1200850).

References

1 A. Pors, O. Albrektsen, I. P. Radko and S. I. Bozhevolnyi, *Sci. Rep.*, 2013, **3**, 2155.
2 S. Sun, K.-Y. Yang, C.-M. Wang, T.-K. Juan, W. T. Chen, C. Y. Liao, Q. He, S. Xiao, W.-T. Kung, G.-Y. Guo, L. Zhou and D. P. Tsai, *Nano Lett.*, 2012, **12**, 6223–6229.
3 X. Ni, N. K. Emani, A. V. Kildishev, A. Boltasseva and V. M. Shalaev, *Science*, 2012, **335**, 427.
4 B. Kanté, A. de Lustrac and J. M. Lourtioz, *Phys. Rev. B: Condens. Matter Mater. Phys.*, 2009, **80**, 035108.
5 G. Dolling, C. Enkrich, M. Wegener, J. F. Zhou, C. M. Soukoulis and S. Linden, *Opt. Lett.*, 2005, **30**, 3198–3200.
6 N. Liu, H. Guo, L. Fu, S. Kaiser, H. Schweizer and H. Giessen, *Nat. Mater.*, 2008, **7**, 31–37.
7 J. Valentine, S. Zhang, T. Zentgraf, E. Ulin-Avila, D. A. Genov, G. Bartal and X. Zhang, *Nature*, 2008, **455**, 376–379.
8 C. M. Soukoulis and M. Wegener, *Nat. Photonics*, 2011, **5**, 523–530.
9 C. L. Holloway, D. C. Love, E. F. Kuester, A. Salandrino and N. Engheta, *Microwaves, Antennas & Propagation, IET*, 2008, **2**(2), 120–129.
10 A. L. Koh, A. I. Fernández-Domínguez, D. W. McComb, S. A. Maier and J. K. W. Yang, *Nano Lett.*, 2011, **11**, 1323–1330.
11 M. Bresin, N. Nehru and J. T. Hastings, *SPIE MOEMS-MEMS. International Society for Optics and Photonics*, 2013, 861306–861306.
12 Y. Xia, J. A. Rogers, K. E. Paul and G. M. Whitesides, *Chem. Rev.*, 1999, **99**, 1823–1848.
13 R. Jin, Y. Cao, C. A. Mirkin, K. L. Kelly, G. C. Schatz and J. G. Zheng, *Science*, 2001, **294**, 1901–1903.
14 B. Nikoobakht and M. A. El-Sayed, *Chem. Mater.*, 2003, **15**, 1957–1962.
15 A. Tao, P. Sinsermsuksakul and P. Yang, *Angew. Chem., Int. Ed.*, 2006, **45**, 4597–4601.
16 Y. Xia, Y. Xiong, B. Lim and S. E. Skrabalak, *Angew. Chem., Int. Ed.*, 2009, **48**, 60–103.
17 J. J. Mock, M. Barbic, D. R. Smith, D. A. Schultz and S. Schultz, *J. Chem. Phys.*, 2002, **116**, 6755–6759.
18 Z. Nie, A. Petukhova and E. Kumacheva, *Nat. Nanotechnol.*, 2009, **5**, 15–25.
19 G. M. Whitesides and B. Grzybowski, *Science*, 2002, **295**, 2418–2421.
20 C. Ciracì, R. T. Hill, J. J. Mock, Y. Urzhumov, A. I. Fernández-Domínguez, S. A. Maier, J. B. Pendry, A. Chilkoti and D. R. Smith, *Science*, 2012, **337**, 1072–1074.
21 B. Gao, M. J. Rozin and A. R. Tao, *Nanoscale*, 2013, **5**, 5677–5691.
22 M. Rycenga, C. M. Cobley, J. Zeng, W. Li, C. H. Moran, Q. Zhang, D. Qin and Y. Xia, *Chem. Rev.*, 2011, **111**, 3669–3712.
23 N. I. Landy, S. Sajuyigbe, J. J. Mock, D. R. Smith and W. J. Padilla, *Phys. Rev. Lett.*, 2008, **100**, 207402.

24 A. V. Kildishev, A. Boltasseva and V. M. Shalaev, *Science*, 2013, **339**, 1232009.
25 D. Nykypanchuk, M. M. Maye, D. van der Lelie and O. Gang, *Nature*, 2008, **451**, 549–552.
26 S. Y. Park, A. K. R. Lytton-Jean, B. Lee, S. Weigand, G. C. Schatz and C. A. Mirkin, *Nature*, 2008, **451**, 553–556.
27 P. Akcora, H. Liu, S. K. Kumar, J. Moll, Y. Li, B. C. Benicewicz, L. S. Schadler, D. Acehan, A. Z. Panagiotopoulos, V. Pryamitsyn, V. Ganesan, J. Ilavsky, P. Thiyagarajan, R. H. Colby and J. F. Douglas, *Nat. Mater.*, 2009, **8**, 354–359.
28 R. Shenhar, T. B. Norsten and V. M. Rotello, *Adv. Mater.*, 2005, **17**, 657–669.
29 K. K. Caswell, J. N. Wilson, U. H. F. Bunz and C. J. Murphy, *J. Am. Chem. Soc.*, 2003, **125**, 13914–13915.
30 G. A. DeVries, M. Brunnbauer, Y. Hu, A. M. Jackson, B. Long, B. T. Neltner, O. Uzun, B. H. Wunsch and F. Stellacci, *Science*, 2007, **315**, 358–361.
31 C.-L. Chen, P. Zhang and N. L. Rosi, *J. Am. Chem. Soc.*, 2008, **130**, 13555–13557.
32 Z. Nie, D. Fava, E. Kumacheva, S. Zou, G. C. Walker and M. Rubinstein, *Nat. Mater.*, 2007, **6**, 609–614.
33 T. Ming, X. Kou, H. Chen, T. Wang, H.-L. Tam, K.-W. Cheah, J.-Y. Chen and J. Wang, *Angew. Chem., Int. Ed.*, 2008, **47**, 9685–9690.
34 F. Li, W. C. Yoo, M. B. Beernink and A. Stein, *J. Am. Chem. Soc.*, 2009, **131**, 18548–18555.
35 V. E. Ferry, L. A. Sweatlock, D. Pacifici and H. A. Atwater, *Nano Lett.*, 2008, **8**, 4391–4397.
36 R. A. Pala, J. White, E. Barnard, J. Liu and M. L. Brongersma, *Adv. Mater.*, 2009, **21**, 3504–3509.
37 D. R. Smith, W. J. Padilla, D. C. Vier, S. C. Nemat-Nasser and S. Schultz, *Phys. Rev. Lett.*, 2000, **84**, 4184.
38 V. M. Shalaev, W. Cai, U. K. Chettiar, H.-K. Yuan, A. K. Sarychev, V. P. Drachev and A. V. Kildishev, *Opt. Lett.*, 2005, **30**, 3356–3358.
39 V. M. Shalaev, *Nat. Photonics*, 2007, **1**, 41–48.
40 B. Gao, G. Arya and A. R. Tao, *Nat. Nanotechnol.*, 2012, **7**, 433–437.
41 B. Gao, Y. Alvi, V. Li and A. R. Tao, *CrystEngComm*, 2014, **16**, 9434–9440.
42 Y. Sun and Y. Xia, *Science*, 2002, **298**, 2176–2179.
43 M. E. Mackay, A. Tuteja, P. M. Duxbury, C. J. Hawker, B. Van Horn, Z. Guan, G. Chen and R. S. Krishnan, *Science*, 2006, **311**, 1740–1743.
44 Y. Xiong, *Chem. Commun.*, 2011, **47**, 1580–1582.

Faraday Discussions

DISCUSSIONS

Applications to Soft Matter: general discussion

Vinothan N. Manoharan, Guruswamy Kumaraswamy, Madivala G. Basavaraj, Sanat Kumar, Siddharth Kulkarni, Ranjini Bandyopadhyay, Sudeep Punnathanam, Ajeet Srivastav, Lola Gonzalez-Garcia, G V Pavan Kumar, Daan Frenkel, Radhika Poojari, Priyadarshi Roy Chowdhury, Gourav Shrivastav, Mukta Tripathy, Vimala Sridurai, Alison Edwards, Madhura Som, B. L. V Prasad, Lynn Walker, Neena S. John, Yon Ju-Nam, Yogesh M. Joshi, Nirmalya Bachhar, Charusita Chakravarty, Rajdip Bandyopadhyaya, Zakiya Shireen, Nicholas Kotov, Oleg Gang, Alamgir Karim, Mario Tagliazucchi, Erika Eiser and Andrea Tao

DOI: 10.1039/C6FD90005D

Gourav Shrivastav communicated:† The authors have shown the self assembly of the particles *via* oscillatory forces by periodical change in pH. 1. How does the rate of ionization of acids and bases affect the result?

2. Does slow oscillation break the dimer formed and therefore is a fast pH required to discriminate between the dimers and free particles ?

Mario Tagliazucchi communicated in reply: 1. We don't believe that the rate of ionization of weak acids or bases will impact the dynamics of our system. In our previous work[1] we have shown that the pH oscillations should have a timescale ranging from microseconds to seconds (depending on the size of the colloids) in order to obtain self-assembled structures. Acid–base neutralization is usually faster than this timescale.

2. Slow oscillations will indeed break the dimers into free particles. For that reason, fast oscillations (*i.e.* oscillations faster than the diffusional timescale of the particles) are required to obtain the dimers as well as the other self-assembled morphologies.

1 M. Tagliazucchi, E. A. Weiss and Igal Szleifer, *PNAS*, 2014, **111**, 9751–9756.

Charusita Chakravarty communicated: The authors use Brownian dynamics simulations of charged particles with weak acidic and basic character and oscillating pH to generate interesting self-assembled structures under far from

† The paper by I. Szleifer was not presented at the meeting, but questions and responses have been provided afterwards.

equilibrium conditions. What would be the possible experimental realizations of this simulation?

Could varying the salt concentration in the colloidal clay systems studied in the articles by Joshi and Bandhyopadhyay be used for this type of non-equilibrium self-assembly?

Ranjini Bandyopadhyay communicated in reply: There are, indeed, important similarities between the system discussed in paper 7909 and the clay suspensions studied experimentally in papers 8549 (by Ali and Bandyopadhyay) and 8570 (by Jatav and Joshi). For example, in both the clay suspensions (Sodium Montmorillonite, Na-MMT, and Laponite) discussed in 8549 and 8570, structures form spontaneously due to the details of the time-dependent interactions between the constituent particles/particle stacks (tactoids). The sensitivity of the assembled structures to the inter-particle potentials has been discussed in detail in paper 7909. There also exist important differences between the system studied in 7909 and the clay suspensions studied in 8549 and 8570. Clay tactoids (stacks), which invariably exist in aqueous suspensions, break up spontaneously and release, during this exfoliation process, sodium ions into the bulk medium. Since tactoid exfoliation is a largely uncontrolled process, controlling the details of the inter-particle potentials in clay suspensions is a difficult task. Furthermore, in the clay suspensions in papers 8549 and 8570, strong inter-particle interactions drive the systems toward kinetic arrest, with the suspensions forming attractive gels in the presence of salt in the bulk medium, or glassy suspensions that are dominated by inter-particle repulsive interactions in the absence of externally added salt. This unavoidable process of kinetic arrest is seen in both clay suspensions and is absent in the model system studied in paper 7909. It is also to be noted that the presence of strong charge polydispersity in the Laponite system studied in paper 8570 and the presence of charge, size and shape polydispersities in the Na-MMT system studied in paper 8549, rule out the possibility of the formation of ordered structures such as those reported in paper 7909.

Mario Tagliazucchi communicated in reply: The possible experimental realizations of this system were discussed in our recent published work.[1] Briefly, we first estimated the timescale of the pH oscillations required for dissipative self-assembly. This timescale is in the second–microsecond range for colloids in the nanometer–micrometer size range. We believe that pH oscillations can be achieved in this timescale using photoacids[2] or electrochemical pH control.[3] The main challenge in such experiments would be to achieve and maintain a sufficiently low ionic strength (in our system, the Debye length of the solution should be similar to the size of the particles in order to obtain the self-assembled structures). Oscillating the salt concentration of the clay systems of Joshi and Bandhyopadhyay may lead to dissipative structures if the timescale of the oscillations is faster than the diffusional timescale of the particles $\sim d^{2/D}$, where d is the size of the colloid and D is its diffusion coefficient. The structures that may form will be different from ours, since the system is different.

1 M. Tagliazucchi, E. A. Weiss and Igal Szleifer, *PNAS*, 2014, **111**, 9751–9756.
2 R. M. D. Nunes, M. Pineiro and L. G. Arnaut, *J. Am. Chem. Soc.*, 2009, **131**, 9456–9462.
3 R. Shacham, D. Avnir Daniel Mandler, *Adv. Mater.*, 1999, **11**, 384–388.

Priyadarshi Roy Chowdhury communicated: "The colloidal molecules are dissipative structures that cannot be obtained in the absence of pH oscillations". Could you discuss this comment in more detail.

Mario Tagliazucchi communicated in reply: In the conditions of composition and density where we observe colloidal molecules, well-defined colloidal molecules are absent in fixed pH simulations. Thus, the colloidal molecules only form in the presence of pH oscillations and they are dissipative structures. This result does not preclude, of course, that the same colloidal molecules can form for other systems (*i.e.* particles interacting through other type of potentials) or even in the same system for other conditions.

Lynn Walker opened the discussion on the paper by Madivala G. Basavaraj: The mechanism for adsorption of colloidal particles at fluid interfaces is one of pinning due to formation of a solid–fluid contact line. This should not decrease the interfacial tension at low coverages. So why is the drop in interfacial tension with time interpreted as adsorption of particles in one case (one pH condition) and not in another (another pH condition). I wonder if the interfacial tension is actually responding to the adsorption of some other species in the system?

Madivala G. Basavaraj replied: If we compare the dynamic surface tension data for a pure water–decane interface and pH=2 case, it can be argued that the ionic species in the suspension (at pH=2), even if they adsorb on to the interface, do not reduce the surface tension. In our experiments, the drop in surface tension is observed when the pendant hematite suspension drops at pH of 4.0 and 6.5 are present in decane. We argue that this drop is due to particle adsorption to the suspension drop–decane interface. Since there is no change in surface tension with time for pH=2, we infer no particle adsorption. We refer to our recent work,[1] where we observe similar results for the adsorption of silica nanoparticles, where the electrostatic double layer interactions are screened by the addition of sodium chloride. We have not looked at the adsorption of other ionic species present in the system at this point. However, in the silica nanoparticles system, if NaCl present in the suspension drop was to adsorb at the interface, then the interfacial tension should have increased.[2] As the decrease in surface tension is observed in two independent studies this is a result of adsorption of particles due to low surface charge density.

1 V. R. Dugyala, J. S. Muthukuru, E. Mani and M. G. Basavaraj, *Phys. Chem. Chem. Phys.*, 2016, **18**, 5499.
2 reference: Chapter 3, Arthur W. Adamson and Alice P. Gast, Physical Chemistry of Surfaces, Wiley-Blackwell.

Vinothan N. Manoharan addressed Madivala G. Basavaraj and Lynn Walker: Following up on Prof. Walker's question on the observed time-dependence of the dynamic surface tension: ellipsoids adsorbed at the interface must distort the shape of the interface, and as a result there are capillary interactions between adsorbed particles. When the particles are close enough, their capillary distortions overlap, reducing the surface energy. Could the change in the surface tension as a function of time reflect a transition between the adsorption of isolated ellipsoids (where their capillary distortions do not overlap) and the adsorption of ellipsoids close to other particles, such that they interact with one another?

Lynn Walker replied: I wonder if this would be a way to estimate a surface concentration? Since the geometry of the particles is known, is a calculation of the surface concentration needed to have overlap of the capillary distortion possible? It would certainly be an estimate, but it could back up the dynamics.

Madivala G. Basavaraj responded: In the dynamic surface tension data, initially the decrease is relatively steep followed by a more gradual change. If the transition from the adsorption of isolated ellipsoids to multiple ellipsoids (close to each other and associated interface deformation overlap) were to cause the change in surface tension with time, then the surface tension should start to change after there are enough ellipsoids at the interface, which is not the case. Therefore throughout the dynamic surface tension measurements there is particle adsorption, however, the rate of adsorption depends on several factors which are discussed in the manuscript.

Lynn Walker remarked: As a follow up to my earlier question, I think that the dynamics of the interfacial tension might contain more information than is being considered. The dynamics are quite rich. If this is particle adsorption as Basa argues, then what can be learned from the timescales (the reaction-diffusion equation with an EOS like Volmer might be appropriate)? If this is adsorption of another species or even replacement of counterions near the interface, then that could also be interesting in understanding the phenomenon.

Madivala G. Basavaraj responded: I completely agree. It would be very interesting to look at the dynamics in more detail – especially – contact line dynamics, possible adsorption of ionic species, and accumulation of ions close to the interface. This would add to a much better and complete understanding of the adsorption of particles to fluid interfaces.

Neena S. John commented: In the case of isotropic particles, the effective radius is considered to calculate the interfacial free energy factor. For the present case of ellipsoids, do you consider the 'aspect ratio' for the calculations?

Also, will the ellipsoids assemble at a planar interface of decane–water in a similar manner?

Madivala G. Basavaraj answered: In the calculation of interfacial free energy, the area that the particle occupies at the interface is considered, that is how the radius of spherical particle figures in these calculations. The area is typically the area of intersection of the sphere and the interface plane. As long as this area is known, such calculations can be made for particles of any regular or arbitrary shape.

For ellipsoids, there are shape induced interface deformations, therefore, this area of intersection is not a perfect ellipse, but an undulated ellipse. Therefore, one has to consider this effect. Such deformations have been measured for ellipsoids and rod-shaped particles. Considering these effects in interface free energy calculations will involve more tedious calculations. Moreover, for complex shaped particles there are no analytical expressions for the calculation of the area of intersection. The behaviour of ellipsoids at interfaces is known to be very different from that of spheres. Ellipsoids that are homogeneous (not Janus) are know to position themselves flat at the interface – with their major axis parallel to the interface plane. For ellipsoids, due to long range capillary forces, both the structure and interfacial properties are known to be aspect ratio dependent.

Guruswamy Kumaraswamy asked: Could you please comment on the change in wettability of your haematite particles with a change in pH? Could this be sufficient to account for their pH-dependent segregation at the oil–water interface, without having to resort to the image charge argument?

Madivala G. Basavaraj responded: The idea of a possible change in wettability of hematite particles with a change in pH is an interesting one. I wonder if there are any tools to measure such wettability changes in solution. We have measured the contact angle of micron sized hematite particles (of different shapes – such as spherocylinders, cubes, dumbbells) at the water–decane and water–air interface by considering an aqueous sub-phase maintained at pH=6.5. The contact angles are obtained from the analysis of microscopy images obtained *via* a gel trapping technique. The values of the three-phase contact angles obtained irrespective of particle shape are ~50 and ~40 degrees, respectively. This suggests that the particles are still partially hydrophilic. It would be interesting to carry out wetting studies by considering aqueous phases at different pH values to see if there is any significant change in particle wettability.

B. L. V Prasad asked two questions: 1. Are the particles that are left to equilibrate in the system for 1 hr assuming any Janus character? Can this phenomenon explain the observations? 2. Please comment on why 1 hr is needed? What could be happening in this 1 hr?

Madivala G. Basavaraj answered: In our drop compression experiments, 1 hr of equilibration time is used to provide enough time for the particles to diffuse and eventually adsorb to the interface. This time period helped us to clearly demonstrate the buckling phenomena in the suspension drops at pH=6.5. In principle, buckling is expected to occur even if the equilibration time is lower, however, at a higher compression ratio. There is no evidence to support the hypothesis that the particles would assume Janus character. Since there is no chemical or physical charge to the particle during our experiments, particles assuming Janus character is very much unlikely.

Vinothan N. Manoharan commented: Another reason that the surface tension might evolve in time is contact-line pinning. Experiments from my group[1] show that micrometer-scale particles can take months to years to reach their equilibrium position at a fluid–fluid interface. We attribute the slow relaxation to pinning of the contact line on defects on the particle surfaces. In a dense monolayer of such particles, the surface energy might gradually decrease over time.

1 D. M. Kaz, R. McGorty, M. Mani, M. P. Brenner and V. N. Manoharan, *Nature Mat.*, 2012, **11**, 138–142.

Madivala G. Basavaraj responded: In the present study we use ellipsoids that are about 250 nm in length (major axis) and 50 nm in diameter (minor axis). It is clear from our experiments and earlier reports that the orientation of homogenous (in terms of surface chemistry) ellipsoids is such that their major axis is parallel to the water–decane interface. Thus, the relevant lengthscale is the length of the minor axis of ellipsoids and associated contact line. The lengthscale and timescale associated with contact-line pinning and equilibration period for the particles used in the present study are clearly much smaller than that for micron-sized particles. The timescale of the experiment which is of the order of hours seems much longer, however, the measurement of contact line dynamics for smaller particles will help conclude if surface tension evolution due to contact line pinning is indeed the reason for the decrease in surface tension observed. Since the adsorption of particles is energetically favored there is strong reason to believe that the adsorption of particles does decrease the interfacial tension. This in our study is confirmed from the buckling experiments and SEM micrographs.

Ajeet Srivastav remarked: How would the shape of the nanoparticles (from spherical to cylindrical), in particular the nanowires, affect the observations with the higher interfacial area available to interact with?

Madivala G. Basavaraj answered: In general, interfaces coated with anisotropic particles such as ellipsoids or graphene sheets are shown to be solid-like and show interesting buckling behavior. There are reports on the compression of planar monolayers of rod-like particles, however, we are not aware of any pendant drop buckling experiments.

Ranjini Bandyopadhyay opened the discussion on the paper by Lynn M. Walker: Since the amphiphilicity of the middle blocks of the triblock copolymers are temperature-dependent, have you looked at the temperature-dependence of nanoparticle transport in your experiments? Also, would you comment on the release of the nanoparticles at specified targets?

Lynn Walker answered: Temperature is an important variable in these systems. For this work, we chose to use concentration to vary the structure since temperature would alter the PEO/water environment and add another complication. Isothermal experiments allowed us to compare to polymer theories for diffusion in semidilute polymer solutions. We have done other work with temperature, but this adds in new micelle transitions (from spheres to cylinders) and phase boundaries. This would be the next stage of the work and we have not yet designed the appropriate study. To get the nanoparticulate phase out of the nanostructured material, we could either cool the system below the CMT ($\sim$15 °C) or dilute with solvent (water). This would result in a polymer/nanoparticle dispersion which could be separated with centrifugation or filtration. It is worth noting that the block polymer chosen is FDA approved for many uses inside the body, so separation might not be necessary.

G V Pavan Kumar asked: Can you arrest the nanoparticles and further transfer them on to a substrate ?

Lynn Walker replied: It should be possible. However, we have purposely focused on tools to allow for structuring nanoparticles in three dimensions. I think that there are probably better ways (lithography and directed assembly) to do things on a substrate. Here, we wanted to find ways to use self assembly in three dimensions.

Alison Edwards commented: It is self-evident that the materials you are discussing which give clear diffraction patterns that you have shown here are crystalline under the conditions of your experiment. Your patterns are reminiscent of those obtained both by experiment and in analysis of diffuse scattering from 'hard' materials – especially those published by Richard Welberry. I would note that it is the long range order which gives rise to the sharp diffraction spots definitive of what is considered to be crystallinity, whereas the 'disorder' at short lengthscales gives the intensity outside the Bragg peaks – of which you have a great deal. Intriguingly, diffuse scatter does not drop away with 2-theta – do your patterns continue out in reciprocal space (or Q if that is your preferred measure)? The concepts developed by Prof. Welberry could be extended to make use of the data you have available in these patterns and there are some quite old modelling methods which can also be applied to such data to extract models for such materials.

Lynn Walker answered: We will follow up and thank you for the reference to Welberry's work. We come to this field from the structured fluid world and are realizing the importance of approaching these as crystalline materials (albeit 'soft' ones with significant form factor effects). Our patterns do continue in q space but only for 3–4 peaks. Using the proper conditions, we are able to align the power with shear to see 'single crystal' patterns. For some of these, we see higher order peaks. However, the peak height drops off quickly and the incoherent background comes up quickly. We usually have 3 orders of magnitude in intensity and that is often our limiting factor in peak resolution. As we have determined the conditions for 'annealing' the crystals, our scattering patterns are becoming even more crystalline and we will definitely approach these from the crystalline point of view!

Yogesh M. Joshi said: At 25°C how sharp is the liquid to crystalline transition for pluronic F127 when the concentration is increased? Is there any biphasic region where both crystal and liquid states coexist? Depending upon how fast the temperature of 25°C is attained, could there be a possibility of having a disordered glassy structure in Pluronic F127?

Lynn Walker replied: If you look at Fig. 1 in our manuscript, you will note that we avoided samples at the C_{gl} point (the concentration at 25°C of the transition). This is because the results are much more time dependent in this region. For all of our work, we were careful to slowly warm the samples from the liquid to the solid state and then allow for significant annealing time. So, we are confident that the results reported are for materials that are crystalline. It is possible to trap these materials in a glassy state by rapid heating from the liquid to the solid. It is also possible to create metastable structures by cooling/heating rapidly through the OOT seen in many of these systems. Our paper[1] discusses these issues with structural measurements. We have not yet addressed the issue of nanoparticle diffusion in a kinetically trapped system.

1 T. A. Lafollette and L. M. Walker, *Polymers*, 2011, **3**, 281.

Sudeep Punnathanam asked: In cases where the mobility of the nanoparticles is severely limited, are you able to crystalize structures that are similar to those observed in substitutionally ordered alloy systems?

Lynn Walker responded: This is a good question and one for which I do not have an answer. There is no reason that this would not work and we are interested in investigating it. The nanoparticles in the hindered cases are held in the interstitial pockets by steric forces, so we lack the specific interactions in colloidal aggregates. I suspect that we would be able to form clusters of a specific size (set by the block copolymer geometry), but I am not sure that we would generate the highly ordered structures reported in oppositely charged nanoparticle or colloidal crystals and alloys.

Siddharth Kulkarni asked: By increasing the particle size above the critical micelle size (characteristic repeat distance), is it possible to see the particles at the grain boundaries? Further, is it possible that such particles partitioned at the boundaries can form a network and enhance the viscoelastic properties of the triblock copolymer matrix?

Lynn Walker replied: We did look at different particle sizes. If the particle size is significantly larger than the unit cell size (micelle size), then the particles sample the bulk behavior of the material. We see no templating or location of the particles coincident to the block copolymer matrix or any grain boundaries.[1] Our earlier templating work does not suggest partitioning to grain boundaries. We do see the spacing of the nanoparticles picking up the crystal structure of the block copolymer micelles. In this work, we sheared one sample to form a single crystal and saw no difference in diffusivity of the nanoparticles. This is indirect evidence, but adds to our body of work that suggests that particles do not partition to grain boundaries in this system. I should also note that we have not been able to determine a domain size for these materials; it requires very little shear to align these into near single crystals.

One difference to note is that this is a highly solvent swollen system. Most work on particles dispersed in block polymers has focused on melts. In a melt system, the energy difference between a grain boundary and the organized micelles is probably considerably more than in this solvent swollen system.

1 D. C. Pozzo, K. R. Hollabaugh and L. M. Walker, *J. Rheol.*, 2005, **49**, 759–782.

Guruswamy Kumaraswamy queried: Could you please comment on the diffusivity of the fluorescent probes in your ordered micellar systems relative to semidilute PEO solutions where the mesh size is comparable to PEO in the micellar phase?

Lynn Walker answered: We thought about this comparison a great deal. Fig. 8 in the manuscript has a prediction for the diffusivity for PEO solutions equivalent to the coronas in the block copolymer solution. We used the molecular weight of the PEO block in the copolymer and calculated concentration using the volume available without the PPO cores. What we see is that the diffusivity in the block copolymer is significantly slower than what would be observed in the equivalent PEO solution. We compared to experimental data (Fig. 8) and a theoretical model for diffusion of nonsticky spheres through unentangled polymer solutions (Fig. 9). The comparison differs slightly, but the overall results are similar in that the nanoscale structure of the block copolymer solution causes a decrease in transport beyond just transport through the corona.

Priyadarshi Roy Chowdhury communicated: Does the complexity of the self-assembly of block co-polymers cause any hindrance to the nanoparticle diffusivity?

Lynn Walker communicated in reply: Yes. This was the point of the work to show that the nanoscale structure is what controls mobility of the nanoparticulate phase dispersed in the block copolymer solution. The findings show that it is possible but that tuning the diffusivity requires understanding the details of both the static and dynamic nanoscale structure.

Madivala G. Basavaraj opened the discussion on the paper by Ranjini Bandyopadhyay: In your analysis of clay particles you mention that the particles are polydisperse and often exist as stacks. I am curious if particles in each stack are identical. If they are, would it be possible to separate them based on size and obtain monodisperse particles?

Ranjini Bandyopadhyay responded: It can be seen from the SEM image in Fig. 1 of our paper that the sodium montmorillonite (Na-MMT) clay particles are polydisperse, irregularly shaped and flexible. Na-MMT tactoids can therefore not be treated as monodisperse stacks. One of the aims of our work was to obtain a clearer understanding of the sedimentation of natural clay gels in the context of geophysical phenomena such as the formation of river deltas. Ordered clay stacks are unlikely in natural clay suspensions and therefore not desirable in our experiments. That said, it must be mentioned that Na-MMT tactoids exfoliate into smaller entities due to the hydration of the interlayer sodium ions which push the negatively-charged particles apart, ultimately resulting in an accelerated exfoliation process. Na-MMT can also be exfoliated by the intercalation of polymer chains which screen the electrostatic interaction between the clay particles, triggering exfoliation into smaller tactoids/stacks and ultimately to single entities. However, this process is extremely sensitive to the concentration of clay in the aqueous suspension.[1] Depending on the sample preparation protocol, there can also be a tactoid-reformation process which results in the tactoids growing in size.[2] Clay tactoids are more likely to eventually exfoliate into single clay particles if the suspension is diluted substantially. Once the tactoids have completely exfoliated, it is presumably possible to use some size segregation technique such as centrifugation to separate clay particles according to size. Complete exfoliation of all tactoids followed by size-separation may therefore be an easier way to size-segregate particles than by first making monodisperse stacks/tactoids followed by exfoliation.

1 S. Ali and R. Bandyopadhyay, *Langmuir*, 2013, **29**, 12663–12669.
2 S. Ali and R. Bandyopadhyay, *Appl. Clay Sci.*, 2015, **203**, 85.

Yogesh M. Joshi remarked: Where do the samples studied by you lie on the Na-Montmorillonite–water phase diagram. It is interesting to see that the system studied by you shows sedimentation. Such sedimentation is also observed for the Laponite suspension[1] wherein the average concentration of the clay rich region was independent of Laponite concentration. Do you also observe a similar phenomenon?

1 P. Mongondry, J. F. Tassin and Taco Nicolai, *J. Colloid Interface Sci.*, 2005, **283**, 397–405.

Ranjini Bandyopadhyay replied: The concentration of sodium montmorillonite (Na-MMT) in the samples studied in our paper is far below the concentration required to form a soft glassy sample in the absence of salt. Soft glassy behaviour in aqueous Na-MMT suspensions is first observed at 4 wt% in the absence of salt.[1] In our paper, the concentration of Na-MMT in the Na-MMT gels is fixed at 0.8 wt%. Without salt, therefore, these clay suspensions would have remained in a sol state. There are several similarities in the settling behaviour of Na-MMT gels (reported in our paper) and Laponite gels in the presence of salt.[2] In both cases, sedimentation is observed at low clay concentrations, with the settling speed depending on C_s, the concentration of salt. As mentioned earlier, we are in the low clay concentration (C) regime, and explore the structures and sedimentation behaviours of Na-MMT gels over a range of C_s. We track the location of the settling interface and report that the time duration for the initial quiescent regime (the regime immediately before sedimentation starts) depends strongly on C_s. The sedimentation behaviour observed in our work is also very sensitive to the preparation protocol, *i.e.* whether salt is added before or after adding clay (SAC and SBC samples in our paper). In their work, Mongondry *et al.* speculate that their results do not rule out particle aggregation in the formation of Laponite gels. This picture agrees very well with our conclusions from the cryo-SEM images of Na-MMT gels (Fig. 11 in our paper). The authors have reported the possibility of gel formation even in samples with extremely low concentrations of clay in the absence of salt. Our paper, in contrast, deals exclusively with the structure and sedimentation of clay gels in the presence of substantial amounts of salt. Mongondry *et al.* plot the reduced scattered light intensity (Fig. 9 in their paper) *vs* the wavevector q and define a correlation length ξ, a characteristic lengthscale for structure formation of the clay suspensions studied by them. They find that for the highest clay concentrations investigated, there is no change in the reduced intensity. It is important to note here that these samples do not contain salt and the authors explain their results in terms of repelling unconnected aggregates or an ordered house-of cards (HoC) structure. In the Na-MMT gels studied by us, however, there is substantial salt in the medium and there are connections between tactoids/exfoliated particles in all the samples (Fig. 11). We systematically tracked the evolution of the clay network structure using these cryo-SEM images and explained the observed sedimentation behaviour in terms of the thermal fluctuation-induced particle rearrangements which modify the local yield stresses of the gel. It is clearly evident from Fig. 11 that in contrast to the work by Mongondry *et al.*, we do not have a well ordered house of cards structure. It is also important to note that the structures that form in our gels are of the order of several micrometers. It is therefore unlikely that if we were to perform systematic scattered intensity measurements for our clay gels, we would see a constant q-independent intensity as reported by those authors.

1 S. Ali and R. Bandyopadhyay, *Soft Matter*, 2016, **12**, 414–421.
2 P. Mongondry, J. F. Tassin and Taco Nicolai, *J. Colloid Interface Sci.*, 2005, **283**, 397–405.

Daan Frenkel commented: You mentioned the use of ultrasonic techniques to probe particle size distribution under conditions where light-scattering cannot be used due to the opacity of the sample. Could you explain how this ultra-sound technique works and indicate when it is (and when it is not) the preferred tool to determine particle-size distributions.

Ranjini Bandyopadhyay responded: Ultrasound attenuation spectroscopy (UAS) is a valuable tool to estimate particle size distributions (PSDs) for opaque and non-transparent samples.[1] The technique involves the measurement of attenuation of very low intensity ultrasound passing through a colloidal dispersion. The attenuation coefficient, α, calculated using the Beer–Lambert law, is expressed in terms of a compression complex wave number, k. This compression complex wave number, k, is related to the PSD of the constituents of the dispersion by modelling the attenuation spectrum appropriately and with prior knowledge of the density of the medium, the density of the particulate phase in suspension, the viscosity of the medium at that temperature and the sound speed in the solid medium. In our fits to the acquired attenuation data, only viscous loss was considered. Complete details can be found in one of our publications.[2] We used UAS extensively to calculate the PSD of clay tactoids in aging clay suspensions at clay volume fractions very close to that required for dynamical arrest of the tactoids. In conventional techniques such as DLS, size estimation is performed by relating the decay of the intensity autocorrelation data to the diffusion coefficient which is then related to the particle size using the Stokes Einstein equation. In the volume fraction regime where the suspension is approaching non-ergodic behaviour, the Stokes Einstein equation ceases to be applicable, thereby making the above treatment using UAS extremely powerful in estimating PSDs. However, it is noted here that the PSD estimated using UAS is based on modelling the data appropriately. Some prior knowledge of the sample, as mentioned above, is therefore an essential requirement for successful analysis of the data.

1 A. S. Dukhin and P. J. Goetz, *Characterization of Liquids, Nano-and Microparticulates, and Porous Bodies using Ultrasound*, Elsevier, New York, 2nd edn., 2010.
2 S. Ali and R. Bandyopadhyay, *Langmuir*, 2013, **29**, 12663.

G V Pavan Kumar said: Your technique of using ultrasound is interesting. Adding to the discussion, I would like mention that one may possibly utilize light and sound simultaneously to probe the assembly in liquid. An example would be to use the Raman–Nath effect.

Yogesh M. Joshi remarked: In Fig. 11, the SEM for the suspension after 138 h shows continuous layers of clays. The lateral dimension of a clay particle is mentioned to be around 400 nm. Could such observation of continuous layers over several microns of lengthscale originate from sintering of the clay particles?

Ranjini Bandyopadhyay replied: As Prof. Guruswamy has pointed out, sintering of Laponite or montmorillonite plates is rather unlikely. These minerals are chemically very inert and are not expected to sinter.

Yogesh M. Joshi remarked: A suspension of clays such as bentonite, montmorillonite, Laponite, *etc.* when added to water shows a soft solid like consistency and is out of thermodynamic equilibrium. Such out of equilibrium structures are highly sensitive to preparation protocol. The work by Prof. Bandyopadhyay very clearly shows how an apparently trivial matter such as whether salt is added before or after mixing clay has a profound influence on the resultant structure and properties. It appears that being sensitive to preparation protocols is a characteristic feature of many out of equilibrium soft materials. Consequently, a product with soft solid-like consistency is influenced by not just the ingredients but the preparation protocol.

Madivala G. Basavaraj asked: What is the correct picture that emerges out of your study on the destabilization of clay dispersions – stable dispersions, then aggregation, followed by settling and then consolidation? How do the results of the sedimentation experiments compare to theories that have been developed for the setting of aggregates?

Ranjini Bandyopadhyay replied: The picture that emerges from our experiments can be summarized as follows: when clay is mixed in water with enough salt to partially screen the repulsion between the constituent particles, the clay particles aggregate in 'overlapping coin' or 'house of cards' configurations or in a combination of both configurations. The time required for stable gel formation is strongly dependent on the concentration of salt in the medium. The gel remains stable for a length of time that depends on the concentration of salt in the medium and the preparation protocol (salt added before clay (SBC) or salt added after clay (SAC)). This initial quiescent regime is followed by a transient/delayed collapse of the gel and a slow compaction of the clay particles into a dense sediment at the bottom of the sample cell. The results are understood in terms of thermal fluctuation-driven particle rearrangements that occur in the gel continuously, modifying the mechanical response of the gel and giving rise to the observed settling behaviour. This scenario has been summarized as a cartoon in Fig. 12 of our paper. Incidentally, the regimes of collapse discussed here have been observed previously in the sedimentation of depletion gels and in colloidal suspensions flocculated at their secondary minimum (references 17–23 of the paper). Using simple theoretical arguments, Manley *et al.* (reference 18 in this paper) showed that delayed sedimentation (or transient collapse) observed in colloidal Ludox gels depends on the comparative values of the gravitational stress on the gel and the network yield stress. The same arguments and physical picture apply to the colloidal sodium montmorillonite (Na-MMT) gels studied here. As speculated by us in the conclusions section of the paper, the observed delayed gel collapse and the subsequent densification could be, in principle, modelled theoretically by using a poroelastic model as a starting point with the clay particle/tactoid association rate K_A and the dissociation rate K_D of the clay platelets as input parameters. Unfortunately, given the complex nature of the exfoliation state of Na-MMT, measuring K_A and K_D will clearly not be easy. Therefore, even though general features such as the delayed collapse of gels have already been explained theoretically in the literature, modelling other observations such as the differences in settling behaviour in SAC and SBC gels remain a far greater challenge.

Siddharth Kulkarni addressed Ranjini Bandyopadhyay and Yogesh M. Joshi: Can the gel formation be predicted by looking at the frequency sweep spectra? Is direct visualization more reliable than the rheology data?

Ranjini Bandyopadhyay responded: The frequency sweep spectra of soft glassy materials are characterized by certain ubiquitous features such as the frequency-independent elastic modulus and a weakly frequency-dependent viscous modulus.[1] The rheological signatures of critical gel formation have been discussed by Professor Joshi in paper 8570. In our work, we correlated the honeycomb structures of our clay-in-salt suspensions and their stability behaviours by means of imaging using scanning electron microscopy (SEM) under cryogenic conditions and light transmission experiments. Cryo-SEM allows us to resolve the microscopic details of the structure at very short lengthscales (approximately 100 nanometers), something that bulk rheology measurements could not have achieved. In a previous work,[2] our group found that above a critical salt concentration, there is a microstructural change in clay gels and that the rheological responses of the suspension could be correlated to plate–plate associations of the Laponite particles probed using cryo-SEM. The nature of these plate–plate associations is clearly very important in determining the stability of the honeycomb structures. In the present paper, we systematically investigated the stability of these structures and correlated them with structural details (obtained using cryo-SEM and resolvable down to the scale of 100 nanometers) at a lower clay volume fraction.

1 H. M. Wyss, K. Miyazaki, J. Mattsson, Z. Hu, D. R. Reichman and D. A. Weitz, *Phys. Rev. Lett.*, 2007, **98**, 238303.
2 S. Ali and R. Bandyopadhyay, *Soft Matter*, 2016, **12**, 414.

Yogesh M. Joshi replied: The seminal work of Winter and Chambon[1] proposed that in a gel forming system, tan δ is independent of frequency at the critical gel state, such that G' and G'' show power law dependence on frequency with identical exponent. Consequently, by monitoring evolution of G' and G'' at different frequencies it is possible to infer the gel formation process. On the other hand scattering techniques such as light scattering and SAXS can also be used to obtain the fractal dimension associated with a gel. Interestingly, the fractal dimension obtained from frequency independent tan δ at the critical gel point matches very well with the one obtained from the scattering studies.

1 H. H. Winter and F. Chambon, *J. Rheol.*, 1986, **30**, 367–382.

Guruswamy Kumaraswamy said: Sintering of Laponite or montmorillonite plates is rather unlikely. These are 2 : 1 phyllosilicates with tetrahedral silicate layers that sandwich an octahedral middle layer. These are chemically very inert and are unlikely to sinter. One way to verify that there is no sintering in the aged gel would be to completely dry the gel, crush it into a powder and then redisperse to form a gel again. If the clay has sintered, then the gelation process and the nature of the gel formed after drying will be very different.

Nirmalya Bachhar opened the discussion on the paper by E. Eiser: How does the concentration of the NPs affect formation of the structure? Does it change the physical properties of the system which may affect the final application?

Erika Eiser replied: While the different ratios of colloids to rods did not lead to large changes in the homogenised structures (also apparent from the chord-distribution analysis) we did observe a much bigger change in structure for the rapidly cooled open structures. Indeed the characteristic lengthscales in the open structure shortened for higher gold-nanoparticle fractions. We argue that it depends what structure you aim for. In the slowly cooled homogenised structures we do not see significant changes.

Yon Ju-Nam asked: In paper 8600, the inclusion of thiol groups in the DNA molecules was mentioned as DNA–thiol is required for the ligand exchange carried out in the preparation of the functionalised gold nanoparticles. Could it be possible to know in more detail how these thiol groups are added to the DNA molecule?

Erika Eiser responded: While thiolated DNA can be bought readily from IDT technology, we also performed the thiolation of the 5′-phosphate end of the DNA using a scaled up version of a protocol by Zanocco *et al.*[1].

The latter is advisable if large amounts of thiolated DNA are required.

1 G. T. Hermanson, *Bioconjugate Techniques*, Academic Press, 2nd edition, 2008.

Sanat Kumar asked: Do you get ordered phases as seen for cylinder–spherical mixtures in the limit of hard sphere (packing) interactions alone?

Erika Eiser replied: The phase diagram of the pure fd-virus solutions has been investigated in great detail by Seth Fraden and Zvonomir Dogic (Brandeis University). Fd-viruses are known to form a cholesteric phase. The Brandeis group has shown that the isotropic to cholesteric transition depends on the added-salt concentration, as the side-coat proteins are negatively charged. Hence, in the limit of high added-salt concentrations the viruses behave like hard rods. We also observe a transition to high-density liquid-crystalline phases. However, in the present work we only consider very low virus-concentrations.

Charusita Chakravarty asked: In your earlier work on DNA-colloids, you were able to generate finite aggregates. In this study, you generate open-structured gels. Could you provide some insight into the nature of effective interactions between colloids and mechanisms of self-assembly that lead to these differences?

Erika Eiser replied: The main difference in whether one would obtain finite size clusters or a percolating network is the length of the spacer between the colloid and the ‘active’ or ‘sticky’ part of the DNA. In our previous work[1] we coated 1 micrometer large polystyrene particles with lambda-DNA, which is about 48000 base pairs long and has a radius of gyration of 800 nm. While one end was attached to the colloids the main double-stranded part of the DNA served as steric stabiliser, and only the last 12 bases were in single stranded form for the binding. The lambda-DNA formed a mushroom-monolayer around the colloids. Because of the length of the dsDNA spacer all available ssDNA on one colloid could bind to all available complementary ssDNA on a different colloid. Hence, all bonds could be saturated, preventing the system from forming a percolating structure. We showed that by shortening the polymeric spacer below a certain length two colloids with complementary DNA could no longer saturate all their available bonds thus forming percolating networks.[2] The same holds for nano-colloids. In our present study the DNA spacer was always a few nanometers shorter than the colloids used.

1 T. Schmatko, B. Bozorgui, N. Geerts, D. Frenkel, E. Eiser and W. C. K. Poon, *Soft Matter*, 2007, **3**, 703.
2 N. Geerts, T. Schmatko and E. Eiser, *Langmuir*, 2008, **24**, 5118.

G V Pavan Kumar remarked: Your structures may be harnessed for random lasing applications. Could you please comment on this further?

Erika Eiser replied: While my group focuses on designing disordered colloidal materials with photonic activity and as porous battery and electrode materials it has been shown that colloidal glasses and dry-pressed powders can act as random lasers. In a classical laser a weakly scattering material is placed in a cavity with two mirrors to provide gain for spontaneous emission. In such a system the cavity determines the frequency and thus the mode of the laser. In a disordered gel or glass made of strongly scattering colloids any incoming light will propagate through multiple scattering – in other words the wavelets undergo a diffusive motion. Letokhov[1] showed that in a diffusive process the gain is proportional to the colloidal volume while the losses are only proportional to the surface. Consequently, under certain circumstances the diffusing light can constructively interfere and produce higher order lasing modes that cannot be found in traditional lasers. How the light is diffusing through a random colloidal material can be influenced by the interaction strength between the colloids, which can be controlled, for instance, by the choice of the DNA attached to the colloids.

An excellent review on random lasers can be found in 'The physics and applications of random lasers' by D. S. Wiersma.[2]

1 V. S. Letokhov, *Zh. Eksp. Teor. Fiz.*, 1967, **53**, 1442.
2 D. S. Wiersma, *Nat. Phys.*, 2008, **4**, 359–367.

Vinothan N. Manoharan said: As I understand, your measurement of the structure factor is done optically. Do you think the gels might have some structure at higher q (that is, smaller lengthscales) than you are able to probe optically?

Erika Eiser replied: The structure factor is obtained *via* a Fourier transform of the microscope image. The maximum q-range that can be probed is determined by the (lateral) resolution of the microscope and the pixel size of the camera (whichever is largest). In our case, the pixel size corresponds to about 200 nm, which is somewhat smaller than the resolution of the microscope (approx. 500 nm). The observed decay of the structure factor for large q is therefore determined by the optical resolution of the microscope. With optical, single-photon imaging, it will not be possible to get a better resolution. Hence, although there may well be structure in $S(q)$ at higher q, we are unable to probe it. With other techniques (*e.g.* SAXS) it should be possible to see structural features due to the correlation in the positions of the nano-particles.

Vinothan N. Manoharan asked further: In your paper, you compare your results for rod–sphere systems to sphere–sphere systems. How do the results compare to rod–rod systems – for example, either systems consisting solely of functionalized virus particles or systems with similar aspect ratios such as carbon nanotubes?

Erika Eiser answered: The big difference with rod-like hard particles without functionalisation is that these tend to form bundles. At high enough concentrations, fd-viruses in solution do show liquid crystalline order. Similarly DNA-functionalized viruses show an isotropic to cholesteric transition, when heated above the hybridisation temperature of the DNA attached. At volume fractions below that transition and below the DNA melt temperature, random aggregates of the viruses develop as they bind wherever they meet, displaying a network reminiscent of the cytoskeleton formed by actin filaments in cells. In an earlier publication[1] we imaged such aggregates with AFM, and measured the aggregation dynamics with dynamic light scattering.

1 R. R. Unwin, R. A. Cabanas, T. Yanagishima, T. R. Blower, H. Takahashi, G. P. C. Salmond, J. M. Edwardson, S. Fraden and E. Eiser, *PCCP*, 2015, **17**, 8194–8202.

Radhika Poojari communicated: What does fd stand for? How can one calculate the number of virions attached to the gold nanoparticles? Can this technique be used in device based applications for biomolecular marker detection in diseases?

Erika Eiser communicated in reply: The name fd derives from the fact that it belongs to the filamentous phage fd. It can be genetically modified to produce viruses with different lengths and charge per unit length in a mono disperse fashion making them a great material for studying liquid crystal behaviour.[1,2] As the fd-viruses are only about 6 nm thick it is impossible to calculate the number of gold colloids per vision from micrographs. In our experiments we used ratios of 2.5, 5 and 10 colloids per virus. However, when allowing DNA hybridisation to take place, a gold-nanoparticle can bind several viruses. An important application of filamentous fd is phage display: phages are viruses that invade bacteria to reproduce. One can genetically insert any type of DNA segment into the virus DNA which is then expressed as a protein in the bacterial host. When the virus escapes the host it incorporates the 'new protein' in the coat protein, hence the word phage display.[3] But filamentous fd has also been used for piezoelectric energy generation.[4]

1 Z. Dogic and S. Fraden, *Curr. Opin. Colloid Interface Sci.*, 2006, **11**, 47.
2 J. Sambrook and D. W. Russell, *Molecular cloning: a laboratory manual*, Cold Spring Harbor Laboratory Press, 3rd edn, 2001.
3 G. P. Smith and V. A. Petrenko, *Chem. Rev.*, 1997, **97**, 391.
4 B. Yang Lee, J. Zhang, C. Zueger, W.-J. Chung, S. Y. Yoo, E. Wang, J. Meyer, R. Ramesh and S.-W. Lee, *Nat. Nanotechnol.*, 3012, 7, 351.

Priyadarshi Roy Chowdhury communicated: What would be the effect of very low temperature and moderately high temperature on the growth of inhomogeneities, as mentioned in this paper?

Erika Eiser communicated in reply: Colloidal gels and crystals driven by DNA binding are different to those formed due to depletion or other chemical interactions. Our structure factor analysis for different quench depths shows that the 'homogenised gels' all have a peak at roughly the same wave vector q. However, a system quenched into the spinodal decomposition region will show a structural peak position that depends on the depth of the quench. In other words, the peak shifts to smaller q-values for lower quenches. In view of this observation we can say that, if quenched not too fast, the DNA binding arrests the spinodally decomposing sample so rapidly that the resulting structure is fixed within the first few degrees. Consequently the depth does not change the system structure no matter how deep we quench. This is very different to for instance supercooled colloidal liquids.

However, if quenched much faster than needed for DNA hybridisation, it can happen that the DNA binding will be stuck in strongly out-of-equilibrium traps leading to a more open system similar to a cluster fluid.

Zakiya Shireen asked: What would happen if you used different shapes such as ellipsoids or cylinders and how would ellipsoid–ellipsoid or for that matter cylinder–cylinder interactions affect the binding of DNA? How would it be different from that of the rod and sphere interaction?

Erika Eiser responded: We have shown that when coating half of the virions with a given single-stranded DNA and the other half with the complementary ssDNA, we obtain random aggregates.[1] For high enough concentrations we expect a continuous, cross-lined network of these semi-flexible polymers similar to the cytoskeleton in cells. It is obvious that also ellipsoids and other shaped colloids can be used. Several techniques have been developed to coat many different materials, whether it is coat proteins for viruses, silica, polystyrene or others. However, different shapes such as ellipsoids may be much more rigid than the semi-flexible rd-viruses. Therefore the resulting disordered gel will have different mechanical properties for similar colloidal volume fractions. If the non-specific van der Waals and Coulomb interactions are suppressed, only the hard shape repulsions and short-ranged DNA attractions are present. So it is only the shape interactions that drive different structure formations.

1 R. R. Unwin, R. A. Cabanas, T. Yanagishima, T. R. Blower, H. Takahashi, G. P. C. Salmond, J. M. Edwardson, S. Fraden and E. Eiser, *PCCP*, 2015, **17**, 8194–8202.

Rajdip Bandyopadhyaya said: This question is in two parts, and concerns how assemblies behave dynamically as a function of time, with implications in real-world applications:

1. During the formation of assembled structures of nanoparticles (either in a dispersion or in a polymer matrix), can it be deliberately made to happen as a function of time? In other words, can the structure be made to follow an arbitrary time-function, in the form of some premeditated timing?

2. Once formed, can such assemblies or structures be made to respond to changes in the environment? If so, how fast can these assemblies respond to imposed changes in the environment? How strong is the coupling between the assembly and the surrounding? Can such assemblies be made to be adaptive?

Erika Eiser responded: In DNA driven aggregation it is important to realise that the process is completely determined by the hybridisation of complementary DNA strands. This process is relatively slow, meaning the two complementary strands need to 'zip-up' correctly. In general only AT or GC base pairs can form in the presence of a monovalent salt. Hence when lowering the temperature of the sample these pairs can form spontaneously, but do not necessarily need to form along the correct position in the sequence. Typical cooling rates for forming correct hybridisation are required to be about 20 s per degree Centigrade. 1. Therefore, if cooled rapidly incorrect hybridisation can take place leading to metastable structures/bonds that will come apart slowly when well below the melt temperature of the sequence used. Thus, very slow arrangements can take place. If cooled along the equilibrium melt curves a much stronger gel can form. To conclude, yes, it is possible to build in a cooling protocol leading to a desired structure. 2. As shown in simulations[1] and in part in experiments (in preparation) DNA can act as a self-healing material because the strands are purely bound *via* hydrogen bonds, which are in essence weak. Consequently, DNA-bound colloidal gels can be sheared and then rejuvenated by heating the sample to around the DNA's melt temperature. Moreover, a gel can be set in any macroscopic shape required.

1 L. Di Michele, D. Fiocco, F. Varrato, S. Sastry, E. Eiser and G. Foffi, *Soft Matter*, 2014, **10**, 3633–3648.

Nicholas Kotov opened the discussion on the paper by Andrea R. Tao: In the last slide you showed interesting spectra of the gold dimers that appear to me exciting and intriguing. Could you go through what's going on with the dimers here?

Andrea Tao answered: What we observe is that our metal nanocube dimers are strongly coupled when assembled in the face-to-face configurations. However, if and only if light is polarized along the dimer axis do we see the appearance of the coupled mode (in the 700–800 nm window). At angles of incidence where the dimer is not 'visible', *i.e.* light only sees the cross-section of one nanocube, the optical response is almost identical to that of a single cube in a polymer.

Oleg Gang said: Since cubes have 3 distinctive plasmonic peaks (related to the length of the edges, and two diagonals), it might be interesting to consider different coupling of those plasmonic modes, depending on how the cubes are oriented relatively to each other.

Andrea Tao responded: I agree – we've only been able to achieve a large yield of dimer formation for the face-to-face assemblies but have looked at edge-to-edge arranged cubes for assemblies consisting of a broad distribution of cluster sizes. Near-field localization is definitely stronger in the edge-to-edge assemblies, but the far-field spectra are composed of largely broad features. Corner-to-corner assemblies would require 3D assembly or a controlled particle rotation, which we haven't been able to achieve just yet.

Vinothan N. Manoharan commented: An alternative approach to metamaterials that does not require long-range order is the metafluid,[1] which is a disordered suspension of plasmonic *meta*-atoms. However, making such a material requires fabricating highly symmetric *meta*-atoms with isotropic optical response.

1 Y. A. Urzhumov, G. Shvets, J. Fan, F. Capasso, D. Brandl and P. Nordlander , *Optics Express*, 2007, **15**, 14129.

Alamgir Karim replied: Thanks! The concept of metafluids and the paper is indeed very interesting! It would be worthwhile trying to get more of those types of plasmonic *meta*-atoms and study more on their colligative properties in the nanopatterned polymeric matrices of the kind we are working on by a simple soft stamping method.

Vimala Sridurai asked: In the reflectance measurements of the *meta*-atoms, the scans are done at 4 degrees and 33 degrees for both horizontal and vertical arrangements.

Why were those two angles chosen and what do you expect in the reflectance peaks for the in between angles?

Andrea Tao answered: These two angles were chosen to be near-incidence and oblique angles to measure reflectance. The coupled mode of the dimers occurs in the 700–800 nm range; we see that for angles and polarizations that are incident with the dimer axis, the coupled mode is maximum. We expect to see the maximum reflectance intensity when the cross-section of the dimer is maximized.

G V Pavan Kumar commented: Your combination of vertical nanocubes is interesting, and may be harnessed as patch optical antennas. Have you tried performing SERS on them?

Andrea Tao responded: Yes, we have. The optical patch antenna works quite well as a SERS substrate, as long as your analyte is located within the optical gap. In some capabilities (*e.g.* if you have no way of binding your analyte to a metal surface), this can be limiting.

Madivala G. Basavaraj asked: During your talk you mentioned there is localisation of light in the tiny gap between the particles because of a linker for the case where cubes are attached face-to-face. If this is true, then can one use roughness and achieve similar localisation? That is – can we assemble two cubes attached face-to-face that have a similar gap now due to particle roughness. Would the effect be the same?

Andrea Tao responded: Surface roughness can certainly lead to smaller gap sizes, but it may not give you the desired effect of increasing the degree of electromagnetic field localization. The scale of your surface roughness will play a significant role in determining this. It is likely that surface roughness with RMS values.

Madhura Som asked: Are there alternate strategies we could use to prevent/slow down lateral diffusion of the nanocubes by engineering the polymer matrix, perhaps something along the lines of Dr Lynn Walker's work on self-assembled block copolymer micelle solutions?

Andrea Tao answered: I think confinement strategies for dictating nanoparticle diffusion would be fascinating, if we could engineer particles to avoid getting pinned at interfaces and still undergo assembly.

Vinothan N. Manoharan addressed Andrea Tao and Erika Eiser: Prof. Tao and Dr Eiser, in both of your papers you describe methods to control the structure of nanoparticle assemblies, with applications toward optical materials. Are there applications in which it would be important to control the dynamics of the assemblies – that is, how the structure changes over time?

Erika Eiser responded: The structures that we present are intrinsically non-equilibium structures. Hence the control of the kinetics is essential for the design of such materials for specific applications.

Andrea Tao replied: In the case of optical materials, potential applications could range from a simple color-based detection scheme where the material changes color upon the presence of a particular chemical species, to optical coatings that dynamically block out particular wavelengths of light. In more advanced optics, one could potentially think of designing a dynamic material for routing or switching capabilities. But such applications are inherently limited by the slow kinetics of nanoparticle assembly.

Rajdip Bandyopadhyaya commented: This is a follow-up on the earlier question of how nanoparticle assemblies behave as a function of time and how that couples the system (particle assembly) to the environment, when such assembly timescales are real-world timescales of minutes in an actual application scenario.

Now the question is that, real-world applications of such pre-programmed structures are supposed to realize some machine-like functions. The problem can come from two sources:

(a) Different components of such a machine (where each component can be a different kind of particle-assembly realizing different functionalities) can interact in an unpredictable manner because of undefined functions in some parameter-space (size, shape, composition *etc.*) of these assemblies.

(b) More importantly, a real-world environment will have many external random noise sources, which will impact how different assemblies interact with each other in an unpredictable way.

So is there a way to make each individual assembled structure in such a way to build-in many redundancies to make the overall collection of assemblies a robust system? All this could lead to a more deterministic response from an inherently mesoscopic-scale assembly of a few hundred nanoparticles.

Lola Gonzalez-Garcia questioned: Transparent conductive materials are gaining interest for the fabrication of devices such as solar cells, OLEDs or touch screens. The use of metal nanoparticle networks is a promising approach, however, a compromise between transparency and conductivity should be achieved. 1. Would it be realistic to use a self-assembly strategy to lower the percolation threshold to get materials with percolating networks as their overall structure?

2. In this case, would it also be possible to address the problems that ligands cause, acting as insulating barriers, for electron transfer between the metal nanoparticles?

Andrea Tao replied: Self-assembly strategies could indeed be used to increase to degree of network formation in similar nanowire or nanoparticle films. One could imagine using this to design ultrathin metallic networks with a specific density of junctions or a specific transport pathway. There are a number of researchers and companies who are actively using colloidal metal nanostructures to address the fabrication of transparent conductive materials. Most of those that I am aware of use metal nanowires, where the nanowires are laid down in a mesh-like film and then welded together using heat or photoactivation. This certainly addresses the difficulties caused by ligands, which could indeed serve as potential barriers for electron conduction between particles or wires.

Erika Eiser answered: Our rod–sphere network demonstrates that one can achieve very low volume fraction samples to achieve percolating networks that might indeed form transparent electrodes. Although the fd-viruses are not conducting, and the DNA-bridges may present a resistance to electric conductivity it has been shown before that a DNA bridge can be metalized. The fd-viruses may be replaced by conducting carbon-nanotubes. Hence, it is possible to use these systems as transparent electrode materials. We are presently working on it.

Eventually, DNA may be exchanged by conducting host–guest molecules or ligands, which can be synthesised on a largescale.

Daan Frenkel opened the discussion on the concluding remarks by Oleg Gang: The possibilities of designing truly complex self-assembling structures is expanding rapidly. Undoubtedly, there will be important applications in the (not too distant future).

However, it is useful to make our aims more explicit. At this stage, we should formulate a few specific 'holy grails': the design of self-assembling materials or devices that can perform functions that cannot be carried out with any existing material. I am thinking of objectives that could inspire the field on a timescale of, say, five years.

Nicholas Kotov replied: My 'holy grail' would be self-replicating nanoparticles. The reason is that this is a system that would help us make quite a few fundamental discoveries. Technological applications of such nanoparticles in biomimetic self-healing catalytic systems, can be thought about as well. Enantioselective self-assembly of helices from nanoparticles under circularly polarized light indicates to me that the origin of homochirality on Earth might be related to self-organization of nanoscale inorganic matter. So, tracing potential mechanisms for how the self-assembly of inorganic nanoparticles can take place at the early age of Earth (*i.e.* Hadean eon) would be my second 'holy grail'.

Nicholas Kotov commented further: Replication of some functions of a microbe would be my other choice. Besides being enigmatic and, to some degree, academically provocative, this Holy Grail should in nanoparticle assembly lead to multiple technological advances in catalysis and energy harvesting.

Oleg Gang responded: I fully agree that some kind of functional, autonomic and self-sufficient small machinery, like a microbe (but not a real one), would be an excellent demonstration of our capabilities in assembly. But we all know it is still far off. We should not try to just make a copy of what works in biology, but should exploit the new properties of nanoparticles. An airplane is inspired by a bird, but it does not duplicate it.

Daan Frenkel said: To give a rough indication of something to aim for: we should try to go beyond 'conventional' smart materials that exhibit one-to-one stimulus-response behaviour. Rather, we could consider materials that perform simple 'logical' operations that allow them to respond in a non-trivial (and certainly non-linear) way to the change in a number of independent external parameters. Mimicking living systems should not be the first objective.

Nicholas Kotov responded: It seems to me that non-trivial and non-linear system behavior is typical of living cells, isn't it?

Madivala G. Basavaraj commented: Over the years there have been exciting ways to assemble nanoparticles and colloids into a range of suprastructures. These assemblies have been put into use for example as gas sensors. One of the challenges in the future would be the sustainable reuse of these assemblies and their application in the real world where there may be dirt and other conditions that can either disrupt or hamper the functionality of these assemblies.

Oleg Gang responded: I agree, for any real-life application usually some "dirt" discrimination mechanism/process is needed.

Daan Frenkel remarked: Many potential applications of complex self-assembly are in chemistry. A nano-structure could be used to implement a multi-step reaction. There are already examples where DNA origami structures can assist in the sequential assembly of complex hetero-polymers (or oligomers). However, the future nano-scale chemical factories need not be DNA-based.

Nirmalya Bachhar questioned: Based on the entire discussion can we yet come to a conclusion on which lengthscale or timescale we should avoid using coarse grained models like DLVO theory? Can there be any rule of thumb for that beside the existing limitations it already has?

Nicholas Kotov answered: In my estimates 20 nm (particle diameter and the interparticle distance) would be a good rule of thumb when non-additivity of particle interactions starts to strongly affect their behavior.

Oleg Gang replied: I think it is highly system specific, moreover, it might be specific to the questions one asks about the systems.

Guruswamy Kumaraswamy addressed Daan Frenkel and Oleg Gang: What is the state-of-the-art in using self-assembly techniques to realize photonic meta-materials in the visible region? For example, is it currently possible to realize periodic assemblies of nanoscale horseshoe-type metal resonators (or metal nanoparticle assemblies with broken symmetry) over macroscopic distances of, say, 1 mm? If this hasn't yet been achieved, do you feel that this is a sufficiently challenging target for the next couple of years?

Oleg Gang answered: As far as I know, no assembly of nanoscale resonators has been achieved even for smaller areas. It does require not only resonator positioning, but also the fabrication and control of their relative orientations. It can be a very good targeted structure. However, it will be important to have optical calculations to support and guide the fabrication of such specific structures.

Daan Frenkel replied: As I am not an expert on photonic *meta*-materials, I do not feel that I can confidently assess the state of the art. I will therefore focus on what DNA-brick structures (or their successors) could enable. As our ability increases to control the growth of periodically repeated DNA-brick structures, we should hopefully be able to make millimetre-sized crystals (but crystal growth is an art). However, once such structures can be made, they can presumably be functionalized with resonators or other components, that can be placed with high precision in non-trivial, functional arrangements (not necessarily photonic).

G V Pavan Kumar questioned: The self assembly of nanoparticles has already made an impact on nano-optics and nano-spectroscopy by providing model systems such as plasmonic dimers with gaps less than 10 nm. The prospects of realizing optical metamaterials using self-assembly of nanoparticles is very interesting. I think it can open up new avenues in nanophotonics.

Vinothan N. Manoharan answered: I agree. In nanophotonics, self-assembly might actually be a more precise fabrication method than top-down techniques such as e-beam lithography. As an example, self-assembled monolayers can control the gap distance between particles to nanometer precision or better.

Mukta Tripathy addressed Vinothan N. Manoharan: During the introductory lecture by Steve Granick, Vinothan Manoharan suggested that we begin to think about time-anisotropy in order to build new functionalities. There are many systems where timescales can be separated, because of the occurrence of both fast and slow processes, such as glassy particles rattling in a cage at short times, and then 'hopping' out of a cage. However, the effects of entropy will dominate at long times, and perhaps to produce a functional material that switches between short-time and long-time behavior, we might have to think of energy-harvesting active matter.

I would like to know more about any specific directions of thought Vinothan has had, or any specific systems he has in mind.

Vinothan N. Manoharan responded: This is a good point. I had in mind materials made from nanoparticles whose interactions change in time. This can be done if the interactions are mediated by functional groups on the particles such as DNA strands. I agree that to make time-anisotropic interactions that are different from what one might observe in a glassy dynamical process, one should focus on active systems (such as dynamic DNA nanotechnology), where there is an energy input.

Sanat Kumar asked Oleg Gang: You mentioned error correction and healing as being important parts of self-assembly but you didn't discuss them in detail in your talk. What do we know now? And beyond what we can think about at the moment do we need to achieve these goals?

Nicholas Kotov further commented: DNA assembly of gold nanoparticles impresses me with the diversity of structures that could be created and complexity of nanoparticle encoding that could be two or three levels deep. When the superstructure starts containing a very large number of particle errors, particle–particle attachment becomes a limiting factor. What is your thinking about this? How could we prevent or 'anneal' them?

Priyadarshi Roy Chowdhury communicated: Could you discuss the necessity of structural error correction associated with complex self assembly?

Oleg Gang answered: We can envision constructing those thermodynamically favorable pathways that allow for more efficient error correction, as for example demonstrated in Prof. Frenkel's talk. Also, depending on the system details, it is possible to consider more active error correction mechanisms. Again, they might just be thermodynamically driven, but might use particles or molecules that diffuse freely in the self-assembled matrix. We have certain examples of such realizations for nanoparticle systems, but they are only in the initial stages.

Priyadarshi Roy Chowdhury communicated: Could you discuss the contrasting properties of nanoparticle assembly, supramolecular assembly and colloids.

Oleg Gang communicated in reply: The most important differences are in the effects of diffusion, interactions, shapes and effect of temperature on the assembly behavior. For example, molecular shapes are significantly more complex than colloidal ones, as a result various unusual packings can be accommodated.

Priyadarshi Roy Chowdhury communicated: Could you discuss the criteria responsible for reorganization of assemblies.

Oleg Gang communicated in reply: One needs to establish the conditions for shape- or interaction-shifting

Priyadarshi Roy Chowdhury communicated: What are the effects of co-ordination number on nucleation in many component self assembly?

Oleg Gang communicated in reply: The answer is not clear at the present time since we know very little about those systems, in particularly from the experimental point of view.

Priyadarshi Roy Chowdhury communicated: What are the effects of colloids undergoing colored interactions?

Oleg Gang communicated in reply: We witness only the beginning of those studies. I believe there are many arrested states for those systems, and we will need to investigate to what degree they can hamper the structure formation, and what energy landscape can help to overcome them.

Faraday Discussions

PAPER

Nanoparticle assembly: from fundamentals to applications: concluding remarks

Oleg Gang

Received 1st March 2016, Accepted 1st March 2016
DOI: 10.1039/c6fd90004f

Nanoparticles, due to their broadly tunable functions, are major building blocks for generating new materials. However, building such materials for practical applications by self-assembly is quite challenging. Following the Faraday Discussion on "Nanoparticle Assembly: from Fundamentals to Applications" we discuss here the current trends in the field of self-assembly, including: understanding the unique interplay of molecular and nanoscale effects, a development of novel approaches for the creation of targeted nanoparticle architectures, advances in controlling dynamic behavior of systems and enabling new functions through specifically formed structures.

Introduction

The great advances in nano-synthesis methods of the last two decades have permitted a fabrication of new functional blocks, nanoparticles (NP), whose properties can be significantly expanded beyond the atomic or macroscale forms of materials with the same chemical compositions. Despite our ability to generate such individual nanoscale blocks, there is an increasing need to establish methods for generating materials and devices that incorporate nanoparticles of different kinds in a unified manner. Moreover, since a material's properties are directly related to the underlying structure, it is essential to control the NP arrangements in these new types of materials. Approaches based on self-assembly offer tremendous cost-advantages and an ease of manufacturing compared to lithographic methods. Furthermore, self-assembly addresses tasks that are intrinsically challenging for conventional lithography processes, such as creating three-dimensional architectures or structures containing pre-fabricated nano-objects of multiple types. On the other hand, the methods of self-assembly typically have significant limitations: they form relatively simple structures from the same type of components, and they rarely permit the rational design of systems. Developing principles and establishing fabrication approaches for organizing nano-components into targeted materials with desired functions is a primary goal of nanoparticle self-assembly. The Faraday Discussion covered the range of topics

Center for Functional Nanomaterials, Brookhaven National Laboratory, Upton, NY 11733, USA. E-mail: ogang@bnl.gov

from synthesis of nanoparticles and their assemblies to advancing theoretical methods for predicting the structure formation, and from structural properties of nanocomposites (inorganic particle and organic material) to novel functionalities of particle systems. The broad range of nanoscale materials, progress in experimental and theoretical approaches, and potential applications were discussed.

Driving forces of nanoparticle self-assembly

Despite a variety of scales and driving forces the general principles of the spontaneous formation of ordered organizations, *i.e.* self-assembly, remain similar: the requirement to comply with the interactions and entropic constraints of the components results in the realization of certain structural arrangements. Sphere packing is one of the earliest and the simplest examples of how objects can be arranged into ordered arrays based on their intrinsic properties, in this case, shape. The question of the 'best' arrangements of simple spheres can be traced back to Sanskrit texts (499 CE), and it is typically well known as the Kepler conjecture: face-centered cubic and hexagonal close lattices of equal spheres exhibit the highest packing density. The mathematical proof of the Kepler conjecture was shown only very recently by Thomas Hales.[1] The notion of forming an ordered structure based on the properties of objects defines the objective of self-assembly.

In the realm of physical systems, even non-interacting objects can form ordered phases from the disordered liquid state. As was shown more than half a century ago by Onsager for rods and by Wood and Alder for hard spheres, entropy increases for crystallized systems due to the larger number of available microscopic states. The phases can be significantly diversified as soon as more complex blocks are introduced. For example, assembling binary mixtures of spherical particles of dissimilar sizes results in a rather rich set of crystallographic symmetries. As soon as an object shape becomes anisotropic the complexity of organization increases drastically. A recent comprehensive simulation study on the assembly of solid shaped blocks showed the richness of phases, the formation of ordered, plastic and quasi-crystalline organizations.[2,3] Real physical systems rarely obey the idealized conditions of non-interacting objects, however the richness of these interactions is a scientists' playground. The interparticle interactions can significantly modulate the assembly behavior leading to phases beyond those determined by shape considerations. For example, many studies have explored how van der Waals,[4] electrostatic,[5] dipolar,[6,7] and magnetic forces[5] between particles affect the structure formation. Moreover, the idealized representation of a particle as a solid block is also far from reality. In this respect, the structure of shells formed by ligands and polymers[8–13] around a particle core has a profound effect on the phase behavior. These effects become more important as soon as we move to nanoscale blocks where particle and molecular dimensions become comparable.

Nanoparticles: the intersection of molecules and colloids

In contrast to typical micron-sized colloidal particles, the sizes of nanoscale colloids, or nanoparticles, approach those of molecules. Since coating a nanoparticle with ligands is needed for particle stabilization and functionality, the organic shells are an integral part of the particle structure. Therefore, the idealized representation of nanoscale objects as describing them only by shape is far from complete. In fact, those organic shells play a crucial role in the way NPs interact and assemble. However, to deduct the relationship between particle assembly behavior and shell properties is often difficult, mainly due to the limited methods for revealing the NP shell structure. Indeed, commonly used electron microscopy and X-ray scattering methods are well positioned to extract information about inorganic particle cores, but not shells.

There are several aspects of organic shells that lead to their special role for nanoparticles. First of all, organic molecules introduce a certain degree of 'softness' to particles and provide interactions through chemical groups and/or charges. Further, novel effects can arise for molecules attached to a surface with a high (and not necessarily constant) curvature. For example, due to entropic reasons molecules can distribute heterogeneously on the surface of a shaped nanoparticle, thus, leading to the formation of a shell, which is not conformal with the underlying core shape. Additionally, a relatively small number of molecules on a nanoparticle might manifest in a discrete character of shells, and that can result in the anisotropic interactions. Hence, soft shells can significantly modulate particle shape, influence interparticle binding motifs, and even contribute to establishing anisotropic interactions. Accordingly, the resulting NP arrays can be quite different from expectations based only on geometrical considerations.[8,15,16]

The directionality provided by organic shells can be used to regulate nanoscale assembly, for instance, *via* electrostatics, steric effects, and molecular bonds established between organic tethers. In addition, related effects can occur when nanoparticles are imbedded in the polymer matrix, and the structure of polymers can be significantly affected by the particles. Apart from the fascinating structural effects, the molecular arrangements can have significant influence on the mechanical properties of nanoparticle-based materials. For example, Michael Bockstaller of Carnegie Mellon University discussed the relationships between the structure of a polymer brush grafted to a nanoparticle, the degree of order in the system and mechanical response of these hybrid materials (DOI: 10.1039/C5FD00121H). Also, Jacques Jestin of Laboratoire Léon Brillouin CEA-CNRS presented a study on the interplay of polymer chain conformations and particle assembly behavior, relevant to industrial problems (DOI: 10.1039/C5FD00130G). As was shown by Andrea Tao from the University of California, San Diego, the polymer distribution on cubic nanoparticles can play an important role in particle placement relative to each other, and consequently, that is reflected in the plasmonic properties (DOI: 10.1039/C5FD00134J).[14]

Complex morphologies through self-assembly

The majority of self-assembly studies in the field are typically concerned with the question: what is the origin of assembly behavior and how can one track back the particle characteristics and interparticle forces to explain the observed phases? These important questions drove the field and resulted in great advances in our understanding of assembly phenomena. However, the demand for transferring self-assembly effects into the territory of concrete material fabrication methods requires answering a different question: how to assemble a particular targeted organization from those specific particles? Given the fact that nanoparticle blocks carry the functions and the final material requires a certain organization of those blocks, the minimum modification of the blocks would be highly beneficial from a practical perspective. On a fundamental level, this required addressing an inverse problem in self-assembly: rather than asking why the structure has formed, one should ask how to drive the system into formation of the targeted structure.

A number of approaches have been proposed in recent years to address that question. For example, an ability to modulate the interactions between particles in an arbitrary way permits inducing the desired organization, as has been shown in computational studies.[15] However, in order to achieve such complex structures long-range interactions (several interparticle distances) with tailored potential shapes are needed. Unfortunately, it is quite challenging to implement these ideas experimentally.

Alternative approaches for controlling particle structures have been considered using anisotropic interactions. For example, one of the broadly used models, known as patchy particles, requires a precise arrangement of binding patches on a particle surface, and that restricts interparticle connections. Thus, through the geometrical arrangement of the patches one can, in principal, control the formation of assembled clusters and lattices. Extensive studies were conducted[2,17–23] and interesting organizations were observed both theoretically and experimentally. However, the formation of ordered lattices from patchy particles might be hampered by defects due to particle orientation freedom and such imperfections are difficult to heal.[24,25] Besides, that approach requires massive high-fidelity particle fabrication, which is quite challenging at the present time, particularly at the nanoscale.[21,26–28]

Recently, a new idea was demonstrated: instead of tailoring the particle anisotropic properties, it was proposed to use an anisotropic connecting block as the connector between spherical particles.[29–31] In this approach, the directionality of interactions is determined by the block's anisotropic characteristics, thus, simply by introducing different types of blocks, different types of organization can be induced without changing the spherical particles. The demonstration of this approach for blocks that interact with spheres *via* facets and *via* vertices was shown experimentally, while DNA was used to regulate the local binding properties. For example, cubic and octahedral NPs were used to coordinate, by their faces, spherical particles into clusters and lattices.[29] Also, using precisely designed and shaped molecular constructs, based on DNA, it was possible to coordinate particles through attachment to the vertices of the constructs and that resulted in the formation of a variety of NP lattice types, including cubic, body-centred-tetragonal, diamond and others, using polyhedral DNA frames as topological interparticle connectors and particle cages.[30,31] While the effect of block shapes and the placement of binding sites on the final lattice symmetry still have to be understood, the approach opens exciting opportunities for building periodic structures in a predictable way.

The idea of anisotropic interactions can be significantly advanced towards the fabrication of complex shaped morphologies if a particle possesses not only anisotropic binding properties but also addressable interactions, as was demonstrated by a computational study by Daan Frenkel of Cambridge University

(DOI: 10.1039/C5FD00135H). In this case using DNA addressability it is possible to differentiate different bonds, and such "coloring" of bonds provides a means to connect a large number of particles with different types of coloring in quite unique architectures.[32–34] Remarkably, the formed structures have programmed structural elements with dimensions of many particle sizes. Erica Eiser of Cambridge University showed the experimental example of these DNA-based methods (DOI: 10.1039/C5FD00120J). It is perceived that approaches based on so-called programmable interactions, for example using DNA specificity of interparticle interactions, might potentially allow for the practical realization of self-assembly by design. An interesting question that arises for the assembly of complex non-periodic architectures is how to provide information about the desired structure to be assembled into the self-assembly process?

In addition to controlling local intercalations alternative approaches are based on inducing particle morphologies, for example, through the media in order to promote the assembly process. Guruswamy Kumaraswamy from CSIR-National Chemical Laboratory at Pune demonstrated the use of ice crystallization for particle aggregation between crystalline ice domains, so-called ice templating (DOI: 10.1039/C5FD00125K). The possibility of using lipid membranes to mediate the formation of complex organizations of curved particles was discussed by Mohamed Laradji of the University of Memphis (DOI: 10.1039/C5FD00144G).

Dynamic self-organization processes

One of the long-standing challenges in self-assembly is the ability to simultaneously control structure formation on different length-scales and its transformation in time or on demand.[35] The structural plasticity of organic molecules forming NP shells and their sensitivity to environmental parameters and molecular stimuli makes NP systems highly suitable for post-assembly transformations. The change in interparticle distance is quite important for modulating functional materials made from NPs and a number of approaches have been demonstrated. On the other side it is even more exciting to regulate a change in the entire type of structural organization on demand. Recent studies show such implementations by employing switchable DNA bonds.[36,37]

Conventional self-assembly approaches typically rely on a system evolving towards its equilibrium state. However, even energetically favorable states can be difficult to realize, owing to the complexity of evolution pathways. It is important to stress that self-assembly, as a process, is non-equilibrium. Depending on the system details, the process might be thermodynamic or driven by external forces. Regardless of the process details, it is never guaranteed that a system will reach a well-ordered structural organization, either because of metastable states or due to the intrinsic characteristics of the system. In fact, a plethora of theoretical and experimental studies has shown that ordered systems only form under well-controlled conditions in which a balance of interactions and mobility of components is achieved. The 'successful' (*i.e.* producing order) assembly is realized only when the assembly and disassembly continuously occur. That implies that interactions are strong enough to drive components toward each other but, at the same time, the components have the ability to explore micro-states, which can only occur if there is no permanent binding. While for thermodynamically driven assembly the interaction energy during assembly should be on the order of kT, for systems driven by external forces much stronger interactions can be allowed.

For instance, arrested states can be overcome more easily for systems far from equilibrium. In non-equilibrium scenarios, additional control is achieved *via* regulated kinetic effects, which often allow the control of structural organization on several scales. A number of experimental approaches have been developed for enhancing order in a system using local heating and mechanical methods. Alamgir Karim of the University of Akron discussed the use of shear forces to promote nanoparticle organization in string-like structures in polymer films (DOI: 10.1039/C5FD00141B). The transient, far from equilibrium, states often have more profound effects on improving the order. The interplay of kinetic and equilibrium effects in this respect is far from being understood. An important related question pertains the characteristic length of a particle's motion and its timescale during the assembly and domain re-arrangements. Obtaining a better understanding of particle transport in highly confined environments is important in this respect. Lynn Walker of Carnegie Mellon University demonstrated NP diffusion in the matrix formed by block copolymer micelles (DOI: 10.1039/C5FD00122F).

Another aspect of dynamic self-assembly is using external fields to induce the athermal mobility of a system's components or alter the interactions between them. It was shown by Steve Granick of the BS Center for Soft and Living Matter in Korea and the University of Illinois at Urbana-Champaign that complex collective motion can be induced, for example on Janus magnetic particles in magnetic fields,[38] and the field controls the character of the motion and dynamic structure (DOI: 10.1039/C6FD00024J). Also, plasmofluidic fields can be efficiently used to drive plasmonic NPs, as was shown by Pavan Kumar of IISER Pune (DOI: 10.1039/C5FD00127G).

Outlook

In recent years, a variety of powerful approaches for particle assembly have been developed using ligands, polymers, biomolecules, and particle shaping. The approaches rely either on a final balance of interactions and entropic factors, on a more deliberate encoding of components, or on the use of patterns and templates for particle positioning. Much progress has been achieved in the creation of clusters and lattices using particles of different sizes, shapes, and materials. While we are far from the complete understanding of these systems, both experimental and theoretical efforts demonstrate significant progress in our ability to predict and to explain the formation of ordered phases. In the near future we may witness a rapid development of approaches that bring the science of self-assembly to a new level; new strategies might allow for the rational self-assembly of desired architectures with and without periodic organization, and with the ability to reconfigure these structures on demand. These advances can be translated into the practical fabrication of novel types of materials from functional particles and such systems might enable the engineering of desired optical, magnetic and bio-related functions.[39] For example, Nick Kotov of the University of Michigan demonstrated that chiroptical activity can be fully engineered into particle clusters and those systems can be applied for gas detection (DOI: 10.1039/C5FD00138B). Rajdip Bandyopadhyaya from IIT Bombay showed a catalytic activity enhancement by NPs imbedded into a silica matrix (DOI: 10.1039/C5FD00126A). The notion of limited practical applicability of self-assembled systems is of high concern, as was discussed by many participants. Thus, there is an increasing need to transfer the development of self-assembly methods into the fabrication of functional materials and devices.

In the upcoming years self-assembly will tackle the range of problems that should establish it as a versatile manufacturing platform at the nanoscale. Moreover, it might enable the creation of materials that mimic biological systems in their adaptive and energy-processing properties, but will still take advantage of inorganic components.

These potentially interesting developments might address the following challenges in building complex systems:

• Reverse engineering of multicomponent self-assembled structures: from arbitrarily prescribed lattices to non-periodic designed morphologies, and to supra-nanoscale architectures.

• Error correction and structure healing in self-assembled structures.

• Systems with regulated dynamic behavior and evolution pathways.

• Incorporation of active components into assembly schemes to overcome the limitations of thermal systems.

• An ability to deliver information and instructions about required processes and structuring to individual components and to a system on a global level.

Merging such capabilities of self-assembly methods with the functions of nanoscale building blocks will open tremendous prospects for creating materials from blueprints, establishing materials with dynamic performance, and inventing material systems that can autonomously manage energy flows for the function-required reorganizations.

Acknowledgements

Research, carried out at the Center for Functional Nanomaterials, Brookhaven National Laboratory, is supported by the U.S. Department of Energy, Office of Basic Energy Sciences, under Contract No. DE-SC0012704.

References

1 T. C. Hales, A proof of the Kepler conjecture, *Ann. Math.*, 2005, **162**, 1065–1185.
2 P. F. Damasceno, M. Engel and S. C. Glotzer, Predictive Self-Assembly of Polyhedra into Complex Structures, *Science*, 2012, **337**, 453–457, DOI: 10.1126/science.1220869.
3 U. Agarwal and F. A. Escobedo, Mesophase behaviour of polyhedral particles, *Nat. Mater.*, 2011, **10**, 230–235, DOI: 10.1038/Nmat2959.
4 Y. S. Xia, *et al.*, Self-assembly of self-limiting monodisperse supraparticles from polydisperse nanoparticles, *Nat. Nanotechnol.*, 2011, **6**, 580–587, DOI: 10.1038/Nnano.2011.121.
5 E. V. Shevchenko, D. V. Talapin, N. A. Kotov, S. O'Brien and C. B. Murray, Structural diversity in binary nanoparticle superlattices, *Nature*, 2006, **439**, 55–59, DOI: 10.1038/Nature04414.
6 Q. Chen, *et al.*, Supracolloidal Reaction Kinetics of Janus Spheres, *Science*, 2011, **331**, 199–202, DOI: 10.1126/science.1197451.

7 Q. Chen, S. C. Bae and S. Granick, Directed self-assembly of a colloidal kagome lattice, *Nature*, 2011, **469**, 381–384, DOI: 10.1038/Nature09713.

8 D. Nykypanchuk, M. M. Maye, D. van der Lelie and O. Gang, DNA-guided crystallization of colloidal nanoparticles, *Nature*, 2008, **451**, 549–552, DOI: 10.1038/Nature06560.

9 R. J. Macfarlane, *et al.*, Nanoparticle Superlattice Engineering with DNA, *Science*, 2011, **334**, 204–208, DOI: 10.1126/science.1210493.

10 Y. G. Zhang, F. Lu, K. G. Yager, D. van der Lelie and O. Gang, A general strategy for the DNA-mediated self-assembly of functional nanoparticles into heterogeneous systems, *Nat. Nanotechnol.*, 2013, **8**, 865–872, DOI: 10.1038/Nnano.2013.209.

11 H. M. Xiong, D. van der Lelie and O. Gang, Phase Behavior of Nanoparticles Assembled by DNA Linkers, *Phys. Rev. Lett.*, 2009, **102**, 015504.

12 M. T. Casey, *et al.*, Driving diffusionless transformations in colloidal crystals using DNA handshaking, *Nat. Commun.*, 2012, **3**, 1209, DOI: 10.1038/Ncomms2206, 2012.

13 F. Varrato, *et al.*, Arrested demixing opens route to bigels, *Proc. Natl. Acad. Sci. U. S. A.*, 2012, **109**, 19155–19160, DOI: 10.1073/pnas.1214971109.

14 A. Tao, P. Sinsermsuksakul and P. Yang, Tunable plasmonic lattices of silver nanocrystals, *Nat. Nanotechnol.*, 2007, **2**, 435–440, DOI: 10.1038/nnano.2007.189.

15 M. C. Rechtsman, F. H. Stillinger and S. Torquato, Synthetic diamond and wurtzite structures self-assemble with isotropic pair interactions, *Phys. Rev. E*, 2007, **75**, 031403.

16 S. Vial, D. Nykypancuk, D. G. Yager, A. V. Tkachenko, and O. Gang, Linear Mesostructures in DNA-Nanorod Self-Assembly, *ACS Nano*, 2013, **7**, 5437.

17 F. Romano, E. Sanz and F. Sciortino, Phase diagram of a tetrahedral patchy particle model for different interaction ranges, *J. Chem. Phys.*, 2010, **132**, 184501.

18 J. Russo, P. Tartaglia and F. Sciortino, Association of limited valence patchy particles in two dimensions, *Soft Matter*, 2010, **6**, 4229–4236.

19 Q. Chen, S. C. Bae and S. Granick, Directed self-assembly of a colloidal kagome lattice, *Nature*, 2012, **469**, 381–384.

20 E. V. Shevchenko, D. V. Talapin, N. A. Kotov, S. O'Brien and C. B. Murray, Structural diversity in binary nanoparticle superlattices, *Nature*, 2006, **439**, 55–59.

21 Y. F. Wang, *et al.*, Colloids with valence and specific directional bonding, *Nature*, 2012, **491**, 51–U61, DOI: 10.1038/Nature11564.

22 F. Lu, K. G. Yager, Y. Zhang, H. Xin and O. Gang, Superlattices assembled through shape-induced directional binding, *Nat. Commun.*, 2015, **6**, 1–10.

23 M. N. O'Brien, M. R. Jones, B. Lee and C. A. Mirkin, Anisotropic nanoparticle complementarity in DNA-mediated co-crystallization, *Nat. Mater.*, 2015, **14**, 833–839.

24 J. Russo, P. Tartaglia and F. Sciortino, Association of limited valence patchy particles in two dimensions, *Soft Matter*, 2010, **6**, 4229–4236, DOI: 10.1039/c0sm00091d.

25 J. M. Tavares, N. G. Almarza and M. M. T. da Gama, Three-dimensional patchy lattice model: Ring formation and phase separation, *J. Chem. Phys.*, 2014, **140**, 044905, DOI: 10.1063/1.4863135.

26 T. G. W. Edwardson, K. M. M. Carneiro, C. K. McLaughlin, C. J. Serpell and H. F. Sleiman, Site-specific positioning of dendritic alkyl chains on DNA cages enables their geometry-dependent self-assembly, *Nat. Chem.*, 2013, **5**, 868–875.

27 K. Suzuki, K. Hosokawa and M. Maeda, Controlling the number and positions of oligonucleotides on gold nanoparticle surfaces, *J. Am. Chem. Soc.*, 2009, **131**, 7518–7519.

28 X. Ye, *et al.*, Shape alloys of nanorods and nanospheres from self-assembly, *Nano Lett.*, 2013, **13**, 4980–4988.

29 F. Lu, K. G. Yager, Y. G. Zhang, H. L. Xin and O. Gang, Superlattices assembled through shape-induced directional binding, *Nat. Commun.*, 2015, **6**, 6912.

30 W. Y. Liu, *et al.*, Diamond family of nanoparticle superlattices, *Science*, 2016, **351**, 582–586, DOI: 10.1126/science.aad2080.

31 Y. Tian, T. Wang, Y. Zhang, H. Li and O. Gang, Lattice Engineering *via* Nanoparticle-DNA Frameworks, *Nat. Mater.*, 2016, DOI: 10.1038/nmat4571.

32 W. M. Jacobs and D. Frenkel, Self-Assembly of Structures with Addressable Complexity, *J. Am. Chem. Soc.*, 2016, **138**, 2457–2467, DOI: 10.1021/jacs.5b11918.

33 J. D. Halverson and A. V. Tkachenko, DNA-programmed mesoscopic architecture, *Phys. Rev. E: Stat., Nonlinear, Soft Matter Phys.*, 2013, **87**, 062310, DOI: 10.1103/PhysRevE.87.062310.

34 Y. G. Ke, L. L. Ong, W. M. Shih and P. Yin, Three-Dimensional Structures Self-Assembled from DNA Bricks, *Science*, 2012, **338**, 1177–1183, DOI: 10.1126/science.1227268.

35 S. Mann, Self-assembly and transformation of hybrid nano-objects and nanostructures under equilibrium and non-equilibrium conditions, *Nat. Mater.*, 2009, **8**, 781–792, DOI: 10.1038/nmat2496.

36 Y. G. Zhang, *et al.*, Selective transformations between nanoparticle superlattices *via* the reprogramming of DNA-mediated interactions, *Nat. Mater.*, 2015, **14**, 840–847, DOI: 10.1038/nmat4296.

37 W. B. Rogers and V. N. Manoharan, Programming colloidal phase transitions with DNA strand displacement, *Science*, 2015, **347**, 639–642, DOI: 10.1126/science.1259762.

38 J. Yan, S. C. Bae and S. Granick, Rotating crystals of magnetic Janus colloids, *Soft Matter*, 2015, **11**, 147–153, DOI: 10.1039/c4sm01962h.

39 L. Xu, *et al.*, Nanoparticle assemblies: dimensional transformation of nanomaterials and scalability, *Chem. Soc. Rev.*, 2013, **42**, 3114–3126, DOI: 10.1039/c3cs35460a.

Poster titles

One-pot sequential nanocomposite synthesis for growing NiWO4 on reduced graphene oxide with enhanced electrochemical performance in supercapacitor, **Aman Agrawal, Rudra Kumar, Rajaram K. Nagarale and Ashutosh Sharma**, *Indian Institute of Technology Kanpur, India*

Thermo- and photo-voltage generation: an innovative use of homeopathic medicine, **Poonam Bandyopadhyay, Sweta Charkraborty, Ruma Basu, Sukhen Das, Durga Sankar Bhar, Khurana Manchand, Rajkumar Noyak, Debadutta Anil Kumar and Papiya Nandy**, *Center for Interdisciplinary Research and Education, India*

Functionalized graphene oxide quantum dots-poly vinyl alcohol hybrid hydrogel: A colorimetric sensor for detection of Fe2+, Co2+ and Cu2+ ions in aqueous media, **Upama Baruah and Devashish Chowdhury**, *Institute of Advanced Study in Science and Technology, India*

Explosive vapour detection using light weight and flexible carbon nanoparticles coated 3D macroporous composite silk scaffolds as green chemiresistor, **Sudesna Chakravarty, Nandana Bhardwaj, Biman B. Mandal and Neelotpal Sen Sarma**, *Institute of Advanced Study in Science and Technology, India*

Templated nano-engineered carbon soot-PDMS superhydrophobic surfaces for enriched 3D ionic crystal growth, **Govind P. Chate, Shashwat S. Banerjee, Bhanudas S. Kuchekar and Jayant K. Khandare**, *MAEER's Maharashtra Institute of Pharmacy, MIT campus, Kothrud, Pune-38, India*

Layer by layer-gold nanoassembly as siRNA carrier for BACE1 gene silencing, **Aparna Chaudhary and Sanjeev Garg**, *IIT Kanpur, India*

Competing directionality in systems of cube-like magnetic colloids, **Joe Donaldson and Sofia Kantorovich**, *University of Vienna, Austria*

Probing the interaction between an energy efficient biomolecule adenosine triphosphate (ATP) and ZnO nanostructures using Raman spectroscopy, **Kartik Ghosh, Austin M. Shearin, Daniel Soden, Priyanka Karnati, Anagh Bhaumik, Ridwan Sakidja, Adam Wanekaya and Robert K. Delong**, *Missouri State University, United States*

Comparison of hydration behaviour and conformational preferences of Trp-cage mini-protein in different rigid body water models, **Madhulika Gupta, Divya Nayar and Charusita Chakravarty**, *Indian Institute of Technology Delhi, India*

Synthesis and characterization of carbon nanotube/hydroxyapatite/clay based hybrid antimicrobial biomaterial for potential tissue engineering application,

Subrata Kar, R. Basu, R. K. Manchanda, A. K. Khorana, D. Nayak and P. Nandy, *Physics Department, Jadavpur University, India*

Self-assembly and rheology of colloidal particles in hexagonal and nematic liquid crystals, **Siddharth Kulkarni and Prachi Thareja**, *Indian Institute of Technology, Gandhinagar, India*

Self-assembly and magnetic field driven crack morphology in film of colloidal ellipsoids, **Hisay Lama, Venkateshwar Rao Dugyala, Dillip K. Satapathy and Basavaraj M. G.**, *Indian Institute of Technology Madras, India*

Au3+ capped paper carbon dots for fluorimetric detection of organic and inorganic sulphur bearing analytes, **Sristi Majumdar, Upama Baruah, Gitanjali Majumdar, Debajit Thakur and Devashish Chowdhury**, *Institute of Advanced Study in Science and Technology, India*

Synthesis and characterization of poly(ethylene glycol) diacrylate hydrogel encapsulated giant unilamellar vesicles for biomedical application, **Mohana Mukherjee and Sammer Jadhav**, *IIT Bombay, India*

A leaf-like branched DNA material by molecular self-assembly of genomic sequences, **Ashok K. Nayak and U. Subudhi**, *CSIR-Institute of Minerals & Materials Technology, India*

Gold nanoparticles passivated with alkylthiols: simulations of solvation in the infinite dilution limit, **Saurav Prasad, Hari O. S. Yadav and Charusita Chakravarty**, *Department of Chemistry, Indian Institute of Technology Delhi, New Delhi-110016, India*

Size dependent quenching of fluorescence of 2-AS by Au nanoparticles: Probing NSET validity, **Soumyadipta Rakshit, Rimi Roy and Subhash Chandra Bhattacharya**, *Department of Chemistry, Jadavpur University, India*

Urea templated porous Mn/Ti layered double hydroxide nanoparticles for highly efficient photocatalytic degradation of organic dye pollutants, **Priyadarshi Roy Chowdhury and Krishna G. Bhattacharyya**, *Gauhati University, Guwahati, Assam, India*

In-situ removal of surfactants from shape-controlled Pt nanaoparticles for electrochemical applications, **Devivaraprasad Ruttala, Tathagata Kar and Manoj Neergat**, *Indian Institute of Technology, Bombay, India*

Optimization of resonance frequency, quality factor of PAN/PVDF modified polymer blend by graphene for electronic application, **Ajay Sharma, Girish Joshi and Pandey Mayank**, *VIT University,Vellore, India*

Electro-sensing base for the oxidation and determination of bioactive molecules at nanoparticle assembly modified electrode, **S. D. Bukkitgar and N. P. Shetti**, *Department of Chemistry, K L E Institute of Technology, India*

Irreversible aggregation of binary sticky spheres, **Zakiya Shireen and Sujin B. Babu**, *Indian Institute of Technology Delhi, India*

Quantitative measures of asymmetry in ligand-shells of nanoparticles, **Gourav Shrivastav, Hari O. S. Yadav and Charusita Chakravarty**, *Department of Chemistry, Indian Institute of Technology Delhi, India*

Self-propelling MnO2 nanoparticle-based paper microjets, **Amit Kumar Singh, Tapas Kumar Mandal and Dipankar Bandyopadhyay**, *Centre for Nanotechnology, IIT Guwahati, India*

Neuron-specific nano-bio interface design for colloidal metallic nanoparticles, **Madhura Som and Andrea R. Tao**, *Department of Nano Engineering, University of California at San Diego, United States*

Temperature dependent magnetization in cobalt nanowires electrodeposited within alumina template using sulphate bath: The role of boric acid as a buffer, **Ajeet K. Srivastav and Rajiv Shekhar**, *Indian Institute of Technology, Kanpur, India*

Entropy driven ordering of polymer-linked nanoparticles, **Mukta Tripathy**, *Department of Chemical Engineering, Indian Institute of Technology Bombay, India*

Electrochemical insight into the Brust-Schiffrin synthesis of Au nanoparticles, **Akihiro Uehara, Samuel G. Booth, Sin Yuen Chang, Sven L. M. Schroeder, Takahito Imai, Teruo Hashimoto, J. Frederick W. Mosselmans and Robert A. W. Dryfe**, *Research Reactor Institute, Kyoto University, Japan*

Creating temperature-independent solvation environment for nanoparticle self-assembly, **Hari O. S. Yadav, Gourav Shrivastav, Manish Agarwal and Charusita Chakravarty**, *Department of Chemistry, Indian Institute of Technology Delhi, India*

Experimental probe of triplet correlation function in simple and tetrahedral fluids, **Debdas Dhabal, Kjartan Thor Wikfeldt, Lawrie Skinner and Charusita Chakravarty**, *Department of Chemistry, Indian Institute of Technology Delhi, India*

Colloidal assembly by ice templating, **Bipul Biswas, Mandan K. Chowdhury and Guruswamy Kumar**, *Polymer Science & Engineering Division, CSIR-National Chemical Laboratory, Pune-411 008, India*

A novel modified screen-printed electrode with triazole surfactant assembled on silver nano-particles for potentiometric determination of uranium, **Tamer Awad Ali and Z. F. Akl**, *Egyptian Petroleum Research Institute (EPRI), Egypt*

Multi-walled carbon nanotube and nanosilica chemically modified screen-printed electrodes for determination of Fe(III) ion in different watersamples, **Tamer Awad Ali, Abeer M. Elsied and Gehad G. Mohamed**, *Analytical Chemistry Department, Faculty of Pharmacy,October 6 University, Giza, Egypt*

Assembly and release of target DNA on polymer substrate based nanostructures of nanorod - linker – protoporphyrine, **Sumana Das, Ramakrishna Vasireddi, Krishna H. Villa and D. Roy Mahapatra**, *IIT Bangalore, India*

Self-assembly of ultrathin gold nanowires for transparent conductive materials, **L. Gonzalez-Garcia, B. Reiser, J. H. M. Maurer, D. Gerstner, I. Kanelidis and T. Kraus**, *INM-Leibniz Institute for New Materials, Germany*

Understanding adsorption behavior of silica nanoparticles over a cellulose surface in an aqueous medium, **Parul Katiyar, Tarak K. Pakra, Jayant K. Singh, D. Y. Sarkar and Eboleena Amitava Pramanik**, *IIT Kanpur, India*

Comparative fluorescence studies on the interaction of Bi2S3 nanorods with serum albumins, **Selvaraj Naveenraj, Ramalinga Viswanathan Mangalaraja, Thangaraj Pandiyarajan and Sambandam Anandan**, *Department of Materials Engineering (DIMAT), University of Concepcion, Chile*

Computational studies on bare and ligated bimetallic gold clusters, **Manohar Pillegowda and Ganga Periyasamy**, *Bangalore University, India*

Facile and rapid synthesis of nanocrystalline vanadium oxide assisted by microwave heating, **Vandana P. Shinde and Pradip P. Patil**, *Department of Physics, North Maharashtra University, Jalgaon, India*

Protein directed synthesis of luminescent gold nanoparticles and nanoclusters: α-helical to β-sheet proteins, **S Deepthi Yarramala and Pulla Rao Chebrolu**, *Indian Institute of Technology Bombay, India*

Os nanostructures on reduced graphene oxide formed at a liquid/liquid interface, **K. Bramhaiah, C. Kavitha and Neena S. John**, *Centre for Nano and Soft Matter Sciences, India*

Electrical Transport in a random quasi-2D assembly of Ag nanoclusters, **C. Bansal and S. G. Praveen**, *School of Physics, University of Hyderabad, India*

The Skinner Prize for the best poster was jointly awarded to Siddharth Kulkarni of Indian Institute of Technology Gandhinagar, India, for his poster on Self-assembly and rheology of colloidal particles in hexagonal and nematic liquid crystals, Hisay Lama of Indian Institute of Technology Madras, India, for his poster on Self-assembly and magnetic field driven crack morphology in film of colloidal ellipsoids and Madhura Som of the University of California at San Diego, United States, for her poster on Neuron-specific nano-bio interface design for colloidal metallic nanoparticles.

List of participants

Mr Aman Agrawal, *Indian Institute of Technology Kanpur India, India*
Dr Tamer Ali, *Egyptian Petroleum Research Institute (EPRI), Egypt*
Dr Vidya Avasare, *Sir Parashurambhau College, India*
Mr Nirmalya Bachhar, *IIT Bombay, India*
Mr Vivekananda Bal, *IIT Bombay, India*
Ms Poonam Bandyopadhyay, *CIRE and Jadavpur University, India*
Dr Ranjini Bandyopadhyay, *Raman Research Institute, India*
Professor Rajdip Bandyopadhyaya, *IIT Bombay, India*
Ms Upama Baruah, *Institute of Advanced Study in Science and Technology, India*
Mr Sreenu Bhanoth, *Defence Institute of Advanced Technology, India*
Mr Sushrut Bhanushali, *IIT Bombay, India*
Mr Bipul Biswas, *NCL Pune, India*
Mr Pritam Biswas, *IIT Bombay, India*
Professor Michael Bockstaller, *Carnegie Mellon University, USA*
Mr Praveen Kumar Bommineni, *Indian Institute of Science, India*
Miss Monali Borthakur, *Gauhati University, India*
Dr Robert Botet, *CNRS, France*
Dr Charusita Chakravarty, *Indian Institute of Technology Delhi, India*
Miss Sudesna Chakravarty, *Institute of Advanced Study in Science and Technology, India*
Mr Govind Chate, *MAEER's Maharashtra Institute of pharmacy, India*
Miss Aparna Chaudhary, *IIT Kanpur, India*
Mr Semen Chervinskii, *University of Eastern Finland / Institute of Photonics, Finland*
Dr Emma Cooper, *Royal Society of Chemistry, United Kingdom*
Dr Marc Couty, *Manufacture Francaise des Pneumatiques MICHELIN, France*
Dr Sumana Das, *Indian Institute of Science, India*
Mr Nipon Deka, *Gauhati University, India*
Mr Debdas Dhabal, *Department of Chemistry, Indian Institute of Technology Delhi, India*
Dr Alison Edwards, *Bragg Institute, Australian Nuclear Science and Technology Organization, Australia*
Dr Erika Eiser, *University of Cambridge, United Kingdom*
Dr Abeer Elsied, *Analytical Chemistry Department, Faculty of Pharmacy, October 6 University, Giza, Egypt. Egypt*
Professor Daan Frenkel, *University of Cambridge, United Kingdom*
Professor Kartik Ghosh, *Missouri State University, USA*
Professor Achille Giacometti, *Dipartimento di Scienze Molecolari e Nanosistemi, Universita Ca' Foscari di Venezia, Campus Scientifico, Italy*
Dr Fiona Gillespie, *Royal Society of Chemistry, United Kingdom*
Miss Neeha Gogoi, *Gauhati University, India*
Dr Lola Gonzalez-Garcia, *INM-Leibniz Institute for New Materials, Germany*
Professor Steve Granick, *IBS Center for Soft and Living Matter, South Korea*
Dr Martyn Green, *MANTIS Deposition, United Kingdom*

Miss Madhulika Gupta, *Indian Institute of Technology Delhi, India*
Miss Nikita Gupta, *Royal Society of Chemistry, India*
Miss Masoom Nasiha Hussain, *GUIST (gauhati university), India*
Ms Shweta Jatav, *Indian Institute of Technology Kanpur, India*
Dr Jacques Jestin, *Laboratoire Léon Brillouin CEA-CNRS, France*
Dr Neena S. John, *Centre for Nano and Soft Matter Sciences, India*
Dr Aparna Joshi, Research Director - *K. D. Joshi Rubber Industries Pvt. Ltd., India*
Professor Yogesh Joshi, *Indian Institute of Technology Kanpur, India*
Dr Yon Ju-Nam, *Swansea University, United Kingdom*
Dr Sofia Kantorovich, *University of Vienna, Austria*
Mr Subrata Kar, *Jadavpur University, India*
Professor Alamgir Karim, *University of Akron, USA*
Ms Parul Katiyar, *IIT Kanpur, India*
Ms Deepa Khandekar, *IIT Bombay, India*
Dr Hima Bindu Kolli, *Dipartimento di Scienze Molecolari e Nanosistemi, Universita Ca' Foscari di Venezia, Campus Scientifico, Italy*
Professor Nicholas Kotov, *University of Michigan, USA*
Dr Ramanan Krishnamoorti, *Department of Chemical and Biomolecular Engineering, University of Houston, USA*
Mr Siddharth Kulkarni, *Indian Institute of Technology, Gandhinagar, India*
Professor G. V. Pavan Kumar, *IISER Pune, India*
Professor Sanat Kumar, *Columbia University, USA*
Professor Sunil Kumar P. B., *IIT Madras, India*
Dr Guruswamy Kumaraswamy, *CSIR-National Chemical Laboratory, India*
Mr Ninad Kumbhojkar, *Institute of Chemical Technology, Mumbai, India*
Mr Hisay Lama, *Indian Institute of Technology Madras, India*
Dr Mohamed Laradji, *University of Memphis, USA*
Dr Basavaraj Madivala, *Indian Institute of Technology Madras, India*
Miss Sristi Majumdar, *Institute of Advanced Study in Science and Technology, India*
Professor Vinothan Manoharan, *Harvard University, USA*
Mr Mayank Misra, *Columbia University, USA*
Miss Mohana Mukherjee, *IIT Bombay, India*
Dr Selvaraj Naveenraj, Post-doctoral Researcher, *Department of Materials Engineering (DIMAT), University of Concepcion, Chile*
Mr Ashok Kumar Nayak, *CSIR-Institute of Minerals and Materials Technology, India*
Dr Vibhuti Patel, *Royal Society of Chemistry, United Kingdom*
Mr Darius Patel, *MANTIS Deposition, United Kingdom*
Mr Manohar Pillegowda, *Bangalore University, India*
Dr Radhika Poojari, *Department of Biosciences and Bioengineering, Indian Institute of Technology Bombay, India*
Dr Abhiram Prabhu, *Royal Society of Chemistry, India*
Dr B. L. V. Prasad, *National Chemical Laboratory, India*
Mr Saurav Prasad, *Department of Chemistry, Indian Institute of Technology Delhi, New Delhi-110016, India, India*
Professor Sudeep Punnathanam, *Department of Chemical Engineering, India*
Mr Anil Rajapantul, *IIT Bombay, India*
Mr Soumyadipta Rakshit, *Department of Chemistry, Jadavpur University, India*
Miss Rimi Roy, *Department of Chemistry, Dinhata College, India*
Mr Priyadarshi Roy Chowdhury, *Gauhati University, Guwahati, Assam, India*
Mr Devivaraprasad Ruttala, *Indian Institute of Technology, Bombay, India*
Dr Praveen S. G. *Indian Institute of Science Education and Research, India*
Mr Namsani Sadanandam, *Indian Institute of Technology Knapur, India*
Dr Pedro A. Sánchez, *University of Vienna, Austria*

Dr Jyoti Seth, *IIT Bombay, India*
Mr Ajay Sharma, *VIT University, India*
Dr Nagaraj Shetti, *Department of chemistry, K L E Institute of Technology, India*
Dr Vandana Shinde, *Department of Physics, North Maharashtra University, Jalgaon, India*
Miss Zakiya Shireen, *Indian Institute of Technology Delhi, India*
Mr Gourav Shrivastav, *Indian Institute of Technology Delhi, India*
Mr Amit Kumar Singh, *Centre for Nanotechnology, IIT Guwahati, India*
Professor Jayant Singh, *IIT Kanpur, India*
Ms Madhura Som, *University of California at San Diego, USA*
Ms Vimala Sridurai, *Centre for Nano and Soft Matter Sciences, India*
Dr Ajeet Srivastav, *Visvesvaraya National Institute of Technology, India*
Dr Igal Szleifer, *Department of Biomedical Engineering, Department of Chemistry and Chemistry of Life Processes Institute, Northwestern University, USA*
Dr Mario Tagliazucchi, *INQUIMAE-CONICET, Ciudad Universitaria, Pabellon 2, Ciudad Autonoma de Buenos Aires, Argentina*
Professor Andrea Tao, *University of California, San Diego, USA*
Professor Mukta Tripathy, *Department of Chemical Engineering, Indian Institute of Technology Bombay, India*
Dr Akihiro Uehara-Kasuno, *Research Reactor Institute, Kyoto University, Japan*
Professor Lynn Walker, *Carnegie Mellon University, USA*
Mr Hari Om Sharanam Yadav, *Department of Chemistry, Indian Institute of Technology Delhi, India*
Ms S. Deepthi Yarramala, *Indian Institute of Technology Bombay, India*